The Biology of Gastric Cancers

Timothy C. Wang • James G. Fox
Andrew S. Giraud

Editors

The Biology of Gastric Cancers

 Springer

Editors
Timothy C. Wang
Silberberg Professor of Medicine
Chair, Division of Digestive and
  Liver Diseases
Columbia University
New York, NY
USA

James G. Fox
Director, Division of Comparative
  Medicine
Professor, Department of Biological
  Engineering
Massachusetts Institute of Technology
Cambridge, MA
USA

Andrew S. Giraud
Professor and Research Theme Director
Infection, Immunity and Environment
Murdoch Childrens Research Institute
Royal Melbourne Hospital
Footscray, Melbourne
Australia

ISBN: 978-0-387-69181-7   e-ISBN: 978-0-387-69182-4
DOI: 10.1007/978-0-387-69182-4

Library of Congress Control Number: 2008923739

springer.com

Foreword

As someone who has spent nearly half his life wondering about the relationship between *Helicobacter* and gastric cancer, I find this textbook on the subject exciting and timely. In fact, I am not aware of any other volume that has been able to distil so much new knowledge into such a comprehensive account of a poorly understood field.

Taking my own view, as a scientist placed in the middle of the spectrum between basic science and clinical medicine, I can see that the editors, Jim Fox, Andy Giraud, and Timothy Wang, provide a broad mix of expertise, which ensures that the subject is treated with the right balance. From clinicopathologic observations in humans, to epidemiology, through animal models, to molecular and cell biology, this team has hit the mark for most readers. Fox is a well-known leader in animal models with broad expertise. He pioneered the field with observations on *Helicobacter* species in animals, from the time when only one spiral gastric bacterium was known, "*Campylobacter pyloridis*." Fox partners with Wang, whose team recently announced a dramatic advance in the field of carcinogenesis—the observation that bone marrow–derived stem cells participate in the changes that become cancer. To this nice mix has been added Andy Giraud from my own country, who brings to the table some remarkable genetic models of gastric cancer based on alterations in the gp130/stat3-signaling pathway.

Examining the structure of the book in more detail, I see that they have set the scene with chapters explaining the core knowledge that the novice must learn. The reason being: these longstanding truths will need to be accounted for in order to seek a unifying etiologic hypothesis for gastric cancer. Correa, a pathologist who is himself Colombian, starts by explaining the clinical patterns and histology noted over a lifetime dealing with gastric cancer from many of the Latin American hotspots. Then, Parsonnet summarizes 40 years of prospective data linking *Helicobacter*, by its causative role in gastritis, as the likely cause of gastric cancer also. But differing gastric cancer incidence in various countries means that *Helicobacter*, at least the main accelerant of carcinogenesis, must also interact with many dietary chemicals that directly touch the gastric mucosa. Updated knowledge of this dietary epidemiology is therefore explained.

The less common gastric tumors with immune, neuroendocrine, and genetic causes are dealt with next. MALT Lymphoma leads the list, providing the first inkling that a lifelong immunologic insult to the gut mucosa is not without peril. These nonmucosal neoplasms must be accurately diagnosed now as their molecular triggers can be specifically treated (GIST tumors). Neuroendocrine neoplasia, always of interest in conditions in which so much therapy relates to chronic acid suppression, is carefully dealt with in two chapters as is the little-known linkage between Epstein-Barr virus and gastric tumors. Some congenital tumors even serve to demonstrate a molecular parallel with normal carcinogenesis. Appropriately, the first group of chapters is rounded off with the controversial relationships among human genetic polymorphisms, acid secretion, *Helicobacter*, and cancer; i.e., the emphasis of host factors rather than bacterial factors as determinants of cancer risk.

The second half of the book contains chapters that draw the reader into basic research and molecular aspects of gastric cancer. In the postgenomic era, as more tumor DNA libraries are collected, cancer genetics are beginning to be understood. Eventually, it is quite likely that automated DNA-based methods (e.g., gene arrays) will replace human-driven tumor diagnosis. Molecular genetics and proteomic observations already show promise as methods of sorting out premalignant and malignant conditions presently labeled rather vaguely as the "metaplasias." As new hypotheses are developed, these need to be tested in appropriate animal models. Although many choices exist, the response of each model to *Helicobacter*, histologically and immunologically, is quite variable and not well understood. If vaccines against *Helicobacter* are ever going to be possible, better understanding of the host–bacterial interaction is necessary. Regardless of the individual deficiencies, investigators can use new animal models to rapidly dissect out disease associations such as the dietary carcinogen/*Helicobacter*/salt connection or the sexual predilection for cancer in male humans and some male mice. Throughout the book, broad chapters on animal models are bolstered by focused chapters on genetically altered mice, other rodents, and even primate models.

The various oncogenes and molecular switches responsible for cancer associations are examined in more detail by experts in these more narrow fields. Developmental processes required for metaplasia again raise the issue of host factors versus bacterial factors as the cause of many cancers. In the bacterial field, CagA is certainly the most studied and probably the most relevant of these, and polymorphisms within its structure might explain its variable strength as a molecular switch. The discovery of the way in which modern molecular techniques were applied to this toxin makes a fascinating story in itself.

Several chapters deal with host immunity, important because lifelong persistence of *Helicobacter* is a unique success story for the bacterium and an apparent failure for the gastrointestinal immune system. Similarly, because inflammation correlates both with increased ulcerogenesis, hyperacidity, and cancer protection in some populations with simultaneous atrophy, hyposecretion, and cancer risk in apparently similar individuals, the fine details of the immune response seem to offer a productive area of research. Again, new molecular markers, oncogenes, and even cytokines are all relevant and worthy of study.

Finally, the role of bone marrow–derived stem cells in gastric cancer is carefully addressed. Because animal models require time to develop cancer as each new hypothesis is tested, understanding of this area serves to show where new studies are leading and how these might relate to diagnosis and prognosis as interventional human studies are planned. These would likely be bacterial eradication primarily, but the dream of most of us: that of a simple vaccine, is also addressed in a futuristic final chapter.

With such a scholarly trio of editors, this book has attracted a superlative team of contributors, each able to bring the reader up to date with the most cutting-edge advances, in chapters that will remain current for many years and continue to provide a solid foundation for study in the field for decades.

January 28, 2008   Barry Marshall

Preface

This book represents the very first volume devoted to the subject of the basic biology of gastric cancer. Although there were major achievements 30–40 years ago in the description of the histopathology of stomach tumors and the preneoplastic conditions that preceded these lesions, it was of course the pioneering (and Nobel prize–winning) studies involving the discovery of *Helicobacter pylori* by Marshall and Warren that set the stage for the more recent insights on the role of chronic inflammation and cancer. As articulated in the chapters of this compendium, written by experts in the field, there have been many impressive advances in our understanding of the pathogenesis of gastric adenocarcinoma, gastric MALT lymphoma, gastrointestinal stromal tumors of the stomach, and gastric carcinoid tumors. The text begins with an overview of these neoplasms and their histopathology, the role of diet and *H. pylori*, and their epidemiology, before moving into the current view and understanding of molecular mechanisms. The book, of course, reflects many of the interests of the editors, with an emphasis on animal models, but also devotes considerable attention to human studies, host genetics, and the role of bacterial factors and other environmental factors influencing the pathogenesis of gastric cancer. The text pays homage to the cutting-edge concepts of cancer stem cells, tumor microenvironment, chronic inflammation, and genetic susceptibility to cancer. Although every potential mechanism contributing to gastric carcinogenesis could not be covered because of space limitations, we have attempted to highlight the important areas that would appeal to a broad readership.

The topics presented in this book should be of interest to clinicians and investigators interested in gastrointestinal cancer, as well as basic investigators in related areas of cancer research. Gastric cancer has in many respects become a paradigm for the broader association between chronic inflammation and cancer, and the association between infection and cancer. At present, it is thought that only 15%–20% of cancers can be directly attributed to infection, but the cause of many cancers is still not known, similar to the situation that existed for gastric cancer before the discovery of *H. pylori*. Thus, we believe that the insights developed from the studies described in these chapters may inform researchers working on other organ-specific neoplasms. In addition, it is our hope that this body of work will stimulate clinical and translational studies that will advance early diagnosis and treatment of gastric tumors.

We extend our appreciation not only to the authors of all the chapters in this book for the tremendous accomplishments, but also to the many reviewers who took the time to critique and edit the manuscripts. The reviewers included John Atherton, Pelayo Correa, Jean Crabtree, Stanley Falkow, JeanMarie Houghton, Peter Isaacson, Robert Jensen, Andrew Leiter, Wai K. Leung, Stephen Meltzer, Christopher A. Moskaluk, Guilleromo Perez-Perez, Arlin Rogers, Massimo Rugge, George Sachs, Linda Samuelson, and Andrea Todisco. We also thank our assistants (Mary Beth Shanahan and Lucy Wilhelm) for their work in formatting and editing the chapters. We acknowledge warmly the patience and attention put into this book by our colleagues at Springer, particularly Rachel Warren and Thomas Brazda. Finally, we especially thank the Funderberg foundation for their creation of the Robert and Sally Funderberg Gastric Cancer Awards, which stimulated interest in this field, and the National Institutes of Health for their support of basic and translational research in gastric cancer.

New York, NY, USA   Timothy C. Wang
Cambridge, MA, USA   James G. Fox
Melbourne, Australia   Andrew S. Giraud

Contents

Contributors

Mohammad S. Alam
Division of Gastroenterology and Hepatology, Department of Medicine,
University of Virginia, Charlottesville, VA, USA

Takeshi Azuma
Department of Gastroenterology, Kobe University School of Medicine, Kobe, Japan

Monica M. Bertagnolli
Department of Surgery, Division of Surgical Oncology, Brigham and Women's
Hospital, Dana-Farber Cancer Center, Boston, MA, USA

Asima Bhattacharyya
Department of Medicine, Division of Gastroenterology and Hepatology,
University of Virginia, Charlottesville, VA, USA

Alex Boussioutas
Department of Medicine, University of Melbourne, Footscray, Melbourne, Australia

M. Constanza Camargo
Division of Gastroenterology, Department of Medicine, Vanderbilt University
Medical School, Nashville, TN, USA

Annie On On Chan
Department of Medicine, Unviersity of Hong Kong, Hong Kong

Tsutomu Chiba
Department of Gastroenterology and Hepatology, Kyoto University Graduate
School of Medicine, Tokyo, Japan

Pelayo Correa
Division of Gastroenterology, Department of Medicine, Vanderbilt University
Medical School, Nashville, TN, USA

Sheila E. Crowe
Division of Gastroenterology and Hepatology, Department of Medicine,
University of Virginia, Charlottesville, VA, USA

Emad M. El-Omar
Department of Medicine and Therapeutics, Institute of Medical Sciences,
Aberdeen University, Foresterhill, Aberdeen, Scotland

Peter B. Ernst
Division of Gastroenterology and Hepatology, Department of Medicine,
University of Virginia, Charlottesville, VA, USA

James G. Fox
Division of Comparative Medicine, Department of Biological Engineering,
Massachusetts Institute of Technology, Cambridge, MA, USA

Hirokazu Fukui
Department of Surgical and Molecular Pathology, Dokkyo University School of
Medicine, Dokkyo, Japan

Roman Galysh, Jr.
Gastroenterology Fellow, Department of Medicine, Division of Gastroenterology/
Hepatology, University of Virginia Health System, Charlottesville, VA, USA

Andrew S. Giraud
Murdoch Childrens Research Institute, Royal Melbourne Hospital, Footscray,
Melbourne, Australia

James R. Goldenring
Nashville Department of Veterans Affairs Medical Center, Nashville, TN, USA

Bjorn I. Gustafsson
Department of Surgery, Yale University School of Medicine, New Haven, CT, USA

Jason L. Hornick
Brigham and Women's Hospital, Department of Pathology, Harvard Medical
School, Dana-Farber Cancer Institute, Boston, MA, USA

JeanMarie Houghton
Department of Medicine and Cancer Biology, University of Massachusetts
Medical School, Worcester, MA, USA

Dawn A. Israel
Division of Gastroenterology, Department of Medicine, Vanderbilt University
Medical Center, Nashville, TN, USA

Louise M. Judd
Department of Pediatrics, University of Melbourne, Royal Children's Hospital,
Parkville, Melbourne, Australia

Mark Kidd
Yale University School of Medicine, New Haven, CT, USA

Yoshikazu Kinoshita
Department of Gastroenterology and Hepatology, Shimane University School of
Medicine, Shimane, Japan

Jaw-Town Lin
Departments of Internal Medicine and Pathology, National Taiwan University
Hospital, Taipei, Taiwan

Chun Liu
Nutrition and Cancer Biology Laboratory, Jean Mayer United States Department
of Agriculture Human Nutrition Research Center on Aging, Tufts University,
Boston, MA, USA

Maximillian V. Malfertheiner
GI Surgical Pathobiology Research Group, Yale University School of Medicine,
New Haven, CT, USA

Juanita L. Merchant
Department of Internal Medicine, University of Michigan, Ann Arbor, MI USA

Irvin M. Madlin
Yale University School of Medicine, New Haven, CT, USA

Pierre Michetti
Service de Gastro-entérologie et d'Hépatologie, Centre Hospitalier Universitaire
Vaudois and University of Lausanne, Lausanne, Switzerland

Anne Müller
Institute of Molecular Cancer Research (Institut für Molekulare Krebsforschung),
University of Zurich, Zurich, Switzerland

Sachiyo Nomura
Department of Gastrointestinal Surgery, Graduate School of Medicine, University
of Tokyo, Tokyo, Japan

Hiroko Oshima
Division of Genetics, Cancer Research Institute, Kanazawa University, Japan

Masanobu Oshima
Division of Genetics, Cancer Research Institute, Kanazawa University, Japan

Julie Parsonnet
Stanford University School of Medicine, Stanford, CA, USA

Richard M. Peek, Jr.
Division of Gastroenterology, Departments of Medicine and Cancer Biology,
Vanderbilt University Medical Center, and Department of Veterans Affairs
Medical Center, Nashville, TN, USA

M. Blanca Piazuelo
Division of Gastroenterology, Department of Medicine, Vanderbilt University
Medical School, Nashville, TN, USA

Steven M. Powell
Division of Gastroenterology/Hepatology, University of Virginia Health System,
Charlottesville, VA, USA

Chandrajit P. Raut
Departments of Surgery and Pathology, Brigham & Women's Hospital, Center for
Sarcoma and Bone Oncology, Dana-Farber Cancer Institute, and Harvard Medical
School, Boston, MA, USA

Guido Rindi
Dipartimento di Patologia e Medicina di Laboratorio, Sezione di Anatomia
Patologica, Universita' degli Studi, Parma, Italy

Arlin B. Rogers
Division of Comparative Medicine, Massachusetts Institute of Technology,
Cambridge, MA, USA

Robert M. Russell
Nutrition and Cancer Biology Laboratory, Jean Mayer United States Department
of Agriculture Human Nutrition Research Center on Aging, Tufts University,
Boston, MA, USA

Chia-Tung Shun
Departments of Internal Medicine and Pathology, National Taiwan University
Hospital, Taipei, Taiwan

Enrico Solcia
Dipartimento di Patologia Umana ed Ereditaria, Sezione di Anatomia Patologica,
Università di Pavia, Pavis, Italy

Makoto Mark Taketo
Department of Pharmacology, Kyoto University Graduate School of Medicine, Japan

Andrea Varro
Department of Physiology, School of Biomedical Sciences, University of
Liverpool, Liverpool, UK

Patrick Tan
Duke-NUS Graduate Medical School, Genome Institute of Singapore, Singapore

Dominique Velin
Service de Gastro-entérologie et d'Hépatologie, Centre Hospitalier Universitaire
Vaudois and University of Lausanne, Lausanne, Switzerland

Jonathan Volk
Department of Medicine, Stanford University School of Medicine, Stanford, CA, USA

Timothy C. Wang
Division of Digestive and Liver Diseases, Columbia University, New York, NY, USA

Xiang-Dong Wang
Nutrition and Cancer Biology Laboratory, Jean Mayer United States Department
of Agriculture Human Nutrition Research Center on Aging, Tufts University,
Boston, MA, USA

Benjamin Chun-Yu Wong
Department of Medicine, University of Hong Kong, Hong Kong

Ming-Shiang Wu
Departments of Internal Medicine and Pathology, National Taiwan University
Hospital, Taipei, Taiwan

Yana Zavros
Department of Molecular and Cellular Physiology. University of Cincinnati,
Cincinnati, OH, USA

Editors' Biographies

Timothy C. Wang, MD

Dr. Wang is the Silberberg Professor of Medicine and Chair of the Division of Digestive and Liver Diseases at Columbia University Medical Center. He is an internationally respected clinician and scientist, specializing in *H. pylori* infection and gastrointestinal cancer. He is a graduate of Columbia College of Physicians and Surgeons. He was a faculty member for ten years at Massachusetts General Hospital and Division Chief at the University of Massachusetts Medical School before returning to Columbia University in 2004 as Chief of Digestive and Liver Diseases and the Silberberg Professor of Medicine. He is a recipient of the AGA Funderberg Gastric Cancer Award and the Victor Mutt medal in Gut Hormone Research, and is a member of American Society for Clinical Investigation (ASCI) and Association of American Physicians (AAP). He is a former and current Associate Editor of Gastroenterology., and Editor in Chief of the journal Therapeutic Advances in Gastroenterology. He is active in the American Gastroenterological Association, currently serving as chair of the Future Trends Committee, and Chair of the GCMB NIH study section. He has authored over 150 original peer-reviewed publications. Dr. Wang's research centers on the role of inflammation and stem cells in the development of gastrointestinal cancers. His main interest has been in developing transgenic mouse models of stomach cancer, and together with Dr. Fox, they have developed the first *Helicobacter*-dependent mouse models of gastric cancer. Dr. Wang's laboratory reported that gastric cancer in mice can originate in part from bone marrow-derived stem cells. As the leader of the Columbia University Tumor Microenvironment Network, he has investigated the importance of the stem cell niche in governing stem cell differentiation, and how this niche is altered in chronic inflammatory states that predispose to cancer. He is currently investigating the role of both fibroblasts and myeloid cells in carcinogenesis.

James C. Fox, MD

Dr. Fox is Professor and Director of the Division of Comparative Medicine and a Professor in the Department of Biological Engineering at the Massachusetts Institute of Technology. He is also an Adjunct Professor at Tufts University School of Veterinary Medicine and the University of Pennsylvania, School of Veterinary Medicine. He is a Diplomate and a past president of the American College of Laboratory Animal Medicine, past president of the Massachusetts Society of Medical Research, past chairman of AAALAC Council, past chairman of the NCCR/NIH Comparative Medicine Study Section, and a member of various other organizations including AAAS, AALAS, AVMA, AAVMC, IDSA and ASM. He also is an elected fellow of the Infectious Disease Society of America, a past member of the Board of Directors of Public Responsibility in Medicine and Research (PRIM&R), ACLAM, AAALAC, MSMR, AAVMC and is currently Chairman of the AVMA and AAVMC Animal Welfare Committee. Professor Fox is the author of over 460 articles, 75 chapters, 4 patents and has edited and authored 12 texts, in the field of in vivo model development and comparative medicine. He has given over 250 invited lectures, consults nationally and internationally with government, academia and industry, has served on the editorial board of several journals, is a past member of the NIH/NCRR Scientific Advisory Council, and a current member of ILAR Council of the National Academy of Sciences.

Recently Professor Fox was elected to the Institute of Medicine of the National Academy of Sciences. He has received numerous scientific awards including the AVMA's Charles River Prize in Comparative Medicine, the AALAS Nathan Brewer Scientific Achievement Award, and the AVMA/ASLAP Excellence in Research Award. In 2006, Dr. Fox received the Distinguished Alumni Award from Colorado State University and in 2007 was selected as the inaugural recipient of the American College of Laboratory Animal Medicine Comparative Medicine Scientist Award. He has been studying infectious diseases of the gastrointestinal tract for the past 30 years and has focused on the pathogenesis of *Campylobacter* spp. and *Helicobacter* spp. infection in humans and animals. His laboratory developed the ferret as a model for both campylobacter and helicobacter associated disease as well as the first rodent model to study helicobacter associated gastric disease, including gastric cancer. Dr. Fox is considered an international authority on the epidemiology and pathogenesis of enterohepatic helicobacters in humans and animals. He is largely responsible for identifying, naming, and describing many of the diseases attributed to various *Helicobacter* species; most notably their association with hepatitis, liver tumors, inflammatory bowel disease and colon cancer. His laboratory most recently has described the pivotal role that *Helicobacter* spp. play in the development of cholesterol gallstones in mice fed a lithogenic diet; thus linking this finding to his earlier description of Helicobacter spp. associated chronic cholecystitis and gallstones in Chilean women, a population at high risk of developing gallbladder cancer. He also has had a long-standing interest in zoonotic diseases as well as biosafety issues associated with *in vivo* models.

His past and current research has been funded by NIH and NCI, as well as by private industrial sources, for the past 30 years. Dr. Fox ranks in the top 5% of all NIH awardees during the past 25 years. He has been the principal investigator of an NIH postdoctoral training grant for veterinarians for the past 20 years and has trained 45 veterinarians for careers in biomedical research. He also has a NIH training grant to introduce veterinary students to careers in biomedical research. He recently chaired a committee for the National Academy of Sciences which published a report entitled "National Need and Priorities for Veterinarians in Biomedical Research" which highlights the urgent need for increased numbers of veterinarians involved in the biomedical research arena. He currently is a member of NAS "Committee to Assess the Current and Future Workforce Needs in Veterinary Medicine."

Andrew S. Giraud

Dr. Giraud is currently Professor and Research Theme Director, Infection, Immunity and Environment at Murdoch Childrens Research Institute in Melbourne, Australia. His research interests are: (1) establishing mouse genetic models of gastric cancer in order to better understand how initiating factors promote inflammation, metaplasia and neoplasia. (2) translation of outcomes from mouse models in order to better understand, detect and treat human gastric adenocarcinoma. For instance we have found that in the mouse and in human gastric cancer the transcription factor STAT3 is continuously activated and promotes tumor growth. We are currently testing new small molecule STAT3 inhibitors as anticancer agents. (3) investigating the biology and mechanisms of activation of the gastrokine and trefoil peptide families, members of which are stomach-specific tumor suppressor genes.

Dr. Giraud is a member of the American Gastroenterology Association and the Gastroenterological Society of Australia. He has published widely in gastroenterological science including in a number of papers in *Gastroenterology* and *Nature Medicine* in the last six years.

(*continued*)

Currently Professor and Research Theme Director, Infection, Immunity and Environment at Murdoch Childrens Research Institute in Melbourne, Australia. Research interests are;

(i)   establishing mouse genetic models of gastric cancer in order to better understand how initiating factors promote inflammation, metaplasia and neoplasia.

(ii) translation of outcomes from mouse models in order to better understand, detect and treat human gastric adenocarcinoma. For instance we have found that in the mouse and in human gastric cancer the transcription factor STAT3 is continuously activated and promotes tumor growth. We are currently testing new small molecule STAT3 inhibitors as anti-cancer agents.

(iii) investigating the biology and mechanisms of activation of the gastrokine and trefoil peptide families, members of which are stomach-specific tumor suppressor genes.

Member of the American Gastroenterology Association and the Gastroenterological Society of Australia.

Published widely in gastroenterological science including in a number of papers in *Gastroenterology* and *Nature Medicine* in the last 6 years.

Chapter 1
Overview and Pathology of Gastric Cancer

Pelayo Correa, M. Constanza Camargo, and M. Blanca Piazuelo

Overview

For most of the twentieth century, gastric cancer was the major cancer burden worldwide. Its etiology and pathogenesis were obscure. Several events have changed that outlook, and currently, it ranks in the second place of mortality from cancer, after lung cancer. A gradual decline in the incidence of gastric cancer has been taking place for several decades, first seen in more affluent societies and then in other countries. The reasons for this decline are not entirely clear. However, it has coincided with several societal changes. Improvement in economic parameters in many populations was reflected in better home sanitation and better general nutrition. Fresh fruits and vegetables became available year round because of new technologies in home refrigeration and transportation. Other changes in dietary practices, such as the decrease in the intake of salt may also be linked to the decrease in gastric cancer incidence. Another relevant event was the development of the technology of the flexible fiberoptic endoscopy and its gradual generalization in the practice of medicine. This instrument allowed the scrutiny and early management of precancerous lesions identified in gastric biopsies. The major precancerous lesions had been previously identified by pathologists, based on autopsy studies and gastrectomy specimens. These observations clearly established the fact that there were changes in the gastric mucosa that preceded the clinical diagnosis of cancer by several decades.

Another major event was the gradual recognition that an infectious agent, namely, *Helicobacter pylori*, is intimately associated with gastric cancer development. In 1994, the International Agency for Research on Cancer (IARC) classified *H. pylori* infection as a class I carcinogen (IARC, 1994). That decision was based entirely on epidemiologic observations, without support in experimental studies. Soon after that, cancer development was documented in Mongolian gerbils infected with *H. pylori* (Watanabe et al. 1998). Since then, the importance of infectious agents as potential carcinogens has been recognized. Agents that induce chronic active inflammation are of special interest. A major research focus at the present time explores the role of inflammation in cancer causation. This research opens the way to the identification of events that may either lead to or prevent a neoplastic outcome, possibly leading to cancer prevention strategies.

T.C. Wang et al. (eds.) *The Biology of Gastric Cancers*,
© Springer Science+Business Media, LLC 2009

Epidemiology

In 2002, there were 933,937 new cases and 700,349 deaths from gastric cancer world-wide (Parkin et al. 2005). In the United States, the number of cases has remained around 20,000 for several years (Jemal et al. 2007), probably reflecting immigration from countries with high gastric cancer risk and aging of the population.

The geographic distribution of gastric cancer is spotty. Areas of highest risk have traditionally been Japan, Korea, China, Eastern Europe, and the Andean regions of the Americas. In contrast, Australia, Africa, the coastal regions of the Americas, and Southern Asia have traditionally been areas of low risk. Western Europe and North America, with considerably higher risk several decades ago, have experienced a marked decrease since then, and at the present time are considered areas of low risk (Ferlay et al. 2004). Table 1.1 shows that the risk is more than 10 times greater in high-risk compared with low-risk countries. This contrast, together with

Table 1.1 Gastric cancer incidence and mortality in males in selected countries, 2002. Globocan, IARC (Ferlay et al. 2004)

Country	Incidence		Mortality	
	Cases	ASR	Deaths	ASR
Republic of Korea	15,912	69.7	8,322	37.1
Japan	73,785	62.0	35,338	28.7
Chile	3,206	46.1	2,257	32.5
China	264,460	41.4	206,632	32.7
Costa Rica	633	41.2	459	30.1
Ecuador	1,636	37.8	1,355	31.0
Peru	3,459	37.5	2,692	29.5
Russian Federation	29,633	36.7	25,728	31.8
Colombia	5,159	36.0	3,929	27.8
Portugal	2,092	27.6	1,601	20.3
Poland	4,962	20.7	4,017	16.6
Italy	9,850	18.8	6,903	12.6
Spain	5,161	15.7	3,928	11.4
Germany	10,558	15.1	7,395	10.4
Argentina	2,888	14.6	2,076	10.3
Mexico	4,502	13.1	3,370	9.9
United Kingdom	6,311	12.5	4,464	8.7
France	4,925	10.4	3,417	7.0
Australia	1,370	9.8	800	5.7
Finland	387	9.6	326	7.9
Canada	2,027	9.1	1,331	5.9
Philippines	1,944	8.9	1,743	8.1
Sweden	659	8.0	571	6.8
Denmark	342	7.9	241	5.5
United States of America	13,710	7.2	7,761	4.0
Cuba	498	7.1	491	6.9
India	22,650	5.7	19,346	4.9

Abbreviations: ASR: age-standardized rate (cases per 100,000 persons per year).

the decreasing rates observed in some populations, suggests a strong etiologic role for environmental factors. The gradual decrease in gastric cancer rates has followed a birth cohort pattern, first shown by Haenszel in the 1950s for the United States. Age-specific incidence rates for each new generation are lower than those of the previous generation (Haenszel and Kurihara 1968).

Studies of migrant Japanese populations from their birth place to their adopted location in the United States (mostly Hawaii and California) showed that the first generation of immigrants (Nissei) had high cancer rates very similar to those of Japan. But the second, U.S. born generation (Issei) had much lower rates, similar to those of their adopted country (Haenszel and Kurihara 1968; Kolonel et al. 1981). The conclusion drawn by Haenszel was that the experience of childhood determined the cancer risk for life. It has later been shown that this "Haenszel phenomenon" is attributable to infection with *H. pylori* during childhood, which induces a life-long chronic active gastritis. The childhood environment of the Issei did not favor such infection. These observations fit very well the birth control pattern of the declining cancer rates. Younger generations have lower rates of *H. pylori* infection (Sipponen et al. 1994; Sipponen 1995).

The Etiology of Gastric Cancer

After the 1994 IARC report, the causative role of *H. pylori* infection has been supported by epidemiologic and experimental evidence. Case-control studies examining the association between *H. pylori* infection and diagnosis of gastric cancer produced discrepant results. Such inconsistencies were the result of ignoring the temporal relationship between the infection, starting in childhood, and the clinical diagnosis of gastric cancer several decades later. Three separate cohort studies, two in the United States (Hawaii and California) and one in England, assessed anti-*Helicobacter* antibodies in serum samples taken years before clinical diagnosis of gastric cancer (Forman et al. 1991; Nomura et al. 1991; Parsonnet et al. 1997). The undeniable association helped explain previous inconsistent findings. It has been shown that advanced precancerous gastric lesions such as extensive atrophy and intestinal metaplasia are not favorable niches for *H. pylori* colonization. Studies in Japan have shown that gastric cancer risk is higher in patients with negative serology for *H. pylori* infection (probably lost), but marked gastric atrophy reflected in low pepsinogen I levels (Ohata et al. 2004). Case-control studies addressing the temporality issue have shown that gastric cancer risk in younger patients and/or small tumors have a clear association with *H. pylori* infection. The association was not seen in large tumors or older patients, probably indicating loss of a longstanding infection where the gastric microenvironment was no longer favorable to *H. pylori* colonization (Fukuda et al. 1995).

It has been estimated that half of the world population is infected with *H. pylori*, but only a small minority develops gastric cancer: approximately 3 per 10,000 infected subjects per year (estimated by the authors). It has long been recognized that patients with peptic duodenal ulcer, in most cases caused by *H. pylori* infection, are not at increased gastric cancer risk (Hansson et al. 1996; Uemura et al. 2001). But patients with peptic gastric ulcer, also caused by *H. pylori* infection, do have an increased risk of cancer (Hansson et al. 1996; Ogura et al. 2006). Duodenal ulcer is usually accompanied by longstanding antral predominant nonatrophic gastritis. Gastric ulcer, however, is a sequel of multifocal atrophic gastritis. It does seem that *H. pylori* infection is the driving force in gastric carcinogenesis. But the outcome of the infection is modulated by the genetic susceptibility of the infected individual and the external environment. Figure 1.1 is a diagrammatic representation of a model of gastric carcinogenesis. Multiple forces interact and determine if the infection will lead to nonatrophic gastritis or multifocal atrophic gastritis. The latter is the first step in the precancerous cascade. A brief description of the main etiologic forces follows. They are described in more detail in other chapters.

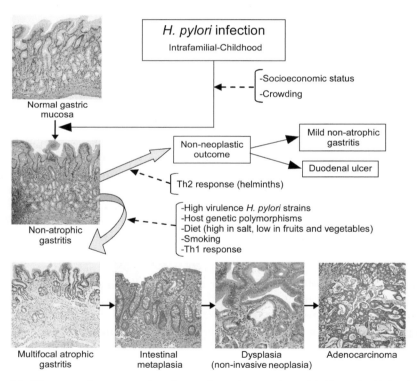

Fig. 1.1 Diagrammatic representation of a gastric carcinogenesis model. The main driving force is the infection with *Helicobacter pylori*. The interaction of multiple factors (human, bacterial, and environmental) determine the type of gastritis, either nonatrophic, which does not increase gastric cancer risk, or multifocal atrophic, which may represent the first step in the gastric carcinogenic sequence. (Reproduced and adapted from Correa et al. 2006, with permission.)

The Bacterial Agent

In 2005, the Nobel Prize in Medicine or Physiology was awarded to Robin Warren and Barry Marshall for their pioneer work linking *H. pylori* infection to peptic ulcers and gastric carcinoma. Spiral bacteria had previously been described in the gastric mucosa by Giulio Bizzozero (Bizzozero 1892) in Pavia one century before the classic letters to the editor of Lancet by Warren and Marshall (Warren and Marshall 1983). *H. pylori* colonizes the gastric mucosa and is mostly localized extracellularly (Figure 1.2). However, it has been recently shown that the bacteria may find their way to the cytoplasm of the gastric epithelial cells and the underlying lamina propria (Dubois and Boren 2007; Necchi et al. 2007; Semino-Mora et al. 2003). The capacity of *H. pylori* to induce gastric cancer has been linked to its virulence, largely determined by its cytotoxin CagA and vacuolating toxin VacA. *cagA* positive and *vacAs1m1* bacterial genotypes have been linked to severe gastritis, atrophy, intestinal metaplasia, and gastric cancer. *cagA*-negative, *vacAs2m2* genotypes are less virulent, induce a milder, nonatrophic type of gastritis, and do not always lead to a neoplastic outcome (Cittelly et al. 2002; Kato et al. 2006a; Quiroga et al. 2005; Tham et al. 2001). The CagA protein is injected into the cytoplasm of the epithelial cell by a type IV secretion system. Once injected, CagA is phosphorylated and starts a process leading to changes in the cellular morphology and the immunologic response to the infection (Peek and Crabtree 2006).

Host Susceptibility

Genetic susceptibility is another important determinant in the gastric carcinogenic process. Familial clustering is observed in approximately 10% of the cases and 1%–3% is hereditary. In the latter group, a well-characterized syndrome, the

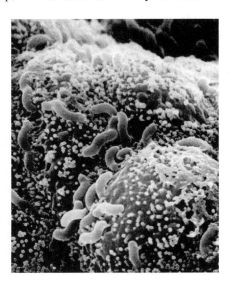

Fig. 1.2 Scanning electron microphotograph of gastric mucosa infected with *Helicobacter pylori*. The bacteria concentrate at the intercellular junctions. (Photograph courtesy of Dr. Francisco Hernandez, Electronic Microscopy Department, University of Costa Rica, Costa Rica.)

hereditary diffuse gastric cancer, is associated with germline mutations in the epithelial cadherin (E-cadherin) gene in approximately 30% of the families (Carneiro et al. 2007). This syndrome was initially described by Guilford and colleagues in 1998 in three New Zealand Maori families with early onset diffuse gastric cancer (Guilford et al. 1998). Since then, at least 151 families have been reported in different countries (Carneiro et al. 2007; Kaurah et al. 2007; Medina-Franco et al. 2007a). The autosomal dominant mutation with a penetrance of 70% leads to death from gastric cancer at an early age.

Besides the described syndrome, a large body of evidence demonstrates the influence of genetic variants in gastric cancer risk as part of the multifactorial etiology of the sporadic cases. The genetic variants that have received more attention are those located in genes encoding pro- and antiinflammatory cytokines involved in the response against *H. pylori* infection. Based on two recent reviews (Gonzalez et al. 2002; Hamajima et al. 2006) and some original articles (Camargo et al. 2006; Goto et al. 2006; Hou et al. 2007; Jin et al. 2007a, 2007b; Kaise et al. 2007; Kamangar et al. 2006; Kato et al. 2006a; Larsson et al. 2006; Lee et al. 2005; Liu et al. 2006; Lu et al. 2005a, 2005b; Medina-Franco et al. 2007b; Pinto-Correia et al. 2006; Savage et al. 2004; Sicinschi et al. 2006; Silva et al. 2001; Silva et al. 2003; Tang et al. 2007; Tatemichi et al. 2005; Wang et al. 2007a, 2007b; Zhou et al. 2007; Zintzaras 2006), a summary of the principal polymorphisms studied for association with diffuse and intestinal types of gastric cancer is presented in Table 1.2. The genetic variants are grouped according to the processes relevant to carcinogenesis in which their encoded proteins intervene. Biologic mechanisms that explain direct and inverse associations among those polymorphisms and gastric cancer risk need further investigation. Incorporation of gene–gene and gene–environment interactions into future association studies may reveal the complex etiology of this neoplasia. Design of ethnic-specific studies is recommended because of the documented differential distribution of genetic variants among ethnic groups (Hassan et al. 2003; Zabaleta et al. 2008). Some meta-analyses have suggested that risk markers are specific by ethnicity (Camargo et al. 2006; Zintzaras 2006). Genetic susceptibility is discussed in detail in Chapter 9.

External Environment

Several epidemiologic studies of gastric cancer conducted before the recognition of the role of *H. pylori* infection had identified several environmental risk factors. Two dietary modalities were consistently shown to influence cancer risk in diverse populations. The first one was based on the observation that in all populations at high cancer risk the consumption of salt was very high. The linear correlation between salt intake and gastric cancer risk was abundantly demonstrated (Joossens et al. 1996; Tsugane 2005). The second observation was provided mainly by case-control studies showing a protective effect of adequate intake of fresh fruits and vegetables. This effect was first reported by Hirayama in Japan (Hirayama 1986), and then corroborated in many different countries, as summarized by Block et al. (1992), and recently by others (Lunet et al. 2007; Tsugane and Sasazuki 2007).

Table 1.2 Polymorphisms in genes related to processes relevant to gastric carcinogenesis

Process	Gene and variant
Protection of gastric mucosa against damaging agents	*MUC1*
	MUC6
	TFF2
Inflammatory response	*IL1B (−31T > C, −511C > T, +3954C > T)*
	IL1RN*2
	IL10 (−1082G > A, −819C > T)
	IL8 (−51T > A, +396T > G, +781C > T)
	IL8RB (+785T > C, +1208C > T, +1440G > A)
	IL4-590 C > T
	TNF (−308G > A, −238G > A)
	COX-2 (PTG52-1195G > A, -765G > C)
	iNOS (NOS2A 150C > T, -2445C > G)
	CD14-260C > T
	IFNGR2 Ex7-128 C > T
	HLA class I gene
	HLA class II gene
Ability to detoxify carcinogens (Phase I and II enzymes)	*CYP2E-1053C > T (RsaI)*
	GSTM1 null
	GSTT1 null
	GSTM3*B (IVS6del3)
	GSTP1 1578A > G (l105V)
	NAT1*10 (1088T > A, 1095C > A)
	NAT2
Protection against oxidative damage and other inductors of DNA damage	Oxidative damage:
	MTHFR (677C > T and 1298A > C)
	MTHFD (1958G > A and 401T > C)
	DNA repair:
	XRCC1 26304C > T (R194W) 28152G > A (R399Q)
	OGG1 S326C
	Oncogenes and tumor suppressor genes:
	MYCL1 (previous symbol: L-myc) EcoRI
	TP53 codon 72
Cell proliferation ability/cell differentiation/ cell homeostasis	*Cyclin D1 870A > G*
	PGC
	CDH1-160C > A
	EGF promoter *(61A > G, −1380A > G, −1744G > A)*
	TGFB1-509C > T
	TGFBR2-875A > G
	JWA-76G > C

Abbreviations: TFF: trefoil peptides; TNF: tumor necrosis factor; COX2: cyclooxygenase 2; iNOS: inducible nitric oxide synthase; IFNGR2: interferon gamma receptor 2; GST: glutathione S-transferase; NAT: N-acetyl transferase; MTHFR: 5,10-methylenetetrahydrofolate reductase; MTHFD: methylenetetrahydrofolate dehydrogenase; XRCC1: x-ray repair cross complementing group 1; OGG1: 7,8-dihydro-8-oxoguanine-DNA glycosylase; TP53: tumor protein 53; PGC: pepsinogen C; CDH1: E-cadherin; EGF: epidermal growth factor; TGFB: transforming growth factor beta; TGFBR2: TGFB receptor II.

Although socioeconomic status is not a causal risk factor, it surrogates other varia-bles that have been inversely associated with gastric cancer risk, such as sanitary con-ditions and nutrition. Socioeconomic status determines the age and the prevalence of *H. pylori* infection. Previous studies have suggested that the high prevalence observed in developing countries is related to family crowding during childhood and the appar-ent transmission from older to younger siblings (Goodman and Correa 2000). Another external environmental factor is smoking (IARC 2002; Nishino et al. 2006). Tobacco contains a wide variety of carcinogens, including N-nitroso compounds, which have been associated with gastric carcinogenesis (Mirvish 1995). Finally, comorbid condi-tions may also determine gastric cancer risk. It is known that concomitant infection with helminths can bias the common Th1-type immune response against *H. pylori* to a less deleterious Th2-type response and decrease the development of gastric atrophy (Whary and Fox 2004; Whary et al. 2005). These findings suggest the possible role of helminths in the prevention of cancer development.

Models of Gastric Carcinogenesis

It is not clear by which mechanisms the *H. pylori* infection drives the genetic altera-tions that ultimately replace the normal gastric epithelium with a neoplastic one. Ample evidence supports the notion that oxidative stress induces cellular changes that may start and maintain the carcinogenesis process. The inflammatory cells attracted to the gastric mucosa by the infection results in the expression of inducible nitric oxide synthase (iNOS) (Mannick et al. 1996), which can lead to high concentrations of nitric oxides, especially in the lipid-containing cellular membranes (Liu et al. 1998). iNOS is also present in the cytoplasm of metaplastic, dysplastic, and epithelial cancer cells (Piazuelo et al. 2004; Pignatelli et al. 2001). Accumulation of reactive oxygen and nitrogen species can lead to oxidation, nitration, and nitrosation of molecules leading to DNA and protein alterations. The mutagenic effects of nitric oxides apparently are counteracted by DNA-protecting enzymes, also found in the inflammatory infiltrate, such as superoxide dismutase and catalase (Pignatelli et al. 1998). Subjects infected with *cagA*-positive, *vacAs1m1*, and *iceA1 H. pylori* genotype showed higher levels of oxidative DNA damage than uninfected patients or patients infected with *cagA*-negative, *vacAs2 m2*, and *iceA2 H. pylori* genotype (Ladeira et al. 2007).

Several models of carcinogenesis have been devised. One of them proposes that the gastric precancerous cascade represents a sequential accumulation of stable mutations and other molecular abnormalities similar to the Vogelstein model of colon carcinogenesis (Fearon and Vogelstein 1990; Smith et al. 2006; Yasui et al. 2006). In support of this model, multiple reports have described alterations in pro-tooncogenes, tumor suppressor genes, cell cycle regulators, and growth factors. The multitude of reported abnormalities and the lack of convincing evidence of their sequential accumulation cast some doubts on the relevance of such a model for gastric carcinogenesis. Another weakness of this model is that the described abnor-malities do not account for the invasive capacity of the affected epithelial cells.

A second proposed model, called the dedifferentiation/redifferentiation model, does account for the invasive capacity of cancer cells. That activity is linked to the ability of some agents to transform epithelial cells into mesenchymal cells. Epithelial–mesenchymal transition (EMT), a normal process in embryogenesis, has been well studied in the sea urchin and in colorectal cancer (Kirchner and Brabletz 2000). This process is driven by the β-catenin/Wnt signaling pathway. In the absence of Wnt signaling, intracellular levels of β-catenin are regulated by a multi-protein complex that promotes its degradation. On the contrary, in the presence of Wnt ligand, coactivation of receptors leads to inhibition of the degradation complex and the intracellular accumulation of β-catenin that results in nuclear translocation. In the nucleus, β-catenin binds to the TCF/LEF family of transcription factors, modulating expression of a broad range of target genes. In human carcinomas, the better differentiated glands of the tumor display β-catenin in the apical cytoplasm and cell membrane, indicating its involvement in the cell–cell adhesion. In the invasive edge, the tumor cells lose their epithelial phenotype and gain a dedifferentiated, mesenchyme-like phenotype, with nuclear β-catenin accumulation. The main consequence of EMT is the acquisition of motility and invasiveness. The observation that metastases of many carcinomas recapitulate the epithelial differentiation and morphology of the primary tumor indicates a redifferentiation toward the epithelial phenotype of the primary tumor (Brabletz et al. 2002; Brabletz et al. 2005). In gastric carcinogenesis, the oncogenic rodent-adapted 7.13 strain of *H. pylori* activates β-catenin in epithelial cells, a process dependent on translocation of CagA into host epithelial cells. Nuclear β-catenin is increased in gastric epithelium from gerbils infected with the *H. pylori* carcinogenic strain as well as from persons carrying *cagA*-positive versus *cagA*-negative strains or uninfected persons. These results indicate that activation of β-catenin by *H. pylori* may be an important early event that precedes malignant transformation (Franco et al. 2005).

As discussed in detail in another chapter, there is new evidence supporting the origin of the gastric neoplastic and preneoplastic cells from bone marrow–derived cells (BMDCs) homing to the gastric mucosa as a result of *H. pylori* infection (Houghton et al. 2004; Li et al. 2006). A hallmark of BMDCs is the absence of β-catenin in cell junction complexes and its nuclear localization. From this point of view, it would seem that the BMDC model and the Wnt/β-catenin-driven model of dedifferentiation/redifferentiation are complementary.

Pathology

Approximately 90% of gastric cancers are adenocarcinomas, which are divided into two main histologic types (intestinal and diffuse) according to the classification by the Finnish pathologists Jarvi and Lauren (Jarvi and Lauren 1951; Lauren 1965). There are major differences in their histopathology, epidemiology, and molecular pathogenesis. Intestinal-type adenocarcinomas are the most common tumors in populations at high gastric cancer risk, occur most frequently in elderly men, and are associated with relatively better survival. Diffuse-type adenocarcinomas are relatively more frequent in women and individuals under age 50, and usually have a less favorable prognosis. The intestinal type has been declining slowly in incidence for several decades to the point that in low-risk populations its incidence is similar to that of diffuse-type carcinomas. Early gastric cancer is defined as adenocarcinoma confined to the mucosa or submucosa, irrespective of the presence of regional lymph node metastases. In Japan, early gastric cancer accounts for approximately 50% of all gastric cancers, whereas in Western countries the proportion is approximately 5%–10% (Sano and Hollowood 2006).

Most gastric adenocarcinomas arise in the antrum and the lesser curvature. During the last decades, the proportion of carcinomas involving the cardia has increased (Blot et al. 1991; Devesa et al. 1998; Powell and McConkey 1992), a phenomenon called "cranialization-trend in gastric cancer location." Cardia adenocarcinomas share epidemiologic and pathologic characteristics with esophageal adenocarcinomas, such as predominance in white males, presence of hiatal hernia, and lack of association with *H. pylori* infection. It is not always possible to differentiate esophageal from cardia adenocarcinomas arising in the gastroesophageal region. Several topographic classifications have been proposed (Hamilton and Aaltonen 2000; Siewert and Stein 1996, 1998).

Intestinal-Type Adenocarcinomas

Macroscopically, intestinal-type adenocarcinomas can display several patterns: polypoid, fungating, ulcerated, and infiltrative. Figure 1.3 shows a gastrectomy specimen with an ulcerated tumor in the area of the incisura angularis. The specimen was stained for alkaline phosphatase, shown in red. This digestive enzyme is normally present in the intestine, as seen in the duodenal portion of the specimen. It is abnormally present in the stomach as a result of intestinal metaplasia, as seen in the photograph surrounding the ulcerated carcinoma. The hallmark of intestinal-type gastric carcinomas is the fact that the neoplastic cells are connected to each other and are able to form tubules and glands, in the same way seen in the normal intestinal mucosa. This cohesion preserves the cell polarity and is attributed to the well-developed adhesion complexes that bind neighboring cells. These protein complexes contain β-catenin and E-cadherin. Figure 1.4A shows a histologic section from an intestinal-type adenocarcinoma immunostained for E-cadherin. Well-formed tubular structures show strong staining for E-cadherin in the cellular membranes.

Diffuse-Type Adenocarcinomas

Grossly, diffuse-type adenocarcinomas are characterized by tumor infiltration and thickening of the gastric wall, expanding with time and eventually transforming the

Fig. 1.3 Photograph of a gastrectomy specimen opened along the greater curvature and stained for alkaline phosphatase. A large ulcerated carcinoma occupies the lesser middle curvature and is surrounded by mucosa with intestinal metaplasia, stained in red. Note similar staining at the distal end of the specimen, corresponding to duodenum. (Reproduced from Correa and Stemmermann 2005, with permission.) (*See Color Plates*)

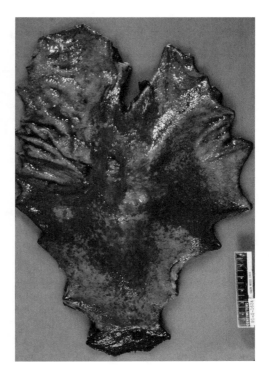

stomach into a rigid tube, which has been called "linitis plastica." Microscopically, the neoplastic cells are usually round and individually invade the gastric wall. Nuclei are frequently compressed against the cell membrane by the abundant intracytoplasmic mucin, resulting in the characteristic signet-ring appearance. The pattern of invasion of the diffuse-type adenocarcinoma results from the absence of cohesion among tumor cells because of lack of intercellular adhesion molecules. E-cadherin immunostaining is absent in the membranes of tumor cells and present in the original gastric glands shown in the periphery of the field in Figure 1.4B. The infiltrative growth pattern results in a desmoplastic reaction, and occasionally invades the neighboring structures of the esophagus and the duodenum.

Precancerous Lesions

Intestinal-type carcinomas are preceded by a series of lesions appearing sequentially as the carcinogenic process advances. The stages of this precancerous cascade are well defined: normal → chronic active nonatrophic gastritis → multifocal atrophic gastritis → intestinal metaplasia (complete, then incomplete) → dysplasia → invasive

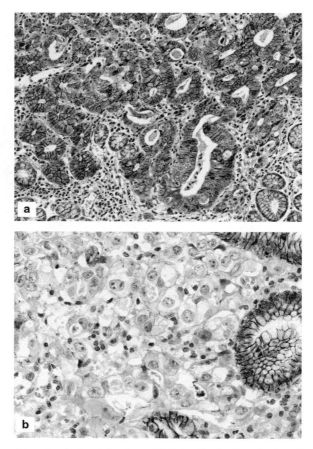

Fig. 1.4 Immunohistochemical staining for epithelial cadherin (E-cadherin). (**A**) Intestinal-type gastric adenocarcinoma shows E-cadherin localized in the cellular membranes. (**B**) Diffuse-type adenocarcinoma lacks expression of E-cadherin. The normal gastric glands seen at the right side of the figure show normal E-cadherin staining

carcinoma (Correa et al. 1975). Chronic, active nonatrophic gastritis displays interstitial infiltration of the gastric mucosa by chronic inflammatory cells: lymphocytes, plasma cells, and macrophages. The "active" designation refers to the presence of polymorphonuclear neutrophils, often seen intraepithelially, and predominantly affecting the gland necks. Up to this point, the lesions do not necessarily go into the precancerous cascade with an eventual neoplastic outcome. As mentioned, patients with duodenal ulcer generally have nonatrophic antral gastritis, but are not at increased gastric cancer risk. The next stage in the cascade is characterized by loss of the original gastric glands (atrophy) in a multifocal pattern. Atrophic glands may be replaced by glands with intestinal phenotype. These cells may have the full set of digestive enzymes such as alkaline phosphatase, sucrase, and trehalase. Figure 1.5

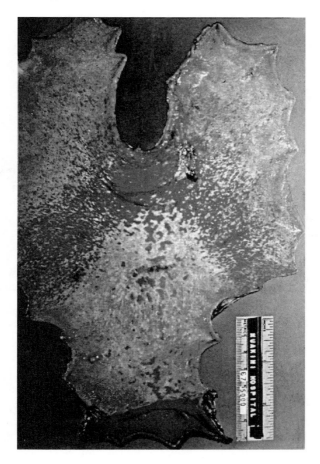

Fig. 1.5 Photograph of a gastrectomy specimen opened along the greater curvature and stained for alkaline phosphatase. Foci of intestinal metaplasia (stained in red) coalesce at the antrum–corpus junction, especially in the incisura angularis. (Reproduced from Correa and Stemmermann 2005, with permission.) (*See Color Plates*)

is a gross photograph of a gastrectomy specimen stained red for alkaline phosphatase. As can be seen, the foci of intestinal metaplasia are more concentrated in the antrum–corpus junction, especially in the incisura angularis, where they coalesce. The complete or small intestinal phenotype is characterized by absorptive enterocytes with a prominent brush border alternating with well-developed mucous goblet cells and sometimes Paneth cells (Figure 1.6A). As the process advances, the intestinalized cells change their phenotype to the incomplete type that resembles colonic mucosa, lined only by irregular goblet cells of various sizes, without a brush border (Figure 1.6B). In the next step of the precancerous process, the mature phenotype of the metaplastic cells is lost and the glands become lined by epithelium

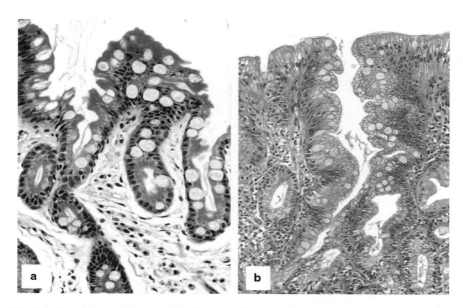

Fig. 1.6 Gastric intestinal metaplasia (hematoxylin and eosin stain). **(A)** Complete-type metaplasia consisting of eosinophilic absorptive enterocytes (brush border) and well-formed goblet cells at regular intervals. **(B)** Incomplete-type metaplasia consisting of goblet cells with mucus vacuoles of varying sizes and absence of brush border

with neoplastic phenotype, characterized by cellular crowding, large, hyperchromatic nuclei, and increase in the mitotic activity. As long as the neoplastic cells remain bound within the original epithelial compartment and do not trespass the basement membrane, they are classified as dysplasia. Because of increasing evidence that dysplastic lesions may show molecular alterations similar to those of invasive adenocarcinomas, and the high risk of progression of dysplastic lesions into invasive cancer, the term noninvasive neoplasia (synonym: intraepithelial neoplasia) has been recommended (Rugge et al. 2000; Rugge et al. 2005). The same lesion is called "adenoma" in Japanese literature. In low-grade dysplasia, the nuclei are generally basally located and conserve their polarity. Figure 1.7 illustrates glands with low-grade dysplasia, extending to the surface epithelium. High-grade dysplasia shows severely distorted glandular architecture and glandular crowding, marked nuclear polymorphism and stratification with nuclei reaching the luminal surface, and loss of cell polarity. Mitoses are numerous and occasional abnormal mitotic figures may be observed.

A different type of metaplasia has been emphasized recently as described in detail by Goldenring and Nomura in Chapter 13. It has long been recognized that in humans the oxyntic mucosa may be partially replaced by glands with antral phenotype, so-called "antralization of the gastric corpus" or "pseudopyloric metaplasia" (Tarpila et al. 1969). In Japanese patients, this type of metaplasia has

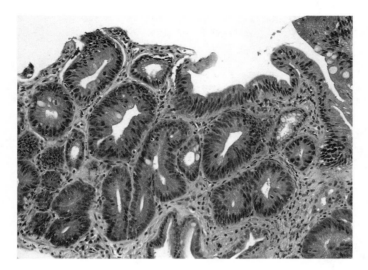

Fig. 1.7 Low-grade gastric dysplasia (noninvasive neoplasia). Irregular glands with large, hyperchromatic, elongated nuclei and frequent mitosis. The nuclei tend to preserve their basal arrangement. The dysplastic cells extend to the surface epithelium

been observed surrounding early carcinomas and adenocarcinomas of the gastric stump (remnant) (El-Zimaity et al. 2002; Yamaguchi et al. 2002). In the latter group, spasmolytic polypeptide-expressing metaplasia was identified in the metaplastic mucosa surrounding the remnant cancers. Spasmolytic polypeptide-expressing metaplasia is the predominant cancer precursor in experimental rodent models (Goldenring and Nomura 2006).

The precancerous cascade does not precede most diffuse-type carcinomas. No clear precancerous lesions have been reported for sporadic diffuse carcinomas. Prophylactic gastrectomies after chromoendoscopic surveillance has revealed multiple small signet-ring cell carcinomas distributed throughout the gastric mucosa (Norton et al. 2007). In one gastrectomy specimen, illustrated in Figure 1.8, 318 independent microscopic foci of signet-ring cell carcinomas were found, as shown by black dots (Charlton et al. 2004). These microscopic foci show independent neoplastic signet-ring cells extending beneath the foveolar or glandular epithelium in the gastric mucosa in a "pagetoid" manner (Figure 1.9) (Carneiro and Sobrinho-Simoes 2005), before they invade the lamina propria.

Cancer Control

The prognosis of gastric cancer once it has invaded the muscularis propria is dismal. In the United States, approximately two-thirds of the cases are diagnosed at that stage and the overall 5-year survival rate is 24% (Jemal et al. 2007). In Japan,

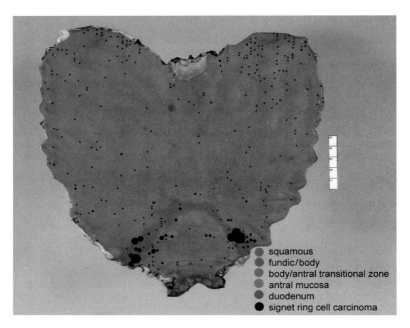

squamous
fundic/body
body/antral transitional zone
antral mucosa
duodenum
signet ring cell carcinoma

Fig. 1.8 Gastrectomy specimen from a 15-year-old Maori adolescent girl with hereditary diffuse adenocarcinoma. Every focus of in situ carcinoma identified grossly is depicted by a black dot. A total of 318 in situ carcinomas were identified in this specimen. Tumors around the antrum–corpus junction were larger. (Photograph courtesy of Dr. Amanda Charlton, Department of Pathology, Middlemore Hospital, Auckland, New Zealand. Reproduced from Charlton et al. 2004, with permission.) (*See Color Plates*)

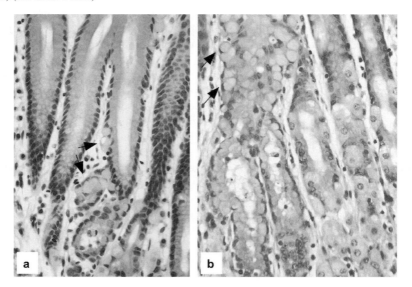

Fig. 1.9 Microphotograph of gastric mucosa from a Portuguese patient with hereditary diffuse adenocarcinoma. (**A**) Neoplastic goblet cells are seen at the base of the foveolar and glandular epithelium. (**B**) Oxyntic mucosa showing pagetoid spread of malignant neoplastic goblet cells (arrows). (Photograph courtesy of Dr. Fatima Carneiro, Institute of Molecular Pathology and Immunology of the University of Porto, Portugal.)

a country with one of the highest gastric cancer incidence rates, the search for "early" cancers is intense. A mass screening program has been conducted since 1960. The category "early gastric cancer" includes primary tumors present in the mucosa or submucosa but not the muscularis propria. Occasionally, they metastasize to the adjacent lymph nodes. Early carcinomas are being discovered with increasing frequency in Japan, in some series reaching 40%–60% (Oda et al. 2006). The 5-year survival rate for early cancer is more than 90% (Tan and Fielding 2006). The Japanese model of early detection is difficult to replicate in other populations. In populations at low risk, screening the general population is not cost effective. In developing countries at high gastric cancer risk, economic and logistic circumstances prevent the adoption of such a practice. Therefore, an attractive proposal to reduce the incidence and mortality of gastric cancer is to identify subjects at high risk, who may benefit from targeted screening, and other prophylactic measures to prevent cancer development.

There is now evidence that polymorphic variation in a number of genes may help to identify susceptible subjects, those carriers of genetic risk-associated alleles and infected with virulent strains of *H. pylori*. A study in Portugal reported that subjects carrying the polymorphism *IL1B-511T* and infected with *cagA*-positive bacteria have a relative risk 25 times greater than those without the T allele infected with *cagA*-negative bacteria. For subjects carrying the polymorphism *IL1B-511T* infected with *H. pylori vacAs1*, the risk is 87 times greater than those carrying less virulent genotypes (*vacAs2*) and *IL1B-511CC* (Figueiredo et al. 2002). It is now possible to diagnose *H. pylori*, cagA and vacA status by serologic techniques (Blaser et al. 1995; Camorlinga-Ponce et al. 1998; Ghose et al. 2007). Presence of several risk-associated alleles may facilitate the identification of subjects at high risk.

The prolonged precancerous process offers opportunities for intervention to prevent its progression. Several chemoprevention trials have reported that *H. pylori* eradication prevents progression of the precancerous lesions (Fuccio et al. 2007). Current evidence suggests that curing the infection early in the process, before atrophy and metaplasia develop, prevents gastric cancer (Wong et al. 2004). Additional evidence from a large-scale retrospective multicentric study in Japan, conducted in patients who underwent upper gastrointestinal endoscopy, showed that *H. pylori* eradication decreased gastric cancer incidence (Kato et al. 2006b). Because in most high-risk countries more than 70% of the subjects are infected, massive anti-*H. pylori* therapy is unrealistic and not suitable.

In summary, prevention must be the strategy to control gastric cancer. Sound prevention measures differ by population. Special attention should be given to tumors with familial aggregation and high-risk pockets in low gastric cancer risk countries. Identifying individuals at the highest risk in the general population can be achieved by testing blood samples for genetic susceptibility markers, as well as infection with highly virulent strains of *H. pylori*. Identified subjects must enter surveillance programs based on endoscopic monitoring aimed to detect and manage dysplasias and early carcinomas.

Several public health measures to reduce gastric cancer risk should include treatment and eradication of *H. pylori* gastritis, reduce salt intake, and promote adequate consumption of fresh fruits and vegetables. It should be expected that the secular trend of decreasing gastric cancer rates will continue. Such steady decrease could be accelerated by adapting the above-mentioned prevention strategies.

References

Bizzozero, G. (1892). Sulle ghiandole tubulari del tubo gastro-enterico e sui rapporti del loro epitelio coll'epitelio di rivestimento della mucosa. Atti della Reale Accademia delle Scienze di Torino. 28:233–51.

Blaser, M.J., Perez-Perez, G.I., Kleanthous, H., Cover, T.L., Peek, R.M., Chyou, P.H., Stemmermann, G.N., and Nomura, A. (1995). Infection with Helicobacter pylori strains possessing cagA is associated with an increased risk of developing adenocarcinoma of the stomach. Cancer Res. 55(10):2111–5.

Block, G., Patterson, B., and Subar, A. (1992). Fruit, vegetables, and cancer prevention: a review of the epidemiological evidence. Nutr Cancer. 18(1):1–29.

Blot, W.J., Devesa, S.S., Kneller, R.W., and Fraumeni, J.F., Jr. (1991). Rising incidence of adenocarcinoma of the esophagus and gastric cardia. JAMA. 265(10):1287–9.

Brabletz, T., Jung, A., and Kirchner, T. (2002). Beta-catenin and the morphogenesis of colorectal cancer. Virchows Arch. 441(1):1–11.

Brabletz, T., Jung, A., Spaderna, S., Hlubek, F., and Kirchner, T. (2005). Opinion: migrating cancer stem cells—an integrated concept of malignant tumour progression. Nat Rev Cancer. 5(9):744–9.

Camargo, M.C., Mera, R., Correa, P., Peek, R.M., Jr, Fontham, E.T., Goodman, K.J., Piazuelo, M.B., Sicinschi, L., Zabaleta, J., and Schneider, B.G. (2006). Interleukin-1beta and interleukin-1 receptor antagonist gene polymorphisms and gastric cancer: a meta-analysis. Cancer Epidemiol Biomarkers Prev. 15(9):1674–87.

Camorlinga-Ponce, M., Torres, J., Perez-Perez, G., Leal-Herrera, Y., Gonzalez-Ortiz, B., Madrazo de la Garza, A., Gomez, A., and Munoz, O. (1998). Validation of a serologic test for the diagnosis of Helicobacter pylori infection and the immune response to urease and CagA in children. Am J Gastroenterol. 93(8):1264–70.

Carneiro, F., Oliveira, C., Suriano, G., and Seruca, R. (2007). Molecular pathology of familial gastric cancer. J Clin Pathol. May 18 [Epub ahead of print].

Carneiro, F., and Sobrinho-Simoes, M. (2005). Hereditary diffuse gastric cancer: lessons from histopatholgy. Adv Anat Pathol. 12(3):151–2.

Charlton, A., Blair, V., Shaw, D., Parry, S., Guilford, P., and Martin, I.G. (2004). Hereditary diffuse gastric cancer: predominance of multiple foci of signet ring cell carcinoma in distal stomach and transitional zone. Gut. 53(6):814–20.

Cittelly, D.M., Huertas, M.G., Martinez, J.D., Oliveros, R., Posso, H., Bravo, M.M., and Orozco, O. (2002). Helicobacter pylori genotypes in non atrophic gastritis are different of the found in peptic ulcer, premalignant lesions and gastric cancer in Colombia. Rev Med Chil. 130(2):143–51. Article in Spanish.

Correa, P., Haenszel, W., Cuello, C., Tannenbaum, S., and Archer, M. (1975). A model for gastric cancer epidemiology. Lancet. 2(7924):58–60.

Correa, P., Piazuelo, M.B., and Camargo, M.C. (2006). Etiopathogenesis of gastric cancer. Scand J Surg. 95(4):218–24.

Correa, P., and Stemmermann, G.N. (2005). The saga of gastric corpus atrophy. Pathologica. 97(5):319–22.

Devesa, S.S., Blot, W.J., and Fraumeni, J.F., Jr. (1998). Changing patterns in the incidence of esophageal and gastric carcinoma in the United States. Cancer. 83(10):2049–53.

Dubois, A., and Boren, T. (2007). Helicobacter pylori is invasive and it may be a facultative intracellular organism. Cell Microbiol. 9(5):1108–16.

El-Zimaity, H.M., Ota, H., Graham, D.Y., Akamatsu, T., and Katsuyama, T. (2002). Patterns of gastric atrophy in intestinal type gastric carcinoma. Cancer. 94(5):1428–36.

Fearon, E.R., and Vogelstein, B. (1990). A genetic model for colorectal tumorigenesis. Cell. 61(5):759–67.

Ferlay, J., Bray, F., Pisani, P., and Parkin, D.M. (2004). GLOBOCAN 2002: cancer incidence, mortality and prevalence worldwide. IARC Cancer Base No. 5, version 2.0. Lyon: IARC*Press*.

Figueiredo, C., Machado, J.C., Pharoah, P., Seruca, R., Sousa, S., Carvalho, R., Capelinha, A.F., Quint, W., Caldas, C., van Doorn, L.J., Carneiro, F., and Sobrinho-Simoes, M. (2002). Helicobacter pylori and interleukin-1 genotyping: an opportunity to identify high-risk individuals for gastric carcinoma. J Natl Cancer Inst. 94(22):1680–7.

Forman, D., Newell, D.G., Fullerton, F., Yarnell, J.W., Stacey, A.R., Wald, N., and Sitas, F. (1991). Association between infection with Helicobacter pylori and risk of gastric cancer: evidence from a prospective investigation. BMJ. 302(6788):1302–5.

Franco, A.T., Israel, D.A., Washington, M.K., Krishna, U., and Fox, J.G., Rogers, A.B., Neish, A.S., Collier-Hyams, L., Perez-Perez, G.I., Hatakeyama, M., Whitehead, R., Gaus, K., O'Brien, D.P., Romero-Gallo, J., and Peek, R.M., Jr. (2005). Activation of beta-catenin by carcinogenic Helicobacter pylori. Proc Natl Acad Sci USA. 102(30):10646–51.

Fuccio, L., Zagari, R.M., Minardi, M.E., and Bazzoli, F. (2007). Systematic review: Helicobacter pylori eradication for the prevention of gastric cancer. Aliment Pharmacol Ther. 25(2):133–41.

Fukuda, H., Saito, D., Hayashi, S., Hisai, H., Ono, H., Yoshida, S., Oguro, Y., Noda, T., Sato, T., and Katoh, M. (1995). Helicobacter pylori infection, serum pepsinogen level and gastric cancer: a case-control study in Japan. Jpn J Cancer Res. 86(1):64–71.

Ghose, C., Perez-Perez, G.I., Torres, V.J., Crosatti, M., Nomura, A., Peek, R.M., Jr., Cover, T.L., Francois, F., and Blaser, M.J. (2007). Serological assays for identification of human gastric colonization by Helicobacter pylori strains expressing VacA m1 or m2. Clin Vaccine Immunol. 14(4):442–50.

Goldenring, J.R., and Nomura, S. (2006). Differentiation of the gastric mucosa III. Animal models of oxyntic atrophy and metaplasia. Am J Physiol Gastrointest Liver Physiol. 291(6):G999–1004.

Gonzalez, C.A., Sala, N., and Capella, G. (2002). Genetic susceptibility and gastric cancer risk. Int J Cancer. 100(3):249–60.

Goodman, K.J., and Correa, P. (2000). Transmission of Helicobacter pylori among siblings. Lancet. 355(9201):358–62.

Goto, Y., Ando, T., Naito, M., Goto, H., and Hamajima, N. (2006). Inducible nitric oxide synthase polymorphism is associated with the increased risk of differentiated gastric cancer in a Japanese population. World J Gastroenterol. 12(39):6361–5.

Guilford, P., Hopkins, J., Harraway, J., McLeod, M., McLeod, N., Harawira, P., Taite, H., Scoular, R., Miller, A., and Reeve, A.E. (1998). E-cadherin germline mutations in familial gastric cancer. Nature. 392(6674):402–5.

Haenszel, W., and Kurihara, M. (1968). Studies of Japanese migrants. I. Mortality from cancer and other diseases among Japanese in the United States. J Natl Cancer Inst. 40(1):43–68.

Hamajima, N., Naito, M., Kondo, T., and Goto, Y. (2006). Genetic factors involved in the development of Helicobacter pylori-related gastric cancer. Cancer Sci. 97(11):1129–38.

Hamilton, S.R., Aaltonen, L.A., editors. (2000). WHO classification of tumours. Pathology and genetics of tumours of the digestive system. Lyon: IARC*Press*.

Hansson, L.E., Nyren, O., Hsing, A.W., Bergstrom, R., Josefsson, S., Chow, W.H., Fraumeni, J.F., Jr., and Adami, H.O. (1996). The risk of stomach cancer in patients with gastric or duodenal ulcer disease. N Engl J Med. 335(4):242–9.

Hassan, M.I., Aschner, Y., Manning, C.H., Xu, J., and Aschner, J.L. (2003). Racial differences in selected cytokine allelic and genotypic frequencies among healthy, pregnant women in North Carolina. Cytokine. 21(1):10–6.

Hirayama, T. (1986). Nutrition and cancer: a large scale cohort study. Prog Clin Biol Res. 206:299–311.

Hou, L., El-Omar, E.M., Chen, J., Grillo, P., Rabkin, C.S., Baccarelli, A., Yeager, M., Chanock, S.J., Zatonski, W., Sobin, L.H., Lissowska, J., Fraumeni, J.F., Jr., and Chow, W.H. (2007). Polymorphisms in Th1-type cell-mediated response genes and risk of gastric cancer. Carcinogenesis. 28(1):118–23.

Houghton, J., Stoicov, C., Nomura, S., Rogers, A.B., Carlson, J., Li, H., Cai, X., Fox, J.G., Goldenring, J.R., and Wang, T.C. (2004). Gastric cancer originating from bone marrow-derived cells. Science. 306(5701):1568–71.

IARC. (1994). Monographs on the evaluation of carcinogenic risks to humans. Schistosomes, liver flukes and Helicobacter pylori. Vol. 61. Lyon: International Agency for Research on Cancer; pp. 177–240.

IARC. (2002). Monographs on the evaluation of carcinogenic risk to humans: tobacco smoke and involuntary smoking. Vol. 83. Lyon: International Agency for Research on Cancer.

Jarvi, O., and Lauren, P. (1951). On the role of heterotopias of the intestinal epithelium in the pathogenesis of gastric cancer. Acta Pathol Microbiol Scand. 29:26–44.

Jemal, A., Siegel, R., Ward, E., Murray, T., Xu, J., and Thun, M.J. (2007). Cancer statistics, 2007. CA Cancer J Clin. 57(1):43–66.

Jin, G., Miao, R., Deng, Y., Hu, Z., Zhou, Y., Tan, Y., Wang, J., Hua, Z., Ding, W., Wang, L., Chen, W., Shen, J., Wang, X., Xu, Y., and Shen, H. (2007a). Variant genotypes and haplotypes of the epidermal growth factor gene promoter are associated with a decreased risk of gastric cancer in a high-risk Chinese population. Cancer Sci. 98(6):864–8.

Jin, G., Wang, L., Chen, W., Hu, Z., Zhou, Y., Tan, Y., Wang, J., Hua, Z., Ding, W., Shen, J., Zhang, Z., Wang, X., Xu, Y., and Shen, H. (2007b). Variant alleles of TGFB1 and TGFBR2 are associated with a decreased risk of gastric cancer in a Chinese population. Int J Cancer. 120(6):1330–5.

Joossens, J.V., Hill, M.J., Elliott, P., Stamler, R., Lesaffre, E., Dyer, A., Nichols, R., and Kesteloot, H. (1996). Dietary salt, nitrate and stomach cancer mortality in 24 countries. European Cancer Prevention (ECP) and the INTERSALT Cooperative Research Group. Int J Epidemiol. 25(3):494–504.

Kaise, M., Miwa, J., Suzuki, N., Mishiro, S., Ohta, Y., Yamasaki, T., and Tajiri, H. (2007). Inducible nitric oxide synthase gene promoter polymorphism is associated with increased gastric mRNA expression of inducible nitric oxide synthase and increased risk of gastric carcinoma. Eur J Gastroenterol Hepatol. 19(2):139–45.

Kamangar, F., Cheng, C., Abnet, C.C., and Rabkin, C.S. (2006). Interleukin-1B polymorphisms and gastric cancer risk: a meta-analysis. Cancer Epidemiol Biomarkers Prev. 15(10):1920–8.

Kato, I., van Doorn, L.J., Canzian, F., Plummer, M., Franceschi, S., Vivas, J., Lopez, G., Lu, Y., Gioia-Patricola, L., Severson, R.K., Schwartz, A.G., and Munoz, N. (2006a). Host-bacterial interaction in the development of gastric precancerous lesions in a high risk population for gastric cancer in Venezuela. Int J Cancer. 119(7):1666–71.

Kato, M., Asaka, M., Nakamura, T., Azuma, T., Tomita, E., Kamoshida, T., Sato, K., Inaba, T., Shirasaka, D., Okamoto, S., Takahashi, S., Terao, S., Suwaki, K., Isomoto, H., Yamagata, H., Nomura, H., Yagi, K., Sone, Y., Urabe, T., Akamatsu, T., Ohara, S., Takagi, A., Miwa, J., and Inatsuchi, S. (2006b). Helicobacter pylori eradication prevents the development of gastric cancer: results of a long-term retrospective study in Japan. Aliment Pharmacol Ther. 24(Suppl 4):203–6.

Kaurah, P., MacMillan, A., Boyd, N., Senz, J., De Luca, A., Chun, N., Suriano, G., Zaor, S., Van Manen, L., Gilpin, C., Nikkel, S., Connolly-Wilson, M., Weissman, S., Rubinstein, W.S., Sebold, C., Greenstein, R., Stroop, J., Yim, D., Panzini, B., McKinnon, W., Greenblatt, M., Wirtzfeld, D., Fontaine, D., Coit, D., Yoon, S., Chung, D., Lauwers, G., Pizzuti, A., Vaccaro, C., Redal, M.A., Oliveira, C., Tischkowitz, M., Olschwang, S., Gallinger, S., Lynch, H., Green, J., Ford, J., Pharoah, P., Fernandez, B., and Huntsman, D. (2007). Founder and recurrent CDH1 mutations in families with hereditary diffuse gastric cancer. JAMA. 297(21):2360–72.

Kirchner, T., and Brabletz, T. (2000). Patterning and nuclear beta-catenin expression in the colonic adenoma-carcinoma sequence. Analogies with embryonic gastrulation. Am J Pathol. 157(4):1113–21.

Kolonel, L.N., Nomura, A.M., Hirohata, T., Hankin, J.H., and Hinds, M.W. (1981). Association of diet and place of birth with stomach cancer incidence in Hawaii Japanese and Caucasians. Am J Clin Nutr. 34(11):2478–85.

Ladeira, M.S., Bueno, R.C., Dos Santos, B.F., Pinto, C.L., Prado, R.P., Silveira, M.G., Rodrigues, M.A., Bartchewsky, W., Jr., Pedrazzoli, J., Jr., Ribeiro, M.L., and Salvadori, D.M. (2007). Relationship among oxidative DNA damage, gastric mucosal density and the relevance of cagA, vacA and iceA genotypes of Helicobacter pylori. Dig Dis Sci. May 23 [Epub ahead of print].

Larsson, S.C., Giovannucci, E., and Wolk, A. (2006). Folate intake, MTHFR polymorphisms, and risk of esophageal, gastric, and pancreatic cancer: a meta-analysis. Gastroenterology. 131(4):1271–83.

Lauren, P. (1965). The two histological main types of gastric carcinoma: diffuse and so-called intestinal-type carcinoma. An attempt at a histo-clinical classification. Acta Pathol Microbiol Scand. 64:31–49.

Lee, W.P., Tai, D.I., Lan, K.H., Li, A.F., Hsu, H.C., Lin, E.J., Lin, Y.P., Sheu, M.L., Li, C.P., Chang, F.Y., Chao, Y., Yen, S.H., and Lee, S.D. (2005). The −251T allele of the interleukin-8 promoter is associated with increased risk of gastric carcinoma featuring diffuse-type histopathology in Chinese population. Clin Cancer Res. 11(18):6431–41.

Li, H.C., Stoicov, C., Rogers, A.B., and Houghton, J. (2006). Stem cells and cancer: evidence for bone marrow stem cells in epithelial cancers. World J Gastroenterol. 12(3):363–71.

Liu, X., Miller, M.J., Joshi, M.S, Thomas, D.D., and Lancaster, J.R., Jr. (1998). Accelerated reaction of nitric oxide with O2 within the hydrophobic interior of biological membranes. Proc Natl Acad Sci USA. 95(5):2175–9.

Liu, F., Pan, K., Zhang, X., Zhang, Y., Zhang, L., Ma, J., Dong, C., Shen, L., Li, J., Deng, D., Lin, D., and You, W. (2006). Genetic variants in cyclooxygenase-2: expression and risk of gastric cancer and its precursors in a Chinese population. Gastroenterology. 130(7):1975–84.

Lu, W., Pan, K., Zhang, L., Lin, D., Miao, X., and You, W. (2005a). Genetic polymorphisms of interleukin (IL)-1B, IL-1RN, IL-8, IL-10 and tumor necrosis factor {alpha} and risk of gastric cancer in a Chinese population. Carcinogenesis. 26(3):631–6.

Lu, Y., Xu, Y.C., Shen, J., Yu, R.B., Niu, J.Y., Guo, J.T., Hu, X., and Shen, H.B. (2005b). E-cadherin gene C-160A promoter polymorphism and risk of non-cardia gastric cancer in a Chinese population. World J Gastroenterol. 11(1):56–60.

Lunet, N., Valbuena, C., Vieira, A.L., Lopes, C., Lopes, C., David, L., Carneiro, F., and Barros, H. (2007). Fruit and vegetable consumption and gastric cancer by location and histological type: case-control and meta-analysis. Eur J Cancer Prev. 16(4):312–27.

Mannick, E.E., Bravo, L.E., Zarama, G., Realpe, J.L., Zhang, X.J., Ruiz, B., Fontham, E.T., Mera, R., Miller, M.J., and Correa, P. (1996). Inducible nitric oxide synthase, nitrotyrosine, and apoptosis in Helicobacter pylori gastritis: effect of antibiotics and antioxidants. Cancer Res. 56(14):3238–43.

Medina-Franco, H., Barreto-Zuniga, R., and Garcia-Alvarez, M.N. (2007a). Preemptive total gastrectomy for hereditary gastric cancer. J Gastrointest Surg. 11(3):314–7.

Medina-Franco, H., Medina, A.R., Vizcaino, G., and Medina-Franco, J.L. (2007b). Single nucleotide polymorphisms in the promoter region of the E-cadherin gene in gastric cancer: case-control study in a young Mexican population. Ann Surg Oncol. 14(8):2246–9.

Mirvish, S.S. (1995). Role of N-nitroso compounds (NOC) and N-nitrosation in etiology of gastric, esophageal, nasopharyngeal and bladder cancer and contribution to cancer of known exposures to NOC. Cancer Lett. 93(1):17–48.

Necchi, V., Candusso, M.E., Tava, F., Luinetti, O., Ventura, U., Fiocca, R., Ricci, V., and Solcia, E. (2007). Intracellular, intercellular, and stromal invasion of gastric mucosa, preneoplastic lesions, and cancer by Helicobacter pylori. Gastroenterology. 132(3):1009–23.

Nishino, Y., Inoue, M., Tsuji, I., Wakai, K., Nagata, C., Mizoue, T., Tanaka, K., and Tsugane, S., and Research Group for the Development and Evaluation of Cancer Prevention Strategies in Japan. (2006). Tobacco smoking and gastric cancer risk: an evaluation based on a systematic review of epidemiologic evidence among the Japanese population. Jpn J Clin Oncol. 36(12):800–7.

Nomura, A., Stemmermann, G.N., Chyou, P.H., Kato, I., Perez-Perez, G.I., and Blaser, M.J. (1991). Helicobacter pylori infection and gastric carcinoma among Japanese Americans in Hawaii. N Engl J Med. 325(16):1132–6.

Norton, J.A., Ham, C.M., Dam, J.V., Jeffrey, R.B., Longacre, T.A., Huntsman, D.G., Chun, N., Kurian, A.W., and Ford, J.M. (2007). CDH1 truncating mutations in the E-cadherin gene: an indication for total gastrectomy to treat hereditary diffuse gastric cancer. Ann Surg. 245(6):873–9.

Oda, I., Saito, D., Tada, M., Iishi, H., Tanabe, S., Oyama, T., Doi, T., Otani, Y., Fujisaki, J., Ajioka, Y., Hamada, T., Inoue, H., Gotoda, T., and Yoshida, S. (2006). A multicenter retrospective study of endoscopic resection for early gastric cancer. Gastric Cancer. 9(4):262–70.

Ogura, M., Yamaji, Y., Hikiba, Y., Maeda, S., Matsumura, M., Okano, K., Sassa, R., Yoshida, H., Kawabe, T., and Omata, M. (2006). Gastric cancer among peptic ulcer patients: retrospective, long-term follow-up. Dig Liver Dis. 38(11):811–4.

Ohata, H., Kitauchi, S., Yoshimura, N., Mugitani, K., Iwane, M., Nakamura, H., Yoshikawa, A., Yanaoka, K., Arii, K., Tamai, H., Shimizu, Y., Takeshita, T., Mohara, O., and Ichinose, M. (2004). Progression of chronic atrophic gastritis associated with Helicobacter pylori infection increases risk of gastric cancer. Int J Cancer. 109(1):138–43.

Parkin, D.M., Bray, F., Ferlay, J., and Pisani, P. (2005). Global cancer statistics, 2002. CA Cancer J Clin. 55(2):74–108.

Parsonnet, J., Friedman, G.D., Orentreich, N., and Vogelman, H. (1997). Risk for gastric cancer in people with CagA positive or CagA negative Helicobacter pylori infection. Gut. 40(3):297–301.

Peek, R.M., Jr., and Crabtree, J.E. (2006). Helicobacter infection and gastric neoplasia. J Pathol. 208(2):233–48.

Piazuelo, M.B., Haque, S., Delgado, A., Du, J.X., Rodriguez, F., and Correa, P. (2004). Phenotypic differences between esophageal and gastric intestinal metaplasia. Mod Pathol. 17(1):62–74.

Pignatelli, B., Bancel, B., Esteve, J., Malaveille, C., Calmels, S., Correa, P., Patricot, L.M., Laval, M., Lyandrat, N., and Ohshima, H. (1998). Inducible nitric oxide synthase, anti-oxidant enzymes and Helicobacter pylori infection in gastritis and gastric precancerous lesions in humans. Eur J Cancer Prev. 7(6):439–47.

Pignatelli, B., Bancel, B., Plummer, M., Toyokuni, S., Patricot, L.M., and Ohshima, H. (2001). Helicobacter pylori eradication attenuates oxidative stress in human gastric mucosa. Am J Gastroenterol. 96(6):1758–66.

Pinto-Correia, A.L., Sousa, H., Fragoso, M., Moreira-Dias, L., Lopes, C., Medeiros, R., and Dinis-Ribeiro, M. (2006). Gastric cancer in a Caucasian population: role of pepsinogen C genetic variants. World J Gastroenterol. 12(31):5033–6.

Powell, J., and McConkey, C.C. (1992). The rising trend in esophageal adenocarcinoma and gastric cardia. Eur J Cancer Prev. 1(3):265–9.

Quiroga, A.J., Cittelly, D.M., and Bravo, M.M. (2005). BabA2, oipA and cagE Helicobacter pylori genotypes in Colombian patients with gastroduodenal diseases. Biomedica. 25(3):325–34. Article in Spanish.

Rugge, M., Correa, P., Dixon, M.F., Hattori, T., Leandro, G., Lewin, K., Riddell, R.H., Sipponen, P., and Watanabe, H. (2000). Gastric dysplasia: the Padova international classification. Am J Surg Pathol. 24(2):167–76.

Rugge, M., Nitti, D., Farinati, F., di Mario, F., and Genta, R.M. (2005). Non-invasive neoplasia of the stomach. Eur J Gastroenterol Hepatol. 17(11):1191–6.

Sano, T., and Hollowood, A. (2006). Early gastric cancer: diagnosis and less invasive treatments. Scand J Surg. 95(4):249–55.

Savage, S.A., Abnet, C.C., Mark, S.D., Qiao, Y.L., Dong, Z.W., Dawsey, S.M., Taylor, P.R., and Chanock, S.J. (2004). Variants of the IL8 and IL8RB genes and risk for gastric cardia adenocarcinoma and esophageal squamous cell carcinoma. Cancer Epidemiol Biomarkers Prev. 13(12):2251–7.

Semino-Mora, C., Doi, S.Q., Marty, A., Simko, V., Carlstedt, I., and Dubois, A. (2003). Intracellular and interstitial expression of Helicobacter pylori virulence genes in gastric precancerous intestinal metaplasia and adenocarcinoma. J Infect Dis. 187(8):1165–77.

Sicinschi, L.A., Lopez-Carrillo, L., Camargo, M.C., Correa, P., Sierra, R.A., Henry, R.R., Chen, J., Zabaleta, J., Piazuelo, M.B., and Schneider, B.G. (2006). Gastric cancer risk in a Mexican population: role of Helicobacter pylori CagA positive infection and polymorphisms in interleukin-1 and -10 genes. Int J Cancer. 118(3):649–57.

Siewert, J.R., and Stein, H.J. (1996). Adenocarcinoma of the gastroesophageal junction. Classification, pathology and extent of resection. Dis Esophagus. 9:173–182.

Siewert, J.R., and Stein, H.J. (1998). Classification of adenocarcinoma of the oesophagogastric junction. Br J Surg. 85(11):1457–9.

Silva, F., Carvalho, F., Peixoto, A., Seixas, M., Almeida, R., Carneiro, F., Mesquita, P., Figueiredo, C., Nogueira, C., Swallow, D.M., Amorim, A., and David, L. (2001). MUC1 gene polymorphism in the gastric carcinogenesis pathway. Eur J Hum Genet. 9(7):548–52.

Silva, F., Carvalho, F., Peixoto, A., Teixeira, A., Almeida, R., Reis, C., Bravo, L.E., Realpe, L., Correa, P., and David, L. (2003). MUC1 polymorphism confers increased risk for intestinal metaplasia in a Colombian population with chronic gastritis. Eur J Hum Genet. 11(5):380–4.

Sipponen, P. (1995). Helicobacter pylori: a cohort phenomenon. Am J Surg Pathol. 19(Suppl 1): S30–6.

Sipponen, P., Helske, T., Jarvinen, P., Hyvarinen, H., Seppala, K., and Siurala, M. (1994). Fall in the prevalence of chronic gastritis over 15 years: analysis of outpatient series in Finland from 1977, 1985, and 1992. Gut. 35(9):1167–71.

Smith, M.G., Hold, G.L., Tahara, E., and El-Omar, E.M. (2006). Cellular and molecular aspects of gastric cancer. World J Gastroenterol. 12(19):2979–90.

Tan, Y.K., and Fielding, J.W. (2006). Early diagnosis of early gastric cancer. Eur J Gastroenterol Hepatol. 18(8):821–9.

Tang, W.Y., Wang, L., Li, C., Hu, Z.B., Chen, R., Zhu, Y.J., Shen, H.B., Wei, Q.Y., and Zhou, J.W. (2007). Identification and functional characterization of JWA polymorphisms and their association with risk of gastric cancer and esophageal squamous cell carcinoma in a Chinese population. J Toxicol Environ Health A. 70(11):885–94.

Tarpila, S., Telkka, A., and Siurala, M. (1969). Ultrastructure of various metaplasias of the stomach. Acta Pathol Microbiol Scand. 77(2):187–95.

Tatemichi, M., Sawa, T., Gilibert, I., Tazawa, H., Katoh, T., and Ohshima, H. (2005). Increased risk of intestinal type of gastric adenocarcinoma in Japanese women associated with long forms of CCTTT pentanucleotide repeat in the inducible nitric oxide synthase promoter. Cancer Lett. 217(2):197–202.

Tham, K.T., Peek, R.M., Jr., Atherton, J.C., Cover, T.L., Perez-Perez, G.I., Shyr, Y., and Blaser, M.J. (2001). Helicobacter pylori genotypes, host factors, and gastric mucosal histopathology in peptic ulcer disease. Hum Pathol. 32(3):264–73.

Tsugane, S. (2005). Salt, salted food intake, and risk of gastric cancer: epidemiologic evidence. Cancer Sci. 96(1):1–6.

Tsugane, S, and Sasazukiv, S. (2007). Diet and the risk of gastric cancer: review of epidemiological evidence. Gastric Cancer. 10(2):75–83.

Uemura, N., Okamoto, S., Yamamoto, S., Matsumura, N., Yamaguchi, S., Yamakido, M., Taniyama, K., Sasaki, N., and Schlemper, R.J. (2001). Helicobacter pylori infection and the development of gastric cancer. N Engl J Med. 345(11):784–9.

Wang, L., Ke, Q., Chen, W., Wang, J., Tan, Y., Zhou, Y., Hua, Z., Ding, W., Niu, J., Shen, J., Zhang, Z., Wang, X., Xu, Y., and Shen, H. (2007a). Polymorphisms of MTHFD, plasma homocysteine levels, and risk of gastric cancer in a high-risk Chinese population. Clin Cancer Res. 13(8):2526–32.

Wang, P., Xia, H.H., Zhang, J.Y., Dai, L.P., Xu, X.Q., and Wang, K.J. (2007b). Association of interleukin-1 gene polymorphisms with gastric cancer: a meta-analysis. Int J Cancer. 120(3):552–62.

Warren, J.R., and Marshall, B.J. (1983). Unidentified curved bacilli on gastric epithelium in active chronic gastritis. Lancet. (8336):1273–5.

Watanabe, T., Tada, M., Nagai, H., Sasaki, S., and Nakao, M. (1998). Helicobacter pylori infection induces gastric cancer in mongolian gerbils. Gastroenterology. 115(3):642–8.

Whary, M.T., and Fox, J.G. (2004). Th1-mediated pathology in mouse models of human disease is ameliorated by concurrent Th2 responses to parasite antigens. Curr Top Med Chem. 4(5):531–8.

Whary, M.T., Sundina, N., Bravo, L.E., Correa, P., Quinones, F., Caro, F., and Fox, J.G. (2005). Intestinal helminthiasis in Colombian children promotes a Th2 response to Helicobacter pylori: possible implications for gastric carcinogenesis. Cancer Epidemiol Biomarkers Prev. 14(6):1464–9.

Wong, B.C., Lam, S.K., Wong, W.M., Chen, J.S., Zheng, T.T., Feng, R.E., Lai, K.C., Hu, W.H., Yuen, S.T., Leung, S.Y., Fong, D.Y., Ho, J., Ching, C.K., Chen, J.S., and China Gastric Cancer Study Group. (2004). Helicobacter pylori eradication to prevent gastric cancer in a high-risk region of China: a randomized controlled trial. JAMA. 291(2):187–94.

Yamaguchi, H., Goldenring, J.R., Kaminishi, M., and Lee, J.R. (2002). Identification of spasmolytic polypeptide expressing metaplasia (SPEM) in remnant gastric cancer and surveillance postgastrectomy biopsies. Dig Dis Sci. 47(3):573–8.

Yasui, W., Sentani, K., Motoshita, J., and Nakayama, H. (2006). Molecular pathobiology of gastric cancer. Scand J Surg. 95(4):225–31.

Zabaleta, J., Schneider, B.G., Ryckman, K., Hooper, P.F., Camargo, M.C., Piazuelo, M.B., Sierra, R.A., Fontham, E.T.H., Correa, P., Williams, S.M., and Ochoa, A.C. (2008). Ethnic differences in cytokine gene polymorphisms: potential implications for cancer development. Cancer Immunol Immunother. 57(1):107–14.

Zhou, Y., Li, N., Zhuang, W., Liu, G.J., Wu, T.X., Yao, X., Du, L., Wei, M.L., and Wu, X.T. (2007). P53 codon 72 polymorphism and gastric cancer: a meta-analysis of the literature. Int J Cancer. 121(7):1481–6.

Zintzaras, E. (2006). Association of methylenetetrahydrofolate reductase (MTHFR) polymorphisms with genetic susceptibility to gastric cancer: a meta-analysis. J Hum Genet. 51(7):618–24.

Chapter 2
Epidemiology of Gastric Cancer and *Helicobacter pylori*

Jonathan Volk and Julie Parsonnet

Introduction

In 1994, the International Agency for Research on Cancer (IARC) declared *Helicobacter pylori* to be a type I carcinogen, or a definite cause of cancer in humans (Humans 1994). This landmark decision—based almost exclusively on epidemiologic evidence—was immediately controversial. Some argued that *H. pylori* could not be considered a "cause" of cancer but only a "risk factor" or "cofactor" (although the difference between these two is largely semantic). Others maintained that the absence of experimental evidence in animals was a critical flaw in the IARC's arguments and left the possibility of residual confounding. Others straddled the fence in anticipation of randomized clinical trials looking at whether *H. pylori* eradication prevented cancer.

Now, almost 15 years after the IARC's declaration, there is broad consensus that *H. pylori* causes gastric adenocarcinoma and lymphoma despite the failure of the only completed randomized treatment trial to demonstrate adenocarcinoma prevention (Wong et al. 2004). Although the biology presented in this book—including animal models—provides plausibility for this acausal association between *H. pylori* and cancer (Pritchard and Przemeck 2004), it is the depth and breadth of epidemiologic evidence that remains the foundation for the scientific conviction. Herein, we present these epidemiologic findings.

Gastric Cancer Epidemiology

Gastric cancers remain a leading cause of cancer morbidity and mortality worldwide, with adenocarcinomas arising from gastric glands accounting for 90% of incident cases (Aaltonen et al. 2000; Coleman et al. 1993). Other less-common epithelial stomach cancers include squamous cell carcinoma, adenosquamous carcinoma, small cell, and carcinoid tumors. Nonepithelial stomach cancers such as leiomyomas, stromal tumors, and malignant lymphomas can also be found in the stomach (Aaltonen et al. 2000). For this chapter, we limit discussion to adenocarcinomas.

T.C. Wang et al. (eds.) *The Biology of Gastric Cancers*,
© Springer Science + Business Media, LLC 2009

Classifications of Adenocarcinomas of the Stomach

Adenocarcinomas—malignant tumors of glandular epithelium—can be classified as proximal tumors, which originate in the stomach cardia and gastroesophageal junction, or distal tumors, which originate in the fundus, body, or antrum of the stomach (Brown and Devesa 2002). The location of the tumor is important, because incidence, trends over time, risk factors, and mortality rates differ between distal and proximal adenocarcinomas.

In 1965, gastric adenocarcinomas were further subdivided into two distinct histologic subtypes, intestinal and diffuse (Lauren 1965), known as the Lauren classification. Intestinal-type adenocarcinomas form recognizable glands whereas diffuse adenocarcinomas consist of poorly cohesive cells that diffusely infiltrate the gastric wall and form no recognizable glands (Aaltonen et al. 2000). The Lauren classification has proven useful for observing the natural history and epidemiology of gastric cancer. Even as early as 1965, Lauren observed a male predominance and older age at diagnosis for patients with intestinal-type compared with diffuse-type gastric cancer (Lauren 1965).

Importantly, preneoplastic lesions are only associated with the later development of intestinal-type adenocarcinoma. In a model proposed by Correa (Correa 1992; Correa et al. 1975), this multistep process for progression of normal gastric cells to adeno-carcinomas starts with chronic gastritis. Chronic gastritis later develops into intestinal metaplasia and dysplasia, with the partial or complete loss of cell differentiation (Aaltonen et al. 2000; Correa 2002). Ultimately, these areas of dysplasia are thought to progress to adenocarcinoma. However, because individual lesions are not followed over time without intervention, it is impossible to prove that a specific preneoplastic lesion has progressed to gastric adenocarcinoma. Others have suggested that intestinal metaplasia does not progress to carcinoma but is instead an alternate endpoint from a shared cause (Tatematsu et al. 2003). In contrast to intestinal-type adenocarcinoma, there are no clearly defined precursors for diffuse-type adenocarcinoma.

Incidence of Gastric Cancer

Based on information from the United States National Cancer Institute's Surveillance Epidemiology and End Results Program (SEER), which collected data from 17 sites between 2002–2004, 0.90% (1/111) of men and women born today in the United States will be diagnosed with stomach cancer during their lifetime. SEER data from 2000–2004 indicate current age-adjusted gastric cancer incidence rates of 11.4 per 100,000 men and 5.6 per 100,000 women annually. The median age of diagnosis is 71 years, and 70% of cases are diagnosed between ages 55–84. As of January 2004, 34,708 men and 25,592 women with gastric cancer were living in the United States (Ries et al. 2007, based on November 2006 SEER data submission, posted to the SEER Web site, 2007 #2571).

Geographical Distribution

Gastric cancer incidence rates vary significantly in different countries and regions of the world. Japan has the highest rates of stomach cancer with 77.9 cases per 100,000 males and 33.3 per 100,000 females annually. Other countries in eastern Asia (including China and Korea), eastern Europe, and tropical South America also have high gastric cancer incidence rates (Ferlay et al. 2004). In contrast, lower incidence rates of gastric cancer have been observed in eastern and northern Africa, North America, and south and southeast Asia, with average age-standardized incidence rates of 5.9–9.0 per 100,000 men and 2.6–5.3 per 100,000 women. Diagnostic and surveillance deficiencies may account for some of the reported low rates in some of these countries. Although incidence differs throughout the world, men consistently manifest incidence higher than women (Parkin et al. 1999b).

Data from England and Wales demonstrate that country of birth is a stronger predictor of stomach cancer risk than the country of current residence (Coggon et al. 1990). Immigrants who migrate from regions at high risk for gastric cancer to regions at lower risk have an intermediate risk level. A study of immigrants from Japan to Hawaii found a lower age-adjusted rate of stomach cancer among the immigrants compared with the remainder of the Japanese population still living in Japan. An even lower rate was observed among second-generation immigrants (Kolonel et al. 1981). A similar study of people from higher-risk populations in England, Scotland, Ireland, Poland, the former Yugoslavia, Greece, and Italy who moved to Australia—a low-risk population for gastric cancer—found a risk reduction that increased with time spent in Australia (McMichael et al. 1980).

Gastric cancer incidence rates also differ by ethnicity. In the United States between 2000–2004, the highest annual incidence rates were observed among Asian/Pacific Islanders (18.9 per 100,000 men and 10.8 per 100,000 women), followed by blacks (17.5 per 100,000 men and 9.1 per 100,000 women), Hispanics (16.0 per 100,000 men and 9.6 per 100,000 women), American Indians and Alaska Natives (16.3 per 100,000 men and 7.9 per 100,000 women), and whites (10.2 per 100,000 men and 4.7 per 100,000 women) (Ries et al. 2007). Interestingly, the trends observed among ethnic groups are observed for both men and women.

Trends Over Time

The incidence of gastric cancer has decreased over time. In the United States from 1975 to 2004, there was a statistically significant decrease in gastric cancer observed for both men and women (Ries et al. 2007). Outside of the United States, the incidence of stomach cancer is also decreasing (Coleman et al. 1993; Parkin et al. 1999b), even in higher-risk countries (Parsonnet 1995). In 1990, there were only 6% more gastric cancer cases compared with 1985 despite an aging population and an increase in total number of people in the world (Parkin et al. 1999a).

Differences in Epidemiology of Various Tumor Types

Changes in incidence rate described above do not apply to all types of gastric cancer. Distal (fundus, body, and antrum), intestinal-type gastric cancer incidence has declined, whereas the incidence of proximal (cardia and gastroesophageal junction) adenocarcinomas has increased. This decrease in intestinal-type tumors accounts for the overall declining trend in gastric cancer incidence (Correa and Chen 1994; Howson et al. 1986).

In contrast to more distal tumors, the incidence of proximal tumors of the gastric cardia has risen steadily from 1974 to 1994 among men, and to a lesser extent, among women. The age-standardized incidence rates increased from 2.1 to 3.3 per 100,000 among whites and 1.0 to 1.9 per 100,000 among blacks (Devesa et al. 1998). These proximal tumors are more common in developed countries, especially among whites and those of higher socioeconomic status. Obesity and gastroesophageal reflux disease (GERD) both have important pathogenic roles for this gastric cancer subtype (Crew and Neugut 2006).

Mortality and Case Fatality

Worldwide, an estimated 850,000 people died from gastric cancer in 2001 (522,000 men and 328,000 women), making gastric cancer the second leading cause of cancer death after lung cancer (Parkin et al. 1999a). As of 1990, the 5-year mortality rate for stomach cancer remained poor, only better than lung and pancreatic cancer in most regions of the world (Parkin et al. 1999a). A notable exception to this dismal prognosis is in Japan, where intensive screening efforts have resulted in the earlier diagnosis of gastric cancer and a much higher 5-year survival rate (57%). In the United States between 1992 and 1998, the overall 5-year survival for gastric cancer for all disease stages was 22% (Jemal et al. 2003); localized disease had much higher 5-year survival rates (59%) than metastatic disease (2%).

According to SEER data collected between 2000 and 2004, the median age of death from gastric cancer was 74 years, with an age-adjusted death rate of 4.2 per 100,000 people (5.9 per 100,000 men and 3.0 per 100,000 women) (Ries et al. 2007). Age-adjusted mortality rates increase dramatically with age. The mortality rate for individuals older than age 85 is 46.1 per 100,000 people, compared with 12.9 per 100,000 for people aged 65–69 and 1.3 per 100,000 for those aged 40–44. Notably, deaths from gastric cancer differed significantly among different ethnic groups and paralleled the trends observed for gastric cancer incidence. Rates were highest among black Americans (11.9 per 100,000 men and 5.8 per 100,000 women), followed by Asian/Pacific Islanders (10.5 per 100,000 men and 6.2 per 100,000 women), American Indian/Alaska native (9.6 per 100,000 men and 5.5 per 100,000 women), Hispanic (9.1 per 100,000 men and 5.1 per 100,000 women), and white Americans (5.2 per 100,000 men and 2.6 per 100,000 women) (Ries et al. 2007).

Until the 1940s, gastric cancer was the most common cause of cancer death among men (Howson et al. 1986). Decreases in stomach cancer mortality started as early as 1926. Joint point analysis from the SEER database reveals statistically significant decreases in gastric cancer mortality in the United States. Between 1990 and 2004, there was a −3% annual percentage change in gastric cancer mortality, with a −3.5% decrease observed for men between the years 1991 and 2004 and a −2.6% decrease observed among women during the same time period (Ries et al. 2007). A statistically significant decrease in gastric cancer mortality rates was also noted between 1975 and 1987 and a nonsignificant decrease between 1987 and 1990 (Ries et al. 2007). According to estimates from the National Center for Health Statistics(American Cancer Society), an estimated 21,260 Americans (13,000 men and 8,260 women) will be diagnosed with stomach cancer in 2007, with 11,210 deaths. This number represents a decrease from the 12,100 deaths seen in 2003.

The decrease in mortality rates seen in the last several decades likely reflects a decreasing incidence of gastric cancer, because the case-fatality rates have changed little. An exception to this, however, may be evident in regions where gastric cancer screening has been undertaken. By identifying early cancers, surgical intervention can be applied when it has the highest probability of benefit. Motivated by high rates of gastric cancer incidence and mortality, Japan introduced a mass screening program in the 1960s (Fukao et al. 1995). Double-contrast barium x-rays were offered in all municipalities of Japan by 1975, and more recently, people have been screened with endoscopy, serum pepsinogen concentrations, and anti-*H. pylori* antibody tests. Using observational data, including cohort studies, case-control studies, and ecologic analyses from different municipalities with varying levels of screening participation, these screening efforts are thought to have caused a decrease in stomach cancer mortality in Japan (Fukao et al. 1995; Miyamoto et al. 2007). No randomized trials of screening have been conducted and it is not clear whether similar screening efforts would yield a benefit in populations with lower rates of gastric cancer. Currently, neither the American Cancer Society (Smith et al. 2003) nor the National Cancer Institute (National Cancer Institute 2007) recommends stomach cancer screening in the United States.

Risk Factors for Gastric Cancer, Excluding Helicobacter pylori

Risk factors for gastric cancer that have undergone extensive investigation include genetic susceptibility, use of cigarettes, poor nutrition, alcohol abuse, obesity, gastroesophageal reflux, pernicious anemia, radiation exposure, and partial gastrecto-mies. When examining the relationships among these risk factors and the subsequent development of gastric cancer, it is important to control for *H. pylori*, an important confounding variable that is discussed in detail in its own section. Unfortunately, many studies of gastric cancer risk factors do not adequately control for *H. pylori* and other potential confounders. For example, although higher stomach cancer rates are seen among individuals with low socioeconomic status (Neugut et al.

1996), this depressed socioeconomic status is confounded by increased usage of tobacco, decreased intake of fruits and vegetables, poorer sanitary conditions that may and probable increased *H. pylori* transmission.

Genetic Factors

Although most familial gastric cancers results from concordant *H. pylori* infection within families (Lugli et al. 2007), there are some families with extraordinarily high rates of diffuse-type gastric cancer occurring at unusually young ages. Genetic susceptibility has proved to have a role in these families. The most well-studied familial genetic defect is a germline mutation in a cell-adhesion protein, E-cadherin. First observed in a kindred group of Maori ethnicity (Guilford et al. 1998), germline mutations of *e-cadherin* have been identified in familial cancers in Europe (Gayther et al. 1998; Guilford et al. 1999), Japan (Shinmura et al. 1999; Yabuta et al. 2002), the United States (Oliveira et al. 2002), and Korea (Yoon et al. 1999). An individual who possesses one of the germline *e-cadherin* gene mutations listed in the International Gastric Cancer Linkage Consortium has a cumulative risk of gastric cancer before 80 years of age of 67% [95% confidence interval (CI), 33%–99%] for men and 83% (95% CI, 58%–99%) for women (Pharoah et al. 2001). Fortunately, *e-cadherin* mutations are rare and are thought to account for less than 3% of gastric tumors (Stone et al. 1999).

Other genetic factors besides *e-cadherin* have been linked to malignancy. Gastric tumors are sometimes observed in hereditary cancer syndromes such as hereditary nonpolyposis colorectal cancer, Li-Fraumeni syndrome, familial adenomatous polyposis, and Peutz-Jeghers syndrome (Caldas et al. 1999). Since the first report in 1953, numerous studies have demonstrated that individuals with blood type A (El Hajj et al. 2007) have higher rates of intestinal-type gastric cancer, chronic atrophic gastritis, intestinal metaplasia, and dysplasia (El Hajj et al. 2007; Kneller et al. 1992) It is postulated that *H. pylori* expressing the BabA adhesin are better able to adhere to the gastric epithelium on individuals expressing blood type A antigens (Gerhard et al. 1999; Ilver et al. 1998), enhancing the persistence of *H. pylori* infection and subsequent gastric cancer development.

Environmental Factors

Familial clusters of cancer do not always indicate inherited genes, because smoking, environmental exposures, alcohol, and especially *H. pylori* also aggregate in families. Cigarette smoking increases the risk of proximal and distal stomach cancers independent of *H. pylori* infection. Although smoking and *H. pylori* often go hand-in-hand, an IARC working group on Tobacco Smoke and Involuntary Smoking concluded that smoking tobacco independently increased gastric cancer risk (IARC 2004) and that confounding by *H. pylori* could be "reasonably ruled out." They also noted a dose-response relationship with tobacco exposure and cancer risk, with decreased risk observed with increased duration of tobacco cessation. A large study by Chao et al. that followed 467,788 men and 588,053 women for 14 years concluded that 28% of gastric cancers in men and 14% in women could be attributed to smoking (Chao et al. 2002); unfortunately, *H. pylori* was not directly evaluated as a confounder in this study.

In 1997, the World Cancer Research Fund (WCRF) and the American Institute for Cancer Research (AICR) conducted a thorough review exploring the relationship between food and nutrition and cancer risk (World Cancer Research Fund and American Institute for Cancer Research 1997). Using data from case-control and cohort studies, the panel concluded that diets high in fruits and vegetables conferred a protective benefit for stomach cancer. In the European Prospective Investigation into Cancer and Nutrition cohort (EPIC), a large study of 521,457 men and women aged 35–70 years, increased intake of total meat, red meat, and processed meat was associated with distal gastric cancers, although not with proximal cancers. The relationship between diet and distal cancers was particularly significant striking among individuals also infected with *H. pylori* (Gonzalez et al. 2006). This study also demonstrated an inverse relationship between gastric cancer risk and high levels of plasma vitamin C, particularly among individuals with high intakes of red meat and other processed meats.

Salt was found to be a risk factor for gastric tumors by the WCRF/AICR panel (World Cancer Research Fund and American Institute for Cancer Research 1997). One proposed mechanism for this deleterious effect is that salt results in mucosal damage in the stomach, and consequently, an inflammatory regenerative response with increased DNA synthesis and cell proliferation (Bergin et al. 2003). This increased cell proliferation in turn increases the risk for tumorigenesis. The use of refrigeration has been found to decrease stomach cancer risk, likely the result of less salt being used in refrigerated food, as well as the decreased risk of food contamination with carcinogenic compounds.

Proximal gastric cancers have a different pathogenesis and different risk factors than from tumors of the distal stomach. For example, although alcohol does not affect overall gastric cancer rates, heavy drinking may increase rates of proximal tumors (Terry et al. 2002). A significant association between body mass index and proximal tumors—but not distal tumors—has been observed (Chow et al. 1999; Ji et al. 1997; Terry et al. 2002). Although GERD is a clear risk factor for the development of adenocarcinoma of the esophagus (Farrow et al. 2000), the evidence in support of GERD as a risk factor for gastric tumors is mixed. Some studies have concluded that reflux is only weakly associated with proximal adenocarcinomas of the gastric cardia (Mayne and Navarro 2002). Other studies demonstrated increased proximal stomach cancers with increased GERD symptom severity (Lagergren et al. 1999). However, other studies have found no relationship between GERD and proximal gastric cancers (Farrow et al. 2000).

Other risk factors for gastric cancer are rarely encountered. Pernicious anemia, which may result in severe atrophic gastritis and intrinsic factor deficiency, increases risk for both gastric cancer and carcinoid tumors (Hsing et al. 1993; Kokkola et al. 1998). Exposure to radiation also increases gastric cancer risk. After the bombing of Hiroshima and Nagasaki, a greater than expected number of gastric tumors occurred to those exposed in childhood (Kai et al. 1997).

Partial gastrectomy is a long-recognized predisposing factor for cancer. Although peptic ulcer disease and its surgical treatment have become increasingly rare in the modern era of acid inhibition, *H. pylori* eradication, and aggressive endoscopy (Higham et al. 2002; Targownik and Nabalamba 2006), some patients still undergo partial gastrectomy for gastrointestinal bleeding. A review of 58 studies showed that individuals who had partial gastrectomies and survived more than 15 years after surgery had a two- to fourfold increase in stomach cancer risk (Stalnikowicz and Benbassat 1990). A metaanalysis also done in 1990 (Tersmette et al. 1990) concluded that this increased cancer risk was limited to a subset of patients who had partial gastrectomies for gastric ulcers (relative risk = 2.12; 95% CI, 1.73–2.59) as compared with duodenal ulcers (relative risk = 0.84; 95% CI, 0.66–1.05). In support of this finding, individuals with a history of duodenal ulcer seem to be at lower risk for gastric malignancy than those without (Hansson et al. 1996).

The impact of proton pump inhibitors and histamine-2 antagonists in gastric carcinogenesis is unclear. In animals, these medications have been linked to gastric carcinoid tumors (Gillen and McColl 2001). Although there have been no longitudinal studies exploring whether proton pump inhibitors and histamine antagonists increase gastric cancer risk in humans, a postmarketing surveillance report of cimetidine indicated no increase in gastric cancer incidence with its use (Colin-Jones et al. 1992). Despite this, some researchers suggest eradicating *H. pylori* before the initiation of these medications because hypochlorhydria promotes extension of *H. pylori* throughout the stomach and development of atrophic gastritis (Kuipers et al. 1996).

Epstein-Barr Virus–Related Tumors

Fewer than 1% of gastric cancers are lymphoepithelioma-like carcinomas (LELCs) — epithelial tumors with extensive lymphoid infiltration into the stroma. These tumors are histopathologically similar to nasopharyngeal carcinomas and contain monoclonally integrated Epstein-Barr virus (EBV) (Herrmann and Niedobitek 2003; Wu et al. 2000). LELC gastric tumors have distinctive oncogene expression, such as p53 overexpression and underexpression of c-erb2 and E-cadherin that likely have causal roles for this gastric cancer subtype (Wu et al. 2000). EBV-associated tumors occur more frequently in males and in younger patients. They are also more often located in the gastric body or cardia rather than in the antrum (Takada 2000). EBV may also have a role in non-LELC gastric tumors, although the evidence remains inconclusive.

Helicobacter pylori Epidemiology

Spiral-shaped bacteria were first observed in the stomachs of humans well over a century ago (Weisse 1996). Although early research linked these spiral-shaped bacteria to gastric inflammation and other upper gastrointestinal disorders (Doenges

1938; Kreinitz 1906), later studies would deem these bacteria contaminants (Palmer 1954) and the field stagnated.

The subsequent history of *H. pylori*'s "discovery" and the ultimate Nobel prize to Marshall and Warren is now widely known. Together, Marshall and Warren showed that "unidentified curved bacilli" in gastric biopsies were associated with "active, chronic gastritis" and with ulcers (Marshall and Warren 1984; Marshall et al. 1985b). In the face of the world's skepticism, Marshall ultimately swallowed an inoculum of *H. pylori* to prove its pathogenicity (Marshall et al. 1985a). Although they were barraged with criticism, Marshall and Warren persisted in their insistence that the organism caused gastritis and ulcer disease and finally a cascade of data from investigators around the world proved their theory correct.

Research reveals that humans have been infected with *H. pylori* for at least 58,000 years, before human migration from Africa (Linz et al., Nature, 2007). The investigators, using a large data set of *H. pylori* strains, found a decrease in bacterial genetic diversity with increasing distance from east Africa. In fact, human migration in modern times can be predicted from phylogenetic bacterial models (Falush et al. 2003; Kersulyte et al. 2000).

Prevalence and Incidence by Region

Approximately half of the world's population is infected with *H. pylori*, although prevalence rates differ tremendously in different regions of the world. Prevalence of *H. pylori* infection increases with age, and is higher in developing than in developed countries (Brown 2000). An exception to this increasing *H. pylori* prevalence with age is the lower prevalence often seen in the very elderly (Taylor et al. 1995). It is likely that advancing gastric atrophy and intestinal metaplasia with age sometimes causes loss of infection with advancing age in the elderly (Ohata et al. 2004).

In developing countries, infection can be so common as to be almost universal in adults, although there are some notable exceptions. For example, Indonesia and Papua New Guinea have reported disproportionately low prevalences of *H. pylori* in some regions (Mitchell et al. 1988; Tokudome et al. 2005). The pattern of higher prevalence in developing countries is also seen among children, with prevalence rates estimated as low as 1.2% in a sample of 2–4 year-old children from the Netherlands to as high as 70–80% in some developing countries (Magalhaes Queiroz and Luzza 2006; Mourad-Baars et al. 2007).

H. pylori incidence is more difficult to determine than prevalence because the initial infection invariably goes unnoticed and undiagnosed. Based on changes in prevalence with age, the incidence of *H. pylori* has been estimated to be 1% per year among white Americans and as high as 3% per year among African Americans (Graham et al. 1991; Parsonnet 1995). The incidence of *H. pylori* in developing countries is much higher, with yearly incidence rates as high as 3%–10% (Parsonnet

1995). Incidence of infection is highest in early childhood; the majority of infections in high prevalence areas occur before age 5 (Granstrom et al. 1997; Malaty et al. 2002; Rothenbacher et al. 2002; Rowland et al. 2006). In studies of children younger than 2 years, there have been reports of children who initially test positive for *H. pylori* but on repeat breath test are seronegative negative (Goodman et al. 2005; Klein et al. 1994; Rothenbacher et al. 2002; Thomas et al. 1999). It has not yet been determined whether the conversion reversion from *H. pylori* seropositive positive to seronegative negative in children reflects transient infections or, rather, false positives in a population with low seroprevalence infection prevalence (Nurgalieva et al. 2006; Perry and Parsonnet 2005; Rosenstock et al. 2000). For adults in developed countries throughout the world, the incidence is low, estimated at 0.5% of susceptible adults becoming infected yearly.

In recent years, there has been a decrease in *H. pylori* prevalence worldwide that has coincided with improved hygiene and socioeconomic status. For example, a study of healthy adults from southern China showed a significant decrease in *H. pylori* seroprevalence (62.5% to 49.3%) from 1993 to 2003 (Chen et al. 2007). Because *H. pylori* infection typically remains in the stomach for life, a decrease in incidence over time is eventually manifest by disproportionately higher rates of *H. pylori* in the elderly than in the young (a birth cohort effect). This birth cohort effect has been documented in Europe and the United States, with a 10% decrease in incidence per decade (Banatvala et al. 1993; Parsonnet et al. 1992; Roosendaal et al. 1997).

Risk Factors for Infection

Lower socioeconomic status, often measured indirectly using level of education, household crowding, sharing of beds, plumbing, and water sanitation, has been consistently identified as a risk factor for *H. pylori* infection (Brown 2000). A large study of 3,194 people from 17 countries showed an inverse relationship between education and *H. pylori* seroprevalence in 11 of 17 populations studied; the remaining 6 populations also showed this relationship, but without reaching clinical significance (Eurogast Study 1993). Similarly, a study of children from northeastern Brazil found that children from lower socioeconomic status had much higher *H. pylori* seroprevalence rates (55%) compared with the wealthier children (16.4%) (Parente et al. 2006). Socioeconomic status in childhood, rather than later in life, is most important in determining *H. pylori* infection (Torres et al. 1998; Ueda et al. 2003). This finding is substantiated by studies of immigrants. Adult immigrants from countries with high prevalence of *H. pylori* infection have prevalence that parallels their country of origin (Perez-Perez et al. 2005; Tsai et al. 2005). Children of immigrants have a prevalence closer to that of the new country, especially after controlling for such variables as household crowding and parents' level of education (O'Rourke et al. 2003; Tsai et al. 2005).

Large differences in *H. pylori* seroprevalence also exist among ethnic groups even within the same country or region. These differences are likely the result, at least in part, of socioeconomic differences. In the United States, non-Hispanic whites have lower *H. pylori* infection rates than African Americans and Hispanic populations (Graham et al. 1991; Hopkins et al. 1990). In a large cross-sectional study of adult Americans from 1988 to 1991, the overall *H. pylori* seroprevalence was 33%, with the highest rates seen in Mexican Americans (62%) followed by African Americans (53%) and non-Hispanic whites (26%) (Everhart et al. 2000).

Similar differences between ethnic groups have been observed outside of the Americas. In a study done in New Zealand, indigenous Maori have significantly higher rates of infection (39%–70%) than white New Zealanders (15%) (Morris et al. 1986). Similarly, Aborigines from western Australia have 2–3 times higher rates of *H. pylori* infection than nonindigenous Australians (Windsor et al. 2005). Other studies have also found ethnic differences in Belgium (Blecker et al. 1993) and Malaysia (Goh and Parasakthi 2001).

Higher *H. pylori* infection rates have been documented in communities where hygiene is poor, such as institutions for the disabled and in orphanages (Brown 2000). Other risks include bed sharing among children, which is strongly correlated with transmission (Mendall et al. 1992; Perry et al. 2005). In contrast, attending daycare (Wizla-Derambure et al. 2001) and school (Tindberg et al. 2001) with infected classmates was not found to increase risk of infection.

Mechanisms of Transmission

New *H. pylori* infections go undetected unless they are iatrogenic or induced experimentally. As a result, the mechanism of *H. pylori* transmission has been difficult to determine definitively, and the preponderance of evidence is circumstantial. Nevertheless, these data suggest that person-to-person transmission is the most important or perhaps only means of *H. pylori* transmission.

The primary source of evidence to support person-to-person transmission comes from cross-sectional data from families. Numerous studies have found clustering of *H. pylori* infections within families, with infection rates highest among first-degree relatives of infected individuals (Drumm et al. 1990; Kivi et al. 2005; Nguyen et al. 2006). Mothers seem to be more important sources for *H. pylori* transmission than fathers. Studies in Japan and Germany found that children with *H. pylori*–infected mothers were significantly more likely to be infected than children with *H. pylori*–seronegative mothers; no associations were found between *H. pylori* infection in fathers and their children (Fujimoto et al. 2007; Weyermann et al. 2006). A prospective study done in Japan with 9 years of follow-up supports these findings with *H. pylori* seroconversion occurring only in children with *H. pylori*–infected mothers (Malaty et al. 2000).

DNA fingerprinting studies also support person-to-person transmission within the family unit. *H. pylori* strains are more similar within families than among unrelated individuals (Bamford et al. 1993; Raymond et al. 2004). Similar fingerprints are particularly common among siblings (Wang et al. 1993); in one study, 81% of siblings shared the same strain of *H. pylori*, compared with 56% of mothers and their children and 0% of fathers and their children (Kivi et al. 2003).

Three possible mechanisms for person-to-person transmission of *H. pylori* have been proposed: fecal–oral, oral–oral, and gastro–oral (defined here as either transmission by vomitus, or the iatrogenic introduction of *H. pylori* into the stomach through the use of infected instruments). The data in support of fecal–oral transmission are inconclusive. Although viable *H. pylori* have been detected in feces of both children and adults (Kelly et al. 1994), some studies only report culturing *H. pylori* from cathartic rather than normal stools (Parsonnet et al. 1999). In support of fecal–oral transmission, a study of 671 healthcare workers concluded that contact with fecal matter was a significant risk factor for *H. pylori* infection (De Schryver et al. 2006). Additionally, a prospective study found that gastroenteritis in an *H. pylori*–infected household member was associated with a fourfold increased risk of new infection in another household member (Perry et al. 2006). In this study, however, diarrhea was not an independent risk factor for infection if vomiting was not also present. In opposition to the fecal–oral hypothesis, no excess risk for *H. pylori* infection has been observed in sewage workers (Friis et al. 1996; Jeggli et al. 2004).

Fewer data support oral–oral transmission. Although *H. pylori* has been detected by polymerase chain reaction from saliva and dental plaques (Krajden et al. 1989; Mapstone et al. 1993), it has rarely been cultured from the mouth (Ferguson et al. 1993; Krajden et al. 1989; Parsonnet et al. 1999). Dental workers do not have an increased risk for *H. pylori* (Malaty et al. 1992). One would anticipate high rates of transmission of *H. pylori* between married couples if saliva were the source, and this is not consistently observed. Some investigators report little *H. pylori* concordance among married couples (Miyaji et al. 2000; Perez-Perez et al. 1991), whereas others find a high correlation among couples even after controlling for confounding factors (Brown et al. 2002; Singh et al. 1999; Stone et al. 2000). The strongest studies, which examined the genotype of infecting strains within couples, have not found strong evidence of concordance; the great majority of spouses are infected with different strains (Kuo et al. 1999; Luman et al. 2002; Suzuki et al. 1999).

The strongest data support gastric–oral transmission, especially in the setting of gastric intubation. The majority of these iatrogenic cases result from direct inoculation of contaminated gastric contents into the stomach via incompletely cleaned endoscopic equipment (Langenberg et al. 1990). Iatrogenic outbreaks have been documented (Graham et al. 1988; Pardo-Mindan et al. 1989; Ramsey et al. 1979; Tytgat 1995), and the strains of *H. pylori* isolated from patients using the same endoscopy equipment have been identical. Endoscopists and endoscopy nurses are also at increased risk for *H. pylori* infection (Chong et al. 1994; Lin et al. 1994).

Given the ubiquity of *H. pylori* infection throughout the world and the infrequency of endoscopic examinations, direct gastric–oral inoculation cannot be the primary route of transmission. However, indirect gastric–oral transmission seems plausibly important. *H. pylori* has consistently been cultured in large quantities from vomitus (Brown 2000; Leung et al. 1998; Parsonnet et al. 1999). In addition, in a prospective study, exposure to an *H. pylori*–infected individual with vomiting conferred a sixfold increased risk for new infection (Perry et al. 2006). Additional data from observational studies concluded that children exposed to emesis were significantly more likely to be infected than children not exposed to emesis (Ito et al. 2006; Luzza et al. 2000). Vomitus has also been implicated in transmission of *H. pylori* among monkeys (Solnick et al. 2006).

Frequently, infections that are transmitted from person to person can also be transmitted via water or other environmental vectors. Evidence for contaminated water as a source of *H. pylori* transmission, however, is weak. Some studies in developing countries have found increased risk for *H. pylori* infection in individuals using, irrigating with, or swimming in, unclean water (Glynn et al. 2002; Goodman et al. 1996; Hopkins et al. 1993; Hulten et al. 1996; Karita et al. 2003; Klein et al. 1991; Nurgalieva et al. 2002). However, the household-based clustering of *H. pylori* infection in populations with municipal water sources, and the lack of concordance of *H. pylori* with other waterborne diseases raises doubt about the importance of these findings (Egemen et al. 2006; Lin et al. 2005). Although *H. pylori* DNA has often been amplified from untreated water, it has been cultured only once using immunomagnetic separation on raw sewage (Lu et al. 2002). Moreover, when *H. pylori* is exposed to water or when it is under other forms of stress, it loses its classic spiral morphology and becomes coccoid. The coccoid form of *H. pylori* cannot be cultured, and it is still debated in the literature whether it is viable and able to infect (Chen 2004; Delport and van der Merwe 2007; Sorberg et al. 1996).

Food and animal exposure have also been implicated as possible routes of transmission, although none consistently (Brown 2000; Hopkins et al. 1993). *H. pylori* naturally infects monkeys (Drazek et al. 1994; Dubois et al. 1994) and cats (Handt et al. 1995), and has also been found in sheep and their milk (Dore et al. 2001), and on houseflies (Grubel et al. 1997; Osato et al. 1998). Pet ownership, however, is not linked to infection (indeed, it has been found to be protective) (Graham et al. 1991; Webb et al. 1994), and animal exposure more generally is unlikely to explain the extremely high rates of infection worldwide.

Links Between *Helicobacter pylori* and Cancer

Interest in *H. pylori* as a potential cancer-causing agent began soon after the pioneering discoveries of Marshall and Warren in the 1980s. It had been known for many years that gastric adenocarcinomas often arose in areas of gastritis. Because of its induction of chronic gastritis, investigators began almost immediately to take interest in *H. pylori*'s causal role in malignancy.

The first studies to examine this association were ecologic and compared regional *H. pylori* incidence with regional cancer incidence (Eurogast Study 1993; Forman et al. 1990). Many subsequent case-control studies examined the prevalence of *H. pylori* in persons with and without cancer. Now, in retrospect, it is understood that these studies—which in a metaanalysis indicated a 1.8-fold increased risk of cancer (Huang et al. 1998)—underestimated the true risk. The underestimate resulted from the loss of *H. pylori* infection and its serologic response as the stomach progresses toward malignancy (Genta and Graham 1993; Masci et al. 1996; Osawa et al. 1996). Stronger support for a role in cancer comes from nested case-control studies. These studies examine *H. pylori* prevalence in stored sera obtained from cases and matched controls years before the development of cancer. Together, these demonstrate a stronger risk of cancer [odds ratio (OR) = 3.0; 95% CI, 2.3–3.8 (Helicobacter and Cancer Collaborative Group 2001)]. The risk was particularly high when sera were drawn more than 10 years before the development of cancer in the case (OR = 5.9; 95% CI, 3.9–10.3), again suggesting that diagnostic artifact occurs when sera are obtained close to the time of cancer diagnosis. Even more compelling data for an association between *H. pylori* and cancer come from longitudinal cohort studies. In a large prospective trial conducted in Japan, only those infected with *H. pylori* later developed gastric cancer; 36 of 1,246 infected individuals developed gastric cancer compared with none of the 280 uninfected participants (infinite OR) (Uemura et al. 2001). A prospective study of 1,225 Taiwanese patients confirmed this "infinite" OR (p = 0.015) (Hsu et al. 2007). The broad spectrum of strongly supportive studies have led some to speculate that *H. pylori* is a necessary factor in the development of gastric cancer of the distal stomach (Brenner et al. 2004).

As a single agent, *H. pylori* may be responsible for as many as 5.5% of all cancers (Parkin 2006), making it the leading infectious cause of cancer worldwide. This figure, however, derives only from tumors of the gastric antrum and body. The role of *H. pylori* in proximal tumors is more debated, in large part because of the difficulties differentiating adenocarcinomas of the proximal stomach from those of the gastroesophageal junction (Odze 2005; Richter 2007). Overall, *H. pylori* infection seems to be important for inflammation in the proximal stomach but remains inversely related to proximal cancers of the gastric cardia and gastroesophageal junction (Yang and Davis 1988). In a large prospective nested case-control study of 29,133 participants aged 50–69 years, *H. pylori* was strongly associated with distal gastric cancer (OR = 7.9; 95% CI, 3.0–20.9) but inversely proportional to proximal gastric cancers (adjusted OR = 0.31; 95% CI, 0.11–0.89) (Kamangar et al. 2006). This finding confirmed the results of a metaanalysis of previous nested case-control studies that showed no increased risk for the development of proximal gastric cancers among those infected with *H. pylori* (Helicobacter and Cancer Collaborative Group 2001).

Proximal tumors often occur in the setting of GERD, and *H. pylori* infection is less common in patients with these symptoms (OR = 0.60; 95% CI, 0.47–0.78) (Raghunath et al. 2003). Pathophysiologically, *H. pylori* may prevent GERD by decreasing gastric pH. Given the strong causal relationship between GERD and adenocarcinoma of the esophagus (Lagergren et al. 1999), *H. pylori* may confer a protective benefit for cardia tumors as well. Although incidence rates of distal gastric adenocarcinoma have declined with decreased *H. pylori* infection rates, it is likely that these decreasing *H. pylori* infection rates observed in western countries explain the simultaneous increase in proximal adenocarcinomas.

Effect Modifiers for Helicobacter pylori and Malignancy

Most people infected with *H. pylori* will remain free of symptoms and will never develop gastric cancer in their lifetimes. Host genetic factors, bacterial variation, and diet and environmental cofactors all have significant roles in the variable evolution and presentation of *H. pylori* infection.

Genetic Factors

Approximately 10% of gastric cancers cluster in families but only a small portion of these cancers result from known hereditary cancer syndromes. Other genetic factors have been investigated that might influence the consequences.

Notably, the intensity and type of immune response to infection with *H. pylori* is determined by host genetics. In 2000, El-Omar demonstrated that specific polymorphisms of interleukin (IL)-1β, an important inflammatory cytokine and potent inhibitor of gastric acid secretion, contribute to intestinal-type stomach cancer progression (El-Omar et al. 2000). These findings have now been extensively replicated worldwide. Polymorphisms in tumor necrosis factor (TNF)-α (Machado et al. 2003), IL-1 receptor antagonist (El-Omar et al. 2000), and IL-10 (El-Omar et al. 2003) also influence intestinal-type gastric cancer evolution whereas polymorphisms in the IL-8 promotor have been linked to diffuse-type cancer (Lee et al. 2005). Individuals unfortunate enough to possess polymorphisms in IL-1β, TNF-α, and IL-10 have a 27-fold increased risk for the development of gastric cancer (El-Omar et al. 2003).

Human leukocyte antigen (HLA) genotypes have been variably linked to gastric cancer, with different HLA types associated with cancer or cancer protection in different regions of the world (Garza-Gonzalez et al. 2004; Hirata et al. 2007; Li et al. 2005; Perri et al. 2002; Quintero et al. 2005; Watanabe et al. 2006).

Bacterial Factors

Variations in *H. pylori* genes confer different risks of cancer development. *H. pylori* can undergo point mutations and chromosomal rearrangements (Blaser and Berg 2001) and, consequently, there is an impressive degree of genetic diversity, even within a single host (Cooke et al. 2005; Israel et al. 2001; Kim et al. 2004). Approximately half of all strains of *H. pylori* contain a 40-kb DNA virulence cassette known as the pathogenicity island (PAI) (Stein et al. 2002; Yamazaki et al. 2003). This cassette, which is discussed in detail elsewhere in this book, encodes a Type IV secretion system that injects the CagA protein into the host epithelial cell. *H. pylori* possessing this cassette produce greater gastric inflammation and a higher risk of intestinal-type malignancies than strains that do not contain this gene (Chow et al. 1998; Parsonnet et al. 1997). A metaanalysis published in 2003 showed that

CagA is an independent risk factor for distal gastric cancers (OR = 1.64; 95% CI, 1.21–2.24) (Huang et al. 2003).

Certain genotypes (the s1 and m1 genotypes) of the *vacA* gene, a gene that encodes a vacuolating cytotoxin, are associated both with the presence of a viable pathogenicity island and with the development of cancer (Con et al. 2007). Other polymorphic bacterial factors linked to cancer contribute to *H. pylori* adherence to gastric epithelial cells (*babA*), bacterial invasion into the gastric glands, and persistence of infection within the gastric lumen.

From an epidemiologic perspective, global distribution of the more pathogenic genotypes might help to explain disease distribution (Bravo et al. 2002; Kersulyte et al. 2000). For example, in Asia, a region with high gastric cancer incidence, nearly all strains of *H. pylori* possess the PAI, whereas the rates are closer to 50% in the United States and Europe (Covacci et al. 1999). In addition to regional differences in the presence of the PAI, polymorphisms within the PAI vary geographically and may relate to disease pathogenesis (Yamaoka et al. 2000). Also being mapped to determine population effects are the *vacA* genotypes; less virulent strains with *vacA* s2 genotype are extremely rare in Asia, whereas they comprise 20%–40% of strains in North America, northern Europe, and Australia (Van Doorn et al. 1999). Recently, an *H. pylori* genotype database has been developed to assess the breadth of gene sequences in isolates entered worldwide (Ahmed et al. 2007); this database will enable a broader understanding of the genetics of virulence and disease. Such an understanding will be complicated, however, by the existence of multiple genotypes of varying pathogenicity within individual hosts (Matteo et al. 2007).

Environmental Factors

Although diet has been extensively studied in gastric carcinogenesis, it is not well studied in the setting of *H. pylori* infection. A prospective cohort study in Scandinavia demonstrated a protective effect of vitamin C and beta-carotene in individuals infected with *H. pylori* (Ekstrom et al. 2000). This finding was initially supported in a prospective study from Colombia that demonstrated that *H. pylori* eradication and increased dietary vitamin C and beta-carotene independently prevented progression of preneoplastic lesions to cancer (Correa et al. 2000). Long-term follow-up from this trial, however, showed that the benefits of vitamin C and beta-carotene—but not of *H. pylori* eradication—disappeared when participants were followed up for a longer period of time (Mera et al. 2005).

Increased dietary salt may also increase gastric cancer risk. As mentioned above, before the discovery of *H. pylori*, salt was linked to gastric cancer in humans. In animal models, increased dietary salt in the setting of *H. pylori* augments gastric carcinogenesis (Fox et al. 2003). In humans, a study of 2,476 participants followed prospectively for 14 years, the years yielded a significant relationship between increased salt consumption and the development of gastric cancer, but only in subjects who were both infected with *H. pylori* and had atrophic gastritis (Shikata et al. 2006).

Another area of increasing interest is the possibility of coinfection with other organisms influencing the outcome of *H. pylori* in humans. By mitigating the Th1 inflammatory response, helminths could theoretically reduce gastric inflammation and cancer incidence. This reduction of inflammation has been observed in animals coinfected with *H. pylori* felis and helminths (Fox et al. 2000) and has also been suggested in small ecologic studies of humans (Mitchell et al. 2002; Whary et al. 2005). The clinical significance of these findings is now an intense area of investigation.

Unanswered Epidemiologic Questions

Although much has been learned about *H. pylori* and its relationship to gastric cancer since 1982, some critical questions remain unresolved and additional research is needed.

Why Do Males Have Higher Risk for Cancer?

More men than women develop gastric cancer. In fact, distal, noncardia gastric cancers is on average twice as common among men compared with women (Crew and Neugut 2006). This higher incidence of gastric cancer observed among men is partially explained by higher *H. pylori* infection rates. A metaanalysis of large, population-based studies found that male gender was significantly associated with *H. pylori* infection (OR = 1.16; 95% CI, 1.11–1.22), although this difference was not observed in studies of children sufficient to explain the differences in cancer incidence (De Martel and Parsonnet 2006). Other cofactors beyond *H. pylori* infection such as smoking, alcohol, increased dietary salt, or even a protective effect of female hormones have not been demonstrated to explain the differences observed (Ferreccio et al. 2007; Lindblad et al. 2005). Understanding gender differences in gastric cancer would provide insights into carcinogenesis more generally.

Is There an African Enigma?

Although *H. pylori* infection has a significant role in gastric cancer development, higher *H. pylori* infection rates are not always associated with higher gastric cancer rates. As early as 1992, Holcombe noted that despite high rates of *H. pylori* infection and *H. pylori*–associated gastritis in Nigeria, gastric cancer was uncommon; they termed this "the Africa enigma" (Holcombe 1992). Additional research has noted a similar pattern of high *H. pylori* infection rates and low gastric cancer incidence in India, Bangladesh, Pakistan, and Thailand (Miwa et al. 2002; Singh and

Ghoshal 2006). Some have maintained that microbial coinfection with helminths explains these paradoxic *H. pylori* responses (Whary et al. 2005). For example, several human studies have found shifts in *H. pylori* immunoglobulin (Ig)G antibodies in helminth-infected populations to IgG1 rather than IgG2 (Mitchell et al. 2002; Whary et al. 2005), indicating different—and possibly less inflammatory—immunologic response to gastric infection. Others maintain that dietary factors, host genetics, or bacterial factors explain these observations (Ghoshal et al. 2007; Louw et al. 2001; Singh and Ghoshal 2006). Still others aver there is no "African enigma," only low life expectancy among the poorest individuals, deficient cancer reporting, and *H. pylori* diagnostic artifact (Agha and Graham 2005). It is unlikely, however, that anyone would argue that host response to infection does not vary from individual to individual. Understanding this variability across populations could be the key to identifying attainable cancer intervention strategies.

Can Treatment of Helicobacter pylori Prevent Cancer?

Treatment of *H. pylori* infection to prevent gastric cancer is an appealing prevention strategy and numerous studies have indicated it is also likely to be cost effective (Fendrick et al. 1999; Mason et al. 2002; Parsonnet et al. 1996; Roderick et al. 2003). Yet, to date, no randomized, prospective trials have shown that eradication of existing infection prevents cancer. Several studies, however, show tantalizing evidence that such treatment might work. A nonrandomized, nonblinded trial of *H. pylori* eradication in Japanese patients with early gastric cancer showed significantly lower rates of cancer relapse in patients who received eradication therapy than those who did not (Uemura et al. 1997); this finding has been supported by larger retrospective analyses (Nakagawa et al. 2006) but has yet to be supported by a randomized clinical trial. Randomized trials of preneoplastic conditions indicate that *H. pylori* therapy may improve the overall pathology of the stomach, decreasing atrophic gastritis and intestinal metaplasia (Correa et al. 2000; Ley et al. 2004; Sung et al. 2000). Although an improvement in histopathology has been reported in only a minority of subjects, the study with longest follow-up indicates that the differences between treated and untreated subjects may become increasingly evident as years pass (Mera et al. 2005).

The one randomized trial of cancer prevention completed to date—a study conducted in China—was underpowered for the final endpoint of gastric cancer (Parsonnet and Forman 2004). Among a subset of participants, however, eradication did show a benefit; i.e., among participants who had no preneoplastic lesions in their first endoscopy, there was a decreased incidence of gastric cancer after active treatment (Wong et al. 2004). Further randomized trials are underway. Given the difficulties of conducting these trials and the decreasing incidence of gastric cancer worldwide, however, it is possible none will ever be completed successfully. In the absence of clinical trial support, there is yet no policy in place in any country to screen populations for infection and treat all infected individuals. Instead, consensus groups have recommended screening and treating those at highest risk: i.e., those with a family history of cancer, with prior gastric surgery, or with documented atrophic gastritis (Malfertheiner et al. 2007).

What Is the Best Approach to Helicobacter pylori Prevention?

H. pylori and its related diseases are disappearing spontaneously worldwide. Acceleration of the organism's disappearance through primary prevention, however, could save myriad lives. Because the preponderance of evidence supports person-to-person transmission of *H. pylori*, efforts to improve hygiene and handwashing may be effective strategies for gastric cancer prevention. These methods have the added advantage of preventing other enteric infections, and of proven feasibility. Consuming a diet high in vegetables and fruit and low in salt may also be an effective way to reduce stomach cancer rates (World Cancer Research Fund and American Institute for Cancer Research 1997), but effectuating dietary change can be an onerous challenge.

A vaccine for *H. pylori* is another appealing primary prevention strategy for gastric cancer. The precedent for a vaccine to prevent cancer has already been established with the hepatitis B vaccine to prevent hepatomas and HPV vaccine to prevent cervical cancer. Unfortunately, the development of an *H. pylori* vaccine has proved to be more difficult than the hepatitis B and HPV vaccines. Although an enormous amount has been learned about the immune response to *H. pylori* infection (Suerbaum and Michetti 2002), there are as yet no known correlates of protective immunity for infection or reinfection in humans. In mouse models, infection can be prevented with a variety of vaccines (Arora and Czinn 2005). The human response to *H. pylori* is more complicated, however, and no vaccine has yet reached efficacy trials in humans (Kabir 2007). Should a vaccine appear, it is likely to be a cost-effective approach to cancer prevention (Rupnow et al. 1999; Rupnow et al. 2001).

In considering *H. pylori* prevention, however, it is also important to assess whether there may be downsides to the strategy. Recent observational data suggest that *H. pylori* might provide some protection against acute diseases of childhood. Theoretically, by upregulating the systemic Th1 immune response, infection might assist in combating common childhood infections. A small amount of evidence supports this; *H. pylori* infection has been linked to protection from both diarrheal disease (Perry et al. 2004)—a leading cause of death in children worldwide—and tuberculosis (Perry et al. 2007). In addition, some argue that absence of *H. pylori*'s Th1 stimulation in young children has fostered increases in asthma and other allergic diseases (Chen and Blaser 2007; Kosunen et al. 2002; Pessi et al. 2005). Finally, in adults, an ever-increasing amount of data indicates that the absence of *H. pylori* promotes both GERD and its long-term consequence—adenocarcinoma of the esophagus (Cremonini et al. 2003; de Martel et al. 2005).

Although there is no doubt that the areas of the world that have spontaneously lost *H. pylori* from their resident flora experience greater life expectancy and decreased morbidity than those in which it persists, we have, in the past, always "let nature take its course." With a vaccine, we would be in a position to precipitate the loss of *H. pylori* even from regions of the world where acute infectious diseases run rampant. One would be prudent, then, to ask whether the survival of *H. pylori* in these microbial ecologies provides a survival advantage to children that counterbalances the risks of gastric adenocarcinoma later in life.

References

Aaltonen, L.A., Hamilton, S.R., World Health Organization., and International Agency for Research on Cancer. (2000). Pathology and genetics of tumours of the digestive system. Lyon: IARC*Press*.

Agha, A., and Graham, D.Y. (2005). Evidence-based examination of the African enigma in relation to Helicobacter pylori infection. Scand J Gastroenterol. 40:523–9.

Ahmed, N., Majeed, A.A., Ahmed, I., Hussain, M.A., Alvi, A., Devi, S.M., Rizwan, M., Ranjan, A., Sechi, L.A., and Megraud, F. (2007). genoBASE pylori: a genotype search tool and database of the human gastric pathogen Helicobacter pylori. Infect Genet Evol. 7:463–8.

American Cancer Society, Atlanta, GA. (2007). http://www.cancer.org/docroot/MED/content/downloads/MED_1_1x_CFF2007_Estimated_Deaths_Sites_by_State.asp.

Arora, S., and Czinn, S.J. (2005). Vaccination as a method of preventing Helicobacter pylori-associated gastric cancer. Cancer Epidemiol Biomarkers Prev. 14:1890–1.

Bamford, K.B., Bickley, J., Collins, J.S., Johnston, B.T., Potts, S., Boston, V., Owen, R.J., and Sloan, J.M. (1993). Helicobacter pylori: comparison of DNA fingerprints provides evidence for intrafamilial infection. Gut. 34:1348.

Banatvala, N., Mayo, K., Megraud, F., Jennings, R., Deeks, J.J., and Feldman, R.A. (1993). The cohort effect and Helicobacter pylori. J Infect Dis. 168:219.

Bergin, I.L., Sheppard, B.J., and Fox, J.G. (2003). Helicobacter pylori infection and high dietary salt independently induce atrophic gastritis and intestinal metaplasia in commercially available outbred Mongolian gerbils. Dig Dis Sci. 48:475–85.

Blaser, M.J., and Berg, D.E. (2001). Helicobacter pylori genetic diversity and risk of human disease. J Clin Invest. 107:767–73.

Blecker, U., Hauser, B., Lanciers, S., Peeters, S., Suys, B., and Vandenplas, Y. (1993). The prevalence of Helicobacter pylori-positive serology in asymptomatic children. J Pediatr Gastroenterol Nutr. 16:252.

Bravo, L.E., van Doom, L.J., Realpe, J.L., and Correa, P. (2002). Virulence-associated genotypes of Helicobacter pylori: do they explain the African enigma? Am J Gastroenterol. 97:2839–42.

Brenner, H., Arndt, V., Stegmaier, C., Ziegler, H., and Rothenbacher, D. (2004). Is Helicobacter pylori infection a necessary condition for noncardia gastric cancer? Am J Epidemiol. 159:252–8.

Brown, L.M. (2000). Helicobacter pylori: epidemiology and routes of transmission. Epidemiol Rev. 22:283–97.

Brown, L.M., and Devesa, S.S. (2002). Epidemiologic trends in esophageal and gastric cancer in the United States. Surg Oncol Clin North Am. 11:235–56.

Brown, L.M., Thomas, T.L., Ma, J.L., Chang, Y.S., You, W.C., Liu, W.D., Zhang, L., Pee, D., and Gail, M.H. (2002). Helicobacter pylori infection in rural China: demographic, lifestyle and environmental factors. Int J Epidemiol. 31:638–45.

Caldas, C., Carneiro, F., Lynch, H.T., Yokota, J., Wiesner, G.L., Powell, S.M., Lewis, F.R., Huntsman, D.G., Pharoah, P.D., Jankowski, J.A., MacLeod, P., Vogelsang, H., Keller, G., Park, K.G., Richards, F.M., Maher, E.R., Gayther, S.A., Oliveira, C., Grehan, N., Wight, D., Seruca, R.,

Roviello, F., Ponder, B.A., and Jackson, C.E. (1999). Familial gastric cancer: overview and guidelines for management. J Med Genet. 36:873–80.

Chao, A., Thun, M.J., Henley, S.J., Jacobs, E.J., McCullough, M.L., and Calle, E.E. (2002). Cigarette smoking, use of other tobacco products and stomach cancer mortality in US adults: The Cancer Prevention Study II. Int J Cancer. 101, 380–9.

Chen, T.S. (2004). Is the coccoid form of Helicobacter pylori viable and transmissible? J Chin Med Assoc. 67:547–8.

Chen, Y., and Blaser, M.J. (2007). Inverse associations of Helicobacter pylori with asthma and allergy. Arch Intern Med. 167:821–7.

Chen, J., Bu, X.L., Wang, Q.Y., Hu, P.J., and Chen, M.H. (2007). Decreasing seroprevalence of Helicobacter pylori infection during 1993–2003 in Guangzhou, southern China. Helicobacter. 12:164–9.

Chong, J., Marshall, B.J., Barkin, J.S., McCallum, R.W., Reiner, D.K., Hoffman, S.R., and O'Phelan, C. (1994). Occupational exposure to Helicobacter pylori for the endoscopy professional: a sera epidemiological study. Am J Gastroenterol. 89:1987.

Chow, W.H., Blaser, M.J., Blot, W.J., Gammon, M.D., Vaughan, T.L., Risch, R.A., Perez-Perez, G.I., Schoenberg, J.B., Stanford, J.L., Rotterdam, H., West, A.B., and Fraumeni, J.F. (1998). An inverse relation between cagA+ strains of Helicobacter pylori infection and risk of esophageal and gastric cardia adenocarcinoma. Cancer Res. 58:588.

Chow, W.H., Swanson, C.A., Lissowska, J., Groves, F.D., Sobin, L.H., Nasierowska-Guttmejer, A., Radziszewski, J., Regula, J., Hsing, A.W., Jagannatha, S., Zatonski, W., and Blot, W.J. (1999). Risk of stomach cancer in relation to consumption of cigarettes, alcohol, tea and coffee in Warsaw, Poland. Int J Cancer. 81:871–6.

Coggon, D., Osmond, C., and Barker, D.J. (1990). Stomach cancer and migration within England and Wales. Br J Cancer. 61:573–4.

Coleman, M.P., Esteve, J., Damiecki, P., Arslan, A., and Renard, H. (1993). Trends in Cancer incidence and mortality. Lyon: International Agency for Research on Cancer.

Colin-Jones, D.G., Langman, M.J., Lawson, D.H., Logan, R.F., Paterson, K.R., and Vessey, M.P. (1992). Postmarketing surveillance of the safety of cimetidine: 10 year mortality report. Gut. 33:1280.

Con, S.A., Takeuchi, H., Valerin, A.L., Con-Wong, R., Con-Chin, G.R., Con-Chin, V.G., Nishioka, M., Mena, F., Brenes, F., Yasuda, N., Araki, K., and Sugiura, T. (2007). Diversity of Helicobacter pylori cagA and vacA genes in Costa Rica: its relationship with atrophic gastritis and gastric cancer. Helicobacter 12:547–52.

Cooke, C.L., Huff, J.L., and Solnick, J.V. (2005). The role of genome diversity and immune evasion in persistent infection with Helicobacter pylori. FEMS Immunol Med Microbiol. 45:11–23.

Correa, P. (1992). Human gastric carcinogenesis: a multistep and multifactorial process—First American Cancer Society Award lecture on cancer epidemiology and prevention. Cancer Res. 52:6735.

Correa, P. (2002). Gastric neoplasia. Curr Gastroenterol Rep 4:463–70.

Correa, P., and Chen, V.W. (1994). Gastric cancer. Cancer Surv. 19–20:55–76.

Correa, P., Fontham, E.T., Bravo, J.C., Bravo, L.E., Ruiz, B., Zarama, G., Realpe, J.L., Malcom, G.T., Li, D., Johnson, W.D., and Mera, R. (2000). Chemoprevention of gastric dysplasia: randomized trial of antioxidant supplements and anti-Helicobacter pylori therapy. J Natl Cancer Inst. 92:1881.

Correa, P., Haenszel, W., Cuello, C., Tannenbaum, S., and Archer, M. (1975). A model for gastric cancer epidemiology. Lancet. 2:58.

Covacci, A., Telford, J.L., Del Giudice, G., Parsonnet, J., and Rappuoli, R. (1999). Helicobacter pylori virulence and genetic geography. Science. 284:1328–33.

Cremonini, F., Di Caro, S., Delgado-Aros, S., Sepulveda, A., Gasbarrini, G., Gasbarrini, A., and Camilleri, M. (2003). Meta-analysis: the relationship between Helicobacter pylori infection and gastro-oesophageal reflux disease. Aliment Pharmacol Ther. 18:279–89.

Crew, K.D., and Neugut, A.I. (2006). Epidemiology of gastric cancer. World J Gastroenterol. 12:354–62.

Delport, W., and van der Merwe, S.W. (2007). The transmission of Helicobacter pylori: the effects of analysis method and study population on inference. Best Pract Res Clin Gastroenterol. 21:215–36.

de Martel, C., Llosa, A.E., Farr, S.M., Friedman, G.D., Vogelman, J.H., Orentreich, N., Corley, D.A., and Parsonnet, J. (2005). Helicobacter pylori infection and the risk of development of esophageal adenocarcinoma. J Infect Dis. 191:761–7.

De Martel, C., and Parsonnet, J. (2006). Helicobacter pylori infection and gender: a meta-analysis of population-based prevalence surveys. Dig Dis Sci. 51.

De Schryver, A., Van Winckel, M., Cornelis, K., Moens, G., Devlies, G., and De Backer, G. (2006). Helicobacter pylori infection: further evidence for the role of feco-oral transmission. Helicobacter. 11:523–8.

Devesa, S.S., Blot, W.J., and Fraumeni, J.F., Jr. (1998). Changing patterns in the incidence of esophageal and gastric carcinoma in the United States. Cancer. 83:2049–53.

Doenges, J.L. (1938). Spirochaetes in gastric glands of macacus rhesus and humans without defined history of related disease. Proc Soc Exp Biol Med. 38:536.

Dore, M.P., Sepulveda, A.R., El-Zimaity, H., Yamaoka, Y., Osato, M.S., Mototsugu, K., Nieddu, A.M., Realdi, G., and Graham, D.Y. (2001). Isolation of Helicobacter pylori from sheep-implications for transmission to humans. Am J Gastroenterol. 96:1396–401.

Drazek, E.S., Dubois, A., and Holmes, R.K. (1994). Characterization and presumptive identification of Helicobacter pylori isolates from rhesus monkeys. J Clin Microbiol. 32:1799.

Drumm, B., Perez-Perez, G.I., Blaser, M.J., and Sherman, P.M. (1990). Intrafamilial clustering of Helicobacter pylori infection. N Engl J Med. 322:359.

Dubois, A., Fiala, N., Heman-Ackah, L.M., Drazek, E.S., Tarnawski, A., Fishbein, W.N., Perez-Perez, G.I., and Blaser, M.J. (1994). Natural gastric infection with Helicobacter pylori in monkeys: a model for spiral bacteria infection in humans. Gastroenterology. 106:1405.

Egemen, A., Yilmaz, O., Akil, I., and Altuglu, I. (2006). Evaluation of association between hepatitis A and Helicobacter pylori infections and routes of transmission. Turk J Pediatr. 48:135–9.

Ekstrom, A.M., Serafini, M., Nyren, O., Hansson, L.E., Ye, W., and Wolk, A. (2000). Dietary antioxidant intake and the risk of cardia cancer and noncardia cancer of the intestinal and diffuse types: a population-based case-control study in Sweden. Int J Cancer. 87:133–40.

El Hajj, I., Hashash, J.G., Baz, E.M., Abdul-Baki, H., and Sharara, A.I. (2007). ABO blood group and gastric cancer: rekindling an old fire? South Med J. 100:726–7.

El-Omar, E.M., Carrington, M., Chow, W.H., McColl, K.E., Bream, J.H., Young, H.A., Herrera, J., Lissowska, J., Yuan, C.C., Rothman, N., Lanyon, G., Martin, M., Fraumeni, J.F., Jr., and Rabkin, C.S. (2000). Interleukin-1 polymorphisms associated with increased risk of gastric cancer. Nature. 404:398–402.

El-Omar, E.M., Rabkin, C.S., Gammon, M.D., Vaughan, T.L., Risch, H.A., Schoenberg, J.B., Stanford, J.L., Mayne, S.T., Goedert, J., Blot, W.J., Fraumeni, J.F., Jr., and Chow, W.H. (2003). Increased risk of noncardia gastric cancer associated with proinflammatory cytokine gene polymorphisms. Gastroenterology. 124:1193–201.

Eurogast Study Group. (1993). Epidemiology of, and risk factors for, Helicobacter pylori infection among 3194 asymptomatic subjects in 17 populations. Gut. 34:1672.

Everhart, J.E., Kruszon-Moran, D., Perez-Perez, G.I., Tralka, T.S., and McQuillan, G. (2000). Seroprevalence and ethnic differences in Helicobacter pylori infection among adults in the United States. J Infect Dis. 181:1359–63.

Falush, D., Wirth, T., Linz, B., Pritchard, J.K., Stephens, M., Kidd, M., Blaser, M.J., Graham, D.Y., Vacher, S., Perez-Perez, G.I., Yamaoka, Y., Megraud, F., Otto, K., Reichard, U., Katzowitsch, E., Wang, X., Achtman, M., and Suerbaum, S. (2003). Traces of human migrations in Helicobacter pylori populations. Science. 299:1582–5.

Farrow, D.C., Vaughan, T.L., Sweeney, C., Gammon, M.D., Chow, W.H., Risch, H.A., Stanford, J.L., Hansten, P.D., Mayne, S.T., Schoenberg, J.B., Rotterdam, H., Ahsan, H., West, A.B., Dubrow, R., Fraumeni, J.F., Jr., and Blot, W.J. (2000). Gastroesophageal reflux disease, use of H2 receptor antagonists, and risk of esophageal and gastric cancer. Cancer Causes Control. 11:231–8.

Fendrick, A.M., Chernew, M.E., Hirth, R.A., Bloom, B.S., Bandekar, R.R., and Scheiman, J.M. (1999). Clinical and economic effects of population-based Helicobacter pylori screening to prevent gastric cancer. Arch Intern Med. 159:142.

Ferguson, D.A., Jr., Li, C., Patel, N.R., Mayberry, W.R., Chi, D.S., and Thomas, E. (1993). Isolation of Helicobacter pylori from saliva. J Clin Microbiol. 31:2802.

Ferlay, J., Bray, F., Pisani, P., and Parkin, D.M. (2004). GLOBOCAN 2002: cancer incidence, mortality and prevalence worldwide. Lyon: IARC*Press*.

Ferreccio, C., Rollan, A., Harris, P.R., Serrano, C., Gederlini, A., Margozzini, P., Gonzalez, C., Aguilera, X., Venegas, A., and Jara, A. (2007). Gastric cancer is related to early Helicobacter pylori infection in a high-prevalence country. Cancer Epidemiol Biomarkers Prev. 16:662–7.

Forman, D., Sitas, F., Newell, D.G., Stacey, A.R., Boreham, J., Peto, R., Campbell, T.C., Li, J.Y., and Chen, J. (1990). Geographic association of Helicobacter pylori antibody prevalence and gastric cancer mortality in rural China. Int J Cancer. 46:608.

Fox, J.G., Beck, P., Dangler, C.A., Whary, M.T., Wang, T.C., Shi, H.N., and Nagler-Anderson, C. (2000). Concurrent enteric helminth infection modulates inflammation and gastric immune responses and reduces Helicobacter-induced gastric atrophy. Nat Med. 6:536.

Fox, J.G., Rogers, A.B., Ihrig, M., Taylor, N.S., Whary, M.T., Dockray, G., Varro, A., and Wang, T.C. (2003). Helicobacter pylori-associated gastric cancer in INS-GAS mice is gender specific. Cancer Res. 63:942–50.

Friis, L., Engstrand, L., and Edling, C. (1996). Prevalence of Helicobacter pylori infection among sewage workers. Scand J Work Environ Health. 22:364–8.

Fujimoto, Y., Furusyo, N., Toyoda, K., Takeoka, H., Sawayama, Y., and Hayashi, J. (2007). Intrafamilial transmission of Helicobacter pylori among the population of endemic areas in Japan. Helicobacter. 12:170–6.

Fukao, A., Tsubono, Y., Tsuji, I., Hisamichi, S., Sugahara, N., and Takano, A. (1995). The evaluation of screening for gastric cancer in Miyagi Prefecture, Japan: a population-based case-control study. Int J Cancer. 60:45–8.

Garza-Gonzalez, E., Bosques-Padilla, F.J., Perez-Perez, G.I., Flores-Gutierrez, J.P., and Tijerina-Menchaca, R. (2004). Association of gastric cancer, HLA-DQA1, and infection with Helicobacter pylori CagA+ and VacA+ in a Mexican population. J Gastroenterol. 39:1138–42.

Gayther, S.A., Gorringe, K.L., Ramus, S.J., Huntsman, D., Roviello, F., Grehan, N., Machado, J.C., Pinto, E., Seruca, R., Halling, K., MacLeod, P., Powell, S.M., Jackson, C.E., Ponder, B.A., and Caldas, C. (1998). Identification of germ-line E-cadherin mutations in gastric cancer families of European origin. Cancer Res. 58:4086–9.

Genta, R.M., and Graham, D.Y. (1993). Intestinal metaplasia, not atrophy or achlorhydria, creates a hostile environment for Helicobacter pylori. Scand J Gastroenterol. 28:924.

Gerhard, M., Lehn, N., Neumayer, N., Boren, T., Rad, R., Schepp, W., Miehlke, S., Classen, M., and Prinz, C. (1999). Clinical relevance of the Helicobacter pylori gene for blood-group antigen-binding adhesin. Proc Natl Acad Sci USA. 96:12778.

Ghoshal, U.C., Tripathi, S., and Ghoshal, U. (2007). The Indian enigma of frequent H. pylori infection but infrequent gastric cancer: is the magic key in Indian diet, host's genetic make up, or friendly bug? Am J Gastroenterol. 102:2113–4.

Gillen, D., and McColl, K.E. (2001). Problems associated with the clinical use of proton pump inhibitors. Pharmacol Toxicol. 89:281–6.

Glynn, M.K., Friedman, C.R., Gold, B.D., Khanna, B., Hutwagner, L., Iihoshi, N., Revollo, C., and Quick, R. (2002). Seroincidence of Helicobacter pylori infection in a cohort of rural Bolivian children: acquisition and analysis of possible risk factors. Clin Infect Dis. 35:1059–65.

Goh, K.L., and Parasakthi, N. (2001). The racial cohort phenomenon: seroepidemiology of Helicobacter pylori infection in a multiracial South-East Asian country. Eur J Gastroenterol Hepatol. 13:177–83.

Gonzalez, C.A., Jakszyn, P., Pera, G., Agudo, A., Bingham, S., Palli, D., Ferrari, P., Boeing, H., del Giudice, G., Plebani, M., Carneiro, F., Nesi, G., Berrino, F., Sacerdote, C., Tumino, R., Panico, S., Berglund, G., Siman, H., Nyren, O., Hallmans, G., Martinez, C., Dorronsoro, M., Barricarte, A., Navarro, C., Quiros, J.R., Allen, N., Key, T.J., Day, N.E., Linseisen, J., Nagel, G.,

Bergmann, M.M., Overvad, K., Jensen, M.K., Tjonneland, A., Olsen, A., Bueno-de-Mesquita, H.B., Ocke, M., Peeters, P.H., Numans, M.E., Clavel-Chapelon, F., Boutron-Ruault, M.C., Trichopoulou, A., Psaltopoulou, T., Roukos, D., Lund, E., Hemon, B., Kaaks, R., Norat, T., and Riboli, E. (2006). Meat intake and risk of stomach and esophageal adenocarcinoma within the European Prospective Investigation Into Cancer and Nutrition (EPIC). J Natl Cancer Inst. 98:345–54.

Goodman, K.J., Correa, P., Tengana Aux, H.J., Ramirez, H., DeLany, J.P., Guerrero Pepinosa, O., Lopez Quinones, M., and Collazos Parra, T. (1996). Helicobacter pylori infection in the Colombian Andes: a population-based study of transmission pathways. Am J Epidemiol. 144:290–9.

Goodman, K.J., O'Rourke, K., Day, R.S., Wang, C., Nurgalieva, Z., Phillips, C.V., Aragaki, C., Campos, A., and de la Rosa, J.M. (2005). Dynamics of Helicobacter pylori infection in a US-Mexico cohort during the first two years of life. Int J Epidemiol. 34:1348–55.

Graham, D.Y., Alpert, L.C., Smith, J.L., and Yoshimura, H.H. (1988). Iatrogenic Campylobacter pylori infection is a cause of epidemic achlorhydria. Am J Gastroenterol. 83:974.

Graham, D.Y., Malaty, H.M., Evans, D.G., Evans, D.J., Jr., Klein, P.D., and Adam, E. (1991). Epidemiology of Helicobacter pylori in an asymptomatic population in the United States. Effect of age, race, and socioeconomic status. Gastroenterology. 100:1495–501.

Granstrom, M., Tindberg, Y., and Blennow, M. (1997). Seroepidemiology of Helicobacter pylori infection in a cohort of children monitored from 6 months to 11 years of age. J Clin Microbiol. 35:468–70.

Grubel, P., Hoffman, J.S., Chong, F.K., Burstein, N.A., Mepani, C., and Cave, D.R. (1997). Vector potential of houseflies (Musca domestica) for Helicobacter pylori. J Clin Microbiol. 35:1300–3.

Guilford, P.J., Hopkins, J.B., Grady, W.M., Markowitz, S.D., Willis, J., Lynch, H., Rajput, A., Wiesner, G.L., Lindor, N.M., Burgart, L.J., Toro, T.T., Lee, D., Limacher, J.M., Shaw, D.W., Findlay, M.P., and Reeve, A.E. (1999). E-cadherin germline mutations define an inherited cancer syndrome dominated by diffuse gastric cancer. Hum Mutat. 14:249–55.

Guilford, P., Hopkins, J., Harraway, J., McLeod, M., McLeod, N., Harawira, P., Taite, H., Scoular, R., Miller, A., and Reeve, A.E. (1998). E-cadherin germline mutations in familial gastric cancer. Nature. 392:402–5.

Handt, L.K., Fox, J.G., Stalis, I.H., Rufo, R., Lee, G., Linn, J., Li, X., and Kleanthous, H. (1995). Characterization of feline Helicobacter pylori strains and associated gastritis in a colony of domestic cats. J Clin Microbiol. 33:2280–9.

Hansson, L.E., Nyren, O., Hsing, A.W., Bergstrom, R., Josefsson, S., Chow, W., Fraumeni, J.F., and Adami, H. (1996). Risk of stomach cancer in patients with gastric or duodenal ulcer disease. New Engl J Med. 335:242.

Helicobacter and Cancer Collaborative Group (2001). Gastric cancer and Helicobacter pylori: a combined analysis of 12 case control studies nested within prospective cohorts. Gut. 49:347–53.

Herrmann, K., and Niedobitek, G. (2003). Epstein-Barr virus-associated carcinomas: facts and fiction. J Pathol. 199:140–5.

Higham, J., Kang, J.Y., and Majeed, A. (2002). Recent trends in admissions and mortality due to peptic ulcer in England: increasing frequency of haemorrhage among older subjects. Gut. 50:460–4.

Hirata, I., Murano, M., Ishiguro, T., Toshina, K., Wang, F.Y., and Katsu, K. (2007). HLA genotype and development of gastric cancer in patients with Helicobacter pylori infection. Hepatogastroenterology. 54:990–4.

Holcombe, C. (1992). Helicobacter pylori: the African enigma. Gut 33:429.

Hopkins, R.J., Russell, R.G., O'Donnoghue, J.M., Wasserman, S.S., Lefkowitz, A., and Morris, J.G., Jr. (1990). Seroprevalence of Helicobacter pylori in Seventh-Day Adventists and other groups in Maryland. Lack of association with diet. Arch Intern Med. 150:2347.

Hopkins, R.J., Vial, P.A., Ferreccio, C., Ovalle, J., Prado, P., Sotomayor, V., Russell, R.G., Wasserman, S.S., and Morris, J.G., Jr. (1993). Seroprevalence of Helicobacter pylori in Chile: vegetables may serve as one route of transmission. J Infect Dis. 168:222.

Howson, C., Hiyama, T., and Wynder, E. (1986). The decline in gastric cancer: epidemiology of an unplanned triumph. Epidemiol Rev. 8:1.

Hsing, A.W., Hansson, L.E., McLaughlin, J.K., Nyren, O., Blot, W.J., Ekbom, A., and Fraumeni, J.F., Jr. (1993). Pernicious anemia and subsequent cancer. A population-based cohort study. Cancer. 71:745–50.

Hsu, P.I., Lai, K.H., Hsu, P.N., Lo, G.H., Yu, H.C., Chen, W.C., Tsay, F.W., Lin, H.C., Tseng, H. H., Ger, L.P., and Chen, H.C. (2007). Helicobacter pylori infection and the risk of gastric malignancy. Am J Gastroenterol. 102:725–30.

Huang, J.Q., Sridhar, S., Chen, Y., and Hunt, R.H. (1998). Meta-analysis of the relationship between Helicobacter pylori seropositivity and gastric cancer. Gastroenterology. 114:1169–79.

Huang, J.Q., Zheng, G.F., Sumanac, K., Irvine, E.J., and Hunt, R.H. (2003). Meta-analysis of the relationship between cagA seropositivity and gastric cancer. Gastroenterology. 125:1636–44.

Hulten, K., Han, S., Enroth, H., Klein, P., Opekun, A., Gilman, R., Evans, D., Engstrand, L., Graham, D., and El-Zaatari, F. (1996). Helicobacter pylori in the drinking water in Peru. Gastroenterology. 110:1031–1035.

IARC (2004). Tobacco Smoke and Involuntary Smoking. IARC Monographs. 83.

IARC Working Group on the Evaluation of Carcinogenic Risks to Humans. (1994). Helicobacter pylori schistosomes, liver flukes and Helicobacter pylori: views and expert opinions of an IARC Working Group on the Evaluation of Carcinogenic Risks to Humans. Lyon: IARC; pp. 177.

Ilver, D., Arnqvist, A., Ogren, J., Frick, I.M., Kersulyte, D., Incecik, E.T., Berg, D.E., Covacci, A., Engstrand, L., and Boren, T. (1998). Helicobacter pylori adhesin binding fucosylated histo-blood group antigens revealed by retagging. Science. 279(5349):373–7.

Israel, D.A., Salama, N., Krishna, U., Rieger, U.M., Atherton, J.C., Falkow, S., and Peek, R.M., Jr. (2001). Helicobacter pylori genetic diversity within the gastric niche of a single human host. Proc Natl Acad Sci USA. 98:14625–30.

Ito, L.S., Oba-Shinjo, S.M., Shinjo, S.K., Uno, M., Marie, S.K., and Hamajima, N. (2006). Community-based familial study of Helicobacter pylori infection among healthy Japanese Brazilians. Gastric Cancer. 9:208–16.

Jeggli, S., Steiner, D., Joller, H., Tschopp, A., Steffen, R., and Hotz, P. (2004). Hepatitis E, Helicobacter pylori, and gastrointestinal symptoms in workers exposed to waste water. Occup Environ Med. 61:622–7.

Jemal, A., Murray, T., Samuels, A., Ghafoor, A., Ward, E., and Thun, M.J. (2003). Cancer statistics, 2003. CA Cancer J Clin. 53:5–26.

Ji, B.T., Chow, W.H., Yang, G., McLaughlin, J.K., Gao, R.N., Zheng, W., Shu, X.O., Jin, F., Fraumeni, J.F., Jr., and Gao, Y.T. (1997). Body mass index and the risk of cancers of the gastric cardia and distal stomach in Shanghai, China. Cancer Epidemiol Biomarkers Prev. 6:481–5.

Kabir, S. (2007). The current status of Helicobacter pylori vaccines: a review. Helicobacter. 12:89–102.

Kai, M., Luebeck, E.G., and Moolgavkar, S.H. (1997). Analysis of the incidence of solid cancer among atomic bomb survivors using a two-stage model of carcinogenesis. Radiat Res. 148:348–58.

Kamangar, F., Dawsey, S.M., Blaser, M.J., Perez-Perez, G.I., Pietinen, P., Newschaffer, C.J., Abnet, C.C., Albanes, D., Virtamo, J., and Taylor, P.R. (2006). Opposing risks of gastric cardia and noncardia gastric adenocarcinomas associated with Helicobacter pylori seropositivity. J Natl Cancer Inst. 98:1445–52.

Karita, M., Teramukai, S., and Matsumoto, S. (2003). Risk of Helicobacter pylori transmission from drinking well water is higher than that from infected intrafamilial members in Japan. Dig Dis Sci. 48:1062–7.

Kelly, S.M., Pitcher, M.C., Farmery, S.M., and Gibson, G.R. (1994). Isolation of Helicobacter pylori from feces of patients with dyspepsia in the United Kingdom. Gastroenterology. 107:1671.

Kersulyte, D., Mukhopadhyay, A.K., Velapatino, B., Su, W., Pan, Z., Garcia, C., Hernandez, V., Valdez, Y., Mistry, R.S., Gilman, R.H., Yuan, Y., Gao, H., Alarcon, T., Lopez-Brea, M., Balakrish Nair, G., Chowdhury, A., Datta, S., Shirai, M., Nakazawa, T., Ally, R., Segal, I., Wong, B.C., Lam, S.K., Olfat, F.O., Boren, T., Engstrand, L., Torres, O., Schneider, R.,

Thomas, J.E., Czinn, S., and Berg, D.E. (2000). Differences in genotypes of Helicobacter pylori from different human populations. J Bacteriol. 182:3210–8.

Kim, J.W., Kim, J.G., Chae, S.L., Cha, Y.J., and Park, S.M. (2004). High prevalence of multiple strain colonization of Helicobacter pylori in Korean patients: DNA diversity among clinical isolates from the gastric corpus, antrum and duodenum. Korean J Intern Med. 19:1–9.

Kivi, M., Johansson, A.L., Reilly, M., and Tindberg, Y. (2005). Helicobacter pylori status in family members as risk factors for infection in children. Epidemiol Infect. 133:645–52.

Kivi, M., Tindberg, Y., Sorberg, M., Casswall, T.H., Befrits, R., Hellstrom, P.M., Bengtsson, C., Engstrand, L., and Granstrom, M. (2003). Concordance of Helicobacter pylori strains within families. J Clin Microbiol. 41:5604–8.

Klein, P.D., Gilman, R.H., Leon-Barua, R., Diaz, F., Smith, E.O., and Graham, D.Y. (1994). The epidemiology of Helicobacter pylori in Peruvian children between 6 and 30 months of age. Am J Gastroenterol. 89:2196.

Klein, P.D., Graham, D.Y., Gaillour, A., Opekun, A.R., and Smith, E.O. (1991). Water source as risk factor for Helicobacter pylori infection in Peruvian children. Gastrointestinal Physiology Working Group. Lancet. 337:1503–6.

Kneller, R.W., You, W.C., Chang, Y.S., Liu, W.D., Zhang, L., Zhao, L., Xu, G.W., Fraumeni, J.F., Jr., and Blot, W.J. (1992). Cigarette smoking and other risk factors for progression of precancerous stomach lesions. J Natl Cancer Inst. 84:1261.

Kokkola, A., Sjoblom, S.M., Haapiainen, R., Sipponen, P., Puolakkainen, P., and Jarvinen, H. (1998). The risk of gastric carcinoma and carcinoid tumours in patients with pernicious anaemia. A prospective follow-up study. Scand J Gastroenterol. 33:88–92.

Kolonel, L.N., Nomura, A.M., Hirohata, T., Hankin, J.H., and Hinds, M.W. (1981). Association of diet and place of birth with stomach cancer incidence in Hawaii Japanese and Caucasians. Am J Clin Nutr. 34:2478–85.

Kosunen, T.U., Hook-Nikanne, J., Salomaa, A., Sarna, S., Aromaa, A., and Haahtela, T. (2002). Increase of allergen-specific immunoglobulin E antibodies from 1973 to 1994 in a Finnish population and a possible relationship to Helicobacter pylori infections. Clin Exp Allergy. 32:373–8.

Krajden, S., Fuksa, M., Anderson, J., Kempston, J., Boccia, A., Petrea, C., Babida, C., Karmali, M., and Penner, J.L. (1989). Examination of human stomach biopsies, saliva, and dental plaque for Campylobacter pylori. J Clin Microbiol. 27:1397.

Kreinitz, W. (1906). Ueber das Auftreten von Spirochaeten verschiedener Form im Mageninhalt bei Carcinoma ventriculi. Dtsch Med Wochenschr. 32:872.

Kuipers, E.J., Lundell, L., Klinkenberg-Knol, E.C., Havu, N., Festen, H.P., Liedman, B., Lamers, C.B., Jansen, J.B., Dalenback, J., and Snel, P. (1996). Atrophic gastritis and Helicobacter pylori infection in patients with reflux esophagitis treated with omeprazole or fundoplication. N Engl J Med. 334:1018.

Kuo, C.H., Poon, S.K., Su, Y.C., Su, R., Chang, C.S., and Wang, W.C. (1999). Heterogeneous Helicobacter pylori isolates from H. pylori-infected couples in Taiwan. J Infect Dis. 180:2064–8.

Lagergren, J., Bergstrom, R., Lindgren, A., and Nyren, O. (1999). Symptomatic gastroesophageal reflux as a risk factor for esophageal adenocarcinoma. N Engl J Med. 340:825–31.

Langenberg, W., Rauws, E.A., Oudbier, J.H., and Tytgat, G.N. (1990). Patient-to-patient transmission of Campylobacter pylori infection by fiberoptic gastroduodenoscopy and biopsy. J Infect Dis. 161:507.

Lauren, P. (1965). The two histological main types of gastric cancer: diffuse and so-called intestinal type carcinoma. Acta Pathol Microbiol Scand. 64:31.

Lee, W.P., Tai, D.I., Lan, K.H., Li, A.F., Hsu, H.C., Lin, E.J., Lin, Y.P., Sheu, M.L., Li, C.P., Chang, F.Y., Chao, Y., Yen, S.H., and Lee, S.D. (2005). The –251T allele of the interleukin-8 promoter is associated with increased risk of gastric carcinoma featuring diffuse-type histopathology in Chinese population. Clin Cancer Res. 11:6431–41.

Leung, W.K., Sung, J.Y., Siu, K.L.K., Kwok, K.L., Cheng, A.F.B., and Sung, R. (1998). Isolation of H. pylori from vomitus in children. Gastroenterology. 114.

Ley, C., Mohar, A., Guarner, J., Herrera-Goepfert, R., Figueroa, L.S., Halperin, D., Johnstone, I., and Parsonnet, J. (2004). Helicobacter pylori eradication and gastric preneoplastic conditions: a randomized, double-blind, placebo-controlled trial. Cancer Epidemiol Biomarkers Prev. 13:4–10.

Li, Z., Chen, D., Zhang, C., Li, Y., Cao, B., Ning, T., Zhao, Y., You, W., and Ke, Y. (2005). HLA polymorphisms are associated with Helicobacter pylori infected gastric cancer in a high risk population, China. Immunogenetics. 56:781–7.

Lin, H.Y., Chuang, C.K., Lee, H.C., Chiu, N.C., Lin, S.P., and Yeung, C.Y. (2005). A seroepidemiologic study of Helicobacter pylori and hepatitis A virus infection in primary school students in Taipei. J Microbiol Immunol Infect. 38:176–82.

Lin, S.K., Lambert, J.R., Schembri, M.A., Nicholson, L., and Korman, M.G. (1994). Helicobacter pylori prevalence in endoscopy and medical staff. J Gastroenterol Hepatol. 9:319.

Lindblad, M., Rodriguez, L.A., and Lagergren, J. (2005). Body mass, tobacco and alcohol and risk of esophageal, gastric cardia, and gastric non-cardia adenocarcinoma among men and women in a nested case-control study. Cancer Causes Control. 16:285–94.

Linz, B., Balloux, F., Moodley, Y., Manica, A., Liu, H., Roumagnac, P., Falush, D., Stamer, C., Prugnolle, F., van der Merwe, S.W., Yamaoka, Y., Graham, D.Y., Perez-Trallero, E., Wadstrom, T., Suerbaum, S., and Achtman, M. (2007). An African origin for the intimate association between humans and Helicobacter pylori. Nature. 445:915–8.

Louw, J.A., Kidd, M.S., Kummer, A.F., Taylor, K., Kotze, U., and Hanslo, D. (2001). The relationship between Helicobacter pylori infection, the virulence genotypes of the infecting strain and gastric cancer in the African setting. Helicobacter. 6:268–73.

Lu, Y., Redlinger, T.E., Avitia, R., Galindo, A., and Goodman, K. (2002). Isolation and genotyping of Helicobacter pylori from untreated municipal wastewater. Appl Environ Microbiol. 68:1436–9.

Lugli, A., Zlobec, I., Singer, G., Kopp Lugli, A., Terracciano, L.M., and Genta, R.M. (2007). Napoleon Bonaparte's gastric cancer: a clinicopathologic approach to staging, pathogenesis, and etiology. Nat Clin Pract Gastroenterol Hepatol. 4:52–7.

Luman, W., Zhao, Y., Ng, H.S., and Ling, K.L. (2002). Helicobacter pylori infection is unlikely to be transmitted between partners: evidence from genotypic study in partners of infected patients. Eur J Gastroenterol Hepatol. 14:521–8.

Luzza, F., Mancuso, M., Imeneo, M., Contaldo, A., Giancotti, L., Pensabene, L., Doldo, P., Liberto, M.C., Strisciuglio, P., Foca, A., Guandalini, S., and Pallone, F. (2000). Evidence favouring the gastro-oral route in the transmission of Helicobacter pylori infection in children. Eur J Gastroenterol Hepatol. 12:623–7.

Machado, J.C., Figueiredo, C., Canedo, P., Pharoah, P., Carvalho, R., Nabais, S., Castro Alves, C., Campos, M.L., Van Doorn, L.J., Caldas, C., Seruca, R., Carneiro, F., and Sobrinho-Simoes, M. (2003). A proinflammatory genetic profile increases the risk for chronic atrophic gastritis and gastric carcinoma. Gastroenterology. 125:364–71.

Magalhaes Queiroz, D.M., and Luzza, F. (2006). Epidemiology of Helicobacter pylori infection. Helicobacter. 11(Suppl 1):1–5.

Malaty, H.M., El-Kasabany, A., Graham, D.Y., Miller, C.C., Reddy, S.G., Srinivasan, S.R., Yamaoka, Y., and Berenson, G.S. (2002). Age at acquisition of Helicobacter pylori infection: a follow-up study from infancy to adulthood. Lancet. 359:931–5.

Malaty, H.M., Evans, D.J., Jr., Abramovitch, K., Evans, D.G., and Graham, D.Y. (1992). Helicobacter pylori infection in dental workers: a seroepidemiology study. Am J Gastroenterol. 87:1728.

Malaty, H.M., Kumagai, T., Tanaka, E., Ota, H., Kiyosawa, K., Graham, D.Y., and Katsuyama, T. (2000). Evidence from a nine-year birth cohort study in Japan of transmission pathways of Helicobacter pylori infection. J Clin Microbiol. 38:1971–3.

Malfertheiner, P., Megraud, F., O'Morain, C., Bazzoli, F., El-Omar, E., Graham, D., Hunt, R., Rokkas, T., Vakil, N., and Kuipers, E.J. (2007). Current concepts in the management of Helicobacter pylori infection: the Maastricht III Consensus Report. Gut. 56:772–81.

Mapstone, N.P., Lynch, D.A., Lewis, F.A., Axon, A.T., Tompkins, D.S., Dixon, M.F., and Quirke, P. (1993). Identification of Helicobacter pylori DNA in the mouths and stomachs of patients with gastritis using PCR. J Clin Pathol. 46:540.

Marshall, B.J., Armstrong, J.A., McGeche, D.B., and Glancy, R.J. (1985a). Attempt to fulfill Koch's postulates for pyloric Campylobacter. Med J Aust. 142:436.

Marshall, B.J., McGeche, D.B., Rogers, P.A., and Glancy, R.J. (1985b). Pyloric Campylobacter infection and gastroduodenal disease. Med J Aust. 142:439.

Marshall, B.J., and Warren, J.R. (1984). Unidentified curved bacilli in the stomach of patients with gastritis and peptic ulceration. Lancet. 1:1311.

Masci, E., Viale, E., Freschi, M., Porcellati, M., and Tittobello, A. (1996). Precancerous gastric lesions and Helicobacter pylori. Hepatogastroenterology. 43:854.

Mason, J., Axon, A.T., Forman, D., Duffett, S., Drummond, M., Crocombe, W., Feltbower, R., Mason, S., Brown, J., and Moayyedi, P. (2002). The cost-effectiveness of population Helicobacter pylori screening and treatment: a Markov model using economic data from a randomized controlled trial. Aliment Pharmacol Ther. 16:559–68.

Matteo, M.J., Granados, G., Perez, C.V., Olmos, M., Sanchez, C., and Catalano, M. (2007). Helicobacter pylori cag pathogenicity island genotype diversity within the gastric niche of a single host. J Med Microbiol. 56:664–9.

Mayne, S.T., and Navarro, S.A. (2002). Diet, obesity and reflux in the etiology of adenocarcinomas of the esophagus and gastric cardia in humans. J Nutr. 132:3467S–70S.

McMichael, A.J., McCall, M.G., Hartshorne, J.M., and Woodings, T.L. (1980). Patterns of gastro-intestinal cancer in European migrants to Australia: the role of dietary change. Int J Cancer. 25:431.

Mendall, M.A., Goggin, P.M., Molineaux, N., Levy, J., Toosy, T., Strachan, D., and Northfield, T.C. (1992). Childhood living conditions and Helicobacter pylori seropositivity in adult life. Lancet. 339:896.

Mera, R., Fontham, E.T., Bravo, L.E., Bravo, J.C., Piazuelo, M.B., Camargo, M.C., and Correa, P. (2005). Long term follow up of patients treated for Helicobacter pylori infection. Gut. 54:1536–40.

Mitchell, H.M., Ally, R., Wadee, A., Wiseman, M., and Segal, I. (2002). Major differences in the IgG subclass response to Helicobacter pylori in the first and third worlds. Scand J Gastroenterol. 37:517.

Mitchell, H.M., Lee, A., Berkowicz, J., and Borody, T. (1988). The use of serology to diagnose active Campylobacter pylori infection. Med J Aust. 149:604–9.

Miwa, H., Go, M.F., and Sato, N. (2002). H. pylori and gastric cancer: the Asian enigma. Am J Gastroenterol. 97:1106.

Miyaji, H., Azuma, T., Ito, S., Abe, Y., Gejyo, F., Hashimoto, N., Sugimoto, H., Suto, H., Ito, Y., Yamazaki, Y., Kohli, Y., and Kuriyama, M. (2000). Helicobacter pylori infection occurs via close contact with infected individuals in early childhood. J Gastroenterol Hepatol. 15:257–62.

Miyamoto, A., Kuriyama, S., Nishino, Y., Tsubono, Y., Nakaya, N., Ohmori, K., Kurashima, K., Shibuya, D., and Tsuji, I. (2007). Lower risk of death from gastric cancer among participants of gastric cancer screening in Japan: a population-based cohort study. Prev Med. 44:12–9.

Morris, A., Nicholson, G., Lloyd, G., Haines, D., Rogers, A., and Taylor, D. (1986). Seroepidemiology of Campylobacter pyloridis. NZ Med J. 99:657.

Mourad-Baars, P.E., Verspaget, H.W., Mertens, B.J., and Mearin, M.L. (2007). Low prevalence of Helicobacter pylori infection in young children in the Netherlands. Eur J Gastroenterol Hepatol. 19:213–6.

Nakagawa, S., Asaka, M., Kato, M., Nakamura, T., Kato, C., Fujioka, T., Tatsuta, M., Keida, K., Terao, S., Takahashi, S., Uemura, N., Kato, T., Aoyama, N., Saito, D., Suzuki, M., Imamura, A., Sato, K., Miwa, H., Nomura, H., Kaise, M., Oohara, S., Kawai, T., Urabe, K., Sakaki, N., Ito, S., Noda, Y., Yanaka, A., Kusugami, K., Goto, H., Furuta, T., Fujino, M., Kinjyou, F., and Ookusa, T. (2006). Helicobacter pylori eradication and metachronous gastric cancer after endoscopic mucosal resection of early gastric cancer. Aliment Pharmacol Ther. 24(Suppl 4):214–8.

National Cancer Institute, Bethesda. (2007). http://www.cancer.gov/cancertopics/pdq/screening/gastric/healthprofessional/allpages. Accessed March 5, 2007.

Neugut, A.I., Hayek, M., and Howe, G. (1996). Epidemiology of gastric cancer. Semin Oncol. 23:281–91.

Nguyen, V.B., Nguyen, G.K., Phung, D.C., Okrainec, K., Raymond, J., Dupond, C., Kremp, O., Kalach, N., and Vidal-Trecan, G. (2006). Intra-familial transmission of Helicobacter pylori infection in children of households with multiple generations in Vietnam. Eur J Epidemiol. 21:459–63.

Nurgalieva, Z.Z., Malaty, H.M., Graham, D.Y., Almuchambetova, R., Machmudova, A., Kapsultanova, D., Osato, M.S., Hollinger, F.B., and Zhangabylov, A. (2002). Helicobacter pylori infection in Kazakhstan: effect of water source and household hygiene. Am J Trop Med Hyg. 67:201–6.

Nurgalieva, Z.Z., Opekun, A.R., and Graham, D.Y. (2006). Problem of distinguishing false-positive tests from acute or transient Helicobacter pylori infections. Helicobacter. 11:69–74.

Odze, R.D. (2005). Pathology of the gastroesophageal junction. Semin Diagn Pathol. 22:256–65.

Ohata, H., Kitauchi, S., Yoshimura, N., Mugitani, K., Iwane, M., Nakamura, H., Yoshikawa, A., Yanaoka, K., Arii, K., Tamai, H., Shimizu, Y., Takeshita, T., Mohara, O., and Ichinose, M. (2004). Progression of chronic atrophic gastritis associated with Helicobacter pylori infection increases risk of gastric cancer. Int J Cancer. 109:138–43.

Oliveira, C., Bordin, M.C., Grehan, N., Huntsman, D., Suriano, G., Machado, J.C., Kiviluoto, T., Aaltonen, L., Jackson, C.E., Seruca, R., and Caldas, C. (2002). Screening E-cadherin in gastric cancer families reveals germline mutations only in hereditary diffuse gastric cancer kindred. Hum Mutat. 19:510–7.

O'Rourke, K., Goodman, K.J., Grazioplene, M., Redlinger, T., and Day, R.S. (2003). Determinants of geographic variation in Helicobacter pylori infection among children on the US-Mexico border. Am J Epidemiol. 158:816–24.

Osato, M.S., Ayub, K., Le, H.H., Reddy, R., and Graham, D.Y. (1998). Houseflies are an unlikely reservoir or vector for Helicobacter pylori. J Clin Microbiol. 36:2786–8.

Osawa, H., Inoue, F., and Yoshida, Y. (1996). Inverse relation of serum Helicobacter pylori antibody titres and extent of intestinal metaplasia. J Clin Pathol. 49:112–5.

Palmer, E.D. (1954). Investigation of the gastric spirochaetes of the human. Gastroenterology. 27:218.

Pardo-Mindan, F.J., Joly, M., Robledo, C., Sola, J., and Valerdiz, S. (1989). Duodenal ulcer "epidemic" in a pathology department. Lancet. 1:153.

Parente, J.M., da Silva, B.B., Palha-Dias, M.P., Zaterka, S., Nishimura, N.F., and Zeitune, J.M. (2006). Helicobacter pylori infection in children of low and high socioeconomic status in northeastern Brazil. Am J Trop Med Hyg. 75:509–12.

Parkin, D.M. (2006). The global health burden of infection-associated cancers in the year 2002. Int J Cancer. 118:3030–44.

Parkin, D.M., Pisani, P., and Ferlay, J. (1999a). Estimates of the worldwide incidence of 25 major cancers in 1990. Int J Cancer. 80:827–41.

Parkin, D.M., Pisani, P., and Ferlay, J. (1999b). Global cancer statistics. CA Cancer J Clin. 49:33–64.

Parsonnet, J. (1995). The incidence of Helicobacter pylori infection. Aliment Pharmacol Ther. 9(Suppl 2):45–51.

Parsonnet, J., Blaser, M.J., Perez-Perez, G.I., Hargrett-Bean, N., and Tauxe, R.V. (1992). Symptoms and risk factors of Helicobacter pylori infection in a cohort of epidemiologists. Gastroenterology. 102:41.

Parsonnet, J., and Forman, D. (2004). Helicobacter pylori infection and gastric cancer—for want of more outcomes. JAMA. 291:244–5.

Parsonnet, J., Harris, R., Hack, H.M., and Owens, D.K. (1996). Modelling cost effectiveness of Helicobacter pylori screening to prevent gastric cancer: a mandate for clinical trials. Lancet. 348:150–4.

Parsonnet, J., Replogle, M., Yang, S., and Hiatt, R. (1997). Seroprevalence of CagA-positive strains among Helicobacter pylori-infected, healthy young adults. J Infect Dis. 175:1240.

Parsonnet, J., Shmuely, H., and Haggerty, T. (1999). Fecal and oral shedding of Helicobacter pylori from healthy infected adults. JAMA. 282:2240–5.

Perez-Perez, G.I., Olivares, A.Z., Foo, F.Y., Foo, S., Neusy, A.J., Ng, C., Holzman, R.S., Marmor, M., and Blaser, M.J. (2005). Seroprevalence of Helicobacter pylori in New York City populations originating in East Asia. J Urban Health. 82:510–6.

Perez-Perez, G.I., Witkin, S.S., Decker, M.D., and Blaser, M.J. (1991). Seroprevalence of Helicobacter pylori infection in couples. J Clin Microbiol. 29:642–4.

Perri, F., Piepoli, A., Quitadamo, M., Quarticelli, M., Merla, A., and Bisceglia, M. (2002). HLA-DQA1 and -DQB1 genes and Helicobacter pylori infection in Italian patients with gastric adenocarcinoma. Tissue Antigens. 59:55–7.

Perry, S., De Jong, B.C., Hill, P., Adegbola, R., and Parsonnet, J. (2007). Helicobacter pylori and the outcome of M. tuberculosis infection. Infectious Disease Society of America, annual meeting. San Diego, CA: IDSA; pp. LB-22.

Perry, S., de la Luz Sanchez, M., Hurst, P.K., and Parsonnet, J. (2005). Household transmission of gastroenteritis. Emerg Infect Dis. 11:1093–6.

Perry, S., de la Luz Sanchez, M., Yang, S., Haggerty, T.D., Hurst, P., Perez-Perez, G., and Parsonnet, J. (2006). Gastroenteritis and transmission of Helicobacter pylori infection in households. Emerg Infect Dis. 12:1701–8.

Perry, S., and Parsonnet, J. (2005). Commentary: H. pylori infection in early life and the problem of imperfect tests. Int J Epidemiol. 34:1356–8.

Perry, S., Sanchez, L., Yang, S., Haggerty, T.D., Hurst, P., and Parsonnet, J. (2004). Helicobacter pylori and risk of gastroenteritis. J Infect Dis. 190:303–10.

Pessi, T., Virta, M., Adjers, K., Karjalainen, J., Rautelin, H., Kosunen, T.U., and Hurme, M. (2005). Genetic and environmental factors in the immunopathogenesis of atopy: interaction of Helicobacter pylori infection and IL4 genetics. Int Arch Allergy Immunol. 137:282–8.

Pharoah, P.D., Guilford, P., and Caldas, C. (2001). Incidence of gastric cancer and breast cancer in CDH1 (E-cadherin) mutation carriers from hereditary diffuse gastric cancer families. Gastroenterology. 121:1348–53.

Pritchard, D.M., and Przemeck, S.M. (2004). Review article: How useful are the rodent animal models of gastric adenocarcinoma? Aliment Pharmacol Ther. 19:841–59.

Quintero, E., Pizarro, M.A., Rodrigo, L., Pique, J.M., Lanas, A., Ponce, J., Mino, G., Gisbert, J., Jurado, A., Herrero, M.J., Jimenez, A., Torrado, J., Ponte, A., Diaz-de-Rojas, F., and Salido, E. (2005). Association of Helicobacter pylori-related distal gastric cancer with the HLA class II gene DQB10602 and cagA strains in a southern European population. Helicobacter. 10:12–21.

Raghunath, A., Hungin, A.P., Wooff, D., and Childs, S. (2003). Prevalence of Helicobacter pylori in patients with gastro-oesophageal reflux disease: systematic review. BMJ. 326:737.

Ramsey, E.J., Carey, K.V., Peterson, W.L., Jackson, J.J., Murphy, F.K., Read, N.W., Taylor, K.B., Trier, J.S., and Fordtran, J.S. (1979). Epidemic gastritis with hypochlorhydria. Gastroenterology. 76:1449.

Raymond, J., Thiberg, J.M., Chevalier, C., Kalach, N., Bergeret, M., Labigne, A., and Dauga, C. (2004). Genetic and transmission analysis of Helicobacter pylori strains within a family. Emerg Infect Dis. 10:1816–21.

Richter, J.E. (2007). Gastrooesophageal reflux disease. Best Pract Res Clin Gastroenterol. 21:609–31.

Ries, L.A.G., Melbert, D., Krapcho, M., Mariotto, A., Miller, B.A., Feuer, E.J., Clegg, L., Horner, M.J., Howlader, N., Eisner, M.P., Reichman, M., and Edwards, B.K. (based on November 2006 SEER data submission, posted to the SEER Web site, 2007). SEER Cancer Statistics Review, 1975–2004: National Cancer Institute, Bethesda, MD.

Roderick, P., Davies, R., Raftery, J., Crabbe, D., Pearce, R., Patel, P., and Bhandari, P. (2003). Cost-effectiveness of population screening for Helicobacter pylori in preventing gastric cancer and peptic ulcer disease, using simulation. J Med Screen. 10:148–56.

Roosendaal, R., Kuipers, E.J., Buitenwerf, J., van Uffelen, C., Meuwissen, S.G., van Kamp, G.J., and Vandenbroucke-Grauls, C.M. (1997). Helicobacter pylori and the birth cohort effect: evidence of a continuous decrease of infection rates in childhood. Am J Gastroenterol. 92:1480.

Rosenstock, S., Jorgensen, T., Andersen, L., and Bonnevie, O. (2000). Seroconversion and seroreversion in IgG antibodies to Helicobacter pylori: a serology based prospective cohort study. J Epidemiol Community Health. 54:444.

Rothenbacher, D., Bode, G., and Brenner, H. (2002). Dynamics of Helicobacter pylori infection in early childhood in a high-risk group living in Germany: loss of infection higher than acquisition. Aliment Pharmacol Ther 16:1663–8.

Rowland, M., Daly, L., Vaughan, M., Higgins, A., Bourke, B., and Drumm, B. (2006). Age-specific incidence of Helicobacter pylori. Gastroenterology. 130:65–72; quiz 211.

Rupnow, M.F., Owens, D.K., Shachter, R., and Parsonnet, J. (1999). Helicobacter pylori vaccine development and use: a cost-effectiveness analysis using the Institute of Medicine methodology. Helicobacter. 4:272–80.

Rupnow, M.F., Shachter, R.D., Owens, D.K., and Parsonnet, J. (2001). Quantifying the population impact of a prophylactic Helicobacter pylori vaccine. Vaccine. 20:879–85.

Shikata, K., Kiyohara, Y., Kubo, M., Yonemoto, K., Ninomiya, T., Shirota, T., Tanizaki, Y., Doi, Y., Tanaka, K., Oishi, Y., Matsumoto, T., and Iida, M. (2006). A prospective study of dietary salt intake and gastric cancer incidence in a defined Japanese population: the Hisayama study. Int J Cancer. 119:196–201.

Shinmura, K., Kohno, T., Takahashi, M., Sasaki, A., Ochiai, A., Guilford, P., Hunter, A., Reeve, A.E., Sugimura, H., Yamaguchi, N., and Yokota, J. (1999). Familial gastric cancer: clinico-pathological characteristics, RER phenotype and germline p53 and E-cadherin mutations. Carcinogenesis. 20:1127–31.

Singh, K., and Ghoshal, U.C. (2006). Causal role of Helicobacter pylori infection in gastric cancer: an Asian enigma. World J Gastroenterol. 12:1346–51.

Singh, V., Trikha, B., Vaiphei, K., Nain, C.K., Thennarasu, K., and Singh, K. (1999). Helicobacter pylori: evidence for spouse-to-spouse transmission. J Gastroenterol Hepatol. 14:519–22.

Smith, R.A., Cokkinides, V., and Eyre, H.J. (2003). American Cancer Society guidelines for the early detection of cancer, 2003. CA Cancer J Clin. 53:27–43.

Solnick, J.V., Fong, J., Hansen, L.M., Chang, K., Canfield, D.R., and Parsonnet, J. (2006). Acquisition of Helicobacter pylori infection in rhesus macaques is most consistent with oral-oral transmission. J Clin Microbiol. 44:3799–803.

Sorberg, M., Nilsson, M., Hanberger, H., and Nilsson, L.E. (1996). Morphologic conversion of Helicobacter pylori from bacillary to coccoid form. Eur J Clin Microbiol Infect Dis. 15:216–9.

Stalnikowicz, R., and Benbassat, J. (1990). Risk of gastric cancer after gastric surgery for benign disorders. Arch Intern Med. 150:2022–6.

Stein, M., Bagnoli, F., Halenbeck, R., Rappuoli, R., Fantl, W.J., and Covacci, A. (2002). c-Src/Lyn kinases activate Helicobacter pylori CagA through tyrosine phosphorylation of the EPIYA motifs. Mol Microbiol. 43:971–80.

Stone, J., Bevan, S., Cunningham, D., Hill, A., Rahman, N., Peto, J., Marossy, A., and Houlston, R.S. (1999). Low frequency of germline E-cadherin mutations in familial and nonfamilial gastric cancer. Br J Cancer. 79:1935–7.

Stone, M.A., Taub, N., Barnett, D.B., and Mayberry, J.F. (2000). Increased risk of infection with Helicobacter pylori in spouses of infected subjects: observations in a general population sample from the UK. Hepatogastroenterology. 47:433–6.

Suerbaum, S., and Michetti, P. (2002). Helicobacter pylori infection. N Engl J Med. 347:1175–86.

Sung, J.J., Lin, S.R., Ching, J.Y., Zhou, L.Y., To, K.F., Wang, R.T., Leung, W.K., Ng, E.K., Lau, J.Y., Lee, Y.T., Yeung, C.K., Chao, W., and Chung, S.C. (2000). Atrophy and intestinal metaplasia one year after cure of H. pylori infection: a prospective, randomized study. Gastroenterology. 119:7.

Suzuki, J., Muraoka, H., Kobayasi, I., Fujita, T., and Mine, T. (1999). Rare incidence of inter-spousal transmission of Helicobacter pylori in asymptomatic individuals in Japan. J Clin Microbiol. 37:4174–6.

Takada, K. (2000). Epstein-Barr virus and gastric carcinoma. Mol Pathol. 53:255–61.

Targownik, L.E., and Nabalamba, A. (2006). Trends in management and outcomes of acute non-variceal upper gastrointestinal bleeding: 1993–2003. Clin Gastroenterol Hepatol. 4:1459–66.

Tatematsu, M., Nozaki, K., and Tsukamoto, T. (2003). Helicobacter pylori infection and gastric carcinogenesis in animal models. Gastric Cancer. 6:1–7.

Taylor, D.N, Parsonnet, J. (1995). The epidemiology and natural history of Helicobacter pylori infection. In: Blaser, M.J, Smith, P.D, Ravdin, J.I, Grenber, H.B, Guerrant, R.L (eds). Infections of the Gastrointestinal Tract. New York: Raven Press, pp. 551–64.

Terry, M.B., Gaudet, M.M., and Gammon, M.D. (2002). The epidemiology of gastric cancer. Semin Radiat Oncol. 12:111–27.

Tersmette, A.C., Offerhaus, G.J., Tersmette, K.W., Giardiello, F.M., Moore, G.W., Tytgat, G.N., and Vandenbroucke, J.P. (1990). Meta-analysis of the risk of gastric stump cancer: detection of high risk patient subsets for stomach cancer after remote partial gastrectomy for benign conditions. Cancer Res. 50:6486–9.

Thomas, J.E., Dale, A., Harding, M., Coward, W.A., Cole, T.J., and Weaver, L.T. (1999). Helicobacter pylori colonization in early life. Pediatr Res. 45:218–23.

Tindberg, Y., Bengtsson, C., Granath, F., Blennow, M., Nyren, O., and Granstrom, M. (2001). Helicobacter pylori infection in Swedish school children: lack of evidence of child-to-child transmission outside the family. Gastroenterology. 121:310–6.

Tokudome, S., Soeripto, Triningsih, F.X., Ananta, I., Suzuki, S., Kuriki, K., Akasaka, S., Kosaka, H., Ishikawa, H., Azuma, T., and Moore, M.A. (2005). Rare Helicobacter pylori infection as a factor for the very low stomach cancer incidence in Yogyakarta, Indonesia. Cancer Lett. 219:57–61.

Torres, J., Leal-Herrera, Y., Perez-Perez, G., Gomez, A., Camorlinga-Ponce, M., Cedillo-Rivera, R., Tapia-Conyer, R., and Munoz, O. (1998). A community-based seroepidemiologic study of Helicobacter pylori infection in Mexico. J Infect Dis. 178:1089–94.

Tsai, C.J., Perry, S., Sanchez, L., and Parsonnet, J. (2005). Helicobacter pylori infection in different generations of Hispanics in the San Francisco Bay Area. Am J Epidemiol. 162:351–7.

Tytgat, G.N.J. (1995). Endoscopic transmission of Helicobacter pylori infection. Aliment Pharmacol Ther. 9(Suppl 2):105.

Ueda, M., Kikuchi, S., Kasugai, T., Shunichi, T., and Miyake, C. (2003). Helicobacter pylori risk associated with childhood home environment. Cancer Sci. 94:914–8.

Uemura, N., Mukai, T., Okamoto, S., Yamaguchi, S., Mashiba, H., Taniyam, K., Sasaki, N., Haruma, K., Sumii, K., and Kajiyama, G. (1997). Effect of Helicobacter pylori eradication on subsequent development of cancer after endoscopic resection of early gastric cancer. Cancer Epidemiol Biomarkers Prev. 6:639.

Uemura, N., Okamoto, S., Yamamoto, S., Matsumura, N., Yamaguchi, S., Yamakido, M., Taniyama, K., Sasaki, N., and Schlemper, R.J. (2001). Helicobacter pylori infection and the development of gastric cancer. N Engl J Med. 345:784.

Van Doorn, L.J., Figueiredo, C., Megraud, F., Pena, S., Midolo, P., Queiroz, D.M., Carneiro, F., Vanderborght, B., Pegado, M.D., Sanna, R., De Boer, W., Schneeberger, P.M., Correa, P., Ng, E.K., Atherton, J., Blaser, M.J., and Quint, W.G. (1999). Geographic distribution of vacA allelic types of Helicobacter pylori. Gastroenterology. 116:823–30.

Wang, J.T., Sheu, J.C., Lin, J.T., Wang, T.H., and Wu, M.S. (1993). Direct DNA amplification and restriction pattern analysis of Helicobacter pylori in patients with duodenal ulcer and their families. J Infect Dis. 168:1544.

Watanabe, Y., Aoyama, N., Sakai, T., Shirasaka, D., Maekawa, S., Kuroda, K., Wambura, C., Tamura, T., Nose, Y., and Kasuga, M. (2006). HLA-DQB1 locus and gastric cancer in Helicobacter pylori infection. J Gastroenterol Hepatol. 21:420–4.

Webb, P.M., Knight, T., Greaves, S., Wilson, A., Newell, D.G., Elder, J., and Forman, D. (1994). Relation between infection with Helicobacter pylori and living conditions in childhood: evidence for person to person transmission in early life. BMJ. 308:750.

Weisse, A.B. (1996). Barry Marshall and the resurrection of Johannes Fibiger. Hosp Pract. 31:105.

Weyermann, M., Adler, G., Brenner, H., and Rothenbacher, D. (2006). The mother as source of Helicobacter pylori infection. Epidemiology. 17:332–4.

Whary, M.T., Sundina, N., Bravo, L.E., Correa, P., Quinones, F., Caro, F., and Fox, J. (2005). Intestinal helminthiasis in Colombian children promotes a Th2 response to Helicobacter pylori: possible implications for gastric carcinogenesis. Cancer Epidemiol Biomarkers Prev. 14:1464–9.

Windsor, H.M., Abioye-Kuteyi, E.A., Leber, J.M., Morrow, S.D., Bulsara, M.K., and Marshall, B.J. (2005). Prevalence of Helicobacter pylori in Indigenous Western Australians: comparison between urban and remote rural populations. Med J Aust. 182:210–3.

Wizla-Derambure, N., Michaud, L., Ategbo, S., Vincent, P., Ganga-Zandzou, S., Turck, D., and Gottrand, F. (2001). Familial and community environmental risk factors for Helicobacter pylori infection in children and adolescents. J Pediatr Gastroenterol Nutr. 33:58–63.

Wong, B.C., Lam, S.K., Wong, W.M., Chen, J.S., Zheng, T.T., Feng, R.E., Lai, K.C., Hu, W.H., Yuen, S.T., Leung, S.Y., Fong, D.Y., Ho, J., Ching, C.K., and Chen, J.S. (2004). Helicobacter pylori eradication to prevent gastric cancer in a high-risk region of China: a randomized controlled trial. JAMA. 291:187–94.

World Cancer Research Fund, and American Institute for Cancer Research. (1997). Stomach, food, nutrition and the prevention of cancer: a global perspective. Washington, D.C.: American Institute for Cancer Research; pp. 148–75.

Wu, M.S., Shun, C.T., Wu, C.C., Hsu, T.Y., Lin, M.T., Chang, M.C., Wang, H.P., and Lin, J.T. (2000). Epstein-Barr virus-associated gastric carcinomas: relation to H. pylori infection and genetic alterations. Gastroenterology. 118:1031–8.

Yabuta, T., Shinmura, K., Tani, M., Yamaguchi, S., Yoshimura, K., Katai, H., Nakajima, T., Mochiki, E., Tsujinaka, T., Takami, M., Hirose, K., Yamaguchi, A., Takenoshita, S., and Yokota, J. (2002). E-cadherin gene variants in gastric cancer families whose probands are diagnosed with diffuse gastric cancer. Int J Cancer. 101:434–41.

Yamaoka, Y., Osato, M.S., Sepulveda, A.R., Gutierrez, O., Figura, N., Kim, J.G., Kodama, T., Kashima, K., and Graham, D.Y. (2000). Molecular epidemiology of Helicobacter pylori: separation of H. pylori from East Asian and non-Asian countries. Epidemiol Infect. 124:91–6.

Yamazaki, S., Yamakawa, A., Ito, Y., Ohtani, M., Higashi, H., Hatakeyama, M., and Azuma, T. (2003). The CagA protein of Helicobacter pylori is translocated into epithelial cells and binds to SHP-2 in human gastric mucosa. J Infect Dis. 187:334–7.

Yang, P.C., and Davis, S. (1988). Epidemiological characteristics of adenocarcinoma of the gastric cardia and distal stomach in the United States, 1973–1982. Int J Epidemiol. 17:293.

Yoon, K.A., Ku, J.L., Yang, H.K., Kim, W.H., Park, S.Y., and Park, J.G. (1999). Germline mutations of E-cadherin gene in Korean familial gastric cancer patients. J Hum Genet. 44:177–80.

Chapter 3
Diet and Gastric Cancer

Chun Liu, Xiang-Dong Wang, and Robert M. Russell

Introduction

Worldwide, gastric cancer is the fourth most common cancer and the second leading cause of cancer death (Parkin et al. 2005). In 2002, approximately 933,937 men and women were diagnosed with gastric cancer, and 700,349 died from gastric cancer worldwide (Parkin et al. 2005). In the United States, gastric cancer is relatively uncommon, and ranked fourteenth among both incident cancers and cancer deaths with approximately 22,280 new cases and 11,430 deaths in 2006 (Jemal et al. 2006). The death rate from gastric cancer has steadily decreased over the last several decades in the United States (Jemal et al. 2006) and worldwide (Parkin et al. 2005), suggesting that environmental factors such as diet have a critical role in the etiology of this malignancy. Growing evidence indicates that gastric cardia and noncardia cancer differ in etiology (Devesa et al. 1998). For example, *Helicobacter pylori* infection is an established strong risk factor for gastric noncardia cancer, but it is not associated with the risk for gastric cardia cancer (Engel et al. 2003; Helicobacter and Cancer Collaborative Group 2001). In addition, the incidence rate of gastric cardia cancer has increased in the United States (Corley and Kubo 2004; Devesa et al. 1998; Devesa and Fraumeni 1999) and European countries (Botterweck et al. 2000a), whereas the incidence rate of gastric noncardia cancer has steadily decreased (Corley and Kubo 2004). It has been suggested that the increasing rate of obesity is contributing to the increase in gastric cardia cancer (Engel et al. 2003; Kubo and Corley, 2006). Other established risk factors for gastric cancer include age and tobacco smoking (Engel et al. 2003; Freedman et al. 2007; Kelley and Duggan, 2003; Sjodahl et al. 2007).

Fruits and Vegetables

Numerous observational epidemiologic studies have evaluated the association between consumption of fruits and vegetables and risk for gastric cancer. The studies in general suggest an inverse association, particularly for raw and allium

T.C. Wang et al. (eds.) *The Biology of Gastric Cancers*,
© Springer Science+Business Media, LLC 2009

vegetables and citrus fruits (Gonzalez et al. 2006b; International Agency for Research on Cancer 2003; Nouraie et al. 2005; Riboli and Norat 2003).

Two metaanalyses published in 2003 (International Agency for Research on Cancer 2003; Riboli and Norat 2003) and one in 2005 (Lunet et al. 2005) have shown that the inverse association is stronger for fruits than for vegetables, but is weaker in cohort studies than in case-control studies. In the International Agency for Research on Cancer (IARC) report in 2003, the summary relative risks comparing high to low categories for fruits were 0.63 [95% confidence interval (CI), 0.58–0.69] from 37 case-control studies and 0.85 (95% CI, 0.77–0.95) from 11 cohort studies (International Agency for Research on Cancer 2003). Regarding vegetables, the summary relative risks comparing high to low categories were 0.66 (95% CI, 0.61–0.71) from 20 case-control studies and 0.94 (95% CI, 0.84–1.06) from 5 cohort studies (International Agency for Research on Cancer 2003). A few case-control studies included in the metaanalysis also showed that the association between intake of fruits and vegetables and gastric cancer risk was not different by gastric anatomic site (International Agency for Research on Cancer 2003). In the metaanalysis of cohort studies in 2005, the summary relative risks of gastric cancer incidence were 0.82 (95% CI, 0.73–0.93) for fruits and 0.88 (95% CI, 0.69–1.13) for vegetables (Lunet et al. 2005).

Recently, results were reported on two large cohort studies in Europe on fruit and vegetable consumption and gastric cancer risk by anatomic site. In the prospective analysis of the Alpha-Tocopherol, Beta-Carotene (ATBC) Cancer Prevention Study among 29,133 male smokers, high consumption of fruits was associated with reduced risk of gastric noncardia cancer, but not with gastric cardia cancer (Nouraie et al. 2005). However, consumption of vegetables was not associated with risk for either gastric cardia or noncardia cancer in the ATBC Cancer Prevention Study (Nouraie et al. 2005). The results from the European Prospective Investigation into Cancer and Nutrition (EPIC) cohort study among 521,457 men and women living in 10 European countries showed no significant association of intakes of fresh fruits, total vegetables, or specific groups of vegetables with risk for gastric cancer regardless of anatomic site, although a nonsignificant inverse association was found for citrus fruits or onion and garlic and risk for gastric cardia cancer only (Gonzalez et al. 2006b).

Randomized trial data evaluating the efficacy of fruit and vegetable consumption on the risk of gastric cancer are not available. Randomization would minimize confounding that could affect the validity of observational data (i.e., the randomization ensures that all factors except for treatment status that could influence the outcome of interest are distributed equally between the active treatment and placebo groups). Using double-blind and placebo-controlled procedures in randomized trials diminishes selection bias and recall bias (case-control design) that observational studies are prone to. Recall bias in case-control studies is of concern because patients with gastric cancer are more likely to recall less consumption of fruits and vegetables than their healthy controls, resulting in a spurious stronger inverse association between fruit and vegetable consumption and gastric cancer risk as shown in the metaanalysis (International Agency for Research on Cancer 2003). Because gastric cancer is relatively rare with a long latency period, the costs of conducting a large randomized trial of fruit and vegetable consumption and gastric cancer association with long-term intervention and follow-up would be extremely high. It is unlikely that such a trial would ever be conducted. Although available data that suggest an inverse association between consumption of fruits and vegetables and risk for gastric cancer come from observational case-control and cohort studies, the consistency in the association between fruit and vegetable consumption and gastric cancer risk in various populations with wide ranges of intake suggests that certain fruits and vegetables may have a role in reducing gastric cancer risk.

Vitamins and Minerals

There are extensive experimental, epidemiologic, and clinical studies on nutrients with antioxidant properties, including carotenoids, vitamin C, vitamin E, and selenium, and risk for gastric cancer. Recent systematic reviews and metaanalyses of randomized trial data found that antioxidant supplements of β-carotene, vitamin A, and vitamin E, with the potential exception of selenium and vitamin C, had no significant effect on the incidence of gastrointestinal cancers (Bjelakovic et al. 2004a; Bjelakovic et al. 2004b); on the contrary, β-carotene, vitamin A, and vitamin E increased mortality (Bjelakovic et al. 2007). To date, no randomized trials have evaluated the effect of lycopene, α-carotene, lutein/zeaxanthin, or β-cryptoxanthin on the prevention of gastric cancer.

Carotenoids

Carotenoids, lipid-soluble compounds, are rich within fruits and vegetables and are responsible for the color of many fruits and vegetables (Wang 2004). α-Carotene, β-carotene, lycopene, lutein/zeaxanthin, and β-cryptoxanthin are the most abundant carotenoids from the diet and in the circulation of humans (Stahl and Sies 1996; Wang 2004). Certain carotenoids (such as α-carotene, β-carotene, and β-cryptoxanthin) present in fruit and vegetables can be partially metabolized to retinol (Russell 1998; Wang et al. 1994).

Observational Epidemiologic Studies

Observational epidemiologic data, albeit not entirely consistent, have demonstrated that lycopene is associated with reduced risk for gastric cancer (Table 3.1), but the results for other carotenoids and retinol have been mixed. High consumption of tomato and tomato-based products (major sources of lycopene), and high intakes or plasma levels of lycopene are associated with reduced risk for

Table 3.1 Epidemiologic studies of lycopene and gastric cancer risk

Reference	Location	Outcome	No. of cases	Exposure	Comparison	Relative risk (95% CI)
Prospective studies						
Yuan et al. 2004	Shanghai, China	GC	191	Serum lycopene	Quartile 4 vs. 1	0.63 (0.34–1.15)
Jenab et al. 2006a	Europe	GC	244	Plasma lycopene	Quartile 4 vs. 1	0.63 (0.36–1.09)
Botterweck et al. 2000b	Netherlands	GC	282	Dietary lycopene	Quintile 5 vs. 1	1.0 (0.7–1.5)
Nouraie et al. 2005	Finland	GCC	57 (M)	Dietary lycopene	Quartile 4 vs. 1	0.97 (0.41–2.34)
		GNCC	163 (M)	Dietary lycopene	Quartile 4 vs. 1	0.62 (0.37–1.03)
Larsson et al. 2007a	Sweden	GC	139	Dietary lycopene	Quartile 4 vs. 1	0.92 (0.53–1.58)
Case-control studies						
Harrison et al. 1997	New York	GC, intestinal	60	Dietary lycopene	One S.D. increment	0.7 (0.4–1.2)
		GC, diffuse	31	Dietary lycopene	One S.D. increment	0.8 (0.5–1.4)
Garcia-Closas et al. 1999	Spain	GC	354	Dietary lycopene	Quartile 4 vs. 1	1.55 (0.91–2.64)
De Stefani et al. 2000	Uruguay	GC	120	Dietary lycopene	Tertile 3 vs. 1	0.37 (0.19–0.73)
Lissowska et al. 2004	Poland	GC	274	Dietary lycopene	Quartile 4 vs. 1	1.19 (0.77–1.82)
Haenszel et al. 1972	Hawaiian	GC	220	Tomato intake	≥11 vs. <4/mo	0.39, p < 0.05, al
	Japanese					0.31, p <0.05 Issei
						0.49, NS, Nisei
Tajima and Tominaga 1985	Japan	GC	93	Tomato intake	≥4 vs. ≥1/week	1.24, NS
Correa et al. 1985	Louisiana	GC	194	Tomato intake	>vs. ≤ median	0.82 (0.53–1.28), Whites
			197			0.56 (0.34–0.90), Blacks
La Vecchia et al. 1987	Italy	GC	206	Tomato intake	Tertile 3 vs. 1	0.69, NS
Buiatti et al. 1989	Italy	GC	1,016	Tomato intake	Tertile 3 vs. 1	0.70, p for trend <0.001
Graham et al. 1990	New York	GC	186 (M)	Tomato intake		Reduced risk, p <0.05
			107 (F)			Reduced risk, NS
Gonzalez et al. 1991	Spain	GC	354	Tomato intake	Quartile 4 vs. 1	0.9 (0.5–1.5)

Boeing et al. 1991	Poland	GC	741	Tomato intake	Tertile 3 vs. 1	0.77, p for trend <0.05
Tuyns et al. 1992	Belgium	GC	449	Cooked tomatoes	>0 vs. 0 g/week	0.12, NS
				Raw tomatoes	>10 vs. 0 g/week	0.74, NS
Hansson et al. 1993	Sweden	GC	338	Tomato intake, adolescence	>2.9 vs. 0 times/mo	0.36 (0.23–0.58)
				20 years early	>15 vs. <2 times/mo	0.72 (0.47–1.11)
Franceschi et al. 1994	Italy	GC	723	Tomato intake	Quartile 4 vs. 1	0.43 (0.33–0.55)
Ekstrom et al. 2000	Sweden	GCC	74	Tomato intake	≥4/week vs. ≤1–3/mo	0.6 (0.3–1.1)
		GNCC	406	Tomato intake	≥4/week vs. ≤1–3/mo	0.8 (0.6–1.2)

Abbreviations: CI: confidence interval; GC: gastric cancer; GCC: gastric cardia cancer; GNCC: gastric noncardia cancer; NS: nonsignificant; M: male; F: female; mo: month.

gastric cancer in various populations (De Stefani et al. 2000; Ekstrom et al. 2000; Giovannucci 1999; Jenab et al. 2006a; Yuan et al. 2004). In the Shanghai cohort study, high serum levels of lycopene, α-carotene, and β-carotene were associated with a significantly reduced risk of developing gastric cancer, but no significant association was observed for serum levels of lutein/zeaxanthin, β-cryptoxanthin, and retinol (Yuan et al. 2004). In the EPIC cohort, plasma lycopene, zeaxanthin, β-cryptoxanthin, and retinol, but not α-carotene, β-carotene, and lutein, were inversely associated with risk for gastric cancer (Jenab et al. 2006a). In the prospective analysis of data from the General Population Trial of Linxian, China, among poorly nourished and high-risk populations, serum levels of β-carotene at baseline were not associated with either gastric cardia or noncardia cancer (Abnet et al. 2003). In that study, the observed inverse association for serum β-cryptoxanthin and retinol was different by different anatomic sites of the stomach; retinol was inversely associated with risk for gastric cardia cancer, but not with gastric noncardia cancer, whereas β-cryptoxanthin was not associated with gastric cardia cancer, but inversely associated with risk for gastric noncardia cancer (Abnet et al. 2003). In addition, high serum levels of lutein/zeaxanthin had no association with gastric cardia cancer, but were associated with increased risk for gastric noncardia cancer (Abnet et al. 2003). In the cohort analysis of the ATBC Cancer Prevention Study, lycopene intake was not associated with gastric cardia cancer, but inversely associated with risk for gastric noncardia cancer, whereas retinol intake was inversely associated with risk for gastric cardia cancer but not with gastric noncardia cancer (Nouraie et al. 2005). Both intake and serum level of β-carotene were not associated with risk of any anatomic site (Nouraie et al. 2005). In the recent report from the Swedish Mammography Cohort and the Cohort of Swedish Men, high intakes of α-carotene and β-carotene and retinol were associated with a lower risk of gastric cancer, but no significant association was found for intakes of lycopene, lutein/zeaxanthin, or β-cryptoxanthin (Larsson et al. 2007a). In the Netherlands Cohort Study, intakes of lycopene, α-carotene, lutein/zeaxanthin, and β-cryptoxanthin were not associated with risk, but an increased risk was found for β-carotene and retinol intake (Botterweck et al. 2000b).

Randomized Trials

Randomized trials also have not produced consistent evidence for the efficacy of β-carotene or combined β-carotene and other antioxidants on the prevention of gastric cancer (Blot et al. 1993; Correa et al. 2000; Hennekens et al. 1996; Virtamo et al. 2003; You et al. 2006). Daily β-carotene supplementation (30 mg) for 6 years among individuals with multifocal nonmetaplastic atrophy and/or intestinal meta-plasia (premalignant lesions) significantly increased the rates of regression in the Columbia trial (Correa et al. 2000). In the General Population Trial of Linxian, China, among 29,584 high-risk individuals with suboptimal nutritional status, daily randomized treatment with supplements containing β-carotene (15 mg), α-tocopherol (30 mg), and selenium from yeast (50 μg) for 5.25 years resulted in a nonsignificant reduction in the incidence and mortality from gastric cardia and noncardia cancer (Blot et al. 1993). Two randomized trials in well-nourished Western populations, however, found no beneficial effect of β-carotene on the reduction in risk for gastric cancer (Hennekens et al. 1996; Virtamo et al. 2003). Although the number of gastric cancers was small, 50 mg of β-carotene supplementation on alternate days for 12 years did not affect the risk for gastric cancer among 22,071 apparently healthy U.S. male physicians (Hennekens et al. 1996). In the ATBC Cancer Prevention Study, randomized daily treatment with 20 mg of β-carotene or with combined 20 mg of β-carotene and 50 mg of α-tocopherol for 5–8 years resulted in nonsignificantly increased risk for gastric cancer among 29,133 Finnish male smokers (Virtamo et al. 2003), and had no impact on the occurrence of neoplastic changes of the stomach in male smokers with atrophic gastritis (Varis et al. 1998).

Experimental Studies

Carotenoids (lycopene, lutein, and β-carotene) and retinoids have been shown to inhibit the incidence and growth of chemically induced gastric tumors in laboratory animal studies (Azuine et al. 1992; Bhuvaneswari et al. 2001; Goswami and Sharma 2005; Velmurugan et al. 2002a; Velmurugan et al. 2002b; Velmurugan et al. 2005; Velmurugan and Nagini 2005). Experimental and animal studies suggest several potential mechanisms by which carotenoids affect gastric carcinogenesis. Carotenoids can act as antioxidants to neutralize reactive oxygen species, thereby protecting DNA from oxidative damage (McCall and Frei 1999; Velmurugan et al. 2002a; Velmurugan et al. 2002b; Velmurugan and Nagini 2005); decrease cell proliferation and induce apoptosis (Liu et al. 2006; Velmurugan et al. 2005); modify cell–cell communication (Krinsky and Johnson 2005); enhance host immunologic functions (Kelley and Bendich 1996); and reduce *H. pylori* bacterial load and gastric inflammation by shifting the T-lymphocyte response from a predominant Th1 response dominated by γ-interferon to a Th1/Th2 response characterized by a decrease in γ-interferon but an increase in interleukin (IL)-4 (Bennedsen et al. 1999; Liu and Lee 2003; Wang et al. 2000).

Using a ferret animal model, the effects of lycopene supplementation on cigarette smoke–induced changes in protein levels of the p53 tumor suppressor gene, p53 target genes (p21$^{Waf1/Cip1}$ and Bax-1), cell proliferation, and apoptosis in the gastric mucosa were evaluated (Liu et al. 2006). p21$^{waf1/cip1}$, a CDK inhibitor, is a key component in the cell-cycle arrest in G1, and Bax is a pro-apoptotic member of the Bcl-2 family. p21$^{waf1/cip1}$ and Bax-1 function as mediators to promote p53-dependent apoptosis (Figure 3.1) (el-Deiry 1998; Robles et al. 2002; Rom and Tchou-Wong 2003). In that study, ferrets were assigned to cigarette smoke exposure or to no cigarette smoke exposure and to low-dose, or high-dose lycopene supplementation, or to no lycopene for 9 weeks. Lycopene concentrations were significantly elevated in a dose-dependent manner in the gastric mucosa of ferrets supplemented with lycopene alone, but were markedly reduced in ferrets supplemented with lycopene and exposed to smoke. It was found that total p53 and phosphorylated p53 levels were greater in

Fig. 3.1. Potential effect of lycopene on p53 tumor suppressor pathway in cigarette smoke–induced gastric carcinogenesis

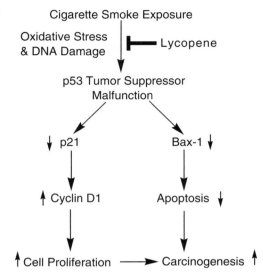

ferrets exposed to smoke alone than in all other groups. However, smoke-elevated total p53 and phosphorylated p53 were markedly attenuated by both doses of lycopene. p21$^{(Waf1/Cip1)}$, Bax-1, and cleaved caspase 3 (an index for apoptosis) were substantially decreased, whereas cell proliferation indices such as cyclin D1 and proliferating cellular nuclear antigen (PCNA) were increased in ferrets exposed to smoke alone, whereas lycopene prevented smoke-induced changes in p21$^{(Waf1/Cip1)}$, Bax-1, cleaved caspase 3, cyclin D1, and PCNA in a dose-dependent manner (Liu et al. 2006). p53 phosphorylation, especially at serine 15, is an early cellular response to various genotoxic carcinogens and stresses that produce reactive oxygen species, and facilitates both the accumulation and functional activation of p53 (She et al. 2000; Tibbetts et al. 1999). Our data indicate that lycopene may prevent smoke exposure–induced changes in p53, p53 phosphorylation, p53 target genes, cell proliferation, and apoptosis in the gastric mucosa of ferrets, suggesting that lycopene may protect against the development of gastric cancer by modulating p53-dependent cell-cycle control and apoptosis.

Evidence from human studies also suggests that lycopene might reduce gastric cancer risk through the inhibition of inflammatory pathways. One small study compared concentrations of plasma lycopene and C-reactive protein (CRP), a marker of systemic inflammation (Ridker 2003), simultaneously between 12 gastrointestinal cancer patients and 12 controls and found higher CRP but lower lycopene levels in gastrointestinal cancer patients than controls (McMillan et al. 2000). In addition, treatment with the antiinflammatory agent ibuprofen among gastrointestinal cancer patients moderately increased plasma lycopene concentrations (McMillan et al. 2000). Some cross-sectional studies also showed that plasma lycopene concentrations are inversely associated with inflammatory markers, including CRP (Boosalis et al. 1996; Chang et al. 2005; Ford et al. 2003; Kritchevsky et al. 2000; McMillan et al. 2002; Rowley et al. 2003; van Herpen-Broekmans et al. 2004), IL-6 (Walston et al. 2006), soluble intercellular adhesion molecule-1 (van Herpen-Broekmans et al. 2004), and adhesion molecule E-selectin (Rowley et al. 2003).

Vitamin C

Vitamin C (ascorbic acid) is a water-soluble antioxidant rich within fruits and vegetables and can regenerate vitamin E from its oxidized form (Frei 1994; Halliwell 2001; Zhang and Farthing 2005).

Observational Epidemiologic Studies

Epidemiologic case-control studies have consistently found an inverse association between vitamin C and gastric cancer risk (Chen et al. 2002; Kono and Hirohata 1996). Most of the prospective cohort studies have also shown an inverse association. In the Basel Study, serum vitamin C levels were inversely associated with mortality from gastric cancer (Stahelin et al. 1991). In the EPIC cohort, plasma vitamin C levels were associated with reduced risk for gastric cancer, and such an inverse association did not differ according to anatomic site (cardia and noncardia), histologic subtype (diffuse and intestinal), or *H. pylori* infection (Jenab et al. 2006b). Use of vitamin C supplements was also associated with reduced mortality from gastric cancer in the American Cancer Society Cancer Prevention Study II cohort (Jacobs et al. 2002). In the cohort analysis of the ATBC Cancer Prevention Study, high vitamin C intake was associated with reduced risk for gastric noncardia cancer but not for gastric cardia cancer (Nouraie et al. 2005). However, vitamin C intake had no association with risk for gastric cancer in the Netherlands Cohort Study (Botterweck et al. 2000b).

Randomized Trials

Findings from a few randomized trials that have evaluated the efficacy of vitamin C in the prevention of gastric cancer are not entirely consistent. A randomized trial in Japan among individuals with chronic gastritis provides evidence that vitamin C slows the progression of gastric mucosal atrophy, a precancerous lesion of gastric cancer (Sasazuki et al. 2003). In that trial, daily treatment with 50 or 500 mg of vitamin C for 5 years significantly reduced the ratio of serum pepsinogen I/II, a marker of gastric atrophy (Sasazuki et al. 2003). In the Italy trial among patients with intestinal metaplasia (premalignant lesions) on the gastric mucosa after *H. pylori* eradication, vitamin C supplementation (500 mg/day) for 6 months significantly increased the regression rate of intestinal metaplasia (Zullo et al. 2000). In the Columbia trial, individuals with multifocal nonmetaplastic atrophy and/or intestinal metaplasia (premalignant lesions) assigned to vitamin C supplements (1 g twice a day) also had significantly increased rates of regression over 6 years of treatment (Correa et al. 2000). However, combined vitamin C (250 mg), vitamin E (100 IU), and selenium from yeast (37.5 µg) supplementation twice daily for 7.3 years did not reduce the prevalence of precancerous gastric lesions and the incidence of gastric cancers in the Linqu trial in China (You et al. 2006). Randomized daily treatment with combined vitamin C (120 mg) and molybdenum supplements also had no significant effect on both incidence and mortality from gastric cardia and noncardia cancer in the General Population Trial of Linxian, China (Blot et al. 1993).

In human studies, treatment with high-dose vitamin C (5 g daily for 4 weeks) eradicated *H. pylori* infection (Jarosz et al. 1998). Vitamin C supplementation (1 g twice per day for 4–12 months) led to reduction in the formation of nitrotyrosine, a nitrating product, among patients with *H. pylori* nonatrophic gastritis (Mannick et al. 1996). However, combined vitamin C (200 mg) and vitamin E (50 mg) supplementation twice daily for 4 weeks failed to reduce reactive oxygen species and lipid peroxidation in the gastric mucosa of individuals with *H. pylori* gastritis (Everett et al. 2002).

Experimental Studies

The effect of vitamin C on experimentally induced gastric cancer in laboratory rodents is equivocal—studies have shown an inhibitory effect (Balansky et al. 1986; Kawasaki et al. 1982; Oliveira et al. 2003), no effect (Shirai et al. 1985), or a promoting effect (Shibata et al. 1993). A few studies reported a promoting effect of vitamin C on forestomach carcinogenesis when it was coadministrated with sodium nitrite after pretreatment with chemical carcinogens (Mirvish et al. 1976; Mirvish et al. 1983; Okazaki et al. 2006; Yoshida et al. 1994), whereas vitamin C itself had no influence on forestomach or glandular stomach carcinogenesis (Okazaki et al. 2006).

Data from experimental and animal studies suggest potential mechanisms by which vitamin C affects gastric carcinogenesis. Vitamin C reduces gastric mucosal oxidative stress and DNA damage (Oliveira et al. 2003; Sun et al. 2005) and gastric inflammation (Sjunnesson et al. 2001; Sun et al. 2005; Wang et al. 2000) by scavenging reactive oxygen species; inhibits gastric nitrosation reaction for the formation of N-nitroso compounds by reducing nitrious acid to nitric oxide and producing dehydroascorbic acid in the stomach (Bartsch et al. 1988; Mirvish 1986; Mirvish 1994; Mirvish 1995; Zhang and Farthing 2005); enhances host immunologic functions (Kelley and Bendich 1996; Zhang and Farthing 2005); has a direct effect on *H. pylori* growth and virulence (Chatterjee et al. 2005; Zhang et al. 1997; Zhang and Farthing 2005); and inhibits gastric cell proliferation and induces apoptosis (Zhang et al. 2002; Zhang and Farthing 2005).

Vitamin E

Observational Epidemiologic Studies

Vitamin E (tocopherol) is a potent lipid-soluble antioxidant (Frei 1994). Findings from observational epidemiologic studies that have examined the association between vitamin E and gastric cancer risk have been inconsistent. High serum α-tocopherol levels were associated with reduced risk for gastric cancer in a Finnish cohort (Knekt et al. 1988) and mortality from gastric cancer in the Basel Study cohort (Stahelin et al. 1991). Plasma α-tocopherol, but not γ-tocopherol, was inversely associated with risk for gastric cancer in the EPIC cohort (Jenab et al. 2006a). However, in the Shanghai cohort study, serum levels of α-tocopherol and γ-tocopherol were not associated with risk for gastric cancer (Yuan et al. 2004). High vitamin E intake also was not associated with gastric cancer risk in the Netherlands Cohort Study (Botterweck et al. 2000b). In addition, use of vitamin E supplements had no association with mortality from gastric cancer in the Cancer Prevention Study II cohort (Jacobs et al. 2002). In the cohort analysis of the ATBC Cancer Prevention Study, the association seemed to differ according to anatomic site—high intakes of α-tocopherol and γ-tocopherol were associated with increased risk for gastric cardia cancer, but with reduced risk for gastric noncardia cancer (Nouraie et al. 2005). In the same study, elevated serum levels of α-tocopherol were associated with increased risk for gastric cardia cancer, but had no association with risk for gastric noncardia cancer (Nouraie et al. 2005).

Randomized Trials

Several randomized trials have evaluated vitamin E as a single agent or combined with other antioxidants in the primary or secondary prevention of gastric cancer (Lee et al. 2005; Virtamo et al. 2003). The results from these trials provide little evidence that vitamin E reduces risk for gastric cancer and may even suggest an increased risk in men (Virtamo et al. 2003). In the General Population Trial of Linxian, China, randomized daily treatment with supplements containing β-carotene (15 mg), vitamin E (30 mg), and selenium from yeast (50 µg) among poorly nourished populations for 5.25 years resulted in a nonsignificant reduction in both incidence and mortality from gastric cardia and noncardia cancer (Blot et al. 1993). In the Women's Health Study, randomized treatment with 600 IU of α-tocopherol every other day over 10 years did not lower the risk for gastric cancer in 39,876 apparently healthy U.S. women (Lee et al. 2005). In the ATBC Cancer Prevention Study, randomized daily treatment with 50 mg of α-tocopherol or with combined 20 mg of β-carotene and 50 mg of α-tocopherol over 5 to 8 years resulted in a nonsignificantly increased risk for gastric cancer among Finnish male smokers (Virtamo et al. 2003). Two trials that have evaluated the effect of vitamin E on premalignant lesions also yielded mixed results. Daily vitamin E supplementation (400 IU) for 1 year among intestinal metaplasia patients increased rates of regression in the Russia trial (Bukin et al. 1997). However, combined vitamin E (100 IU), vitamin C (250 mg), and selenium from yeast (37.5 µg) supplementation twice daily for 7.3 years did not reduce the prevalence of precancerous gastric lesions and the incidence of gastric cancers in the Linqu trial in China (You et al. 2006).

Experimental Studies

In animal studies, vitamin E has been shown to reduce gastric oxidative stress and mucosal cell membranes from lipid peroxidation (Sun et al. 2005); inhibit nitrosa-tion (Bartsch et al. 1988; Mirvish 1986; Sun et al. 2005); and reduce chemically induced gastric tumors, specifically in combination with other antioxidants, in some studies (Balansky et al. 1986; Mirvish 1986; Sjunnesson et al. 2001; Wu et al. 2001), but not in others (Miyauchi et al. 2002; Takahashi et al. 1986a). Vitamin E was ineffective in reducing gastric inflammation and premalignant lesions in Mongolian gerbils (Sun et al. 2005). In a human study, vitamin E (50 mg) combined with vitamin C (200 mg) twice daily for 4 weeks did not decrease reactive oxygen species and lipid peroxidation in gastric mucosa in individuals with *H. pylori* gas-tritis (Everett et al. 2002).

Selenium

Selenium, an essential trace element for humans, is a component of 25 selenopro-teins such as glutathione peroxidases (Brenneisen et al. 2005; Moghadaszadeh and Beggs 2006; Rayman 2000; Rayman 2005). The biologic effects of selenium occur mainly through the function of selenoproteins, which are involved in diverse biologic pathways, including antioxidant defense, which can reduce oxi-dative stress and DNA damage; induction of phase II conjugating enzymes for detoxification of carcinogens and reduction of DNA adduct formation; inhibition of cell proliferation and promotion of DNA repair and apoptosis through p53 tumor suppressor gene; inactivation of protein kinase C (a receptor that is critical for tumor promotion); maintenance of normal DNA methylation; generation of active thyroid hormone; and enhancement of immune function, fertility, and mus-cle movement and function (Brenneisen et al. 2005; Moghadaszadeh and Beggs 2006; Rayman 2000; Rayman 2005). Some of these health effects of selenium are not entirely attributable to its enzymatic functions (Brenneisen et al. 2005; Moghadaszadeh and Beggs 2006; Rayman 2000; Rayman 2005). Limited animal studies reported mixed results on the effect of selenium on chemically induced gastric tumors (Bergman and Slanina 1986; el-Bayoumy 1985; Hu et al. 1992; Kobayashi et al. 1986; Su et al. 2005).

Observational Epidemiologic Studies and Randomized Trials

Data from human epidemiologic studies and one clinical trial suggest an important inverse association between selenium and gastric cancer risk, and the effect seems to be strongest in individuals with the low selenium status (Table 3.2). High toenail selenium levels were associated with a nonsignificantly reduced risk of gastric cancer in the Netherlands Cohort Study (van den Brandt et al. 1993). Elevated serum selenium levels were also associated with reduced gastric cancer risk in a Finnish cohort, a population with low selenium intake (Knekt et al. 1990), but not in a cohort of Japanese Americans in Hawaii (Nomura et al. 1987) and a case-control study in Japan (Kabuto et al. 1994), where populations have relatively adequate selenium intake (Kabuto et al. 1994; Rayman 2005). In the General Population Trial of Linxian, China, populations with poor selenium status, randomized daily treatment with supplements containing β-carotene (15 mg), vitamin E (30 mg), and selenium from yeast (50 μg) for 5.25 years led to a nonsignificant reduction in both incidence and mortality from gastric cardia and noncardia cancer (Blot et al. 1993). In the prospective analysis of the General Population Trial of Linxian, China, baseline serum selenium levels were highly significantly associated with decreased incidence and mortality of gastric cardia cancer, but not with noncardia cancer (Mark et al. 2000; Wei et al. 2004). For gastric cancer incidence, the relative risks comparing the highest to lowest quartile of serum selenium were 0.47 (95% CI, 0.33–0.65) for gastric cardia cancer and 1.07 (95% CI, 0.55–2.08) for noncardia cancer (Mark et al. 2000). The corresponding relative risks for mortality were 0.31 (95% CI, 0.11–0.87) for gastric cardia cancer and 1.64 (95 CI, 0.49–5.48) for gastric noncardia cancer (Wei et al. 2004). However, combined vitamin C (250 mg), vitamin E (100 IU), and selenium from yeast (37.5 μg) supplementation twice daily for 7.3 years did not reduce the prevalence of precancerous gastric lesions and the incidence of gastric cancers in the Linqu trial in China (You et al. 2006).

There are two ongoing randomized trials testing the efficacy of selenium as a single agent on cancer incidence, namely, the Selenium and Vitamin E Cancer Prevention Trial (SELECT) and the Prevention of Cancer by Intervention with Selenium (PRECISE) trial. The SELECT trial is examining the effect of selenium (200 μg/day from L-selenomethionine) and/or vitamin E (400 IU/day of all rac alpha-tocopheryl acetate) supplementation for a minimum of 7 years (maximum of 12 years) on prostate cancer incidence in non–African American men aged 55 years or older and African American men aged 50 years or older (Lippman et al. 2005). The PRECISE trial is going to examine the effect of selenium on the incidence of cancer in the United Kingdom, Sweden, and Denmark, three countries with low selenium status (Rayman 2000). The results from these two randomized trials will provide critical answers to the question of the role of selenium in the primary prevention of cancer.

Albeit data are not entirely consistent, cumulative evidence suggests that antioxidant supplements of β-carotene, vitamin A, and vitamin E, with the potential exception of selenium and vitamin C, have no significant effect on the incidence of gastrointestinal cancers (Bjelakovic et al. 2004a; Bjelakovic et al. 2004b).

Table 3.2 Epidemiologic studies and randomized trials of selenium and gastric cancer risk

Reference	Location	Outcome	No. of cases	Exposure	Comparison	Relative risk (95% CI)
Prospective studies						
Nomura et al. 1987	Hawaiian Japanese	GC	66	Serum Se	Quintile 1 vs. 5	0.9, NS*
Knekt et al. 1990	Finland	GC	58 (M)	Serum Se	Quintile 5 vs. 1	0.26, p < 0.01
			37 (F)			0.59, NS
Mark et al. 2000	Linxian, China	GCC	402	Serum Se	Quartile 4 vs. 1	0.47 (0.33–0.65)
		GNCC	87			1.07 (0.55–2.08)
Wei et al. 2004	Linxian, China	GCC*	36	Serum Se	Quartile 4 vs. 1	0.31 (0.11–0.87)
		GNCC*	24			1.64 (0.49–5.48)
van den Brandt et al. 1993	Netherlands	GC	92	Toenail Se	Quintile 5 vs. 1	0.64 (0.33–1.27)
Case-control studies						
Kabuto et al. 1994	Japan	GC	202	Serum Se	Quartile 1 vs. 4	1.0 (0.5–1.9)
Randomized trials						
Blot et al. 1993	Linxian, China	GC	539	Combined β-carotene (15 mg)/vitamin E (30 mg)/Se (50 μg)	Treatment vs. placebo	0.84 (0.71–1.00)
		GCC	435			0.85 (0.70–1.02)
		GNCC	104			0.82 (0.56–1.20)
Blot et al. 1993	Linxian, China	GC*	331	Combined β-carotene (15 mg)/vitamin E (30 mg)/Se (50 μg)	Treatment vs. placebo	0.79 (0.64–0.99)
		GCC*	253			0.82 (0.64–1.04)
		GNCC*	78			0.72 (0.46–1.14)

You et al. 2006	Linqu, China	Dysplasia or GC	434 in 1999	Combined vitamin C (250 mg)/vitamin E (100 IU)/Se (37.5 µg)	Treatment vs. placebo	1.10 (0.89–1.37) in 1999
			921 in 2003			1.03 (0.87–1.23) in 2003
		GC only	58			1.03 (0.61–1.73)

Abbreviations: CI: confidence interval; Se: selenium; GC: gastric cancer; GCC: gastric cardia cancer; GNCC: gastric noncardia cancer; M: male; and F: female.

*Death from GCC and GNCC.

A number of factors may contribute to the discrepancy in findings on the relationship between antioxidant nutrients and risk of gastric cancer among studies, including different populations at risk, using cancer versus intermediate endpoints in the evaluation, using dietary assessments versus nutrient biomarkers, and duration of nutrient supplementation or intake. For instance, the Physicians' Health Study and the ATBC Cancer Prevention Study, two randomized trials in well-nourished Western populations, showed no beneficial effect of β-carotene on the reduction in risk for gastric cancer (Hennekens et al. 1996; Virtamo et al. 2003), whereas the General Population Trial of Linxian, China, among high-risk individuals with suboptimal nutritional status, revealed a nonsignificant reduction in gastric cancer with supplements containing β-carotene, vitamin E, and selenium from yeast (Blot et al. 1993). Because supplements in the Linxian trial were given in combination, it is difficult to infer the reduction in gastric cancer risk was a result of β-carotene or other components such as selenium in supplements. Using intermediate endpoints has the advantage of enabling investigators to conduct clinical trials in a relatively short period with fewer participants. However, the results from trials using intermediate endpoints are not always consistent with the trials using cancer as the endpoints because intermediate outcomes sometimes can regress to normal status and only a small fraction of intermediate outcomes will develop into cancer. When evaluating the precancerous process, both the duration of intervention and the stage of the precancerous process when the intervention is implemented can contribute to the inconsistency in the trial results (Fontham et al. 2005). Without using nutrient biomarkers (e.g., blood concentrations), a small or moderate association between antioxidant nutrients and gastric cancer risk can be potentially masked by measurement errors caused by imprecise dietary assessments in observational studies.

Nitrate, Nitrite, and Nitrosamines

There are two sources of nitrosamines that humans are exposed to, namely, preformed exogenous nitrosamines and nitrosamines produced endogenously from nitrate and nitrite (Jakszyn and Gonzalez 2006; Mirvish 1995). Preformed nitrosamines are present mainly in nitrite-cured meat and fish and other foods; smoked, pickled, and salty preserved foods; and alcoholic beverages (beer and whiskey) (Jakszyn and Gonzalez 2006; Mirvish 1995). Nitrate, a natural compound, is present in vegetables and drinking water (Jakszyn and Gonzalez 2006; Mirvish 1995) and is used as a food additive in cheese and cured meat (van Loon et al. 1997). N-nitroso compounds are also found in tobacco products, drugs, and industrial materials (Mirvish 1995). Dietary nitrate can be reduced to nitrite by oral bacteria and then to N-nitroso compounds (e.g., nitrosamines) by acid-catalyzed and bacterial nitrosation in the stomach through reaction with compounds such as amines, amides, and amino acids (Mirvish 1995; van Loon et al. 1997). The *in vivo* formation of nitrosamines can also occur via nitric oxide formation during inflammation (Mirvish 1995).

N-nitroso compounds have been shown to be carcinogenic in animal studies (Mirvish 1995). Two nitrosamines (N-nitrosodiethylamine and N-nitrosodimethyl-amine) (Figure 3.2) are classified as probably carcinogenic to humans (group 2A) by the IARC (International Agency for Research on Cancer 2006). Epidemiologic studies suggest a positive association between nitrosamines and gastric cancer risk, but data are still inconclusive (Jakszyn and Gonzalez 2006). Most of the epidemiologic investigations on nitrosamine and related food intake and gastric cancer risk have been case-control investigations, which support a positive association of nitrite, nitrosamine, processed meat and fish, preserved vegetables, and smoked food intake with risk for gastric cancer (Jakszyn and Gonzalez 2006).

The association between intake of nitrate, nitrite, or nitrosamine and risk of gastric cancer has been evaluated in only a few prospective cohorts, and the findings are not entirely consistent. In the Netherlands Cohort Study, intake of nitrate was not associated with gastric cancer risk whereas intake of nitrite was nonsignificantly positively associated with the risk (van Loon et al. 1997; van Loon et al. 1998). There was no association between intakes of nitrate, nitrite, or N-nitrosodimethylamine and risk of gastric cancer in a Finnish cohort study (Knekt et al. 1999). The EPIC cohort study found no association between dietary intake of N-nitrosodimethylamine and gastric cancer risk, but endogenous formation of N-nitroso compounds was significantly associated with risk for gastric noncardia cancer (relative risk, 1.42; 95% CI, 1.14–1.78 for an increase of 40 µg/day) but not with gastric cardia cancer (relative risk, 0.96; 95% CI, 0.69–1.33) (Jakszyn et al. 2006). Data from the EPIC cohort also suggest a possible interaction of endogenous formation of nitroso compounds with *H. pylori* infection or plasma vitamin C levels; the positive association between endogenous formation of nitroso compounds and risk for gastric noncardia cancer was present only in those who were infected with *H. pylori* or had lower plasma vitamin C levels (Jakszyn et al. 2006).

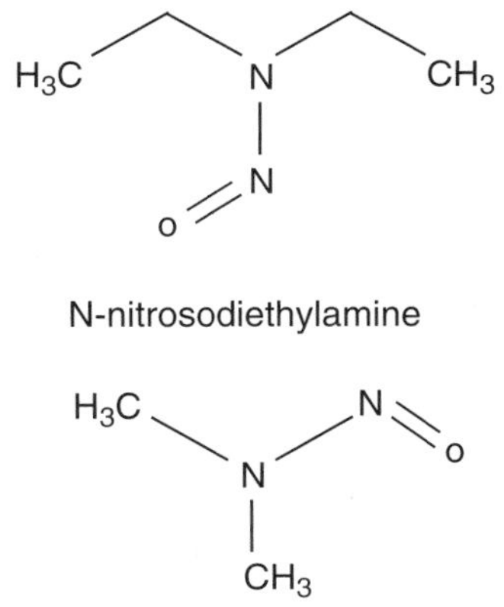

N-nitrosodiethylamine

N-nitrosodimethylamine

Fig. 3.2 Chemical structures of two nitrosamines

Processed meat, an important source of N-nitroso compounds, refers to those preserved by adding nitrate, nitrite, or salt, or by smoking (Larsson et al. 2006). In a metaanalysis that summarized available evidence from six prospective cohort studies and nine case-control studies published from January 1966 through March 2006, the summary relative risks for an increment in processed meat consumption of 30 g/day (approximately half of an average serving) were 1.15 (95% CI, 1.04–1.27) for the cohort studies and 1.38 (95% CI, 1.19–1.60) for the case-control studies (Larsson et al. 2006). In the EPIC cohort, when the association between processed meat and gastric cancer was evaluated by anatomic site, per 50 g/day of increase in processed meat was associated with a significant 2.45-fold increase in gastric noncardia cancer, but not with gastric cardia cancer (Gonzalez et al. 2006a).

In summary, a totality of evidence from animal studies and observational epidemiologic investigations suggests that nitrosamine and related food intake may increase risk for gastric cancer.

Alcohol

The relationship between alcohol consumption and gastric cancer risk remains controversial. An early metaanalysis of 14 case-control studies and two cohort studies showed that alcohol consumption was associated with a modest increase in risk for gastric cancer; the summary relative risk for an increase of 25 g/day of alcohol was 1.07 (95% CI, 1.04–1.10) (Bagnardi et al. 2001). However, recent data from four large prospective cohort studies and a prospective analysis of data from an automated database provide little support for the association between total alcohol consumption and gastric cancer risk (Barstad et al. 2005; Freedman et al. 2007; Larsson et al. 2007b; Lindblad et al. 2005; Sjodahl et al. 2007). Data from the Norwegian cohort suggest that alcohol may interact with smoking to increase gastric cancer risk. In that study, smoking doubled the risk for gastric cancer, and alcohol consumption had no significant association with the risk, but combined high exposure to cigarettes (>20/day) and alcohol (>5 occasions/14 days) increased risk for gastric noncardia cancer 4.9-fold (95% CI: 1.90–12.62), compared with nonusers of both cigarettes and alcohol (Sjodahl et al. 2007).

Findings have been mixed regarding the relation of alcoholic beverage type to gastric cancer risk (Barstad et al. 2005; Freedman et al. 2007; Larsson et al. 2007b). In the Danish cohort of men and women, spirits were positively, but wine was inversely, associated with gastric cancer risk, and no association was found for beer (Barstad et al. 2005). In the NIH–AARP Diet and Health Study cohort among 474,606 men and women, hard liquor was significantly associated with increased risk for gastric cardia cancer, but not for gastric noncardia cancer; no significant association was found for beer and wine (Freedman et al. 2007). By contrast, in the Swedish Mammography Cohort, consumption of medium-strong/strong beer was significantly associated with increased risk; the relative risk for women who consumed more than one serving of medium-strong/strong beer per week was 2.09 (95% CI, 1.11–3.93) compared with nondrinkers, but consumption of light beer, wine, and spirits was not significantly associated with risk (Larsson et al. 2007b). N-nitrosodimethylamine found in beer, whiskeys, and other hard liquor has been suggested to be responsible for the increased risk for gastric cancer observed for beer and spirits consumption (Larsson et al. 2007b).

Animal studies on alcohol and gastric cancer are equivocal. In experimental studies in rodents, 10% ethanol administered in drinking water did not promote the development of gastric tumors induced by N-methyl-N′-nitro-N-nitrosoguanidine (MNNG), a known gastric carcinogen (Takahashi et al. 1986b; Wada et al. 1998; Watanabe et al. 1992). In addition, when wine or 11% ethanol was coadministrated with MNNG, wine or 11% ethanol was found to inhibit MNNG-induced gastroduo-denal carcinoma in rats (Cerar and Pokorn 1996). However, in an early study, the intraperitoneal injection of 20% ethanol in 0.9% sodium chloride increased the incidence and number of the MNNG-induced gastric cancers of glandular stomach in rats (Iishi et al. 1989). Because ethanol and sodium chloride were given together, it is difficult to infer that the increased risk was attributable to ethanol or to sodium chloride, which has been shown to increase MNNG-induced gastric tumors in rats (Watanabe et al. 1992).

Taken together, data from both epidemiologic studies and animal experimental studies provide little support for a harmful effect of alcohol on the development of gastric cancer.

Salt

Numerous studies in laboratory rodents have shown that salt enhances chemically induced gastric tumors (Kato et al. 2006; Newberne et al. 1987; Sorbye et al. 1994c; Takahashi and Hasegawa 1985; Takahashi et al. 1983; Takahashi et al. 1994; Watanabe et al. 1992). A high concentration of salt in the stomach induces gastric inflammation; damages the mucosal protective layer; increases DNA synthesis and cell proliferation; and alters the mucous microenvironment in a dose-dependent manner (Furihata et al. 1984; Kato et al. 2006; Sorbye et al. 1994a, 1994b; Takahashi and Hasegawa 1985). The gastric mucosal damage and increased cell proliferation might increase carcinogenesis and risk for gastric cancer (Takahashi and Hasegawa 1985).

The mucosal damage caused by a high-salt diet enhances persistent *H. pylori* infection in mice (Fox et al. 1999; Kato et al. 2006). A few studies have shown that a high-salt diet also acts synergistically with *H. pylori* infection to promote the development of gastric tumors induced by N-methyl-N-nitrosourea (MNU) in Mongolian gerbils (Kato et al. 2006; Nozaki et al. 2002). However, in those studies, MNU initiation was essential for tumor induction, because no tumors were found in gerbils without MNU treatment despite their *H. pylori* and/or high-salt diet treatments (Kato et al. 2006; Nozaki et al. 2002). A study using outbred Mongolian gerbils found that atrophic gastritis and intestinal metaplasia developed in both animals infected with *H. pylori* and uninfected animals fed with a 2.5% salt diet, but no dysplasia or tumor was observed (Bergin et al. 2003). No synergism was found between *H. pylori* infection and a high-salt diet in relation to gastric tumors in the INS-GAS (Fox et al. 2003) and a wild-type B6129 mouse models (Rogers et al. 2005). The high-salt diet was associated with less-severe gastric lesions than the basal diet in male INS-GAS mice (Fox et al. 2003). In these mouse models, the high-salt diet led to a shift in antimicrobial humoral immunity from a Th1 to a Th2 pattern, which indicates even some protection against tumor progression by the high-salt diet (Fox et al. 2003; Rogers et al. 2005). These data suggest that a high-salt diet might not be sufficient to initiate gastric carcinogenesis by itself and that a high-salt diet might not have a copromoting effect with *H. pylori* infection on gastric cancer.

The association between salt and salted foods and gastric cancer risk has been evaluated in more than 40 epidemiologic case-control studies and several prospective cohort studies. Most case-control studies have found a positive association between intake of salt and high-salt foods such as salted vegetables, fish, and cured meat products and risk of gastric cancer (Tsugane 2005). A high-salt diet and *H. pylori* infection have also been found to act synergistically to increase the risk for gastric cancer in a Korean case-control study (Lee et al. 2003). Prospective data are limited and less consistent. In a U.S. cohort of men, salted fish intake was associated with increased mortality from gastric cancer (Kneller et al. 1991). A Japanese cohort study also observed a nonsignificantly increased mortality from gastric cancer associated with high intake of pickled foods and traditional soups (Ngoan et al. 2002). Researchers in The Netherlands Cohort Study reported that intakes of salt and several types of cured meat were weakly positively associated with gastric cancer risk, whereas salt added to the hot meal or to soup during cooking or use of table salt had no association (van den Brandt et al. 2003). In another large prospective cohort study in Japan, high salt intake was significantly associated with increased risk for gastric cancer in men, but not in women, and highly salted foods such as salted fish roe and salted fish preserves were strongly associated with increased risk in both men and women (Tsugane et al. 2004). However, a prospective study of Japanese men in Hawaii reported no association with intake of table salt/shoyu (Nomura et al. 1990).

High-salt foods such as processed meat products and salted fish are also important sources of nitrites and/or nitrosamines (Jakszyn and Gonzalez 2006; Mirvish 1995). A recent evaluation from a joint World Health Organization/Food and Agriculture Organization Expert Consultation concluded that salt-preserved foods and salt are probable risk factors for gastric cancer (World Health Organization 2003).

Body Weight

Accumulating evidence suggests that increased body weight is associated with increased risk for gastric cardia cancer (Chow et al. 1998; Kubo and Corley 2006; Lagergren et al. 1999; Lindblad et al. 2005; MacInnis et al. 2006; Merry et al. 2007;

Ryan et al. 2006), but not for gastric noncardia cancer (Lindblad et al. 2005; MacInnis et al. 2006). A systematic review of four published U.S. and European studies that have evaluated the association between body mass index (BMI) and risk of gastric cardia adenocarcinoma reported that being overweight (BMI ≥25 to <30 kg/m^2) or obese (BMI ≥30 kg/m^2) was significantly associated with a 1.5-fold increase in risk of gastric cardia adenocarcinoma (Kubo and Corley 2006). Three recent studies also found a positive association between BMI and risk for gastric cardia cancer (MacInnis et al. 2006; Merry et al. 2007; Ryan et al. 2006). In the Netherlands Cohort Study, the relative risks of gastric cardia adenocarcinoma were 1.32 (95% CI, 0.94–1.85) for overweight and 2.73 (95% CI, 1.56–4.79) for obese men and women, compared with individuals with normal weight (BMI ≥20 to <25 kg/m2) (Merry et al. 2007). The parallel increase in both the incidence of gastric cardia cancer and the prevalence of obesity suggests that obesity likely contributes to some of the increasing incidence of gastric cardia adenocarcinoma in the United States (Mayne and Navarro 2002).

Summary

Data from epidemiologic studies and experimental and animal studies suggest that diet has an important role in the etiology of gastric cancer. Available data suggest that high intake of fresh fruit and vegetables, lycopene and lycopene products, and potentially vitamin C and selenium might reduce the risk for gastric cancer. Data also support that high intake of nitrosamines, processed meat products, salt and salted foods, and being overweight or obese might increase the risk. Current data provide little support for an association of β-carotene, vitamin E, and alcohol consumption with risk for gastric cancer.

References

Abnet, C.C., Qiao, Y.L., Dawsey, S.M., Buckman, D.W., Yang, C.S., Blot, W.J., Dong, Z.W., Taylor, P.R., and Mark, S.D. (2003). Prospective study of serum retinol, beta-carotene, beta-cryptoxanthin, and lutein/zeaxanthin and esophageal and gastric cancers in China. Cancer Causes Control. 14:645–55.

Azuine, M.A., Goswami, U.C., Kayal, J.J., and Bhide, S.V. (1992). Antimutagenic and anticarcinogenic effects of carotenoids and dietary palm oil. Nutr Cancer. 17:287–95.

Bagnardi, V., Blangiardo, M., La Vecchia, C., and Corrao, G. (2001). Alcohol consumption and the risk of cancer: a meta-analysis. Alcohol Res Health. 25:263–70.

Balansky, R.M., Blagoeva, P.M., Mircheva, Z.I., Stoitchev, I., and Chernozemski, I. (1986). The effect of antioxidants on MNNG-induced stomach carcinogenesis in rats. J Cancer Res Clin Oncol. 112:272–5.

Barstad, B., Sorensen, T.I., Tjonneland, A., Johansen, D., Becker, U., Andersen, I.B., and Gronbaek, M. (2005). Intake of wine, beer and spirits and risk of gastric cancer. Eur J Cancer Prev. 14:239–43.

Bartsch, H., Ohshima, H., and Pignatelli, B. (1988). Inhibitors of endogenous nitrosation. Mechanisms and implications in human cancer prevention. Mutat Res. 202:307–24.

Bennedsen, M., Wang, X., Willen, R., Wadstrom, T., and Andersen, L.P. (1999). Treatment of H. pylori infected mice with antioxidant astaxanthin reduces gastric inflammation, bacterial load and modulates cytokine release by splenocytes. Immunol Lett. 70:185–9.

Bergin, I.L., Sheppard, B.J., and Fox, J.G. (2003). Helicobacter pylori infection and high dietary salt independently induce atrophic gastritis and intestinal metaplasia in commercially available outbred Mongolian gerbils. Dig Dis Sci. 48:475–85.

Bergman, K., and Slanina, P. (1986). Effects of dietary selenium compounds on benzo (a)-pyrene-induced forestomach tumours and whole-blood glutathione peroxidase activities in C3H mice. Anticancer Res. 6:785–90.

Bhuvaneswari, V., Velmurugan, B., Balasenthil, S., Ramachandran, C.R., and Nagini, S. (2001). Chemopreventive efficacy of lycopene on 7,12-dimethylbenz[a]anthracene-induced hamster buccal pouch carcinogenesis. Fitoterapia. 72:865–74.

Bjelakovic, G., Nikolova, D., Gluud, L.L., Simonetti, R.G., and Gluud, C. (2007). Mortality in randomized trials of antioxidant supplements for primary and secondary prevention: systematic review and meta-analysis. JAMA. 297:842–57.

Bjelakovic, G., Nikolova, D., Simonetti, R.G., and Gluud, C. (2004a). Antioxidant supplements for preventing gastrointestinal cancers. Cochrane Database Syst Rev. CD004183.

Bjelakovic, G., Nikolova, D., Simonetti, R.G., and Gluud, C. (2004b). Antioxidant supplements for prevention of gastrointestinal cancers: a systematic review and meta-analysis. Lancet. 364:1219–28.

Blot, W.J., Li, J.Y., Taylor, P.R., Guo, W., Dawsey, S., Wang, G.Q., Yang, C.S., Zheng, S.F., Gail, M., Li, G.Y., et al. (1993). Nutrition intervention trials in Linxian, China: supplementation with specific vitamin/mineral combinations, cancer incidence, and disease-specific mortality in the general population. J Natl Cancer Inst. 85:1483–92.

Boeing, H., Jedrychowski, W., Wahrendorf, J., Popiela, T., Tobiasz-Adamczyk, B., and Kulig, A. (1991). Dietary risk factors in intestinal and diffuse types of stomach cancer: a multicenter case-control study in Poland. Cancer Causes Control. 2:227–33.

Boosalis, M.G., Snowdon, D.A., Tully, C.L., and Gross, M.D. (1996). Acute phase response and plasma carotenoid concentrations in older women: findings from the nun study. Nutrition. 12:475–8.

Botterweck, A.A., Schouten, L.J., Volovics, A., Dorant, E., and van Den Brandt, P.A. (2000a). Trends in incidence of adenocarcinoma of the oesophagus and gastric cardia in ten European countries. Int J Epidemiol. 29:645–54.

Botterweck, A.A., van den Brandt, P.A., and Goldbohm, R.A. (2000b). Vitamins, carotenoids, dietary fiber, and the risk of gastric carcinoma: results from a prospective study after 6.3 years of follow-up. Cancer. 88:737–48.

Brenneisen, P., Steinbrenner, H., and Sies, H. (2005). Selenium, oxidative stress, and health aspects. Mol Aspects Med. 26:256–67.

Buiatti, E., Palli, D., Decarli, A., Amadori, D., Avellini, C., Bianchi, S., Biserni, R., Cipriani, F., Cocco, P., Giacosa, A., Marubini, E., Puntoni, R., Vindigni, C., Fraumeni, J.F., Jr., and Blot, W.J. (1989). A case-control study of gastric cancer and diet in Italy. Int J Cancer. 44:611–6.

Bukin, Y.V., Draudin-Krylenko, V.A., Kuvshinov, Y.P., Poddubniy, B.K., and Shabanov, M.A. (1997). Decrease of ornithine decarboxylase activity in premalignant gastric mucosa and regression of small intestinal metaplasia in patients supplemented with high doses of vitamin E. Cancer Epidemiol Biomarkers Prev. 6:543–6.

Cerar, A., and Pokorn, D. (1996). Inhibition of MNNG-induced gastroduodenal carcinoma in rats by synchronous application of wine or 11% ethanol. Nutr Cancer. 26:347–52.

Chang, C.Y., Chen, J.Y., Ke, D., and Hu, M.L. (2005). Plasma levels of lipophilic antioxidant vitamins in acute ischemic stroke patients: correlation to inflammation markers and neurological deficits. Nutrition. 21:987–93.

Chatterjee, A., Bagchi, D., Yasmin, T., and Stohs, S.J. (2005). Antimicrobial effects of antioxidants with and without clarithromycin on Helicobacter pylori. Mol Cell Biochem. 270:125–30.

Chen, H., Tucker, K.L., Graubard, B.I., Heineman, E.F., Markin, R.S., Potischman, N.A., Russell, R.M., Weisenburger, D.D., and Ward, M.H. (2002). Nutrient intakes and adenocarcinoma of the esophagus and distal stomach. Nutr Cancer. 42:33–40.

Chow, W.H., Blot, W.J., Vaughan, T.L., Risch, H.A., Gammon, M.D., Stanford, J.L., Dubrow, R., Schoenberg, J.B., Mayne, S.T., Farrow, D.C., Ahsan, H., West, A.B., Rotterdam, H., Niwa, S., and Fraumeni, J.F., Jr. (1998). Body mass index and risk of adenocarcinomas of the esophagus and gastric cardia. J Natl Cancer Inst. 90:150–5.

Corley, D.A., and Kubo, A. (2004). Influence of site classification on cancer incidence rates: an analysis of gastric cardia carcinomas. J Natl Cancer Inst. 96:1383–7.

Correa, P., Fontham, E.T., Bravo, J.C., Bravo, L.E., Ruiz, B., Zarama, G., Realpe, J.L., Malcom, G.T., Li, D., Johnson, W.D., and Mera, R. (2000). Chemoprevention of gastric dysplasia: randomized trial of antioxidant supplements and anti-helicobacter pylori therapy. J Natl Cancer Inst. 92:1881–8.

Correa, P., Fontham, E., Pickle, L.W., Chen, V., Lin, Y.P., and Haenszel, W. (1985). Dietary determinants of gastric cancer in south Louisiana inhabitants. J Natl Cancer Inst. 75:645–54.

De Stefani, E., Boffetta, P., Brennan, P., Deneo-Pellegrini, H., Carzoglio, J.C., Ronco, A., and Mendilaharsu, M. (2000). Dietary carotenoids and risk of gastric cancer: a case-control study in Uruguay. Eur J Cancer Prev. 9:329–34.

Devesa, S.S., Blot, W.J., and Fraumeni, J.F., Jr. (1998). Changing patterns in the incidence of esophageal and gastric carcinoma in the United States. Cancer. 83:2049–53.

Devesa, S.S., and Fraumeni, J.F., Jr. (1999). The rising incidence of gastric cardia cancer. J Natl Cancer Inst. 91:747–9.

Ekstrom, A.M., Serafini, M., Nyren, O., Hansson, L.E., Ye, W., and Wolk, A. (2000). Dietary antioxidant intake and the risk of cardia cancer and noncardia cancer of the intestinal and diffuse types: a population-based case-control study in Sweden. Int J Cancer. 87:133–40.

el-Bayoumy, K. (1985). Effects of organoselenium compounds on induction of mouse forestomach tumors by benzo(a)pyrene. Cancer Res. 45:3631–5.

el-Deiry, W.S. (1998). Regulation of p53 downstream genes. Semin Cancer Biol. 8:345–57.

Engel, L.S., Chow, W.H., Vaughan, T.L., Gammon, M.D., Risch, H.A., Stanford, J.L., Schoenberg, J.B., Mayne, S.T., Dubrow, R., Rotterdam, H., West, A.B., Blaser, M., Blot, W.J., Gail, M.H., and Fraumeni, J.F., Jr. (2003). Population attributable risks of esophageal and gastric cancers. J Natl Cancer Inst. 95:1404–13.

Everett, S.M., Drake, I.M., White, K.L., Mapstone, N.P., Chalmers, D.M., Schorah, C.J., and Axon, A.T. (2002). Antioxidant vitamin supplements do not reduce reactive oxygen species activity in Helicobacter pylori gastritis in the short term. Br J Nutr. 87:3–11.

Fontham, E.T., Correa, P., Mera, R., Bravo, L.E., Bravo, J.C., Piazuelo, M.B., and Camargo, M.C. (2005). Duration of exposure, a neglected factor in chemoprevention trials. Cancer Epidemiol Biomarkers Prev. 14:2465–6.

Ford, E.S., Liu, S., Mannino, D.M., Giles, W.H., and Smith, S.J. (2003). C-reactive protein concentration and concentrations of blood vitamins, carotenoids, and selenium among United States adults. Eur J Clin Nutr. 57:1157–63.

Fox, J.G., Dangler, C.A., Taylor, N.S., King, A., Koh, T.J., and Wang, T.C. (1999). High-salt diet induces gastric epithelial hyperplasia and parietal cell loss, and enhances Helicobacter pylori colonization in C57BL/6 mice. Cancer Res. 59:4823–8.

Fox, J.G., Rogers, A.B., Ihrig, M., Taylor, N.S., Whary, M.T., Dockray, G., Varro, A., and Wang, T.C. (2003). Helicobacter pylori-associated gastric cancer in INS-GAS mice is gender specific. Cancer Res. 63:942–50.

Franceschi, S., Bidoli, E., La Vecchia, C., Talamini, R., D'Avanzo, B., and Negri, E. (1994). Tomatoes and risk of digestive-tract cancers. Int J Cancer. 59:181–4.

Freedman, N.D., Abnet, C.C., Leitzmann, M.F., Mouw, T., Subar, A.F., Hollenbeck, A.R., and Schatzkin, A. (2007). A prospective study of tobacco, alcohol, and the risk of esophageal and gastric cancer subtypes. Am J Epidemiol. 165:1424–33.

Frei, B. (1994). Reactive oxygen species and antioxidant vitamins: mechanisms of action. Am J Med. 97(Suppl 3A):5S–13S.

Furihata, C., Sato, Y., Hosaka, M., Matsushima, T., Furukawa, F., and Takahashi, M. (1984). NaCl induced ornithine decarboxylase and DNA synthesis in rat stomach mucosa. Biochem Biophys Res Commun. 121:1027–32.

Garcia-Closas, R., Gonzalez, C.A., Agudo, A., and Riboli, E. (1999). Intake of specific carotenoids and flavonoids and the risk of gastric cancer in Spain. Cancer Causes Control. 10:71–5.

Giovannucci, E. (1999). Tomatoes, tomato-based products, lycopene, and cancer: review of the epidemiologic literature. J Natl Cancer Inst. 91:317–31.

Gonzalez, C.A., Jakszyn, P., Pera, G., Agudo, A., Bingham, S., Palli, D., Ferrari, P., Boeing, H., del Giudice, G., Plebani, M., Carneiro, F., Nesi, G., Berrino, F., Sacerdote, C., Tumino, R., Panico, S., Berglund, G., Siman, H., Nyren, O., Hallmans, G., Martinez, C., Dorronsoro, M., Barricarte, A., Navarro, C., Quiros, J.R., Allen, N., Key, T.J., Day, N.E., Linseisen, J., Nagel, G., Bergmann, M.M., Overvad, K., Jensen, M.K., Tjonneland, A., Olsen, A., Bueno-de-Mesquita, H.B., Ocke, M., Peeters, P.H., Numans, M.E., Clavel-Chapelon, F., Boutron-Ruault, M.C., Trichopoulou, A., Psaltopoulou, T., Roukos, D., Lund, E., Hemon, B., Kaaks, R., Norat, T., and Riboli, E. (2006a). Meat intake and risk of stomach and esophageal adenocarcinoma within the European Prospective Investigation into Cancer and Nutrition (EPIC). J Natl Cancer Inst. 98:345–54.

Gonzalez, C.A., Pera, G., Agudo, A., Bueno-de-Mesquita, H.B., Ceroti, M., Boeing, H., Schulz, M., Del Giudice, G., Plebani, M., Carneiro, F., Berrino, F., Sacerdote, C., Tumino, R., Panico, S., Berglund, G., Siman, H., Hallmans, G., Stenling, R., Martinez, C., Dorronsoro, M., Barricarte, A., Navarro, C., Quiros, J.R., Allen, N., Key, T.J., Bingham, S., Day, N.E., Linseisen, J., Nagel, G., Overvad, K., Jensen, M.K., Olsen, A., Tjonneland, A., Buchner, F.L., Peeters, P.H., Numans, M.E., Clavel-Chapelon, F., Boutron-Ruault, M.C., Roukos, D., Trichopoulou, A., Psaltopoulou, T., Lund, E., Casagrande, C., Slimani, N., Jenab, M., and Riboli, E. (2006b). Fruit and vegetable intake and the risk of stomach and oesophagus adenocarcinoma in the European Prospective Investigation into Cancer and Nutrition (EPIC-EURGAST). Int J Cancer. 118:2559–66.

Gonzalez, C.A., Sanz, J.M., Marcos, G., Pita, S., Brullet, E., Saigi, E., Badia, A., and Riboli, E. (1991). Dietary factors and stomach cancer in Spain: a multi-centre case-control study. Int J Cancer. 49:513–9.

Goswami, U.C., and Sharma, N. (2005). Efficiency of a few retinoids and carotenoids in vivo in controlling benzo[a]pyrene-induced forestomach tumour in female Swiss mice. Br J Nutr. 94:540–3.

Graham, S., Haughey, B., Marshall, J., Brasure, J., Zielezny, M., Freudenheim, J., West, D., Nolan, J., and Wilkinson, G. (1990). Diet in the epidemiology of gastric cancer. Nutr Cancer. 13:19–34.

Haenszel, W., Kurihara, M., Segi, M., and Lee, R.K. (1972). Stomach cancer among Japanese in Hawaii. J Natl Cancer Inst. 49:969–88.

Halliwell, B. (2001). Vitamin C and genomic stability. Mutat Res. 475:29–35.

Hansson, L.E., Nyren, O., Bergstrom, R., Wolk, A., Lindgren, A., Baron, J., and Adami, H.O. (1993). Diet and risk of gastric cancer. A population-based case-control study in Sweden. Int J Cancer. 55:181–9.

Harrison, L.E., Zhang, Z.F., Karpeh, M.S., Sun, M., and Kurtz, R.C. (1997). The role of dietary factors in the intestinal and diffuse histologic subtypes of gastric adenocarcinoma: a case-control study in the U.S. Cancer. 80:1021–8.

Helicobacter and Cancer Collaborative Group. (2001). Gastric cancer and Helicobacter pylori: a combined analysis of 12 case control studies nested within prospective cohorts. Gut. 49:347–53.

Hennekens, C.H., Buring, J.E., Manson, J.E., Stampfer, M., Rosner, B., Cook, N.R., Belanger, C., LaMotte, F., Gaziano, J.M., Ridker, P.M., Willett, W., and Peto, R. (1996). Lack of effect of long-term supplementation with beta carotene on the incidence of malignant neoplasms and cardiovascular disease. N Engl J Med. 334:1145–9.

Hu, G., Han, C., Wild, C.P., Hall, J., and Chen, J. (1992). Lack of effects of selenium on N-nitrosomethylbenzylamine-induced tumorigenesis, DNA methylation, and oncogene expression in rats and mice. Nutr Cancer. 18:287–95.

Iishi, H., Tatsuta, M., Baba, M., and Taniguchi, H. (1989). Promotion by ethanol of gastric carcinogenesis induced by N-methyl-N'-nitro-N-nitrosoguanidine in Wistar rats. Br J Cancer. 59:719–21.

International Agency for Research on Cancer. (2003). IARC handbooks of cancer prevention. Vol. 8. Fruits and vegetables. Lyon: IARC*Press*.

International Agency for Research on Cancer. (2006). Overall evaluation of carcinogenicity to humans. IARC monographs. Vol. 1–86. (http://monographs.iarc.fr). Lyon: IARC*Press*.

Jacobs, E.J., Connell, C.J., McCullough, M.L., Chao, A., Jonas, C.R., Rodriguez, C., Calle, E.E., and Thun, M.J. (2002). Vitamin C, vitamin E, and multivitamin supplement use and stomach cancer mortality in the Cancer Prevention Study II cohort. Cancer Epidemiol Biomarkers Prev. 11:35–41.

Jakszyn, P., Bingham, S., Pera, G., Agudo, A., Luben, R., Welch, A., Boeing, H., Del Giudice, G., Palli, D., Saieva, C., Krogh, V., Sacerdote, C., Tumino, R., Panico, S., Berglund, G., Siman, H., Hallmans, G., Sanchez, M.J., Larranaga, N., Barricarte, A., Chirlaque, M.D., Quiros, J.R., Key, T.J., Allen, N., Lund, E., Carneiro, F., Linseisen, J., Nagel, G., Overvad, K., Tjonneland, A., Olsen, A., Bueno-de-Mesquita, H.B., Ocke, M.O., Peeters, P.H., Numans, M.E., Clavel-Chapelon, F., Trichopoulou, A., Fenger, C., Stenling, R., Ferrari, P., Jenab, M., Norat, T., Riboli, E., and Gonzalez, C.A. (2006). Endogenous versus exogenous exposure to N-nitroso compounds and gastric cancer risk in the European Prospective Investigation into Cancer and Nutrition (EPIC-EURGAST) study. Carcinogenesis. 27:1497–501.

Jakszyn, P., and Gonzalez, C.A. (2006). Nitrosamine and related food intake and gastric and oesophageal cancer risk: a systematic review of the epidemiological evidence. World J Gastroenterol. 12:4296–303.

Jarosz, M., Dzieniszewski, J., Dabrowska-Ufniarz, E., Wartanowicz, M., Ziemlanski, S., and Reed, P.I. (1998). Effects of high dose vitamin C treatment on Helicobacter pylori infection and total vitamin C concentration in gastric juice. Eur J Cancer Prev. 7:449–54.

Jemal, A., Siegel, R., Ward, E., Murray, T., Xu, J., Smigal, C., and Thun, M.J. (2006). Cancer statistics, 2006. CA Cancer J Clin. 56:106–30.

Jenab, M., Riboli, E., Ferrari, P., Friesen, M., Sabate, J., Norat, T., Slimani, N., Tjonneland, A., Olsen, A., Overvad, K., Boutron-Ruault, M.C., Clavel-Chapelon, F., Boeing, H., Schulz, M., Linseisen, J., Nagel, G., Trichopoulou, A., Naska, A., Oikonomou, E., Berrino, F., Panico, S., Palli, D., Sacerdote, C., Tumino, R., Peeters, P.H., Numans, M.E., Bueno-de-Mesquita, H.B., Buchner, F.L., Lund, E., Pera, G., Chirlaque, M.D., Sanchez, M.J., Arriola, L., Barricarte, A., Quiros, J.R., Johansson, I., Johansson, A., Berglund, G., Bingham, S., Khaw, K.T., Allen, N., Key, T., Carneiro, F., Save, V., Del Giudice, G., Plebani, M., Kaaks, R., and Gonzalez, C.A. (2006a). Plasma and dietary carotenoid, retinol and tocopherol levels and the risk of gastric adenocarcinomas in the European prospective investigation into cancer and nutrition. Br J Cancer. 95:406–15.

Jenab, M., Riboli, E., Ferrari, P., Sabate, J., Slimani, N., Norat, T., Friesen, M., Tjonneland, A., Olsen, A., Overvad, K., Boutron-Ruault, M.C., Clavel-Chapelon, F., Touvier, M., Boeing, H., Schulz, M., Linseisen, J., Nagel, G., Trichopoulou, A., Naska, A., Oikonomou, E., Krogh, V., Panico, S., Masala, G., Sacerdote, C., Tumino, R., Peeters, P.H., Numans, M.E., Bueno-de-Mesquita, H.B., Buchner, F.L., Lund, E., Pera, G., Sanchez, C.N., Sanchez, M.J., Arriola, L., Barricarte, A., Quiros, J.R., Hallmans, G., Stenling, R., Berglund, G., Bingham, S., Khaw, K.T., Key, T., Allen, N., Carneiro, F., Mahlke, U., Del Giudice, G., Palli, D., Kaaks, R., and Gonzalez, C.A. (2006b). Plasma and dietary vitamin C levels and risk of gastric cancer in the European Prospective Investigation into Cancer and Nutrition (EPIC-EURGAST). Carcinogenesis. 27:2250–7.

Kabuto, M., Imai, H., Yonezawa, C., Neriishi, K., Akiba, S., Kato, H., Suzuki, T., Land, C.E., and Blot, W.J. (1994). Prediagnostic serum selenium and zinc levels and subsequent risk of lung and stomach cancer in Japan. Cancer Epidemiol Biomarkers Prev. 3:465–9.

Kato, S., Tsukamoto, T., Mizoshita, T., Tanaka, H., Kumagai, T., Ota, H., Katsuyama, T., Asaka, M., and Tatematsu, M. (2006). High salt diets dose-dependently promote gastric chemical carcinogenesis in Helicobacter pylori-infected Mongolian gerbils associated with a shift in mucin production from glandular to surface mucous cells. Int J Cancer. 119:1558–66.

Kawasaki, H., Morishige, F., Tanaka, H., and Kimoto, E. (1982). Influence of oral supplementation of ascorbate upon the induction of N-methyl-N'-nitro-N-nitrosoguanidine. Cancer Lett. 16:57–63.

Kelley, D.S., and Bendich, A. (1996). Essential nutrients and immunologic functions. Am J Clin Nutr. 63:994S–6S.

Kelley, J.R., and Duggan, J.M. (2003). Gastric cancer epidemiology and risk factors. J Clin Epidemiol. 56:1–9.

Knekt, P., Aromaa, A., Maatela, J., Alfthan, G., Aaran, R.K., Hakama, M., Hakulinen, T., Peto, R., and Teppo, L. (1990). Serum selenium and subsequent risk of cancer among Finnish men and women. J Natl Cancer Inst. 82:864–8.

Knekt, P., Aromaa, A., Maatela, J., Alfthan, G., Aaran, R.K., Teppo, L., and Hakama, M. (1988). Serum vitamin E, serum selenium and the risk of gastrointestinal cancer. Int J Cancer. 42:846–50.

Knekt, P., Jarvinen, R., Dich, J., and Hakulinen, T. (1999). Risk of colorectal and other gastrointestinal cancers after exposure to nitrate, nitrite and N-nitroso compounds: a follow-up study. Int J Cancer. 80:852–6.

Kneller, R.W., McLaughlin, J.K., Bjelke, E., Schuman, L.M., Blot, W.J., Wacholder, S., Gridley, G., CoChien, H.T., and Fraumeni, J.F., Jr. (1991). A cohort study of stomach cancer in a high-risk American population. Cancer. 68:672–8.

Kobayashi, M., Kogata, M., Yamamura, M., Takada, H., Hioki, K., and Yamamoto, M. (1986). Inhibitory effect of dietary selenium on carcinogenesis in rat glandular stomach induced by N-methyl-N -nitro-N-nitrosoguanidine. Cancer Res. 46:2266–70.

Kono, S., and Hirohata, T. (1996). Nutrition and stomach cancer. Cancer Causes Control. 7:41–55.

Krinsky, N.I., and Johnson, E.J. (2005). Carotenoid actions and their relation to health and disease. Mol Aspects Med. 26:459–516.

Kritchevsky, S.B., Bush, A.J., Pahor, M., and Gross, M.D. (2000). Serum carotenoids and markers of inflammation in nonsmokers. Am J Epidemiol. 152:1065–71.

Kubo, A., and Corley, D.A. (2006). Body mass index and adenocarcinomas of the esophagus or gastric cardia: a systematic review and meta-analysis. Cancer Epidemiol Biomarkers Prev. 15:872–8.

Lagergren, J., Bergstrom, R., and Nyren, O. (1999). Association between body mass and adenocarcinoma of the esophagus and gastric cardia. Ann Intern Med. 130:883–90.

Larsson, S.C., Bergkvist, L., Naslund, I., Rutegard, J., and Wolk, A. (2007a). Vitamin A, retinol, and carotenoids and the risk of gastric cancer: a prospective cohort study. Am J Clin Nutr. 85:497–503.

Larsson, S.C., Giovannucci, E., and Wolk, A. (2007b). Alcoholic beverage consumption and gastric cancer risk: a prospective population-based study in women. Int J Cancer. 120:373–7.

Larsson, S.C., Orsini, N., and Wolk, A. (2006). Processed meat consumption and stomach cancer risk: a meta-analysis. J Natl Cancer Inst. 98:1078–87.

La Vecchia, C., Negri, E., Decarli, A., D'Avanzo, B., and Franceschi, S. (1987). A case-control study of diet and gastric cancer in northern Italy. Int J Cancer. 40:484–9.

Lee, I.M., Cook, N.R., Gaziano, J.M., Gordon, D., Ridker, P.M., Manson, J.E., Hennekens, C.H., and Buring, J.E. (2005). Vitamin E in the primary prevention of cardiovascular disease and cancer: the Women's Health Study—a randomized controlled trial. JAMA. 294:56–65.

Lee, S.A., Kang, D., Shim, K.N., Choe, J.W., Hong, W.S., and Choi, H. (2003). Effect of diet and Helicobacter pylori infection to the risk of early gastric cancer. J Epidemiol. 13:162–8.

Lindblad, M., Rodriguez, L.A., and Lagergren, J. (2005). Body mass, tobacco and alcohol and risk of esophageal, gastric cardia, and gastric non-cardia adenocarcinoma among men and women in a nested case-control study. Cancer Causes Control. 16: 285–94.

Lippman, S.M., Goodman, P.J., Klein, E.A., Parnes, H.L., Thompson, I.M., Jr., Kristal, A.R., Santella, R.M., Probstfield, J.L., Moinpour, C.M., Albanes, D., Taylor, P.R., Minasian, L.M., Hoque, A., Thomas, S.M., Crowley, J.J., Gaziano, J.M., Stanford, J.L., Cook, E.D., Fleshner, N.E., Lieber, M.M., Walther, P.J., Khuri, F.R., Karp, D.D., Schwartz, G.G., Ford, L.G., and

Coltman, C.A., Jr. (2005). Designing the Selenium and Vitamin E Cancer Prevention Trial (SELECT). J Natl Cancer Inst. 97:94–102.

Lissowska, J., Gail, M.H., Pee, D., Groves, F.D., Sobin, L.H., Nasierowska-Guttmejer, A., Sygnowska, E., Zatonski, W., Blot, W.J., and Chow, W.H. (2004). Diet and stomach cancer risk in Warsaw, Poland. Nutr Cancer. 48:149–59.

Liu, B.H., and Lee, Y.K. (2003). Effect of total secondary carotenoids extracts from Chlorococcum sp on Helicobacter pylori-infected BALB/c mice. Int Immunopharmacol. 3:979–86.

Liu, C., Russell, R.M., and Wang, X.D. (2006). Lycopene supplementation prevents smoke-induced changes in p53, p53 phosphorylation, cell proliferation, and apoptosis in the gastric mucosa of ferrets. J Nutr. 136:106–11.

Lunet, N., Lacerda-Vieira, A., and Barros, H. (2005). Fruit and vegetables consumption and gastric cancer: a systematic review and meta-analysis of cohort studies. Nutr Cancer. 53:1–10.

MacInnis, R.J., English, D.R., Hopper, J.L., and Giles, G.G. (2006). Body size and composition and the risk of gastric and oesophageal adenocarcinoma. Int J Cancer. 118:2628–31.

Mannick, E.E., Bravo, L.E., Zarama, G., Realpe, J.L., Zhang, X.J., Ruiz, B., Fontham, E.T., Mera, R., Miller, M.J., and Correa, P. (1996). Inducible nitric oxide synthase, nitrotyrosine, and apoptosis in Helicobacter pylori gastritis: effect of antibiotics and antioxidants. Cancer Res. 56:3238–43.

Mark, S.D., Qiao, Y.L., Dawsey, S.M., Wu, Y.P., Katki, H., Gunter, E.W., Fraumeni, J.F., Jr., Blot, W.J., Dong, Z.W., and Taylor, P.R. (2000). Prospective study of serum selenium levels and incident esophageal and gastric cancers. J Natl Cancer Inst. 92:1753–63.

Mayne, S.T., and Navarro, S.A. (2002). Diet, obesity and reflux in the etiology of adenocarcinomas of the esophagus and gastric cardia in humans. J Nutr. 132:3467S–70S.

McCall, M.R., and Frei, B. (1999). Can antioxidant vitamins materially reduce oxidative damage in humans? Free Radic Biol Med. 26:1034–53.

McMillan, D.C., Sattar, N., Talwar, D., O'Reilly, D.S., and McArdle, C.S. (2000). Changes in micronutrient concentrations following anti-inflammatory treatment in patients with gastrointestinal cancer. Nutrition. 16:425–8.

McMillan, D.C., Talwar, D., Sattar, N., Underwood, M., O'Reilly, D.S., and McArdle, C. (2002). The relationship between reduced vitamin antioxidant concentrations and the systemic inflammatory response in patients with common solid tumours. Clin Nutr. 21:161–4.

Merry, A.H., Schouten, L.J., Goldbohm, R.A., and van den Brandt, P.A. (2007). Body mass index, height and risk of adenocarcinoma of the oesophagus and gastric cardia: a prospective cohort study. Gut. 56:1503–11.

Mirvish, S.S. (1986). Effects of vitamins C and E on N-nitroso compound formation, carcinogenesis, and cancer. Cancer. 58:1842–50.

Mirvish, S.S. (1994). Experimental evidence for inhibition of N-nitroso compound formation as a factor in the negative correlation between vitamin C consumption and the incidence of certain cancers. Cancer Res. 54:1948s–51s.

Mirvish, S.S. (1995). Role of N-nitroso compounds (NOC) and N-nitrosation in etiology of gastric, esophageal, nasopharyngeal and bladder cancer and contribution to cancer of known exposures to NOC. Cancer Lett. 93:17–48.

Mirvish, S.S., Pelfrene, A.F., Garcia, H., and Shubik, P. (1976). Effect of sodium ascorbate on tumor induction in rats treated with morpholine and sodium nitrite, and with nitrosomorpholine. Cancer Lett. 2:101–8.

Mirvish, S.S., Salmasi, S., Cohen, S.M., Patil, K., and Mahboubi, E. (1983). Liver and forestomach tumors and other forestomach lesions in rats treated with morpholine and sodium nitrite, with and without sodium ascorbate. J Natl Cancer Inst. 71:81–5.

Miyauchi, M., Nakamura, H., Furukawa, F., Son, H.Y., Nishikawa, A., and Hirose, M. (2002). Promoting effects of combined antioxidant and sodium nitrite treatment on forestomach carcinogenesis in rats after initiation with N-methyl-N'-nitro-N-nitrosoguanidine. Cancer Lett. 178:19–24.

Moghadaszadeh, B., and Beggs, A.H. (2006). Selenoproteins and their impact on human health through diverse physiological pathways. Physiology (Bethesda). 21:307–15.

Newberne, P.M., Charnley, G., Adams, K., Cantor, M., Suphakarn, V., Roth, D., and Schrager, T.F. (1987). Gastric carcinogenesis: a model for the identification of risk factors. Cancer Lett. 38:149–63.

Ngoan, L.T., Mizoue, T., Fujino, Y., Tokui, N., and Yoshimura, T. (2002). Dietary factors and stomach cancer mortality. Br J Cancer. 87:37–42.

Nomura, A., Grove, J.S., Stemmermann, G.N., and Severson, R.K. (1990). A prospective study of stomach cancer and its relation to diet, cigarettes, and alcohol consumption. Cancer Res. 50:627–31.

Nomura, A., Heilbrun, L.K., Morris, J.S., and Stemmermann, G.N. (1987). Serum selenium and the risk of cancer, by specific sites: case-control analysis of prospective data. J Natl Cancer Inst. 79:103–8.

Nouraie, M., Pietinen, P., Kamangar, F., Dawsey, S.M., Abnet, C.C., Albanes, D., Virtamo, J., and Taylor, P.R. (2005). Fruits, vegetables, and antioxidants and risk of gastric cancer among male smokers. Cancer Epidemiol Biomarkers Prev. 14:2087–92.

Nozaki, K., Shimizu, N., Inada, K., Tsukamoto, T., Inoue, M., Kumagai, T., Sugiyama, A., Mizoshita, T., Kaminishi, M., and Tatematsu, M. (2002). Synergistic promoting effects of Helicobacter pylori infection and high-salt diet on gastric carcinogenesis in Mongolian gerbils. Jpn J Cancer Res. 93:1083–9.

Okazaki, K., Ishii, Y., Kitamura, Y., Maruyama, S., Umemura, T., Miyauchi, M., Yamagishi, M., Imazawa, T., Nishikawa, A., Yoshimura, Y., Nakazawa, H., and Hirose, M. (2006). Dose-dependent promotion of rat forestomach carcinogenesis by combined treatment with sodium nitrite and ascorbic acid after initiation with N-methyl-N′-nitro-N-nitrosoguanidine: possible contribution of nitric oxide-associated oxidative DNA damage. Cancer Sci. 97:175–82.

Oliveira, C.P., Kassab, P., Lopasso, F.P., Souza, H.P., Janiszewski, M., Laurindo, F.R., Iriya, K., and Laudanna, A.A. (2003). Protective effect of ascorbic acid in experimental gastric cancer: reduction of oxidative stress. World J Gastroenterol. 9:446–8.

Parkin, D.M., Bray, F., Ferlay, J., and Pisani, P. (2005). Global cancer statistics, 2002. CA Cancer J Clin. 55:74–108.

Rayman, M.P. (2000). The importance of selenium to human health. Lancet. 356:233–41.

Rayman, M.P. (2005). Selenium in cancer prevention: a review of the evidence and mechanism of action. Proc Nutr Soc. 64:527–42.

Riboli, E., and Norat, T. (2003). Epidemiologic evidence of the protective effect of fruit and vegetables on cancer risk. Am J Clin Nutr. 78:559S–69S.

Ridker, P.M. (2003). Clinical application of C-reactive protein for cardiovascular disease detection and prevention. Circulation. 107:363–9.

Robles, A.I., Linke, S.P., and Harris, C.C. (2002). The p53 network in lung carcinogenesis. Oncogene. 21:6898–907.

Rogers, A.B., Taylor, N.S., Whary, M.T., Stefanich, E.D., Wang, T.C., and Fox, J.G. (2005). Helicobacter pylori but not high salt induces gastric intraepithelial neoplasia in B6129 mice. Cancer Res. 65:10709–15.

Rom, W.N., and Tchou-Wong, K.M. (2003). Functional genomics in lung cancer and biomarker detection. Am J Respir Cell Mol Biol. 29:153–6.

Rowley, K., Walker, K.Z., Cohen, J., Jenkins, A.J., O'Neal, D., Su, Q., Best, J.D., and O'Dea, K. (2003). Inflammation and vascular endothelial activation in an Aboriginal population: relationships to coronary disease risk factors and nutritional markers. Med J Aust. 178:495–500.

Russell, R.M. (1998). Physiological and clinical significance of carotenoids. Int J Vitam Nutr Res. 68:349–53.

Ryan, A.M., Rowley, S.P., Fitzgerald, A.P., Ravi, N., and Reynolds, J.V. (2006). Adenocarcinoma of the oesophagus and gastric cardia: male preponderance in association with obesity. Eur J Cancer. 42:1151–8.

Sasazuki, S., Sasaki, S., Tsubono, Y., Okubo, S., Hayashi, M., Kakizoe, T., and Tsugane, S. (2003). The effect of 5-year vitamin C supplementation on serum pepsinogen level and Helicobacter pylori infection. Cancer Sci. 94:378–82.

She, Q.B., Chen, N., and Dong, Z. (2000). ERKs and p38 kinase phosphorylate p53 protein at serine 15 in response to UV radiation. J Biol Chem. 275:20444–9.

Shibata, M.A., Hirose, M., Kagawa, M., Boonyaphiphat, P., and Ito, N. (1993). Enhancing effect of concomitant L-ascorbic acid administration on BHA-induced forestomach carcinogenesis in rats. Carcinogenesis. 14:275–80.

Shirai, T., Masuda, A., Fukushima, S., Hosoda, K., and Ito, N. (1985). Effects of sodium L-ascorbate and related compounds on rat stomach carcinogenesis initiated by N-methyl-N′-nitro-N-nitro-soguanidine. Cancer Lett. 29:283–8.

Sjodahl, K., Lu, Y., Nilsen, T.I., Ye, W., Hveem, K., Vatten, L., and Lagergren, J. (2007). Smoking and alcohol drinking in relation to risk of gastric cancer: a population-based, prospective cohort study. Int J Cancer. 120:128–32.

Sjunnesson, H., Sturegard, E., Willen, R., and Wadstrom, T. (2001). High intake of selenium, beta-carotene, and vitamins A, C, and E reduces growth of Helicobacter pylori in the guinea pig. Comp Med. 51:418–23.

Sorbye, H., Gislason, H., Kvinnsland, S., and Svanes, K. (1994a). Effect of salt on cell proliferation and N-methyl-N′-nitro-N-nitrosoguanidine penetration to proliferative cells in the forestomach of rats. J Cancer Res Clin Oncol. 120:465–70.

Sorbye, H., Kvinnsland, S., and Svanes, K. (1994b). Effect of salt-induced mucosal damage and healing on penetration of N-methyl-N′-nitro-N-nitrosoguanidine to proliferative cells in the gastric mucosa of rats. Carcinogenesis. 15:673–9.

Sorbye, H., Maaartmann-Moe, H., and Svanes, K. (1994c). Gastric carcinogenesis in rats given hypertonic salt at different times before a single dose of N-methyl-N′-nitro-N-nitrosoguanidine. J Cancer Res Clin Oncol. 120:159–63.

Stahelin, H.B., Gey, K.F., Eichholzer, M., Ludin, E., Bernasconi, F., Thurneysen, J., and Brubacher, G. (1991). Plasma antioxidant vitamins and subsequent cancer mortality in the 12-year follow-up of the prospective Basel Study. Am J Epidemiol. 133:766–75.

Stahl, W., and Sies, H. (1996). Lycopene: a biologically important carotenoid for humans? Arch Biochem Biophys. 336:1–9.

Su, Y.P., Tang, J.M., Tang, Y., and Gao, H.Y. (2005). Histological and ultrastructural changes induced by selenium in early experimental gastric carcinogenesis. World J Gastroenterol. 11:4457–60.

Sun, Y.Q., Girgensone, I., Leanderson, P., Petersson, F., and Borch, K. (2005). Effects of antioxidant vitamin supplements on Helicobacter pylori-induced gastritis in Mongolian gerbils. Helicobacter. 10:33–42.

Tajima, K., and Tominaga, S. (1985). Dietary habits and gastro-intestinal cancers: a comparative case-control study of stomach and large intestinal cancers in Nagoya, Japan. Jpn J Cancer Res. 76:705–16.

Takahashi, M., Furukawa, F., Toyoda, K., Sato, H., Hasegawa, R., and Hayashi, Y. (1986a). Effects of four antioxidants on N-methyl-N′-nitro-N-nitrosoguanidine initiated gastric tumor development in rats. Cancer Lett. 30:161–8.

Takahashi, M., and Hasegawa, R. (1985). Enhancing effects of dietary salt on both initiation and promotion stages of rat gastric carcinogenesis. Princess Takamatsu Symp. 16:169–82.

Takahashi, M., Hasegawa, R., Furukawa, F., Toyoda, K., Sato, H., and Hayashi, Y. (1986b). Effects of ethanol, potassium metabisulfite, formaldehyde and hydrogen peroxide on gastric carcinogenesis in rats after initiation with N-methyl-N′-nitro-N-nitrosoguanidine. Jpn J Cancer Res. 77:118–24.

Takahashi, M., Kokubo, T., Furukawa, F., Kurokawa, Y., Tatematsu, M., and Hayashi, Y. (1983). Effect of high salt diet on rat gastric carcinogenesis induced by N-methyl-N′-nitro-N-nitrosoguanidine. Gann. 74:28–34.

Takahashi, M., Nishikawa, A., Furukawa, F., Enami, T., Hasegawa, T., and Hayashi, Y. (1994). Dose-dependent promoting effects of sodium chloride (NaCl) on rat glandular stomach carcinogenesis initiated with N-methyl-N′-nitro-N-nitrosoguanidine. Carcinogenesis. 15:1429–32.

Tibbetts, R.S., Brumbaugh, K.M., Williams, J.M., Sarkaria, J.N., Cliby, W.A., Shieh, S.Y., Taya, Y., Prives, C., and Abraham, R.T. (1999). A role for ATR in the DNA damage-induced phosphorylation of p53. Genes Dev. 13:152–7.

Tsugane, S. (2005). Salt, salted food intake, and risk of gastric cancer: epidemiologic evidence. Cancer Sci. 96:1–6.

Tsugane, S., Sasazuki, S., Kobayashi, M., and Sasaki, S. (2004). Salt and salted food intake and subsequent risk of gastric cancer among middle-aged Japanese men and women. Br J Cancer. 90:128–34.

Tuyns, A.J., Kaaks, R., Haelterman, M., and Riboli, E. (1992). Diet and gastric cancer. A case-control study in Belgium. Int J Cancer. 51:1–6.

van den Brandt, P.A., Botterweck, A.A., and Goldbohm, R.A. (2003). Salt intake, cured meat consumption, refrigerator use and stomach cancer incidence: a prospective cohort study (Netherlands). Cancer Causes Control. 14:427–38.

van den Brandt, P.A., Goldbohm, R.A., van't Veer, P., Bode, P., Dorant, E., Hermus, R.J., and Sturmans, F. (1993). A prospective cohort study on toenail selenium levels and risk of gastrointestinal cancer. J Natl Cancer Inst. 85:224–9.

van Herpen-Broekmans, W.M., Klopping-Ketelaars, I.A., Bots, M.L., Kluft, C., Princen, H., Hendriks, H.F., Tijburg, L.B., van Poppel, G., and Kardinaal, A.F. (2004). Serum carotenoids and vitamins in relation to markers of endothelial function and inflammation. Eur J Epidemiol. 19:915–21.

van Loon, A.J., Botterweck, A.A., Goldbohm, R.A., Brants, H.A., and van den Brandt, P.A. (1997). Nitrate intake and gastric cancer risk: results from the Netherlands cohort study. Cancer Lett. 114:259–61.

van Loon, A.J., Botterweck, A.A., Goldbohm, R.A., Brants, H.A., van Klaveren, J.D., and van den Brandt, P.A. (1998). Intake of nitrate and nitrite and the risk of gastric cancer: a prospective cohort study. Br J Cancer. 78:129–35.

Varis, K., Taylor, P.R., Sipponen, P., Samloff, I.M., Heinonen, O.P., Albanes, D., Harkonen, M., Huttunen, J.K., Laxen, F., and Virtamo, J. (1998). Gastric cancer and premalignant lesions in atrophic gastritis: a controlled trial on the effect of supplementation with alpha-tocopherol and beta-carotene. The Helsinki Gastritis Study Group. Scand J Gastroenterol. 33:294–300.

Velmurugan, B., Bhuvaneswari, V., Burra, U.K., and Nagini, S. (2002a). Prevention of N-methyl-N′-nitro-N-nitrosoguanidine and saturated sodium chloride-induced gastric carcinogenesis in Wistar rats by lycopene. Eur J Cancer Prev. 11:19–26.

Velmurugan, B., Bhuvaneswari, V., and Nagini, S. (2002b). Antiperoxidative effects of lycopene during N-methyl-N′-nitro-N-nitrosoguanidine-induced gastric carcinogenesis. Fitoterapia. 73:604–11.

Velmurugan, B., Mani, A., and Nagini, S. (2005). Combination of S-allylcysteine and lycopene induces apoptosis by modulating Bcl-2, Bax, Bim and caspases during experimental gastric carcinogenesis. Eur J Cancer Prev. 14:387–93.

Velmurugan, B., and Nagini, S. (2005). Combination chemoprevention of experimental gastric carcinogenesis by s-allylcysteine and lycopene: modulatory effects on glutathione redox cycle antioxidants. J Med Food. 8:494–501.

Virtamo, J., Pietinen, P., Huttunen, J.K., Korhonen, P., Malila, N., Virtanen, M.J., Albanes, D., Taylor, P.R., and Albert, P. (2003). Incidence of cancer and mortality following alpha-tocopherol and beta-carotene supplementation: a postintervention follow-up. JAMA. 290:476–85.

Wada, S., Hirose, M., Shichino, Y., Ozaki, K., Hoshiya, T., Kato, K., and Shirai, T. (1998). Effects of catechol, sodium chloride and ethanol either alone or in combination on gastric carcinogenesis in rats pretreated with N-methyl-N′-nitro-N-nitrosoguanidine. Cancer Lett. 123:127–34.

Walston, J., Xue, Q., Semba, R.D., Ferrucci, L., Cappola, A.R., Ricks, M., Guralnik, J., and Fried, L.P. (2006). Serum antioxidants, inflammation, and total mortality in older women. Am J Epidemiol. 163:18–26.

Wang, X.-D. (2004). Carotenoid oxidative/degradative products and their biological activities. In: Krinsky, N.I., Mayne, S.T., Sies, H., editors. Carotenoids in health and disease. New York: Marcel Dekker; pp. 1–35.

Wang, X.-D., Krinsky, N.I., Benotti, P.N., and Russell, R.M. (1994). Biosynthesis of 9-cis-retinoic acid from 9-cis-b-carotene in human intestinal mucosa in vitro. Arch Biochem Biophys. 313:150–5.

Wang, X., Willen, R., and Wadstrom, T. (2000). Astaxanthin-rich algal meal and vitamin C inhibit Helicobacter pylori infection in BALB/cA mice. Antimicrob Agents Chemother. 44:2452–7.

Watanabe, H., Takahashi, T., Okamoto, T., Ogundigie, P.O., and Ito, A. (1992). Effects of sodium chloride and ethanol on stomach tumorigenesis in ACI rats treated with N-methyl-N′-nitro-N-nitrosoguanidine: a quantitative morphometric approach. Jpn J Cancer Res. 83:588–93.

Wei, W.Q., Abnet, C.C., Qiao, Y.L., Dawsey, S.M., Dong, Z.W., Sun, X.D., Fan, J.H., Gunter, E.W., Taylor, P.R., and Mark, S.D. (2004). Prospective study of serum selenium concentrations and esophageal and gastric cardia cancer, heart disease, stroke, and total death. Am J Clin Nutr. 79:80–5.

World Health Organization (2003). Report of a Joint WHO/FAO Expert Consultation. Diet, nutrition and the prevention of chronic diseases. WHO technical report series; 916. Geneva, Switzerland.

Wu, K., Shan, Y.J., Zhao, Y., Yu, J.W., and Liu, B.H. (2001). Inhibitory effects of RRR-alpha-tocopheryl succinate on benzo(a)pyrene (B(a)P)-induced forestomach carcinogenesis in female mice. World J Gastroenterol. 7:60–5.

Yoshida, Y., Hirose, M., Takaba, K., Kimura, J., and Ito, N. (1994). Induction and promotion of forestomach tumors by sodium nitrite in combination with ascorbic acid or sodium ascorbate in rats with or without N-methyl-N′-nitro-N-nitrosoguanidine pre-treatment. Int J Cancer. 56:124–8.

You, W.C., Brown, L.M., Zhang, L., Li, J.Y., Jin, M.L., Chang, Y.S., Ma, J.L., Pan, K.F., Liu, W.D., Hu, Y., Crystal-Mansour, S., Pee, D., Blot, W.J., Fraumeni, J.F., Jr., Xu, G.W., and Gail, M.H. (2006). Randomized double-blind factorial trial of three treatments to reduce the prevalence of precancerous gastric lesions. J Natl Cancer Inst. 98:974–83.

Yuan, J.M., Ross, R.K., Gao, Y.T., Qu, Y.H., Chu, X.D., and Yu, M.C. (2004). Prediagnostic levels of serum micronutrients in relation to risk of gastric cancer in Shanghai, China. Cancer Epidemiol Biomarkers Prev. 13:1772–80.

Zhang, Z.W., Abdullahi, M., and Farthing, M.J. (2002). Effect of physiological concentrations of vitamin C on gastric cancer cells and Helicobacter pylori. Gut. 50:165–9.

Zhang, Z.W., and Farthing, M.J. (2005). The roles of vitamin C in Helicobacter pylori associated gastric carcinogenesis. Chin J Dig Dis. 6:53–8.

Zhang, H.M., Wakisaka, N., Maeda, O., and Yamamoto, T. (1997). Vitamin C inhibits the growth of a bacterial risk factor for gastric carcinoma: Helicobacter pylori. Cancer. 80:1897–903.

Zullo, A., Rinaldi, V., Hassan, C., Diana, F., Winn, S., Castagna, G., and Attili, A.F. (2000). Ascorbic acid and intestinal metaplasia in the stomach: a prospective, randomized study. Aliment Pharmacol Ther. 14:1303–9.

Chapter 4
MALT Lymphoma: Clinicopathologic Features and Molecular Pathogenesis

Anne Müller

Introduction

Extranodal marginal-zone B-cell lymphomas of mucosa-associated tissue (MALT lymphomas) constitute a group of low-grade non-Hodgkin's lymphomas that share an indolent clinical course and unique pathologic and molecular features. They arise in the context of chronic lymphoid proliferation in a wide variety of mucosal sites. The majority of MALT lymphomas are found in the stomach (70%); other tumor sites include the lung (14%), ocular adnexa (12%), thyroid (4%), and small intestine (1%).

MALT lymphomas were first recognized as a discrete entity in 1994 by the Revised European-American Lymphoma (REAL) classification. In the more recent WHO classification system, MALT lymphomas are grouped with splenic and nodal marginal-zone lymphomas based on the normal cell counterpart that these three B-cell neoplasms have in common (the marginal-zone B cell), despite the fact that they are morphologically, phenotypically, and genotypically distinct disease entities. Nodal and extranodal marginal-zone lymphomas further differ significantly with respect to their clinical behavior. In contrast to their nodal counterparts, MALT lymphomas are characterized by a relatively favorable prognosis, late dissemination, and superior overall survival rates.

MALT lymphomas also differ from both splenic and nodal marginal-zone lymphomas in that they typically arise at sites that are normally devoid of organized lymphoid tissue. The lymphoid tissue from which the lymphomatous clones emerge is rather acquired in the setting of chronic infection or autoimmune disorder. In the case of gastric lymphomas, persistent infection with the bacterium *Helicobacter pylori* resulting in a chronic local inflammatory response is now unequivocally recognized as the background in which this type of lymphoma can develop.

MALT lymphomas account for approximately 8% of all non-Hodgkin's lymphomas; they represent the third most common subtype after diffuse large B-cell and follicular lymphoma. The incidence of all lymphomas has increased by 80% in the United States within the last 20 years and similar statistics have been reported in European countries. Interestingly, this increase seems to be higher in extranodal than in nodal forms, and particularly high in the gastrointestinal (GI) tract. However,

T.C. Wang et al. (eds.) *The Biology of Gastric Cancers*,
© Springer Science+Business Media, LLC 2009

because diagnostic procedures have improved significantly over the same period, the real increase in MALT lymphoma incidence is uncertain.

In the stomach, lymphomas represent the second most common malignancy with 5% of cases, following adenocarcinoma. Among the histologic types, MALT lymphomas and diffuse large B-cell lymphomas are most common in the stomach (approximately 60% and 40%, respectively), whereas mantle, follicular, and T-cell lymphomas are rarely observed (<1% each). Although MALT lymphoma can transform to diffuse large B-cell lymphoma and mixed forms are frequently diagnosed, it is not clear whether all diffuse large B-cell lymphomas are derived from previous MALT lymphoma or whether they can also arise *de novo*.

In the past, the favorable clinical course of MALT lymphomas has often resulted in their description as "pseudo-lymphomas." However, several characteristics support their true neoplastic nature, namely: the monoclonal origin of tumors in the majority of cases; the presence of characteristic chromosomal aberrations; their potential to transform to high grade, diffuse large B-cell lymphoma; and their (low) propensity for dissemination to lymph nodes, spleen, and other sites.

Histologic and Immunophenotypic Features of MALT Lymphoma

In 1987, Isaacson and Spencer reported that the histology of MALT lymphoma recapitulates that of the Peyer's patch (Figure 4.1) rather than that of lymph nodes and thereby introduced the MALT lymphoma concept. MALT lymphoma cells show an immunophenotype identical to that of marginal-zone B cells. The marginal zone in lymph nodes, Peyer's patches and the spleen surrounds the germinal center and follicular mantle (Figure 4.1). It consists of memory B cells that can resemble centrocytes (centrocyte-like B cells) or monocytes (monocytoid B cells). Correspondingly, MALT lymphoma cells are positive for surface immunoglobulin (Ig)M and pan-B-cell markers (CD19, CD20, CD79a) and for the marginal-zone markers CD35 and CD21 (Spencer et al. 1985). They are negative for CD5, CD10, CD23, and cyclin D1. MALT lymphomas also contain large numbers of tumor-infiltrating T cells, some of which are CD4+ and some are CD8+ (Knörr et al. 1999; Müller et al. 2005).

The histologic features of low-grade MALT lymphomas are similar regardless of the site of origin. A characteristic feature is the presence of lymphoepithelial lesions (LELs), a term that describes clusters of lymphocytes that invade and partially destroy the gastric epithelium (Figure 4.2A and B). The infiltrate consists of centrocyte-like cells, but large cells (immunoblasts and centroblasts) are usually also present in variable numbers. Plasma-cell differentiation is found in approximately one-third of gastric MALT lymphomas. Nonmalignant reactive follicles can be present, and their germinal centers are sometimes colonized by the lymphoma cells (Isaacson et al. 1991).

High numbers of large cells in a MALT lymphoma are indicative of histologic transformation into diffuse large B-cell lymphoma. This transformation is usually

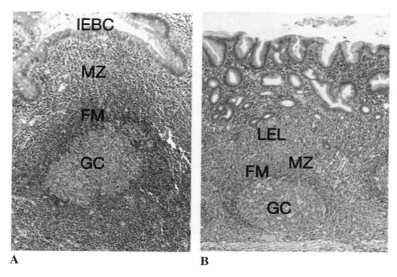

Fig. 4.1 Histologic similarities between a Peyer's patch (**A**) and MALT lymphoma (**B**). (**A**) Peyer's patches contain a germinal center (GC), where B cells (centrocytes and centroblasts) encounter antigens that have diffused across the mucosal surface. In response to this contact, they undergo repeated mutation (the germinal center reaction) until the encoded immunoglobulin is highly specific for the particular antigen. They then transform into larger cells (blasts) that leave the germinal center, enter the circulation, and home back to the intestinal mucosa having differentiated into either memory B cells or plasma cells. GCs are surrounded by the follicular mantle (FM), which comprises B cells that have not passed through a germinal center. The follicular mantle, in turn, is surrounded by the marginal zone (MZ), which contains memory B cells that sometimes resemble centrocytes (centrocyte-like B cells) or monocytes (monocytoid B cells). Marginal-zone B cells can migrate into the germinal center, and some—the intraepithelial marginal-zone B cells (IEBCs)—are present within the epithelium covering the Peyer's patch and form the so-called lymphoepithelium that is a defining feature of MALT. (**B**) MALT lymphomas comprise a reactive B-cell follicle surrounded by neoplastic marginal-zone B cells that invade gastric gland epithelium to form lymphoepithelial lesions (LELs). (Reprinted from Isaacson and Du 2004, with permission from Macmillan Publishers. Copyright 2004.)

defined by the presence of compact aggregates or a sheet-like proliferation of large cells (Chan et al. 1990).

As described above, *H. pylori* infection provides an ongoing antigenic stimulus that can sustain the growth of gastric low-grade lymphoma. Mutation analysis of the V_H genes of lymphoma cells has indeed shown somatic hypermutations with a pattern indicative of antigen-selective pressure, suggesting that the tumor cells have undergone positive selection in germinal centers (Du et al. 1996b; Qin et al. 1995, 1997). This antigen dependency may explain why the tumors remain localized for long periods of time. Intraclonal variation (i.e., ongoing mutations) further suggests that antigenic drive may have a role in the clonal expansion of tumor cells (Du et al. 1996b; Qin et al. 1997; Thiede et al. 1998).

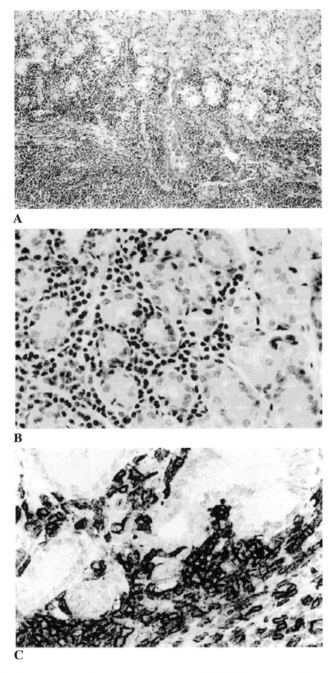

Fig. 4.2 Histologic and immunohistochemical features of MALT lymphoma. (**A**) Lymphocytic infiltrate of gastric glandular epithelium with formation of lymphoepithelial lesions. (**B**) Lymphoepithelial lesion with invasion and partial destruction of mucosal glands by lymphoma cells. (**C**) Infiltrating lymphoma cells stain strongly with an anti-CD20 antibody, confirming their B-cell origin. (From Cohen et al. 2006.)

The Role of *H. pylori* Infection in MALT Lymphomagenesis

Multiple lines of evidence indicate that gastric MALT lymphoma arises from MALT that is acquired in the course of chronic infection with *H. pylori*, in an organ that is normally devoid of organized lymphoid tissue. First, the infection is present in a large majority of cases of MALT lymphoma, with infection rates (as diagnosed by culture or histology) ranging from 60% to more than 90% of cases (Eidt et al. 1994; Nakamura et al. 1997, 1998; Wotherspoon et al. 1991). The rates vary considerably depending on the study; in any case, the prevalence of infection is consistently higher in patients with MALT lymphoma than in the respective general population. Interestingly, the frequency of *H. pylori* carriage at diagnosis is lower in gastric lymphoma than in chronic gastritis or peptic ulcer cases (both >90%) (Nakamura et al. 1997), suggesting that the favorable niche that the bacterium ordinarily inhabits in the stomach disappears as the disease progresses. Indeed, the prevalence of infection is lower in lymphomas invading beyond the submucosa compared with those restricted to the mucosa and submucosa (48% vs. 76%) and in high-grade vs. low-grade tumors (55% vs. 72%) (Nakamura et al. 1997). Independent of the detection of an ongoing infection at the time of diagnosis, a large case control study provided definitive evidence in 1994 that previous or ongoing *H. pylori* infection (measured by serology) was significantly associated with the development of gastric, but not nongastric lymphomas (both high-grade and low-grade; Table 4.1) (Parsonnet et al. 1994). Furthermore, regions with a high incidence of gastric MALT lymphoma (such as the Veneto region of Italy) have an accompanying high prevalence of *H. pylori* infection (Doglioni et al. 1992).

Experimental evidence lends additional support to the role of *H. pylori* infection in MALT lymphomagenesis. In 1993, Isaacson and coworkers showed that unsorted tumor cell suspensions containing accessory cells such as T cells and macrophages could be kept alive and even proliferated upon addition of heat-killed whole-cell extracts of *H. pylori* (Hussell et al. 1993a). This reaction was specific for only one bacterial strain per case of lymphoma. Additional experimental evidence is provided by animal models of *Helicobacter* infection. In the appropriate laboratory mouse strains, infection with either *H. pylori* or its close relatives *H. heilmannii* and *H. felis* induces acquisition of MALT that can give rise to gastric lymphoma after approximately 18–24 months post infection (Enno et al. 1995). The histopathologic and immunologic features of the disease in mice accurately reproduce those observed in patients (discussed in more detail below).

Table 4.1 Odds ratios and 95% confidence intervals for the association of *Helicobacter pylori* infection with gastric and nongastric non-Hodgkin's lymphoma

Type of lymphoma	No. of patients	*H. pylori* infection (in %)		Matched odds ratio	95% confidence interval
		Patients	Matched controls		
Gastric	33	85	55	6.3	2.0–19.9
Nongastric	31	65	59	1.2	0.5–3.0

Source: Parsonnet et al. 1994.

*Infection status was determined by serum enzyme-linked immunosorbent assay.

The most striking evidence in support of *H. pylori*'s role in MALT lymphom-agenesis comes from the observation that, in a majority of patients, the tumor can be treated successfully by eradication of the infection. This concept was first tested in 1993, when Isaacson's group accomplished complete remission of MALT lymphoma in six patients treated for the infection alone (Wotherspoon et al. 1993). Multiple additional studies have since reproduced this result (e.g., Montalban et al. 2001; Neubauer et al. 1997; Thiede et al. 2000); the response rate is typically estimated to be approximately 75%, with the remaining cases showing no response to antibiotics. As a result of this finding, eradication therapy directed against the infection is now generally used as first-line treatment of the disease.

Clinical Aspects: Diagnosis and Treatment

Clinical Presentation, Diagnosis, and Staging of MALT Lymphoma

The common presenting symptoms of MALT lymphomas are epigastric discomfort and dyspepsia. Gastric bleeding, systemic symptoms, and bone marrow involvement are rare. The majority of patients have no abnormal findings upon physical examination, and the endoscopic appearance (Figure 4.3A) of low-grade MALT lymphoma can mimic that of benign diseases such as chronic gastritis or a peptic ulcer. The lymphoma is definitively diagnosed by histopathologic examination of biopsy

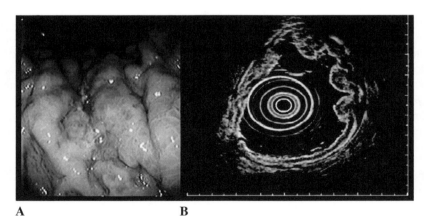

A   B

Fig. 4.3 (**A**) An endoscopic aspect of a mucosa-associated lymphoid tissue lymphoma of the stomach: gastric folds are prominent, unable to be extended, with a small ulcer along the greater curvature of the gastric body. (**B**) Ultrasound-endoscopy investigation of the same MALT patient shows thickening of the wall, with fusion of the normal stratification because of involvement of the mucosa, submucosa, and muscularis-mucosa membranes, whose external margins appear to be reasonably regular. (From Ferrucci and Zucca 2007.)

specimens taken during an esophageal-gastro-duodenoscopy. Diagnostic gastric biopsies are taken from multiple areas of abnormal-appearing tissue; additional samples are taken from macroscopically uninvolved mucosa in each region of the stomach, duodenum, and gastroesophageal junction. Biopsies from gastric low-grade lymphomas typically display the characteristic histologic features, such as LELs and a diffuse infiltrate of marginal-zone "centrocyte-like" B cells. In those cases for which the diagnosis is uncertain, the WHO recommends the demonstration of B-cell monoclonality. This is accomplished by polymerase chain reaction (PCR)-based amplification of rearranged VDJ sequences of the Ig heavy-chain gene. However, monoclonality analysis is questionable because it gives both false-negative and false-positive results. Not all histologically diagnosed MALT lymphomas are monoclonal; the rates lie between 63% and 88%, depending on the study (Montalban et al. 2001; Neubauer et al. 1997; Thiede et al. 2000). This might mostly be attributable to the multifocal nature of gastric MALT lymphomas (Wotherspoon et al. 1992). A monoclonal diagnosis, however, is not necessarily indicative of malignant disease. Multiple studies have shown that simple gastritis can be monoclonal (e.g., Rudolph et al. 1997; Savio et al. 1996; Saxena et al. 2000). This is especially true if lymphoid follicles are present (Wuendisch et al. 2003), as is the case in approximately 50% of gastritis cases associated with *H. pylori* infection. Monoclonality is typically detected in 10%–40% of chronic active gastritis cases, and in 50% or more if lymphoid hyperplasia or reactive follicles are found. Recent studies suggest that melting-curve or GeneScan analysis is more accurate than electrophoresis analysis of PCR products in detecting malignant monoclonal B-cell populations, and that biopsies taken in deeper tissues are less likely to yield false-positive results (Steiff et al. 2006). Interestingly, higher rates of B-cell monoclonality were found in one study in histologically benign biopsies in patients that later developed MALT lymphomas compared with patients whose disease did not progress (79% vs. 21%) (Nakamura et al. 1998), suggesting that monoclonality precedes the formation of frank lymphomas.

In summary, a diagnosis of malignant lymphoma based on the finding of PCR-based B-cell monoclonality is probably not warranted in the absence of clear histologic proof of lymphoma. Interestingly, monoclonality is also frequently observed in posttreatment biopsies of patients in complete histologic remission (Neubauer et al. 1997; Thiede et al. 2001). This observation is discussed in more detail below.

Histologic grading of gastric biopsies is usually performed according to a scoring system developed by Wotherspoon in 1993 (Table 4.2). It is based on the presence of scattered plasma cells only (score 0 = normal), small clusters of lymphocytes (score I = chronic active gastritis), or lymphoid follicles (score II = chronic active gastritis with lymphoid follicles). If the lymphoid follicles are surrounded by lymphocytes that infiltrate diffusely into the lamina propria, a score of III is warranted ("suspicious lymphoid infiltrate, probably reactive"). A score of IV indicates infiltration of small groups of lymphocytes into the epithelium ("suspicious lymphoid infiltrate, probably lymphoma"), and a score of V identifies MALT lymphoma based on the presence of prominent LELs.

The problem of staging a gastric lymphoma is controversial, and various systems are still in use (see comparison in Table 4.3). One frequently used staging system for

Table 4.2 Scoring of mononuclear mucosal infiltrate

Wotherspoon index	Definition	Histologic characteristics
0	Normal mucous membrane	Occasional plasmacytes in lamina propria
1	Active chronic gastritis	Small lymphocytic aggregates in the lamina propria, without lymphoid follicles or lymphoepithelial lesions
2	Active chronic gastritis with lymphoid follicles	Prominent lymphoid follicles surrounded by a mantle zone and plasmacytes, without lymphoepithelial lesions
3	Suspected lymphoid infiltration, probably reactive	Lymphoid follicles surrounded by small lymphocytes that widely infiltrate the lamina propria and, occasionally, the epithelium
4	Suspected lymphoid infiltration, probably lymphoma	Lymphoid follicles surrounded by centrocytic-like cells that widely infiltrate the lamina propria and the epithelium
5	MALT lymphoma	Dense and widespread infiltration of centrocytic-like cells in the lamina propria, with prominent lymphoepithelial lesions

Source: Adapted from Wotherspoon et al., 1993.

GI-tract lymphomas was proposed by Blackledge et al. in 1979 and was modified by the Lugano workshop in 1994 (Table 4.3, last column) (Rohatiner et al. 1994). It takes into consideration whether the tumor is confined to the GI tract (where it can be limited to the mucosa or involve the muscularis propria and serosa), or extends into the abdomen with spread to local or distant lymph nodes, or shows adjacent organ/supradiaphragmatic involvement. In recent years, the tumor-node-metastasis (TNM) staging system has more frequently been applied to gastric lymphomas (Table 4.3, second column) (Steinbach et al. 1999). This has become possible because endoscopic ultrasound is now routinely used to measure the depth of infiltration of the gastric wall (Figure 4.3B). This parameter determines the T stage and correlates strongly with lymph node involvement and therefore has prognostic relevance (Steinbach et al. 1999).

Almost 90% of MALT lymphomas are localized to the stomach at diagnosis; dissemination is usually seen only to other parts of the GI tract or the splenic marginal zone (Du et al. 1997). Gastric MALT lymphoma cells highly express the homing receptor $\alpha_4\beta_7$ integrin. Its ligand, the vascular addressin molecule MAdCAM-1 is expressed in endothelial cells of the gastrointestinal mucosa (Briskin et al. 1997) and the splenic marginal zone (Kraal et al. 1995), which explains the preferential dissemination to these organs.

In clinical practice, the staging evaluation of gastric MALT lymphoma requires several routine examinations: endoscopy and biopsy (for histopathology and *H. pylori* staining), examination of Waldeyer's ring, endoscopic ultrasound, computed tomography scan of the chest, abdomen, and pelvis, a bone marrow biopsy, and serum quantification of lactate dehydrogenase and β2-microglobulin. Past or present infection with *H. pylori* is assessed by serology.

Table 4.3 Comparison of several frequently used MALT lymphoma classification systems

Lymphoma extension	TNM modifications in the Paris system (Ruskone et al. 2003)	Adapted TNM staging system (Steinbach et al. 1999)	Musshoff staging (Musshoff 1977)	Modified Blackledge system (Lugano staging, Rohatiner et al. 1994)
Mucosa	T1m N0 M0	T1 N0 M0	I_{E1}	Stage I = confined to GI tract (single primary or multiple, noncontiguous)
Submucosa	T1sm N0 M0		I_{E2}	
Muscularis propria	T2 N0 M0	T2 N0 M0		
serosa	T3 N0 M0	T3 N0 M0		
Perigastric lymph nodes	T1–3 N1 M0	T1–3 N1 M0	II_{E1}	Stage II = extending into abdomen (Stage II_1 = local nodal involvement; Stage II_2 = distant nodal involvement)
More distant regional nodes	T1–3 N2 M0	T1–3 N2 M0	II_{E2}	
Extraabdominal nodes	T1–3 N3 M0			
Invasion of adjacent tissues		T4 N0 M0	I_E	Stage II_E = penetration of serosa to involving adjacent organs or tissues
Lymph nodes on both sides of the diaphragm, and/or additional extranodal sites with noncontinuous involvement of separate GI sites	T1–4 N3 M0	T1–4 N3 M0	III_E	
Or noncontinuous involvement of non-GI sites	T1–4 N0–3 M1 T1–4 N0–3 M2	T1–4 N0–3 M1	IV_E	Stage IV = disseminated extranodal involvement or concomitant supradiaphragmatic nodal involvement
Bone marrow not assessed	T1–4 N0–3 M0–2 BX			
Bone marrow not involved	T1–4 N0–3 M0–2 B0			
Bone marrow involvement	T1–4 N0–3 M2 B1			

Source: Ferrucci and Zucca 2007.

Abbreviations: TNM: tumor-node-metastasis; GI: gastrointestinal.

Treatment Options for MALT Lymphoma

The recognition of the association of *H. pylori* infection with gastric MALT lymphoma has revolutionized the treatment of this disease. Whereas surgery was historically considered to be the gold standard, it has now been replaced completely by stomach-preserving approaches. The treatment recommendations vary considerably depending on *H. pylori* status, presence of chromosomal translocations, staging at time of diagnosis, involvement of superficial vs. deeper tissues, and the degree of high-grade transformation (see Figure 4.4 for a typical treatment

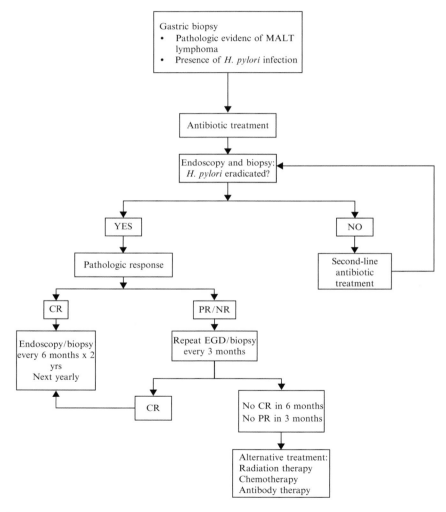

Fig. 4.4 Gastric mucosa-associated lymphoid tissue (MALT) lymphoma treatment algorithm. CR: complete response; EGD: esophagogastroduodenoscopy; NR: no response; PR: partial response. (From Cohen et al. 2006.)

algorithm). Therefore, treatment options for localized, *H. pylori*-positive disease are discussed separately from those for disseminated, *H. pylori*-negative or relapsed lymphomas and from high-grade lymphomas.

Treatment of Early, H. pylori–Associated Gastric MALT Lymphoma

The vast majority of patients with gastric MALT lymphoma are diagnosed with localized disease. Therefore, localized treatment regimens targeting the stomach have historically been the treatments of choice. Surgery (total gastrectomy or partial resection) was chosen as first-line treatment until the early 1990s even for patients presenting with early-stage disease, either alone or followed by radiotherapy or postoperative chemotherapy. These regimens resulted in 5-year survival rates as high as 90%. However, surgical treatment is associated with several major drawbacks. MALT lymphoma is a multifocal disease; therefore, partial resection does not always prevent local recurrence. Total gastrectomy has a higher potential for cure, but can lead to an impaired quality of life. The surgical approach to treatment of gastric lymphoma was first questioned in 1991, when a large prospective study showed that advanced-stage, aggressive lymphomas (including those originating primarily in the stomach) could be treated by intensive chemotherapy alone (Salles et al. 1991). The effectiveness of chemotherapy in advanced stages prompted reexamination of the surgical role in localized stages as well. All surgical treatments have now been abandoned in favor of eradication therapy, radiotherapy, and/or chemo- and immunotherapy even in patients failing primary therapy and are now reserved for complications such as obstruction, perforation, or hemorrhage.

Eradication therapy as a first-line treatment option was introduced in 1993 in a report by Isaacson and coworkers, who had successfully treated a small group of patients with antibiotic therapy (Wotherspoon et al. 1993). This treatment resulted in eradication of *H. pylori* infection and regression of the tumors in the majority of patients. The initial results have now been reproduced in multiple independent studies (e.g., Fischbach et al. 2004; Montalban et al. 2001; Neubauer et al. 1997; Savio et al. 1996; Thiede et al. 2000; Wuendisch et al. 2005), with complete remission typically achieved in more than 70% of patients presenting with localized disease (the range is 55%–95%, depending on the study). Antibiotic treatment consisting of a combination of amoxicillin, clarithromycin, tetracycline, and/or metronidazole with a proton pump inhibitor therefore is now considered the standard first-line treatment for *H. pylori*-positive MALT lymphoma. It is particularly successful in patients whose tumors are confined to the mucosa and submucosa, but typically fails in patients with involvement of the stomach wall or perigastric lymph nodes. Lymphomas that harbor the t(11;18)(q21;q21) chromosomal translocation are also generally refractory to eradication therapy. *H. pylori*-negative patients never respond to eradication therapy.

Eradication therapy is associated with enormous benefits to the patient over all other treatment options. There are virtually no side effects, and the tumors regress quickly, typically within 3–6 months of successful eradication. However, it is now generally accepted that eradication of *H. pylori* does not offer a complete cure, but

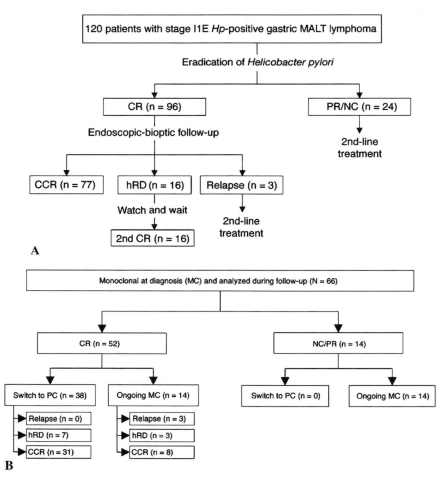

Fig. 4.5 Long-term follow-up of gastric MALT lymphoma after *Helicobacter pylori* eradication. **(A)** Response and follow-up of 120 patients included in a long-term study, on the basis of macroscopic and histologic findings. **(B)** Analysis of B-cell clonality of gastric MALT lymphoma during follow-up. Of 120 patients, 91 were analyzed at diagnosis. Twenty-two patients were polyclonal (PC), and 69 patients were monoclonal (MC). CR: complete remission; NC: no change; PR: partial remission; hRD: histologic residual disease; CCR: continuous complete remission. (Reprinted from Wundisch at. 2005, with permission from the American Society of Clinical Oncology.)

suppresses rather than eliminates the lymphomatous clones. In virtually all studies that have examined "molecular" vs. "histologic remission" by PCR screening for B-cell monoclonality or other molecular techniques, persistence of the lymphoma clones was demonstrated in a significant subset of patients long after complete histologic remission was achieved (Montalban et al. 2001; Neubauer et al. 1997; Thiede et al. 2001). For example, one of the first long-term studies (with a median follow-up of more than 6 years; see Figure 4.5A) that was conducted to measure the efficacy of eradication therapy reported that 27% of patients in complete remission had ongoing monoclonality and this subset was at significantly higher risk of relapse, histologic residual disease, or an only partial response (Figure 4.5B) (Wuendisch et al. 2005). In contrast, none of the patients who had shifted to a polyclonal B-cell pattern in the stomach relapsed. The authors concluded that clonality analysis might be useful in identifying a patient population requiring less intense follow-up with fewer endoscopies. In general, relapse rates are fairly high in patients treated for the infection alone (1%–20% depending on the study) (Fischbach et al. 2004; Nobre-Leitao et al. 1998; Papa et al. 2000; Tursi et al. 1997). Tumor relapses have been attributed to either *H. pylori* reinfection or the emergence of lymphoma clones that have gained independence of antigenic drive. Indeed, several case reports indicate that reinfection can induce an extremely rapid tumor recurrence because of persistence and rapid reactivation of tumor cells despite histologic remission (Cammarota et al. 1995; Horstmann et al. 1994). In clinical practice, the discrepancy between "histologic" and "molecular" remission translates into the need for close endoscopic follow-up of patients who have received eradication therapy alone. It is now recommended that endoscopic examinations be repeated every 6 months for 2 years after complete remission is achieved, and on a yearly basis thereafter. In an effort to identify novel regimens allowing for higher rates of molecular remission, an international prospective trial is now underway that compares the alkylating agent chlorambucil with observation after *H. pylori* eradication. In any case, it remains to be seen whether long-lasting molecular remission will translate into a higher cure fraction and lower relapse rates.

Treatment of Helicobacter pylori–Negative, Advanced, or Relapsed Gastric MALT Lymphoma

No definitive guidelines exist for those patients who fail to respond to eradication therapy, are *H. pylori* negative at diagnosis, or relapse after successful tumor regression. Prospective randomized trials comparing different therapeutic procedures have not been published to date. However, several recent studies have reported complete remission of *H. pylori*–negative or antibiotic-refractory lymphomas after radiotherapy on the stomach and perigastric lymph nodes, with 5-year survival rates of more than 90% (Schechter et al. 1998; Tsang et al. 2003; Yahalom 2001). Radiotherapy has since become the therapy of choice for this subset of patients.

In patients in whom the disease has spread, systemic therapy such as chemo- or immunotherapy or a combination of both must be considered. Alkylating agents (chlorambucil or cyclophosphamide) or purine analogs achieve response rates of up to 100% (with 75% and 100% complete remission, respectively) in patients with disseminated MALT lymphoma (Hammel et al. 1995; Thieblemont 2005). Among the combination regimens investigated so far, fludarabine plus mitoxantrone or mitoxantrone, chlorambucil, and prednisone were shown to be highly active (Wöhrer et al. 2003; Zinzani et al. 2004). The anti-CD20 antibody Rituximab has been shown to be effective with response rates of approximately 70% when given alone (Martinelli et al. 2005). A retrospective study published in 2006 reported that the combination of Rituximab with cyclophosphamide, doxorubicin/mitoxantrone, vincristine, and prednisone was highly effective with a response rate of 100% (77% complete remission) even in patients that were heavily pretreated with other chemotherapeutic regimens or had relapsed after *H. pylori* eradication (Raderer et al. 2006). An intriguing new option for patients with disseminated disease may be the delivery of targeted radiation using an anti-CD20 radioimmunoconjugate. Initial experimental treatment of very few patients indeed suggests that this approach might prove to be effective even in stage IV patients who have previously received both chemo- and radiation therapy (Witzig et al. 2001).

Because of the indolent nature of many cases of MALT lymphoma and its low propensity for spread, it is important to point out that some patients with persistent disease may be managed using a watch-and-wait strategy without active therapy. This approach may be considered because the prognosis of MALT lymphomas is generally very favorable, with 5-year survival rates better than 80%, regardless of the type of treatment.

Treatment of High-Grade Gastric Lymphoma

Gastric high-grade/diffuse large B-cell lymphoma treatment has changed as radically as that of low-grade lymphoma, as surgical resection has been abandoned for more conservative approaches. The guidelines for localized stages now recommend front-line chemo-immunotherapy with Rituximab plus cyclophosphamide, doxorubicin, vincristine, and prednisone followed by "involved field" radiotherapy. Advanced-stage patients usually undergo more cycles of the same regimen, but third-line regimens additionally including bleomycin, adriamycin, methotrexate, and oncovin are also used, generally with no differences in terms of complete remission induction and 5-year survival rates.

High-grade lymphomas are generally considered to be *H. pylori* independent, and therefore resistant to antibiotic therapy. Indeed, some cases of low-grade lymphoma that did not respond to eradication therapy and therefore required gastrectomy were later shown to contain a significant high-grade component. Several studies using molecular biology and immunohistochemical techniques have suggested that low- and high-grade components found in one patient are typically derived from the same clone (Peng et al. 1997). The view that high-grade disease is not treatable by antibiotic therapy was recently challenged by several small studies reporting that indeed aggressive high-grade gastric lymphomas can regress upon eradication therapy alone, implying that these tumors have somehow retained their dependence on antigenic drive (Chen et al. 2001). These results must, however, be validated by larger, prospective studies before treatment recommendations can be changed.

Susceptibility to MALT Lymphoma: Genetics of the Host and Pathogen

Host Genetic Susceptibility to Gastric MALT Lymphoma

Very little solid evidence is available about germline mutations that influence the risk of MALT lymphoma development in *H. pylori*–positive subjects over other disease outcomes, especially in comparison to the large body of literature that is now available on genetic polymorphisms and the risk of developing gastric adenocarcinoma. However, several reports have recently addressed the role of polymorphisms in genes encoding proinflammatory cytokines, sensor molecules of the innate immune system, and T-cell regulatory molecules in MALT lymphoma risk.

Polymorphisms possibly influencing MALT lymphoma risk were found in the *TNF*-α gene and the interleukin (IL)-1 cluster. A polymorphism upstream of the *TNF*-α gene (*TNF*-α −857 C/T) was identified to influence MALT lymphoma risk in a Taiwanese patient population (Wu et al. 2004). The rare −857 T allele, which had previously also been negatively linked to Crohn's disease, was shown to be underrepresented in MALT lymphoma patients compared with healthy controls, conferring a threefold decreased risk (Wu et al. 2004). −857 T is associated with decreased TNF-α expression upon lipopolysaccharide (LPS) stimulation, probably because of binding of the transcription factor OCT1, which in turn blocks the transactivating effect of NF-κB. Interestingly, the same allele seems to confer a 1.8-fold increased risk of MALT lymphoma in Caucasians, as shown for a group of 144 European lymphoma patients (Hellmig et al. 2005a). This was particularly true for high-grade lymphoma cases.

The IL-1RN 2/2 genotype, which is well known to affect gastric cancer risk, was also linked to MALT lymphoma risk in a group of 66 cases from northern England (Rollinson et al. 2003). This association could not be reproduced, however, in a group of 153 German patients (Hellmig et al. 2004). In summary, whether polymorphisms in proinflammatory genes or their receptors truly mediate genetic susceptibility to gastric MALT lymphoma remains unclear despite multiple large-scale studies.

Several investigators have searched for a link between MALT lymphoma and polymorphisms affecting the activity of sensor molecules of the innate immune system that detect pathogen-associated molecular patterns such as bacterial lipoproteins, LPS, and peptidoglycan. One study found a negative association between MALT lymphoma and the rare mutation Asp299Gly in toll-like receptor (TLR) 4, the innate immune sensor for LPS (Hellmig et al. 2005b), whereas another study found a positive association (Nieters et al. 2006). This mutation attenuates TLR-4 signaling and diminishes the inflammatory response. Interestingly, the same allele is overrepresented in gastric cancer patients vs. controls, suggesting that nonfunctional TLR-4 is a risk factor for gastric cancer. Another group of recognition molecules, the NOD proteins, detect intracellular rather than extracellular pathogen-associated molecules. A study comparing 83 MALT lymphoma patients with 428 infected control subjects demonstrated that MALT lymphoma risk is increased in carriers of the rare R702W mutation in the NOD2 protein (Rosenstiel et al. 2006). This mutation, which is also frequently found in patients with Crohn's disease, diminishes the NF-κB response to NOD protein activation.

Finally, polymorphisms affecting surface localization of the T-cell inhibitory protein CTLA-4 were reported to affect MALT lymphoma risk (Cheng et al. 2006). In this study, which involved 60 lymphoma patients and 250 controls in Taiwan, the allele that is less efficient in inhibiting T-cell activation was linked to a significantly higher MALT lymphoma risk.

Characteristics of Helicobacter pylori Strains Associated with MALT Lymphoma

The disease outcome associated with chronic *H. pylori* infection differs greatly depending on the colonizing strain. In general, severe disease outcomes such as gastroduodenal ulceration and gastric adenocarcinoma are closely linked to strains that harbor the cag-pathogenicity island (cag-PAI) along with its secreted effector molecule CagA, as well as certain variants of the vacuolating cytotoxin (VacA), and IceA. The picture is less clear for MALT lymphoma strains. Research to date has focused on whether the *cagA* gene is prevalent in strains isolated from MALT lymphoma patients more than in strains associated with other disease outcomes or asymptomatic infection. An initial report published in 1997 suggested that strains isolated from gastric lymphoma patients were more likely than those isolated from a control group to harbor the *cagA* gene (Eck et al. 1997). Subsequent studies that stratified the patient population according to low- vs. high-grade disease found that indeed only strains isolated from patients whose lymphomas had progressed to high grade were *cagA* positive, whereas the prevalence of *cagA* was similar in low-grade lymphoma and asymptomatic gastritis (Delchier et al. 2001; Peng et al. 1998). This finding suggests that high-grade transformation may be more likely to occur after infection by *cagA*-positive strains of *H. pylori*.

Pathogenesis of MALT Lymphoma

Early MALT lymphomas are usually *H. pylori* dependent and can be treated by eradication of the bacteria through antibiotic therapy (see above). In contrast, chromosomal translocations conferring autonomous growth can occur as the lymphomas progress and can render them independent of environmental signals. Therefore, the mechanisms driving tumor cell proliferation are probably quite different in early and late gastric MALT lymphomas and are therefore discussed separately.

Pathogenesis of Early, Helicobacter pylori–Dependent MALT Lymphoma: Role of T-Cell–Derived Costimulatory Signals

Several lines of evidence suggest that "early" MALT lymphoma cells do not grow autonomously, but require T-helper cell-derived activation and proliferation signals. Numerous studies have shown that tumor-infiltrating T cells are abundant in MALT lymphoma masses (Figure 4.6C) (Knörr et al. 1999; Müller et al. 2005). Both CD4[+] and CD8[+] T cells are found (Figure 4.6C), with a typical ratio of approximately 4:1. Flow cytometric and immunohistochemical analysis of human biopsies as well as of material from experimentally infected mice further demonstrated that infiltrating T cells are activated and express the activation marker CD69 (Knörr et al. 1999; Müller et al. 2005). Antigen-specific activation of infiltrating T cells is also demonstrated by their expression of CD40 ligand, a surface marker that mediates T-/B-cell interaction via CD40 on the B-cell surface (Knörr et al. 1999). Tumor-infiltrating T cells in MALT lymphomas of humans and MALT lymphoma-like lesions in mice also express the immunocompetence marker CD28 (Knörr et al. 1999; Müller et al. 2005), another surface molecule of T cells, which interacts with the B7-related surface protein CD80 on antigen-presenting cells and transmits the costimulatory signals required for full activation of T cells upon antigen contact.

These observations have raised the possibility that *H. pylori*–specific, activated, tumor-infiltrating T cells might provide growth signals to the tumor B cells. Indeed, several *in vitro* studies using cultures of explanted, unsorted tumor cells showed that these could be kept alive and were induced to proliferate by addition of heat-killed *H. pylori* extract (Hussell et al. 1993a, 1996). Depletion of T cells from the cultures abrogated proliferation, suggesting that *H. pylori*–specific T cells have an important role in driving tumor cell proliferation (Hussell et al. 1996). Interestingly, ligation of the CD40 receptor by an agonistic antibody can replace T cells in this scenario, suggesting that the direct interaction of T and B cells via CD40 is required for tumor growth (Greiner et al. 1997). This coculture system further demonstrated that T-cell–derived cytokines typical of the Th2 subset have an important role in B-cell proliferation, because addition of IL-4 and IL-10, but not addition of the Th1 cytokines IFN-γ or IL-2, enhanced the effect of CD40 ligation on tumor cell survival and proliferation (Figure 4.6A and B) (Greiner et al. 1997). Indeed, MALT lymphomas in humans and MALT lymphoma–like lesions in mice were found to

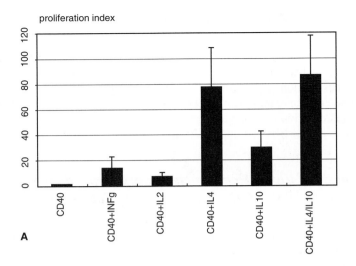

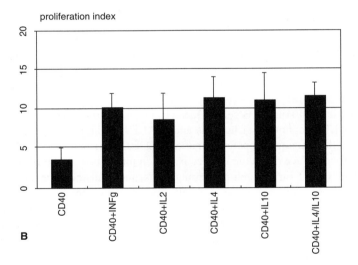

Fig. 4.6 Role of Th2 cytokines in MALT lymphoma pathogenesis. (**A**) and (**B**), Summary of low-grade [(**A**) n = 6] and high-grade [(**B**) n = 5] MALT lymphoma B-cell proliferation that is induced by anti-CD40 and cytokines. A total of 5×10^4 purified B cells were cultured in triplicate with or without cytokines and anti-CD40 as indicated. B-cell proliferation was determined at day 5 of culture by the addition of ^{3}H thymidine present during the last 16 hours of the culture period. The proliferation index is expressed as cpm assay/cpm medium. The results give the cumulative mean value of all patients calculated from the means of three independent proliferation tests performed for each patient. Each error bar gives the standard deviation of the respective cumulative mean. (**C**) Phenotypic characterization of tumor-infiltrating immune cells in murine MALT lymphomas. Serial cryosections were stained in blue-gray with monoclonal antibodies against the indicated antigens and counterstained with nuclear fast red. A hematoxylin and eosin–stained section is included for orientation. Lymphoid aggregates are colonized by CD3+ T cells, but not

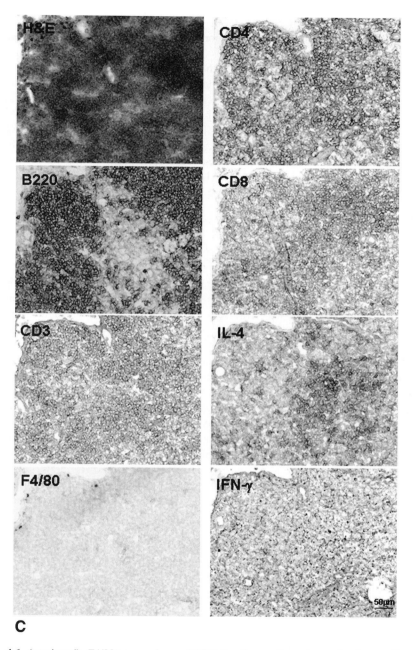

C

Fig. 4.6 (continued) F4/80+ macrophages. B220+ B cells constitute the predominant cell type. Tumor-infiltrating T cells are mostly CD4+ and produce interleukin (IL)-4, but not interferon (IFN)-γ. A scale bar is included in the bottom right panel. (**A** and **B** reprinted from Greiner 1997, with permission from the American Society for Investigative Pathology; **C** reprinted from Müller et al. 2005, with permission from the American Society for Investigative Pathology.)

express high levels of Th2 cytokines *in vivo* (Figure 4.6C) (Knörr et al. 1999; Müller et al. 2005), lending further support to the role of a Th2 helper response in MALT lymphomagenesis.

It is of interest to note in this context that only mouse strains with a genetic predisposition toward Th2 polarization, such as BALB/c, develop MALT lymphomas upon persistent experimental infection with *Helicobacter* species. Strains with a Th1 disposition such as C57/Bl6 develop *Helicobacter*-associated pathologies in line with a requirement for Th1-biased T-cell responses, such as chronic active gastritis, gastric ulceration, and gastric adenocarcinoma, but are protected from MALT lymphoma. In summary, Th2-polarized T-cells translate *H. pylori*–specific signals into growth signals for early MALT lymphoma B cells, which themselves appear not to be *Helicobacter* specific (see below).

Specificity of MALT Lymphoma–Derived Immunoglobulin

MALT lymphoma cells carry rearranged and somatically mutated Ig genes, implying that they are derived from antigen-activated memory B cells. The replacement vs. silent mutation ratios in the framework regions of the IgV_H genes are significantly < 1.5, implying that, despite the high mutation frequencies, selective forces preserve the B-cell receptor in MALT lymphomas. Intraclonal variations caused by ongoing somatic mutations and/or replacement of a part of the variable heavy segment (receptor revision) have been reported (Du et al. 1996a, 1996b; Qin et al. 1997; Thiede et al. 1998). Because MALT lymphoma cells express functional Ig, their specificity is of great interest. Somewhat unexpectedly, multiple studies have shown that tumor Ig (which is typically generated by hybridoma technology) consistently fails to recognize *H. pylori* antigen in Western blots or immunohisto-chemically stained sections. In contrast, binding of tumor cell immunoglobulins to various structures of normal human tissues (follicular dendritic cells, venules, epithelial cells, connective tissue) has been described (Greiner et al. 1994; Hussell et al. 1993b). Consistent with this observed autoreactivity of tumor immunoglobulins, preferential use of certain V_H family genes is detected, which seem to be frequently involved in autoantibody production (Du et al. 1996a; Qin et al. 1995).

In a recent study comparing the structure of the antigen receptors of a large panel of mature B non-Hodgkin's lymphomas, MALT lymphomas were shown to be unique in that a significant proportion expressed B-cell antigen receptors with strong CDR3 homology to rheumatoid factors (RFs) (Bende et al. 2005). RFs are antibodies against the Fc portion of human IgG found in patients with rheumatoid arthritis; RFs and IgG form immune complexes, which are part of the disease process of various autoimmune diseases. The RF-CDR3 homology of MALT lymphoma immunoglobulins without exception included N-region–encoded residues in the hypermutated IgV_H genes, indicating that they were stringently selected for reactivity with auto-IgG. Using 10 MALT lymphoma–derived antibodies, it was further shown by *in vitro* binding studies that many antibodies with RF-CDR3 homology indeed possess strong RF activity (Figure 4.7A) (Bende et al. 2005). None of the MALT lymphomas with the translocation (11;18) possess RF-CDR3 homology or bind IgG *in vitro*, confirming that MALT lymphomas harboring t(11;18) do not depend on BCR-mediated NF-κB activation for survival and proliferation.

A recent study has cast some doubt on the general applicability of the autoimmune hypothesis (Lenze et al. 2006). Of seven analyzed single-chain antibodies cloned from MALT lymphomas, only one reacted with an autoantigen, the plasma cell–specific protein Ufc1 (Figure 4.7B and C). The others did not show any reactivity in screens with comprehensive expression and peptide libraries. It remains to be seen in larger studies which proportion of MALT lymphomas express autoreactive antibodies, and which additional (auto)-antigens can be identified.

Pathogenesis of Late, Helicobacter pylori–Independent MALT Lymphoma: Effect of Chromosomal Translocations

Non-Hodgkin's lymphomas have characteristic chromosomal alterations, which are usually subtype specific. The most notorious is t(14;18)(q32;q21) leading to overexpression of Bcl-2 in follicular lymphoma. In MALT lymphoma, three chromosomal translocations have been identified that all affect different genes and occur with varying frequencies in specific anatomic sites. It is noteworthy that all three translocations affect the NF-κB signaling pathway, which in B cells regulates antigen-dependent activation, proliferation, survival, and induction of effector

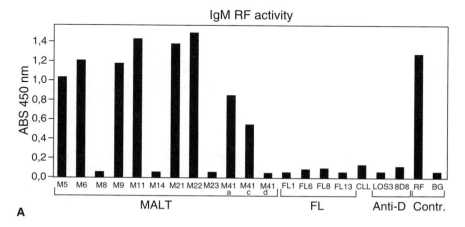

Fig. 4.7 Autoreactivity of tumor-derived immunoglobulins. (**A**) Rheumatoid factor (RF) activity of lymphoma-idiotype–derived antibodies (LIDAs). Binding activity of immunoglobulin (Ig)M LIDAs, RF control serum, and anti-Rh(D) control IgM Abs (*LOS3 and 8D8*) in the IgM-RF

A. Müller

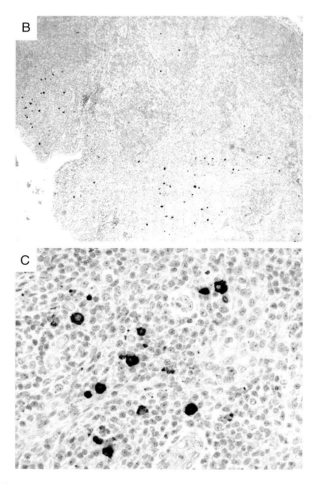

Fig. 4.7 (continued) ELISA. All samples were tested at a stratified concentration of 500 ng/mL IgM. The ABS 450 nm is plotted. (**B**) and (**C**) Immunohistochemical staining of human tonsil with a patient single-chain fragment variable antibody. Overview (**B**) and detail (**C**) of the distribution pattern and morphology of the positive cells. (**A:** Reproduced from Bende et al. Rockefeller University Press. Copyright 2005. **B** and **C:** This research was originally published in Lenze et al. 2006. © The American Society of Hematology.)

functions. The net result of all three translocations is increased proliferation and inhibition of apoptosis, thereby conferring a growth advantage to affected cells.

The translocation **t(11;18)(q21;q21)** fuses the apoptosis inhibitor 2 (*API2*) gene on chromosome 11q21 to the *MALT1* gene on chromosome 18q21, resulting in the novel chimeric API2-MALT1 protein (Figure 4.8A) (Dierlamm et al. 1999). It is the most frequently observed chromosomal translocation in MALT lymphomas, and is usually diagnosed by fluorescence *in situ* hybridization (Figure 4.8B) or reverse transcriptase (RT)-PCR. This translocation is particularly common in

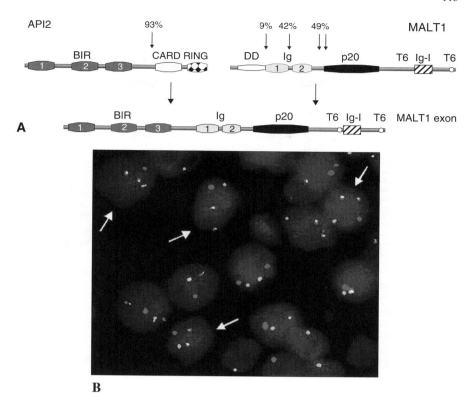

Fig. 4.8 (**A**) Structure of the *API2* and *MALT1* genes. The arrowheads indicate known breakpoints and their frequencies. BIR: baculovirus inhibitor of apoptosis repeat; CARD: caspase recruitment domain; DD: death domain; RING: really interesting new gene; p20: caspase-like p20 domain: Ig-I: immunoglobulin-like domain; T6: TRAF6 binding site. (**B**) Interphase fluorescence *in situ* hybridization (FISH) pattern of (11;18)/*API2-MALT1*. FISH with an *API2/MALT1* dual-color, dual-fusion translocation probe shows colocalization of the red and green signals (arrows). (**A:** Reprinted from Sagaert et al. 2007, with permission from Macmillan Publishers. Copyright 2007. **B:** Reprinted from Nakamura et al. 2007, with permission from the BMJ Publishing Group.) (*See Color Plates*)

gastric and pulmonary MALT lymphomas and is almost never found in MALT lymphomas of other anatomic sites. Reported rates vary from 24% to 48% of gastric MALT lymphomas, depending on the study. Lymphomas with this translocation typically do not respond to *H. pylori* eradication therapy and are more likely to disseminate to the local lymph nodes or distant sites (Liu et al. 2001a, 2001b). However, t(11;18)(q21;q21)-positive MALT lymphomas are not associated with high-grade transformation and are genetically stable, i.e., they rarely harbor additional clonal aberrations (Figure 4.9) (Starostik et al. 2002). In contrast, t(11;18)(q21;q21)-negative tumors often show trisomy of chromosomes 3, 12, and 18, as well as other aberrations (Figure 4.9). The API2-MALT1 fusion protein contains the N-terminal region of API2, a caspase inhibitor, with three consecutive

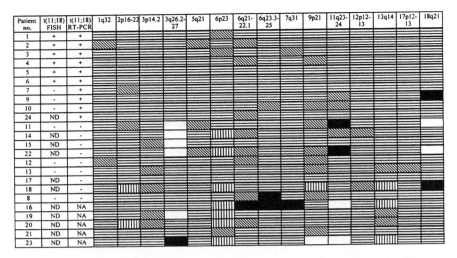

Fig. 4.9 Pattern of allelic imbalances in gastric MALT lymphoma and their relationship to the t(11;18). Chromosomal regions analyzed are shown at the top of each column. Patients are listed in the first column with their t(11;18) status as detected by fluorescence *in situ* hybridization (FISH) (ND: not done) or reverse transcriptase-polymerase chain reaction (RT-PCR) (NA: no amplificate) given in the second and third columns, respectively. Status of each locus is indicated: horizontal stripes, retention of heterozygosity; diagonal stripes, not informative; vertical stripes, no amplificate; empty bars, genomic DNA amplification; and black bars, LOH. (This research was originally published in Starostik et al. 2002. © The American Society of Hematology.)

BIR (baculovirus IAP repeat) domains as well as the caspase-like domain of the C-terminus of MALT1. The full-length MALT1 protein is an essential component of the antigen-receptor–dependent NF-κB activation in lymphocytes. The fusion protein is believed to derive its oncogenic properties through oligomerization via its BIR domains (McAllister-Lucas et al. 2001).

The translocation **t(1;14)(p22;q32)** also occurs in advanced and eradication therapy–resistant lymphomas (Ye et al. 2003). It affects only about 5% of all MALT lymphomas. The translocation fuses the *BCL10* gene to the Ig locus, thereby bringing Bcl-10 expression under the control of the strong Ig heavy-chain gene enhancer (Willis et al. 1999; Zhang et al. 1999). This results in Bcl-10 over-expression and nuclear localization. Studies in *BCL10* knockout mice have shown that the protein is essential in the development and function of lymphocytes, where it is part of the antigen-receptor signaling pathway activating NF-κB (Ruland et al. 2001; Xue et al. 2003). The knockout mice are markedly immunodeficient because of defective NF-κB activation. In contrast, mice that express transgenic Bcl-10 under the control of the IgH enhancer are characterized by expansion of the splenic marginal zone, in line with a role for Bcl-10 in regulating the proliferation of marginal-zone B cells (Morris et al. 2001).

The translocation **t(14;18)(q32;q21)** juxtaposes the *MALT1* gene to the Ig heavy-chain enhancer, thereby causing its overexpression (Sanchez-Izquierdo et al. 2003; Streubel et al. 2003). This translocation affects approximately 20% of MALT lymphomas and occurs more often in non-GI MALT lymphomas. In lymphoma cells harboring t(14;18)(q32;q21), both MALT1 and Bcl-10 are overexpressed in the cytoplasm (Sanchez-Izquierdo et al. 2003).

Studies in knockout mice have found synergistic roles for Bcl-10 and MALT1 in the activation of NF-κB after crosslinking the antigen-receptor in T cells, and it is likely that similar mechanisms exist in B cells (see model, Figure 4.10; Isaacson and Du 2004). According to this model, the lipid-raft-located adaptor CARD11 forms a ternary complex with Bcl-10 and MALT1 upon antigen recognition of the T-cell receptor. This interaction induces oligomerization of MALT1, which in turn can bind to and induce the oligomerization and activation of TNF receptor associated factor 6 (TRAF6). Activated TRAF6 interacts with a ubiquitin-conjugating enzyme (E2) and mediates polyubiquitylation of NEMO/IKKγ. Multiubiquitylated NEMO activates IKKs α and β, which phosphorylate and cause the degradation of IκB and the release of NF-κB.

If Bcl-10 is overexpressed in B cells as a result of the t(1;14)(p22;q32) translocation, it oligomerizes through its CARD domain and shortcuts the signaling pathway by triggering MALT1 oligomerization in an antigen-independent manner. MALT1 in contrast cannot oligomerize by itself because it lacks the required structural domains. In cells that harbor the t(14;18)(q32;q21) translocation (and therefore overexpress MALT1), high levels of both MALT1 and Bcl-10 are observed in the cytoplasm (Sanchez-Izquierdo et al. 2003), suggesting that MALT1 can bind to and stabilize Bcl-10. This interaction can activate NF-κB in the absence of antigen binding. In the case of the t(11;18)(q21;q21) translocation producing the API2-MALT1 fusion protein, the BIR domains contributed by the API2 N-terminus are believed to induce oligomerization of the fusion protein and eliminate the need for Bcl-10 interaction (McAllister-Lucas et al. 2001). In addition to its (normal) cytoplasmic localization, Bcl-10 is found in the nuclei of t(1;14)(p22;q32)-positive and t(11;18)(q21;q21)-positive MALT lymphoma cells, but the oncogenic potential of nuclear Bcl-10 remains unclear.

Molecular Mechanisms of High-Grade Transformation

Low-grade MALT lymphoma can transform into (high-grade) diffuse large B-cell lymphoma (DLBCL), which is pathogenetically different from its nodal counterpart in several ways. For example, *BCL2* rearrangements occur frequently in nodal DLBCL, but are lacking in gastric DLBCL (Cogliatti et al. 2000). Conversely, a high incidence of *MYC* rearrangements is observed in gastric DLBCL, whereas they are rare in the nodal counterpart. It is not clear whether all gastric DLBCLs originate from low-grade MALT lymphoma, or whether other pathogenetic mechanisms exist. The best-studied factor believed to be associated with high-grade transformation is Bcl-6. Bcl-6 normally functions as a transcriptional switch controlling germinal center formation and is downregulated as lymphocytes within the GC differentiate into memory B or plasma cells. As a result, Bcl-6 is highly expressed in normal GC cells as well as in neoplasms derived from GC cells, such

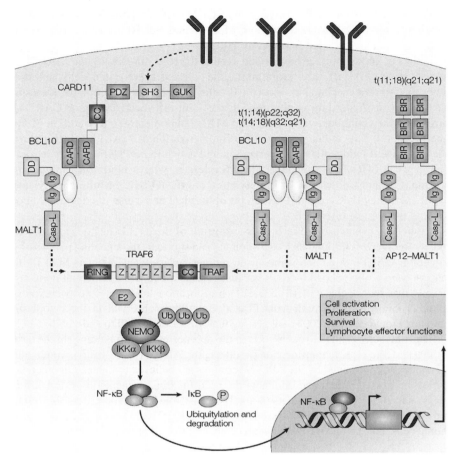

Fig. 4.10 The different chromosomal translocations involved in MALT lymphoma affect a common molecular pathway. After antigen-receptor stimulation, CARD11 is activated to recruit BCL10 through caspase recruitment domain (CARD)–CARD interactions, and this induces BCL10 oligomerization (for simplicity, oligomerization events are not shown). BCL10 then binds the immunoglobulin (Ig)-like domain of mucosa-associated lymphoid tissue lymphoma translocation protein 1 (MALT1) through a short region downstream of its CARD and induces MALT1 oligomerization. Oligomerized MALT1 binds to tumour-necrosis factor receptor associated factor 6 (TRAF6) and induces its oligomerization, resulting in the activation of TRAF6 ubiquitin-ligase activity 109. Activated TRAF6 interacts with a ubiquitin-conjugating enzyme (E2) and mediates polyubiquitylation of nuclear factor-κB (NF-κB) essential modulator (NEMO), which is also known as IκB kinase-γ (IKKγ). By mechanisms that are incompletely understood, multiubiquit-ylated NEMO induces the activation of IKKα and IKKβ, which causes phosphorylation and degradation of IκB and the release of NF-κB. NF-κB then translocates to the nucleus and trans-activates genes important for cellular activation, proliferation, and survival, and induction of effector functions of lymphocytes. In MALT lymphomas involving t(1;14)(p22;q32), *BCL10* is placed under the regulation of the Ig heavy-chain gene enhancers and is overexpressed. In these lymphomas, BCL10 is thought to form oligomers through its CARD domain without the need of upstream signals, leading to constitutive NF-κB activation. In MALT lymphomas with

as follicular lymphoma. In contrast, low-grade MALT lymphoma is usually negative for Bcl-6 as well as another GC marker, CD10, which is in line with its derivation from marginal-zone B cells. Thus, immunostaining for Bcl-6 and CD10 has an important role in the differential diagnosis of B-cell malignancies with different normal cell counterparts (Dogan et al. 2000).

High-grade gastric (and nongastric) DLBCLs are heterogeneous in terms of Bcl-6 expression. Indeed, immunostaining for Bcl-6 as well as CD10 and MUM1 can be used to classify DLBCL into GC B-cell (GCB)-like and non-GCB subgroups with prognostic significance (Chen et al. 2006). Cases with high Bcl-6 expression generally have a more favorable prognosis. Bcl-6 expression can be deregulated through at least three different mechanisms. On the one hand, mutations in the two "*BCL6* binding motifs" BSE1A and BSE1B can prevent Bcl-6 from binding to its own promoter, thus disrupting its negative autoregulation. These mutations are always associated with high Bcl-6 expression and are found in approximately 25% of DLBCL cases (Chen et al. 2006). On the other hand, chromosomal translocations involving the *BCL6* gene juxtapose its 5' regulatory region to the promoter region of an Ig gene; fusions to other constitutively expressed, non-Ig genes have also been found. Approximately 40% of DLBCLs carry *BCL6* translocations, which also are typically associated with high expression of the gene product (Chen et al. 2006). Neither deregulating mutations nor translocations involving Bcl-6 are found at high frequency in low-grade MALT lymphomas. It has therefore been speculated that low- to high-grade transformation is brought about by mutations or translocations of the *BCL6* gene. The translocations involving Ig genes in particular could arise during erroneous Ig isotype class switching, which has been shown to occur in a proportion of gastric MALT lymphomas. The presence of many reactive follicles in MALT lymphoma may provide a setting for both types of genetic alterations of *BCL6*, thus resulting in high-grade transformation.

The third mechanism of Bcl-6 deregulation involves amplification of the region 3q26.2-3q27. This region on chromosome 3 harbors the genes coding for Bcl-6 and the phosphatidylinositol-3 kinase catalytic subunit, which has been implicated as an oncogene in ovarian cancer. In a study comparing the genetic abnormalities in gastric MALT lymphoma and gastric DLBCL by microsatellite screening, amplification of the 3q26.2-3q27 region was shown to be the most frequent allelic imbalance affecting both disease entities (approximately 20%) (Starostik et al. 2002). This and

Fig. 4.10 (continued) t(14;18)(q32;q21), MALT1 is placed under the control of the Ig heavy-chain gene enhancer and is overexpressed. In these tumors, the oligomerization and activation of MALT1 is thought to be dependent on BCL10. In MALT lymphomas with t(11;18)(q21;q21), the resulting API2-MALT1 fusion product is believed to self-oligomerize through the baculovirus IAP repeat (BIR) domain of the API2 molecule, therefore leading to constitutive NF-κB activity. Casp-L: caspase-like domain; CC: coiled coil; DD: death domain; GUK: guanylate-kinase-like domain; SH3: Src homology 3 domain; Ub: ubiquitin; Z: zinc finger. (Reprinted from Isaacson and Du 2004, with permission from Macmillan Publishers. Copyright 2004.)

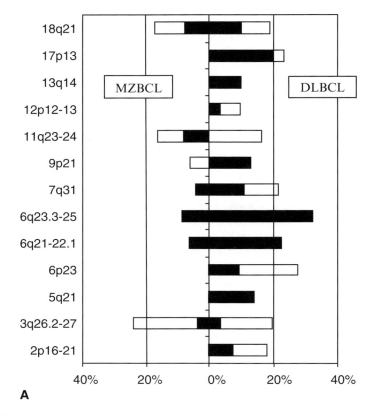

Fig. 4.11 **(A)** Comparison of the allelotypes of low-grade MALT lymphoma (marginal zone B-cell lymphoma; MZBCL) and high-grade DLBCL (diffuse large B-cell lymphoma). Frequency of allelic imbalance (percentage of informative analyses) in individual regions is expressed as a

several other allelic imbalances were shared by t(11;18)(q21;q21)-negative MALT lymphomas and gastric DLBCL (Figure 4.11), whereas t(11;18)(q21;q21)-positive MALT lymphoma cases were genetically stable, i.e., did not acquire additional genetic aberrations (Figures 4.9, 11A). Overall, the frequency of aberrations was much lower in the MALT lymphoma cases compared with the DLBCL. A distinct subgroup of allelic imbalances is unique to the DLBCL cases and never found in any of the MALT lymphoma cases. The genes affected by these aberrations are good additional candidates for a role in low- to high-grade transition. Among the aberrations in this group, LOH is seen in the regions 5q21 (*APC* gene locus), 9p21 (*INK4A/ARF*), 13q14 (*RB*), and 17p13 (*p53*) (Starostik et al. 2002). Indeed, complete inactivation of the p53 tumor suppressor by mutation or deletion has been observed in DLBCL also in independent studies (e.g. Du et al. 1995), as has inacti-

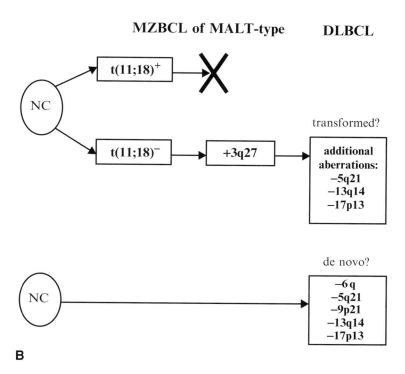

Fig. 4.11 (continued) bar diagram: open bar, amplification; black bar, LOH. The left-hand side shows results for MZBCL; the right-hand side, results for DLBCL. (**B**) Two pathways of gastric MALT lymphoma development from normal cells (NC). The first one is characterized by the t(11;18); peculiarly, these lymphomas only rarely accumulate additional genetic aberrations and do not seem to transform into DLBCL. On the other side, the t(11;18)-negative tumors acquire various genetic aberrations exemplified by the 3q26.2-27 amplification, and some of them may eventually transform into high-grade DLBCL (listed are some of the most common additional aberrations). The 6q aberration displaying cases might be primary DLBCL. (This research was originally published in Starostik et al. 2002. © the American Society of Hematology.)

vation of the *INK4A* gene encoding a cyclin-dependent kinase inhibitor and negative regulator of the cell cycle (Neumeister et al. 1997).

Based on the similarities and differences in the allelic imbalances found in both groups of tumors, it has been proposed that gastric lymphomas develop along two distinct pathways (see model in Figure 4.11B) (Starostik et al. 2002). One group of tumors develops along the pathway determined by the dysregulation of the *API2* and *MALT1* genes brought about by the t(11;18). These tumors do not accumulate enough secondary genetic aberrations to transform into DLBCL and remain in the MALT lymphoma stage. In contrast, MALT lymphomas characterized by the absence of t(11;18) and increased accumulation of various genetic abnormalities, most frequently the 3q26.2-27 amplification, could be the source of tumors that eventually do transform into high-grade DLBCL (Starostik et al. 2002). Based on the additional finding that many DLBCLs do not harbor any of the shared abnormalities, it was further speculated that DLBCLs can be categorized into two groups: one that derives from a low-grade counterpart (characterized by the 3q27 and other shared aberrations) and one that evolves *de novo* (Starostik et al. 2002).

Animal Models of MALT Lymphoma

Models of Infection-Induced MALT Lymphoma

In 1995, Adrian Lee and coworkers published a first report showing that chronic infection of BALB/c mice with *Helicobacter* mimics the histopathologic features of human MALT lymphoma (Enno et al. 1995). The mice were experimentally infected with a *Helicobacter* strain isolated from cats, *H. felis*. *H. felis* colonizes murine stomachs at high densities, and shows preference for the antral region of the stomach as well as the junction between squamous and glandular epithelium. In mice that were persistently infected for 22 months or more, the development of lymphoid follicles (MALT) was frequently observed (Enno et al. 1995). These contained germinal centers composed of centroblasts with scattered macrophages and indistinct mantle zones. LELs, a hallmark of MALT lymphomas in humans, were frequently seen in areas of dense bacterial colonization. Early LELs with small aggregates of centrocyte-like cells infiltrating the foveolar epithelium were shown to progress to late lesions with destruction of both the glandular and foveolar epithelium (Figure 4.12; Enno et al. 1995). The development of late lesions was accompanied by loss of germinal centers and a more monomorphic appearance of the lymphoid infiltrate. The infiltrates extended across the mucosa and submucosa and involvement of the muscularis propria was occasionally observed. Thus, these lesions have histopathologic characteristics that meet the criteria for MALT lymphoma and can be considered to be the murine counterpart of the human disease. Depending on the bacterial strain used (see below), between 50% and 90% of infected animals develop LELs by 22 months postinfection; frank MALT lymphomas are diagnosed in 4%–25% of mice (O'Rourke et al. 2004).

A subsequent study using the same model showed that eradication of the bacteria at 20 months postinfection with a combination therapy consisting of metronidazole, tetracycline, and bismuth induced regression of these MALT lymphoma–like lesions (Enno et al. 1998). By 4 months posttreatment, all lymphoid follicles had dissolved and LELs were no longer detected in the majority of mice. In this and other subsequent studies, 10%–25% of animals have proven to be resistant to antibiotic therapy; i.e., *Helicobacter* eradication is successful but the lymphomas nevertheless fail to regress (Enno et al. 1998; Müller et al. 2005). These rates are comparable to resistance rates seen in patients receiving similar antibiotic regi-

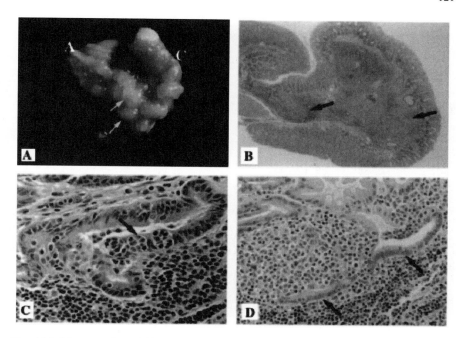

Fig. 4.12 Macroscopic and histologic features of MALT lymphoma-like lesions in mice. (**A**) Nodular gastric mucosa (arrows) of *H. felis*–infected SPF mouse; A = antrum; C = corpus mucosa (original magnification, ×5). (**B**) Section through nodular mucosa showing lymphoid hyperplasia in the submucosa. Poorly demarcated germinal centers (arrows) are apparent within the lymphoid tissue [hematoxylin and eosin (H&E); original magnification, ×13]. (**C**) Small aggregates of centrocyte-like cells (arrows) infiltrate hyperplastic foveolar epithelium in this relatively early lymphoepithelial (LE) lesion (H&E; original magnification, ×100). (**D**) Fully developed LE lesion showing destruction of glandular and foveolar epithelium with a central epithelial remnant (arrows) (H&E; original magnification, ×63). (Reprinted from Enno et al. 2003, with permission from the American Society for Investigative Pathology.)

mens. In addition, in mice as well as in patients, "histologic" remission is not equivalent to "molecular" remission. Transcriptional profiling of whole stomach extracts demonstrated that numerous transcripts indicative of lymphoma cells (i.e., their "molecular signature") were still present in virtually all mice many months after successful eradication and concomitant lymphoma regression (Figure 4.13) (Müller et al. 2005). It is remarkable that experimental reinfection of such "successfully treated" mice generally results in very rapid tumor relapse, suggesting that the lymphomatous clones are still present and can be reactivated very rapidly after the reintroduction of *Helicobacter* antigen (Müller et al. 2005).

It seems extraordinary in this context that protective immunity against *Helicobacter* is not acquired by the host animal, even after lifelong infection, and despite a vigorous and sustained immune response. In keeping with the observation that experimental reinfection of mice is usually successful, the global reinfection

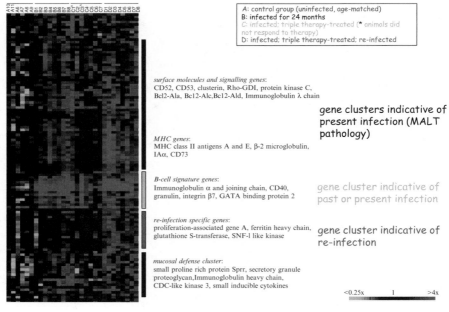

A: control group (uninfected, age-matched)
B: infected for 24 months
C: infected; triple therapy-treated (* animals did not respond to therapy)
D: infected; triple therapy-treated; re-infected

surface molecules and signalling genes:
CD52, CD53, clusterin, Rho-GDI, protein kinase C,
Bcl2-Ala, Bcl2-Alc,Bcl2-Ald, Immunoglobulin λ chain

gene clusters indicative of present infection (MALT pathology)

MHC genes:
MHC class II antigens A and E, β-2 microglobulin,
IAα, CD73

B-cell signature genes:
Immunoglobulin α and joining chain, CD40,
granulin, integrin β7, GATA binding protein 2

gene cluster indicative of past or present infection

re-infection specific genes:
proliferation-associated gene A, ferritin heavy chain,
glutathione S-transferase, SNF-1 like kinase

gene cluster indicative of re-infection

mucosal defense cluster:
small proline rich protein Sprr, secretory granule
proteoglycan,Immunoglobulin heavy chain,
CDC-like kinase 3, small inducible cytokines

<0.25x   1   >4x

Fig. 4.13 Transcriptional profiling reveals clusters of genes whose expression patterns correlate with treatment outcome. RNA was prepared from whole stomach tissue of several animals per treatment group, reverse transcribed, and hybridized to 38,000 element murine cDNA arrays. Data were filtered with respect to spot quality (spots with regression correlations <0.6 were omitted) and data distribution (genes whose $\log_2$ of red-to-green normalized ratio is more than 1.5 standard deviations away from the mean in at least three arrays were selected) before clustering of the resulting 125 genes. Only genes for which information was available for >70% of arrays were included. Clusters of genes that are differentially regulated between treatment groups are annotated, and representative genes are listed for every cluster. (Reprinted from Müller et al., 2005, with permission from the American Society for Investigative Pathology.) (*See Color Plates*)

rate with *H. pylori* was shown to be just as high as or only slightly lower than the primary infection rate (Parsonnet 2003). Recurrence of MALT lymphomas as a result of reinfection is relatively uncommon, affecting only between 1% and 20% of successfully treated patients, depending on the study (Fischbach et al. 2004; Nobre-Leitao et al. 1998; Papa et al. 2000; Tursi et al. 1997). Nevertheless, in light of the finding that 97% of spouses of patients with *H. pylori*–positive MALT lymphomas are also seropositive (far exceeding the seroprevalence rates in the general population) (Fischbach et al. 2000), it has been suggested that spouses receive eradication therapy along with the patients to eliminate this common source of reinfection.

In mice as well as in humans, MALT lymphomas can transform to diffuse large B-cell lymphoma; the tumors can consist of both high- and low-grade components and typically affect all layers of the stomach including the stomach wall (Figure

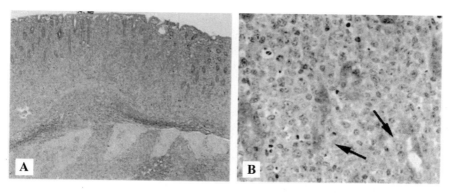

Fig. 4.14 (**A**) High-grade (large B-cell) lymphoma in a mouse infected for 22 months. Diffuse infiltrate involving all of stomach wall. Original magnification, ×50. (**B**) High-power view of high-grade lymphoma with tumor cells destroying gastric glands (arrows). Original magnification, ×312.5. (Reprinted from Enno et al. 1998, with permission from the American Society for Investigative Pathology.)

4.14) (Enno et al. 1998). LELs can frequently be detected in murine gastric DLBCL, even in the absence of low-grade components, suggesting their origin in MALT lymphoma. In addition, dissemination of the gastric lymphomatous clone to the spleen is occasionally observed in mice, suggesting that systemic dissemination can occur also in this species, albeit rarely.

Larger studies that also included mice infected with *H. pylori* as well as another closely related *Helicobacter*, *H. heilmannii*, indicated that lymphomas develop in response to all three species, albeit with different rates and time courses (Müller et al. 2003a). *H. heilmannii* isolates from monkeys seem to be particularly virulent in the mouse model, showing an early onset of lymphoid follicle formation at 6 months postinfection and frank lymphomas at 18 months postinfection (O'Rourke et al. 2004). *H. heilmannii* has been associated with the whole spectrum of human gastric diseases, including peptic ulceration, gastric adenocarcinoma, and MALT lymphoma.

A major drawback of this model is the need for long-term infection of the mice, which makes these types of experiments prohibitively expensive in most institutions. Interestingly, a more rapid course of lymphomagenesis is observed in BALB/c mice that have been thymectomized as neonates (Fukui et al. 2004). Thymectomy induces spontaneous autoimmune gastritis through a mechanism involving depletion of CD25$^+$ regulatory T cells, which, in collaboration with *H. pylori* infection, apparently leads to the rapid acquisition of lymphoid follicles in the stomach 2 months postinfection. LELs develop gradually but much more rapidly than in normal BALB/c mice, with 100% of mice showing LELs at 12 months postinfection (Fukui et al. 2004). In this model as well as the slower, spontaneous MALT lymphoma model, experimental evidence suggests that expression of the anti-apoptotic factor Bcl-X$_L$ may at least in part be responsible for tumor cell survival (Fukui et al. 2004; Morgner et al. 2001). Lymphoma cells cultivated *ex vivo* in the presence of *H. pylori* whole cell sonicate survived longer than unstimulated cells, and survival was associated with an induction of Bcl-X$_L$ mRNA.

The development of MALT lymphoma can effectively be prevented by prophylactic immunization of mice with an *H. felis* whole-cell sonicate vaccine administered orally along with cholera toxin as adjuvant (Müller et al. 2003b; Sutton et al. 2004). This vaccine reduces bacterial densities upon challenge, but does not completely protect mice from colonization with the vaccine strain. The remaining bacterial burden induces the acquisition of MALT in 33% of immunized mice, but apparently is not sufficient to drive formation of LELs or progression to lymphoma (Müller et al. 2003b; Sutton et al. 2004).

Other animal models have only sporadically been used to study MALT lymphoma. In 1997, natural infection of ferrets with *H. mustelae* resulted in MALT lymphoma and diffuse large B-cell lymphoma in a small number of animals, again with histologic characteristics similar to the respective human diseases (Erdman et al. 1997). Both types of lymphomas had arisen from areas of chronic inflammation in the antral mucosa, which typically appears upon natural infection with *H. mustelae* after weaning and increases in extent and severity as the animals grow older. All lymphomas in ferrets were monoclonal as indicated by light-chain restriction (Erdman et al. 1997). Immunohistochemical evaluation of clonality as a means of discerning lymphoma from severe gastritis is possible in ferrets (as well as in humans), because they have an approximately balanced κ/λ ratio. Mice in contrast have a disparate κ/λ ratio of 20:1, making those comparisons difficult.

Transgenic Mouse Models of MALT Lymphoma

The biologic effects of two of the three common chromosomal translocations occurring in advanced MALT lymphoma have been investigated in transgenic mice expressing the resulting fusion products and have shed light on the common pathway affected by the translocations. As noted earlier, the t(11;18)(q21;q21) translocation generates the API2-MALT1 fusion protein. API2-MALT1 expressed under the control of the Ig heavy-chain enhancer triggers the specific expansion of splenic marginal-zone B cells (Figure 4.15) (Baens et al. 2006), the normal cell counterpart of neoplastic MALT lymphoma cells. In splenic lymphocytes expressing API2-MALT, increased NF-κB activation is observed and leads to the survival of B cells *ex vivo* (Figure 4.16; Baens et al. 2006), supporting the hypothesis that the fusion protein acts by constitutively activating this pathway. Expression of the API2-MALT1 protein is not sufficient to produce lymphoma masses, neither in the spleen nor in other tissues. However, if the transgenic mice are subjected to strong antigenic stimulation by injection of Freund's adjuvant, the spleens exhibit extensive lymphoid hyperplasia that resembles splenic mar-

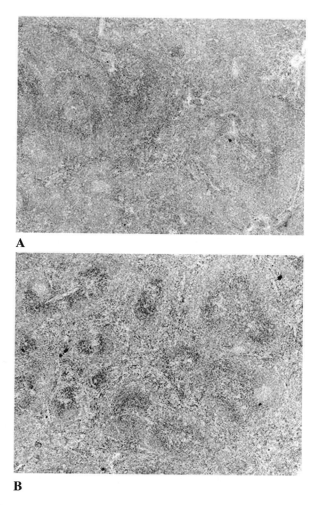

Fig. 4.15 Spleen morphology of API2-MALT1 transgenic mice and WT littermates at age 20 weeks. Transgenic mice show a reduction of red pulpa with an increase of white pulpa as the result of an expanded marginal zone (**A**) compared with their WT littermates (**B**). (Reprinted from Baens et al. 2006, with permission from the American Association for Cancer Research.)

ginal-zone lymphoma morphologically and immunophenotypically (Sagaert et al. 2006). This finding suggests that antigenic stimulation and the expression of API2-MALT1 can act synergistically to induce NF-κB–dependent hyperproliferation of B cells.

A similar phenotype is observed in transgenic mice expressing Bcl-10 under the control of the Ig heavy-chain enhancer, thereby mimicking the effect of the t(1;14)(p22;q32) translocation (Morris 2001). These mice also show a marked and

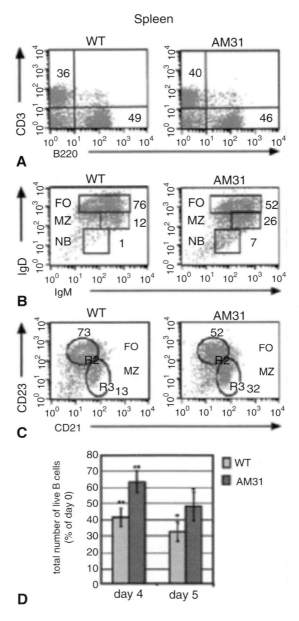

Fig. 4.16 API2-MALT1 promotes B-cell survival. FACS analyses for splenic B cells from 12-week-old WT and API2-MALT1-transgenic (AM31) mice labeled with the indicated antibodies. Percentages of positive cells are shown. (**A**) B-cell and T-cell populations (CD3[+]/B220[+]) in spleen (lymphocyte gate). (**B**) Surface IgM/IgD. FO: follicular B cells; MZ: marginal-zone B cells; NB: naive B cells. (**C**) CD21/CD23 expression on B220-gated splenic B lymphocytes. (**D**) Splenic CD19[+] B cells from AM31 or WT mice were cultured *in vitro*, and the number of surviving B cells was expressed as a proportion of the number present at day 0. Columns: mean values from four mice (12 weeks) of each group; bars: standard deviation. p = 0.0012 and p = 0.013 at days 4 and 5, respectively. (Reprinted from Baens et al. 2006, with permission from the American Association for Cancer Research.)

specific expansion of the splenic marginal zone reminiscent of human marginal-zone lymphoma.

References

Baens, M., Fevery, S., Sagaert, X., Noels, H., Hagens, S., Broeckx, V., Billiau, A. D., De Wolf-Peeters, C., and Marynen, P. (2006). Selective expansion of marginal zone B cells in Emicro-API2-MALT1 mice is linked to enhanced IkappaB kinase gamma polyubiquitination. Cancer Res. 66:5270–7.

Bende, R.J., Aarts, W.M., Riedl, R.G., de Jong, D., Pals, S.T., and van Noesel, C.J. (2005). Among B cell non-Hodgkin's lymphomas, MALT lymphomas express a unique antibody repertoire with frequent rheumatoid factor activity. J Exp Med. 201:1229–41.

Blackledge, G., Bush, H., Dodge, O.G., and Crowther, D. (1979). A study of gastro-intestinal lymphoma. J Clin Oncol. 5:209–19.

Briskin, M., Winsor-Hines, D., Shyjan, A., Cochran, N., Bloom, S., Wilson, J., McEvoy, L.M., Butcher, E.C., Kassam, N., Mackay, C.R., Newman, W., and Ringler, D.J. (1997). Human mucosal addressin cell adhesion molecule-1 is preferentially expressed in intestinal tract and associated lymphoid tissue. Am J Pathol. 151:97–110.

Cammarota, G., Montalto, M., Tursi, A., Vecchio, F.M., Fedeli, G., and Gasbarrini, G. (1995). Helicobacter pylori reinfection and rapid relapse of low-grade B-cell gastric lymphoma. Lancet. 345:192.

Chan, J.K., Ng, C.S., and Isaacson, P.G. (1990). Relationship between high-grade lymphoma and low-grade B-cell mucosa-associated lymphoid tissue lymphoma (MALToma) of the stomach. Am J Pathol. 136:1153–64.

Chen, Y.W., Hu, X.T., Liang, A.C., Au, W.Y., So, C.C., Wong, M.L., Shen, L., Tao, Q., Chu, K.M., Kwong, Y.L., Liang, R.H., and Srivastava, G. (2006). High BCL6 expression predicts better prognosis, independent of BCL6 translocation status, translocation partner, or BCL6-deregulating mutations, in gastric lymphoma. Blood. 108:2373–83.

Chen, L.T., Lin, J.T., Shyu, R.Y., Jan, C.M., Chen, C.L., Chiang, I.P., Liu, S.M., Su, I.J., and Cheng, A.L. (2001). Prospective study of Helicobacter pylori eradication therapy in stage I(E) high-grade mucosa-associated lymphoid tissue lymphoma of the stomach. J Clin Oncol. 19:4245–51.

Cheng, T.Y., Lin, J.T., Chen, L.T., Shun, C.T., Wang, H.P., Lin, M.T., Wang, T.E., Cheng, A.L., and Wu, M.S. (2006). Association of T-cell regulatory gene polymorphisms with susceptibility to gastric mucosa-associated lymphoid tissue lymphoma. J Clin Oncol. 24:3483–9.

Cogliatti, S.B., Griesser, H., Peng, H., Du, M.Q., Isaacson, P.G., Zimmermann, D.R., Maibach, R.C., and Schmid, U. (2000). Significantly different bcl-2 expression profiles in gastric and non-gastric primary extranodal high-grade B-cell lymphomas. J Pathol. 192:470–8.

Cohen, S.M., Petryk, M., Varma, M., Kozuch, P.S., Ames, E.D., and Grossbard, M.L. (2006). Non-Hodgkin's lymphoma of MALT. Oncologist. 11:1100–17.

Delchier, J.C., Lamarque, D., Levy, M., Tkoub, E.M., Copie-Bergman, C., Deforges, L., Chaumette, M.T., and Haioun, C. (2001). Helicobacter pylori and gastric lymphoma: high seroprevalence of CagA in diffuse large B-cell lymphoma but not in low-grade lymphoma of mucosa-associated lymphoid tissue type. Am J Gastroenterol. 96:2324–8.

Dierlamm, J., Baens, M., Wlodarska, I., Stefanova-Ouzounova, M., Hernandez, J.M., Hossfeld, D.K., De Wolf-Peeters, C., Hagemeijer, A., Van den Berghe, H., and Marynen, P. (1999). The apoptosis inhibitor gene API2 and a novel 18q gene, MLT, are recurrently rearranged in the t(11;18)(q21;q21) associated with mucosa-associated lymphoid tissue lymphomas. Blood. 93:3601–9.

Dogan, A., Bagdi, E., Munson, P., and Isaacson, P.G. (2000). CD10 and BCL-6 expression in paraffin sections of normal lymphoid tissue and B-cell lymphomas. Am J Surg Pathol. 24:846–52.

Doglioni, C., Wotherspoon, A.C., Moschini, A., de Boni, M., and Isaacson, P.G. (1992). High incidence of primary gastric lymphoma in northeastern Italy. Lancet. 339:834–5.

Du, M., Diss, T.C., Xu, C., Peng, H., Isaacson, P.G., and Pan, L. (1996b). Ongoing mutation in MALT lymphoma immunoglobulin gene suggests that antigen stimulation plays a role in the clonal expansion. Leukemia. 10:1190–7.

Du, M.Q., Peng, H.Z., Dogan, A., Diss, T.C., Liu, H., Pan, L.X., Moseley, R.P., Briskin, M.J., Chan, J.K., and Isaacson, P.G. (1997). Preferential dissemination of B-cell gastric mucosa-associated lymphoid tissue (MALT) lymphoma to the splenic marginal zone. Blood. 90:4071–7.

Du, M., Peng, H., Singh, N., Isaacson, P.G., and Pan, L. (1995). The accumulation of p53 abnormalities is associated with progression of mucosa-associated lymphoid tissue lymphoma. Blood. 86:4587–93.

Du, M.Q., Xu, C.F., Diss, T.C., Peng, H.Z., Wotherspoon, A.C., Isaacson, P.G., and Pan, L.X. (1996a). Intestinal dissemination of gastric mucosa-associated lymphoid tissue lymphoma. Blood. 88:4445–4451.

Eck, M., Schmausser, B., Haas, R., Greiner, A., Czub, S., and Müller-Hermelink, H.K. (1997). MALT-type lymphoma of the stomach is associated with Helicobacter pylori strains expressing the CagA protein. Gastroenterology. 112:1482–6.

Eidt, S., Stolte, M., and Fischer, R. (1994). Helicobacter pylori gastritis and primary gastric non-Hodgkin's lymphomas. J Clin Pathol. 47:436–9.

Enno, A., O'Rourke, J., Braye, S., Howlett, R. and Lee, A. (1998). Antigen-dependent progression of mucosa-associated lymphoid tissue (MALT)-type lymphoma in the stomach: effects of antimicrobial therapy on gastric MALT lymphoma in mice. Am J Pathol. 152:1625–32.

Enno, A., O'Rourke, J.L., Howlett, C.R., Jack, A., Dixon, M.F., and Lee, A. (1995). MALToma-like lesions in the murine gastric mucosa after long-term infection with Helicobacter felis: a mouse model of Helicobacter pylori induced gastric lymphoma. Am J Pathol. 147:217–22.

Erdman, S.E., Correa, P., Coleman, L.A., Schrenzel, M.D., Li, X., and Fox, J.G. (1997). Helicobacter mustelae-associated gastric MALT lymphoma in ferrets. Am J Pathol. 151:273–80.

Ferrucci, P.F., and Zucca, E. (2007). Primary gastric lymphoma pathogenesis and treatment: what has changed over the past 10 years? Br J Haematol. 136:521–38.

Fischbach, W., Goebeler-Kolve, M.E., Dragosics, B., Greiner, A., Stolte, M. (2004). Long term outcome of patients with gastric marginal zone B cell lymphoma of mucosa associated lymphoid tissue (MALT) following exclusive Helicobacter pylori eradication therapy: experience from a large prospective series. Gut. 53:34–7.

Fischbach, W., Jung, T., Goebeler-Kolve, M., and Eck, M. (2000). Comparative analysis of the Helicobacter pylori status in patients with gastric MALTtype lymphoma and their respective spouses. Z Gastroenterol. 38:627–30.

Fukui, T., Okazaki, K., Tamaki, H., Kawasaki, K., Matsuura, M., Asada, M., Nishi, T., Uchida, K., Iwano, M., Ohana, M., Hiai, H., and Chiba, T. (2004). Immunogenetic analysis of gastric MALT lymphoma-like lesions induced by Helicobacter pylori infection in neonatally thymectomized mice. Lab Invest. 84:485–92.

Greiner, A., Knoerr, C., Yufen, Q., Sebald, W., Schimpl, A., Banchereau, J., and Mueller-Hermelink, H.K. (1997). Low grade B cell lymphomas of MALT require CD40 mediated signaling and Th2 type cytokines for in vitro growth and differentiation. Am J Pathol. 150:1583–93.

Greiner, A., Marx, A., Hessemann, J., Leebmann, J., Schmausser, B., and Mueller-Hermelink, H.K. (1994). Idiotype identity in a MALT type lymphoma and B cells in Helicobacter associated chronic gastritis. Lab Invest. 70:572–8.

Hammel, P., Haioun, C., Chaumette, M.T., Gaulard, P., Divine, M., Reyes, F., and Delchier, J.C. (1995). Efficacy of single agent chemotherapy in low grade B cell MALT lymphoma with prominent gastric expression. J Clin Oncol. 13:2524–9.

Hellmig, S., Fischbach, W., Goebeler-Kolve, M.E., Fölsch, U.R., Hampe, J., and Schreiber, S. (2005a). A functional promotor polymorphism of TNF-alpha is associated with primary gastric B-Cell lymphoma. Am J Gastroenterol. 100:2644–9.

Hellmig, S., Fischbach, W., Goebeler-Kolve, M.E., Fölsch, U.R., Hampe, J., and Schreiber, S. (2005b). Association study of a functional Toll-like receptor 4 polymorphism with susceptibility to gastric mucosa-associated lymphoid tissue lymphoma. Leuk Lymphoma. 46:869–72.

Hellmig, S., Vollenberg, S., Goebeler-Kolve, M.E., Fischbach, W., Hampe, J., Fölsch, U.R., and Schreiber, S. (2004). IL-1 gene cluster polymorphisms and development of primary gastric B-cell lymphoma in Helicobacter pylori infection. Blood. 104:2994–5.

Horstmann, M., Erttmann, R., and Winkler, K. (1994). Relapse of MALT lymphoma associated with Helicobacter pylori after antibiotic treatment. Lancet. 343:1098–9.

Hussell, T., Isaacson, P.G., Crabtree, J.E., and Spencer, J. (1993a). The response of cells from low grade B cell gastric lymphomas of MALT to Helicobacter pylori. Lancet. 342:571–4.

Hussell, T., Isaacson, P.G., Crabtree, J.E., Dogan, A., and Spencer, J. (1993b). Immunoglobulin specificity of low grade gastrointestinal lymphoma of MALT type. Am J Pathol. 142:285–92.

Hussell, T., Isaacson, P.G., Crabtree, J.E., and Spencer, J. (1996). Helicobacter-specific tumor infiltrating T-cells provide contact dependent help for the growth of malignant B-cells in low grade gastric lymphomas of MALT. J Pathol. 178:122–7.

Isaacson, P.G., and Du, M.-Q. (2004). MALT lymphoma: from morphology to molecules. Nat Rev Cancer. 4:644–53.

Isaacson, P.G., and Spencer, J. (1987). Malignant lymphoma of mucosa-associated lymphoid tissue. Histopathology. 11:445–62.

Isaacson, P.G., Wotherspoon, A.C., Diss, T., and Pan, L.X. (1991). Follicular colonization in B-cell lymphoma of mucosaassociated lymphoid tissue. Am J Surg Pathol. 15:819–28.

Knörr, C., Amrehn, C., Seeberger, H., Rosenwald, A., Stilgenbauer, S., Ott, G., Müller Hermelink, H.K., Greiner, A. (1999). Expression of costimulatory molecules in low-grade mucosa-associated lymphoid tissue-type lymphomas in vivo. Am J Pathol. 155:2019–27.

Kraal, G., Schornagel, K., Streeter, P.R., Holzmann, B., and Butcher, E.C. (1995). Expression of the mucosal vascular addressin, MAdCAM-1, on sinus-lining cells in the spleen. Am J Pathol. 147:763–71.

Lenze, D., Berg, E., Volkmar, R., Weiser, A.A, Greiner, A., Knoerr C., Anagnastopoulos, I., Stein, H., and Hummel, M. (2006). Influence of antigen on the development of MALT lymphoma. Blood. 107:1141–8.

Liu, H., Ruskon-Fourmestraux, A., Lavergne-Slove, A., Ye, H., Molina, T., Bouhnik, Y., Hamoudi, R.A., Diss, T.C., Dogan, A., Megraud, F., Rambaud, J.C., Du, M.Q., and Isaacson, P.G. (2001a). Resistance of t(11;18) positive gastric mucosa-associated lymphoid tissue lymphoma to Helicobacter pylori eradication therapy. Lancet. 357:39–40.

Liu, H., Ye, H., Dogan, A., Ranaldi, R., Hamoudi, R.A., Bearzi, I., Isaacson, P.G., and Du, M.Q. (2001b). T(11;18)(q21;q21) is associated with advanced mucosa-associated lymphoid tissue lymphoma that expresses nuclear BCL10. Blood. 98:1182–7.

Martinelli, G., Laszlo, D., Ferreri, A.J., Pruneri, G., Ponzoni, M., Conconi, A., Crosta, C., Pedrinis, E., Bertoni, F., Calabrese, L., and Zucca, E. (2005). Clinical activity of rituximab in gastric marginal zone non-Hodgkin's lymphoma resistant to or not eligible for anti-Helicobacter pylori therapy. J Clin Oncol. 23:1979–83.

McAllister-Lucas, L.M., Inohara, N., Lucas, P.C., Ruland, J., Benito, A., Li, Q., Chen, S., Chen, F.F., Yamaoka, S., Verma, I.M., Mak, T.W., and Núñez, G. (2001). Bimp1, a MAGUK family member linking protein kinase C activation to Bcl10-mediated NF-κB induction. J Biol Chem. 276:30589–97.

Montalban, C., Santon, A., Boixeda, D., Redondo, C., Alvarez, I., Calleja, J.L., de Argila, C.M., and Bellas, C. (2001). Treatment of low grade gastric mucosa-associated lymphoid tissue lymphoma in stage I with Helicobacter pylori eradication: long-term results after sequential histologic and molecular follow-up. Haematologica. 86:609–17.

Morgner, A., Sutton, P., O'Rourke, J.L., Enno, A., Dixon, M.F., and Lee, A. (2001). Helicobacter-induced expression of Bcl-X(L) in B lymphocytes in the mouse model: a possible step in the development of gastric mucosa-associated lymphoid tissue (MALT) lymphoma. Int J Cancer. 92:634–40.

Morris, S.W. (2001). American Society of Hematology Education program book. Washington, D.C.: American Society of Hematology; pp. 191–204.

Müller, A., Grimm, J., O'Rourke, J., Lee, A., Guillemin, K., Dixon, M.F., and Falkow, S. (2003a). Distinct gene expression profiles characterize the histopathological stages of disease in Helicobacter-induced mucosa-associated lymphoid tissue lymphoma. Proc Natl Acad Sci. 100:1292–7.

Müller, A., O'Rourke, J., Chu, P., Kim, C.C., Sutton, P., Lee, A., and Falkow, S. (2003b). Protective immunity against Helicobacter is characterized by a unique transcriptional signature. Proc Natl Acad Sci. 100:12289–94.

Müller, A., O'Rourke, J., Chu, P., Lee, A., and Falkow, S. (2005). The role of Helicobacter pylori and H. pylori-specific T-cells in MALT lymphoma pathogenesis. Am J Pathol. 167:797–812.

Musshoff K. (1977). Clinical staging classification of non-Hodgkin's lymphomas. Strahlentherapie. 153:218–21.

Nakamura, S., Aoyagi, K., Furuse, M., Suekane, H., Matsumoto, T., Yao, T., Sakai, Y., Fuchigami, T., Yamamoto, I., Tsuneyoshi, M., and Fujishima, M. (1998). B-cell monoclonality precedes the development of gastric MALT lymphoma in Helicobacter pylori-associated chronic gastritis. Am J Pathol. 152:1271–9.

Nakamura, S., Yao, T., Aoyagi, K., Iida, M., Fujishima, M., and Tsuneyoshi, M. (1997). Helicobacter pylori and primary gastric lymphoma. A histopathologic and immunohistochemical analysis of 237 patients. Cancer. 79:3–11.

Nakamura, S., Ye, H., Bacon, C.M., Goatly, A., Liu, H., Banham, A.H., Ventura, R., Matsumoto, T., Iida, M., Ohji, Y., Yao, T., Tsuneyoshi, M., and Du, M.Q. (2007). Clinical impact of genetic aberrations in gastric MALT lymphoma: a comprehensive analysis using interphase fluorescence in situ hybridisation. Gut. 56:1358–63.

Neubauer, A., Thiede, C., Morgner, A., Alpen, B., Ritter, M., Neubauer, B., Wündisch, T., Ehninger, G., Stolte, M., and Bayerdörffer, E. (1997). Cure of Helicobacter pylori infection and duration of remission of low-grade gastric mucosa-associated lymphoid tissue lymphoma. J Natl Cancer Inst. 89:1350–5.

Neumeister, P., Hoefler, G., Beham-Schmid, C., Schmidt, H., Apfelbeck, U., Schaider, H., Linkesch, W., and Sill, H. (1997). Deletion analysis of the p16 tumor suppressor gene in gastrointestinal mucosa-associated lymphoid tissue lymphomas. Gastroenterology. 112:1871–5.

Nieters, A., Beckmann, L., Deeg, E., and Becker, N. (2006). Gene polymorphisms in Toll-like receptors, interleukin-10, and interleukin-10 receptor alpha and lymphoma risk. Genes Immunol. 7:615–24.

Nobre-Leitao, C., Lage, P., Cravo, M., Cabecadas, J., Chaves, P., Alberto-Santos, A., Correia, J., Soares, J., and Costa-Mira, F. (1998). Treatment of gastric MALT lymphoma by Helicobacter pylori eradication: a study controlled by endoscopic ultrasonography. Am J Gastroenterol. 93:732–6.

O'Rourke, J.L., Dixon, M.F., Jack, A., Enno, A., and Lee, A. (2004). Gastric B-cell mucosa-associated lymphoid tissue (MALT) lymphoma in an animal model of 'Helicobacter heilmannii' infection. J Pathol. 203:896–903.

Papa, A., Cammarota, G., Tursi, A., Gasbarrini, A., and Gasbarrini, G. (2000). Helicobacter pylori eradication and remission of low-grade gastric mucosa associated lymphoid tissue lymphoma: a long-term follow-up study. J Clin Gastroenterol. 31:169–71.

Parsonnet, J. (2003). What is the Helicobacter pylori global reinfection rate? Can J Gastroenterol. 17:46B–8B.

Parsonnet, J., Hansen, S., Rodriguez, L., Gelb, A.B., Warnke, R.A., Jellum, E., Orentreich, N., Vogelman, J.H., and Friedman, G.D. (1994). Helicobacter pylori infection and gastric lymphoma. N Engl J Med. 330:1267–71.

Peng, H., Du, M., Diss, T.C., Isaacson, P.G., and Pan, L. (1997). Genetic evidence for a clonal link between low and high-grade components in gastric MALT B-cell lymphoma. Histopathology. 30:425–9.

Peng, H., Ranaldi, R., Diss, T.C., Isaacson, P.G., Bearzi, I., and Pan, L. (1998). High frequency of CagA+ Helicobacter pylori infection in high-grade gastric MALT B-cell lymphomas. J Pathol. 185:409–12.

Qin, Y., Greiner, A., Hallas, C., Haedicke, W., and Muller-Hermelink, H.K. (1997). Intraclonal offspring expansion of gastric low-grade MALT-type lymphoma: evidence for the role of antigen-driven high-affinity mutation in lymphomagenesis. Lab Invest. 76:477–85.

Qin, Y., Greiner, A., Trunk, M.J., Schmausser, B., Ott, M.M., Müller-Hermelink, H.K. (1995). Somatic hypermutation in low-grade mucosa-associated lymphoid tissue-type B-cell lymphoma. Blood. 86:3528–34.

Raderer, M., Wohrer, S., Streubel, B., Drach, J., Jager, U., Turetschek, K., Troch, M., Puspok, A., Zielinski, C.C., and Chott, A. (2006). Activity of rituximab plus cyclophosphamide, doxorubicin/mitoxantrone, vincristine and prednisone in patients with relapsed MALT lymphoma. Oncology. 70:411–7.

Rohatiner, A., d'Amore, F., Coiffier, B., Crowther, D., Gospodarowicz, M., Isaacson, P., Lister, T.A., Norton, A., Salem, P., Shipp, M., et al. (1994). Report on a workshop convened to discuss the pathological and staging classifications of gastrointestinal tract lymphoma. Ann Oncol. 5:397–400.

Rollinson, S., Levene, A.P., Mensah, F.K., Roddam, P.L., Allan, J.M., Diss, T.C., Roman, E., Jack, A., MacLennan, K., Dixon, M.F., and Morgan, G.J. (2003). Gastric marginal zone lymphoma is associated with polymorphisms in genes involved in inflammatory response and antioxidative capacity. Blood. 102:1007–11.

Rosenstiel, P., Hellmig, S., Hampe, J., Ott, S., Till, A., Fischbach, W., Sahly, H., Lucius, R., Fölsch, U.R., Philpott, D., and Schreiber, S. (2006). Influence of polymorphisms in the NOD1/CARD4 and NOD2/CARD15 genes on the clinical outcome of Helicobacter pylori infection. Cell Microbiol. 8:1188–98.

Rudolph, B., Bayerdörffer, E., Ritter, M., Müller, S., Thiede, C., Neubauer, B., Lehn, N., Seifert, E., Otto, P., Hatz, R., Stolte, M., and Neubauer, A. (1997). Is the polymerase chain reaction or cure of Helicobacter pylori infection of help in the differential diagnosis of early gastric mucosa-associated lymphatic tissue lymphoma? J Clin Oncol. 15:1104–9.

Ruland, J., Duncan, G.S., Elia, A., del Barco Barrantes, I., Nguyen, L., Plyte, S., Millar, D.G., Bouchard, D., Wakeham, A., Ohashi, P.S., and Mak, T.W. (2001). Bcl10 is a positive regulator of antigen receptor-induced activation of NF-κB and neural tube closure. Cell. 104:33–42.

Ruskoné-Fourmestraux, A., Dragosics, B., Morgner, A., Wotherspoon, A., and De Jong, D. (2003). Paris staging system for primary gastrointestinal lymphomas. Gur. 52(6):912–3.

Ruskoné-Fourmestraux, A., Lavergne, A., Aegerter, P.H., Megraud, F., Palazzo, L., de Mascarel, A., Molina, T., and Rambaud, J.L. (2001). Predictive factors for regression of gastric MALT lymphoma after anti-Helicobacter pylori treatment. Gut. 48:297–303.

Sagaert, X., De Wolf-Peeters, C., Noels, H., and Baens, M. (2007). The pathogenesis of MALT lymphomas: where do we stand? Leukemia. 21:389–96.

Sagaert, X., Theys, T., De Wolf-Peeters, C., Marynen, P., and Baens, M. (2006). Splenic marginal zone lymphoma-like features in API2-MALT1 transgenic mice that are exposed to antigenic stimulation. Haematologica. 91:1693–6.

Salles, G., Herbrecht, R., Tilly, H., Berger, F., Brousse, N., Gisselbrecht, C., and Coiffier, B. (1991). Aggressive primary gastrointestinal lymphomas: review of 91 patients treated with the LNH-84 regimen. A study of the Groupe d'Etude des Lymphomes Agressifs. Am J Med. 90:77–84.

Sanchez-Izquierdo, D., Buchonnet, G., Siebert, R., Gascoyne, R.D., Climent, J., Karran, L., Marin, M., Blesa, D., Horsman, D., Rosenwald, A., Staudt, L.M., Albertson, D.G., Du, M.Q., Ye, H., Marynen, P., Garcia-Conde, J., Pinkel, D., Dyer, M.J., and Martinez-Climent, J.A. (2003). MALT1 is deregulated by both chromosomal translocation and amplification in B-cell non-Hodgkin lymphoma. Blood. 101:4539–46.

Savio, A., Franzin, G., Wotherspoon, A.C., Zamboni, G., Negrini, R., Buffoli, F., Diss, T.C., Pan, L., and Isaacson, P.G. (1996). Diagnosis and posttreatment follow-up of Helicobacter pylori-positive gastric lymphoma of mucosa-associated lymphoid tissue: histology, polymerase chain reaction, or both? Blood. 87:1255–60.

Saxena, A., Moshynska, O., Kanthan, R., Bhutani, M., Maksymiuk, A.W., and Lukie, B.E. (2000). Distinct B-cell clonal bands in Helicobacter pylori gastritis with lymphoid hyperplasia. J Pathol. 190:47–54.

Schechter, N.R., Portlock, C.S., and Yahalom, J. (1998). Treatment of MALT lymphoma of the stomach with radiation alone. J Clin Oncol. 16:1916–21.

Spencer, J., Finn, T., Pulford, K.A., Mason, D.Y., and Isaacson, P.G. (1985). The human gut contains a novel population of B lymphocytes which resemble marginal zone cells. Clin. Exp. Immunol. 62(3):607–12.

Starostik, P., Patzner, J., Greiner, A., Schwarz, S., Kalla, J., Ott, G., Müller-Hermelink, H.K. (2002). Gastric marginal zone B-cell lymphomas of MALT type develop along 2 distinct pathogenetic pathways. Blood. 99:3–9.

Steiff, J.N., Neubauer, A., Stolte, M., and Wündisch, T. (2006). Clonality analyses in gastric MALT (mucosa-associated lymphoid tissue). Pathol Res Pract. 202:503–7.

Steinbach, G., Ford, R., Glober, G., Sample, D., Hagemeister, F.B., Lynch, P.M., McLaughlin, P.W., Rodriguez, M.A., Romaguera, J.E., Sarris, A.H., Younes, A., Luthra, R., Manning, J.T., Johnson, C.M., Lahoti, S., Shen, Y., Lee, J.E., Winn, R.J., Genta, R.M., Graham, D.Y., and Cabanillas, F.F. (1999). Antibiotic treatment of gastric lymphoma of mucosa-associated lymphoid tissue. An uncontrolled trial. Ann Intern Med. 131:88–95.

Streubel, B., Lamprecht, A., Dierlamm, J., Cerroni, L., Stolte, M., Ott, G., Raderer, M., and Chott, A. (2003). T(14;18)(q32;q21) involving IGH and MALT1 is a frequent chromosomal aberration in MALT lymphoma. Blood. 101:2335–9.

Sutton, P., O'Rourke, J., Wilson, J., Dixon, M.F., and Lee, A. (2004). Immunisation against Helicobacter felis infection protects against the development of gastric MALT lymphoma. Vaccine. 22:2541–6.

Thieblemont, C. (2005). Clinical presentation and management of marginal zone lymphomas. Hematology. 307–13.

Thiede, C., Alpen, B., Morgner, A., Schmidt, M., Ritter, M., Ehninger, G., Stolte, M., Bayerdörffer, E., and Neubauer, A. (1998). Ongoing somatic mutations and clonal expansions after cure of Helicobacter pylori infection in gastric mucosa-associated lymphoid tissue B-cell lymphoma. J Clin Oncol. 16:3822–31.

Thiede, C., Wündisch, T., Alpen, B., Neubauer, B., Morgner, A., Schmitz, M., Ehninger, G., Stolte, M., Bayerdörffer, E., Neubauer, A, and German MALT Lymphoma Study Group. (2001). Long-term persistence of monoclonal B cells after cure of Helicobacter pylori infection and complete histologic remission in gastric mucosa-associated lymphoid tissue B-cell lymphoma. J. Clin. Oncol. 19(6):1600–9.

Thiede, C., Wündisch, T., Neubauer, B., Alpen, B., Morgner, A., Ritter, M., Ehninger, G., Stolte, M., Bayerdörffer, E., and Neubauer, A. (2000). Eradication of Helicobacter pylori and stability of remissions in low-grade gastric B-cell lymphomas of the mucosa-associated lymphoid tissue: results of an ongoing multicenter trial. Recent Results Cancer Res. 156:125–33.

Tsang, R.W., Gospodarowicz, M.K., Pintilie, M., Wells, W., Hodgson, D.C. Sun, A., Crump, M., and Patterson, B.J. (2003). Localized MALT lymphoma treated with radiation therapy has excellent clinical outcome. J Clin Oncol. 21:4157–64.

Tursi, A., Cammarota, G., Papa, A., Cuoco, L., Fedeli, G., and Gasbarrini, G. (1997). Long-term follow-up of disappearance of gastric mucosa-associated lymphoid tissue after anti Helicobacter pylori therapy. Am J Gastroenterol. 92:1849–52.

Willis, T.G., Jadayel, D.M., Du, M.Q., Peng, H., Perry, A.R., Abdul-Rauf, M., Price, H., Karran, L., Majekodunmi, O., Wlodarska, I., Pan, L., Crook, T., Hamoudi, R., Isaacson, P.G., and Dyer, M.J. (1999). Bcl10 is involved in t(1;14)(p22;q32) of MALT B-cell lymphoma and mutated in multiple tumor types. Cell. 96:35–45.

Witzig, T., Gordon, L., Emmanouilides, C., et al. (2001). Safety and efficacy of Zevalin in four patients with mucosa associated lymphoid tissue (MALT) lymphoma. Blood. 98:254b.

Wöhrer, S., Drach, J., Hejna, M., Scheithauer, W., Dirisamer, A., Püspök, A., Chott, A., and Raderer, M. (2003). Treatment of extranodal marginal zone B-cell lymphoma of mucosa-associated lymphoid tissue (MALT lymphoma) with mitoxantrone, chlorambucil and prednisone (MCP). Ann Oncol. 14:1758–61.

Wotherspoon, A.C., Doglioni, C., Diss, T.C., Pan, L., Moschini, A., de Boni, M., and Isaacson, P.G. (1993). Regression of primary low grade B-cell gastric lymphoma of mucosa associated lymphoid tissue type after eradication of Helicobacter pylori. Lancet. 342:575–7.

Wotherspoon, A.C., Doglioni, C., and Isaacson, P.G. (1992). Low grade gastric B-cell lymphoma of mucosa-associated lymphoid tissue (MALT): a multifocal disease. Histopathology. 20:29–34.

Wotherspoon, A.C., Ortiz Hidalgo, C., Falzon, M.R., and Isaacson, P.G. (1991). Helicobacter pylori-associated gastritis and primary B-cell gastric lymphoma. Lancet. 338:1175–6.

Wu, M.S., Chen, L.T., Shun, C.T., Huang, S.P., Chiu, H.M., Wang, H.P., Lin, M.T., Cheng, A.L., and Lin, J.T. (2004). Promoter polymorphisms of tumor necrosis factor-alpha are associated with risk of gastric mucosa-associated lymphoid tissue lymphoma. Int J Cancer. 110:695–700.

Wuendisch, T., Neubauer, A., Stolte, M., Ritter, M., and Thiede, C. (2003). B-cell monoclonality is associated with lymphoid follicles in gastritis. J Surg Pathol. 27:882–7.

Wuendisch, T., Thiede, C., Morgner, A., Dempfle, A., Günther, A., Liu, H., Ye, H., Du, M., Kim, T., Bayerdörffer, E., Stolte, M., and Neubauer, A. (2005). Long-Term follow-up of gastric MALT lymphoma after Helicobacter pylori eradication. J Clin Oncol. 31:8018–24.

Xue, L., Morris, S.W., Orihuela, C., Tuomanen, E., Cui, X., Wen, R., and Wang, D. (2003). Defective development and function of Bcl10-deficient follicular, marginal zone and B1 B cells. Nat Immunol. 4:857–65.

Yahalom, J. (2001). MALT lymphomas: a radiation oncology viewpoint. Ann Hematol. 80: B100–5.

Ye, H., Liu, H., Attygalle, A., Wotherspoon, A.C., Nicholson, A.G., Charlotte, F., Leblond, V., Speight, P., Goodlad, J., Lavergne-Slove, A., Martin-Subero, J.I., Siebert, R., Dogan, A., Isaacson, P.G., and Du, M.Q. (2003). Variable frequencies of t(11;18)(q21;q21) in MALT lymphomas of different sites: significant association with CagA strains of H. pylori in gastric MALT lymphoma. Blood. 102:1012–8.

Zhang, Q., Siebert, R., Yan, M., Hinzmann, B., Cui, X., Xue, L., Rakestraw, K.M., Naeve, C.W., Beckmann, G., Weisenburger, D.D., Sanger, W.G., Nowotny, H., Vesely, M., Callet-Bauchu, E., Salles, G., Dixit, V.M., Rosenthal, A., Schlegelberger, B., Morris, S.W. (1999). Inactivating mutations and overexpression of BCL10, a caspase recruitment domain-containing gene, in MALT lymphoma with t(1;14)(p22;q32). Nat Genet. 22:63–8.

Zinzani, P.L., Stefoni, V., Musuraca, G., Tani, M., Alinari, L., Gabriele, A., Marchi, E., Pileri, S., and Baccarani, M. (2004). Fludarabine-containing chemotherapy as frontline treatment of nongastrointestinal mucosa-associated lymphoid tissue lymphoma. Cancer. 100:2190–4.

Chapter 5
Gastrointestinal Stromal Tumors of Gastric Origin

Chandrajit P. Raut, Jason L. Hornick, and Monica M. Bertagnolli

Introduction

Gastrointestinal stromal tumors (GISTs) are rare neoplasms. Although they account for only 0.1%–3% of all gastrointestinal (GI) malignancies (Crosby et al. 2001; DeMatteo et al. 2000; Lewis and Brennan 1996; Nishida and Hirota 2000), they represent 80% of GI mesenchymal tumors (Miettinen and Lasota 2001). GISTs may arise from anywhere along the GI tract and exhibit a broad spectrum of clinical behavior. Many are asymptomatic, discovered incidentally during imaging, endoscopy, or laparotomy for unrelated reasons. GISTs exhibit a broad spectrum of clinical activity. Some remain stable for years, whereas others progress rapidly (DeMatteo et al. 2000; Fletcher et al. 2002). Metastases may be identified in 15%–50% of GISTs at the time of diagnosis (Bauer et al. 2003; Roberts and Eisenberg 2002).

Histologic and Molecular Classification

Until recently, the diagnostic criteria for GISTs were poorly defined because of their variable natural history, heterogeneous immunophenotype, and the lack of a specific marker for this tumor type. The term "GIST" was initially a descriptive, generic term, first used in 1983 by Mazur and Clark (Mazur and Clark 1983) to define intraabdominal nonepithelial tumors composed of cells that failed to exhibit either the ultrastructural or immunohistochemical features of smooth muscle or neural tissues. Based on their histologic and immunohistochemical features, GISTs are believed to arise from the interstitial cells of Cajal (ICC), which are myoepithelial components of the intestinal autonomic nervous system that serve as pacemakers regulating intestinal peristalsis (Kindblom et al. 1998).

T.C. Wang et al. (eds.) *The Biology of Gastric Cancers*,
© Springer Science + Business Media, LLC 2009

Histology

Hematoxylin and eosin staining of typical examples of GISTs are shown in Figure 5.1. Three microscopic GIST morphologies have been identified: spindle cell type (70%), epithelioid type (20%), or a mixed type reflecting a combination of both (10%) (Fletcher et al. 2002). The most common spindle cell GISTs contain short fascicles of uniform, bland spindle cells with tapering nuclei and palely eosinophilic, fibrillary cytoplasm. A subset of gastric spindle cell GISTs show distinctive paranuclear vacuoles. Epithelioid GISTs are composed of round cells with abundant clear or eosinophilic cytoplasm, arranged in a sheet-like or nested architecture. GISTs of either type are notable for their homogeneous cytomorphology, although a minority of GISTs, in particular those of epithelioid type, show pleomorphism.

Pathologists have long recognized that GISTs show considerable variability in expression of differentiation antigens used as markers for muscle cells (smooth muscle actin) and nerve sheath cells (S-100). Based on the patterns of expression of these cell lineage markers, these tumors were originally variably classified as leiomyomas, "leiomyoblastomas," leiomyosarcomas, Schwannomas, or "GI autonomic nerve tumors" (Miettinen et al. 2002). It is likely that in the past several different types of epithelioid and spindle cell tumors were included in the diagnostic category of GIST, whereas many true GIST cases were inadvertently placed into other categories. In a population-based review of patients diagnosed with GIST before 2000, investigators in Sweden found that two-thirds of these tumors had been misclassified as smooth muscle tumors (leiomyomas, "leiomyoblastomas," or leiomyosarcomas), and only 28% were correctly identified as GISTs (Nilsson et al. 2005). As a result, clinical reports for GIST published before 2000 are difficult to interpret.

KIT Receptor Tyrosine Kinase Mutations

In 1998, gain-of-function mutations in the c-*KIT* protooncogene were first identified in GISTs (Hirota et al. 1998). *KIT* encodes the KIT receptor tyrosine kinase (RTK), a transmembrane protein that regulates intracellular signal transduction pathways when activated via binding by its ligand, stem cell factor, or steel factor (Savage and Antman 2002). KIT has a critical role in the development and maintenance of

---

Fig. 5.1 Histology of gastrointestinal stromal tumors (GISTs). Hematoxylin and eosin (H&E)-stained sections of (**A**) spindle cell GIST, composed of short fascicles of elongated cells with palely eosinophilic, fibrillary cytoplasm, and (**B**) epithelioid GIST, composed of sheets of polygonal cells with clear to eosinophilic cytoplasm. Both variants are notable for their uniform cytology. Spindle cell GISTs show similar histologic features as leiomyosarcomas (**C**), although the latter contain longer fascicles of spindle cells with brightly eosinophilic cytoplasm (*See Color Plates*)

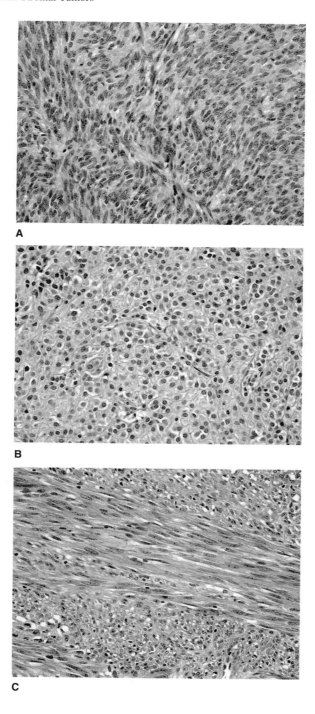

A

B

C

hematopoiesis, germ cells, and intestinal pacemaker cells (Rubin et al. 2001; Tian et al. 1999; Ward et al. 1994). Constitutive activation of KIT by oncogenic mutations has been identified not only in GISTs, but also in neoplasms derived from these other lineages, including systemic mastocytosis, acute myeloid leukemia, and seminoma (Rubin et al. 2001).

Approximately 85% of GISTs harbor *KIT* mutations, resulting in constitutive KIT RTK activation (Rubin et al. 2001). These mutations occur in exon 11 (in 57%–71% of cases), exon 9 (10%–18%), exon 13 (1%–4%), or exon 17 (1%–4%) (Corless et al. 2004; De Giorgi and Verweij 2005; Heinrich et al. 2002, 2003b). *KIT* mutations usually affect the juxtamembrane region and TK domain (Hirota et al. 1998). This results in ligand-independent tyrosine autophosphorylation of KIT, receptor dimerization, and stimulation of downstream signal transduction with subsequent unregulated cell growth and malignant transformation (Hirota et al. 1998).

Immunohistochemistry

KIT expression can be detected by immunohistochemical staining through reactivity with the CD117 antigen, a part of the receptor, providing a specific and reliable diagnostic marker (Figure 5.2). CD117 is expressed in approximately 95% of GISTs, but not in other GI smooth muscle tumors, which typically express high levels of desmin and smooth muscle actin (Fletcher et al. 2002; Rubin et al. 2001). Immunohistochemical analysis for KIT is routinely performed to confirm a suspected diagnosis of GIST. Other (nonspecific) markers frequently expressed in GIST include CD34 (60%–70%), smooth muscle actin (30%–40%), and S-100 (5%). Approximately 5% of GISTs are negative for KIT expression, so a diagnosis of GIST based on other clinicopathologic features should not be excluded based solely on the absence of KIT staining (Medeiros et al. 2004).

Additional Molecular Factors Relevant to Diagnosis and Treatment

A small proportion of GISTs lacking *KIT* mutations nevertheless exhibit constitutively activated KIT protein (Rubin et al. 2001). Furthermore, tumors negative for KIT by immunohistochemistry may still harbor *KIT* mutations (Emile et al. 2004). Approximately 5% of GISTs have activating mutations in the gene encoding a related RTK, *platelet-derived growth factor receptor alpha* (*PDGFRA*) (Heinrich et al. 2003b; Hirota et al. 2003; Medeiros et al. 2004). *PDGFRA* mutations have been identified in exon 18 (2%–6% of tumors), exon 12 (1%–2%), and exon 14

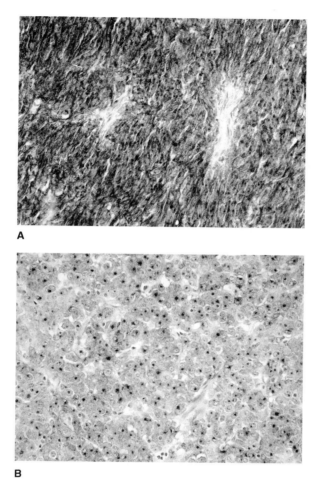

Fig. 5.2 KIT expression in gastrointestinal stromal tumors (GISTs). By immunohistochemistry, 95% of GISTs are positive for KIT (CD117), which is usually predominantly cytoplasmic in distribution (**A**), but may also show paranuclear dot-like staining (**B**) (*See Color Plates*)

(<1%) (Figure 5.3) (Corless et al. 2005; Heinrich et al. 2003b). Finally, 5%–10% of GISTs lack both *KIT* and *PDGFRA* mutations.

Tumor progression studies suggest that, in most cases, mutation of *KIT* or *PDGFRA* is the earliest alteration in a series of genetic losses leading to GIST formation. Animal models and studies of familial GIST kindreds suggest that mutation of *KIT* or *PDGFRA* alone is not sufficient for tumor formation, although this condition does produce ICC hyperplasia (Corless et al. 2004). The transition from ICC hyperplasia to low-risk GIST is often accompanied by 14q deletion (Corless et al. 2004),

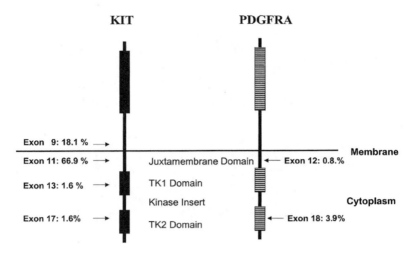

Fig. 5.3 KIT and platelet-derived growth factor receptor alpha (PDGFRA) mutations in GIST. KIT and PDGFRA mutations in GIST produce constitutive ligand-independent receptor activation. Response to tyrosine kinase inhibitors correlates with the location of the activating mutation, with best response in patients whose tumors contain mutations in KIT exon 11. (Reprinted with permission from ASCO. Heinrich et al. 2003a)

and additional loss of 22q is frequently observed in high-risk or metastatic GIST (Corless et al. 2004).

Several components of signaling downstream of activated KIT RTK have been identified. These include phosphorylation of the Grb and PI_3K binding sites of KIT and activation of mitogen-activated protein kinase and Akt (Heinrich et al. 2003a; Hirota et al. 2003). Inhibitor studies have also identified KIT-associated activation of mammalian target of rapamycin signaling (Maki 2004).

Epidemiology

Incidence

Epidemiologic data from the Surveillance, Epidemiology, and End Results (SEER) program from the National Cancer Institute are difficult to interpret because many GI mesenchymal tumors currently classified as GISTs were previously classified as GI smooth muscle neoplasms (Perez et al. 2006). Recently published data provide some new insights concerning the true incidence of GIST. Analysis of the current SEER data (April 2005 release) suggests a near doubling of the incidence of all GI mesenchymal tumors, from 0.17 per 100,000 in 1992 to 0.31 per 100,000 in 2002

(Perez et al. 2006). An estimated 80% of these tumors are GISTs. Experience from ongoing clinical trials for patients with GIST also suggests a higher incidence of 4,500–6,000 new cases per year in the United States (15–20 cases per million population) (Fletcher et al. 2002; Kindblom et al. 2003). Similar values have emerged from several European population–based studies, which include GISTs detected at autopsy. Reevaluation of all potential cases of GIST in one province in Sweden (1983–2000), in Girona, Spain (1994–2001), and in the entire country of Iceland (1990–2003) identified annual incidence rates of 14.5, 10.9, and 11 cases per million population, respectively (Nilsson et al. 2005; Rubio et al. 2007; Tryggvason et al. 2005). In data covering through the year 2003, a Dutch study retrieved records of all tumors diagnosed as GIST from a national pathology registry and reported that the annual incidence increased from 2.1 cases per million inhabitants in 1995 to 12.7 cases per million inhabitants (Goettsch et al. 2005). This change was attributed to both a better understanding of GIST pathobiology and the availability of CD117 staining. These studies imply that GISTs are underrecognized. Two recent studies demonstrated that small, subcentimeter GISTs are quite common in the adult population. Careful analysis of 100 total gastrectomy specimens in patients treated for gastric cancer in Japan uncovered 50 small GISTs in 35 of the patients (Kawanowa et al. 2006). Similarly, in a German series of consecutive autopsies performed on individuals older than 50 years, subcentimeter GISTs were identified in 22.5% (Agaimy et al. 2007). These tumors were KIT positive and often contained either *KIT* or *PDGFRA* mutations (Agaimy et al. 2007). Thus, GISTs are much more prevalent than previously expected, but the majority are very small tumors that are unlikely to become clinically significant.

Familial Gastrointestinal Stromal Tumors

Germline *KIT* mutations are exceedingly rare. Individuals carrying this defect demonstrate an autosomal dominant predilection for developing multiple GISTs (Beghini et al. 2001; Carney and Stratakis 2002; Hirota et al. 2002; Isozaki et al. 2000; Li et al. 2005; Maeyama et al. 2001; Nishida et al. 1998; Robson et al. 2004). Affected individuals are younger than patients with sporadic GISTs, but metastases are more uncommon than in sporadic cases (Robson et al. 2004). Specific *KIT* mutations in these kindreds vary, but usually involve the *KIT* juxtamembrane region (exon 11) (Corless et al. 2004; Isozaki et al. 2000; Li et al. 2005; Maeyama et al. 2001). A GIST kindred with a germline mutation in *PDGFRA* has also been described (Chompret et al. 2004).

Carney's triad is the association of paragangliomas, pulmonary chondromas, and gastric lesions that were classified as "leiomyosarcomas" in the initial report of this condition in seven unrelated young women (Carney 1999). Carney's triad occurs predominantly in women, and affected individuals are generally diagnosed with tumors before age 30 years. The gastric tumors characteristic of this syndrome are now known to represent epithelioid GISTs, based on their morphology and the presence of KIT immunostaining. Mutational analyses, however, have failed to document the presence of activating *KIT* or *PDGFRA* mutations in patients with Carney's triad (Amieux 2004).

Multifocal GISTs are detected in approximately 7% of individuals with von Recklinghausen's neurofibromatosis (NF1), with the small intestine being the most common site (Miettinen et al. 2006; Shinomura et al. 2005; Zoller et al. 1997). In the past, these tumors were frequently misclassified as leiomyomas or autonomic nerve tumors. The vast majority of NF1-associated GISTs are wild-type for both *KIT* and *PDGFRA*; point mutations in *KIT* and *PDGFRA* have been reported in only 8% and 6% of GISTs, respectively, from NF1 patients (Takazawa et al. 2005). In contrast, mutations of the *NF1* gene have not been identified in GISTs in non-NF1 individuals (Kinoshita et al. 2004). The molecular pathogenesis and natural history of GISTs in patients with NF1 remain unknown.

Clinical Presentation

GISTs usually occur in adults, with a median age at diagnosis of 60 years (range 40–80 years) (DeMatteo et al. 2000; Nilsson et al. 2005). Men are affected slightly more often than women (van der Zwan and DeMatteo 2005). These tumors are very rare in children, but can be seen in young patients with familial GIST or as part of Carney's triad (Miettinen and Lasota 2003).

Primary GISTs may arise throughout the GI tract, but are most common in the stomach (50%–70%), followed by small bowel (25%–35%), colon and rectum (5%–10%), mesentery or omentum (7%), and esophagus (<5%) (Emory et al. 1999; Miettinen et al. 2002). Individuals with familial GISTs may have skin hyperpigmentation, dysphagia, multicentric paragangliomas, and diffuse hyperplasia of the intestinal myenteric plexus (Beghini et al. 2001; Carney and Stratakis 2002; Hirota et al. 2002; Isozaki et al. 2000; Li et al. 2005; Maeyama et al. 2001; Nishida et al. 1998; Robson et al. 2004). Small GISTs (≤2 cm) may be asymptomatic, and are often detected incidentally on radiographic studies, endoscopy, or laparotomy. In a study of 288 primary GISTs, 69% of tumors were symptomatic (median tumor size 8.9 cm), 21% were incidentally discovered at laparotomy (median tumor size 2.7 cm), and 10% were identified at autopsy (median tumor size 3.4 cm) (Nilsson et al. 2005). GISTs can be highly vascular and friable, and even small lesions can erode through the intestinal mucosa and produce significant GI bleeding. This is a relatively common presentation, leading to diagnosis of otherwise asymptomatic small GISTs. Like all nonmucosal tumors, GISTs generally expand into the abdominal cavity or retroperitoneum and rarely compromise the intestinal lumen. As a result, they can reach a very large size before producing symptoms. Nonspecific symptoms that may lead to further investigation include nausea, vomiting, abdominal discomfort or pain, or distention (Figure 5.4A). Intraabdominal or intraluminal hemorrhage can occur, and rupture into the peritoneal cavity is possible. This life-threatening complication also carries a high risk of dissemination by peritoneal

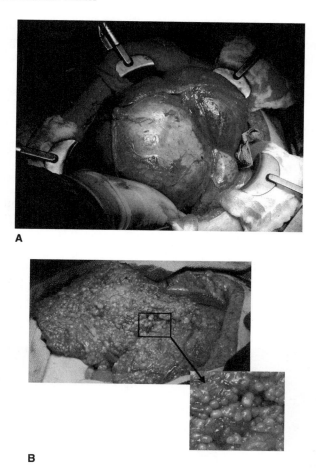

Fig. 5.4 Manifestations of metastatic gastrointestinal stromal tumor (GIST). Photographs taken during debulking surgery show (**A**) complete replacement of the right hepatic lobe by a single metastatic lesion; and (**B**) peritoneal carcinomatosis with complete replacement of the omentum by innumerable GIST nodules (*See Color Plates*)

seeding of the tumor. Obstruction of the GI tract is occasionally a presenting condition, and can lead to perforation.

Between 15% and 50% of GISTs present with overtly metastatic disease, usually to the liver and peritoneum (Bauer et al. 2003; Roberts and Eisenberg 2002). The pattern of metastatic spread is almost entirely intraabdominal, with fewer than 5% of patients developing pulmonary metastases. GISTs virtually never spread to regional lymph nodes, although they may invade adjacent organs such as intestine, liver, or bladder. Diffuse peritoneal spread is not uncommon, and may manifest as innumerable small tumor nodules essentially replacing the omentum, studding the diaphragmatic surface and peritoneal lining, or covering the serosal surfaces of the bowel (Figure 5.4B). In these cases, tumor-associated ascites is typical. It is not unusual to encounter tumor nodules in abdominal incisions, growing through the weakened tissue of a previous incision site. Large, necrotic tumors occasionally form fistulas with bowel, biliary tree, or skin.

Diagnostic Studies

Endoscopy, Fine-Needle Aspiration, and Biopsy

On endoscopy, GISTs can appear as submucosal masses with or without ulceration. These lesions are visually indistinguishable from other GI tumors of smooth muscle origin (Figure 5.5). Size and extent of disease may be determined by endoscopic ultrasound (EUS). Endoscopic biopsies might not obtain sufficient tissue to establish a definitive diagnosis, capturing only normal

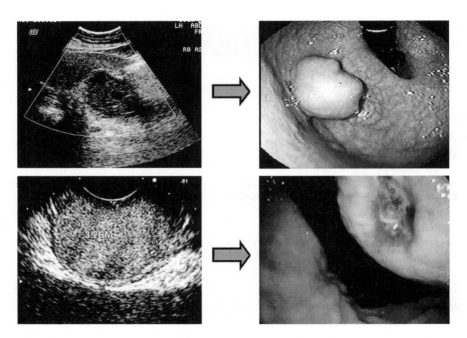

Fig. 5.5 Endoscopic examination of primary gastric gastrointestinal stromal tumor (GIST). GISTs often present as submucosal tumors in the wall of the gastrointestinal (GI) tract. Central necrosis of high-risk tumors or erosion into blood vessels can produce local inflammation or significant GI bleeding (bottom, right) (*See Color Plates*)

mucosa without sampling the underlying submucosal lesion. EUS-guided fine-needle aspiration (FNA) might also fail to yield enough tissue for diagnosis, but might be helpful to exclude other submucosal lesions (Demetri and Morgan 2005). However, cytologic morphology combined with immunohistochemistry performed on cell block material may establish a diagnosis from an EUS–FNA specimen (Rader et al. 2001).

Preoperative biopsy to establish a diagnosis is not always necessary, and the decision to proceed with biopsy should be approached judiciously. If a suspected GIST presents as a solitary lesion and appears resectable, surgery may be undertaken without confirmatory biopsy. However, if the specific histologic diagnosis would impact treatment (for instance, distinguishing a gastric GIST from gastric lymphoma), or if unresectable or metastatic disease is present, an endoscopic or computed tomography (CT)- or ultrasound-guided biopsy is warranted. If endoscopic or image-guided attempts at biopsy are nondiagnostic, laparoscopic or open biopsy or resection may be necessary to guide further therapy.

Radiographic Studies

Contrast-enhanced CT is the imaging modality of choice for the initial evaluation of an abdominal mass or abdominal symptoms. CT scans are critical in determining the anatomic extent of a tumor, assisting with operative planning, and identifying recurrent or metastatic disease during routine surveillance after resection (Demetri and Delaney 2001). CT findings vary with the size of the lesions (Figure 5.6A–D). Primary GISTs are typically well-circumscribed, brightly enhancing masses arising from the GI tract as extraluminal, occasionally exophytic masses. Small GISTs have sharp margins, an intraluminal growth pattern, and are of homogeneous density on both unenhanced and contrast-enhanced scans (Ghanem et al. 2003). In contrast, larger tumors may have irregular margins, extraluminal growth patterns, and a more heterogeneous appearance caused by necrosis, hemorrhage, or areas of degeneration (Demetri and Morgan 2005). The origin of a large mass may be difficult to identify (Demetri and Morgan 2005). Radiographic signs corresponding to a more aggressive GIST include ulceration, necrosis, fistula formation, metastasis, ascites, and signs of infiltration of local tissues. Unfortunately, CT is unable to differentiate between inflammatory adhesions and malignant involvement of adjoining organs. It is also unlikely to identify any peritoneal metastases smaller than 2 cm in diameter.

[^{18}F]fluoro-2-deoxy-D-glucose (^{18}FDG) positron emission tomography (PET) is a functional imaging approach, which reveals metabolic activity within tumor cells via accumulation of ^{18}FDG. PET can serve as a useful adjunct to CT for the initial detection of the extent of disease and the characterization of ambiguous masses. Magnetic resonance imaging is not routinely used for the evaluation of suspected GIST.

Prognostic Factors

Risk Stratification

GISTs exhibit a wide spectrum of clinical behavior. Small lesions may remain stable for years, but some lesions rapidly progress to metastatic disease. A risk stratification schema predictive of progressive disease was initially assembled by consensus during a National Institute of Health/National Cancer Institute–sponsored workshop in 2001 (Fletcher et al. 2002). Tumors were categorized as "very low," "low," "intermediate," and "high" based on tumor size and mitotic count (per 50 high-power field). GISTs

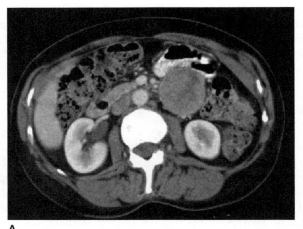

A

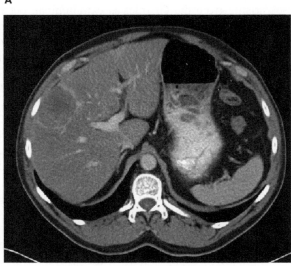

B

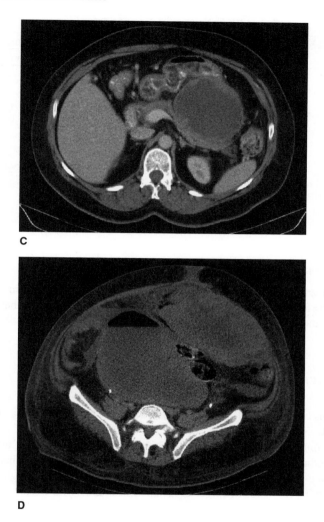

Fig. 5.6 Computed tomography characteristics of a gastrointestinal stromal tumor (GIST). Various manifestations of GIST include: (**A**) heterogeneous primary tumor; (**B**) hepatic metastasis; (**C**) cystic lesion; and (**D**) large necrotic GIST with enterocutaneous fistula

in the high-risk group demonstrate a high rate of recurrence (83%) and a high rate of tumor-related deaths (63%). Lower-risk GISTs are also associated with tumor-related deaths in 1.2% of patients with lower-risk tumors (Nilsson et al. 2005). Furthermore, up to 20% of small GISTs (<5 cm) exhibit metastatic behavior (Rossi et al. 2003). Thus, it is difficult to label a particular GIST as either benign or malignant based on histology alone. All GISTs are now regarded as having malignant potential; use of the term "benign GIST" is discouraged.

Recent studies have further clarified risk of progression. In a series of 1,055 gastric, 629 small intestinal, 144 duodenal, and 111 rectal GISTs from the pre-imatinib era reported from the Armed Forces Institute of Pathology, the risk of recurrence

Table 5.1 Revised risk stratification for gastrointestinal stromal tumors

Mitotic rate	Size (cm)	Risk for progressive disease (%), based on site of origin			
		Stomach	Duodenum	Jejunum/ileum	Rectum
≤5 per 50 HPF	≤2	0 0	0	0	
	>2, ≤5	1.9	4.3	8.3	8.5
	>5, ≤10	3.6	24	—*	—*
	>10	10	52	34	57
≤5 per 50 HPF	≤2	—*	—*	—*	54
	>2, ≤5	16	73	50	52
	>5, ≤10	55	85	—*	—*
	>10	86	90	86	71

Source: Adapted from Miettinen and Lasota (2006).
Abbreviations: HPF: high-power field.
*Insufficient data.

depended not only on tumor size and mitotic index, but also site of origin at the time of primary presentation (Miettinen and Lasota 2006). Gastric GISTs have a more favorable prognosis than small intestinal GISTs of comparable size (Table 5.1).

KIT Mutations

KIT mutations correlate with clinical behavior (Ernst et al. 1998; Lasota et al. 1999; Li et al. 2000; Singer et al. 2002; Taniguchi et al. 1999). In a series of 275 GISTs, the exon 11 mutation frequency was 87% among low-risk tumors (Corless et al. 2004). In another study, patients with missense mutations in exon 11 had a 5-year recurrence-free survival (RFS) rate of 89% versus 40% for those having other types of *KIT* mutations (p = 0.03) (Singer et al. 2002). Patients with exon 13 mutations had a 5-year RFS of 0% versus 51% for all other patients (p = 0.001) (Singer et al. 2002). There is little evidence that progression of GISTs from localized to meta-static disease arises from accumulation of additional *KIT* mutations. In an analysis of 127 "malignant" GISTs, none harbored more than a single mutation in *KIT* exons 9, 11, 13, or 17 (Heinrich et al. 2003). Specific subtypes of exon 11 mutations may also have different risks of malignant behavior. Tumors with exon 11 deletions involving codons 562 to 579 have been associated with a higher rate of metastases than those with deletions in codons 550 to 561 (Emile et al. 2004). Another group reported that GISTs with deletions of exon 11 codons 557 to 558 had a worse RSF (23%) compared with those in which those codons were unaffected (74%) (Martin et al. 2005).

At present, tumor size, mitotic rate, and site of origin are known to have prog-nostic value. The value of using factors such as gender, histology, and *KIT* mutation is unknown. Expression of the tumor suppressors p53 (Al-Bozom 2001) and p16^{INK4} (Schneider-Stock et al. 2003, 2005), the Ki-67 proliferation marker (Yan et al. 2003), and vascular endothelial growth factor (Takahashi et al. 2003) may each have prognostic value, but the utility of these potential prognostic factors awaits confirmation through further study.

Treatment

Historically, GISTs have been treated by the three traditional cancer therapeutic modalities: surgery, chemotherapy, and radiation therapy. Surgery is effective for patients with limited resectable disease, but with this single modality, disease can recur in upwards of 50% of individuals based on the risk stratification described above. Standard cytotoxic chemotherapy and radiotherapy have shown remarkably little efficacy in this disease (Crosby et al. 2001; DeMatteo et al. 2000). The identification of activating *KIT* mutations led to the investigation of specific agents inhibiting its RTK activity. Therapy with tyrosine kinase inhibitors (TKIs), such as imatinib mesylate (STI571 or Gleevec, Novartis Pharmaceuticals) or sunitinib malate (SU11248 or Sutent, Pfizer, Inc.), has proven to be highly effective against unresectable, metastatic, and recurrent disease. These agents are illustrated in Figure 5.7 and discussed in more detail below.

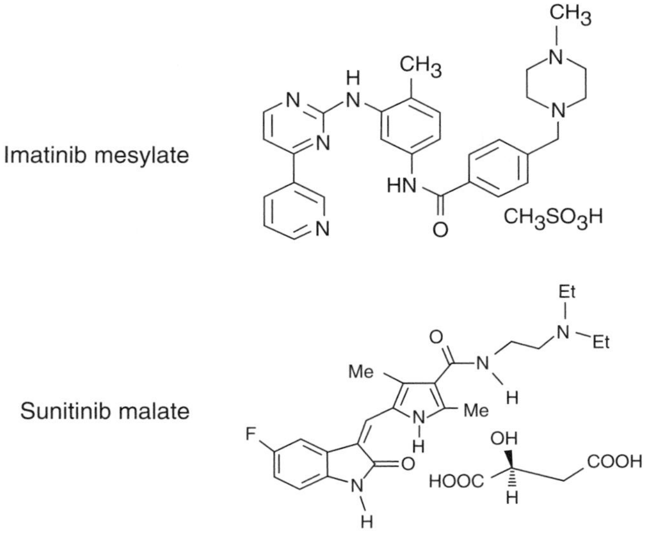

Fig. 5.7 Chemical structures of imatinib mesylate (Gleevec) and sunitinib malate (Sutent)

Surgery

Surgery remains the standard therapy for all patients with resectable, localized GIST, when the operation can be performed without excess morbidity or functional compromise for the patient. A summary of clinical management for primary GIST is shown in Figure 5.8. Complete macroscopic tumor resection is the optimal treatment for patients without evidence of metastatic disease. The goal of surgery should be complete resection of the tumor with an intact pseudocapsule wherever possible. Tumor rupture or violation of the tumor pseudocapsule is associated with an increased risk of recurrence, including dissemination of tumor throughout the peritoneum. Meticulous surgical technique is mandated to avoid tumor rupture, because GISTs are soft and fragile. The optimum width of margin is unknown; unlike adenocarcinomas, staple line or anastomotic recurrences are rare. GISTs are generally well circumscribed and do not exhibit infiltrative borders; thus, modest 1-cm margins are a standard goal. Consensus meetings under the auspices of the National Comprehensive Cancer Network (NCCN) in the United States and the European Society of Medical Oncology (ESMO) in Europe released statements discouraging laparoscopic resection of GISTs because of the higher risk of tumor rupture and subsequent tumor seeding (Blay et al. 2005; Demetri et al. 2004). However, as experience with laparoscopic resections for GIST has accumulated, its role has been reevaluated. The principles of complete macroscopic resection and avoidance of tumor rupture observed during laparotomy apply to laparoscopy.

Lymphadenectomy is not indicated unless regional lymphadenopathy is noted. Primary tumors typically project extraluminally, displacing rather than invading adjacent organs. Nevertheless, *en bloc* multiorgan resection may be necessary to

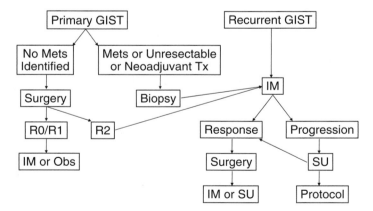

Fig. 5.8 Management of primary and recurrent gastrointestinal stromal tumors (GISTs). Mets: metastases; R0: macroscopically complete resection with microscopically negative margins; R1: macroscopically complete resection with microscopically positive margins; R2: macroscopically incomplete resection; Obs: observation; IM: imatinib; SU: sunitinib

achieve negative microscopic margins and avoid tumor spillage, particularly in the setting of recurrent disease. Inflammatory adhesions should also be removed, because they frequently contain tumor cells. Patients with GISTs in difficult locations or with multiple small tumors should be considered for neoadjuvant TKI therapy, because they may otherwise require more extensive and potentially more morbid surgery. Unresectable primary tumors might demonstrate enough of a favorable response to TKI therapy to render them amenable for surgical extirpation.

Controversy exists concerning which GISTs can be followed without intervention. Smaller tumors (<1 cm), particularly in the esophagus and stomach, have often been followed without resection with evaluations at 6-month intervals. However, with the understanding that all GISTs, no matter how small, have malignant potential (discussed below), the rationale for observation is called into question (Blay et al. 2005; Fletcher et al. 2002). Otani et al. (2006) reported no local or distant recurrences in patients with GISTs <4 cm in size resected either laparoscopically or via laparotomy with a median follow-up of 53 months. However, in this study, 14% of tumors <3 cm in size were still of intermediate-risk based on mitotic rate (Otani et al. 2006). In contrast, thorough examination of 100 consecutive whole gastric specimens identified subcentimeter GISTs in 35% of the specimens (Kawanowa et al. 2006). Miettinen and Lasota reported that no patients with GISTs <2 cm in size with a low mitotic rate, irrespective of site of origin, developed recurrent disease (Miettinen and Lasota 2006).

Chemotherapy and Radiation Therapy

For many patients with GIST, surgery is not curative. Recurrence usually occurs initially in the abdomen (liver, peritoneum) and frequently is asymptomatic. In a study of 200 patients with a likely diagnosis of GIST treated at Memorial Sloan-Kettering Cancer Center in the pre-imatinib era, complete resection was possible in 80 of 93 patients (86%) who presented with nonmetastatic primary disease (DeMatteo et al. 2000). The 5-year disease-free survival rate was 54%, with a median overall survival (OS) of 66 months. Other studies have reported similar results, with a 5-year OS rate of 63% for patients with low-risk tumors (<5 cm in diameter) compared with 34% for those with high-risk tumors (>10 cm) (Besana-Ciani et al. 2003). In another review of patients treated before imatinib approval, the 5-year RFS was 76% ± 9% for patients undergoing an R0 resection, compared with 15% ± 8% in patients undergoing R1 or R2 resections (p = 0.0001) (Singer et al. 2002). Long-term follow-up indicated that 80% of recurrences occurred within 2 years of surgery, with a median OS of 24 months (Samiian et al. 2004). The prognosis is much worse for patients with metastatic disease. In the Memorial Sloan-Kettering study cited above, in the pre-TKI era, median OS was only 19 months for patients with metastatic disease. Less than one-third of patients underwent complete surgical resection, and even those who did generally experienced rapid relapses (DeMatteo et al. 2000). Correlations between *KIT* or *PDGFRA* mutational status and clinical outcome after resection of localized disease are an active area of investigation. As described above, the frequency of *KIT* exon 11 mutations in low-risk localized primary tumors was 87% (Corless et al. 2004). *KIT* exon 9 mutations were found in 17% of the high-risk tumors, but only 3.0% and 2.5% of the intermediate- and low-risk tumors, respectively. The relationship between surgical outcome and the presence of *PDGFRA* and wild-type *KIT* mutations has not been established for localized, resectable disease.

Traditional cytotoxic chemotherapy and radiation therapy have little to offer in the management of GIST. Standard sarcoma regimens, such as those using doxorubicin, taxanes, gemcitabine, ifosfamide, temozolomide, ecteinascidin (ET-743), and hydroxyurea, achieved response rates of generally less than 5%, with median survival rates of 14–18 months, in patients with metastatic disease. Two of the last trials of cytotoxic chemotherapy (one with dacarbazine, mitomycin C, doxorubicin, and cisplatin; the second with temozolomide) in patients with advanced disease in the pre-imatinib era reported response rates from 0% to 5% (Edmonson et al. 2002; Trent et al. 2003). Similarly, these regimens have no role in adjuvant therapy. Nonstandard regimens, such as those using intraperitoneal mitoxantrone and chemoembolization, also achieved marginal, if any, benefit. Reports of radiation therapy have consisted of limited studies with few patients and have demonstrated limited efficacy (Blanke 2003). Thus, neither radiation therapy nor cytotoxic chemotherapy seems to have a role in the treatment of GIST.

Targeted Molecular Therapy

Imatinib Mesylate

A dramatic change in the treatment of GIST occurred after the recognition that activated KIT provided the driving force for GIST growth. Under physiologic conditions, binding of ligand to KIT causes dimerization of the receptor, leading to phosphorylation of intracellular proteins and transmission of a signal resulting in cell growth. Oncogenic mutations of *KIT* result in receptor dimerization and constitutive activation of the kinase (Hirota et al. 1998).

The clinical application of the novel oral targeted TKI imatinib mesylate for the treatment of GIST represented a major advance in therapy. Imatinib is a small molecule originally developed to treat chronic myelogenous leukemia. This orally administered drug also inhibits the KIT and PDGFRA protein RTKs (Druker et al. 1996, 2001; Heinrich et al. 2000, 2003). Imatinib lodges in an adenosine 5′-triphosphate (ATP)-binding pocket that forms upon receptor dimerization. With this association, KIT is unable to bind ATP, the source of the phosphate group required for phosphorylation, and the kinase is rendered inactive (Figure 5.7). The clinical potential of imatinib in this disease was first illustrated when a Finnish patient with GIST metastatic to the liver was treated with a daily oral dose of 400 mg. This patient experienced a dramatic, rapid, and sustained partial antitumor response, with resolution of all cancer-related symptoms in a few weeks (Joensuu et al. 2001). In a pivotal multicenter phase II trial, 147 patients with advanced and/or metastatic KIT-positive GIST were randomized to receive 400 or 600 mg imatinib daily (Demetri et al. 2002). Prior therapy included surgery (98% of patients), chemotherapy for metastatic disease (51%), and radiation therapy (15%). Most patients presented with advanced, bulky disease. Early analysis after a minimum follow-up of 6 months documented partial responses in 54%, stable disease in 28%, and progressive disease in 14%. No complete responses were observed. Among patients with partial responses, reduction in tumor bulk ranged from 50% to 96%. There were no significant differences in response rate or duration of response between the two dose levels. The 1-year OS rate for all patients was 88%, much higher than any OS rate reported before the availability of imatinib for this patient population with advanced disease. Responding patients exhibited a demonstrable decrease in tumor uptake of [18]FDG as detected by PET scan as early as 24 hours after treatment with imatinib. Additionally, performance status improved in the overall patient population. Concordant results were noted in a multicenter European trial (van Oosterom et al. 2001).

Imatinib therapy is generally well tolerated. Although most patients experience some treatment-related toxicities, these are usually mild to moderate in severity. Common adverse events (all grades) include edema (74%), nausea (52%), diarrhea (45%), myalgias (40%), fatigue (35%), rash or dermatitis (31%), abdominal pain (26%), and headache (26%) (Demetri et al. 2002). Serious toxicities (grade 3/4) occur in 21% of patients, including GI or intraabdominal hemorrhages in patients with large, bulky tumors (seen in approximately 5% of patients). On the basis of clinical trial efficacy and safety data, imatinib mesylate was approved by the U.S. Food and Drug Administration (FDA) in February 2002 for treatment of metastatic and/or unresectable GIST.

The use of adjuvant imatinib after resection of primary GIST is being evaluated in several multicenter trials. The American College of Surgeons Oncology Group (ACOSOG) is conducting a phase II intergroup trial (Z9000) evaluating the value of imatinib at 400 mg daily for 12 months after R0/R1 resection of "high-risk" GISTs (defined as those ≥10 cm in size, with intraperitoneal rupture or hemorrhage, or multifocal). This trial has completed accrual, and patients are currently being followed for OS. In a second ACOSOG intergroup trial (Z9001), researchers randomized patients with tumors at least 3 cm in size to imatinib 400 mg per day or placebo for 12 months after resection. This trial was closed early in April 2007 after planned interim analysis confirmed significant improvement in RFS in the treatment arm (DeMatteo et al. 2007).

Impact of Tyrosine Kinase Mutations

Approximately one-third of patients with GIST who fail to express KIT but have *PDGFRA* mutations are sensitive to imatinib and should be considered candidates for TKI therapy (Corless et al. 2005). The mutational status of *KIT* was, in one study, the most important predictive factor for clinical response to this agent

(Heinrich et al. 2003). Partial responses on imatinib were more common in patients with mutations in *KIT* exon 11 (83.5%) than in those with exon 9 muta-tions (47.8%). Median survival was also greater in the exon 11 mutation group (23 months vs. 7 months). In a series of 324 KIT-positive GIST patients (Heinrich et al. 2005), patients with exon 11 mutations had a higher response rate to imat-inib and longer time to progression than those with exon 9 mutations or no kinase mutations (67% vs. 40% and 39%; 576 days vs. 308 days and 251 days). These results suggest that *KIT* and *PDGFRA* mutational status may predict clinical response to imatinib in patients with GIST, and that genotyping might eventually serve as a useful tool in assessing treatment decisions. Such mutational analysis must still be considered experimental and currently is not routinely recommended except as a diagnostic confirmation for the small subset of patients with GIST that fails to express KIT. It is likely that mutational analysis will become routine in the near future, enabling treating physicians to more specifically tailor TKI therapy.

Detection of Response to Imatinib

The development and application of novel detection methods has improved the ability to monitor response to imatinib therapy and to more accurately assess disease progression and resistance. For example, the decline in tumor glycolytic activity measured by PET correlated with clinical benefits in response to imatinib treatment (Demetri et al. 2002) and may be a useful predictor of response in patients with recurrent or metastatic disease (Gayed et al. 2004). In patients with imatinib-refractory GIST whose disease subsequently progressed, PET was able to detect rapid increases in tumor glycolytic activity in some GIST lesions, implying that relative imatinib sensitivity may persist even in patients beginning to progress on this therapy (van den Abbeele et al. 2004). Moreover, PET can also be used to define or predict specific patterns of disease progression after initial response to imatinib, such as the "nodule within a nodule" pattern (i.e., a new enhancing focus within a preexisting nonenhancing or faintly enhancing tumor mass) found in some patients with recurrent disease (Dileo et al. 2004).

On conventional imaging modalities such as CT, tumors, although responding to treatment, may actually increase in size as a result of either intratumoral hem-orrhage or myxoid degeneration (Demetri and Morgan 2005). This corresponds to a decrease in tumor density on CT. Once the tumors become hypodense, they may decrease in size, eventually stabilizing (Demetri and Morgan 2005). Objective radiographic responses are seen at a median of 4 months, whereas maximum responses may not be noted until 6 months or later (Verweij et al. 2004). CT response criteria proposed by Choi et al. from M. D. Anderson Cancer Center use both tumor density (in Hounsfield units) and size to assess response of GIST to TKIs (Table 5.2) (Choi et al. 2007). These criteria have not been uni-versally accepted.

Table 5.2 Computed tomography response criteria

Response	RECIST criteria	Modified CT criteria
Complete response	1. Disappearance of all lesions 2. No new lesions	1. Disappearance of all lesions 2. No new lesions
Partial response	1. A decrease ≥30% in the sum of the largest diameter of target lesions 2. No new lesions	1. A decrease in size ≥10% **or** a decrease in tumor density (HU) ≥15% on CT 2. No new lesions 3. No obvious progression of nonmeasurable disease
Stable disease	1. Neither sufficient shrinkage to qualify for partial response nor sufficient increase to qualify for progressive disease	1. Does not meet criteria for complete response, partial response, or progression 2. No symptomatic deterioration attributed to tumor progression
Progressive disease	1. An increase ≥20% in the sum of the largest diameter of target lesions 2. New lesions	1. An increase in tumor size ≥10% **and** does not meet criteria of partial response by, tumor density (HU) on CT 2. New lesions 3. New intratumoral nodules **or** increase in the size of existing intratumoral tumor nodules

Source: Demetri et al. (2007).
Abbreviations: RECIST: Response Evaluation Criteria in Solid Tumors; CT: computed tomography; HU: Hounsfield units.

Resistance to Imatinib

Imatinib is highly effective in patients with GIST, with primary resistance noted in only 12% of patients (Verweij et al. 2004). However, most patients with unresectable disease eventually progress because of the development of resistance. This secondary resistance correlates with the development of additional mutations in *KIT* or *PDGFRA* in tumor cells, allowing alternate signaling pathways to be used (Duensing et al. 2004; Wakai et al. 2004). Fletcher and colleagues analyzed 16 patients who developed resistance early (n = 3) or late (n = 13) after initial response to imatinib (Fletcher et al. 2003). Four mechanisms of resistance were identified: 1. acquisition of new *KIT* or *PDGFRA* point mutations (the dominant mechanism); 2. *KIT* gene amplification, resulting in increased expression of the KIT receptor; 3. activation of an alternate RTK (e.g., PDGFRA); or 4. KIT or

PDGFRA activation with pretreatment *KIT* or *PDGFRA* mutations outside of the juxtamembrane hotspot regions, but without any secondary genomic mutation. These data indicated that resistance to imatinib mesylate occurred through multiple mechanisms, and that targeting one or more different kinases downstream of KIT and PDGFRA may reduce or prevent clinical resistance.

Sunitinib Malate for Treatment of Imatinib-Resistant Gastrointestinal Stromal Tumors

Sunitinib malate (formerly SU11248) is a novel oral agent that inhibits the signaling of multiple RTKs including KIT, PDGFRA, vascular endothelial growth factor receptor, and Flt-3, resulting in both antineoplastic and antiangiogenic activity (Figure 5.2) (Mendel et al. 2003; O'Farrell et al. 2003). Based on encouraging results in phase I/II studies, an international, multicenter, randomized phase III trial of sunitinib was conducted in patients resistant to or intolerant of imatinib mesylate (Demetri et al. 2006). In this study, 312 patients were randomized to receive therapy with sunitinib or placebo. Patients progressing on placebo therapy were crossed over to active drug. Median time to disease progression was 27.3 weeks with sunitinib compared with 6.4 weeks on placebo. Kaplan-Meier analysis revealed improved survival in patients randomized to initial sunitinib therapy, with median survival not yet reached as of the most recent interim analysis. Hypothyroidism was noted as a sunitinib-associated toxicity. Sunitinib was approved by the U.S. FDA for treatment of metastatic or unresectable GIST in January 2006.

Multimodality Treatment of Advanced Disease

Most patients experience disease recurrence despite successful resection of their primary tumor, with a median time to recurrence of 18–24 months after surgery (DeMatteo et al. 2000; Ng et al. 1992). At the time of recurrence, two-thirds of patients have liver metastases and half have peritoneal disease (DeMatteo et al. 2000). Lung and bone metastases are exceedingly rare, but may develop later in the course of disease, particularly with the extended survival seen in the imatinib era.

Imatinib is the first-line treatment for advanced (unresectable primary or metastatic) GIST (Figure 5.8). It is generally started at 400 mg per day orally. In patients who develop progressive disease on 400 mg of imatinib daily, dose escalation may be effective as demonstrated in phase III studies (Rankin et al. 2004; Zalcberg et al. 2005). If patients experience disease progression or unacceptable toxicities on higher doses of imatinib, then second-line sunitinib is started. When sunitinib resistance develops, protocol-based therapies are usually the next option. Surgery alone has limited efficacy in recurrent or metastatic GIST. Surgery may have a role in the treatment of patients whose disease has not yet progressed on medical therapy. In particular, patients whose disease is rendered technically resectable on TKI treatment may achieve increased progression-free survival (PFS) by removal of disease before drug resistance develops. Even in the context of radiographically stable or responsive disease on TKI therapy, the tumor may still harbor viable cells, and complete pathologic responses are rare (Bauer et al. 2005; Bumming et al. 2003; Raut et al. 2006; Scaife et al. 2003).

Recent studies have established that surgery, including complete resection, is possible in previously unresectable GIST after imatinib therapy (Bauer et al. 2005; Rutkowski et al. 2006; Scaife et al. 2003). A study from the Brigham and Women's Hospital and Dana-Farber Cancer Institute detailed the results of surgery in patients with advanced GIST on TKI therapy (either imatinib or sunitinib) (Raut et al. 2006). Three clinical categories were defined based on the response of advanced GIST to TKI therapy: stable disease, limited progression disease, and generalized progression disease. Outcomes of surgery and survival rates correlated with disease status. An R0/R1 resection was achieved in 78%, 25%, and 7% of patients with stable, limited progression, and generalized progression disease, respectively (p <0.0001). The 12-month PFS rates for patients with stable, limited progression, and generalized progression disease were 80%, 33%, and 0%, respectively (p <0.0001). Similarly, the 12-month OS rates were 95%, 86%, and 0%, respectively (p <0.0001). Subsequently, others have also reported successful resection of advanced GIST after imatinib therapy (Andtbacka et al. 2007; Bonvalot et al. 2006; DeMatteo et al. 2007; Gronchi et al. 2007; Rutkowski et al. 2006). Andtbacka et al. from M. D. Anderson Cancer Center recently reported a median 85% decrease in tumor volume in patients with locally advanced GIST treated with neoadjuvant TKIs (Andtbacka et al. 2007). Similarly, others reported that once multiple tumors start to progress in pattern of generalized disease progression, the value of surgery is dubious (DeMatteo et al. 2007).

In summary, patients with stable disease who underwent cytoreductive surgery achieved substantial rates of PFS and OS. In those with drug resistance at limited sites preoperatively, OS rates were prolonged but surgery did not prevent disease recurrence, potentially reflecting more aggressive tumor biology. Finally, in patients with generalized progression disease, surgery offered no survival benefit, as the median time to disease progression was 2.9 months, and the median time to death was only 5.6 months. In this setting, the principal role of surgery is symptom palliation.

Surveillance

The recent update by the NCCN consensus panel provides a standard postoperative follow-up for patients who have successfully undergone surgical resection of a primary GIST (Demetri et al. 2007). Currently, a history and physical examination are recommended every 3–6 months during the first 5 years. CT scans of the abdomen and pelvis with intravenous contrast should be obtained every 3–6 months during the first 3–5 years, then annually thereafter. Full guidelines may be downloaded from the NCCN Web site: www.nccn.org/professionals/physician_gls/default.asp.

Conclusions

New diagnostic methods show that GISTs are relatively common gastric neo-plasms, and a subset of these lesions progress as a result of uncontrolled KIT or PDGFRA activation. Before the development of imatinib, the principal treatment modality for GIST was surgery. Recurrence was common, and survival in the setting of recurrent or metastatic disease was poor. With the advent of therapy with TKIs, first imatinib, and now sunitinib, patient outcomes have improved considerably. Surgery remains the primary mode of treatment for localized, resectable disease, or localized unresectable disease that has been rendered resectable by TKI therapy. Metastatic disease is often effectively controlled by TKI therapy, although eventual disease progression is likely. These tumors provide an interesting example of the significant improvements in both curative and palliative therapy that are possible after development of effective targeted chemotherapeutic agents.

References

Agaimy, A., Wunsch, P.H., et al. (2007). Minute gastric sclerosing stromal tumors (GIST tumor-lets) are common in adults and frequently show c-KIT mutations. Am J Surg Pathol. 31(1):113–20.

Al-Bozom, I.A. (2001). p53 expression in gastrointestinal stromal tumors. Pathol Int. 51(7):519–23.

Amieux, P.S. (2004). Getting the GIST of the Carney Triad: growth factors, rare tumors, and cel-lular respiration. Pediatr Dev Pathol. 7(4):306–8.

Andtbacka, R.H., Ng, C.S., et al. (2007). Surgical resection of gastrointestinal stromal tumors after treatment with imatinib. Ann Surg Oncol. 14(1):14–24.

Bauer, S., Corless, C.L., et al. (2003). Response to imatinib mesylate of a gastrointestinal stromal tumor with very low expression of KIT. Cancer Chemother Pharmacol. 51(3):261–5.

Bauer, S., Hartmann, J.T., et al. (2005). Resection of residual disease in patients with metastatic gastrointestinal stromal tumors responding to treatment with imatinib. Int J Cancer. 117(2):316–25.

Beghini, A., Tibiletti, M.G., et al. (2001). Germline mutation in the juxtamembrane domain of the kit gene in a family with gastrointestinal stromal tumors and urticaria pigmentosa. Cancer. 92(3):657–62.

Besana-Ciani, I., Boni, L., et al. (2003). Outcome and long term results of surgical resection for gastrointestinal stromal tumors (GIST). Scand J Surg 92(3):195–9.

Blanke, C.D. (2003). Therapeutic options for gastrointestinal stromal tumors. Educational book for the 2003 annual meeting of the American Society of Clinical Oncology. 266–72.

Blay, J.Y., Bonvalot, S., et al. (2005). Consensus meeting for the management of gastrointestinal stromal tumors. Report of the GIST Consensus Conference of 20–21 March 2004, under the auspices of ESMO. Ann Oncol. 16(4):566–78.

Bonvalot, S., Eldweny, H., et al. (2006). Impact of surgery on advanced gastrointestinal stromal tumors (GIST) in the imatinib era. Ann Surg Oncol. 13(12):1596–603.

Bumming, P., Andersson, J., et al. (2003). Neoadjuvant, adjuvant and palliative treatment of gas-trointestinal stromal tumours (GIST) with imatinib: a centre-based study of 17 patients. Br J Cancer. 89:460–4.

Carney, J.A. (1999). Gastric stromal sarcoma, pulmonary chondroma, and extra-adrenal paragan-glioma (Carney Triad): natural history, adrenocortical component, and possible familial occur-rence. Mayo Clin Proc. 74(6):543–52.

Carney, J.A., and Stratakis, C.A. (2002). Familial paraganglioma and gastric stromal sarcoma: a new syndrome distinct from the Carney triad. Am J Med Genet. 108(2):132–9.

Choi, H., Charnsangavej, C., et al. (2007). Correlation of computed tomography and positron emission tomography in patients with metastatic gastrointestinal stromal tumor treated at a single institution with imatinib mesylate: proposal of new computed tomography response criteria. J Clin Oncol. 25(13):1753–9.

Chompret, A., Kannengiesser, C., et al. (2004). PDGFRA germline mutation in a family with multiple cases of gastrointestinal stromal tumor. Gastroenterology. 126(1):318–21.

Corless, C.L., Fletcher, J.A., et al. (2004). Biology of gastrointestinal stromal tumors. J Clin Oncol. 22(18):3813–25.

Corless, C.L., Schroeder, A., et al. (2005). PDGFRA mutations in gastrointestinal stromal tumors: frequency, spectrum and in vitro sensitivity to imatinib. J Clin Oncol. 23(23):5357–64.

Crosby, J.A., Catton, C.N., et al. (2001). Malignant gastrointestinal stromal tumors of the small intestine: a review of 50 cases from a prospective database. Ann Surg Oncol. 8(1):50–9.

De Giorgi, U., and Verweij, J. (2005). Imatinib and gastrointestinal stromal tumors: Where do we go from here? Mol Cancer Ther. 4(3):495–501.

DeMatteo, R.P., Lewis, J.J., et al. (2000). Two hundred gastrointestinal stromal tumors: recurrence patterns and prognostic factors for survival. Ann Surg. 231(1):51–8.

DeMatteo, R.P., Maki, R.G., et al. (2007). Results of tyrosine kinase inhibitor therapy followed by surgical resection for metastatic gastrointestinal stromal tumor. Ann Surg. 245(3):347–52.

DeMatteo, R., Owzar, K., et al. (2007). Adjuvant imatinib mesylate increases recurrence free survival (RFS) in patients with completely localized primary gastrointestinal stromal tumor (GIST): North American Intergroup Phase III trial ACOSOG Z9001. Proc Am Soc Clin Oncol. Abstr. 10079.

Demetri, G.D., Benjamin, R.S., et al. (2007). NCCN task force report: optimal management of patients with gastrointestinal stromal tumor (GIST): update of the NCCN clinical practice guidelines. J Natl Compr Canc Netw. 5(S2):S1–S32.

Demetri, G.D., and Delaney, T. (2001). NCCN: Sarcoma. Cancer Control. 8(6 Suppl 2):94–101.

Demetri, G.D., and Morgan, J.A. (2005). Gastrointestinal stromal tumors, leiomyomas, and leiomyosarcomas of the gastrointestinal tract. On line access.

Demetri, G.D., van Oosterom, A.T., et al. (2006). Efficacy and safety of sunitinib in patients with advanced gastrointestinal stromal tumour after failure of imatinib: a randomised controlled trial. Lancet. 368(9544):1329–38.

Demetri, G.D., von Mehren, M., et al. (2002). Efficacy and safety of imatinib mesylate in advanced gastrointestinal stromal tumors. N Engl J Med. 347(7):472–80.

Dileo, P., Randhawa, R., et al. (2004). Safety and efficacy of percutaneous radio-frequency ablation (RFA) in patients with metastatic gastrointestinal stromal tumor with clonal evolution of lesions refractory to imatinib mesylate. J Clin Oncol. 22(14S):9024.

Druker, B.J., Talpaz, M., et al. (2001). Efficacy and safety of a specific inhibitor of the BCR-ABL tyrosine kinase in chronic myeloid leukemia. N Engl J Med. 344(14):1031–7.

Druker, B.J., Tamura, S., et al. (1996). Effects of a selective inhibitor of the Abl tyrosine kinase on the growth of Bcr-Abl positive cells. Nat Med. 2(5):561–6.

Duensing, A., Medeiros, F., et al. (2004). Mechanisms of oncogenic KIT signal transduction in primary gastrointestinal stromal tumors (GISTs). Oncogene. 23(22):3999–4006.

Edmonson, J.H., Marks, R.S., et al. (2002). Contrast of response to dacarbazine, mitomycin, doxorubicin, and cisplatin (DMAP) plus GM-CSF between patients with advanced malignant gastrointestinal stromal tumors and patients with other advanced leiomyosarcomas. Cancer Invest. 20(5–6):605–12.

Emile, J.F., Theou, N., et al. (2004). Clinicopathologic, phenotypic, and genotypic characteristics of gastrointestinal mesenchymal tumors. Clin Gastroenterol Hepatol. 2(7):597–605.

Emory, T.S., Sobin, L.H., et al. (1999). Prognosis of gastrointestinal smooth-muscle (stromal) tumors: dependence on anatomic site. Am J Surg Pathol. 23(1):82–7.

Ernst, S.I., Hubbs, A.E., et al. (1998). KIT mutation portends poor prognosis in gastrointestinal stromal/smooth muscle tumors. Lab Invest. 78(12):1633–6.

Fletcher, C.D., Berman, J.J., et al. (2002a). Diagnosis of gastrointestinal stromal tumors: a consensus approach. Int J Surg Pathol. 10(2):81–9.

Fletcher, C.D., Berman, J.J., et al. (2002b). Diagnosis of gastrointestinal stromal tumors: a consensus approach. Hum Pathol. 33(5):459–65.

Fletcher, J.A., Corless, C.L., et al. (2003). Mechanisms of resistance to imatinib mesylate (IM) in advanced gastrointestinal stromal tumor (GIST). Proc Am Soc Clin Oncol. 22:815.

Gayed, I., Vu, T., et al. (2004). The role of 18F-FDG PET in staging and early prediction of response to therapy of recurrent gastrointestinal stromal tumors. J Nucl Med. 45(1):17–21.

Ghanem, N., Altehoefer, C., et al. (2003). Computed tomography in gastrointestinal stromal tumors. Eur Radiol. 13(7):1669–78.

Goettsch, W.G., Bos, S.D., et al. (2005). Incidence of gastrointestinal stromal tumours is underestimated: results of a nation-wide study. Eur J Cancer. 41(18):2868–72.

Gronchi, A., Fiore, M., et al. (2007). Surgery of residual disease following molecular-targeted therapy with imatinib mesylate in advanced/metastatic GIST. Ann Surg. 245(3):341–6.

Heinrich, M.C., Blanke, C.D., et al. (2002). Inhibition of KIT tyrosine kinase activity: a novel molecular approach to the treatment of KIT-positive malignancies. J Clin Oncol. 20(6):1692–703.

Heinrich, M.C., Corless, C.L., et al. (2003a). Kinase mutations and imatinib response in patients with metastatic gastrointestinal stromal tumor. J Clin Oncol. 21(23):4342–9.

Heinrich, M.C., Corless, C.L., et al. (2003b). PDGFRA activating mutations in gastrointestinal stromal tumors. Science. 299(5607):708–10.

Heinrich, M.C., Griffith, D.J., et al. (2000). Inhibition of c-kit receptor tyrosine kinase activity by STI 571, a selective tyrosine kinase inhibitor. Blood. 96(3):925–32.

Heinrich, M.C., Shoemaker, J.S., et al. (2005). Correlation of target kinase genotype with clinical activity of imatinib mesylate in patients with metastatic GI stromal tumors (GISTs) expressing KIT (KIT+). American Society of Clinical Oncology annual meeting. Abstr.

Hirota, S., Isozaki, K., et al. (1998). Gain-of-function mutations of c-kit in human gastrointestinal stromal tumors. Science. 279(5350):577–80.

Hirota, S., Nishida, T., et al. (2002). Familial gastrointestinal stromal tumors associated with dysphagia and novel type germline mutation of KIT gene. Gastroenterology. 122(5):1493–9.

Hirota, S., Ohashi, A., et al. (2003). Gain-of-function mutations of platelet-derived growth factor receptor alpha gene in gastrointestinal stromal tumors. Gastroenterology. 125(3):660–7.

Isozaki, K., Terris, B., et al. (2000). Germline-activating mutation in the kinase domain of KIT gene in familial gastrointestinal stromal tumors. Am J Pathol. 157(5):1581–5.

Joensuu, H., Roberts, P.J., et al. (2001). Effect of the tyrosine kinase inhibitor STI571 in a patient with a metastatic gastrointestinal stromal tumor. N Engl J Med. 344(14):1052–6.

Kawanowa, K., Sakuma, Y., et al. (2006). High incidence of microscopic gastrointestinal stromal tumors in the stomach. Hum Pathol. 37(12):1527–35.

Kindblom, L.G., Meis-Kindblom, J.M., et al. (2003). Incidence, prevalence, phenotype, and biologic spectrum of gastrointestinal stroma tumors (GIST): a population-based study [abstract]. Ann Oncol. 13:157.

Kindblom, L.G., Remotti, H.E., et al. (1998). Gastrointestinal pacemaker cell tumor (GIPACT): gastrointestinal stromal tumors show phenotypic characteristics of the interstitial cells of Cajal. Am J Pathol. 152(5):1259–69.

Kinoshita, K., Hirota, S., et al. (2004). Absence of c-kit gene mutations in gastrointestinal stromal tumours from neurofibromatosis type 1 patients. J Pathol. 202(1):80–5.

Lasota, J., Jasinski, M., et al. (1999). Mutations in exon 11 of c-Kit occur preferentially in malignant versus benign gastrointestinal stromal tumors and do not occur in leiomyomas or leiomyosarcomas. Am J Pathol. 154(1):53–60.

Lewis, J.J., and Brennan, M.F. (1996). Soft tissue sarcomas. Curr Probl Surg. 33(10):817–72.

Li, F.P., Fletcher, J.A., et al. (2005). Familial gastrointestinal stromal tumor syndrome: phenotypic and molecular features in a kindred. J Clin Oncol. 23(12):2735–43.

Li, S.Q., O'Leary, T.J., et al. (2000). Analysis of KIT mutation and protein expression in fine needle aspirates of gastrointestinal stromal/smooth muscle tumors. Acta Cytol. 44(6):981–6.

Maeyama, H., Hidaka, E., et al. (2001). Familial gastrointestinal stromal tumor with hyperpigmentation: association with a germline mutation of the c-kit gene. Gastroenterology. 120(1):210–5.

Maki, R.G. (2004). Gastrointestinal stromal tumors respond to tyrosine kinase-targeted therapy. Curr Treat Options Gastroenterol. 7(1):13–7.

Martin, J., Poveda, A., et al. (2005). Deletions affecting codons 557–558 of the c-KIT gene indicate a poor prognosis in patients with completely resected gastrointestinal stromal tumors: a study by the Spanish Group for Sarcoma Research (GEIS). J Clin Oncol. 23(25):6190–8.

Mazur, M.T., and Clark, H.B. (1983). Gastric stromal tumors. Reappraisal of histogenesis. Am J Surg Pathol. 7(6):507–19.

Medeiros, F., Corless, C.L., et al. (2004). KIT-negative gastrointestinal stromal tumors: proof of concept and therapeutic implications. Am J Surg Pathol. 28(7):889–94.

Mendel, D.B., Laird, A.D., et al. (2003). In vivo antitumor activity of SU11248, a novel tyrosine kinase inhibitor targeting vascular endothelial growth factor and platelet-derived growth factor receptors: determination of a pharmacokinetic/pharmacodynamic relationship. Clin Cancer Res. 9(1):327–37.

Miettinen, M., Fetsch, J.F., et al. (2006). Gastrointestinal stromal tumors in patients with neurofibromatosis 1: a clinicopathologic and molecular genetic study of 45 cases. Am J Surg Pathol. 30(1):90–6.

Miettinen, M., and Lasota, J. (2001). Gastrointestinal stromal tumors: definition, clinical, histological, immunohistochemical, and molecular genetic features and differential diagnosis. Virchows Arch. 438(1):1–12.

Miettinen, M., and Lasota, J. (2003). Gastrointestinal stromal tumors (GISTs): definition, occurrence, pathology, differential diagnosis and molecular genetics. Pol J Pathol. 54(1):3–24.

Miettinen, M., and Lasota, J. (2006). Gastrointestinal stromal tumors: pathology and prognosis at different sites. Semin Diagn Pathol. 23(2):70–83.

Miettinen, M., Majidi, M., et al. (2002). Pathology and diagnostic criteria of gastrointestinal stromal tumors (GISTs): a review. Eur J Cancer. 38(Suppl 5):S39–51.

Ng, E.H., Pollock, R.E., et al. (1992). Prognostic implications of patterns of failure for gastrointestinal leiomyosarcomas. Cancer. 69(6):1334–41.

Nilsson, B., Bumming, P., et al. (2005). Gastrointestinal stromal tumors: the incidence, prevalence, clinical course, and prognostication in the preimatinib mesylate era—a population-based study in western Sweden. Cancer. 103(4):821–9.

Nishida, T., Hirota, S., et al. (1998). Familial gastrointestinal stromal tumours with germline mutation of the KIT gene. Nat Genet. 19(4):323–4.

Nishida, T., and Hirota, S. (2000). Biological and clinical review of stromal tumors in the gastrointestinal tract. Histol Histopathol 15(4):1293–301.

O'Farrell, A.M., Foran, J.M., et al. (2003). An innovative phase I clinical study demonstrates inhibition of FLT3 phosphorylation by SU11248 in acute myeloid leukemia patients. Clin Cancer Res. 9(15):5465–76.

Otani, Y., Furukawa, T., et al. (2006). Operative indications for relatively small (2–5 cm) gastrointestinal stromal tumor of the stomach based on analysis of 60 operated cases. Surgery. 139(4):484–92.

Perez, E.A., Livingstone, A.S., et al. (2006). Current incidence and outcomes of gastrointestinal mesenchymal tumors including gastrointestinal stromal tumors. J Am Coll Surg. 202(4):623–9.

Rader, A.E., Avery, A., et al. (2001). Fine-needle aspiration biopsy diagnosis of gastrointestinal stromal tumors using morphology, immunocytochemistry, and mutational analysis of c-kit. Cancer. 93(4):269–75.

Rankin, C., Von Mehren, M., et al. (2004). Dose effect of imatinib (IM) in patients with metastatis GIST—Phase III Sarcoma Group Study S0033. American Society of Clinical Oncology.

Raut, C.P., Hornick, J.L., et al. (2006). Advanced gastrointestinal stromal tumor: potential benefits of aggressive surgery combined with targeted tyrosine kinase inhibitor therapy. Am J Hematol Oncol. In press.

Raut, C.P., Posner, M., et al. (2006). Surgical management of advanced gastrointestinal stromal tumors after treatment with targeted systemic therapy using kinase inhibitors. J Clin Oncol. 24(15):2325–31.

Roberts, P.J., and Eisenberg, B. (2002). Clinical presentation of gastrointestinal stromal tumors and treatment of operable disease. Eur J Cancer. 38(Suppl 5):S37–8.

Robson, M.E., Glogowski, E., et al. (2004). Pleomorphic characteristics of a germ-line KIT mutation in a large kindred with gastrointestinal stromal tumors, hyperpigmentation, and dysphagia. Clin Cancer Res. 10(4):1250–4.

Rossi, C.R., Mocellin, S., et al. (2003). Gastrointestinal stromal tumors: from a surgical to a molecular approach. Int J Cancer. 107(2):171–6.

Rubin, B.P., Singer, S., et al. (2001). KIT activation is a ubiquitous feature of gastrointestinal stromal tumors. Cancer Res. 61(22):8118–21.

Rubio, J., Marcos-Gragera, R., et al. (2007). Population-based incidence and survival of gastrointestinal stromal tumours (GIST) in Girona, Spain. Eur J Cancer. 43(1):144–8.

Rutkowski, P., Nowecki, Z., et al. (2006). Surgical treatment of patients with initially inoperable and/or metastatic gastrointestinal stromal tumors (GIST) during therapy with imatinib mesylate. J Surg Oncol. 93(4):304–11.

Samiian, L., Weaver, M., et al. (2004). Evaluation of gastrointestinal stromal tumors for recurrence rates and patterns of long-term follow-up. Am Surg. 70(3):187–91; discussion 191–2.

Savage, D.G., and Antman, K.H. (2002). Imatinib mesylate: a new oral targeted therapy. N Engl J Med. 346(9):683–93.

Scaife, C.L., Hunt, K.K., et al. (2003). Is there a role for surgery in patients with unresectable cKIT+ gastrointestinal stromal tumors treated with imatinib mesylate? Am J Surg. 186(6):665–9.

Schneider-Stock, R., Boltze, C., et al. (2003). High prognostic value of p16INK4 alterations in gastrointestinal stromal tumors. J Clin Oncol. 21(9):1688–97.

Schneider-Stock, R., Boltze, C., et al. (2005). Loss of p16 protein defines high-risk patients with gastrointestinal stromal tumors: a tissue microarray study. Clin Cancer Res. 11(2 Pt 1):638–45.

Shinomura, Y., Kinoshita, K., et al. (2005). Pathophysiology, diagnosis, and treatment of gastrointestinal stromal tumors. J Gastroenterol. 40(8):775–80.

Singer, S., Rubin, B.P., et al. (2002). Prognostic value of KIT mutation type, mitotic activity, and histologic subtype in gastrointestinal stromal tumors. J Clin Oncol. 20(18):3898–905.

Takahashi, R., Tanaka, S., et al. (2003). Expression of vascular endothelial growth factor and angiogenesis in gastrointestinal stromal tumor of the stomach. Oncology. 64(3):266–74.

Takazawa, Y., Sakurai, S., et al. (2005). Gastrointestinal stromal tumors of neurofibromatosis type I (von Recklinghausen's disease). Am J Surg Pathol. 29(6):755–63.

Taniguchi, M., Nishida, T., et al. (1999). Effect of c-kit mutation on prognosis of gastrointestinal stromal tumors. Cancer Res. 59(17):4297–300.

Tian, Q., Frierson, H.F., Jr., et al. (1999). Activating c-kit gene mutations in human germ cell tumors. Am J Pathol. 154(6):1643–7.

Trent, J.C., Beach, J., et al. (2003). A two-arm phase II study of temozolomide in patients with advanced gastrointestinal stromal tumors and other soft tissue sarcomas. Cancer. 98(12):2693–9.

Tryggvason, G., Gislason, H.G., et al. (2005). Gastrointestinal stromal tumors in Iceland, 1990–2003: the Icelandic GIST study, a population-based incidence and pathologic risk stratification study. Int J Cancer. 117(2):289–93.

van den Abbeele, A.D., Badawi, R.D., et al. (2004). Effects of cessation of imatinib mesylate (IM) therapy in patients (pts) with IM-refractory gastrointestinal stromal tumors (GIST) as visualized by FDG-PET scanning. Proc Am Soc Clin Oncol. 23:198. Abstr. 3012.

van der Zwan, S.M., and DeMatteo, R.P. (2005). Gastrointestinal stromal tumor: 5 years later. Cancer. 104(9):1781–8.

van Oosterom, A.T., Judson, I., et al. (2001). Safety and efficacy of imatinib (STI571) in metastatic gastrointestinal stromal tumours: a phase I study. Lancet. 358(9291):1421–3.

Verweij, J., Casali, P.G., et al. (2004). Progression-free survival in gastrointestinal stromal tumours with high-dose imatinib: randomised trial. Lancet. 364(9440):1127–34.

Wakai, T., Kanda, T., et al. (2004). Late resistance to imatinib therapy in a metastatic gastrointestinal stromal tumour is associated with a second KIT mutation. Br J Cancer. 90(11):2059–61.

Ward, S.M., Burns, A.J., et al. (1994). Mutation of the proto-oncogene c-kit blocks development of interstitial cells and electrical rhythmicity in murine intestine. J Physiol. 480(Pt 1):91–7.

Yan, H., Marchettini, P., et al. (2003). Prognostic assessment of gastrointestinal stromal tumor. Am J Clin Oncol. 26(3):221–8.

Zalcberg, J.R., Verweij, J., et al. (2005). Outcome of patients with advanced gastro-intestinal stromal tumours crossing over to a daily imatinib dose of 800 mg after progression on 400 mg. Eur J Cancer. 41(12):1751–7.

Zoller, M.E., Rembeck, B., et al. (1997). Malignant and benign tumors in patients with neurofibromatosis type 1 in a defined Swedish population. Cancer. 79(11):2125–31.

Chapter 6
Human Gastric Neuroendocrine Neoplasia: Current Pathologic Status

Guido Rindi and Enrico Solcia

Introduction

This chapter provides an overview of nonneoplastic (hyperplasia and dysplasia) and neoplastic endocrine cell lesions of the stomach [the carcinoid, or well-differentiated (neuro)endocrine tumor/carcinoma, and the poorly differentiated endocrine carcinoma, (PDEC)]. The normal endocrine cell counterparts are also briefly described.

The adjective "neuroendocrine" is widely used for this type of tumor and is considered a synonym of "endocrine," as defined by the World Health Organization (WHO) which also maintains the traditional term "carcinoid." All of these various tumor-connoting definitions are used synonymously throughout this chapter.

The Gastric Endocrine Cells

Five different endocrine cell types are described in the gastric mucosa of humans (Table 6.1) (Rindi et al. 2002a; Simonsson et al. 1988; Solcia et al. 1975, 2000a). The most represented endocrine cell type of the oxyntic mucosa is the histamine-producing, enterochromaffin-like (ECL) cell (Figure 6.1A). However, the gastrin-producing G cell is predominant in the antrum (Figure 6.2A). Serotonin-producing enterochromaffin (EC) cells and somatostatin D cells represent relatively minor populations scattered in both the acidopeptic and antral mucosa. Ghrelin-producing cells (Kojima et al. 1999) have been recently described in humans as corresponding to the previously described P/D$_1$ cells and accounting for approximately 15% of all gastric endocrine cell types (Kojima et al. 1999; Rindi et al. 2002a). In theory, all the above cell types are a potential source of proliferating lesions; however, only some of them have been actually recognized in such lesions.

Nonneoplastic Lesions

Nonneoplastic lesions are classified as hyperplasia and dysplasia. Cells with features of ECL and G cells were described as the dominant cell component in nonneoplastic endocrine lesions of the corpus and antrum, respectively. Nonneoplastic endocrine

Table 6.1 Main features of human gastric endocrine cells

Type	Product	Site	%	Stain	Secretory granule ultrastructure		
					Size (nm)	Shape	Inner structure
ECL	Histamine	Oxyntic	40–60	Grim, CgA	160–300	Round	Vesicular, coarsely granular core
D	Somatostatin	Oxyntic	15–25	H-H	200–400	Round	Poorly osmiophilic, homogeneous
EC	Serotonin	Oxyntic	10–20	Grim, CgA	150–350	Pleomorphic	Heavily osmiophilic
P/D$_1$	Ghrelin	Oxyntic	10–15	Grim, CgA	100–200	Round	Thin-haloed
G	Gastrin	Antral	40–60	Grim, CgA	150–350	Round	Vesicular, flocculent core

Source: Solcia et al. 2000a, 2000b; Rindi et al. 2002a.
Abbreviations: Grim: Grimelius silver impregnation; H-H: Hellman-Hellerstrom silver impregnation; M-F: Masson-Fontana silver impregnation; CgA: chromogranin A immunoreactivity.

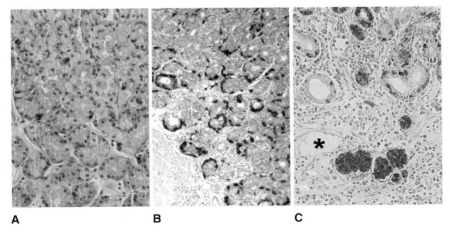

A   **B**   **C**

Fig. 6.1 (**A**) Normal endocrine cell population as assessed by chromogranin A immunohisto-chemistry, in the normotrophic oxyntic mucosa of a normogastrinemic man; the photo shows the central portion of the gland; immunoperoxidase, hematoxylin counterstain, original magnification ×100. (**B**) Diffuse and linear enterochromaffin-like (ECL) cell hyperplasia in a hypergastrinemic patient with MEN1/ZES (multiple endocrine neoplasia syndrome type 1/Zollinger-Ellison syndrome) as assessed by chromogranin A immunohistochemistry; note the normal/hypertrophic oxyntic mucosa with mild inflammation of the lamina propria (lower left corner); immunoperoxidase, hematoxylin counterstain, original magnification ×80. (**C**) Complex ECL cell hyperplasia, linear, chain-forming, and micronodular in a patient with A-CAG as assessed by immunohistochemistry for the vesicular monoamine transporter 2; note the micronodules in the lamina propria and close to a lymphatic vessel (asterisk); immunoperoxidase, hematoxylin counterstain, original magnification ×100

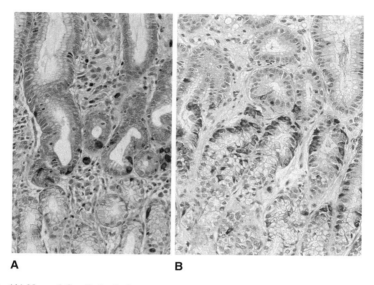

A   **B**

Fig. 6.2 (**A**) Normal G cells in the human antrum; note the distribution at the gland neck; immunoperoxidase, hematoxylin counterstain, original magnification ×140. (**B**) G-cell hyperplasia in a patient with chronic atrophic gastritis of the acidopeptic mucosa; note that G cells form palisades extending in isolated cells toward the upper and lower parts of the antral gland; immunoperoxidase, hematoxylin counterstain, original magnification ×140

cell growths are increasingly reported in routine pathology practice. This finding parallels the reported incremental trend in incidence of gastric endocrine tumors (carcinoids) in surgical and/or endoscopic series. Increased endoscopic procedures, together with increased clinical awareness of such lesions, may be the reasons for this incremental trend. A facilitating role of acid-inhibitory agents through the mild hypergastrinemia they elicit has been considered (Hodgson et al. 2005; Jensen 2006).

Corpus-Fundus

Nonantral endocrine cell lesions have been defined in terms of severity and capacity of progression toward well-differentiated tumors for ECL cells only (Figure 6.1B, C and Table 6.2) (Solcia et al. 1988). Other endocrine cells, for instance EC or ghrelin cells, may contribute to micronodular lesions (Bordi et al. 1974, 1978, 1986, 1991; Solcia et al. 1979, 1980, 1986; Srivastava et al. 2004). ECL cell hyperplasia is one morphologic counterpart of hypergastrinemia. Hypergastrinemia is reputed to promote ECL cell proliferation from hyperplasia to dysplasia and neoplasia via a multistep process (Bordi et al. 1995; Creutzfeldt 1988; D'Adda et al. 1990; Jensen 2006; Solcia et al. 1988).

Table 6.2 Proliferative lesions of nonantral stomach

Definition		Size limits
Normal pattern		Not applicable
Hyperplasia		
Simple		2 SD vs. normal
Linear (chain-forming)		≤5 cells; 2 chains/mm
Micronodular	→	≥5 cells; size 30–150 µm
Adenomatoid		
Dysplasia		
Dysplastic (precarcinoid) lesions	→	<500 mm
Enlarging micronodule		
Fusing micronodule		
Microinvasive lesion		
Nodule with newly formed stroma		
Neoplasia		
Intramucosal tumor (carcinoid)	→	≥500 µm
Microcarcinoid, microcarcinoidosis		
Invasive tumor/carcinoma		
(carcinoid/malignant carcinoid)		

Source: Solcia et al. 1988.

Enterochromaffin-Like Cell Hyperplasia

These lesions lacking neoplastic potential are classified as hyperplastic changes (Solcia et al. 1988, 1991b, 1995) and defined as: 1. *simple (diffuse) hyperplasia*, with increased number (more than twice standard deviation above normal values) of endocrine cells retaining their discrete scattered distribution; 2. *linear or chain-forming hyperplasia*, made by a minimum of five cells, along the basal membrane, in linear sequence, with a minimum of two chains per linear millimeter of mucosa; 3. *micronodular*, clusters of five or more cells (30–150 µm in size), either within the basal membrane of glands, or in the lamina propria, and a minimum of one micronodule per linear millimeter of mucosa; 4. *adenomatoid*, located within the lamina propria and composed of a minimum of five adjacent micronodules with intervening basal membrane. Notably, diffuse and linear ECL cell hyperplasia are often observed in patients treated long-term with proton pump inhibitors (PPI) (Nishi et al. 2005) (Figure 6.1B, C).

Enterochromaffin-Like Cell Dysplasia

Defined as preneoplastic (precarcinod) changes, these lesions are larger in size (150–500 µm) and all are located in the lamina propria. They are defined as: 1. *enlarging micronodules*, or cell clusters >150 µm in size; 2. *fusing micronodules*, apparently resulting from the disappearance of the basal membranes between adjacent micronodules; 3. *microinvasive lesions*, deeply infiltrating the lamina propria and filling the space between glands; 4. *nodule with newly formed stroma*, displaying a microlobular or trabecular structure.

Mechanisms

The most common conditions underlying ECL cell changes caused by longstanding hypergastrinemia are: 1. achlorhydria in nonantral (often autoimmune) chronic atrophic gastritis (A-CAG) (Bordi et al. 1978; Feyrter and Klima 1952; Rubin 1973); 2. functioning gastrin cell tumors (gastrinoma) in Zollinger-Ellison syndrome (ZES) (Bordi et al. 1974, Solcia et al. 1970), with or without multiple endocrine neoplasia syndrome type 1 (MEN1) (Solcia et al. 1990); and 3. long-term PPI treatment (Eissele et al. 1997; Jensen 2006; Lamberts et al. 1993). ECL cell tumors (carcinoids) develop in 13%–43% of MEN1 patients, but are reported in only <1% of sporadic ZES patients (Jensen 2006). Extensive endoscopic sampling of gastric mucosa allows precise histologic assessment of background conditions on which the proliferative lesions arise as well as the assignment of patients to different tumor risk groups. In patients undergoing PPI treatment or with MEN1/ZES, the gastric mucosa is usually normotrophic or hypertrophic, with variable signs of inflammation. In contrast, the mucosa of CAG patients is severely atrophic.

Evidence indicates a central role for gastrin in ECL cell proliferation. In CAG patients, hypergastrinemia induces increased ECL cell-volume density and change of granule ultrastructure (Bordi et al. 1995). After antrectomy, the withdrawal of hypergastrinemia evokes a dramatic shrinkage of actively proliferating ECL cells, sometimes with complete disappearance of hyperplastic changes (Hirschowitz et al. 1992; Richards et al. 1987).

However, besides the low frequency of gastric carcinoid in sporadic ZES patients, further data support the idea that the sole gastrin is insufficient to determine tumor development. Despite persistent gastrin stimulus, the substantial absence of mitotic activity in micronodular hyperplasia (Solcia et al. 1991b) and follow-up studies in CAG patients (Lamberts et al. 1993; Roucayrol and Cattan 1990; Solcia et al. 1992) indicate that micronodular and adenomatoid hyperplastic lesions do not progress to tumors. Such endocrine cell aggregates would mark the remaining of oxyntic glands as destroyed by inflammation. This is similar to what is observed in the colonic mucosa of patients with graft versus host colitis, where endocrine cells are spared by the autoimmune attack and solely mark the gland remnants as gland "graveyard" (Lampert et al. 1985).

However, the association with gastric carcinoids in CAG or MEN/ZES patients, indicates the malignant potential of dysplastic lesions (Rindi et al. 1993, 1999; Solcia et al. 1990). The severity of ECL cell lesions was proven to parallel the accuracy and extension of endoscopic sampling (Bordi et al. 2000), suggesting that the actual rate of dysplastic lesions and carcinoids in hypergastrinemic patients is likely to be underestimated. Accurate follow-up endoscopic strategies are thus advocated in hypergastrinemic patients who are undergoing lifelong PPI treatment (Jensen 2006).

Antrum

Antral gastrin cell hyperplasia (Figure 6.2B) occurs in achlorhydric conditions (Solcia et al. 1998). No preneoplastic potential is attributed to such a lesion. As a morphologic counterpart of hypergastrinemia, G-cell hyperplasia is invariably observed in patients with nonantral gastric atrophy and hypochlorhydria (Arnold et al. 1982, 1992). The lesion is characterized by G-cell palisading toward the upper and lower parts of the antral gland. High numbers of G cells (140–250 per linear millimeter of mucosa vs. 40–90 in controls) are typically associated with reduced numbers of D cells, resulting in an abnormal antral G/D cell ratio (Arnold et al. 1982; Friesen and Tomita 1981; Keuppens et al. 1980; Polak et al. 1972).

Likely as a result of abnormal G (or D) cell sensitivity to acid, increased gastric acid output, G-cell hyperplasia/hyperfunction and increased G/D cell ratio have also occasionally been observed in children either with or without *Helicobacter pylori* infection (Oderda et al. 1993; Rindi et al. 1994). Similar findings were reported in adults with *H. pylori* gastritis (Liu et al. 2005). D-cell hyperplasia was described in the stomach and duodenum of a patient with dwarfism, obesity, dryness of the mouth, and goiter (Holle et al. 1986). Focal, deep, linear, or micronodular hyperplasia of argentaffin EC cells was described in atrophic antral mucosa (Solcia et al. 1970). No dysplastic lesion of antral endocrine cells has been described so far.

Tumors

WHO developed an updated classification of gastric endocrine tumors embracing tumor morphology, behavior-predicting parameters, and clinical syndrome (Table 6.3) (Capella et al. 1995; Hamilton and Aaltonen 2000; Solcia et al. 2000b). Based on the differentiation status of tumor cells, endocrine tumors are currently classified as well- and poorly differentiated lesions (Rindi et al. 1993, 1996, 1999). Well-differentiated tumors/carcinomas are also defined as carcinoids (Figures 6.3–6.5). In the following paragraphs, both terms are used to indicate the same tumor disease. The definition of well-differentiated endocrine carcinoma is preferred to outline proven aggressiveness (i.e., deep-wall invasion and/or metastatic disease).

In most pathology practices, the largest fraction of gastric endocrine tumors is represented by well-differentiated lesions, whereas only a minority are PDECs. In well-differentiated tumors, the dominant cell component is a tumor cell with features of an ECL cell, more rarely a ghrelin, G or EC cells. Additionally, in most instances, all other gastric endocrine cell types may be observed as a minor tumor cell component (Bordi et al. 1991; Borch et al. 1985; Carney et al. 1983; Papotti et al. 2001; Quinonez et al. 1988; Rindi et al. 1993, 2002b; Srivastava et al. 2004;

Table 6.3 General neuroendocrine tumor categories

1. Well-differentiated endocrine tumor
2. Well-differentiated endocrine carcinoma
3. Poorly differentiated endocrine (small-cell) carcinoma
4. Mixed exocrine-endocrine tumor
5. Tumor-like lesions

Source: Solcia et al. 2000b.

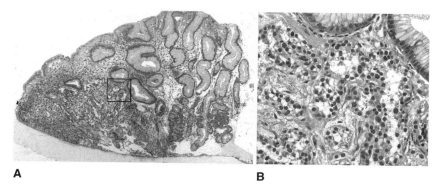

A   B

Fig. 6.3 (**A**) Histology of type I (G1) carcinoid in a small endoscopic biopsy of the corpus mucosa; hematoxylin and eosin (H&E), original magnification ×40. (**B**) Enlargement of the square in **A** detailing the bland tumor histology; H&E, ×200 original magnification ×40

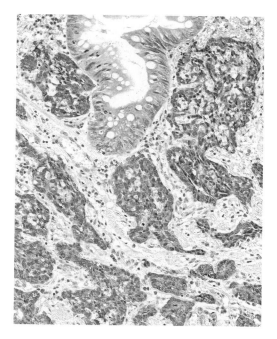

Fig. 6.4 Type 1 gastric carcinoid diffusely immunoreactive for the vesicular monoamine transporter 2; note the intestinal metaplasia in the residual gland (top of the micrograph); immunoperoxidase, hematoxylin counterstain, original magnification ×100

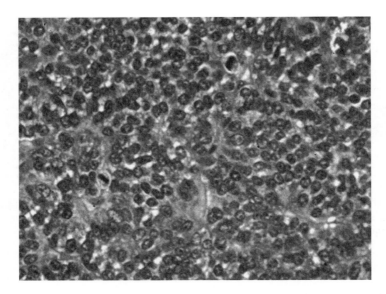

Fig. 6.5 G2 histology in a type III (sporadic) malignant enterochromaffin-like cell carcinoma; note the solid structure with delicate stroma, atypical cells with vesicular nuclei, focally prominent nucleoli, and two evident mitoses; hematoxylin and eosin, original magnification ×200

Tsolakis et al. 2004). In a series of 205 gastric neuroendocrine tumors, 193 (94%) were well-differentiated whereas only 12 were PDECs (Rindi et al. 1996), and 191 of 193 (98%) were mainly composed of ECL cells, only 2 representing G-cell tumors (Rindi et al. 1996). Rare EC cell tumors (Quinonez et al. 1988) and, more recently, one "ghrelinoma" have also been described (Tsolakis et al. 2004).

A higher incidence of gastric endocrine tumors was reported in more recent series, both surgical and endoscopic. In three U.S. large cancer databases covering the years between 1950–1999 and mainly from surgical series, gastric well-differentiated endocrine tumors (carcinoids) showed an increase from 0.5% to 1.77% of all gastric malignancies (which are, however, sharply decreasing in absolute figures) and from 2.4% to 8.7% of all gastrointestinal carcinoids (Modlin et al. 2004). A similar increase was shown in a recent survey of the cancer registries of Florida in the years 1981–2000 (Hodgson et al. 2005). In recent, mainly endoscopic series, higher incidences of gastric carcinoids were reported ranging from 11% to 41% of all gastrointestinal endocrine tumors (Mizuma et al. 1983; Rindi et al. 1993; Sjoblom et al. 1988; Solcia et al. 1991a). The increased incidence of gastric endocrine tumors needs to be weighed within the incremental trend observed for all types of gastrointestinal carcinoids in recent years (Hemminki and Li 2001; Maggard et al. 2004; Modlin et al. 2003, 2005). Nonetheless, aspects peculiar to the stomach also need to be considered. Similar to nonneoplastic ECL cell changes, the increased figures of gastric endocrine tumors may be explained by increased endoscopic practice and clinical awareness of such lesions. The potential facilitating role of potent acid-inhibitory agents (PPIs) has been also proposed and, although yet to be ascertained, this cannot be ruled out (Hodgson et al. 2005; Jensen 2006). To date, there have been no studies that have demonstrated any clear association between PPI use and the occurrence of gastric neuroendocrine tumors.

Histology and Grading

The study of clinicopathologic features of 102 gastric endocrine tumors allowed the development of a malignancy-related histopathology grading system (Rindi et al. 1999). Three histologic grades (G1–3) were proposed: grade 1 (G1) with monomorph structure characterized by solid nests and tubules, mild cellular atypia, and almost absent or very few [0–2/10 high-power field (HPF)] typical mitoses (Figure 6.3). In the common practice, the largest fraction of gastric endocrine tumors fits this category, as reported in 81 of 102 cases in this study.(Rindi et al. 1999) G2 histology displays prevalent solid aggregates, scant punctate necrosis, and relatively elevated mitotic count (≥7/10 HPF) (Figure 6.5). In clinical practice, G2 tumors are rare, essentially correspond to a fraction of sporadic cases, and represented just 5 of 102 in this study (Rindi et al. 1999). G3 histology is characterized by solid structure, sometimes organoid, with abundant "geographical chart" necrosis, overt atypia, and high number of mitoses, often atypical, thus identifying PDECs (Figure 6.6). PDECs are rare, accounting for 16 of 102 cases in this series

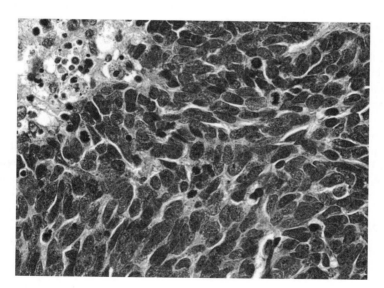

Fig. 6.6 G3 histology of a gastric poorly differentiated endocrine carcinoma; note the solid structure with focal necrosis (upper left corner) and the spindle-shaped, severely atypical carcinoma cells with scant cytoplasm, granular chromatin of the nuclei in the absence or relevant nucleoli, and the brisk mitotic count; hematoxylin and eosin, original magnification ×400

(Rindi et al. 1999). The high-grade, "undifferentiated" histology together with the relative unawareness of such entity can lead to its misdiagnosis as poorly differentiated adenocarcinoma, which likely contributes to the relative rarity of reported gastric PDECs (Brenner et al. 2004, 2007). The grading system proposed for gastric endocrine tumors (Rindi et al. 1999) was the working basis for a recent grading proposal for endocrine tumors of foregut origin (stomach, duodenum, upper jejunum, and pancreas) (Table 6.4) (Rindi et al. 2006).

Clinicopathologic Typing and Prognosis

Carcinoids

The majority of gastric well-differentiated endocrine tumors/carcinomas are not associated with hormone hyperproduction. Gastric carcinoids causing a clinically overt endocrine syndrome are rare. Few ECL cell tumors associated with the "atypical" carcinoid syndrome and exceedingly rare tumors determining other endocrine syndromes have been reported (Larsson et al. 1973; Solcia et al. 1998; Tartaglia et al. 2005).

The most represented well-differentiated tumors of the stomach, and the most common occurrence in clinical practice, are nonfunctioning ECL cell tumors (carcinoids). They are usually located in the corpus-fundus or in the adjacent mucosa extending in the antrum (Bordi et al. 2001). The histologic diagnosis of ECL cell tumor is *bona fide* inferred by: 1. strong argyrophilia (Grimelius and Sevier-Munger silver stains); 2. absence of reactivity for the argentaffin or diazonium tests for serotonin; 3) strong immunoreactivity for chromogranin A, the vesicular monoamine transporter 2 (VMAT2) (Rindi et al. 2000) (Figure 6.4) and scant or occasional immunoreactivity for other gastric-type hormones (ghrelin, somatostatin, gastrin, and serotonin). Transmission electron microscopy is an effective diagnostic tool, which allows identification of electro-dense granules (or large dense core vesicles) with irregular to round cores and adherent to detached, smooth to weavy membranes. Such aspects may variably correspond to those described for typical ECL cell granules (Borch et al. 1985; Capella et al. 1991; Carney et al. 1983; Håkanson et al. 1984; Solcia et al. 1975). Nonetheless, despite its tremendous

Table 6.4 Proposed grading of foregut neuroendocrine tumors

Grade	Mitotic count (10 HPF)*	Ki-67 index (%)†
G1	<2	≤2
G2	2–20	3–20
G3	>20	>20

Source: Rindi et al. 2006.
*10 HPF: high-power field = $2\,mm^2$, at least 40 fields (at 40× magnification) evaluated in areas of highest mitotic density.
†MIB1 antibody; % of 2,000 tumor cells in areas of highest nuclear labeling.

potential utility, the use of transmission electron microscopy is currently largely restricted to research purposes or to address specific diagnostic issues, given the often unavailable technical skills and high costs.

Three clinicopathologic subtypes of ECL cell tumors have been recognized (Rappel et al. 1995; Rindi et al. 1993, 1996, 1999) and proven useful in clinical management and treatment (Borch et al. 2005; Hou and Schubert 2007; Ruszniewski et al. 2006; Safatle-Ribeiro et al. 2007). Type I ECL tumors are associated with diffuse, corpus-restricted CAG (Figures 6.3 and 6.4); type II are associated with MEN1/ZES and hypertrophic gastropathy; and type III, or sporadic, are not associated with any distinctive gastric pathology (Figure 6.5).

The common feature of hypergastrinemia is associated with type I and type II tumors, linking them to the development of nonneoplastic and preneoplastic lesions (see previous paragraph). In contrast, the type III tumor constitutes a neoplastic disease overtly independent from hormonal imbalance or inflammation.

In addition, the relatively recent description of two cases of multiple carcinoids associated with achlorhydria, hypergastrinemia, and corpus mucosa hypertrophy and hyperplasia, in the absence of MEN1/ZES, raised the possibility of a potentially novel type IV gastrin-dependent tumor setting (Abraham et al. 2005; Ooi et al. 1995). In these cases, hypergastrinemia is attributable to gastrin-cell hyperplasia, which in turn is the result of a retained physiologic corpus-antrum axis responding to reduced acid output. In contrast, the presence of cystic and engorged oxyntic glands represents the morphologic counterpart of ineffective acid production by parietal cells.

Type I ECL cell carcinoids comprise the largest fraction and most frequently diagnosed (Figures 6.3 and 6.4). They are prevalent in aged female patients and are associated with antral gastrin cell hyperplasia and other ECL cell nonneoplastic lesions. G1 histology is almost exclusive of type I carcinoids. Type I tumors are often multiple and multicentric, small in size (usually < 1 cm), and invade the mucosa or submucosa. In exceedingly rare instances, larger carcinomas may invade the muscularis propria. Metastases are exceptional and survival is generally excellent (Rindi et al. 1999), although occasional lethal cases have been reported (Wangberg et al. 1990).

Type II ECL cell carcinoids are rare, and in a large study accounted for 12 of 191 cases (6%) (Rindi et al. 1996). Tumors develop in a context of hypertrophic, hypersecretory gastropathy and ECL cell hyperplasia. They tend to occur in adult patients without gender prevalence. G1 histology is the most represented in type II carcinoids. Often multiple and multicentric, type II carcinoids are small in size, limited to the mucosa/submucosa, and display bland histology with no significant mitoses. Survival is usually excellent, although metastases to local lymph nodes have been reported. Tumor-related death is rare; however, two unusually aggressive tumors with mixed G1 and G2/G3 histology have been described (Bordi et al. 1997).

Type III ECL cell carcinoids are usually single lesions that develop in a stomach void of notable pathology. Accounting for 27 of 191 cases (14%), they are preferentially observed in males at a relatively younger age (sixth decade of life), in the absence of hypergastrinemia and gastrin-dependent ECL cell hyperplasia (Rindi et al. 1993, 1996, 1999). Histopathologic analysis typically reveals type III carcinoids to be relatively bland, although solid aggregates and broader trabeculae are often seen with moderate cellular atypia and mitoses. A minority of cases (5 of 17) show a G2 pattern characterized by a more solid structure, round to spindle cells with large nuclei and prominent nucleoli, and high mitotic count (median 9, range 7–18) with atypical mitotic figures (Figure 6.5). The average tumor size is significantly larger compared with type I and II cases (mean 3.2 vs. 0.7 and 1.2 cm, respectively) (Rindi et al. 1996), with deep wall invasion and metastases (70% and 58% of 17 cases studied, respectively) (Rindi et al. 1999). Tumor-related death has been reported in 7 of 26 patients at follow-up (Rindi et al. 1996).

In a large retrospective series, 13 variables (among which were clinical type, tumor grading, size, mitoses number, proliferative fraction by Ki67 expression, P53 hyperexpression/accumulation, lympho/angio-invasion, deep wall invasion, and metastases) were investigated and proven statistically significant predictors of malignancy and survival (Rindi et al. 1999). The most aggressive carcinomas were characterized by G2 and G3 histology, size ≥3 cm, mitotic index of ≥9 in 10 HPF, ≥300 Ki67 index in 10 HPF. However, G1 tumors displayed Ki67 index below 150/10 HPF, size < 1 cm, and growth limited to the mucosa. These data indicate that both histopathologic grading and tumor size are equally important prognostic factors and higher-grade tumors are most often larger in size. Four models of multivariate logistic regression analysis identified the cutoffs for several variables (including mitotic count or Ki67 index) that allowed estimating the potential malignancy of individual cases under study (Rindi et al. 1999). Overall histologic grade, with special reference to the proliferative fraction, proved to be a key predictor of malignancy and outcome.

Poorly Differentiated Endocrine Carcinomas

PDECs display G3 histology, as defined by WHO criteria (Figure 6.6) (Solcia et al. 2000b). PDECs are composed of epithelial cells of small to intermediate size, with poorly developed electron-dense granules (Rindi et al. 1999), tentatively defined as protoendocrine. The low differentiation status of PDECs and other histopathologic features are often shared by poorly differentiated adenocarcinomas. The expression of markers of neuroendocrine differentiation (in the absence of proven mucus production) in the majority of cancer cells is needed to support a diagnosis of PDECs. Contrary to well-differentiated endocrine tumors, chromogranin A and tissue-specific hormones are generally absent or poorly represented in PDECs (Rindi et al. 1993, 1999). Both hormones and chromogranin A are stored within large, dense core vesicles, which are rarely observed in gastrointestinal and gastric PDECs (Gould et al. 1984; Rindi et al. 1999; Sweeney and McDonnell 1980).

In general, gastric PDECs are not associated with an overt endocrine syndrome, and develop in patients of the seventh decade, in the absence of significant sex prevalence (Rindi et al. 1996). Usually large lesions (>4 cm), PDECs may arise in any part of the stomach, are invariably in an advanced stage with diffuse metastasis at diagnosis, leading most patients to death within a few months (Rindi et al. 1996). Most PDECs develop in nonhypergastrinemic patients similar to type III carcinoids, thus implying a distinct pathophysiologic origin to type I and type II carcinoids. Nonetheless, the report of rare cases with coexisting well- and poorly differentiated endocrine tumors, sometimes arising in a CAG or MEN1/ZES background (Bordi et al. 1997; Gould et al. 1984; Rindi et al. 1993, 1999; Sweeney and McDonnell 1980) suggests the potential occurrence of PDEC progression from carcinoid, especially type III with G2 histology.

Tumor-Node-Metastasis Staging

Because of the need for standards in patient stratification and management, a novel tumor-node-metastasis (TNM) staging classification system has been developed for gastric endocrine tumors (Table 6.5) (Rindi et al. 2006). Validation on large series is needed.

Table 6.5 Proposed tumor-node-metastasis (TNM) classification and disease staging for gastric neuroendocrine tumors

TNM
T—primary tumor
TX   primary tumor cannot be assessed
T0   no evidence of primary tumor
Tis   *in situ* tumor/dysplasia (<0.5 mm)
T1   tumor invades lamina propria or submucosa and <1 cm
T2   tumor invades muscularis propria or subserosa or >1 cm
T3   tumor penetrates serosa
T4   tumor invades adjacent structures
Any T add (**m**) for multiple tumors.
N—regional lymph nodes
NX   regional lymph nodes cannot be assessed
N0   no regional lymph node metastasis
N1   regional lymph node metastasis
M—distant metastasis
MX distant metastasis cannot be assessed
M0   no distant metastases
M1* distant metastasis
*M1 specific sites defined according to Sobin and Wittekind (2002).

Stage			
	Disease stages		
Stage 0	Tis	N0	M0
Stage I	T1	N0	M0
Stage IIa	T2	N0	M0
IIb	T3	N0	M0
Stage IIIa	T4	N0	M0
IIIb	Any T	N1	M0
Stage IV	Any T	Any N	M1

Source: Rindi et al. 2006.

Frequency of Enterochromaffin-Like Changes and Carcinoids

The two conditions most frequently associated with gastric endocrine cell growths are A-CAG and ZES. ZES patients, despite long-term hypergastrinemia, usually do not develop high-grade ECL cell lesions and carcinoids, unless in MEN1 background (Cadiot et al. 1993; Lehy et al. 1989, 1992; Peghini et al. 2002). Indeed, only rare gastric carcinoids were reported in ZES patients (Cadiot et al. 1995; Feurle 1994). However, pernicious anemia and A-CAG are well-known risk factors for both gastric adenocarcinoma and carcinoid (Hsing et al. 1993; Ye and Nyren 2003). In different endoscopic series, the frequency of gastric carcinoid in pernicious anemia/A-CAG patients ranged between 1.6% and 10% (Annibale et al. 2001; Kokkola et al. 1998; Sjoblom et al. 1988). Additionally, the frequency of ECL dysplasia matched that of carcinoid (Annibale et al. 2001).

Technical Notes

To assess gastritis, it is recommended to take biopsies from the antrum (two biopsies), incisura angularis (one biopsy), the oxyntic mucosa (two biopsies), and from any other abnormalities (Dixon et al. 1996). This strategy would normally allow the eventual detection of endocrine cell changes in residual islands of oxyntic glands within an otherwise atrophic gastric mucosa.

To specifically assess endocrine cell changes, it is recommended to take biopsy samples from the antrum (two biopsies) and the oxyntic mucosa (four biopsies from fundus) (Ruszniewski et al. 2006). However, the number and the severity of the endocrine cell lesions at histology directly correlate with the number of biopsies taken, especially from the greater curvature, thus justifying a more extensive sampling if a carcinoid is suspected and in patients with ZES (Bordi et al. 2000).

For the assessment of nonantral endocrine cells and derived growths on biopsy samples, immunohistochemistry for chromogranin A and synaptophysin is the minimum recommended (Ruszniewski et al. 2006). For carcinoids, the mitotic count and the Ki67 index per $2\,mm^2$ area (10 HPF) are needed to meet recommended standards (Rindi et al. 2006; Ruszniewski et al. 2006).

Final Remarks

Hyperplastic and dysplastic changes in the stomach have been clearly identified and categorized. The classification system was developed upon the specific clinical demand for neoplastic risk assessment of well-known proliferating lesions.

The largest and most clinically relevant fraction of gastric carcinoids is made by well-differentiated ECL cell tumors/carcinomas. As for nonneoplastic lesions, gastrin has a pivotal role in the genesis of the majority of them, the type I (or A-CAG related) and type II (or MEN1/ZES related) ECL cell carcinoids, which can thus be defined as gastrin-dependent tumors. Their proliferation rate is low and, in most cases, their clinical behavior is substantially benign. In contrast, type III ECL cell carcinoids and most PDECs develop in normogastrinemic patients, and in the absence of an overt pathogenetic background, i.e., a genetic defect or severe atrophy. These clinically relevant features mark the substantial divergence existing between the gastrin-dependent type I or II carcinoids and type III carcinoids or PDECs.

Relevant questions remain open, namely, aiming to define the cell of origin of endocrine tumors, either well differentiated or poorly differentiated. Different genetic backgrounds are likely to be involved in the genesis of these different forms of neoplasia. Finally, the role of gastrin as potent gastric mucosa growth factor will likely need further investigation with a multiple-biopsy approach in chronic hypergastrinemic conditions.

References

Abraham, S.C., Carney, J.A., Ooi, A., Choti, M.A., and Argani, P. (2005). Achlorhydria, parietal cell hyperplasia, and multiple gastric carcinoids: a new disorder. Am J Surg Pathol. 29:969–75.

Annibale, B., Azzoni, C., Corleto, V. D., di Giulio, E., Caruana, P., D'Ambra, G., Bordi, C., and Delle Fave, G. (2001). Atrophic body gastritis patients with enterochromaffin-like cell dysplasia are at increased risk for the development of type I gastric carcinoid. Eur J Gastroenterol Hepatol. 13:1449–56.

Arnold, R., Frank, M., Simon, B., Eissele, R., and Koop, H. (1992). Adaptation and renewal of the endocrine stomach. Scand J Gastroenterol Suppl. 193:20–7.

Arnold, R., Hulst, M.V., Neuhof, C.H., Schwarting, H., Becker, H.D., and Creutzfeldt, W. (1982). Antral gastrin-producing G-cells and somatostatin-producing D-cells in different states of gastric acid secretion. Gut. 23:285–91.

Borch, K., Ahren, B., Ahlman, H., Falkmer, S., Granerus, G., and Grimelius, L. (2005). Gastric carcinoids: biologic behavior and prognosis after differentiated treatment in relation to type. Ann Surg. 242:64–73.

Borch, K., Renvall, H., and Liedberg, G. (1985). Gastric endocrine cell hyperplasia and carcinoid tumors in pernicious anemia. Gastroenterology. 88:638–48.

Bordi, C., Azzoni, C., Ferraro, G., Corleto, V.D., Gibril, F., Delle Fave, G., Lubensky, I.A., Venzon, D.J., and Jensen, R.T. (2000). Sampling strategies for analysis of enterochromaffin-like cell changes in Zollinger-Ellison syndrome. Am J Clin Pathol. 114:419–25.

Bordi, C., Cocconi, G., Togni, R., Vezzadini, P., and Missale, G. (1974). Gastric endocrine cell proliferation. Association with Zollinger-Ellison syndrome. Arch Pathol. 98:274–8.

Bordi, C., Corleto, V.D., Azzoni, C., Pizzi, S., Ferraro, G., Gibril, F., Delle Fave, G., and Jensen, R.T. (2001). The antral mucosa as a new site for endocrine tumors in multiple endocrine neoplasia type 1 and Zollinger-Ellison syndromes. J Clin Endocrinol Metab. 86:2236–42.

Bordi, C., D'Adda, T., Azzoni, C., Pilato, F.P., and Caruana, P. (1995). Hypergastrinemia and gastric enterochromaffin-like cells. Am J Surg Pathol. 19 Suppl 1:S8–19.

Bordi, C., Falchetti, A., Azzoni, C., D'Adda, T., Canavese, G., Guariglia, A., Santini, D., Tomassetti, P., and Brandi, M.L. (1997). Aggressive forms of gastric neuroendocrine tumors in multiple endocrine neoplasia type I. Am J Surg Pathol. 21:1075–82.

Bordi, C., Ferrari, C., D'Adda, T., Pilato, F., Carfagna, G., Bertele, A., and Missale, G. (1986). Ultrastructural characterization of fundic endocrine cell hyperplasia associated with atrophic gastritis and hypergastrinaemia. Virchows Arch A Pathol Anat Histopathol. 409:335–47.

Bordi, C., Gabrielli, M., and Missale, G. (1978). Pathologic changes of endocrine cells in chronic atrophic gastritis. An ultrastructural study on peroral gastric biopsy specimens. Arch Pathol Lab Med. 102:129–35.

Bordi, C., Yu, J.Y., Baggi, M.T., Davoli, C., Pilato, F.P., Baruzzi, G., Gardini, G., Zamboni, G., Franzin, G., Papotti, M., and Bussolati, G. (1991). Gastric carcinoids and their precursor lesions. A histologic and immunohistochemical study of 23 cases. Cancer. 67:663–72.

Brenner, B., Tang, L.H., Klimstra, D.S., and Kelsen, D.P. (2004). Small-cell carcinomas of the gastrointestinal tract: a review. J Clin Oncol. 22:2730–9.

Brenner, B., Tang, L.H., Shia, J., Klimstra, D.S., and Kelsen, D.P. (2007). Small cell carcinomas of the gastrointestinal tract: clinicopathological features and treatment approach. Semin Oncol. 34:43–50.

Cadiot, G., Lehy, T., Ruszniewski, P., Bonfils, S., and Mignon, M. (1993). Gastric endocrine cell evolution in patients with Zollinger-Ellison syndrome. Influence of gastrinoma growth and long-term omeprazole treatment. Dig Dis Sci. 38:1307–17.

Cadiot, G., Vissuzaine, C., Potet, F., and Mignon, M. (1995). Fundic argyrophil carcinoid tumor in a patient with sporadic-type Zollinger-Ellison syndrome. Dig Dis Sci. 40:1275–8.

Capella, C., Finzi, G., Cornaggia, M., Usellini, L., Luinetti, O., and Buffa, R. (1991). Ultrastructural typing of gastric endocrine cells. In The stomach as an endocrine organ, R. Håkanson and F. Sundler eds, 27–51. Amsterdam: Elsevier.

Capella, C., Heitz, P.U., Hofler, H., Solcia, E., and Kloppel, G. (1995). Revised classification of neuroendocrine tumours of the lung, pancreas and gut. Virchows Arch. 425:547–60.

Carney, J.A., Go, V.L., Fairbanks, V.F., Moore, S.B., Alport, E.C., and Nora, F.E. (1983). The syndrome of gastric argyrophil carcinoid tumors and nonantral gastric atrophy. Ann Intern Med. 99:761–6.

Creutzfeldt, W. (1988). The achlorhydria-carcinoid sequence: role of gastrin. Digestion. 39:61–79.

D'Adda, T., Corleto, V., Pilato, F.P., Baggi, M.T., Robutti, F., Delle Fave, G., and Bordi, C. (1990). Quantitative ultrastructure of endocrine cells of oxyntic mucosa in Zollinger-Ellison syndrome. Correspondence with light microscopic findings. Gastroenterology. 99:17–26.

Dixon, M.F., Genta, R.M., Yardley, J.H., and Correa, P. (1996). Classification and grading of gastritis. The updated Sydney System. International Workshop on the Histopathology of Gastritis, Houston 1994. Am J Surg Pathol. 20:1161–81.

Eissele, R., Brunner, G., Simon, B., Solcia, E., and Arnold, R. (1997). Gastric mucosa during treatment with lansoprazole: Helicobacter pylori is a risk factor for argyrophil cell hyperplasia. Gastroenterology. 112:707–17.

Feurle, G.E. (1994). Argyrophil cell hyperplasia and a carcinoid tumour in the stomach of a patient with sporadic Zollinger-Ellison syndrome. Gut. 35:275–7.

Feyrter, F., and Klima, R. (1952). Uber die Histopathologie der Magenveranderungen bei der Anaemia perniciosa. Munch Med Wochenschr. 94:145–53.

Friesen, S.R., and Tomita, T. (1981). Pseudo-Zollinger-Ellison syndrome: hypergastrinemia, hyperchlorhydria without tumor. Ann Surg. 194:481–93.

Gould, V.E., Jao, W., Chejfec, G., Banner, B.F., and Bonomi, P. (1984). Neuroendocrine carcinomas of the gastrointestinal tract. Semin Diagn Pathol. 1:13–18.

Håkanson, E., Ekelund, M., and Sundler, F. (1984). Activation and proliferation of gastric endocrine cells. In volution and tumor pathology of the neuroendocrine system, S. Falkmer, R. Hakanson and F. Sundler eds, 371–98. Amsterdam: Elsevier.

Hamilton, S.R., and Aaltonen, L.A. (eds). (2000). World Health Organization Classification of Tumours, Pathology and Genetics of Tumours of the Digestive System. IARC Press, Lyon.

Hemminki, K., and Li, X. (2001). Incidence trends and risk factors of carcinoid tumors: a nationwide epidemiologic study from Sweden. Cancer. 92:2204–10.

Hirschowitz, B.I., Griffith, J., Pellegrin, D., and Cummings, O.W. (1992). Rapid regression of enterochromaffinlike cell gastric carcinoids in pernicious anemia after antrectomy. Gastroenterology. 102:1409–18.

Hodgson, N., Koniaris, L.G., Livingstone, A.S., and Franceschi, D. (2005). Gastric carcinoids: a temporal increase with proton pump introduction. Surg Endosc. 19:1610–2.

Holle, G.E., Spann, W., Eisenmenger, W., Riedel, J., and Pradayrol, L. (1986). Diffuse somatostatin-immunoreactive D-cell hyperplasia in the stomach and duodenum. Gastroenterology. 91:733–9.

Hou, W., and Schubert, M.L. (2007). Treatment of gastric carcinoids. Curr Treat Options Gastroenterol. 10:123–33.

Hsing, A.W., Hansson, L.E., McLaughlin, J.K., Nyren, O., Blot, W.J., Ekbom, A., and Fraumeni, J.F., Jr. (1993). Pernicious anemia and subsequent cancer. A population-based cohort study. Cancer. 71:745–50.

Jensen, R.T. (2006). Consequences of long-term proton pump blockade: insights from studies of patients with gastrinomas. Basic Clin Pharmacol Toxicol. 98:4–19.

Keuppens, F., Willems, G., De Graef, J., and Woussen-Colle, M.C. (1980). Antral gastrin cell hyperplasia in patients with peptic ulcer. Ann Surg. 191:276–81.

Kojima, M., Hosoda, H., Date, Y., Nakazato, M., Matsuo, H., and Kangawa, K. (1999). Ghrelin is a growth-hormone-releasing acylated peptide from stomach. Nature. 402:656–60.

Kokkola, A., Sjoblom, S.M., Haapiainen, R., Sipponen, P., Puolakkainen, P., and Jarvinen, H. (1998). The risk of gastric carcinoma and carcinoid tumours in patients with pernicious anaemia. A prospective follow-up study. Scand J Gastroenterol. 33:88–92.

Lamberts, R., Creutzfeldt, W., Struber, H.G., Brunner, G., and Solcia, E. (1993). Long-term omeprazole therapy in peptic ulcer disease: gastrin, endocrine cell growth, and gastritis. Gastroenterology. 104:1356–70.

Lampert, I.A., Thorpe, P., Van Noorden, S., Marsh, J., Goldman, J.M., Gordon-Smith, E.C., and Evans, D.J. (1985). Selective sparing of enterochromaffin cells in graft versus host disease affecting the colonic mucosa. Histopathology. 9:875–866.

Larsson, L.I., Ljungberg, O., Sundler, F., Hakanson, R., Svensson, S.O., Rehfeld, J., Stadil, R., and Holst, J. (1973). Antor-pyloric gastrinoma associated with pancreatic nesidioblastosis and proliferation of islets. Virchows Arch A Pathol Pathol Anat. 360:305–14.

Lehy, T., Cadiot, G., Mignon, M., Ruszniewski, P., and Bonfils, S. (1992). Influence of multiple endocrine neoplasia type 1 on gastric endocrine cells in patients with the Zollinger-Ellison syndrome. Gut. 33:1275–9.

Lehy, T., Mignon, M., Cadiot, G., Elouaer-Blanc, L., Ruszniewski, P., Lewin, M.J., and Bonfils, S. (1989). Gastric endocrine cell behavior in Zollinger-Ellison patients upon long-term potent antisecretory treatment. Gastroenterology. 96:1029–40.

Liu, Y., Vosmaer, G.D., Tytgat, G.N., Xiao, S.D., and Ten Kate, F.J. (2005). Gastrin (G) cells and somatostatin (D) cells in patients with dyspeptic symptoms: Helicobacter pylori associated and non-associated gastritis. J Clin Pathol. 58:927–31.

Maggard, M.A., O'Connell, J.B., and Ko, C.Y. (2004). Updated population-based review of carcinoid tumors. Ann Surg. 240:117–22.

Mizuma, K., Shibuya, H., Totsuka, M., and Hayasaka, H. (1983). Carcinoid of the stomach: a case report and review of 100 cases reported in Japan. Ann Chir Gynaecol. 72:23–7.

Modlin, I.M., Kidd, M., Latich, I., Zikusoka, M.N., and Shapiro, M.D. (2005). Current status of gastrointestinal carcinoids. Gastroenterology. 128:1717–51.

Modlin, I.M., Lye, K.D., and Kidd, M. (2003). A 5-decade analysis of 13,715 carcinoid tumors. Cancer. 97:934–59.

Modlin, I.M., Lye, K.D., and Kidd, M. (2004). A 50-year analysis of 562 gastric carcinoids: small tumor or larger problem? Am J Gastroenterol. 99:23–32.

Nishi, T., Makuuchi, H., and Weinstein, W.M. (2005). Changes in gastric ECL cells and parietal cells after long-term administration of high-dose omeprazole to patients with Barrett's esophagus. Tokai J Exp Clin Med. 30:117–21.

Oderda, G., Fiocca, R., Villani, L., Altare, F., Morra, I., and Ansaldi, N.S. (1993). Gastrin cell hyperplasia in childood Helicobacter Pylori gastritis. Eur J Gastroenterol Hepatol. 5:13–6.

Ooi, A., Ota, M., Katsuda, S., Nakanishi, I., Sugawara, H., and Takahashi, I. (1995). An Unusual Case of Multiple Gastric Carcinoids Associated with Diffuse Endocrine Cell Hyperplasia and Parietal Cell Hypertrophy. Endocr Pathol. 6:229–237.

Papotti, M., Cassoni, P., Volante, M., Deghenghi, R., Muccioli, G., and Ghigo, E. (2001). Ghrelin-producing endocrine tumors of the stomach and intestine. J Clin Endocrinol Metab. 86:5052–9.

Peghini, P.L., Annibale, B., Azzoni, C., Milione, M., Corleto, V.D., Gibril, F., Venzon, D.J., Delle Fave, G., Bordi, C., and Jensen, R.T. (2002). Effect of chronic hypergastrinemia on human enterochromaffin-like cells: insights from patients with sporadic gastrinomas. Gastroenterology. 123:68–85.

Polak, J.M., Stagg, B., and Pearse, A.G. (1972). Two types of Zollinger-Ellison syndrome: immunofluorescent, cytochemical and ultrastructural studies of the antral and pancreatic gastrin cells in different clinical states. Gut. 13:501–12.

Quinonez, G., Ragbeer, M.S., and Simon, G.T. (1988). A carcinoid tumor of the stomach with features of a midgut tumor. Arch Pathol Lab Med. 112:838–41.

Rappel, S., Altendorf-Hofmann, A., and Stolte, M. (1995). Prognosis of gastric carcinoid tumours. Digestion. 56:455–62.

Richards, A.T., Hinder, R.A., and Harrison, A.C. (1987). Gastric carcinoid tumours associated with hypergastrinaemia and pernicious anaemia–regression of tumors by antrectomy. A case report. S Afr Med J. 72:51–3.

Rindi, G., Annibale, B., Bonamico, M., Corleto, V., Delle Fave, G., and Solcia, E. (1994). Helicobacter pylori infection in children with antral gastrin cell hyperfunction. J Pediatr Gastroenterol Nutr. 18:152–8.

Rindi, G., Azzoni, C., La Rosa, S., Klersy, C., Paolotti, D., Rappel, S., Stolte, M., Capella, C., Bordi, C., and Solcia, E. (1999). ECL cell tumor and poorly differentiated endocrine carcinoma of the stomach: prognostic evaluation by pathological analysis. Gastroenterology. 116:532–42.

Rindi, G., Bordi, C., Rappel, S., La Rosa, S., Stolte, M., and Solcia, E. (1996). Gastric carcinoids and neuroendocrine carcinomas: pathogenesis, pathology, and behavior. World J Surg. 20:168–72.

Rindi, G., Kloppel, G., Alhman, H., Caplin, M., Couvelard, A., de Herder, W.W., Erikssson, B., Falchetti, A., Falconi, M., Komminoth, P., Korner, M., Lopes, J.M., McNicol, A.M., Nilsson, O., Perren, A., Scarpa, A., Scoazec, J.Y., and Wiedenmann, B. (2006). TNM staging of foregut (neuro)endocrine tumors: a consensus proposal including a grading system. Virchows Arch. 449:395–401.

Rindi, G., Luinetti, O., Cornaggia, M., Capella, C., and Solcia, E. (1993). Three subtypes of gastric argyrophil carcinoid and the gastric neuroendocrine carcinoma: a clinicopathologic study. Gastroenterology. 104:994–1006.

Rindi, G., Necchi, V., Savio, A., Torsello, A., Zoli, M., Locatelli, V., Raimondo, F., Cocchi, D., and Solcia, E. (2002a). Characterisation of gastric ghrelin cells in man and other mammals: studies in adult and fetal tissues. Histochem Cell Biol. 117:511–9.

Rindi, G., Paolotti, D., Fiocca, R., Wiedenmann, B., Henry, J.P., and Solcia, E. (2000). Vesicular monoamine transporter 2 as a marker of gastric enterochromaffin-like cell tumors. Virchows Arch. 436:217–23.

Rindi, G., Savio, A., Torsello, A., Zoli, M., Locatelli, V., Cocchi, D., Paolotti, D., and Solcia, E. (2002b). Ghrelin expression in gut endocrine growths. Histochem Cell Biol. 117:521–5.

Roucayrol, A.M., and Cattan, D. (1990). Evolution of fundic argyrophil cell hyperplasia in non-antral atrophic gastritis. Gastroenterology. 99:1307–14.

Rubin, W. (1973). A fine structural characterization of the proliferated endocrine cells in atrophic gastric mucosa. Am J Pathol. 70:109–18.

Ruszniewski, P., Delle Fave, G., Cadiot, G., Komminoth, P., Chung, D., Kos-Kudla, B., Kianmanesh, R., Hochhauser, D., Arnold, R., Ahlman, H., Pauwels, S., Kwekkeboom, D.J., and Rindi, G. (2006). Well-differentiated gastric tumors/carcinomas. Neuroendocrinology. 84:158–64.

Safatle-Ribeiro, A.V., Ribeiro, U., Jr., Corbett, C.E., Iriya, K., Kobata, C.H., Sakai, P., Yagi, O.K., Pinto, P.E., Jr., Zilberstein, B., and Gama-Rodrigues, J. (2007). Prognostic value of immunohistochemistry in gastric neuroendocrine (carcinoid) tumors. Eur J Gastroenterol Hepatol. 19:21–8.

Simonsson, M., Eriksson, S., Hakanson, R., Lind, T., Lonroth, H., Lundell, L., O'Connor, D.T., and Sundler, F. (1988). Endocrine cells in the human oxyntic mucosa. A histochemical study. Scand J Gastroenterol. 23:1089–99.

Sjoblom, S.M., Sipponen, P., Miettinen, M., Karonen, S.L., and Jrvinen, H.J. (1988). Gastroscopic screening for gastric carcinoids and carcinoma in pernicious anemia. Endoscopy. 20:52–6.

Sobin, L.H., and Wittekind, C. (eds). (2002). TNM Classification of malignant tumours. Wiley-Liss, New York-Toronto.

Solcia, E., Bordi, C., Creutzfeldt, W., Dayal, Y., Dayan, A.D., Falkmer, S., Grimelius, L., and Havu, N. (1988). Histopathological classification of nonantral gastric endocrine growths in man. Digestion. 41:185–200.

Solcia, E., Capella, C., and Vassallo, G. (1970). Endocrine cells of the stomach and pancreas in states of gastric hypersecretion. Rendic R Gastroenterol. 2:147–158.

Solcia, E., Capella, C., Buffa, R., Fiocca, R., Frigerio, B., and Usellini, L. (1980). Identification, ultrastructure and classification of gut endocrine cells and related growths. Invest Cell Pathol. 3:37–49.

Solcia, E., Capella, C., Buffa, R., Usellini, L., Frigerio, B., and Fontana, P. (1979). Endocrine cells of the gastrointestinal tract and related tumors. Pathobiol Annu. 9:163–204.

Solcia, E., Capella, C., Fiocca, R., Rindi, G., and Rosai, J. (1990). Gastric argyrophil carcinoidosis in patients with Zollinger-Ellison syndrome due to type 1 multiple endocrine neoplasia. A newly recognized association. Am J Surg Pathol. 14:503–13.

Solcia, E., Capella, C., Fiocca, R., Sessa, F., LaRosa, S., and Rindi, G. (1998). Disorders of the endocrine system. In Pathology of the gastrointestinal tract., S. C. Ming and H. Goldman eds, 295–322. Philadelphia: Williams and Wilkins.

Solcia, E., Capella, C., Sessa, F., Rindi, G., Cornaggia, M., Riva, C., and Villani, L. (1986). Gastric carcinoids and related endocrine growths. Digestion. 35 Suppl 1:3–22.

Solcia, E., Capella, C., Vassallo, G., and Buffa, R. (1975). Endocrine cells of the gastric mucosa. Int Rev Cytol. 42:223–86.

Solcia, E., Fiocca, R., Sessa, F., Rindi, G., Gianatti, A., Cornaggia, M., and Capella, C. (1991a). Morphology and natural history of gastric endocrine tumors. In The stomach as an endocrine organ, R. Håkanson and F. Sundler eds, 473–498. Amsterdam: Elsevier.

Solcia, E., Fiocca, R., Villani, L., Gianatti, A., Cornaggia, M., Chiaravalli, A., Curzio, M., and Capella, C. (1991b). Morphology and pathogenesis of endocrine hyperplasias, precarcinoid lesions, and carcinoids arising in chronic atrophic gastritis. Scand J Gastroenterol Suppl. 180:146–59.

Solcia, E., Fiocca, R., Villani, L., Luinetti, O., and Capella, C. (1995). Hyperplastic, dysplastic, and neoplastic enterochromaffin-like-cell proliferations of the gastric mucosa. Classification and histogenesis. Am J Surg Pathol. 19 Suppl 1:S1–7.

Solcia, E., Klöppel, G., and Sobin, L. H. (2000b). Histological typing of endocrine tumours. Springer-Verlag, New York.

Solcia, E., Rindi, G., Buffa, R., Fiocca, R., and Capella, C. (2000a). Gastric endocrine cells: types, function and growth. Regul Pept. 93:31–5.

Solcia, E., Rindi, G., Fiocca, R., Villani, L., Buffa, R., Ambrosiani, L., and Capella, C. (1992). Distinct patterns of chronic gastritis associated with carcinoid and cancer and their role in tumorigenesis. Yale J Biol Med. 65:793–804; discussion 827–9.

Srivastava, A., Kamath, A., Barry, S.A., and Dayal, Y. (2004). Ghrelin expression in hyperplastic and neoplastic proliferations of the enterochromaffin-like (ECL) cells. Endocr Pathol. 15:47–54.

Sweeney, E.C., and McDonnell, L.M. (1980). Atypical gastric carcinoids. Histopathology. 4:215–24.

Tartaglia, A., Vezzadini, C., Bianchini, S., and Vezzadini, P. (2005). Gastrinoma of the stomach: a case report. Int J Gastrointest Cancer. 35:211–6.

Tsolakis, A.V., Portela-Gomes, G.M., Stridsberg, M., Grimelius, L., Sundin, A., Eriksson, B.K., Oberg, K.E., and Janson, E.T. (2004). Malignant gastric ghrelinoma with hyperghrelinemia. J Clin Endocrinol Metab. 89:3739–44.

Wangberg, B., Grimelius, L., Granerus, G., Conradi, N., Jansson, S., and Ahlman, H. (1990). The role of gastric resection in the management of multicentric argyrophil gastric carcinoids. Surgery. 108:851–7.

Ye, W., and Nyren, O. (2003). Risk of cancers of the oesophagus and stomach by histology or subsite in patients hospitalised for pernicious anaemia. Gut. 52:938–41.

Chapter 7
Gastric Neuroendocrine Neoplasia

Irvin M. Modlin, Mark Kidd, Maximillian V. Malfertheiner, and Bjorn I. Gustafsson

Introduction

The enterochromaffin-like (ECL) cell is the best-characterized neuroendocrine cell in the gastric mucosa[1] and constitutes one type of at least seven different neuroendocrine cell types (approximately 1%–2% of total mucosal cells) that comprise the fundic mucosa. The ECL cell, however, is the predominant endocrine cell population of the gastric fundus and accounts for approximately 35%–65% of the gastric endocrine cell mass[2] (Figure 7.1). The ECL cell is a pivotal regulator of acid secretion via a mechanism that involves its activation by circulating gastrin produced by antral G cells. Activation of the ECL CCK2R elicits the release of histamine which acts in a paracrine manner to initiate parietal cell secretion of protons into the gastric lumen (Figure 7.2).[3,4] Low acid states engendered by acid suppression or loss of parietal cell mass (atrophic or autoimmune gastritis) results in diminution of acid secretion, elevated luminal pH, activation of antral G cells, hypergastrinemia, and endoscopic and histopathologic evidence of ECL cell hyperplasia and neoplasia variously recognized under the terminology of gastric neuroendocrine tumors (NETs) or gastric carcinoids.[5,6] The latter term should be discarded as archaic and confusing because it embraces at least three different tumor types and does not indicate the specific cell type or malignant phenotype of the lesion.

Gastric NETs have, in recent times, become the subject of substantial clinical and investigative interest. This reflects global concerns regarding the consequences of prolonged hypochlorhydria, longstanding hypergastrinemia (increased use of acid-suppressive pharmacotherapeutic agents), as well as the proposed putative relationship between gastric adenocarcinoma and gastric NETs.[5,7–10] These tumors were previously considered rare lesions,[11] overall representing less than 2% of all gastrointestinal NETs and less than 1% of all gastric neoplasms. The misconception of rarity is redundant because current cancer databases indicate that gastric NETs are increasing in incidence/prevalence and that the current figures are closer to 5%.[12,13] Whether this represents increased clinical awareness, more accurate pathologic identification, or more thorough endoscopic surveillance is debatable, but nevertheless, provides a far larger group of patients whose disease requires

T.C. Wang et al. (eds.) *The Biology of Gastric Cancers*,
© Springer Science+Business Media, LLC 2009

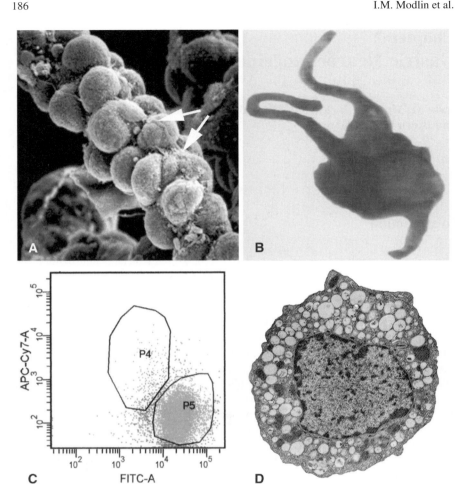

Fig. 7.1 Enterochromaffin-like (ECL) cell topography and isolation. (**A**) A scanning electron microscopic view of the basal third of a gastric gland. Numerous protruding parietal cells are visible in addition to a lesser number of small, compact ECL cells with dendritic processes (arrow) that embrace the basolateral surfaces of the parietal cells. (**B**) Photomicrograph of an HDC-labeled, formalin-fixed, dispersed human rat ECL cell demonstrates the typical morphology with sinuous basal projections exhibiting bulbar terminals containing accumulations of histamine-filled granules. (**C**) Micrograph of FACS-sorted human gastric mucosal cell preparations. ECL cells (P5) show mainly emission at 532 nm (green—FITC-A) whereas red-labeled parietal cells (P4) show 757 nm (red—APC-Cy7-A) and 532-nm emission. Green-labeled ECL cells can be gated and FACS sorted to produce cell suspensions with ~98% ± 2.3% ECL cell purity. (**D**) EM of an isolated ECL cell demonstrating a large central nucleus surrounded by a granular cytoplasm containing a typical admixture of dense peptide/amine containing granules and electroluscent empty vesicles (4,800× magnification). FACS, fluorescent activated cell sorting; FITC, fluorescein isothiocyanate, EM, electron microscopy; HDC, histidine decarboxylase

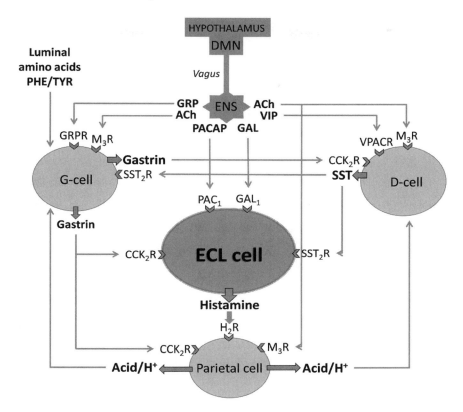

Fig. 7.2 The central role of the enterochromaffin-like (ECL) cell in the regulation of acid secretion. Central regulation occurs via vagal outflow from the dorsal motor nucleus (DMN) of the hypothalamus which is activated by smell, taste, and cortical input. The vagal response is transmitted to gastric enteric neurons [enteric nervous system (ENS)] in the wall of the stomach that activate postganglionic fibers that impinge on a variety of neuroendocrine cells. Thus, PACAP release targets the PAC_1 receptor on the ECL cell and VPAC on the D cell, whereas acetylcholine (M3 receptor) and GRP (GRP receptor) stimulate the gastrin cell. The G cell is, in addition, stimulated to release gastrin by luminal aromatic amino acids including phenylalanine and tyrosine. Negative regulation occurs through galanin-mediated inhibition of the GAL_1 receptor on the ECL cell and M3 receptor mediated inhibition of D and parietal cells. Circulating gastrin stimulates the ECL cell to release histamine which activates the H_2 receptor of the parietal cell to produce acid. Paracrine release of gastrin stimulates somatostatin release from the D cell and provides a local feedback regulation of gastrin release because the D cell is also responsive to luminal pH. Although there is a gastrin (CCK2) receptor on the parietal cell, its role in the stimulation of acid secretion is not well understood. Green arrows indicate stimulation and red arrows inhibition

management. Delineation of the regulation of ECL cell proliferation, characterization of its degree of transformation, and determination of its malignant potential are necessary adjuncts for the development of a rational strategy for clinical management. As a result of the above factors, an intense clinical and scientific scrutiny of gastric neuroendocrine ECL cell tumors has developed.[12]

This chapter examines the biology of the ECL cell, the role of animal models in the investigation of gastric NET pathobiology, and reviews current information regarding human gastric NETs in terms of their diagnosis, management, and outcome.

Animal Models of Enterochromaffin-Like Cell Neoplasia

Mastomys

Mastomys (*Praomys natalensis*) is a sub-Saharan African muroid rodent phylogenetically related to the mouse,[14] that spontaneously develops gastric NETs; 20%–50% of these animals exhibit such lesions by 2 years of age.[15] Serum gastrin levels in these animals are normal[16] and the development of normogastrinemic ECL cell tumors is likely attributable to a gastrin receptor mutant that shows ligand-independent activity.[17] Studies have demonstrated that the *Mastomys* CCK2 receptor, when expressed in COS-7 cells, differs from human, canine, and rat receptor homologs by its ability to constitutively activate inositol phosphate formation.[17] Functional characterization has revealed that three amino acids from the *Mastomys* transmembrane domain VI through the C-terminal end are sufficient to confer constitutive activity. Mutagenesis studies using a combination of ^{344}L, ^{353}I, and ^{407}D confer a level of comparable ligand-independent signaling when introduced into the human receptor.[17]

Although multiple naturally occurring amino acid polymorphisms and/or mutations may result in an enhanced basal level of CCK2 receptor activity, endogenous gastrin, however, is required for ECL cell tumor development in the *Mastomys*.[18] Thus, drug-induced hypergastrinemia consequent upon acid suppression, e.g., after oral ingestion of the histamine H2 receptor blockers (loxtidine, cimetidine) or omeprazole [proton pump inhibitor (PPI) class of agents] significantly accelerates the development of ECL cell tumors (Figure 7.3).[19]

Gastrin

Gastrin-mediated growth regulation is the end result of a ligand-receptor activated signal transduction cascade (usually the MAP kinase pathway[20]) and induction of the AP-1 activator protein-1 (AP-1) complex (a fos/jun-mer) transcription factor[21] which regulates genes necessary for cell-cycle progression (e.g., cyclin genes).[22] Normal ECL cell proliferation is associated with activation of *fos/jun* transcription by the MAPK pathway (ERK1/2) after gastrin-mediated Ras activation.[23] In these normal cells, gastrin activates the Ras-MAPK pathway and mediates cell growth via upstream activation of phosphatidylinositol 3-kinase (PI3K) and the protein kinase B (PKB)/protein kinase C (PKC) pathways. In addition,

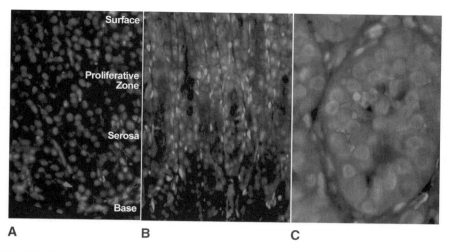

Fig. 7.3 Gastrin-stimulated enterochromaffin-like (ECL) cell proliferation in *Mastomys*. (**A**) Immunostaining of ECL cells (red cells) in normal mucosa with anti-HDC antibodies demonstrating the predominantly basal location and relatively low numbers of ECL cells. Cy5 anti-HDC expression: red and nuclei: DAPI—blue (150× magnification). (**B**) ECL cell distribution in hyperplastic gastric mucosa after 8 weeks of acid suppression demonstrating an increase in ECL cell number (hyperplasia-red cells). Cy5 anti-HDC expression: red and nuclei: DAPI—blue (150× magnification). (**C**) ECL cell neoplasm after16 weeks of acid suppression demonstrating nodule formation. Cy5 anti-HDC expression: red and nuclei: DAPI—blue (400× magnification). HDC, histidine decarboxylase; DAPI, 4′6-diamidino-2-phenylindole

there is evidence that gastrin also upregulates epidermal growth factor (EGF)/ transforming growth factor (TGF)α and the EGF receptor in neoplastic ECL cells.[24] It is thus likely that perturbations in growth factor production and/or responsiveness are implicated in increased ECL cell proliferation, and ultimately NET neoplasia.[24]

AP-1 and Menin

Mechanisms by which gastrin mediates ECL tumorigenesis include a decrease in expression of the negative regulators (JunD and menin) of AP-1 which regulates cell-cycle progression via cyclin D1 expression.[19] Cloning and sequencing the *MEN* gene from *Mastomys* tumors has identified one nucleotide alteration in a second codon position: GGG→GCG resulting in a Gly→Ala$_{511}$ amino acid change in menin.[19] This resulted in a novel missense mutation but no frameshift mutations were noted. No alterations were noted in the JunD binding region or the nuclear localization signals within *Mastomys* menin. The functionality of the missense at codon 511 is unknown, but these data suggest that alterations in the menin pathway, particularly related to AP-1 function, may be implicated in the genesis of gastrin-mediated ECL cell neoplasia.

Histamine 1 Receptor

Whole genome analysis of the *Mastomys* ECL cell tumors has identified compositional changes in AP-1 and upregulation of the histamine H1 receptor in the ECL cell.[19] Blockade of the H1 receptor with the specific receptor antagonist, terfenadine, inhibited DNA synthesis in cultured tumor cells demonstrating that the H1 receptor has a significant influence on ECL cell proliferation.[19]

CCN2/Connective Tissue Growth Factor

Microarray techniques have demonstrated overexpression of the growth factor CCN2 in the *Mastomys* ECL cell tumor model[25] and a transgenic mouse model of metastatic gastric neuroendocrine cancer.[26] CCN2 or connective tissue growth factor (CTGF) is a prototypic member of the CCN family of proteins,[27] and is involved in diverse biologic processes including fibrosis, regulation of cell division, differentiation, embryogenesis, chemotaxis, apoptosis, adhesion, motility, and ion transport.[28] It also has an important and ubiquitous role in neoplasia but its precise role requires further elucidation.[29–32]

Investigations of CCN2 expression in pure preparations (~98%) of *Mastomys* ECL cells demonstrated transcript expression in tumor cells but an absence in normal cells.[25] Of note was the observation that recombinant CCN2 (comprising the C-terminal 98 amino acids containing an EGF binding domain) failed to stimulate normal ECL cell proliferation but caused a 17-fold increase in *Mastomys* tumor ECL cell proliferation [median effective concentration $(EC_{50}) = 0.01$ ng/mL].[25] This proliferative response was augmented by EGF (>26-fold, $EC_{50} = 6$ pg/mL), was mediated through activation of the ERK1/2 pathway, and could be inhibited by the MEK inhibitor, PD98059, but not by the PI3K inhibitor, wortmannin (AKT pathway).[25] Thus, growth factor (CCN2)-mediated growth regulation in neoplastic ECL cells occurs principally through the MAPK pathway (ERK1/2) after CCN2-mediated receptor activation. This is similar to gastrin-mediated growth regulation of normal cells but differs because the PI3K pathway is not activated (Table 7.1). CCN2, which is expressed in gastric rodent NET cells, therefore functions as a proliferative agent during NET neoplasia (Figure 7.4).

Cotton Rat

Female hispid cotton rats (*Sigmodon hispidus*) spontaneously develop gastric carcinomas by approximately 10–16 months of age that appear to have a neuroendocrine cell component.[33] As in the *Mastomys*, ECL cell–derived tumors can be accelerated by pharmacologic acid suppression but this usually takes longer than 6 months.[34] A proportion of the female cotton rats also develop spontaneous gastric hypoacidity and hypergastrinemia.[35] However, unlike the *Mastomys*, these animals

Table 7.1 Ligand-induced growth-mediated signal transduction events in normal and neoplastic enterochromaffin-like cells

	Normal	Neoplasia
Ligand	Gastrin	CCN2/EGF
Receptor	CCK2	LRP/EGF$_R$
Signal transduction-1	PKB/C activation and PI3K production	—*
Signal transduction-2	RAS activation	ND
Signal transduction-3	MAPK(ERK1/2) phosphorylation	MAPK(ERK1/2) phosphorylation
Gene target-1	FOS/JUN ⇧	FOS/JUN⇧ JUN D, MENIN⇩
Gene target-2	Cyclins	Cyclin D1⇧ H$_1$R/EG F$_R$⇧, CCN2⇧

ND: no data.
*No effect.

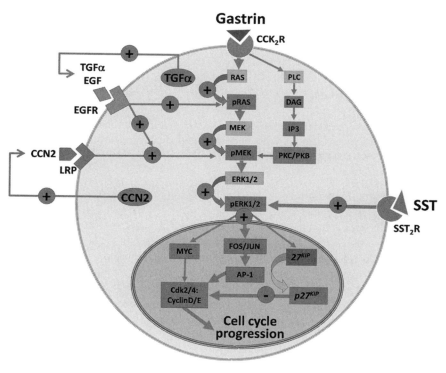

Fig. 7.4 A cartoon of the mechanisms governing cell-cycle regulation in ECL cells. Gastrin regulates the cell cycle via MAPK pathway (Ras-MEK-ERK1/2) activated fos/jun expression and AP-1–regulated cyclin transcription. This is complemented by IP3-mediated activation of the PKC/PKB pathway which coactivates the MAPK pathway. Somatostatin (SST) inhibits proliferation via MAPK-mediated phosphorylation of the cell-cycle inhibitor, P27^{KIP1}, which when activated, inhibits the cyclin E complex. In the transformed ECL cell, the growth factors, CCN2 and TGFα are upregulated and exert an autoregulatory proliferative effect via activation of similar cell-cycle regulatory pathways (MAPK-AP-1 and MAPK-MYC-cyclin). The effects of CCN2 may be directly augmented by EGF-mediated MAPK activation

develop adenocarcinomas rather than NETs. The tumors appear to develop from ECL cells of the oxyntic mucosa with hyperplasia of chromogranin A (CgA) immunoreactive cells, whereas a proportion of the adenocarcinoma tumor cells are both CgA and Sevier-Munger positive.[35,36] Carcinoma development can be prevented by a gastrin receptor (CCK2) antagonist YF476 indicating that gastrin has an important role in tumorigenesis.[36] It is currently unclear, however, whether this lesion reflects an aggressive NET with an adenocarcinomatous phenotype, an adenocarcinoma with a neuroendocrine phenotype, or a "true" neuroendocrine carcinoma (NEC). It is likely that the resolution of this issue requires identification of the neuroendocrine cell of origin and delineation of the specific determinants of cell lineage.

Genetically Engineered Models

A number of receptor knockout and transgenic models have been developed that provide insight into the relationship of receptor activation and ECL cell proliferation (Table 7.2).

H$_2$ Receptor Knockouts

Histamine H2 receptor–deficient mice are characterized by an increase in ECL cell numbers, which is associated with both elevated histamine levels (+300%) and gastrin levels (+400%).[37]

CCK2 Receptor Knockouts

CCK2 receptor knockout mice, in contrast, exhibit a >50% decrease in ECL cell numbers despite a 1,000% increase in circulating gastrin levels.[38] In the absence of CCK2 expression, ECL cells do not proliferate and tumors do not develop.

Gastrin and histamine receptor activation therefore are required for ECL cell development and although these studies do not examine neoplasia *per se*, they suggest that ECL cell differentiation in the gastric niche can be affected by activation of specific (gastrin and histamine) G-protein coupled receptors.

Table 7.2 Receptor knockout and transgenic mouse models used to study enterochromaffin-like (ECL) cell biology

Knockout/ gene target	Physiologic/ cellular effect	Effect on ECL cell numbers	Effect on serum gastrin
Histamine 2 receptor	H$_2$ receptor lost	⇧	⇧
CCK2 receptor	CCK2 receptor lost	⇩	⇧
MEN1 transgenic	Menin, nonfunctional	⇔	ND
SV40 large T antigen	*atp4b* gene, nonfunction	⇧	ND
Reg	Reg overexpressed	⇔	ND

ND: no data.

MEN1 Transgenic Mouse

A mouse model heterozygous for deletion in the MEN1 locus[39] developed a variety of tumors consistent with human MEN1 syndrome, but only one animal (3.7%) developed a gastric NET. These animals exhibited hyperinsulinemia; gastrin levels, however, were not documented. The development of gastric NETs in humans with type II lesions requires a combination of a MEN1 mutation as well as elevated gastrin levels.[40]

SV40 Transgenic Mouse

A transgenic mouse carrying the human SV40 large T antigen in the *atp4b* gene[26] develops metastatic gastric NECs that are not derived from the ECL cell but from increases in the progenitor oxynto-peptic cell lineage.[26] This suggests that a neuroendocrine cell might not necessarily be the precursor of gastric neuroendocrine tumors and that NECs might be derived from a different progenitor cell pool. A microarray analysis of this experimentally induced tumor identified overexpression of ~200 genes, one of which, CCN2, is also overexpressed in *Mastomys* tumors.[25]

Reg Transgenic Mouse

Reg (regenerating gene product) is an ECL cell product and transgenic mice that overexpress Reg exhibit a marked increase in the thickness of the fundic mucosa.[41] This expansion reflects increased chief and parietal cell populations but no alteration in ECL cell numbers. It seems that ECL cell–derived Reg has a growth-promoting effect on gastric progenitor cells and may direct the differentiation of the cells into chief cell and parietal cell lineages but does not seem to be associated with ECL cell tumorigenesis.

The menin and Reg gene products, when mutated or overexpressed, however, are not associated with alterations in the ECL cell phenotype.[39,41] In contrast, increasing a preprogenitor cell lineage (preparietal cells), is associated with the development of ECL cell neoplasia,[26] suggesting, at least in murine models, that mutations in genes directing progenitor cell populations may result in ECL cell neoplasia.

Helicobacter pylori *Models*

Helicobacter pylori inoculation of transgenic mice overexpressing amidated gastrin (INS-GAS)[42] or Mongolian gerbils[43] results in the development of gastric adenocarcinoma. Infected INS-GAS mice have no evidence of ECL cell hyperplasia,[44] but male Mongolian gerbils infected with CagA-positive *H. pylori* for longer than 24 months developed neuroendocrine cell dysplasia and NETs with marked atrophic gastritis of the oxyntic mucosa.[45] Serum gastrin levels were increased ~1,000% in these animals. This suggests that *H. pylori* infection is capable of producing not only gastric carcinomas but also ECL cell tumors in Mongolian gerbils, and that hypergastrinemia has a substantial role.

Although the mechanism by which *H. pylori* causes NET development is considered to be through alterations in gastrin levels, bacterial cell wall lipopolysaccharides are bioactive and potentially mitogenic.[46] In the *Mastomys* model, *H. pylori* lipopolysaccharide induced tumor ECL cell proliferation *in vitro* by activation of the CD14 receptor, ornithine decarboxylase activity, and polyamine biosynthesis.[47] Thus, alternative *H. pylori*–related mechanisms may be involved in ECL cell proliferation.

Low Acid States/Hypergastrinemia

Protracted sustained acid suppression is well known to stimulate ECL cell growth in the gastric mucosa of both mice and rats.[48,49] The mechanism for the development of ECL cell hyperplasia and neoplasia under these circumstances is an increase in serum gastrin levels with a concomitant direct proliferative effect on the ECL cells. However, in other models in which hypergastrinemia is engendered, e.g., after total parietal cell atrophy following oral administration of the protonophore and antiinflammatory agent, the neutrophil elastase inhibitor, DMP 777,[50] ECL cell hyperplasia does not occur. Ciprofibrate, a common lipid-lowering drug that is also a ligand for and modulator of the peroxisome proliferator-activated receptor transcription factor, also causes parietal cell loss. This, however, is associated with both hypergastrinemia and secondary ECL cell hyperplasia.[51] The reasons for these differences are unclear but gastric inflammation was not evident in the male rats treated with DMP-777[50] suggesting that a combination of hypergastrinemia and gastric bacterial colonization may be required for ECL cell development.

Human Gastric Neuroendocrine Tumors

Prevalence

Gastric NETs were previously considered to be extremely rare lesions (Figure 7.5).[11] In the preendoscopic era, they comprised ~0.3% of all gastric tumors and 1.9% of all gastrointestinal NETs.[11,52] The widespread use of endoscopy, increased awareness, and the advance of histopathology and immunohistochemical techniques have led to an increase in the identification of such lesions.[53] More recent studies

Fig. 7.5 The first description of a gastric carcinoid tumor. Max Askanazy (1865–1940) (top right), professor of pathology at Geneva, provided the first description of gastric carcinoids in 1923 (top). His report documented two instances of gastric carcinoid tumors (bottom left—histology) identified at autopsy in elderly patients who had succumbed from pneumonia and the sequelae of prostate hypertrophy, respectively. Both this observation and his initial prescient identification of the large acidophilic parafollicular cells in canine thyroids in 1898 (subsequently erroneously attributed to "Hurthle") have been, for the most part, overlooked or in the latter instance misattributed

indicate that as many as 10%–30% of all NETs may occur in the stomach.[54] Gastric NETs may be divided into a group associated with hypergastrinemia and a second group that occurs under normogastrinemic circumstances. The former exhibit an increased incidence in individuals with atrophic gastritis, pernicious anemia, a variety of autoimmune diseases, and MEN1-associated gastrinoma.[55]

Gastric Neuroendocrine Tumors Incidence: Surveillance, Epidemiology, and End Results Database Evaluation

A 50-year analysis of 562 gastric NETs from the National Cancer Institute (NCI) database indicates an increase in these tumors among all gastric malignancies from 0.3% to 1.8%.[12,13] In addition, the proportion of gastric NETs among all gastrointestinal NETs has increased from 2.4% to 7.1%.[13] Data from the current Surveillance, Epidemiology, and End Results (SEER) database

(2003) identifies an annual increase in gastric NETs of 6% since 1985 with a current calculated incidence of the disease in the U.S. population of ~0.33/100,000.[13] Age-adjusted incidence rates among male, female, black, and white population subsets have all increased since 1973, with the greatest increase (800%) evident in white females and a decrease in the male/female ratio from 0.90 to 0.54.[13]

In the SEER database, 67.5% of gastric NETs were localized, 3.1% regionalized, 6.5% had distant metastases, and 22.9% were unstaged,[12,13] and an association with other malignant neoplasms was evident in 7.8% of cases. The occurrence of synchronous or metachronous non-NET tumors with gastric NETs has decreased by 26% during the course of SEER data collection (1973–2003). The average age at diagnosis over the last 30 years has remained stable (~63 years). The 5-year survival rate for gastric NETs overall has increased from 51% to 63% during the same time period.[13] This presumably reflects the impact of endoscopy on earlier diagnosis of the disease.

The U.S. epidemiologic data demonstrate that gastric NETs have increased in incidence over the last three decades (Figure 7.6), and the different increases across gender and race divisions might reflect genetic-based propensities (or protection) among certain ethnic populations. The clinical and pathologic recognition of the relationship among ECL cell hyperplasia, elevated gastric pH, and hypergastrinemia has increased the level of awareness of gastric NETs and increased biopsy identification of the disease. Although the SEER data analysis cannot assess the role of acid-suppressive medications, some studies have suggested a correlation between the temporal increase in the incidence of gastric NETs and the introduction and widespread use of acid-suppressive medication since the late 1980s.[56]

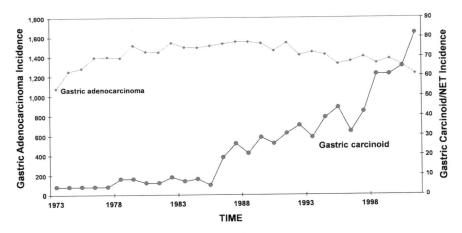

Fig. 7.6 Incidence of gastric carcinoids and gastric adenocarcinomas in U.S. patients. Yearly fold increase in gastric carcinoids (1973–2003) demonstrates >6% annual increase since 1985. (Adapted from the NCI–SEER database, 2005.[53])

General Clinical Presentation

Gastric NETs present with nonspecific symptoms and signs of a gastric mucosal lesion. These include pain (69%), vomiting, upper gastrointestinal (GI) bleeding, dyspepsia, anemia (72%), heme-positive stools, and gastric polyps at endoscopy.[57] Presentation with symptoms of classic carcinoid syndrome (flushing, sweating, itching, lacrimation, and bronchospasm) is unusual (<10%); rather symptoms are often related to the ingestion of food or alcohol. In such instances, the lesion is usually a "sporadic" gastric carcinoid tumor with a serotonin-producing EC cell component that exhibits aggressive behavior more reflective of a NEC or gastric adenocarcinoma. The clinical presentation of the hypergastrinemic group of gastric NETs often occur in association with a variety of autoimmune and endocrine abnormalities including atrophic gastritis (67%), pernicious anemia (58%), hypothyroidism (39%), diabetes (19%), Addison's disease (6%), and hyperparathy-roidism (6%).[58] In general, hypergastrinemia-associated NETs occur more frequently in women and 70%–80% are diagnosed between the fifth and seventh decades.[6,59] The normogastrinemic-associated tumors occur more often in men older than 50 years of age.[6,59]

Gastrin-Sensitive (Type I/II) Neuroendocrine Tumors

The presence of a gastrin-related lesion may be inferred if a history of an autoim-mune disease such as pernicious anemia is present or previous studies indicate underlying chronic atrophic gastritis associated with achlorhydria. Serum gastrin levels in both type I and type II NETs are usually significantly elevated. A secretin test or measurement of acid secretion levels may suggest the gastrinoma origin of the type II–associated lesion. In addition, genetic screening for the *MEN1* gene may facilitate the early diagnosis of hereditary NETs.[40] However, clinical presentations may exhibit wide variations, and include diverse symptomatology indistinguishable from peptic ulcer, gastric polyps, or even carcinoma. These include vomiting, diarrhea, gastrointestinal bleeding, or intermittent gastric outlet obstruction. A clas-sic carcinoid syndrome or atypical carcinoid syndrome attributed to release of his-tamine, or bradykinin-related peptides, is extremely rare in type I and II tumors.[60]

Gastrin-Autonomous (Type III) Neuroendocrine Tumors

These tumors may present as gastric adenocarcinomas with loss of appetite and weight as well as anemia and evidence of local and advanced metastatic disease.[6,53] The atypical carcinoid syndrome may also be evident in some type III lesions and is characterized by cutaneous flushing, profound itching, bronchospasm, and lacrima-tion. This usually reflects the sequelae of unregulated ECL cell histamine production (although other agents, e.g., tachykinins, have been reported) and is often provoked by ingestion of certain kinds of foods, particularly cheeses (tyramine) and wine.[61]

Biologic Relationships

Low Acid States

Low acid states, induced either by endogenous parietal cell destruction (autoimmune disease) or by exogenous pharmacotherapeutic agents such as H_2RAs or PPIs, result in G-cell hypersecretion and culminate in hypergastrinemia. The association between low acid states and gastric neoplasia is well documented.[6,62] Similarly, the gastrin elevation noted in atrophic gastritis and the trophic effect of gastrin on ECL cells is consistent with the hypothesis that a low acid state with elevated plasma gastrin levels drives ECL proliferation.[10] Although it is likely that other trophic regulatory agents (TGFα, β FGF, CCN2, Reg) are implicated, gastrin seems to be the dominant effector.[6,53]

Immune Disease

Autoimmune gastritis is associated with ECL hyperplasia.[63] The postulated mechanism underlying this association is that ECL cell hyperplasia develops in response to glandular damage and the resultant increased serum gastrin. Glandular damage is caused by deep or diffuse lymphocytoplasmatic infiltration within the lamina propria which results in epithelial metaplasia and parietal cell loss and consequently, hypergastrinemia. Longstanding pernicious anemia is also associated with hypergastrinemia and the development of multiple gastric NETs with a background of diffuse ECL cell hyperplasia.[64] These data indicate that, as in animal models, low acid states and/or chronic inflammation of the oxyntic mucosa and atrophy of the oxyntic glands is associated with the development of achlorhydria and hypergastrinemia and results in ECL cell hyperplasia with progression to neoplasia.

Pathology and Pathobiology

Histopathology and Histopathogenesis

In humans, ECL cell lesions have been classified as pseudohyperplasia (cell clustering unassociated with cell proliferation), hyperplasia (diffuse, linear, micronodular, adenomatoid), dysplasia (enlarged, adenomatous or fused micronodules, microinfiltration, nodular growth), and neoplasia (intramucosal or invasive carcinoids). The entire spectrum of ECL cell proliferation, from hyperplasia to dysplasia and neoplasia, has been observed in type I and II NETs. Hyperplastic and pseudohyperplastic changes also occur with some frequency in *H. pylori*–related chronic gastritis associated with ulcer disease or dyspepsia.[65] Controversy, however, exists as to whether human ECL cells actually proliferate.

The current concept of ECL cell histogenesis is based on rodent models of cell transformation from a single self-replicating ECL cell under the proliferative control of gastrin that progresses from hyperplasia to dysplasia and then develops into a NET. In the *Mastomys* model, ECL cells express the cell-cycle marker Ki-67 through all histologic stages of ECL cell transformation from normal through hyperplasia to ECL cell NET.[66] This suggests that the *Mastomys* ECL cell is self-replicating and is transformed by hypergastrinemia into neoplasia. In contrast, in human samples, no replicating (Ki-67–positive) ECL cells have been identified in either normal mucosa or in the different stages of hyperplasia (diffuse, linear, or micronodular).[66] Type III NETs, in contrast, demonstrate positive staining for Ki-67. These data suggest that nontransformed and potentially "gastrin"-regulatable ECL cells (in humans) do not enter the cell cycle. Any increase in ECL cell numbers under conditions of hypergastrinemia or chronic atrophic gastritis (CAG) may instead reflect cell clustering unassociated with cell proliferation. CAG results in wholesale destruction of gastric crypts and it is possible that hyperplasia of ECL cells reflect aggregation of a cell type resistant to immune attack.[66,67]

General Pathologic Classification

Based on the distinct pathobiologic behavior of gastric neuroendocrine tumors, three NET tumor types have been proposed in humans: type I—associated with type A CAG (CAG/A); type II—associated with a Zollinger-Ellison syndrome (ZES)/MEN1; and type III—sporadic gastric carcinoids (Figure 7.7).[57] Type I and II lesions are associated with hypergastrinemia and tumors consist mainly of ECL cells. Type III lesions, although consisting predominantly of ECL cells, may also contain serotonin-, somatostatin-, or even gastrin-positive cells.[68] It has been proposed that human ECL cell lesions may give rise to a significant proportion of the "diffuse-type" gastric carcinomas in a manner analogous to cotton rats.[69] This hypothesis, although intriguing, has received little support.[70]

Type I Neuroendocrine Tumors

Type I is the most frequent tumor type and comprises ~65% of all gastric NETs.[71] The lesion is localized in atrophic oxyntic mucosa in the fundus in individuals with CAG/A with or without pernicious anemia. Characteristically, the lesions are multicentric, small and polypoid, and exhibit little propensity to metastasize (<2%–3%). They tend to behave less aggressively than sporadic (type III) gastric NETs.[1] In the histologic classification of carcinoids (Munich, 1994), small (<1.0 cm) gastric NETs associated with CAG/A were considered well differentiated, limited to the mucosa-submucosa, without angioinvasion, and benign in behavior. The larger-size (1–2 cm) tumors may exhibit low-grade malignant

I.M. Modlin et al.

Gastric Carcinoid/NET Tumor Types

NOMENCLATURE	ENDOSCOPIC APPEARANCE	TUMOR CHARACTERISTICS	HISTOLOGY	HYPER-GASTRINEMIA	BIOLOGICAL BEHAVIOR
TYPE I	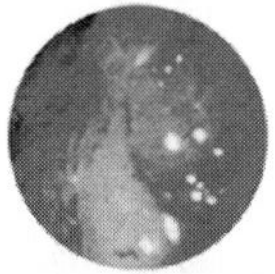	Generally small (<1cm) and multiple; often nodular/polypoid	ECL cell lesion. Stages of ECL cell hyperplasia, dysplasia, and neoplasia present in adjacent mucosa	Present	Slow growth, regional or distant metastases extremely rare (<5%) 5-year survival >95%
TYPE II		Generally small (<1cm) and multiple	ECL cell lesion. Stages of ECL cell hyperplasia, dysplasia, and neoplasia present in adjacent mucosa	Present	Slow growth, may metastasize more often (7-12%) than CAG-associated lesions. 5-year survival high (70-90%) but dependent on gastrinoma prognosis.
TYPE III	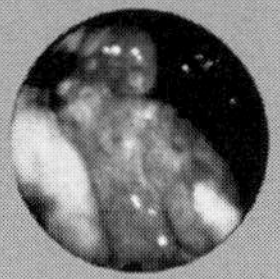	Solitary, often large (>1 cm)	ECL, EC, or X cells. Tumor formation w/o evidence of hyperplasia or precarcinoid dysplasia in adjacent mucosa	Absent	Relatively aggressive growth, frequent metastases to regional nodes (55%) and liver (24%). 5-year survival <35%.

Fig. 7.7 The clinical, biochemical, and pathologic characteristics of gastric carcinoids. Type III represents a heterogeneous group of lesions that require further delineation

behavior, with or without angioinvasion. Tumors in the latter group may be either single or multiple, exhibit a low rate of lymph node invasion (3%–8%), and rarely (~2%) are associated with distant metastases.[72]

Type II Neuroendocrine Tumors

Type II lesions, similar to type I NETs, consist mainly of ECL cells,[73] and exhibit argyrophil cell hyperplasia/dysplasia throughout the oxyntic mucosa which, in contrast to CAG/A, is usually rugose. These tumors, associated with ZES/MEN1, are usually multiple, small (<1 cm), and predominantly of the ECL cell type, although some lesions contain heterogeneous cell populations.[1] The cytologic characteristics of the lesion are similar to the type I, gastrin-dependent NETs. The clinicopathologic behavior of type II NETs occupies an intermediate position between that of the aggressive (gastrin-independent) type III sporadic lesions and the more benign type I NETs. In contrast to the latter, local infiltration may occur at mucosal and submucosal levels and metastases occur in ~12%.[57]

Type III Neuroendocrine Tumors

Type III (sporadic NETs) are less frequent (21%) and display a moderately aggressive behavior with invasive growth and a high incidence (24%–55%) of metastasis.[57,71] They are usually large, solitary lesions that evolve in normal gastric mucosa with normal plasma gastrin levels. Type III lesions often display markedly aggressive local behavior, and metastasize. Although the tendency to metastasize correlates with tumor size, minute tumors have been reported with spread.[74] Factors that predict aggressive behavior include cellular atypia, two or more mitoses per 10 high-powered fields, angioinvasion, and transmural invasion.[57,75]

Sporadic gastric NETs display a fairly uniform light microscopic appearance with typical "carcinoid" histopathologic features. Their growth pattern may be trabecular or gyriform, medullary or solid, glandular or rosette-like, or a combination of any of these types.[61] Overall, type III NETs exhibit a greater similarity to NECs than to NETs *per se*. In this respect, their biologic behavior is aggressive, and local invasive growth and distant metastases are predictable features of their evolution. At diagnosis, local spread is present in approximately 15% of patients and hepatic metastases in ~50%.[53] The 5-year survival rate is significantly higher for localized disease (64.3%) and for lesions with regional metastases (29.9%) than for lesions with distant metastases (10%).[12,13]

A subtype of the type III grouping, which is particularly aggressive in its behavior, is the "atypical" of sporadic carcinoid.[53] This lesion is associated with rapid local gastric progression, distant metastases, and early death. These may be identified by an increased mitotic count with nuclear polymorphism, hypochromasia, and prominent nucleoli.[6] Whether this tumor is any different from a NEC is not clear.

Some classifications include NECs whose appearance and behavior are indistinguishable from those of an adenocarcinoma except that varying percentages of neuroendocrine cells (CgA positive) can be identified within the tumor matrix.[6] NECs, previously known as "atypical carcinoids," represent an aggressive neuroendocrine neoplasm that bears a greater resemblance to type III NETs than to gastrin-associated tumors. These lesions display invasive growth, metastasize with great frequency, and progress rapidly with a prognosis indistinguishable from gastric adenocarcinoma.[57]

The World Health Organization Pathologic Classification

In an attempt to reconcile the high degree of morphologic and biologic heterogeneity of gastrointestinal NETs, the World Health Organization (WHO), in 2000, proposed an updated pathologic schema based on the localization, biology, and prognosis of individual NETs.[76] In regard to gastric NETs, a distinction was made between well-differentiated NETs, which exhibit benign behavior or uncertain malignant potential (WHO type 1); well-differentiated NECs, which are characterized by low-grade malignancy (WHO type 2); and poorly differentiated (usually small cell) NECs of high-grade malignancy (WHO type 3).[77]

WHO type 1 tumors include the traditional type I and type II carcinoids and are characterized as being small in size (1–2 cm), nonfunctioning, confined to the mucosa–submucosa, and of benign or low-grade malignant potential. WHO type 2 tumors include all lesions, even the traditionally annotated type III carcinoids and are characterized as follows: >2 cm in size, have invaded the muscularis propria with or without metastases, and are either nonfunctional or are secretory tumors. WHO type 3 tumors include the traditionally annotated "atypical" or "sporadic" gastric NETs. These have a high grade of malignancy and are considered poorly differentiated NECs. The WHO classification is not generally used, and to avoid confusion, this chapter uses the currently accepted general pathologic classification (types I, II, and III) to distinguish the gastric carcinoid subtypes.

Relationship to Multiple Endocrine Neoplasia Mutations

Multiple endocrine neoplasia syndrome type I (*MEN1*)[78] is an autosomal dominant disorder associated with the gene locus *MEN1* located on 11q13. Its protein product (menin) is involved in transcriptional regulation and genome stability.[78] Somatic mutations, loss of heterozygosity (LOH), or deletion of the wild-type allele results in loss of tumor suppressor function of the *MEN1* gene. Studies demonstrate that one-third of individuals with MEN1 develop gastric NETs, and loss of heterozygosity at the 11q13 location occurs in 75% of ZES/MEN1 NETs and in 41% of MEN1 gastrinomas.[79]

LOH in the 11q13-14 region is also frequently found in type I NETs but infrequently in type III lesions.[80] This suggests a potential involvement for the *MEN1* gene and/or a more telomeric tumor-suppressor gene in the pathogenesis of type I NETs. A low rate (~25%) of LOH at 11q13-14 in type III NETs suggests that the MEN locus might not have a role in the etiopathogenesis of the gastrin-autonomous tumor type.

Molecular Biologic Classification

Hierarchical cluster analysis of gastric molecular signatures demonstrates that type I/II NETs cluster with normal mucosa whereas type III and NECs cocluster.[81] This indicates that gene expression levels are not usually shared by these tumor types and that the etiopathogenesis of gastrin-dependent NETs is different than gastrin-autonomous tumors. An examination of genes that were differently expressed in these samples demonstrates that type I NETs exhibit altered regulation of immunologic genes consistent with the evidence of histologic atrophy and the presence of gastritis.[81] Type II NETs had elevated *fos* genes (more than fourfold, p < 0.001). Type III NETs are characterized by a downregulation of factors in the AP-1 pathway, *fos* and *junD* (more than twofold, p < 0.001). This suggests that alterations in AP-1 occur in human tumors and that differential regulation of this transcriptional machinery (and/or its regulators) may have a role in human ECL cell malignancy. The role of the H1 receptor in mediating ECL cell proliferation is unknown but might be of relevance given its overexpression in type III NETS in the microarray studies.[81]

Malignant Signatures and Candidate Genes

Approximately 270 candidate genes are differentially altered in human type III NETs and NECs compared with type I/II NETs.[81] Using a supervised selection approach to identify biologically relevant genes, candidate markers including CgA (+6.2, p = 0.012); NALP1 (−1 to +2.3; p = 0.017); MAGE-D2 (0.5 to 2.3; p = 0.05), MTA1 (0.6 to 2.1; p = 0.05), and CCN2 (+1.2 to +2.3; p = 0.023) were identified in malignant ECL cell tumors.[82] An evaluation of individual marker transcript expression in gastric NETs, gastric adenocarcinomas, and GISTs indicated that CgA differentiated all three types of NETs from other gastric tumors that do not express this message. In addition, type III lesions had significantly higher levels of CgA mRNA than type I.[83] Expression of the adhesin, MAGE-D2, also differentiated malignant NETs (high expression) from type I tumors (low expression) whereas the regulator of histone deacetylase, MTA1, and the growth factor CCN2, were similarly overexpressed in the type III tumors.[25,82]

The transcript studies were confirmed using a tissue microarray approach and automated quantitative analyses of immunohistochemical staining of the candidate genes. CgA protein was overexpressed in NETs compared with other gastric neoplasia.[83] MTA1, although overexpressed in all gastric neoplasia, was significantly overexpressed in gastrin-autonomous carcinoids compared with type I/II NETs,[82] whereas CCN2 protein expression differentiated type I from type III NETs.[25] The latter result is similar to findings in *Mastomys* tumors and suggests that CCN2 expression is related to autonomous (nongastrin-responsive) tumor growth. Differential expression of these candidate factors may ultimately be developed into a molecular marker panel that can be used to differentiate between and distinguish the malignant potential of the different types of human neoplastic gastric ECL cell tumors.

Management Strategy

Diagnosis

Upper GI endoscopy with biopsy provides direct evaluation of NET size, number, and extent, and can facilitate a precise histologic diagnosis (Figure 7.8). Lesions of the type I or II NETs can be single or multiple, but are usually small in size (<1cm). They are often polypoid, yellowish in color, and on occasion might even exhibit superficial ulceration.[6] It is noteworthy that such lesions can be submucosally located and thus not amenable to luminal inspection. The diagnosis of these lesions is usually undertaken by endoscopic ultrasound and confirmed by histologic examination of fine-needle aspirates.

Other studies may be of some use in the identification of gastric NETs, but for most, endoscopy is adequate.[84] Barium contrast studies are of limited utility in identifying submucosal NETs and are most useful in the detection of polypoid lesions but their discriminate index is low for lesions smaller than 1 cm in size.[6] Endoscopic ultrasound is of value for preoperative staging, particularly with respect to identifying the precise location and extent of submucosal lesions. Computerized tomography (CT) rarely detects the primary lesion but together with magnetic resonance imaging (MRI) may identify lymph node involvement or liver metastases. Somatostatin receptor scintigraphy (SRS) using [111]indium-labeled octreotide may help in determining the location and extent of tumors expressing somatostatin receptors and, in particular, their metastases to lymph nodes, liver, and bone.[85] Although the majority of primary lesions are often too small to be detected by this modality, a baseline study is important to facilitate follow-up, particularly in those individuals who require surgery.[6] Histopathologic evaluation using standard hematoxylin and eosin staining is of use in providing a general morphologic evaluation. For the precise staging of ECL cell proliferation into the various gradations of hyperplastic, dysplastic, or neoplastic, CgA immunocytochemistry is the best diagnostic marker.[53] The histologic classification of ECL cell proliferation is of importance for determining the most appropriate management strategy. In this respect, biopsies of apparently normal mucosa adjacent to, and distant from, the lesion of interest may determine the degree of involvement of the remainder of the mucosa.

The Ki67 index provides an assessment of proliferative rate. According to the WHO 2000 classification,[76,77] a Ki67 index <2% identifies tumors with benign or low-grade malignancy, whereas an index >30% is prognostic for high-grade malignant behavior. An intermediate index of 2%–30% in tumors >2 cm in size is considered indicative of WHO type 2 tumors of low-grade malignancy.[59]

Although a number of other peptides and amines, e.g., pancreastatin, have been identified in the ECL cell, antibodies against many of these agents are often not clinically available and their identification has yielded little further information of either the biologic behavior of the tumor or any long-term outcome. Electron microscopic visualization of ECL cell histamine-containing granules, although definitive in terms of confirming the cell of origin, is difficult and not usually clinically practical.[6]

Plasma and Urine Markers

CgA can be used as a diagnostic marker for gastric NETs,[84] and although plasma CgA levels are increased in all NET patients, they are higher (p < 0.01) in patients with type III NETs than those with type I disease.[86] False positives, however, may occur in patients who are undergoing PPI treatment or in patients with renal failure,[87] and it is evident that even short-term treatment with low-dose omeprazole may be enough to significantly increase CgA levels.[88] Nonetheless, biochemically, CgA is the most sensitive marker for detection. Gastrin levels should also be measured to identify atrophic gastritis and any associated secondary hypergastrinemia. Because ECL cells do not secrete serotonin (5HT), the measurement of plasma 5HT or its degradative product urinary 5-hydroxyindole acetic acid (5-HIAA) levels as markers is limited to individuals with type III lesions with mixed NE cell populations who exhibit evidence of the carcinoid syndrome such as flushing and diarrhea.

Therapy

Medical Therapy

Because the vast majority of gastric carcinoids do not exhibit carcinoid symptomatology, symptomatic therapy is rarely required. If symptoms are an issue (usually type III), the long-acting depot formulations of somatostatin analogs (Somatuline Autogel® and Sandostatin LAR®) are the principal agents prescribed.[53] These analogs decrease the gastric endocrine cell mass in patients with ZES (type II tumors)[89,90] but there are insufficient data to adequately predict the antiproliferative efficacy of this drug in either type I or II lesions,[91] and it is currently not recommended in this group of patients.[59] Other modalities including histamine 1 receptor blockade (Fexofenadine®, Loratadine®, Terfenadine®, or Diphenhydramine®) may also be of benefit in suppressing skin rashes, particularly in histamine-secreting gastric NETs (functional type III tumors).

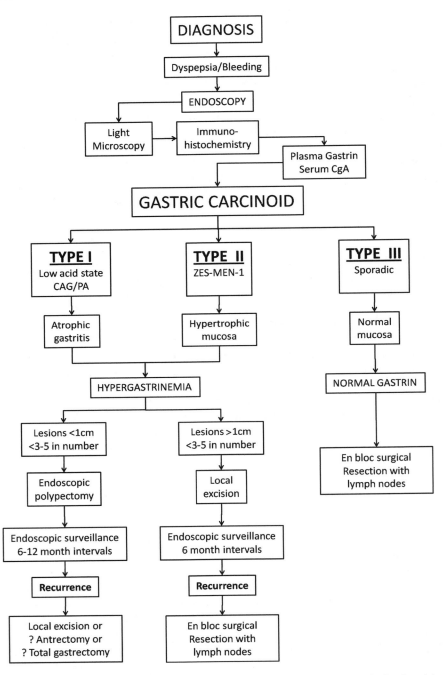

Fig. 7.8 A management algorithm for gastric carcinoids. The decision to undertake local excision of lesions should be determined after endoscopic ultrasound. It is preferable to undertake endoscopic resection if the lesions are superficial or not excessive in number. Although antrectomy is

Intravenous cytotoxic therapy is not indicated for either type I or II tumors but may be used if metastatic lesions (type III or NECs) are evident. Its usage is influenced by the malignancy of the tumor.[59] In selected patients, hepatic artery perfusion, focal embolization (or chemoembolization), cryoablation, or radiofrequency ablation may be considered with or without synchronous chemotherapy, which may be beneficial,[92] whereas the combination of cisplatin and etoposide should be considered for anaplastic NECs.[89,93,94] Peptide receptor radionuclide therapy (PRRT) may be considered as a treatment option but a positive SRS is a prerequisite for using this modality.[59] It should be noted that neither cytotoxics nor PRRT is associated with substantial objective tumor responses (~20%) but may increase progress-free survival.[92]

Endoscopic Therapy

Endoscopy has an important role in the management of gastric NETs given its utility to identify, biopsy, and resect gastric lesions.[95] Endoscopic resection, using endoscopic mucosal resection techniques, is usually successful in type I NETs (lesions < 1 cm and 1–2 tumors present).[96] However, a surgical excision of the lesion should be undertaken if the lesion is larger than 1 cm, more than five lesions are present, or there has been a recurrence of endoscopic polypectomy.[92]

An endoscopic surveillance approach with extensive sampling of both the lesser and greater curvatures is recommended early in individuals at risk (atrophic gastritis, autoimmune disease, MEN1) for dysplasia or gastric NETs.[97] Limited sampling in the greater curvature, however, is adequate in those without an increased risk for undetected development of NETs.[97]

Surgical Therapy

The elimination of sustained hypergastrinemia after antrectomy is associated with a reduction of volume, density, cross-sectional area, and number of endocrine cells (>75% decrease) in the remaining nonantral mucosa.[98,99] The effect of an antrectomy, however, may be unpredictable if ECL cell lesions have become gastrin autonomous.[6] Thus, surgical management of gastric NETs depends on the type and extent of the lesion.

Fig. 7.8 (continued) a theoretically attractive surgical option, because it is based on the biologic premise that the type I and II lesions are gastrin-driven, its application to individual patients is limited because it is not possible to determine whether enterochromaffin-like cells are gastrin dependent or autonomous. In the latter circumstance (~20% of cases), antral resection to remove gastrin-producing cells will not alter disease progression. Antral resection should not be considered in type II lesions because the dominant gastrin source is the gastrinoma.

Type I and II Neuroendocrine Tumors

Type I and II NETs can be managed by repeated endoscopic resection or local wedge excision, unless the lesions are excessive in number or there is evidence of invasion.[6,92] Large lesions that ulcerate or bleed might require more extensive surgical resection, particularly if the patient is young and evidence of diffuse gastric micro carcinoidosis is evident. In general, surgery is effective in ~80% of type I tumors.[59,100]

Type III Neuroendocrine Tumors

Type III lesions should be managed as for gastric adenocarcinoma. Thus, the presence of solitary, large (>1 cm), or invasive tumors mandates an attempt at surgical cure if not palliation.[60,74] In patients whose general condition is consistent with an acceptable operative risk, complete or partial gastrectomy (en bloc resection) with local lymph node resection is appropriate.[101] Lesions larger than 2 cm associated with local invasion require subtotal gastrectomy or extended local resection.[102]

Neuroendocrine Carcinomas

In the presence of metastasis, extensive surgery is rarely indicated, although gastric resection is often necessary to obviate complications such as bleeding, obstruction, or perforation.[6,53]

Follow-Up Studies

No clinical study has been undertaken to determine how often patients with type I and II gastric NETs should undergo endoscopic surveillance. The currently accepted recommendation[59] is that surveillance should be performed every 2 years for type I tumors and every year for type 2 lesions. In general, endoscopic mucosal resection should be undertaken when polyps are >1 cm in size. In type I tumors, the risk of gastric adenocarcinoma developing from intestinal metaplasia in CAG justifies undertaking biopsies of the mucosa. The approach to follow-up of type III tumors depends on the tumor subtype. In well-differentiated tumors and after curative resection, imaging (SRS/CT/MRI) and serum CgA should be performed at 6-month intervals for the first 2 years, and then yearly for 3 additional years. In well-differentiated metastatic tumors, follow-up investigations (SRS/CT/MRI) should be done every 3 months.

Prognostic Features and Outcome Data

The 5-year survival rate for all gastric NETs is 48.6%.[6,12] For those classed as localized lesions, the 5-year survival rate is 64.3%.[103] The prognosis of gastric NETs, however, varies widely depending on tumor type and is, for the most part, dependent on the subtype of the lesion involved as well as any associated or underlying disease process. The outcome of the type I lesion is of particular relevance because it constitutes the most common (~80%) group of gastric NETs encountered in clinical practice.

Type I Neuroendocrine Tumors

Rappel et al. in 1995 reported an observed 78% and an age-corrected 100% survival rate (Kaplan-Meier) in 88 patients with type I lesions.[104] More recently, 5- and 10-year crude survival rates were estimated at 96.1% and 73.9% (not different from the general population) for type I tumors.[100] These data reflect a generally benign course for this disease. Although a good prognosis may be predicted in small, non-metastatic lesions with minimal invasive management in the type I NETs, the potential mortality/morbidity rate of repeated endoscopy resection or even surgical removal (antrectomy/gastrectomy) requires consideration. Furthermore, the possibility of recurrent lesions, which may require additional intervention also, needs to be considered in the discussion of outcome. In general with a type I NET the prediction of a normal life expectancy can be predicted in the vast majority of individuals.

Type II Neuroendocrine Tumors

Although the metastatic rate (3%–12%) of ZES/MEN1 (type II) lesions is reported to be higher than in type I lesions, the short-term prognosis is similar.[105] The long-term prognosis of type II disease, however, ultimately reflects the course of the MEN1 gastrinoma which itself has a 5-year survival rate of 60%–75%.[106] Overall, the survival rate of type II lesions is thus worse than type I given the increased propensity for metastasis of the gastric lesion and the additional morbidity conferred by the natural history of MEN1.

Type III Neuroendocrine Tumors

Type III lesions often display markedly aggressive local behavior, and metastasize. This poor prognosis is directly related to the high metastatic rate, and the overall 5-year survival rate is usually <50%.[11,57,60,103] The 5-year survival rate is significantly higher for localized disease (64.3%) and for lesions with regional metastases (29.9%) than for lesions with distant metastases (10%). Although the tendency to metastasize correlates with tumor size, minute tumors have been reported with spread.[74] Factors that predict aggressive behavior include cellular atypia, two or more mitoses per 10 high-powered fields, Ki67 index >2%, angioinvasion, and transmural invasion.[57,75]

Neuroendocrine Carcinomas

NECs, previously known as "atypical carcinoids," represent an aggressive neuroendocrine neoplasm that bears greater resemblance to "sporadic carcinoids" than to hypergastrinemia-associated tumors. These lesions display invasive growth, metastasize with great frequency, and progress rapidly with a prognosis indistinguishable from gastric adenocarcinoma.[57]

Summary

Gastrointestinal neuroendocrine cells have been recognized since the latter part of the nineteenth century,[107] but it is only within the last three decades that an appreciation of the biology and neoplastic potential of the gastric ECL cell has become apparent. This cell represents a unique model because an intrinsic physiologic agent—gastrin—whose regulatory system has been well-defined, can initiate histamine secretion, cell proliferation, and subsequent neoplastic transformation. Because the ECL cell also contains numerous other biologic agents, e.g., growth factors, it is apparent that the ECL cell has a critical role in the maintenance of gastric mucosal integrity as well as the regulation of acid secretion.

In this context, it is of interest to consider that the ECL cell may be implicated in the genesis of gastric carcinoma either as a determinant of lineage fate (through growth factor production) or as a direct cellular antecedent. In this respect, attention has been drawn to the increasing frequency with which mixed (composite) glandular endocrine cell carcinomas and their precursor lesions are identified.[108] It is likely that investigation of this gastric neuroendocrine cell will yield insight not only into the biology of the neuroendocrine regulatory system of the stomach but also the biogenesis and evolution of other gastric neoplasia.

The difficulties in determining whether a focus of proliferating ECL cells under gastrin regulation represents a cancer with a sinister prognosis mandates the development of a prognostically based gene-marker panel to define the pathobiology of these lesions. In addition, it is clear that optimal therapy cannot be achieved by aggregating all gastric carcinoid lesions into broad groups,[100] and appropriate classification of patients based on the biologic behavior of the tumor remains to be developed.

Conclusion

Gastric NETs are not as rare as previously thought and remain a prominent focus of clinical attention. This reflects both an increased awareness of the entity and the recognition of its relationship to low acid states and hypergastrinemia. Although a number of studies have, as yet, revealed no irreversible pathologic sequelae that might be related to ECL cell neoplasia in patients undergoing profound and sustained PPI-induced acid inhibition,[109-112] the long-term outcome of ECL cell hyperplasia in prolonged low acid states in humans remains to be fully characterized. Continued study of the pathobiology of human ECL cell neoplasia will yield valuable insights into the regulation of gastric mucosal cells, and the commonality of the association of the low acid state to gastric carcinoid development and gastric adenocarcinoma is worthy of further investigation.

References

1. Modlin IM, Tang LH. The gastric enterochromaffin-like cell: an enigmatic cellular link. Gastroenterology 1996;111(3):783–810.
2. Bordi C, D'Adda T, Azzoni C, Ferraro G. Classification of gastric endocrine cells at the light and electron microscopical levels. Microsc Res Tech 2000;48(5):258–71.
3. Prinz C, Zanner R, Gratzl M. Physiology of gastric enterochromaffin-like cells. Annu Rev Physiol 2003;65:371–82.
4. Lindstrom E, Chen D, Norlen P, Andersson K, Hakanson R. Control of gastric acid secretion: the gastrin-ECL cell-parietal cell axis. Comp Biochem Physiol A Mol Integr Physiol 2001;128(3):505–14.
5. Modlin IM, Lawton GP, Miu K, Kidd M, Luque EA, Sandor A, et al. Pathophysiology of the fundic enterochromaffin-like (ECL) cell and gastric carcinoid tumours. Ann R Coll Surg Engl 1996;78(2):133–8.
6. Modlin IM, Lye KD, Kidd M. Carcinoid tumors of the stomach. Surg Oncol 2003; 12(2):153–72.
7. Smith AM, Watson SA, Caplin M, Clarke P, Griffin N, Varro A, et al. Gastric carcinoid expresses the gastrin autocrine pathway. Br J Surg 1998;85(9):1285–9.
8. Kitago M, Inada T, Igarashi S, Mizutani S, Ogata Y, Kubota T. Multiple gastric carcinoid tumors with type A gastritis concomitant with gastric cancer: a case report. Oncol Rep 2001;8(2):343–6.
9. Waldum HL, Haugen OA, Isaksen C, Mecsei R, Sandvik AK. Enterochromaffin-like tumour cells in the diffuse but not the intestinal type of gastric carcinomas. Scand J Gastroenterol Suppl 1991;180:165–9.
10. Modlin IM, Goldenring JR, Lawton GP, Hunt R. Aspects of the theoretical basis and clinical relevance of low acid states. Am J Gastroenterol 1994;89(3):308–18.
11. Godwin JD, 2nd. Carcinoid tumors. An analysis of 2,837 cases. Cancer 1975;36(2):560–9.
12. Modlin IM, Lye KD, Kidd M. A 5-decade analysis of 13,715 carcinoid tumors. Cancer 2003;97(4):934–59.
13. Modlin IM, Lye KD, Kidd M. A 50-year analysis of 562 gastric carcinoids: small tumor or larger problem? Am J Gastroenterol 2004;99(1):23–32.
14. Jansa SA, Weksler M. Phylogeny of muroid rodents: relationships within and among major lineages as determined by IRBP gene sequences. Mol Phylogenet Evol 2004;31(1):256–76.
15. Modlin IM, Lawton GP, Tang LH, Geibel J, Abraham R, Darr U. The mastomys gastric carcinoid: aspects of enterochromaffin-like cell function. Digestion 1994;55(Suppl 3):31–7.
16. Modlin IM, Esterline W, Kim H, Goldenring JR. Enterochromaffin-like cells and gastric argyrophil carcinoidosis. Acta Oncol 1991;30(4):493–8.
17. Schaffer K, McBride EW, Beinborn M, Kopin AS. Interspecies polymorphisms confer constitutive activity to the Mastomys cholecystokinin-B/gastrin receptor. J Biol Chem 1998;273(44):28779–84.
18. Bilchik AJ, Nilsson O, Modlin IM, Sussman J, Zucker KA, Adrian TE. H2-receptor blockade induces peptide YY and enteroglucagon-secreting gastric carcinoids in mastomys. Surgery 1989;106(6):1119–26; discussion 1126–7.

19. Kidd M, Hinoue T, Eick G, Lye KD, Mane SM, Wen Y, et al. Global expression analysis of ECL cells in Mastomys natalensis gastric mucosa identifies alterations in the AP-1 pathway induced by gastrin-mediated transformation. Physiol Genomics 2004;20(1):131–42.
20. Rozengurt E, Walsh JH. Gastrin, CCK, signaling, and cancer. Annu Rev Physiol 2001;63:49–76.
21. Chalmers CJ, Gilley R, March HN, Balmanno K, Cook SJ. The duration of ERK1/2 activity determines the activation of c-Fos and Fra-1 and the composition and quantitative transcriptional output of AP-1. Cell Signal 2007;19(4):695–704.
22. Treinies I, Paterson HF, Hooper S, Wilson R, Marshall CJ. Activated MEK stimulates expression of AP-1 components independently of phosphatidylinositol 3-kinase (PI3-kinase) but requires a PI3-kinase signal To stimulate DNA synthesis. Mol Cell Biol 1999;19(1):321–9.
23. Kinoshita Y, Nakata H, Kishi K, Kawanami C, Sawada M, Chiba T. Comparison of the signal transduction pathways activated by gastrin in enterochromaffin-like and parietal cells. Gastroenterology 1998;115(1):93–100.
24. Tang LH, Modlin IM, Lawton GP, Kidd M, Chinery R. The role of transforming growth factor alpha in the enterochromaffin-like cell tumor autonomy in an African rodent mastomys. Gastroenterology 1996;111(5):1212–23.
25. Kidd M, Modlin IM, Eick GN, Camp RL, Mane SM. Role of CCN2/CTGF in the proliferation of Mastomys enterochromaffin-like cells and gastric carcinoid development. Am J Physiol Gastrointest Liver Physiol 2007;292(1):G191–200.
26. Syder AJ, Karam SM, Mills JC, Ippolito JE, Ansari HR, Farook V, et al. A transgenic mouse model of metastatic carcinoma involving transdifferentiation of a gastric epithelial lineage progenitor to a neuroendocrine phenotype. Proc Natl Acad Sci USA 2004;101(13):4471–6.
27. Bradham DM, Igarashi A, Potter RL, Grotendorst GR. Connective tissue growth factor: a cysteine-rich mitogen secreted by human vascular endothelial cells is related to the SRC-induced immediate early gene product CEF-10. J Cell Biol 1991;114(6):1285–94.
28. Brigstock DR. The CCN family: a new stimulus package. J Endocrinol 2003;178(2):169–75.
29. Kang Y, Siegel PM, Shu W, Drobnjak M, Kakonen SM, Cordon-Cardo C, et al. A multigenic program mediating breast cancer metastasis to bone. Cancer Cell 2003;3(6):537–49.
30. Pan LH, Beppu T, Kurose A, Yamauchi K, Sugawara A, Suzuki M, et al. Neoplastic cells and proliferating endothelial cells express connective tissue growth factor (CTGF) in glioblastoma. Neurol Res 2002;24(7):677–83.
31. Igarashi A, Hayashi N, Nashiro K, Takehara K. Differential expression of connective tissue growth factor gene in cutaneous fibrohistiocytic and vascular tumors. J Cutan Pathol 1998;25(3):143–8.
32. Shakunaga T, Ozaki T, Ohara N, Asaumi K, Doi T, Nishida K, et al. Expression of connective tissue growth factor in cartilaginous tumors. Cancer 2000;89(7):1466–73.
33. Kawase S, Ishikura H. Female-predominant occurrence of spontaneous gastric adenocarcinoma in cotton rats. Lab Anim Sci 1995;45(3):244–8.
34. Fossmark R, Martinsen TC, Bakkelund KE, Kawase S, Waldum HL. ECL-cell derived gastric cancer in male cotton rats dosed with the H2-blocker loxtidine. Cancer Res 2004;64(10):3687–93.
35. Waldum HL, Rorvik H, Falkmer S, Kawase S. Neuroendocrine (ECL cell) differentiation of spontaneous gastric carcinomas of cotton rats (Sigmodon hispidus). Lab Anim Sci 1999;49(3):241–7.
36. Martinsen TC, Kawase S, Hakanson R, Torp SH, Fossmark R, Qvigstad G, et al. Spontaneous ECL cell carcinomas in cotton rats: natural course and prevention by a gastrin receptor antagonist. Carcinogenesis 2003;24(12):1887–96.
37. Fukushima Y, Matsui T, Saitoh T, Ichinose M, Tateishi K, Shindo T, et al. Unique roles of G protein-coupled histamine H2 and gastrin receptors in growth and differentiation of gastric mucosa. Eur J Pharmacol 2004;502(3):243–52.
38. Chen D, Zhao CM, Al-Haider W, Hakanson R, Rehfeld JF, Kopin AS. Differentiation of gastric ECL cells is altered in CCK(2) receptor-deficient mice. Gastroenterology 2002;123(2):577–85.

39. Crabtree JS, Scacheri PC, Ward JM, Garrett-Beal L, Emmert-Buck MR, Edgemon KA, et al. A mouse model of multiple endocrine neoplasia, type 1, develops multiple endocrine tumors. Proc Natl Acad Sci USA 2001;98(3):1118–23.

40. Debelenko LV, Emmert-Buck MR, Zhuang Z, Epshteyn E, Moskaluk CA, Jensen RT, et al. The multiple endocrine neoplasia type I gene locus is involved in the pathogenesis of type II gastric carcinoids. Gastroenterology 1997;113(3):773–81.

41. Miyaoka Y, Kadowaki Y, Ishihara S, Ose T, Fukuhara H, Kazumori H, et al. Transgenic over-expression of Reg protein caused gastric cell proliferation and differentiation along parietal cell and chief cell lineages. Oncogene 2004;23(20):3572–9.

42. Fox JG, Rogers AB, Ihrig M, Taylor NS, Whary MT, Dockray G, et al. Helicobacter pylori-associated gastric cancer in INS-GAS mice is gender specific. Cancer Res 2003;63(5):942–50.

43. Cao X, Tsukamoto T, Nozaki K, Tanaka H, Cao L, Toyoda T, et al. Severity of gastritis deter-mines glandular stomach carcinogenesis in Helicobacter pylori-infected Mongolian gerbils. Cancer Sci 2007;98(4):478–83.

44. Wang TC, Dangler CA, Chen D, Goldenring JR, Koh T, Raychowdhury R, et al. Synergistic interaction between hypergastrinemia and Helicobacter infection in a mouse model of gastric cancer. Gastroenterology 2000;118(1):36–47.

45. Kagawa J, Honda S, Kodama M, Sato R, Murakami K, Fujioka T. Enterochromaffin-like cell tumor induced by Helicobacter pylori infection in Mongolian gerbils. Helicobacter 2002;7(6):390–7.

46. Rudnicka W, Jarosinska A, Bak-Romaniszyn L, Moran A, Planeta-Malecka I, Wadstrom T, et al. Helicobacter pylori lipopolysaccharide in the IL-2 milieu activates lymphocytes from dys-peptic children. FEMS Immunol Med Microbiol 2003;36(3):141–5.

47. Kidd M, Tang LH, Schmid S, Lauffer J, Louw JA, Modlin IM. Helicobacter pylori lipopoly-saccharide alters ECL cell DNA synthesis via a CD14 receptor and polyamine pathway in mastomys. Digestion 2000;62(4):217–24.

48. Betton GR, Dormer CS, Wells T, Pert P, Price CA, Buckley P. Gastric ECL-cell hyperplasia and carcinoids in rodents following chronic administration of H2-antagonists SK&F 93479 and oxmetidine and omeprazole. Toxicol Pathol 1988;16(2):288–98.

49. Poynter D, Selway SA. Neuroendocrine cell hyperplasia and neuroendocrine carcinoma of the rodent fundic stomach. Mutat Res 1991;248(2):303–19.

50. Goldenring JR, Ray GS, Coffey RJ, Meunier PC, Haley PJ, Barnes TB, et al. Reversible drug-induced oxyntic atrophy in rats. Gastroenterology 2000;118(6):1080–93.

51. Bakke I, Hammer TA, Sandvik AK, Waldum HL. PPAR alpha stimulates the rat gastrin-producing cell. Mol Cell Endocrinol 2002;195(1–2):89–97.

52. McDonald R. A study of 356 carcinoids of the gastrointestinal tract. Am J Med 1956;21:867–72.

53. Modlin IM, Kidd M, Latich I. Current status of gastrointestinal carcinoids. Gastroenterology 2005;128(6):1717–51.

54. Borch K. Atrophic gastritis and gastric carcinoid tumours. Ann Med 1989;21(4):291–7.

55. Cadiot G, Laurent-Puig P, Thuille B, Lehy T, Mignon M, Olschwang S. Is the multiple endo-crine neoplasia type 1 gene a suppressor for fundic argyrophil tumors in the Zollinger-Ellison syndrome? Gastroenterology 1993;105(2):579–82.

56. Hodgson N, Koniaris LG, Livingstone AS, Franceschi D. Gastric carcinoids: a temporal increase with proton pump introduction. Surg Endosc 2005;19(12):1610–2.

57. Rindi G, Luinetti O, Cornaggia M, Capella C, Solcia E. Three subtypes of gastric argyrophil carcinoid and the gastric neuroendocrine carcinoma: a clinicopathologic study. Gastroenterology 1993;104(4):994–1006.

58. Gough DB, Thompson GB, Crotty TB, Donohue JH, Kvols LK, Carney JA, et al. Diverse clinical and pathologic features of gastric carcinoid and the relevance of hypergastrinemia. World J Surg 1994;18(4):473–9; discussion 479–80.

59. Ruszniewski P, Delle Fave G, Cadiot G, Komminoth P, Chung D, Kos-Kudla B, et al. Well-differentiated gastric tumors/carcinomas. Neuroendocrinology 2006;84(3):158–64.

60. Gilligan CJ, Lawton GP, Tang LH, West AB, Modlin IM. Gastric carcinoid tumors: the biology and therapy of an enigmatic and controversial lesion. Am J Gastroenterol 1995;90(3):338–52.

61. Creutzfeldt W, Stockmann F. Carcinoids and carcinoid syndrome. Am J Med 1987;82(5B): 4–16.
62. McCloy RF, Arnold R, Bardhan KD, Cattan D, Klinkenberg-Knol E, Maton PN, et al. Pathophysiological effects of long-term acid suppression in man. Dig Dis Sci 1995;40(2 Suppl):96S–120S.
63. Torbenson M, Abraham SC, Boitnott J, Yardley JH, Wu TT. Autoimmune gastritis: distinct histological and immunohistochemical findings before complete loss of oxyntic glands. Mod Pathol 2002;15(2):102–9.
64. Hodges JR, Isaacson P, Wright R. Diffuse enterochromaffin-like (ECL) cell hyperplasia and multiple gastric carcinoids: a complication of pernicious anaemia. Gut 1981;22(3):237–41.
65. Solcia E, Fiocca R, Villani L, Luinetti O, Capella C. Hyperplastic, dysplastic, and neoplastic enterochromaffin-like-cell proliferations of the gastric mucosa. Classification and histogenesis. Am J Surg Pathol 1995;19(Suppl 1):S1–7.
66. Rashid N, Modlin I, Tang L, Wright N. The proliferative status of enterochromaffin-like (ECL) cells in the gastric mucosa. The plot thickens. Gastroenterology 1997;112:A644.
67. Annibale B, Azzoni C, Corleto VD, di Giulio E, Caruana P, D'Ambra G, et al. Atrophic body gastritis patients with enterochromaffin-like cell dysplasia are at increased risk for the development of type I gastric carcinoid. Eur J Gastroenterol Hepatol 2001;13(12):1449–56.
68. Bordi C. Gastric carcinoids. Ital J Gastroenterol Hepatol 1999;31(Suppl 2):S94–7.
69. Waldum HL, Brenna E. Personal review: is profound acid inhibition safe? Aliment Pharmacol Ther 2000;14(1):15–22.
70. Laine L, Ahnen D, McClain C, Solcia E, Walsh JH. Review article: potential gastrointestinal effects of long-term acid suppression with proton pump inhibitors. Aliment Pharmacol Ther 2000;14(6):651–68.
71. Solcia E, Fiocca R, Rindi G, Villani L, Cornaggia M, Capella C. The pathology of the gastrointestinal endocrine system. Endocrinol Metab Clin North Am 1993;22(4):795–821.
72. Carcinoid tumors. International symposium on pathology, epidemiology, clinical aspects and therapy. January 13–16, 1994, Munich, Germany. Proceedings. Digestion 1994;55(Suppl 3):1–113.
73. Lehy T, Mignon M, Cadiot G, Elouaer-Blanc L, Ruszniewski P, Lewin MJ, et al. Gastric endocrine cell behavior in Zollinger-Ellison patients upon long-term potent antisecretory treatment. Gastroenterology 1989;96(4):1029–40.
74. Kumashiro R, Naitoh H, Teshima K, Sakai T, Inutsuka S. Minute gastric carcinoid tumor with regional lymph node metastasis. Int Surg 1989;74(3):198–200.
75. Moesta KT, Schlag P. Proposal for a new carcinoid tumour staging system based on tumour tissue infiltration and primary metastasis; a prospective multicentre carcinoid tumour evaluation study. West German Surgical Oncologists' Group. Eur J Surg Oncol 1990;16(4):280–8.
76. Solica E, Kloppel G, Sobin L. Histological typing of the endocrine tumors. Berlin: Springer Verlag; 2000.
77. Kloppel G, Perren A, Heitz PU. The gastroenteropancreatic neuroendocrine cell system and its tumors: the WHO classification. Ann NY Acad Sci 2004;1014:13–27.
78. Calender A. Molecular genetics of neuroendocrine tumors. Digestion 2000;62(Suppl 1):3–18.
79. Debelenko LV, Zhuang Z, Emmert-Buck MR, Chandrasekharappa SC, Manickam P, Guru SC, et al. Allelic deletions on chromosome 11q13 in multiple endocrine neoplasia type 1-associated and sporadic gastrinomas and pancreatic endocrine tumors. Cancer Res 1997;57(11): 2238–43.
80. D'Adda T, Keller G, Bordi C, Hofler H. Loss of heterozygosity in 11q13-14 regions in gastric neuroendocrine tumors not associated with multiple endocrine neoplasia type 1 syndrome. Lab Invest 1999;79(6):671–7.
81. Kidd M, Shapiro MD, Mane SM, Modlin IM. The utility of AP-1 gene analysis of human gastric ecl cell carcinoid tumors to define gastrin-autonomy and determine tumor types. Gastroenterology 2004;126(Suppl 2):T1184.

82. Kidd M, Modlin IM, Mane SM, Camp RL, Eick GN, Latich I, et al. Utility of molecular genetic signatures in the delineation of gastric neoplasia. Cancer 2006;106(7):1480–8.
83. Kidd M, Modlin IM, Mane SM, Camp RL, Shapiro MD. Q RT-PCR detection of chromogranin A: a new standard in the identification of neuroendocrine tumor disease. Ann Surg 2006; 243(2):273–80.
84. Modlin IM, Latich I, Zikusoka M, Kidd M, Eick G, Chan AK. Gastrointestinal carcinoids: the evolution of diagnostic strategies. J Clin Gastroenterol 2006;40(7):572–82.
85. Krenning EP, Kooij PP, Pauwels S, Breeman WA, Postema PT, De Herder WW, et al. Somatostatin receptor: scintigraphy and radionuclide therapy. Digestion 1996;57(Suppl 1):57–61.
86. Granberg D, Wilander E, Stridsberg M, Granerus G, Skogseid B, Oberg K. Clinical symptoms, hormone profiles, treatment, and prognosis in patients with gastric carcinoids. Gut 1998;43(2):223–8.
87. Sanduleanu S, De Bruine A, Stridsberg M, Jonkers D, Biemond I, Hameeteman W, et al. Serum chromogranin A as a screening test for gastric enterochromaffin-like cell hyperplasia during acid-suppressive therapy. Eur J Clin Invest 2001;31(9):802–11.
88. Giusti M, Sidoti M, Augeri C, Rabitti C, Minuto F. Effect of short-term treatment with low dosages of the proton-pump inhibitor omeprazole on serum chromogranin A levels in man. Eur J Endocrinol 2004;150(3):299–303.
89. Ruszniewski P, Ramdani A, Cadiot G, Lehy T, Mignon M, Bonfils S. Long-term treatment with octreotide in patients with the Zollinger-Ellison syndrome. Eur J Clin Invest 1993;23(5):296–301.
90. Annibale B, Delle Fave G, Azzoni C, Corleto V, Camboni G, D'Ambra G, et al. Three months of octreotide treatment decreases gastric acid secretion and argyrophil cell density in patients with Zollinger-Ellison syndrome and antral G-cell hyperfunction. Aliment Pharmacol Ther 1994;8(1):95–104.
91. Fykse V, Sandvik AK, Waldum HL. One-year follow-up study of patients with enterochromaffin-like cell carcinoids after treatment with octreotide long-acting release. Scand J Gastroenterol 2005;40(11):1269–74.
92. Modlin IM, Latich I, Kidd M, Zikusoka M, Eick G. Therapeutic options for gastrointestinal carcinoids. Clin Gastroenterol Hepatol 2006;4(5):526–47.
93. Jensen R. Carcinoid tumors and the carcinoid syndrome. In: Vincent T. DeVita, Jr., Samuel Hellman, and Steven A. Rosenberg, editors. Cancer: principles and practice of oncology. Philadelphia: Lippincott-Raven Publishers; 1997:1704–23.
94. Moertel CG, Kvols LK, O'Connell MJ, Rubin J. Treatment of neuroendocrine carcinomas with combined etoposide and cisplatin. Evidence of major therapeutic activity in the anaplastic variants of these neoplasms. Cancer 1991;68(2):227–32.
95. Lachter J, Chemtob J. EUS may have limited impact on the endoscopic management of gastric carcinoids. Int J Gastrointest Cancer 2002;31(1–3):181–3.
96. Ichikawa J, Tanabe S, Koizumi W, Kida Y, Imaizumi H, Kida M, et al. Endoscopic mucosal resection in the management of gastric carcinoid tumors. Endoscopy 2003;35(3): 203–6.
97. Bordi C, Azzoni C, Ferraro G, Corleto VD, Gibril F, Delle Fave G, et al. Sampling strategies for analysis of enterochromaffin-like cell changes in Zollinger-Ellison syndrome. Am J Clin Pathol 2000;114(3):419–25.
98. D'Adda T, Pilato FP, Sivelli R, Azzoni C, Sianesi M, Bordi C. Gastric carcinoid tumor and its precursor lesions. Ultrastructural study of a case before and after antrectomy. Arch Pathol Lab Med 1994;118(6):658–63.
99. Kern SE, Yardley JH, Lazenby AJ, Boitnott JK, Yang VW, Bayless TM, et al. Reversal by antrectomy of endocrine cell hyperplasia in the gastric body in pernicious anemia: a morphometric study. Mod Pathol 1990;3(5):561–6.
100. Borch K, Ahren B, Ahlman H, Falkmer S, Granerus G, Grimelius L. Gastric carcinoids: biologic behavior and prognosis after differentiated treatment in relation to type. Ann Surg 2005;242(1):64–73.

101. De Vries EG, Kema IP, Slooff MJ, Verschueren RC, Kleibeuker JH, Mulder NH, et al. Recent developments in diagnosis and treatment of metastatic carcinoid tumours. Scand J Gastroenterol Suppl 1993;200:87–93.
102. Davies MG, O'Dowd G, McEntee GP, Hennessy TP. Primary gastric carcinoids: a view on management. Br J Surg 1990;77(9):1013–4.
103. Modlin IM, Sandor A. An analysis of 8305 cases of carcinoid tumors. Cancer 1997;79(4):813–29.
104. Rappel S, Altendorf-Hofmann A, Stolte M. Prognosis of gastric carcinoid tumours. Digestion 1995;56(6):455–62.
105. Jensen RT. Management of the Zollinger-Ellison syndrome in patients with multiple endocrine neoplasia type 1. J Intern Med 1998;243(6):477–88.
106. Meko JB, Norton JA. Management of patients with Zollinger-Ellison syndrome. Annu Rev Med 1995;46:395–411.
107. Heidenhain R. Untersuchungen uber den Bau der Labdrusen. Arch Mikro Anat 1870;6:368–406.
108. Waldum HL, Aase S, Kvetnoi I, Brenna E, Sandvik AK, Syversen U, et al. Neuroendocrine differentiation in human gastric carcinoma. Cancer 1998;83(3):435–44.
109. Lamberts R, Creutzfeldt W, Stockmann F, Jacubaschke U, Maas S, Brunner G. Long-term omeprazole treatment in man: effects on gastric endocrine cell populations. Digestion 1988;39(2):126–35.
110. Lamberts R, Creutzfeldt W, Struber HG, Brunner G, Solcia E. Long-term omeprazole therapy in peptic ulcer disease: gastrin, endocrine cell growth, and gastritis. Gastroenterology 1993;104(5):1356–70.
111. Solcia E, Rindi G, Havu N, Elm G. Qualitative studies of gastric endocrine cells in patients treated long-term with omeprazole. Scand J Gastroenterol Suppl 1989;166:129–37; discussion 138–9.
112. Solcia E, Villani L, Luinetti O, Fiocca R. Proton pump inhibitors, enterochromaffin-like cell growth and Helicobacter pylori gastritis. Aliment Pharmacol Ther 1993;7(Suppl 1):25–8; discussion 29–31.

Chapter 8
Epstein-Barr Virus and Gastric Cancer

Ming-Shiang Wu, Chia-Tung Shun, and Jaw-Town Lin

Introduction

Chronic inflammation, triggered by infectious agents, has a key role in the development of malignancies (Aggarwal et al. 2006). The estimated total of infection-attributable cancer in the year 2002 is 1.9 million cases, or 17.8% of the global cancer burden (Parkin 2006). Among the principal agents, *Helicobacter pylori* and Epstein-Barr virus (EBV) are two common and widely disseminated pathogens throughout the world, and account for 5.5% and 1% of total malignancies, respectively (Parkin 2006).

H. pylori has been classified by the International Agency for Research on Cancer (IARC) as a class I carcinogen, i.e., a clear and unquestionable causative agent for gastric cancer (GC). EBV was isolated in 1964 from Burkitt's lymphoma (Epstein et al. 1964), which was first described in 1958 to be unique endemic jaw tumors in equatorial African (Burkitt 1958). Shortly after that, Old and colleagues discovered the link between EBV and epithelial malignancies. They incidentally observed that patients with nasopharyngeal cancer (NPC), a prevalent epithelial malignancy in southern China, are associated with a significantly elevated antibody titer to EBV (Old et al. 1966). Later studies demonstrated that EBV is closely linked to carcinogenicity in Burkitt's lymphoma, non-Hodgkin lymphoma in immunosuppressed subjects, sinonasal angiocentric T-cell lymphoma, Hodgkin lymphoma, and NPC. EBV is now considered to be a group I carcinogen for the above-mentioned cancers by the IARC (IARC 1997). The first hint of an association between EBV and GC came in 1990 from a case of undifferentiated lymphoepithelioma-like gastric carcinoma (LELC), in which EBV was detected using polymerase chain reaction (PCR) (Burke et al. 1990). Subsequently, EBV has been found in most cases of the rare tumor subtype (LELC) and a considerable portion of common gastric adenocarcinomas (Osato and Imai 1996; Rugge and Genta 1999; Takada 2000). Unique morphologic and phenotypic features have been reported for EBV-associated GC (EBVaGC). These divergent clinicopathologic characteristics raise the possibility that EBV-positive and -negative GCs adopt different carcinogenetic pathways. However, the evidence concerning the link between EBV and GC, unlike the situation for *H. pylori*, is considered to be

T.C. Wang et al. (eds.) *The Biology of Gastric Cancers*,
© Springer Science+Business Media, LLC 2009

inconclusive at the present time by the IARC (Parkin 2006). This chapter summarizes current knowledge of EBVaGC, including virology, detection methods, epidemiology, histopathology, pathogenesis, and clinical implication.

Virology

EBV belongs to the gamma subfamily of herpesviruses. The EBV genome consists of linear double-stranded DNA, containing approximately 172×10^3 base pairs (Baer et al. 1984). Like other human herpesviruses, EBV causes latent or lytic infection. EBV preferentially binds the CD21 receptor on the surface of B cells and can establish a lifelong latent infection in the B lymphocytes of its host. Infection of other cell types (e.g., epithelial cells) is much less efficient and occurs through separate, as yet poorly defined, pathways (Young and Rickinson 2004). Virions are only produced in lytic infection, and activation to a full lytic cycle is observed only in a minority of infected cells in culture, and is apparently rare *in vivo*. The lytic genes include immediate early genes *BZLF1* and *BRLF1*, early genes *BARF1* and *BHRF1*, and late genes *BCLF1* and *BLLF1*. In latent infection, viral DNA exists as an extrachromosomal circular molecule (episome) in the nucleus. It is noteworthy that infection of naïve B cells with EBV in culture results in transformation of resting B cells into the permanent latently infected lymphoblastoid cell line (LCL) (Nilsson et al. 1971). The LCL provides an *in vitro* system that is invaluable to study the function of the EBV-encoded proteins in the latently infected cells.

Latent episomal EBV genomes are subject to host cell–dependent epigenetic modification. The distinct viral epigenotypes are associated with the distinct EBV latency type, i.e., cell type–specific usage of latent EBV promoters controlling the expression of latent, growth transformation–associated genes (Minarovits 2006). Thus, the expression of EBV-encoded proteins differs depending on the type, differentiation, and activation status of the target cells. Expression of the full set of proteins detected in LCLs is designated as "growth pattern" and is referred to as type III latency. In these cells, six EBV nuclear antigens (EBNAs 1, 2, 3A, 3B, 3C, and lead protein—LP) and three latent membrane proteins (LMPs 1, 2A and 2B) are translated. Among them, LMP1, EBNA2, EBNA3A, and EBNA3C are essential for B-cell immortalization. LMP1 induces activation markers and costimulatory molecules and contributes to the immunogenicity of B cells. Therefore, cells expressing type III latency can exist only during the acute phase of primary infection before the EBV-specific cytotoxic T-lymphocyte (CTL) response develops, and in patients with impaired immune functions, such as transplant recipients who are treated with immunosuppressive agents. Thus, type III latency can be observed in immunocompromised patients with lymphoproliferative disorders, in EBV-infected B cells in the acute phase of infectious mononucleosis, and in LCL. LMP1 can interact with several tumor necrosis factor receptor associated factors (TRAF). The interaction of LMP1 with these

TRAF results in the high expression of NF-κB. LMP1 also upregulates the expression of several antiapoptotic and adhesion genes, including *A20*, *bcl2*, and *ICAM1*. Additionally, it activates the expression of interferon regulatory factor 7, matrix metalloprotease-9, and fibroblast growth factor-2. In epithelial cells, LMP1 has been shown to inhibit differentiation and induce cell proliferation through the activation of the phosphoinositol-3-kinase/Akt pathway. Other viral genes that encode transforming potential include *EBNA2* and *EBNA3*. *EBNA2* is a promiscuous transcriptional activator, activating the promoters of both viral and cellular genes. *EBNA3C* has been shown to cooperate with the protooncogene *Ras* to immortalize and transform rodent fibroblasts. It can directly interact with the Rb tumor suppressor protein, rendering it inactive and promoting tumor progression (Klein et al. 2007).

Besides the above-mentioned nine proteins, two small nonpolyadenylated RNAs (EBER-1,2) encoded by EBV DNA are consistently expressed in all EBV-infected cells. In latency I, only one nuclear protein, i.e., EBNA1 transcribed from Q promoter, is found in EBV-infected cells. Such type program does not induce proliferation and the phenotype corresponds to nonactivated B cells. This type of latency is also seen in EBV-positive Burkitt's lymphoma and gastric carcinoma. Because EBNA1 is not the target of EBV-CTL, neoplastic cells in type I latency may evade immune surveillance because no target antigens of EBV-CTL are expressed. In latency II, EBNA1, EBERs, and LMPs are expressed. Cells from EBV-positive NPC, Hodgkin disease, and T-cell lymphoma display type II latency. Collectively, these different types of latencies might determine the fate of EBV-carrying cells and could be related to the development of different EBV-associated diseases (Table 8.1) (Klein et al. 2007). During the acute or covalent phase of a primary infection, cells expressing EBNA2, -3A, -3C, and LMP-1 are highly susceptible to the action of EBV-CTL. However, there are other accessory defense mechanisms, inducing natural killer (NK) cells, neutralizing antibodies, antibody-dependent cell-mediated cytokines, and cytokines such as interferon.

In general, EBV infection is asymptomatic in childhood, but frequently results in infectious mononucleosis during adolescence (Evans 1972). Antibodies specific for other EBV-encoded proteins expressed in cells with lytic and latent infection appear subsequently in regular sequence. The antibody profile was shown to be characteristic for the different EBV-associated diseases. These antibodies are helpful in distinguishing infectious mononucleosis from other diseases with lymphocytosis.

Table 8.1 Detection of the Epstein-Barr virus and characterization of latency

Latency	Expression	Examples
Type I	EBNA1	Burkitt lymphoma (Ig-myc), gastric cancer
Type II	EBNA1 LMP1	Hodgkin lymphoma, NK/T-cell lymphoma
Type III	EBNA1–6 LMP1	Infectious mononucleosis, after transplant lymphoproliferative disease

Detection Methods

The association of GC with EBV infection has been demonstrated with a variety of techniques. Detection of EBV genome or related proteins in affected tumor tissue is highly recommended for diagnosis of EBV-associated malignancies. PCR, Southern blotting, and *in situ* hybridization (ISH) techniques are available for the detection of DNA. ISH is useful for the detection of RNA. Immunofluorescence, immunohistochemical procedures, and Western blotting are used for the detection of EBV-related proteins. Because of their abundance, the EBERs represent ideal targets for ISH using radiolabeled or nonradioactive probes. ISH for the detection of EBERs has thus become the standard method for the detection of EBV latent infection. Notably, detection of EBV genome in extracts of tumor tissue by PCR does not permit precise conclusion of the cellular source of the virus, because EBV-carrying B cells are present in the peripheral blood as well as in lymphoid and nonlymphoid tissues of most individuals (Niedobitek and Herbst 2006). Definitive designation of a tumor as "EBV-associated" should require unequivocal demonstration of the EBV genome or virus gene product within the tumor cell population. Therefore, it is crucial to use morphology-based analyses to confirm the presence of virus in cancer cells. Furthermore, individual infection events lead to episome with different numbers of terminal repetitive sequences, and analysis of terminal repeat region by Southern blot hybridization can provide evidence regarding the clonality of the viral genome (Hermann and Niedobitek 2003). Analyses of EBV expression products in GC have yielded latency I and suggested that some lytic genes (e.g., *BARF1* and *BHRF1*) may be expressed in EBV-associated GC (Hoshikawa et al. 2002; Luo et al. 2005). Recently, quantitative analysis of circulating EBV DNA by real-time quantitative PCR has demonstrated a positive correlation between the disease stage and clinical events, and is of prognostic significance in NPC (Lo 2001a). For GC, serum DNA may reflect tumoral EBER status (Lo et al. 2001b), and the potential role of monitoring EBV viral load on EBVaGC could be elaborated as in the case of NPC.

Measuring antibodies against major proteins produced by EBV such as EBNA, early antigen (EA), and viral capsid antigen (VCA) enables distinction between a susceptible individual and one with a primary, past, or activated infection. Detection of immunoglobulin (Ig)M antibodies against the VCA and of IgM and IgA antibodies against the EA expressed in lytically infected cells is the diagnostic sign of newly acquired infection. One report has demonstrated that seroprevalence rates of VCA-IgA and EBV-EA IgG were higher in EBV-positive gastric carcinoma cases than in EBV-negative carcinoma cases, with an odds ratio of 3.4 (95% confidence interval: 1.3–8.8) and 6.6 (95% confidence interval: 2.7–16.3), respectively (Shinkura et al. 2000). Another large, prospective, cohort study even revealed that VCA-IgA and neutralizing antibodies against EBV DNase are predictive of the development of NPC (Chien et al. 2001). The clinical usefulness of these serologic markers in EBVaGC warrants further elucidation.

Epidemiology

Since Burke et al. showed that EBV-positive GC resembled undifferentiated NPC (Burke et al. 1990), cumulative data have suggested a role in the development of a subset of GC. In contrast to Burkitt's lymphoma and NPC, which are endemic in equatorial Africa and southern Asia, respectively, EBVaGC is a nonendemic disease distributed throughout the world (Takada 2000). However, there are some regional differences in the incidence of EBVaGC. The highest average reported incidence (16%) was from the United States, whereas the lowest reported incidence (1.7%) was from the United Kingdom (Table 8.2). In Japan, GC is a common disease and the reported incidence of EBVaGC as a proportion of all cases of GC varied greatly from 3.1% to 20.1% (Ojima et al. 1996; Tokunaga et al. 1993). One report showed that the highest incidence (10.3%) was observed in Okinawa, which has the lowest GC mortality rate in Japan. Paradoxically, the lowest incidence (3.1%) was found in Niigata, which has the highest GC mortality rate (Tokunaga et al. 1993). Reports from other Asian countries revealed 4.3%–20.7% in Chinese (Hao et al. 2002; Harn et al. 1995; Hsieh et al. 1998; Yuen et al. 1994;) and 3.4%–13.5% in Koreans (Chang et al. 2000). One migration study showed that EBVaGC is present in 4.7% of Japanese Brazilians, 11.2% non-Japanese Brazilians, and 6.2% Japanese (Koriyama et al. 2001a). Collectively, geographic differences in the incidence of EBVaGC reflect the influence of environmental and cultural factors (Qiu et al. 1997).

Apart from geographic factors, different studies enrolling various cases and using diverse methods may influence the reported incidence of EBVaGC. EBV

Table 8.2 Geographic distribution of prevalence of Epstein-Barr–positive gastric adenocarcinoma

Country	Prevalence (%)	Reference
Asia		
Japan	3.1–20.1	Tokunaga 1993; Ojima 1996
Korea	3.4–13.5	Chang 2000; Takada 2000
Taiwan	9.3–20.7	Harn 1995; Hsieh 1998
Hong Kong	8.4	Yuen 1994
China	4.3–9	Takada 2000; Hao 2002
India	5	Kattoor 2002
Europe		
France	7.7	Chapel 2000
Germany	8.5–18	Ott 1994; Tanaka 2000
Italy	4.6	Leonicini 1993
UK	1.7	Burgess 2002
Latin America		
Peru	3.9	Yoshiwara 2005
Mexico	8.2	Herrera-Goepfert 1999
Chile	16.8	Corvaian 2001
Brazil	10.6	Takada 2000
United States	16	Shibata 1992
Russia	8.7	Galetsky 1997

positivity determined by PCR alone is simple but could be hampered by picking up EBV genome in tumor-infiltrating lymphocytes. Furthermore, the incidence of EBV also varies with histologic subtypes. One report even showed EBV is absent in conventional GC, but is present in more than 80% in LELCs (Ohfuji et al. 1996). Compared with GC, EBV infection has no etiologic role in other digestive cancers, such as esophagus, small bowel, colon, and pancreas (Cho et al. 2001; Kijima et al. 2001; von Rahden et al. 2006).

Because only a small fraction of EBV-infected subjects ever develop GC, additional risk factors are likely to be important determinants of disease risk. One case-control study in Japan showed high salt intake and exposure to wood dust and/or iron fillings might be associated with the development of EBVaGC (Koriyama et al. 2005). Along with environmental and lifestyle factors, host genetic and strain variation of EBV are likely to have a role in the outcomes of EBV infection (Gorzer et al. 2007). In EBV subtyping, attention has been focused on variations in genes with known or presumed oncogenic functions, as well as on CTL epitope variation. *LMP1* is one of the most intensively investigated genes in this regard and a 30-bp deletion at the LMP1 C-terminal region was detected in approximately 90% of EBVaGCs (Hayashi et al. 1998; Lee et al. 2004). Although it has been suggested that the 30-bp deletion variant has a role in tumor progression of EBV-associated Hodgkin's disease, this deletion may not be relevant to the pathogenesis of EBVaGC. Similarly, the results concerning LMP1 polymorphisms in EBVaGC remain controversial (Corvalan et al. 2006). Apart from the LMP1 genotype, other viral gene polymorphisms or mutations in EBNAs have also been found, but their effects on EBVaGC are largely unknown (Chu et al. 1999).

Recently, it has been proposed that the variability of generating host cytokines could be another reason for the variety of clinical outcomes after exposure to microbial pathogens. New evidence has emerged that EBVaGC is predominantly a result of inappropriately regulated gastric immune response to the infection. In other words, the interaction between microorganisms and the host immune system would dictate the disease outcome. Therefore, different cytokine responses determined at the genetic (transcription) level may also explain why there is a variety of individual susceptibilities to a given microbial infection. Among various cytokines, perturbation in the level and activity of interleukin (IL)-1B, IL-10, and tumor necrosis factor (TNF)-α has been reported as the key modulating factor in EBV-related malignancies (Chong et al. 2002; Ouburg et al. 2005). In EBVaGC, IL-1B may act as an autocrine growth factor, but two studies found IL-1B genotypes (−511 C/T and +3194*C/T) did not confer risk in EBVaGC (Sakuma et al. 2005; zur Hausen et al. 2003). One study from Taiwan investigated whether genetic differences in TNF-α and IL-10 production effect various outcomes after exposure to EBV infection. Researchers showed that the frequency of the high-producer allele (−308A) in the TNF-α gene was significantly higher among EBVaGCs, whereas the frequency of the high-producer allele (−1082G) in the IL-10 gene was significantly higher among EBV-negative GC patients (Wu et al. 2002). However, human leukocyte antigen (HLA) molecules are responsible for the presentation of foreign antigens to the immune system. They have a central role in the immune recognition and subsequent clearance of virally infected cells. Because a key feature of HLA genes is their high degree of polymorphism, the potential for identifying individual risks for EBVaGC is plausible. Researchers in Japan performed serologic typing for HLA class I and II and demonstrated that a deficiency of HLA-DR11 and the presence of HLA-DQ3 may be genetic markers for a population at greater risk of EBVaGC (Koriyama et al. 2001b). Taken together, these data support the notion that genetic factors may modify the outcomes of infectious diseases through different cytokine-producing or different antigen-presentation capabilities. Further association studies in different ethnic populations and functional studies of the underlying mechanisms are warranted to clarify the causative role of host genetic in EBVaGC because these findings are still exploratory.

Pathogenesis

EBV has evolved a strategy to survive within cells and avoid host responses that would otherwise lead to strong immune responses resulting in cell growth arrest and apoptosis. EBVaGC usually elicits strong lymphocyte infiltration. These tumor-infiltrating lymphocytes are predominantly HLA class I restricted CD8-positive cytotoxic lymphocytes (Kuzushima et al. 1999). However, the establishment of EBV latent infection in gastric tissue allows the malignant cells to avoid the immune surveillance of both CTL and NK cells by regulating differential expression of HLA class I molecules or Fas/FasL (Dutta et al. 2006; Kume et al. 1999). These data suggest that some EBV cellular proteins might be involved in the strong T-cell response of EBVaGC, but such responses are not effective in killing EBV-infected carcinoma cells (Kuzushima et al. 1999). Immune suppression of the host by various causes has also been implicated in the development of EBVaGC. Recently, an interesting case report described EBVaGC development in a patient after stem cell transplantation for multiple myeloma. The development of GC in this patient was preceded by severe graft-versus-host disease (GVHD) necessitating strong immunosuppression, which resulted in intense reactivation of the EBV infection. Three sequential gastric biopsies performed at 100, 130, and 150 days after hematopoietic stem cell transplantation showed gastritis, dysplasia, and adenocarcinoma, respectively. There was no evidence of *H. pylori* infection. ISH for EBER showed the absence of EBV in the gastritis specimen, but the presence of EBV in dysplasia and carcinoma specimens. This study indicates that mucosal damage caused by GVHD, immunosuppression, and EBV reactivation combined to lead to EBV infection of gastric cells and rapid initiation of carcinogenesis (Au et al. 2005).

The EBV genomes are maintained episomally through the tethering of the viral genome to the host chromosomal DNA by viral latent proteins such as EBNA1. In EBVaGC, EBV genome is present in some precancerous lesions and most cancer cells in a monoclonal episomal form, suggesting that EBV infects a cell before neoplastic transformation (Yanai et al. 1997b). Although EBV genomes do not integrate into the host chromosomal DNA during or before the neoplastic event, it encodes a diverse array of proteins that help contribute to the multistep process of carcinogenesis. Among the encoded proteins, LMP1, EBNA2, and EBNA3 have transformation potentials. In contrast to two other EBV-related malignancies, NPC and Hodgkin's disease, in which LMP1 and EBNA2 and EBNA3 were frequently detected, these proteins were not expressed in EBVaGC, raising doubts about the importance of EBV in gastric carcinogenesis. However, another EBV gene, *BRAF1*, has been shown to have transforming and immortalizing capacities and might offer an alternative way for EBV-mediated oncogenesis other than LMP1 (zur Hausen et al. 2000). In EBVaGC, the BRAF1 transcript was observed and could provide a protective role against apoptosis through an increased Bcl-2 to Bax ratio, thus promoting cancer-cell survival (Wang et al. 2006). Recently, the EBV-encoded small RNA (EBER) has also been shown to induce the expression of insulin-like growth factor-I (IGF-I). IGF-I can act as an autocrine growth factor and promote growth of tumor cells in EBVaGC (Iwarkiri et al. 2003). Furthermore, microRNA (miRNA), 19- to 25-nucleotide-long single-stranded RNAs with functions in cellular differentiation, proliferation, and apoptosis, was investigated in EBVaGC. It was found that BART miRNAs were expressed in EBV-infected gastric carcinoma cell lines and tumor tissues from patients as well as the animal model. These EBV miRNAs might also have crucial roles in the tumorigenesis of EBVaGC (Kim et al. 2007).

These striking epidemiologic and clinical differences indicate important underlying differences in the etiology and behavior of EBV-positive and -negative GCs. Increasing evidence has revealed that accumulations of multiple genetic alterations accompanies the multistep event proceeding from normal to preneoplastic lesions to highly malignant tumors in GC development. Therefore, delineating the involved genes may reflect the variability in the biologic characteristics of GC and elucidate the heterogeneity of their causes and histologic subtypes. Several studies have documented that promoter hypermethylation of various tumor-related genes is extremely frequent in EBVaGC, suggesting that viral oncogenesis might involve DNA hypermethylation (Chang et al. 2003, 2006; Etoh et al. 2004; Kang et al. 2002; Kusano et al. 2006; Sakuma et al. 2004; Sudo et al. 2004; van Rees et al. 2002). EBVaGC may comprise a pathogenetically distinct subgroup in high CpG island methylator phenotype (CIMP-H) GC. In addition to hypermethylation of methylated marker genes (e.g., LOX, HRAS, FLNc, HAND1, and TM), such CIMP-H tumors might display a high frequency of overexpression of DNMT1 and Bcl-2, and loss of E-cadherin, p14, p16, FHIT, and Smad4, but rare occurrence of microsatellite instability, and no mutation of p53 or K-ras. By comparative genomic hybridization and/or fluorescence ISH, loss of chromosomes 4p, 11p, and 18q, and gains in chromosome 11 (trisomy or polysomy) were more frequent in EBVaGC than in EBV-negative cancers (Chan et al. 2001; zur Hausen et al. 2001).Collectively, the association of EBV with GC is further strengthened by the molecular evidence that EBVaGC has distinct genetic alterations from EBV-negative GC.

Although the association of EBV with GC has long been established, unraveling the precise role of EBV in gastric carcinogenesis is still hampered by the common gastric pathogen, *H. pylori*. A recent estimate suggests that the attributable fraction of *H. pylori* in noncardia GC is 74% in developed countries and 78% in developing counties. This represents 63.4% of all stomach cancers (Parkin 2006). In general, EBV infection was observed in 10% of GC and might at least have a more important role in the so-called *H. pylori*–negative GC or cardia GC. Currently, chronic atrophic gastritis and subsequent intestinal metaplasia are thought to have a pivotal role in multistep carcinogenesis of GC. In recent years, abundant data have documented that chronic gastritis attributable to *H. pylori* infection may progress to intestinal metaplasia and even GC. In contrast to EBVaGC, *H. pylori* infection is linked to all histologic subtypes and distal location but not to cardia cancer. It was assumed that these two pathogens are affected by each other or together have important roles directly or indirectly in the pathogenesis of gastric diseases (Shinohara et al. 1998). The background mucosa of GC would provide some valuable clues about gastric carcinogenesis. The majority of GCs associated with *H. pylori* and environmental factors arises from multifocal atrophic gastritis. Therefore, an independent oncogenetic mechanism caused by EBV would be plausible if EBV-positive GC did not have coexistent atrophic gastritis. However, it would be difficult to identify the role of *H. pylori* infection in EBVaGC with coexistent atrophic gastritis or intestinal metaplasia (Osato and Imai 1996). In their preliminary report, Levine et al. (1995) documented that antibody levels in EBV-negative GC were significantly higher than those of EBV-positive GC and controls. It was thus inferred that an inverse relationship between antibody levels of *H. pylori* and EBV status existed. Nevertheless, the other study reported that the rate of *H. pylori* infection was higher in EBV-positive GC patients by histology (Yanai et al. 1999). Compared with the 54.4% of background *H. pylori* infection rate, another study showed 36.4%, 68.4%, and 68.3% of *H. pylori* seropositivity in LELC, EBV-positive non-LELC, and EBV-negative GC, respectively (Wu et al. 2000). This finding is intriguing and suggests that there is a tendency for gastric LELC, with the high frequency of EBV and predominant proximal location, to have less association with *H. pylori* infection than the other two groups. This result is consistent with the notion that the infection route of EBV follows the contiguous spread from the nasopharynx to the proximal stomach. Furthermore, these data suggest that EBV may have a role in *H. pylori*–seronegative GCs, especially those located at the proximal stomach. The detailed mechanism involved in the significantly lower rate of *H. pylori* seropositivity in gastric LELC compared with the other two groups is unknown. Possible reasons are the differences in genetic susceptibility of the infections, or that the gastric milieu in EBV-associated LELC are unable to support *H. pylori*. Considering equal distribution of intestinal metaplasia among EBV-positive and -negative GCs (Yanai et al. 1999; Wu et al. 2000), the latter possibility is more favored. Because mucosal immune or physiologic mechanisms activated by one infection may protect or aggravate the second, more studies are needed to ascertain the relative contributory role of EBV and *H. pylori* in different subsets of GC.

To investigate the association of EBV with GC, several epithelial cell lines derived from patients infected with EBV have been successfully cultured. These include G38, G39, HSC-39, and SNU-719 (Luo et al. 2004; Oh et al. 2004; Tajima et al. 1998). Moreover, an animal model with severe combined immunodeficiency mice was established (Iwasaki et al. 1998). These cell lines and the animal model are useful for studies of the interaction of EBV with epithelial cells and the molecular mechanisms of EBVaGC.

Clinicopathologic Features of Epstein-Barr–Associated Gastric Cancer

EBV involvement in GC varies by sex, histologic type, and tumor location (Tokunaga and Land 1998). One specific subset of GC, GC with rich lymphoid stroma (GCLS), had been described years before its recognition as one of the EBV-related tumors (Watanabe et al. 1976). GCLS, showing morphologic features similar to undifferentiated NPC (also called lymphoepithelioma), has been extensively investigated for an association with EBV. This particular type of tumor, depending on the strictness of the diagnostic criteria, accounts for 1%–4% of all GCs. In most reports, these rare gastric lymphoepitheliomas are characterized by an apparently high prevalence of EBV (80%–100%), predominant proximal location, low prevalence of lymph node metastasis, and a favorable prognosis (Koriyama et al. 2002; van Beek et al. 2004; Wu et al. 2000; Yanai et al. 1997a). A representative example of a lymphoepithelioma-like GC is provided in Figure 8.1. As for EBVaGC with conventional histology, the following unique features have been reported: superficial depressed or ulcerated lesions in the upper part of the stomach, moderately differentiated tubular and poorly differentiated solid types, and greater correlations in male patients and in those with gastritis cystitis polyposa of remnant GC (Kaizaki et al. 2005; Nishikawa et al. 2002; Yanai et al. 1997a). However, the prevalence of EBV is similar in diffuse and intestinal type GC.

To determine the cell lineage of the epithelial cell giving rise to EBVaGC, gastric marker mucins (MUC5AC and MUC6) and intestinal marker molecules (MUC2 and CD10) have been used. The results indicate that EBV infection may occur in the epithelial cells of a null or gastric phenotype, which may be devoid of transdifferentiation potential toward an intestinal phenotype (Barua et al. 2006). Studies of cytokeratin (CK) expression profiles showed that a decreased amount of CK7 is a characteristic of EBVaGC (Kim et al. 2004). Regarding the relationship between EBV infection and precursor lesions, the majority of the surrounding gastric mucosa in EBVaGC showed moderate to severe atrophy and carcinomas were located near the mucosa atrophic border (Yanai et al. 1999). DNA-ISH revealed EBNA1 in epithelial cells of chronic atrophic gastritis and LMP1 in areas of intestinal metaplasia, whereas EBER-1 expression was limited to carcinoma cells (Yanai et al. 1997b).

Clinical Implications

Considering the relatively small proportion of EBVaGC compared with conventional GC, it will rarely be necessary in everyday diagnostic practice to investigate whether a given GC is EBV associated. An exception to this rule may be the study

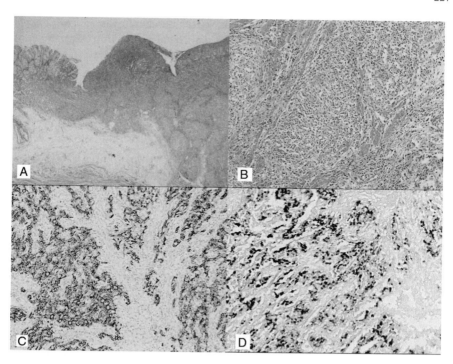

Fig. 8.1 A representative example of microscopic features and *in situ* hybridization of lymphoepithelioma-like carcinoma of stomach. (**A**) Hematoxylin and eosin (H&E) stain, original magnification 40×. (**B**) Pleomorphic cancer cell nests separated by intervening lymphoid stroma (H&E, original magnification 200×). (**C**) Cytokeratin staining shows immunoreactive tumor cells. (**D**) Positive EBER-1 signals in the nucleus of glandular and undifferentiated carcinomas (*in situ* hybridization, original magnification 200×) (*See Color Plates*)

of metastatic undifferentiated carcinoma of an unknown primary site, in particular if it shows morphologic features reminiscent of nonkeratinizing NPC. Detection of EBV in such a situation could suggest the nasopharynx or stomach as a possible primary site (Hermann et al. 2003). In addition, the consistent presence of the EBV genome in tumors offers the potential for novel EBV-directed therapies. Switching the latent form of EBV infection present in most EBV-positive tumor cells into the cytolytic form could be clinically useful because lytic EBV infection leads to host-cell destruction and very few normal cells contain the EBV genome (Westphal et al. 2000). Ganciclovir (GCV) or 3-azido-3deoxythymidine (AZT) can be converted to their active cytotoxic forms in cells that contain EBV-encoded lytic proteins. Studies have demonstrated that GCV or AZT treatment could potentially enhance the therapeutic efficacy of chemotherapy or radiation therapy for EBV-positive cancers when chemotherapeutic agents or radiation induces EBV into lytic cycles (Feng et al. 2002a; Jung et al. 2007; Westphal et al. 2000). Other experimental therapies included induction of the lytic form of EBV infection by adenovirus vector expressing BZLF1 and BRLF1, leading to the destruction and death of

EBV-carrying cancer cells (Feng et al. 2002b). Immunotherapy using LMP2 as the target could also effectively induce CTL against EBVaGC *in vitro* (Okugawa et al. 2004). However, whether these promising *in vitro* results could be translated into clinical practice remains to be further investigated.

Conclusions and Future Prospects

Most lymphoepithelial carcinomas of the stomach are EBV associated, regardless of the individual's geographic or ethnic origin. A substantial portion of conventional GCs carry the EBV virus and the proportion has geographic variations. In general, EBVaGC is responsible for 10% of GCs. This specific subset of patients tends to have proximal location, strong lymphocyte infiltration, more favorable prognosis, and distinct genetic and epigenetic alteration. These divergent clinicopathologic characteristics raise the possibility that EBV-positive and -negative GCs adopt different carcinogenetic pathways. However, the evidence concerning the link between EBV and GC, unlike the situation for *H. pylori*, is inconclusive at the present time. Further studies are mandatory to ascertain the relative contributory role of EBV and *H. pylori* in different subsets of GC and elucidate the precise pathogenetic mechanisms of EBV in gastric carcinogenesis. Moreover, research of EBV in animal models, monitoring the viral load for disease occurrence and treatment response, and development of novel therapy relevant to EBVaGC, are also needed.

References

Aggarwal BB, Shishodia S, Sandur SK, Pandey MK, and Sethi G. 2006. Inflammation and cancer: how hot is the link? Biochem Pharmacol 72:1605–21.

Au WY, Pang A, Chan EC, Chu KM, Shek TWH, and Kwong YL. 2005. Epstein-Barr virus-related gastric adenocarcinoma: an early secondary cancer post hemopoietic stem cell transplantation. Gastroenterology 129:2058–63.

Baer R, Bankier AT, Biggin MD, et al. 1984. DNA sequence and expression of the B95–8 Epstein-Barr virus genome. Nature 310:207–11.

Barua RR, Uozaki H, Chong JM, Ushiku T, Hino R, Chang MS, Nagai H, and Fukayama M. 2006. Phenotype analysis by MUC2, MUC5AC, MUC6, and CD10 expression in Epstein-Barr virus-associated gastric carcinoma. J Gastroenterol 41:733–9.

Burgess DE, Woodman CB, Flavell KJ, Rowlands DC, Crocker J, Scott K, Biddulph JP, Young LS, and Murrray PG. 2002. Low prevalence of Epstein-Barr virus in incident gastric adenocarcinomas from United Kingdom. Br J Cancer 86:702–4.

Burke AP, Yen TS, Shekitka KM, and Sobin LH. 1990. Lymphoepithelial carcinoma of the stomach with Epstein-Barr virus demonstrated by polymerase chain reaction. Mod Pathol 3:377–80.

Burkitt D. 1958. A sarcoma involving the jaws in African children. Br J Surg 46:218–24.

Chan WY, Chan AB, Liu AY, Chow JH, Ng EK, and Chung SS. 2001. Chromosome 11 copy number gains and Epstein-Barr virus-associated malignancies. Diagn Mol Pathol 10:223–7.

Chang MS, Kim WH, Kim CW, and Kim YJ. 2000. Epstein-Barr virus in gastric carcinoma with lymphoid stroma. Histopathology 37:309–15.

Chang MS, Lee HS, Kim HS, Kim SH, Choi SI, Lee BL, Kim CW, Kim YI, Yang M, and Kim WH. 2003. Epstein-Barr virus and microsatellite instability in gastric carcinogenesis. J Pathol 199:447–52.

Chang MS, Uozaki M, Chong JM, Ushiku T, Sakuma K, Ishikawa S, Hino R, Barua RR, Iwasaki Y, Arai K, Fukii H, Nagai H, and Fukayama M. 2006. CpG island methylation status in gastric carcinoma with and without infection of Epstein-Barr virus. Clin Cancer Res 12:2995–3002.

Chapel J, Fabiani B, Davi F, Raphel M, Tepper M, Champault G, and Guettier C. 2000. Epstein-Barr virus and gastric carcinoma in Western patients: comparison of pathological parameters and p53 expression in EBV-positive and negative tumors. Histopathology 36:252–61.

Chien YC, Chen JY, Liu MY, Yang HI, Hsu MM, Chen CJ, and Yang CS. 2001. Serologic markers of Epstein-Barr virus infection and nasopharyngeal carcinoma in Taiwanese men. N Engl J Med 345:1877–82.

Cho YJ, Chang MS, Park SH, Kim HS, and Kim WH. 2001. In situ hybridization of Epstein-Barr virus in tumor cells and tumor-infiltrating lymphocytes of the gastrointestinal tract. Hum Pathol 32:297–301.

Chong JM, Sakuma K, Sudo M, Osawa T, Ohara E, Uozaki H, Shibahara J, Kuroiwa K, Tominaga S, Hippo Y, Aburatani H, Funata N, and Fukayama M. 2002. Interleukin-1beta expression in human gastric carcinoma with Epstein-Barr virus infection. J Virol 76:6825–31.

Chu PG, Chang KL, Chen WG, Chen YY, Shibata D, Hayashi K, Bacchi C, Bacchi M, and Weiss M. 1999. Epstein-Barr virus (EBV) nuclear antigen (ENBA)-4 mutations in EBV-associated malignancies in three different populations. Am J Pathol 155:941–7.

Corvaian A, Koriyama C, Akiba S, Eizuru Y, Backhouse C, Palma M, Argandona J, and Tokunaga M. 2001. Epstein-Barr virus in gastric carcinoma is associated with location in the cardia and with a diffuse histology: a study in one area of Chile. Int J Cancer 94: 527–30.

Corvalan A, Ding S, Koriyama C, Carrasquilla G, Backhouse C, Urzua L, Argandona J, Palma M, Eizuru Y, and Akiba S. 2006. Association of a distinctive strain of Epstein-Barr virus with gastric cancer. Int J Cancer 118:1736–42.

Dutta N, Gupta A, Mazumder DN, and Banerjee S. 2006. Down-regulation of locus-specific human lymphocyte antigen class I expression in Epstein-Barr virus-associated gastric cancer: implication for viral-induced immune evasion. Cancer 106:1685–93.

Epstein MA, Achong BC, and Barr YM. 1964. Virus particles in cultured lymphoblasts from Burkitt's lymphoma. Lancet i:702–3.

Etoh T, Kanai Y, Ushijima S, Nakagawa T, Nakanashi Y, Sasako M, Kitano S, and Hirohashi S. 2004. Increased DNA methyltransferase 1 (DNMT1) protein expression correlates significantly with poor tumor differentiation and frequent DNA hypermethylation of multiple CpG islands in gastric cancer. Am J Pathol 154:689–99.

Evans AS. 1972. Clinical syndromes associated with EB virus infection. Ann Intern Med 18:77–93.

Feng WH, Israel B, Raab-Traub N, Busson P, and Kenney SC. 2002a. Chemotherapy induces lytic EBV replication and confers ganciclovir susceptibility to EBV-positive epithelial cell tumors. Cancer Res 62:1920–6.

Feng WH, Westphal E, Mauser A, Raab-Traub N, Gulley ML, Busson P, and Kenney SC. 2002b. Use of adenovirus vectors expressing Epstein-Barr virus (EBV) immediate-early protein BZRF1 or BRLF1 to treat EBV-positive tumors. J Virol 76:10951–9.

Galetsky SA, Tsvetnov VV, Land CE, Afanasieva TA, Petrovichev NN, Gurtsevitch VE, and Tokunaga M. 1997. Epstein-Barr virus-associated gastric cancer in Russia. Int J Cancer 73:786–9.

Gorzer I, Puchhammer-Stockl E, van Esser JWJ, Niester HGM, and Cornelissen JJ. 2007. Associations among Epstein-Barr virus subtypes, human leukocyte antigen class I alleles and the development of posttransplantation lymphoproliferative disorder in bone marrow transplant recipients. Clin Infect Dis 44:693–5.

Hao Z, Koriyama C, Akiba S, Li J, Luo X, Itoh T, Eizuru Y, and Zou J. 2002. The Epstein-Barr virus-associated gastric carcinoma in Southern and Northern China. Oncol Rep 9:1293–8.

Harn HJ, Chang JY, Wang MW, Ho LI, Lee HS, Chiang JH, and Lee WH. 1995. Epstein-Barr virus-associated gastric adenocarcinomas in Taiwan. Hum Pathol 26:267–71.

Hayashi K, Chen WG, Chen YY, Murakami I, Chen HL, Ohara N, Nose S, Hamaya K, Matsui S, Bacchi MM, Bacchi CE, Chang KL, and Weiss LM. 1998. Deletion of Epstein-Barr virus latent membrane protein 1 gene in Japanese and Brazilian gastric carcinomas, metastatic lesions, and reactive lymphocytes. Am J Pathol 152:191–8.

Hermann K, and Niedobitek G. 2003. Epstein-Barr virus–associated carcinomas: facts and fiction. J Pathol 199:140–5.

Herrera-Goepfert R, Reyes E, Hernadez-Avila M, Mohar A, Shinkura R, Fujiyama C, Akiba S, Eiruru Y, Harada Y, and Tokunaga M. 1999. Epstein-Barr virus-associated gastric carcinoma in Mexico: analysis of 135 consecutive gastrectomies in two hospitals. Mod Pathol 12:873–8.

Hoshikawa Y, Satoh Y, Murakami M, Maeta M, Kaibara N, Ito M, Kurata T, and Sairenji T. 2002. Evidence of the lytic infection of Epstein-Barr virus (EBV) in EBV-positive gastric carci-noma. J Med Virol 66:351–9.

Hsieh LL, Lin PJ, Chen TC, and Ou JT. 1998. Frequency of Epstein-Barr virus-associated gastric adenocarcinoma in Taiwan. Cancer Lett 129:125–9.

IARC monographs on the evaluation of carcinogenic risks to humans. Infections with Epstein-Barr virus and human herpes viruses. Vol.70. Lyon: IARC; 1997.

Iwarkiri D, Eizuru Y, Tokunaga M, and Takada K. 2003. Autocrine growth of Epstein-Barr virus-positive gastric carcinoma cells mediated by an Epstein-Barr virus-encoded small RNA. Cancer Res 63:7062–7.

Iwasaki Y, Chong JM, Hayashi Y, Ikeno R, Arai K, Kitamura M, Koike M, Hirai K, and Fukayama M. 1998. Establishment and characterization of a human Epstein-Barr virus-associated gastric carcinoma in SCID mice. J Virol 72:8321–6.

Jung EJ, Lee YM, Lee BL, Chang MS, and Kim HW. 2007. Ganciclovir augments the lytic induc-tion and apoptosis induced by chemotherapeutic agents in an Epstein-Barr virus-infected gas-tric carcinoma cell line. Anticancer Drugs 18:79–85.

Kaizaki Y, Hosokawa O, Sakurai S, and Fukayama M. 2005. Epstein-Barr virus-associated gastric carcinoma in the remnant stomach: de novo and metachronous gastric remnant carcinoma. J Gastroenterol 40:570–7.

Kang GH, Lee S, Kim WH, Lee HW, Kim JC, Rhyu MG, and Ro JY. 2002. Epstein-Barr virus-positive gastric carcinoma demonstrates frequent aberrant methylation of multiple genes and contributes CpG island methylator phenotype-positive gastric carcinoma. Am J Pathol 160:787–94.

Kattoor J, Koriyama C, Akiba S, Itoh T, Ding S, Eizuru Y, Abraham EK, and Chandralekha B. 2002. Epstein-Barr virus-associated gastric carcinoma in southern India: a comparison with a large-scale Japanese series. J Med Virol 68:384–9.

Kijima Y, Hokita S, Takao S, Baba M, Natsugoe S, Yoshinaka H, Aridome K, Otsuji T, Itoh T, Tokunaga M, Eizuru Y, and Aikou T. 2001. Epstein-Barr virus involvement is mainly restricted to lymphoepithelial type of gastric carcinoma among various epithelial neoplasms. J Med Virol 64:513–8.

Kim DN, Chae HS, Oh ST, Kang JH, Park CH, Park WS, Takada K, Lee JM, Lee WK, and Lee SK. 2007. Expression of viral microRNAs in Epstein-Barr virus-associated gastric carcinoma. J Virol 81:1033–6.

Kim MA, Lee HS, Yang HK, and Kim WH. 2004. Cytokeratin expression profile in gastric carci-noma. Hum Pathol 35:576–81.

Klein E, Kis LL, and Klein G. 2007. Epstein-Barr virus infection in humans: from harmless to life endangering virus-lymphocyte interaction. Oncogene 26:1297–305.

Koriyama C, Akiba S, Iriya K, Yamaguti T, Hamada GS, Itoh T, Eizuru Y, Aikou T, Watanabe S, Tsugane S, and Tokunaga M. 2001a. Epstein-Barr virus-associated gastric carcinoma in Japanese Brazilians and non-Japanese Brazilians in San Paulo. Jpn J Cancer Res 92:911–7.

Koriyama C, Shinkura R, Hamasaki Y, Fujiyoshi T, Eizuru Y, and Tokunaga M. 2001b. Human leukocyte antigen related to Epstein-Barr virus-associated gastric carcinoma in Japanese patients. Eur J Cancer Prev 10:69–75.

Koriyama C, Akiba S, Itoh T, Kijima Y, Sueyoshi K, Corvaian A, Herrer-Goepfer R, and Eieuru Y. 2002. Prognostic significance of Epstein-Barr virus involvement in gastric carcinoma in Japan. Int J Mol Med 10:635–9.

Koriyama C, Akiba S, Minakami Y, and Eizuru Y. 2005. Environmental factors related to Epstein-Barr virus-associated gastric cancer in Japan. J Exp Clin Cancer Res 24:547–53.

Kume T, Oshima K, Yamasshita Y, Shirakusa T, and Kikuchi M. 1999. Relationship between Fas-ligand expression on carcinoma cell and cytotoxic T-lymphocyte response in lymphoepithelioma-like cancer of the stomach. Int J Cancer 84:339–43.

Kusano M, Toyota M, Suzuki H, Akino K, Aoki F, Fujita M, Hosokawa M, Shinomura Y, Imai K, and Tokino T. 2006. Genetic, epigenetic, and clinicopathologic features of gastric carcinoma with the CpG island methylator phenotype and an association with Epstein-Barr virus. Cancer 106:1467–79.

Kuzushima K, Nakamura S, Nakamura T, Yamamura Y, Yokoyama N, Fujita M, Kiyono T, and Tsurumi T. 1999. Increased frequency of antigen-specific CD8(+) cytotoxic T lymphocytes infiltrating an Epstein-Barr virus-associated gastric carcinoma. J Clin Invest 104:163–71.

Lee HS, Chang MS, Yang HK, Lee BL, and Woo HK. 2004. Epstein-Barr virus-positive gastric carcinoma has a distinct protein expression profile in comparison with Epstein-Barr virus-negative carcinoma. Clin Cancer Res 10:1698–705.

Leonicini L, Vindigni C, Megha T, Funto I, Pacenti I, Musaro M, Renieri A, Seri M, Anagnostopoulos J, and Tosi P. 1993. Epstein-Barr virus and gastric cancer: data and unanswered questions. Int J Cancer 53:898–901.

Levine PH, Stemmermann G, Lennette ET, Hidesheim A, Shibata D, Nomura H, Okita K, Miura O, Shimizu N, and Takada K. 1995. Elevated antibody titers to Epstein-Barr virus prior to the diagnosis of Epstein-Barr virus-associated gastric adenocarcinoma. Int J Cancer 60:642–4.

Lo YM. 2001a. Quantitative analysis of Epstein-Barr virus DNA in plasma and serum: applications in tumor detection and monitoring. Ann NY Acad Sci 945:68–72.

Lo YM, Chan WY, Ng EK, Chan LY, Lai PB, Tam JS, and Chung SC. 2001b. Circulating Epstein-Barr virus DNA in the serum of patients with gastric carcinoma. Clin Cancer Res 7:1856–9.

Luo B, Murakami M, Fukuda M, Fujioka A, Yanagihara K, and Sairenji T. 2004. Characterization of Epstein-Barr virus infection in a human signet ring cell gastric carcinoma cell line, HSC-39. Microbes Infect 6:429–39.

Luo B, Wang Y, Wang XF, Liang H, Yan LP, Huang BH, and Zhao P. 2005. Expression of Epstein-Barr virus genes in EBV-associated gastric carcinomas. World J Gastroenterol 11:629–33.

Minarovits J. 2006. Epigenotypes of latent herpesvirus genomes. Curr Top Microbiol Immunol 310:61–80.

Niedobitek G, and Herbst M. 2006. In situ detection of Epstein-Barr virus and phenotype determination of EBV-infected cells. Methods Mol Biol 326:115–37.

Nilsson K, Klein G, Henle, W, and Henle G. 1971. The establishment of lymphoblastoid lines from adult and fetal human lymphoid tissue and its dependence on EBV. Int J Cancer 81:443–50.

Nishikawa J, Yanai H, Hirano A, Okamoto T, Nakamura H, Matsusaki K, Kawano T, Miura O, and Okita K. 2002. High prevalence of Epstein-Barr virus in gastric remnant carcinomas after Billroth-II reconstruction. Scand J Gastroenterol 37:825–9.

Oh ST, Seo JS, Moon UY, Kang KH, Shin DJ, Yoon SK, Kim WH, Park JG, and Lee SK. 2004. A naturally derived gastric cancer cell line shows latency I Epstein-Barr virus infection closely resembling EBV-associated gastric cancer. Virology 320:330–6.

Ohfuji S, Osaki M, Tsujitani S, Ikeguchi M, Sairenji T, and Ito H. 1996. Low frequency of apoptosis in Epstein-Barr virus-associated gastric carcinoma with lymphoid stroma. Int J Cancer 68:710–5.

Ojima H, Fukuda T, Nakajima T, Takenoshita S, and Nagamachi Y. 1996. Discrepancy between clinical and pathological lymph node evaluation in Epstein-Barr virus-associated gastric cancers. Anticancer Res 16:3081–4.

Okugawa K, Itoh T, Kawashima I, Takesako K, Mazda O, Nukaya I, Yano Y, Yamamoto Y, Yamagishi H, and Ueda Y. 2004. Recognition of Epstein-Barr virus-associated gastric carcinoma cells by cytotoxic T lymphocyte induced in vitro with autologous lymphoblastoid cell line and LMP2-derived HLA-A24-restricted 9-mer peptide. Oncol Rep 12:725–31.

Old LJ, Boyse EA, Oettgen HF, Harven ED, Geering, Williamson B, and Clifford P. 1966. Precipitating antibody in human serum to an antigen present in cultured Burkitt's lymphoma cells. Proc Natl Acad Sci USA 56:1699–704.

Osato T, and Imai S. 1996. Epstein-Barr virus and gastric carcinoma. Semin Cancer Biol 7:175–82.

Ott G, Kirchiner T, and Muller-Hermelink HK. 1994. Monoclonal Epstein-Barr virus genomes but lack of EBV related protein expression in different types of gastric carcinoma. Histopathology 25:323–9.

Ouburg S, Bart A, Crusius J, Klinkenberg-Knol EC, Mulder CJ, Salvador Pena A, and Morre SA. 2005. A candidate gene approach of immune mediators effecting the susceptibility to and severity of upper gastrointestinal diseases in relation to Helicobacter and Epstein-Barr virus infection. Eur J Gatroenterol Hepatol 17:1213–24.

Parkin DM. 2006. The global health burden of infection-associated cancers in the year 2002. Int J Cancer 118:3030–44.

Qiu K, Tomita Y, Hashimoto M, Ohsawa M, Kawano K, Wu DM, and Aozasa K. 1997. Epstein-Barr virus in gastric carcinoma in Suzhou, China and Osaka, Japan: association with clinico-pathologic factors and HLA-subtypes. Int J Cancer 71:155–8.

Rugge M, and Genta R. 1999. Epstein-Barr virus: a possible accomplice in gastric oncogenesis. J Clin Gastroenterol 29:3–5.

Sakuma K, Chong JM, Sudo M, Ushiku T, Inoue Y, Shibahara J, Uozaki M, Nagai H, and Fukayama M. 2004. High-density methylation of p14ARF and p16INK4A in Epstein-Barr virus-associated gastric carcinoma. Int J Cancer 112:273–8.

Sakuma K, Uozaki H, Chong Jm, Hironaka M, Sudo M, Ushiku T, Nagai H, and Fukayama M. 2005. Cancer risk to the gastric corpus in Japanese, its correlation with interleukin-1beta gene polymorphism (+3953"T) and Epstein-Barr virus infection. Int J Cancer 115:93–7.

Shibata D, and Weiss LM. 1992. Epstein-Barr virus-associated gastric adenocarcinoma. Am J Pathol 140:769–74.

Shinkura R, Yamamoto N, Koriyama C, Shinmura Y, Eizuru Y, and Tokunaga M. 2000. Epstein-Barr virus-specific antibodies in Epstein-Barr virus-positive and -negative gastric carcinoma cases in Japan. J Med Virol 60:411–6.

Shinohara K, Miyazaki K, Noda N, Satoh D, Terada M, and Wakasugi H. 1998. Gastric diseases related to Helicobacter pylori and Epstein-Barr virus infection. Microbiol Immunol 42:415–21.

Sudo M, Chong JM, Sakuma K, Ushiku T, Uozaki H, Nagai H, Funata N, Matsumoto Y, and Fukayama M. 2004. Promoter hypermethylation of E-cadherin and its abnormal expression in Epstein-Barr virus-associated gastric carcinoma. Int J Cancer 109:194–9.

Tajima M, Komuro M, and Okinaga K. 1998. Establishment of Epstein-Barr virus-positive human gastric epithelial lines. Jpn J Cancer Res 89:262–8.

Takada K. 2000. Epstein-Barr virus and gastric carcinoma. J Clin Pathol Mol Pathol 53:255–61.

Tokunaga M, Uemura Y, Tokudome T, et al. 1993. Epstein-Barr virus-related gastric cancer in Japan: a molecular patho-epidemiological study. Acta Pathol Jpn 43:574–81.

Tokunaga M, and Land CE. 1998. Epstein-Barr virus involvement in gastric cancer: biomarker for lymph node metastasis. Cancer Epidemiol Biomarkers Prev 7:449–50.

van Beek J, zur Hausen A, Klein KE, van de Velde CJ, Middeldorp JM, van den Brule AJ, Meijer CJ, and Bloemena E. 2004. EBV-positive gastric adenocarcinomas: a distinct clinicopatho-logic entity with a low frequency of lymph node involvement. J Clin Oncol 22:664–70.

van Rees BP, Caspers E, zur Hausen A, van den Brule A, Drilenburg P, Weterman MA, and Offerhaus GJ. 2002. Different pattern of allelic loss in Epstein-Barr virus-positive gastric cancer with emphasis on the p53 tumor suppressor pathway. Am J Pathol 161:1207–13.

von Rahden BH, Langner C, Brucher BL, Stein HJ, and Sarbia M. 2006. No association of primary adenocarcinomas of the small bowel with Epstein-Barr virus infection. Mol Carcinog 45:349–52.

Wang Q, Tsao SW, Ooka T, Nicholls Jm, Cheung HW, Fu S, Wong YC, and Wang X. 2006. Anti-apoptotic role of BARF1 in gastric cancer cells. Cancer Lett 238:90–103.

Watanabe H, Enjoji M, and Imai T. 1976. Gastric carcinoma with lymphoid stroma: its morphologic characteristics and prognostic correlation. Cancer 38:232–43.

Westphal EM, Blackstock W, Feng W, Israel B, and Kenney SC. 2000. Activation of lytic Epstein-Barr virus (EBV) infection by radiation and sodium butyrate in vitro and in vivo: a potential method for treating EBV-positive malignancies. Cancer Res 60:5781–8.

Wu MS, Shun CT, Wu CC, Hsu TY, Lin MT, Chang MC, Wang HP, and Lin JT. 2000. Epstein-Barr virus-associated gastric carcinomas: relation to H. pylori infection and genetic alterations. Gastroenterology 118:1031–8.

Wu MS, Huang SP, Chang YT, Shun CT, Chang MC, Lin MT, Wang HP, and Lin JT. 2002. Tumor necrosis factor-a and interleukin-10 promoter polymorphisms in Epstein-Barr virus-associated gastric carcinoma. J Infect Dis 185:106–9.

Yanai H, Murakami T, Yoshiyama H, Takeuchi H, Nishikawa J, Nakamura H, Okita K, Miura O, Shimizu N, and Takada K. 1999. Epstein-Barr virus-associated gastric carcinoma and atrophic gastritis. J Clin Gastroenterol 29:39–43.

Yanai H, Nishikawa J, Mizugaki Y, Shimizu, N, Takada K, Matsusaki K, Toda T, Matsumoto Y, Tada M, and Okita K. 1997a. Endoscopic and pathologic features of Epstein-Barr virus-associated gastric carcinoma. Gastrointest Endosc 45:236–42.

Yanai H, Takada K, Shimizu N, Mizugaki Y, Tada M, and Okita K. 1997b. Epstein-Barr virus infection in non-carcinomatous epithelium. J Pathol 183:293–8.

Yoshiwara E, Koriyama C, Akiba S, Itoh T, Minakami Y, Chirinos JL, Watanabe J, Takano J, Miyagui J, Hidalgo H, Chacon P, Linares V, and Eizuru Y. 2005. Epstein-Barr virus-associated gastric carcinoma in Lima, Peru. J Exp Clin Cancer Res 24:49–54.

Young LS, and Rickinson AB. 2004. Epstein-Barr virus:40 years on. Nature Rev Cancer 4:757–68.

Yuen ST, Chung LP, Leung SY, Luk IS, Chan SY, and Ho J. 1994. In situ hybridization of Epstein-Barr virus in gastric and colorectal adenocarcinomas. Am J Surg Pathol 18:1158–63.

zur Hausen A, Brink AA, Craanen ME, Middeldorp JM, Meijer CJ, and van den Brule AJ. 2000. Unique transcription pattern of Epstein-Barr virus (EBV) in EBV-carrying gastric adenocarcinomas: expression of the transforming BARF1 gene. Cancer Res 60:2745–8.

zur Hausen A, Crusus JBA, Murillo LS, Alizadeh BZ, Morre SA, Meijer CJLM, van dan Brule AJC, and Pena AS. 2003. IL-1B promoter polymorphism and Epstein-Barr virus in Dutch patients with gastric carcinoma. Int J Cancer 107:866–7.

zur Hausen A, van Grieken NC, Meijer GA, Hermsen MA, Bloemena E, Meuwissen SG, Baak JP, Meijer CJ, Kuipers EJ, and van den Brule AJ. 2001. Distinct chromosomal abnormalities in Epstein-Barr virus-carrying gastric carcinomas tested by comparative genomic hybridization. Gastroenterology 121:612–8.

Color Plates

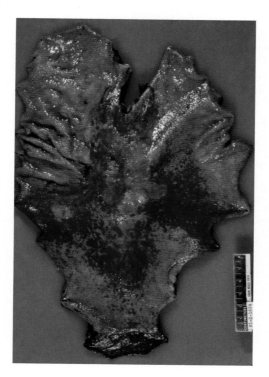

Fig. 1.3 Photograph of a gastrectomy specimen opened along the greater curvature and stained for alkaline phosphatase. A large ulcerated carcinoma occupies the lesser middle curvature and is surrounded by mucosa with intestinal metaplasia, stained in red. Note similar staining at the distal end of the specimen, corresponding to duodenum. (Reproduced from Correa and Stemmermann 2005, with permission.)

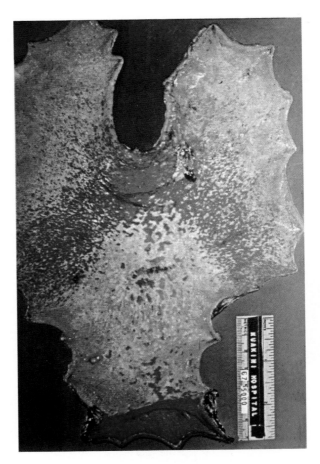

Fig. 1.5 Photograph of a gastrectomy specimen opened along the greater curvature and stained for alkaline phosphatase. Foci of intestinal metaplasia (stained in red) coalesce at the antrum–corpus junction, especially in the incisura angularis. (Reproduced from Correa and Stemmermann 2005, with permission.)

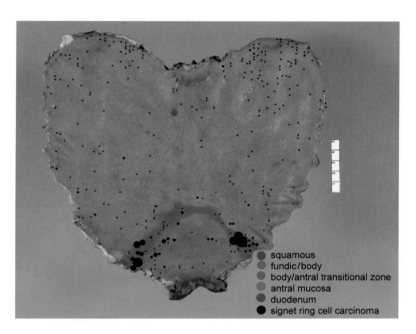

squamous
fundic/body
body/antral transitional zone
antral mucosa
duodenum
signet ring cell carcinoma

Fig. 1.8 Gastrectomy specimen from a 15-year-old Maori adolescent girl with hereditary diffuse adenocarcinoma. Every focus of in situ carcinoma identified grossly is depicted by a black dot. A total of 318 in situ carcinomas were identified in this specimen. Tumors around the antrum–corpus junction were larger. (Photograph courtesy of Dr. Amanda Charlton, Department of Pathology, Middlemore Hospital, Auckland, New Zealand. Reproduced from Charlton et al. 2004, with permission.)

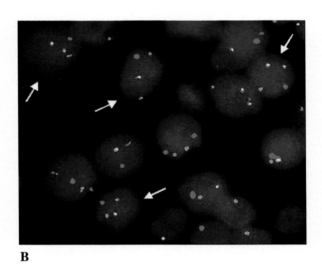

B

Fig. 4.8 (B) Interphase fluorescence *in situ* hybridization (FISH) pattern of (11;18)/*API2-MALT1*. FISH with an *API2/MALT1* dual-color, dual-fusion translocation probe shows colocalization of the red and green signals (arrows). (**A**: Reprinted from Sagaert et al. 2007, with permission from Macmillan Publishers. Copyright 2007. **B**: Reprinted from Nakamura et al. 2007, with permission from the BMJ Publishing Group.)

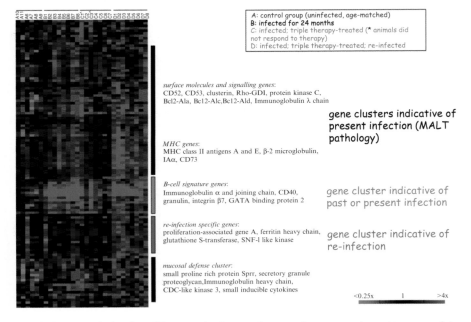

A: control group (uninfected, age-matched)
B: infected for 24 months
C: infected; triple therapy-treated (* animals did not respond to therapy)
D: infected; triple therapy-treated; re-infected

surface molecules and signalling genes:
CD52, CD53, clusterin, Rho-GDI, protein kinase C, Bcl2-Ala, Bcl2-Alc, Bcl2-Ald, Immunoglobulin λ chain

gene clusters indicative of present infection (MALT pathology)

MHC genes:
MHC class II antigens A and E, β-2 microglobulin, IAα, CD73

B-cell signature genes:
Immunoglobulin α and joining chain, CD40, granulin, integrin β7, GATA binding protein 2

gene cluster indicative of past or present infection

re-infection specific genes:
proliferation-associated gene A, ferritin heavy chain, glutathione S-transferase, SNF-l like kinase

gene cluster indicative of re-infection

mucosal defense cluster:
small proline rich protein Sprr, secretory granule proteoglycan, Immunoglobulin heavy chain, CDC-like kinase 3, small inducible cytokines

<0.25x   1   >4x

Fig. 4.13 Transcriptional profiling reveals clusters of genes whose expression patterns correlate with treatment outcome. RNA was prepared from whole stomach tissue of several animals per treatment group, reverse transcribed, and hybridized to 38,000 element murine cDNA arrays. Data were filtered with respect to spot quality (spots with regression correlations <0.6 were omitted) and data distribution (genes whose $\log_2$ of red-to-green normalized ratio is more than 1.5 standard deviations away from the mean in at least three arrays were selected) before clustering of the resulting 125 genes. Only genes for which information was available for >70% of arrays were included. Clusters of genes that are differentially regulated between treatment groups are annotated, and representative genes are listed for every cluster. (Reprinted from Müller et al., with permission from the American Society for Investigative Pathology.)

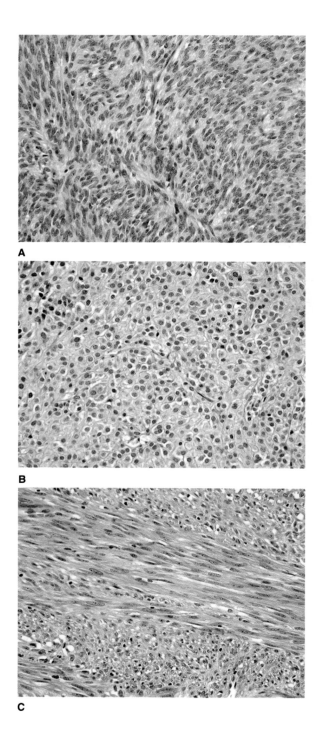

A

B

C

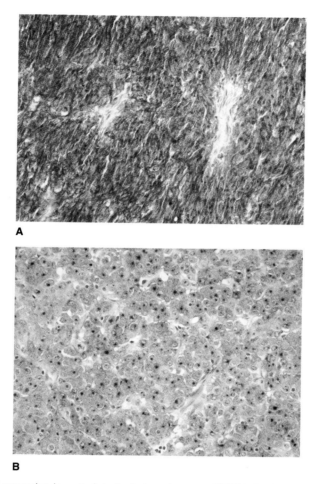

A

B

Fig. 5.2 KIT expression in gastrointestinal stromal tumors (GISTs). By immunohistochemistry, 95% of GISTs are positive for KIT (CD117), which is usually predominantly cytoplasmic in distribution (**A**), but may also show paranuclear dot-like staining (**B**)

Fig. 5.1 Histology of gastrointestinal stromal tumors (GISTs). Hematoxylin and eosin (H&E)-stained sections of (**A**) spindle cell GIST, composed of short fascicles of elongated cells with palely eosinophilic, fibrillary cytoplasm, and (**B**) epithelioid GIST, composed of sheets of polygonal cells with clear to eosinophilic cytoplasm. Both variants are notable for their uniform cytology. Spindle cell GISTs show similar histologic features as leiomyosarcomas (**C**), although the latter contain longer fascicles of spindle cells with brightly eosinophilic cytoplasm

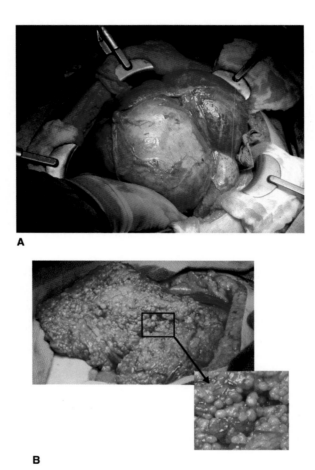

Fig. 5.4 Manifestations of metastatic gastrointestinal stromal tumor (GIST). Photographs taken during debulking surgery show (**A**) complete replacement of the right hepatic lobe by a single metastatic lesion; and (**B**) peritoneal carcinomatosis with complete replacement of the omentum by innumerable GIST nodules

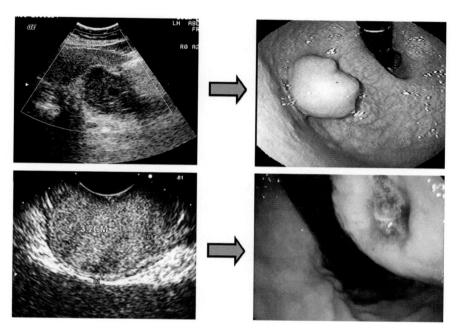

Fig. 5.5 Endoscopic examination of primary gastric gastrointestinal stromal tumor (GIST). GISTs often present as submucosal tumors in the wall of the gastrointestinal (GI) tract. Central necrosis of high-risk tumors or erosion into blood vessels can produce local inflammation or significant GI bleeding (bottom, right)

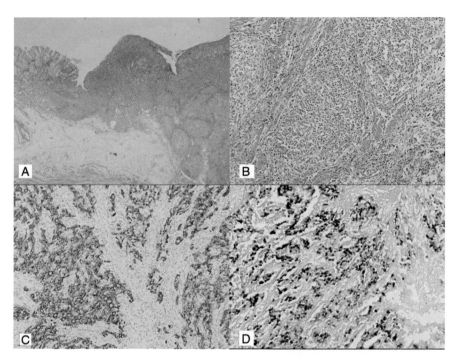

Fig. 8.1 A representative example of microscopic features and *in situ* hybridization of lymphoepithelioma-like carcinoma of stomach. (**A**) Hematoxylin and eosin (H&E) stain, original magnification 40×. (**B**) Pleomorphic cancer cell nests separated by intervening lymphoid stroma (H&E, original magnification 200×). (**C**) Cytokeratin staining shows immunoreactive tumor cells. (**D**) Positive EBER-1 signals in the nucleus of glandular and undifferentiated carcinomas (*in situ* hybridization, original magnification 200×)

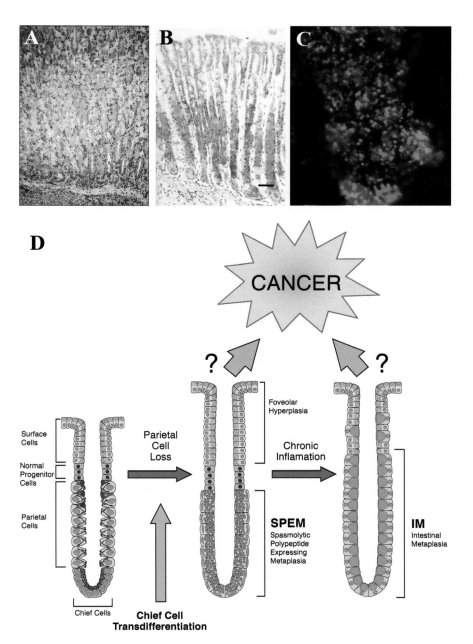

Fig. 13.2 Spasmolytic polypeptide-expressing metaplasia (SPEM) in mice. (**A**) Trefoil factor 2 (TFF2) staining reveals near complete replacement of fundic glands with SPEM in a C57BL/6 mouse infected with *Helicobacter felis* for 9 months. (**B**) TFF2 staining shows prominent SPEM in a mouse treated with DMP-777 for 14 days. (**C**) Dual immunofluorescence staining for intrinsic factor (green) and TFF2 (red) demonstrates dual-staining SPEM cells at the bases of glands from a mouse treated with DMP-777 for 14 days. Note that the intrinsic factor and TFF2 are present in separate granule populations in the SPEM cells. (**D**) A unified hypothesis for the origin of gastric metaplasias. Our studies suggest that SPEM arises from transdifferentiation of chief cells after parietal cell loss. In addition, we hypothesize that, in the presence of chronic inflammation in humans, intestinal metaplasia emerges from further differentiation of SPEM. (Adapted from Goldenring and Nomura 2006.)

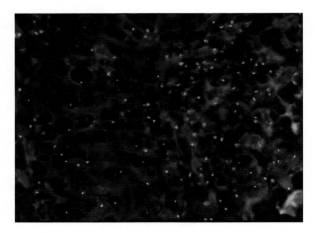

Fig. 22.1 Female tumor cells are present in solid tumors in male patients. The photomicrograph shows a laryngeal squamous cell carcinoma from a male patient that had undergone a bone marrow transplant 7 years earlier from a female donor. The tissue was stained for Y chromosome (red dots) and X chromosomes (green dots) by fluorescent *in situ* hybridization. Nuclei were counterstained in blue with DAPI stain. The tumor appears to be polyclonal, with the red circle on the left indicating a region that is largely XY and of recipient (male) origin, whereas the green circle on the left shows an area that is largely XX and of donor (female) origin. (This figure was provided courtesy of I. Avital and R. Downey, MSKCC.)

Chapter 9
Role of Host Genetic Susceptibility in the Pathogenesis of Gastric Cancer

Emad M. El-Omar

Introduction

At the turn of the twentieth century, gastric cancer was the leading cause of cancer-related death in many Western countries but the incidence has steadily decreased over the past 70 years. Globally, gastric cancer is the second most common cause of cancer-related death and, as a result of population aging and growth, the predicted incidence for 2010 is 1.1 million with the majority of this health burden borne by economically lesser-developed countries.[1] In this chapter, we hope to shed some light on the role of host genetic susceptibility in the pathogenesis of gastric cancer. In particular, we attempt to demonstrate how interactions among an infectious agent, host genetic makeup, and environmental factors could influence the pathogenesis of this cancer. The infectious agent in question is *Helicobacter pylori*, the world's most common chronic bacterial infection and the malignancy is gastric cancer, second only to lung cancer in its global incidence and impact. We demonstrate how this gastric infection could be used as a paradigm for gene–environment interactions in human disease, one that could help unravel the unknowns of a multitude of other microbial-induced malignancies.

Gastric Cancer and *Helicobacter pylori* Infection

In 1994, the International Agency for Research on Cancer declared *H. pylori* a Group I (definite) carcinogen.[2] This bold statement was met with considerable skepticism, but in the ensuing 14 years, evidence from epidemiologic and interventional studies in humans as well as experiments in rodents has convinced many that this bacterial infection is indeed the key factor in the initiation of the neoplastic process in the stomach.

H. pylori is a gram-negative, spiral-shaped, microaerophilic, urease-positive bacillus that is known to have chronically infected the stomachs of more than half the world's population. The infection is acquired during childhood, probably via the fecal/oral or gastric/oral route, and if not treated with antibiotics, will persist throughout life. Although the bacteria mainly reside on the surface mucus gel layer with little invasion of the gastric glands, the host responds with an impressive humoral and cell-mediated immune response. This immune response is largely ineffective, however, as most infections become chronically established with little evidence that spontaneous clearance occurs.

H. pylori has to survive in one of the harshest and least hospitable niches in the human body. Gastric acidity acts as a formidable first-line defense against food-borne pathogens and the constant outpouring of gastric secretions coupled with regular peristalsis, ensure that gastric contents, including microbial agents, are constantly flushed away. Despite this, *H. pylori* seems well equipped and adapted for habitation within this harsh environment. Recent studies show that *H. pylori* maintains its periplasmic pH within viable limits through possession of an acid-induced urea channel that regulates intrabacterial urease activity.[3] Essential nourishment for *H. pylori* is drawn from host gastric tissue through the inflammatory exudate it induces. To understand how this bacterium can predispose to gastric cancer, it is necessary to understand the basic pathophysiologic consequences of its presence within the human stomach.

Helicobacter pylori and Chronic Gastric Inflammation

The key pathophysiologic event in *H. pylori* infection is the initiation of an inflammatory response.[4] This response is probably triggered by the bacterium's lipopolysaccharide, urease, and/or cytotoxins and is mediated by cytokines. The cytokine repertoire comprises a multitude of pro- and antiinflammatory mediators whose function is to coordinate an effective immune/inflammatory response against invading pathogens without causing undue damage to the host.

In addition to their pro- or antiinflammatory properties, some *H. pylori*–induced cytokines have direct effects on gastric epithelial cells that have a profound effect on gastric physiology. For example, the proinflammatory cytokine interleukin-1 beta (IL-1β) is the most potent of known agents that are gastric cytoprotective, anti-ulcer, antisecretory, and inhibitory of gastric emptying.[5] Wolfe and Nompleggi[6] estimated that, on a molar basis, IL-1β is 100 times more potent than both prostaglandins and the proton pump inhibitor omeprazole and 6,000 times more potent than cimetidine in inhibiting acid secretion. Another important proinflammatory cytokine that is upregulated by *H. pylori* infection is tumor necrosis factor alpha (TNF-α), which also inhibits gastric acid secretion, but to a lesser extent than IL-1β.[7]

In physiologic terms, the stomach could be divided into two main compartments: an acidic proximal corpus that contains the acid-producing parietal cells, and a less-acidic distal antrum that does not have parietal cells but contains the endocrine cells that control acid secretion. *H. pylori* infection is first established in parts of the stomach that have a higher pH such as the antrum. Likely, this is the bacterium's attempt to preserve energy, for although *H. pylori* is well equipped for survival at low pH, this is achieved at a high cost of energy expenditure. Thus, high acid production by the parietal cells probably protects the corpus mucosa from initial colonization. Both animal and human ingestion studies suggest that successful colonization of the gastric mucosa is best achieved with the aid of acid suppression.[8,9] Furthermore, pharmacologic inhibition of acid secretion in infected subjects leads to redistribution of the infection and its associated gastritis from an antral to a corpus-predominant pattern.[10] Thus, lack of gastric acid extends the area of colonization and also maximizes the tissue damage resultant from this colonization.

Helicobacter pylori Infection and the Gastric Cancer Versus Duodenal Ulcer Phenotypes

There are three main gastric phenotypes that result from chronic *H. pylori* infection: 1. the most common by far is a mild pangastritis that does not affect gastric physiology and is not associated with significant human disease; 2. a corpus-predominant gastritis associated with multifocal gastric atrophy, hypochlorhydria, and increased risk of gastric cancer (the gastric cancer phenotype)[11]; and 3. an antral-predominant gastritis associated with high gastric acid secretion and increased risk of duodenal ulcer disease (the DU phenotype).[12]

There is accumulating evidence that acid secretory capacity is crucial in determining the distribution and natural history of *H. pylori* infection.[13] In hosts with low secretory capacity (genetically determined or secondary to pharmacologic inhibition), the organism is capable of colonizing a wider niche than would be possible in the presence of high volumes of acid. Colonization of a wider niche, including the corpus mucosa, leads to corpus gastritis with resultant functional inhibition of acid secretion. This inhibition is mediated by *H. pylori*–induced inflammatory cytokines (such as IL-1β and TNF-α) and the net effect is the establishment of a more aggressive gastritis that accelerates the development of gastric atrophy. Once atrophy develops, acid secretion is not only attenuated by the functional inhibition caused by inflammatory mediators but by a more permanent morphologic change that is more difficult to reverse. This situation is very relevant to the subgroup of humans who develop the gastric cancer phenotype in the presence of chronic *H. pylori* infection.

In contrast to subjects who have an increased risk of gastric cancer, subjects who develop duodenal ulcer disease are known to have a large parietal cell mass that is relatively free of *H. pylori*–induced inflammatory activity. This pattern of antral-predominant gastritis with high acid output characterizes the duodenal ulcer diathesis.

The effect of acid secretion on changing the distribution of *H. pylori* colonization and gastritis is most markedly exposed in individuals in whom acid secretion is manipulated by pharmacologic means. Thus, *H. pylori*–infected individuals on long-term proton pump inhibitors undergo a shift in the pattern of gastritis from antral to corpus-predominant, and they have a higher risk of developing gastric atrophy, a precursor lesion for gastric neoplasia.[10] This observation provided a clue as to the role of potential endogenous substances that could also inhibit acid secretion, such as IL-1β and TNF-α. These two cytokines were therefore prime candidates as host genetic factors that may increase risk of gastric cancer.

Genetic Polymorphisms in the *Interleukin-1* Gene Cluster Increase the Risk of Gastric Cancer and Its Precursors

One of the paradoxes of *H. pylori* infection is its association with mutually exclusive clinical outcomes such as gastric cancer and duodenal ulcer disease. A large volume of research has focused on the role of bacterial virulence factors (e.g., *cagA, vacA, BabA, OipA*) in the pathogenesis of these diseases and although these factors undoubtedly contribute to the degree of tissue damage, they do not readily distinguish between the two key outcomes.[14] This prompted some researchers to consider the host genetic factors that may be relevant to this process. Crucially, the search for the appropriate candidate genes had to stem from a profound understanding of gastric physiology and how this is disrupted by *H. pylori* infection. Because *H. pylori* achieves most of its damage through induction of chronic inflammation, it was reasonable to consider genes that control this process as appropriate candidates.

The *IL-1* gene cluster on chromosome 2q contains three related genes within a 430 kb region—*IL-1A, IL-1B*, and *IL-1RN*—which encode for the proinflammatory cytokines IL-1α and IL-1β as well as their endogenous receptor antagonist IL-1ra, respectively.[15] IL-1β is upregulated in the presence of *H. pylori* and has a central role in initiating and amplifying the inflammatory response to this infection.[16] As mentioned above, IL-1β is also an extremely potent inhibitor of gastric acid secretion. Three diallelic polymorphisms in *IL-1B* have been reported, all representing C–T or T–C transitions, at positions −511, −31, and +3954 bp from the transcriptional start site.[17] The *IL-1RN* gene has a penta-allelic 86-bp tandem repeat (VNTR) in intron 2, of which the less common allele 2 (*IL-1RN*2*) is associated with a wide range of chronic inflammatory and autoimmune conditions.[17] The presence of such highly prevalent and functional genetic polymorphisms provided an ideal opportunity to design the appropriate epidemiologic studies to test the role of these candidate loci.

We first studied the correlation of these high IL-1β genotypes (two polymorphisms in the *IL-1B* and *IL-1RN* genes) with hypochlorhydria and gastric atrophy in a Caucasian population of gastric cancer relatives from Scotland. These relatives are known to be at increased risk of developing the same cancer and have a higher prevalence of the precancerous abnormalities but only in the presence of *H. pylori* infection. We found that the high IL-1β genetic markers significantly increase the risk of these precancerous conditions. In a logistic regression model including both genotypes, the estimated age-adjusted odds ratios (ORs) for *IL-1B-511*T/-31*C* and *IL-1RN*2/*2* were 7.5 [95% confidence intervals (CI): 1.8–31] and 2.1 (95% CI: 0.7–6.3), respectively.[18] We proceeded to examine the association between the same IL-1β genetic polymorphisms and gastric cancer itself using another Caucasian case-control study comprising 366 gastric cancer patients and 429 population controls from Poland. We confirmed the same positive association between these genotypes and gastric cancer. In a logistic regression model including both genotypes, the estimated ORs for IL-1B-511*T/-31*T and IL-1RN*2/*2 were 1.6 (95% CI: 1.2–2.2) and 2.9 (95% CI: 1.9–4.4), respectively.[18] The initial observations were confirmed using another Caucasian population-based case-control study from the United States.[19] In this study, the proinflammatory IL-1 genotypes (IL-1B-511*T and IL-1RN) conferred similar ORs for noncardia gastric adenocarcinoma to the Polish study.

In the above studies, the proinflammatory IL-1 genotypes increased the risk of both intestinal and diffuse types of gastric cancer but the risk was restricted to the noncardia subsite. Indeed, the IL-1 markers had no effect on risk of cardia gastric adenocarcinoma, esophageal adenocarcinoma, or esophageal squamous cell carcinoma.[19] The latter findings are entirely in keeping with the proposed mechanism for the effect of these polymorphisms in gastric cancer, namely, reduction of gastric acid secretion. Thus, a high IL-1β genotype increases the risk of noncardia gastric cancer, a disease characterized by hypochlorhydria, whereas it has no effect on cancers associated with high acid exposure such as esophageal adenocarcinoma and some cardia cancers.

The association between IL-1 gene cluster polymorphisms and gastric cancer and its precursors has been confirmed independently by other groups covering Caucasian, Asian, and Hispanic populations.[20–26] Machado et al.[20] were the first to confirm the association between IL-1 markers and gastric cancer in Caucasians and reported similar ORs to those reported by our group. Furthermore, the same group subsequently reported on the combined effects of proinflammatory IL-1 genotypes and H. pylori bacterial virulence factors (cagA positive, VacA s1, and VacA m1). They showed that for each combination of bacterial/host genotype, the odds of having gastric carcinoma were greatest in those with both bacterial and host high-risk genotypes.[21] This highlights the important interaction between host and bacterium in the pathogenesis of gastric cancer.

Unlike the studies mentioned above, some reports, particularly from Asian countries, failed to find an association between IL-1 markers and gastric cancer risk. A number of these studies were underpowered and others used inappropriate controls. However, even excluding the weaker studies, there is still the impression that not all Asian or Caucasian populations have demonstrated a predisposition for gastric cancer in association with "published" proinflammatory IL-1 polymorphisms. In some instances, studies found that there was a positive association but with novel markers of the IL-1B gene.[27] Other studies pointed to the importance of the background prevalence of gastric cancer in the population, with the positive associations being easier to demonstrate in low-incidence compared with high-incidence areas.[26] Finally, some studies reported a positive association with IL-1 markers but this was confined to the opposite alleles from those reported by the majority of studies.[28,29] In one study, the authors confirmed that the opposite alleles (previously found to associate with reduced IL-1β production) were in fact the high IL-1β alleles in their population.[28]

Other potential explanations for lack of association with the IL-1 markers in some populations should also be considered. One very interesting concept that has emerged recently is that of haplotype context. Chen et al.[30] studied the significance of haplotype structure in gene regulation by testing whether individual single nucleotide polymorphisms (SNPs) within a gene promoter region (*IL-1B* in this case) might affect promoter function and, if so, whether function was dependent on haplotype context. They showed very elegantly that significant interactions between SNPs according to haplotype had a profound influence on the functionality of the gene promoter. Because haplotype context is influenced by ethnic background and past selective pressures that are unique to different populations, this may explain why some studies might demonstrate a positive association with a particular marker in some but not all populations.

Lee et al.[31] recently reported on another potential explanation for variation in association with *IL-1B* and gastric cancer. They studied the interaction between SNPs in *IL-1B* and general transcription factor (*GTF2A1*) genes. They found an association between carriage of the *IL1B-31C* allele and gastric cancer among Koreans, which was observed only in subjects with *GTF2A1* GG genotype suggesting synergy between the two genetic markers. Taken at face value, these findings still point to IL-1β as being a crucial cytokine in the pathogenesis of *H. pylori*–induced gastric cancer and its precursors and variations in its gene act as host genetic factors that mediate this effect.

A crucial piece of evidence that confirmed the unique role of IL-1β in *H. pylori*–induced gastric carcinogenesis came from a transgenic mouse model in which IL-1β overproduction was targeted to the stomach by the H+/K+ adenosine triphosphatase beta promoter. With overexpression of IL-1β confined to the stomach, these transgenic mice had a thickened gastric mucosa, produced lower amounts of gastric acid, and developed severe gastritis followed by atrophy, intestinal metaplasia, dysplasia, and adenocarcinoma. Crucially, these IL-1β transgenic mice proceeded through a multistage process that mimicked human gastric neoplasia. These changes occurred even in the absence of *H. pylori* infection, which when introduced led to an acceleration of these abnormalities.[32]

Role of Other Cytokine Gene Polymorphisms

Soon after the *IL-1* gene cluster polymorphisms were identified as risk factors for gastric cancer, the proinflammatory genotypes of TNF-α (*TNF-A*) and *IL-10* were reported as independent additional risk factors for noncardia gastric cancer.[19] TNF-α is another powerful proinflammatory cytokine that is produced in the gastric mucosa in response to *H. pylori* infection. Similar to IL-1β, it has an acid inhibitory effect, albeit much weaker.[7] The *TNF-A*-308 G >A polymorphism is known to be involved in a number of inflammatory conditions. Carriage of the proinflammatory A allele increased the OR for noncardia gastric cancer to 2.2 (95% CI: 1.4–3.7). The role of the *TNF-A*-308 G >A polymorphism in gastric cancer was independently confirmed by a study from Machado et al.[22] IL-10 is an antiinflammatory cytokine that down-regulates IL-1β, TNF-α, interferon-γ, and other proinflammatory cytokines. Relative deficiency of IL-10 may result in a T helper-1 (Th-1)-driven hyperinflammatory response to *H. pylori* with greater damage to the gastric mucosa. We reported that homozygosity for the low-IL-10 *ATA* haplotype (based on three promoter polymorphisms at positions −592, −819 and −1082) increased the risk of noncardia gastric cancer with an OR of 2.5 (95% CI: 1.1–5.7).

We have studied the effect of having an increasing number of proinflammatory genotypes (*IL-1B*-511*T, *IL-1RN*2*2, *TNF-A*-308*A, and *IL-10* ATA/ATA) on the risk of nongastric cancer. The risk increased progressively so that by the time 3–4 of these polymorphisms were present, the OR for gastric cancer was increased to 27-fold (Table 9.1).[19] The fact that *H. pylori* is a prerequisite for the association of these polymorphisms with malignancy demonstrates that, in this situation, inflammation is indeed driving carcinogenesis.

Another important cytokine that has a noted role in the pathogenesis of *H. pylori*–induced diseases is IL-8. This chemokines belongs to the CXC family and is a potent chemoattractant for neutrophils and lymphocytes. It also has effects on cell proliferation, migration, and tumor angiogenesis. The gene has a well-established promoter polymorphism at position −251 (*IL-8*-251 T > A). The A allele is associated with increased production of IL-8 in *H. pylori*–infected gastric mucosa.[33] It was also found to increase the risk of severe inflammation and precancerous gastric abnormalities in Caucasian[33] and Asian populations.[34] However, the same polymorphism was only found to increase risk of gastric cancer in some Asian populations[34–37] with no apparent effect in Caucasians.[38]

It is likely that other proinflammatory cytokine gene polymorphisms will be relevant to gastric cancer initiation and progression. This exciting field has expanded greatly over the past few years and there are now attempts at defining the full complement of risk genotypes that dictate an individual's likelihood of developing cancer.

But how do these proinflammatory cytokine polymorphisms explain the divergent outcome to *H. pylori* infection? We speculate that the effect of these polymorphisms operates early in the disease process and requires the presence of *H. pylori* infection. When *H. pylori* infection challenges the gastric mucosa, a vigorous inflammatory response with a high IL-1β/TNF-α component may seem to be beneficial, but it has the unfortunate effect of switching acid secretion off, thus

Table 9.1 Frequencies and age, sex, and race-adjusted odds ratios (and Cornfield 95% confidence intervals) for the association of one to four proinflammatory polymorphisms in *IL-1B, IL-1RN, IL-10,* and *TNF-A* with noncardia gastric cancer[19]

Polymorphisms	Cases (N = 188)	Controls (N = 210)	Odds ratio (95% confidence interval)
0	22	75	(referent)
1	74	85	2.8 (1.6–5.1)
2	62	46	5.4 (2.7–10.6)
3	28	4	26.3 (7.1–97.1)
4	2	0	∞ (undefined)

allowing the infection to extend its colonization and damaging inflammation to the corpus mucosa, an area that is usually well protected by secretion of acid. A decreased flow of acid will also undermine attempts to flush out these toxic substances causing further damage to the mucosa. More inflammation in the corpus leads to more inhibition of acid secretion and a continuing cycle that accelerates glandular loss and onset of gastric atrophy. It is apparent that this vicious cycle ultimately succeeds in driving the infection out, but at a very high price for the host. This is amply demonstrated by the finding that *H. pylori* density becomes progressively lower with the progression from mild gastritis through severe gastritis, atrophy, and intestinal metaplasia. Indeed, by the time gastric cancer develops, it is extremely difficult to demonstrate any evidence of the infection.[39]

Role of Polymorphisms in the Innate Immune Response Genes

Genetic polymorphisms of the cytokines and chemokines discussed above clearly have an important role in the risk of *H. pylori*–induced gastric adenocarcinoma. However, *H. pylori* is initially handled by the innate immune response and it is conceivable that functionally relevant polymorphisms in genes of this arm of the immune system could affect the magnitude and subsequent direction of the host's response against the infection. The majority of *H. pylori* cells do not invade the gastric mucosa but the inflammatory response against it is triggered through attachment of *H. pylori* to the gastric epithelia.[40] Toll-like receptor 4 (TLR4), the lipopolysaccharide (LPS) receptor, was initially identified as the potential signaling receptor for *H. pylori* on gastric epithelial cells.[41] TLR4 belongs to a family of pattern recognition receptors, of which there are currently 11 members, that activate proinflammatory signaling pathways in response to microbes or pathogen-associated molecular patterns.[42] TLR4, in conjunction with CD14 and MD-2, transduces signals through MyD88, Toll/IL-1 receptor domain, and TRAF6. This promotes transcription of genes, which are involved in immune activation including the transcription factor NF-κB and MAP kinase pathways.[43]

Arbour et al.[44] described a functional polymorphism at position +896 in exon 4 of the *TLR4* gene (dbSNP ID: rs4986790). This A >G transition results in replacement of a conserved aspartic acid residue with glycine at amino acid 299 (Asp299Gly), and alteration in the extracellular domain of the TLR4 receptor. This renders carriers hyporesponsive to LPS challenge by either disrupting transport of TLR4 to the cell membrane or by impairing ligand binding or protein interactions.[44] The mutation has been associated with a variety of inflammatory and infectious conditions including atherosclerosis, myocardial infarction, inflammatory bowel disease, and septic shock.[45–47] Recent work demonstrates that defective signaling through the TLR4 receptor ultimately leads to an exaggerated inflammatory response with severe tissue destruction, even though the initial immune response may be blunted. This is attributed to inadequate production of IL-10–secreting type 1 regulatory cells.[48]

We hypothesized that the *TLR4+896A* >G polymorphism would be associated with an exaggerated and destructive chronic inflammatory phenotype in *H. pylori*–infected subjects. This phenotype would be characterized by gastric atrophy and hypochlorhydria, the hallmarks of subsequent increased risk of gastric cancer. We further hypothesized that the same polymorphism might increase the risk of gastric cancer itself. We proceeded to test the effect of this polymorphism on the *H. pylori*–induced gastric phenotype and the risk of developing premalignant and malignant outcomes. We assessed associations with premalignant gastric changes in relatives of gastric cancer patients, including those with hypochlorhydria and gastric atrophy. We also genotyped two independent Caucasian population–based case-control studies of upper gastrointestinal tract cancer, initially in 312 noncardia gastric cancer cases and 419 controls and then in 184 noncardia gastric cancers, 123 cardia cancers, 159 esophageal cancers, and 211 frequency-matched controls. *TLR4+896G* carriers had a 7.7-fold (95% CI: 1.6–37.6) increased OR for hypochlorhydria; the polymorphism was not associated with gastric acid output in the absence of *H. pylori* infection. Carriers also had significantly more severe gastric atrophy and inflammation.[49] Sixteen percent of gastric cancer patients in the initial study and 15% of the noncardia gastric cancer patients in the replication study had 1 or 2 *TLR4* variant alleles vs. 8% of both control populations (combined OR = 2.4; 95% CI: 1.6–3.4).[49] In contrast, prevalence of *TLR4+896G* was not significantly increased in esophageal squamous cell (2%, OR = 0.4), adenocarcinoma (9%, OR = 0.8), or gastric cardia cancer (11%, OR = 1.2).

The association of *TLR4+896A* >G polymorphism with both gastric cancer and its precursor lesions implies that it is relevant to the entire multistage process of gastric carcinogenesis, which starts with *H. pylori* colonization of the gastric mucosa. Subjects with this polymorphism have an increased risk of severe inflammation and, subsequently, development of hypochlorhydria and gastric atrophy, which are regarded as the most important precancerous abnormalities. This severe inflammation is initiated by *H. pylori* infection but it is entirely feasible that subsequent cocolonization of an achlorhydric stomach by a variety of other bacteria may sustain and enhance the microbial inflammatory stimulus and continue to drive the carcinogenic process. Evidence supporting this concept comes from the work of Sanduleanu et al.[50] who showed that pharmacologic inhibition of acid secretion was associated with a higher prevalence of non-*H. pylori* bacteria. Furthermore, the simultaneous presence of *H. pylori* and non-*H. pylori* bacteria was associated with a markedly increased risk of atrophic gastritis, and with higher circulating levels of IL-1β and IL-8. Supportive evidence also comes from animal studies in which hypochlorhydria was induced in mice either genetically (G⁻/G⁻ gastrin-deficient mice) or pharmacologically (administration of omeprazole). Zavros et al.[51] found that genetic or chemical hypochlorhydria predisposes the stomach to bacterial overgrowth resulting in inflammation, which was not present in the wild-type mice or those not treated with an acid inhibitor.

The potential mechanism by which the *TLR-4* polymorphism increases the risk of gastric cancer and its precursors is intriguing and may lie in the nature of the host's overall response to the *H. pylori* LPS attack. Failure to handle the invasion by appropriately recognizing and activating the necessary pathways may lead to an imbalance of pro- and antiinflammatory mediators. A very elegant demonstration of this phenomenon was recently reported by Higgins et al.[48] The authors infected TLR4-defective C3H/HeJ mice and their wild-type counterparts (C3H/HeN) with an aerosol of *B. pertussis* (a gram-negative bacterium that causes whooping cough) and monitored the course of the infection and its consequences over several weeks. The course of the infection was more severe in the TLR4-defective than the wild-type mice and this was associated with enhanced inflammatory cytokine production, cellular infiltration, and severe pathologic changes in the lungs. Interestingly, Higgins et al. showed that signaling through TLR4 in response to bacterial infection activated IL-10 production, which promoted IL-10–producing T cells and controlled inflammatory pathology during infection in normal, but not TLR4-defective, mice. It is therefore likely that the severe tissue damage observed in the TLR4-defective mice is attributable to the deficiency of the antiinflammatory IL-10, which in turn accentuated the proinflammatory destructive tissue response.

Overall Contribution of a Host Proinflammatory Genetic Makeup to Pathogenesis of Gastric Cancer

We propose that subjects with a proinflammatory genetic makeup based on a combination of markers from cytokine/chemokine genes (e.g., IL-1β, TNF-α, IL-10, IL-8) and the innate immune response (e.g., TLR4), respond to *H. pylori* infection by creating an environment within the stomach that is chronically inflamed with reduced acidity (Figure 9.1). This environment is conducive to the growth of other bacteria within the gastric milieu, leading to sustained inflammation and oxidative/genotoxic stress. Subjects with the same proinflammatory polymorphisms may respond in the same exaggerated manner to these non-*H. pylori* bacteria, thus maintaining the proneoplastic drive. This may explain why *H. pylori* is not required in the latter stages of gastric carcinogenesis and why it is often absent from gastric tumor tissue.

Role of Human Leukocyte Antigen Polymorphisms in Gastric Cancer

Several studies have examined the role of human leukocyte antigen (HLA) class I and II alleles in gastric cancer, in Caucasian and non-Caucasian populations. Lee et al.[52] found that the HLA class II allele *DQB1*0301* was more common in Caucasian patients with gastric adenocarcinoma than noncancer controls. The mechanism linking HLA-*DQB1*0301* with gastric adenocarcinoma was not clear but was thought to be independent of increased susceptibility to *H. pylori* infection. Azuma et al.[53] reported that the allele frequency of *DQA1*0102* was significantly lower in the *H. pylori*–infected subjects with atrophic gastritis compared with infected subjects without atrophy and uninfected subjects. In addition, the allele frequency

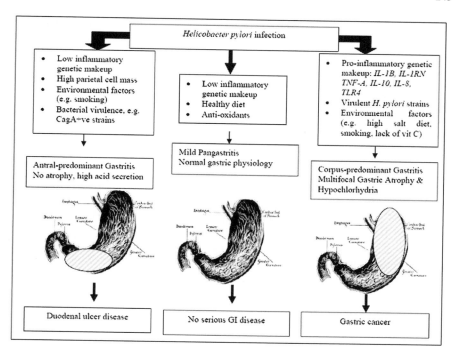

Fig. 9.1 Contribution of host genetic factors to distribution of gastritis and clinical outcome of *Helicobacter pylori* infection including gastric cancer. (Adapted from El Omar.60)

of *DQA1*0102* was also significantly lower in the infected intestinal type gastric adenocarcinoma patients compared with all controls (infected and uninfected). The authors concluded that the HLA *DQA1*0102* allele was protective against gastric atrophy and intestinal type gastric adenocarcinoma. Magnusson et al.[54] studied the effect of HLA class II alleles on risk of *H. pylori* infection and gastric cancer. They confirmed Azuma et al.'s finding that the *DQA1*0102* allele was associated with decreased risk of *H. pylori* infection but they found no protective effect on risk of gastric cancer. They also showed that the *DRB1*1601* allele was significantly associated with an increased gastric cancer risk with an OR of 8.7 (95% CI: 2.7–28.0). The effect of *1601* was more pronounced among Hp-negative subjects, and the association was stronger with the diffuse, rather than with the intestinal, histologic type of gastric cancer.[54] Interestingly, they failed to show any association between gastric cancer and the *DQB1*0301* reported by Lee et al.

These three studies highlight several problems pertinent to the study of HLA disease associations. The HLA system is highly polymorphic and the genetic variation is very much dependent on ethnicity and its background selective heritage. As such, HLA disease associations in certain ethnic groups are unlikely to be relevant to others, simply because certain alleles may be under/overrepresented in these groups independent of the disease under question. The other relevant point here is the power required for such association studies. If one is dealing with the most highly polymorphic genetic system, it follows that the power of the studies has to match the complexity of this system. The study by Lee et al. was based on 52 gastric cancer cases and 260 noncancer controls, whereas Azuma et al. studied 82 cancer cases and 167 controls and Magnusson et al. 130 cancer cases and 263 population controls. It could be argued that all three studies were grossly underpowered to look at associations between gastric cancer and HLA markers. Indeed, the sample size required will likely be in the thousands to get closer to any true associations.

Role of Environmental Factors in Gastric Carcinogenesis

Why do only a few *H. pylori*–infected subjects with these polymorphisms develop gastric cancer? The answer lies in the polygenic and multifactorial nature of most complex human diseases. These genetic factors operate only in the presence of an infectious agent and lead to the development of an atrophic phenotype. Progression of atrophy toward cancer depends on other components of the host genetic constitution acting epistatically, as well as by dietary and other factors in the environment. For example, it is known that men are twice as likely to develop distal gastric adenocarcinoma compared with women. The gender difference in risk raises the interesting possibility that either hormonal factors such as estrogens or perhaps the lower body content of iron in women (carcinogenic in other tissues such as liver) may explain the difference. Furthermore, although *H. pylori* infection and host genetics interact to initiate a hypochlorhydric and atrophic phenotype, environmental cofactors may mediate subsequent neoplastic transformation, even after disappearance of the infection. Diet could be particularly relevant, with greater consumption of fresh fruits and vegetables shown to protect against risk of gastric as well as several other cancers (see Chapter 3). Dietary vitamin C reduces the formation of N-nitroso-compounds and scavenges mutagenic reactive oxygen metabolites generated by gastric inflammation[55] and supplemental vitamin C is associated with significantly lower risk of noncardia gastric cancer.[56] Furthermore, vitamin C concentrations and bioavailability are reduced in the presence of *H. pylori* infection.[57] Another important cofactor is cigarette smoking, which was found to nearly double the risk of transition from atrophic gastritis to dysplasia in a high-risk population.[58] Stenström et al.[59] recently showed that Swedish moist snuff (a form of oral smokeless tobacco) accelerated gastric cancer development in *H. pylori*–infected wild-type and gastrin transgenic mice. Overall, therefore, cytokine gene polymorphisms represent only one component of a complex interplay among host, pathogen, and environmental factors involved in gastric carcinogenesis (Figure 9.1).

Conclusion

Sporadic gastric cancer is a common cancer with a grave prognosis, particularly in the West. A major advance in the fight against this global killer came with the recognition of the role of *H. pylori* infection in its pathogenesis. The cancer represents a classic example of an inflammation-induced malignancy. Host genetic factors, interacting with bacterial virulence and environmental factors, have an important role in the pathogenesis of this cancer (Figure 9.1). In particular, genetic polymorphisms in the adaptive and innate immune response genes seem to increase the risk of cancer, largely through induction of severe gastritis, which progresses to atrophy and hypochlorhydria. The proinflammatory host genetic makeup is only relevant in the presence of infection, initially *H. pylori*, but later could be other bacteria that thrive in an achlorhydric environment. Future research must focus on defining a more comprehensive genetic profile that better predicts the clinical outcome of *H. pylori* infection, including gastric cancer. This may be achieved through advances in technology. For example, it is now possible to scan the entire human genome for SNPs and the so-called whole genome association studies (WGAS) are starting to appear in the literature. One can predict that in the near future we will be able to define a gastric cancer–specific genetic risk profile. Genetic profiling will need to be combined with testing for *H. pylori* infection, which is the most important factor that allows this genetic risk profile to exert its harmful effect. Our aim should be to use the *H. pylori* testing and genetic profiling to deliver eradication therapy to those at risk and to avoid giving unnecessary antibiotic therapy to those who are not at risk of cancer. Such a preventative strategy is the only way of defeating this global killer.

References

1. Parkin DM, Bray F, Ferlay J, Pisani P. Global cancer statistics, 2002. CA Cancer J Clin 2005;55:74–108.
2. IARC monographs on the evaluation of carcinogenic risks to humans. Vol. 61. Schistosomes, liver flukes and Helicobacter pylori. Lyon: IARC; 1994:177–240.
3. Scott DR, Weeks D, Hong C, Postius S, Melchers K, Sachs G. The role of internal urease in acid resistance of Helicobacter pylori. Gastroenterology 1998;114:58–70.
4. Israel DA, Peek RM. Pathogenesis of Helicobacter pylori-induced gastric inflammation. Aliment Pharmacol Ther 2001;15:1271–90.
5. Robert A, Olafsson AS, Lancaster C, Zhang WR. Interleukin-1 is cytoprotective, antisecretory, stimulates PGE2 synthesis by the stomach, and retards gastric emptying. Life Sci 1991;48:123–34.
6. Wolfe MM, Nompleggi DJ. Cytokine inhibition of gastric acid secretion: a little goes a long way. Gastroenterology 1992;102:2177–8.
7. Beales IL, Calam J. Interleukin 1 beta and tumour necrosis factor alpha inhibit acid secretion in cultured rabbit parietal cells by multiple pathways. Gut 1998;42:227–34.
8. Graham DY, Opekun AR, Osato MS, El-Zimaity HMT, Lee CK, Yamaoka Y, Qureshi WA, Cadoz M, Monath TP. Challenge model for Helicobacter pylori infection in human volunteers. Gut 2004;53:1235–43.
9. Danon SJ, O'Rourke JL, Moss ND, Lee A. The importance of local acid production in the distribution of Helicobacter felis in the mouse stomach. Gastroenterology 1995;108:1386–95.
10. Kuipers EJ, Lundell L, Klinkenberg-Knol EC, Havu N, Festen HP, Liedman B, Lamers CB, Jansen JB, Dalenback J, Snel P, Nelis GF, Meuwissen SG. Atrophic gastritis and Helicobacter pylori infection in patients with reflux esophagitis treated with omeprazole or fundoplication. N Engl J Med 1996;334:1018–22.

11. El-Omar EM, Oien K, El Nujumi A, Gillen D, Wirz A, Dahill S, Williams C, Ardill JE, McColl KE. Helicobacter pylori infection and chronic gastric acid hyposecretion. Gastroenterology 1997;113:15–24.
12. El-Omar EM, Penman ID, Ardill JE, Chittajallu RS, Howie C, McColl KE. Helicobacter pylori infection and abnormalities of acid secretion in patients with duodenal ulcer disease. Gastroenterology 1995;109:681–91.
13. McColl KE, El-Omar E, Gillen D. Helicobacter pylori gastritis and gastric physiology. Gastroenterol Clin North Am 2000;29:687–703.
14. Graham DY, Yamaoka Y. Disease-specific Helicobacter pylori virulence factors: the unfulfilled promise. Helicobacter 2000;5(Suppl 1):S3–S9.
15. Dinarello CA. Biologic basis for interleukin-1 in disease. Blood 1996;87:2095–147.
16. El-Omar EM. The importance of interleukin 1beta in Helicobacter pylori associated disease. Gut 2001;48:743–7.
17. Bidwell J, Keen L, Gallagher G, Kimberly R, Huizinga T, McDermott MF, Oksenberg J, McNicholl J, Pociot F, Hardt C, D'Alfonso S. Cytokine gene polymorphism in human disease: on-line databases, supplement 1. Genes Immun 2001;2:61–70.
18. El-Omar EM, Carrington M, Chow WH, McColl KE, Bream JH, Young HA, Herrera J, Lissowska J, Yuan CC, Rothman N, Lanyon G, Martin M, Fraumeni JF, Jr., Rabkin CS. Interleukin-1 polymorphisms associated with increased risk of gastric cancer. Nature 2000;404:398–402.
19. El-Omar EM, Rabkin CS, Gammon MD, Vaughan TL, Risch HA, Schoenberg JB, Stanford JL, Mayne ST, Goedert J, Blot WJ, Fraumeni JF, Jr., Chow WH. Increased risk of noncardia gastric cancer associated with proinflammatory cytokine gene polymorphisms. Gastroenterology 2003;124:1193–201.
20. Machado JC, Pharoah P, Sousa S, Carvalho R, Oliveira C, Figueiredo C, Amorim A, Seruca R, Caldas C, Carneiro F, Sobrinho-Simoes M. Interleukin 1B and interleukin 1RN polymorphisms are associated with increased risk of gastric carcinoma. Gastroenterology 2001;121:823–9.
21. Figueiredo C, Machado JC, Pharoah P, Seruca R, Sousa S, Carvalho R, Capelinha AF, Quint W, Caldas C, Van Doorn LJ, Carneiro F, Sobrinho-Simoes M. Helicobacter pylori and interleukin 1 genotyping: an opportunity to identify high-risk individuals for gastric carcinoma. J Natl Cancer Inst 2002;94:1680–7.
22. Machado JC, Figueiredo C, Canedo P, Pharoah P, Carvalho R, Nabais S, Castro AC, Campos ML, Van Doorn LJ, Caldas C, Seruca R, Carneiro F, Sobrinho-Simoes M. A proinflammatory genetic profile increases the risk for chronic atrophic gastritis and gastric carcinoma. Gastroenterology 2003;125:364–71.
23. Rad R, Prinz C, Neu B, Neuhofer M, Zeitner M, Voland P, Becker I, Schepp W, Gerhard M. Synergistic effect of Helicobacter pylori virulence factors and interleukin-1 polymorphisms for the development of severe histological changes in the gastric mucosa. J Infect Dis 2003;188:272–81.
24. Palli D, Saieva C, Luzzi I, Masala G, Topa S, Sera F, Gemma S, Zanna I, D'Errico M, Zini E, Guidotti S, Valeri A, Fabbrucci P, Moretti R, Testai E, del Giudice G, Ottini L, Matullo G, Dogliotti E, Gomez-Miguel MJ. Interleukin-1 gene polymorphisms and gastric cancer risk in a high-risk Italian population. Am J Gastroenterol 2005;100:1941–8.
25. Furuta T, El-Omar EM, Xiao F, Shirai N, Takashima M, Sugimurra H. Interleukin 1beta polymorphisms increase risk of hypochlorhydria and atrophic gastritis and reduce risk of duodenal ulcer recurrence in Japan. Gastroenterology 2002;123:92–105.
26. Zeng ZR, Hu PJ, Hu S, Pang RP, Chen MH, Ng M, Sung JJY. Association of interleukin 1B gene polymorphism and gastric cancers in high and low prevalence regions in China. Gut 2003;52:1684–9.
27. Lee KA, Ki CS, Kim HJ, Sohn KM, Kim JW, Kang WK, Rhee JC, Song SY, Sohn TS. Novel interleukin 1beta polymorphism increased the risk of gastric cancer in a Korean population. J Gastroenterol 2004;39:429–33.

28. Chang YW, Jang JY, Kim NH, Lee JW, Lee HJ, Jung WW, Dong SH, Kim HJ, Kim BH, Lee JI, Chang R. Interleukin-1B (IL-1B) polymorphisms and gastric mucosal levels of IL-1beta cytokine in Korean patients with gastric cancer. Int J Cancer 2005;114:465–71.

29. Yang J, Hu Z, Xu Y, Shen J, Niu J, Hu X, Guo J, Wei Q, Wang X, Shen H. Interleukin-1B gene promoter variants are associated with an increased risk of gastric cancer in a Chinese population. Cancer Lett 2004;215:191–8.

30. Chen HM, Wilkins LM, Aziz N, Cannings C, Wyllie DH, Bingle C, Rogus J, Beck JD, Offenbacher S, Cork MJ, Rafie-Kolpin M, Hsieh CM, Kornman KS, Duff GW. Single nucleotide polymorphisms in the human interleukin-1B gene affect transcription according to haplotype context. Hum Mol Genet 2006;15:519–29.

31. Lee KA, Park JH, Sohn TS, Kim S, Rhee JC, Kim JW. Interaction of polymorphisms in the interleukin 1B-31 and general transcription factor 2A1 genes on the susceptibility to gastric cancer. Cytokine 2007;38:96–100.

32. Tu S, Cui G, Takaishi S, Tran AV, Frederick DM, Carlson JE, Kurt-Jones E, Wang TC. Overexpression of human interleukin-1 beta in transgenic mice results in spontaneous gastric inflammation and carcinogenesis. Gastroenterology 2005;128:A62.

33. Smith MG, Hold GL, Rabkin CS, Chow WH, McColl KE, Perez-Perez GI, Mowat NAG, El-Omar EM. The IL-8-251 promoter polymorphism is associated with high IL-8 production, severe inflammation and increased risk of pre-malignant changes in H-pylori positive subjects. 2004;124:A23.

34. Taguchi A, Ohmiya N, Shirai K, Mabuchi N, Itoh A, Hirooka Y, Niwa Y, Goto H. Interleukin-8 promoter polymorphism increases the risk of atrophic gastritis and gastric cancer in Japan. Cancer Epidemiol Biomarkers Prev 2005;14:2487–93.

35. Lu W, Pan K, Zhang L, Lin D, Miao X, You W. Genetic polymorphisms of interleukin (IL)-1B, IL-1RN, IL-8, IL-10 and tumor necrosis factor a and risk of gastric cancer in a Chinese population. Carcinogenesis 2005;26:631–6.

36. Lee WP, Tai DI, Lan KH, Li AF, Hsu HC, Lin EJ, Lin YP, Sheu ML, Li CP, Chang FY, Chao Y, Yen SH, Lee SD. The −251T allele of the interleukin-8 promoter is associated with increased risk of gastric carcinoma featuring diffuse-type histopathology in Chinese population. Clin Cancer Res 2005;11:6431–41.

37. Ohyauchi M, Imatani A, Yonechi M, Asano N, Miura A, Iijima K, Koike T, Sekine H, Ohara S, Shimosegawa T. The polymorphism interleukin 8 −251 A/T influences the susceptibility of Helicobacter pylori related gastric diseases in the Japanese population. Gut 2005;54:330–5.

38. Smith MG, Hold GL, Rabkin CS, Chow WH, Fraumeni JF, Jr., Mowat NAG, Ando T, Goto H, El-Omar EM. The interleukin-8-251 promoter polymorphism and risk of gastric cancer in Caucasian and Japanese populations. 2006; 128:A-19.

39. Kuipers EJ. Review article: exploring the link between Helicobacter pylori and gastric cancer. Aliment Pharmacol Ther 1999;13(Suppl 1):3–11.

40. Segal ED, Lange C, Covacci A, Tompkins LS, Falkow S. Induction of host signal transduction pathways by Helicobacter pylori. Proc Natl Acad Sci USA 1997;94:7595–9.

41. Su B, Ceponis PJ, Lebel S, Huynh H, Sherman PM. Helicobacter pylori activates Toll-like receptor 4 expression in gastrointestinal epithelial cells. Infect Immun 2003;71:3496–502.

42. Pasare C, Medzhitov R. Toll-like receptors: linking innate and adaptive immunity. Adv Exp Med Biol 2005;560:11–8.

43. Takeda K, Akira S. Toll-like receptors in innate immunity. Int Immunol 2005;17:1–14.

44. Arbour NC, Lorenz E, Schutte BC, Zabner J, Kline JN, Jones M, Frees K, Watt JL, Schwartz DA. TLR4 mutations are associated with endotoxin hyporesponsiveness in humans. Nat Genet 2000;25:187–91.

45. Franchimont D, Vermeire S, El Housni H, Pierik M, Van Steen K, Gustot T, Quertinmont E, Abramowicz M, Van Gossum A, Deviere J, Rutgeerts P. Deficient host-bacteria interactions in inflammatory bowel disease? The toll-like receptor (TLR)-4 Asp299gly polymorphism is associated with Crohn's disease and ulcerative colitis. Gut 2004;53:987–92.

46. Lorenz E, Mira JP, Frees KL, Schwartz DA. Relevance of mutations in the TLR4 receptor in patients with gram-negative septic shock. Arch Intern Med 2002;162:1028–32.

47. Kiechl S, Lorenz E, Reindl M, Wiedermann CJ, Oberhollenzer F, Bonora E, Willeit J, Schwartz DA. Toll-like receptor 4 polymorphisms and atherogenesis. N Engl J Med 2002;347:185–92.
48. Higgins SC, Lavelle EC, McCann C, Keogh B, McNeela E, Byrne P, O'Gorman B, Jarnicki A, McGuirk P, Mills KH. Toll-like receptor 4-mediated innate IL-10 activates antigen-specific regulatory T cells and confers resistance to Bordetella pertussis by inhibiting inflammatory pathology. J Immunol 2003;171:3119–27.
49. Hold GL, Rabkin CS, Chow WH, Smith MG, Gammon MD, Risch HA, Vaughan TL, McColl KE, Lissowska J, Zatonski W, Schoenberg JB, Blot WJ, Mowat NA, Fraumeni JF, Jr., El-Omar EM. A functional polymorphism of toll-like receptor 4 gene increases risk of gastric carcinoma and its precursors. Gastroenterology 2007;132:905–12.
50. Sanduleanu S, Jonkers D, De Bruine A, Hameeteman W, Stockbrugger RW. Double gastric infection with Helicobacter pylori and non-Helicobacter pylori bacteria during acid-suppressive therapy: increase of pro-inflammatory cytokines and development of atrophic gastritis. Aliment Pharmacol Ther 2001;15:1163–75.
51. Zavros Y, Rieder G, Ferguson A, Samuelson LC, Merchant JL. Genetic or chemical hypochlorhydria is associated with inflammation that modulates parietal and G-cell populations in mice. Gastroenterology 2002;122:119–33.
52. Lee JE, Lowy AM, Thompson WA, Lu M, Loflin PT, Skibber JM, Evans DB, Curley SA, Mansfield PF, Reveille JD. Association of gastric adenocarcinoma with the HLA class II gene DQB10301. Gastroenterology 1996;111:426–32.
53. Azuma T, Ito S, Sato F, Yamazaki Y, Miyaji H, Ito Y, Suto H, Kuriyama M, Kato T, Kohli Y. The role of the HLA-DQA1 gene in resistance to atrophic gastritis and gastric adenocarcinoma induced by Helicobacter pylori infection. Cancer 1998;82:1013–8.
54. Magnusson PKE, Enroth H, Eriksson I, Held M, Nyren O, Engstrand L, Hansson LE, Gyllensten UB. Gastric cancer and human leukocyte antigen: distinct DQ and DR alleles are associated with development of gastric cancer and infection by Helicobacter pylori. Cancer Res 2001;61:2684–9.
55. Correa P. Human gastric carcinogenesis: a multistep and multifactorial process—First American Cancer Society Award Lecture on Cancer Epidemiology and Prevention. Cancer Res 1992;52:6735–40.
56. Mayne ST, Risch HA, Dubrow R, Chow WH, Gammon MD, Vaughan TL, Farrow DC, Schoenberg JB, Stanford JL, Ahsan H, West AB, Rotterdam H, Blot WJ, Fraumeni JF, Jr. Nutrient intake and risk of subtypes of esophageal and gastric cancer. Cancer Epidemiol Biomarkers Prev 2001;10:1055–62.
57. Banerjee S, Hawksby C, Miller S, Dahill S, Beattie AD, McColl KE. Effect of Helicobacter pylori and its eradication on gastric juice ascorbic acid. Gut 1994;35:317–22.
58. Kneller RW, You WC, Chang YS, Liu WD, Zhang L, Zhao L, Xu GW, Fraumeni JF, Jr., Blot WJ. Cigarette smoking and other risk factors for progression of precancerous stomach lesions. J Natl Cancer Inst 1992;84:1261–6.
59. Stenström B, Zhao CM, Rogers AB, Nilsson HO, Sturegard E, Lundgren S, Fox JG, Wang TC, Wadstrom TM, Chen D. Swedish moist snuff accelerates gastric cancer development in Helicobacter pylori-infected wild-type and gastrin transgenic mice. Carcinogenesis 2007;28:2041–6.
60. El Omar EM. Role of host genes in sporadic gastric cancer. Best Pract Res Clin Gastroenterol 2006;20:675–86.

Chapter 10
Cancer Genetics of Human Gastric Adenocarcinoma

Roman Galysh, Jr. and Steven M. Powell

Introduction

Gastric adenocarcinomas are the most common form of gastric cancer and are usually sporadic in nature with genetic alterations acquired over decades. There are, however, rare inherited gastric cancer predisposition traits such as E-cadherin mutations in familial diffuse gastric cancers as discussed in detail below. The complexity of the genetics involved in human gastric adenocarcinoma is reflected in the temporal, regional, and gender variation in gastric cancer incidence rates. A better understanding of these phenomena through molecular and genetic studies of gastric tumorigenesis will provide important insights into cancer development in general and is anticipated to lead to earlier diagnosis and better management options in combating this devastating cancer. The relative importance of bacterial virulence, environmental, and host factors (e.g., age of acquisition, immune response, acid secretion changes) to the clinical outcome of these infections are currently pressing issues, and molecular studies may help discern the most influential of these factors. In addition, multiple gastric tumor pathologic classification systems have been proposed in effort to identify various subgroups with differing biologic behaviors and prognostic indicators. Molecular markers should help facilitate classification of these subgroups.

Inherited Susceptibility

Familial Clustering

Most cases of gastric cancer seem to occur sporadically, without an obvious hereditary component. It is estimated that up to 8%–10% of gastric cancer cases are familial (La Vecchia et al. 1992). Familial clustering has been observed in 12%–25% of gastric carcinoma cases, with a dominant inheritance pattern observed (Goldgar et al. 1994; Videbaek and Mosbech 1954). Notably, Napoleon Bonaparte apparently suffered from gastric cancer involving most of his stomach and may

T.C. Wang et al. (eds.) *The Biology of Gastric Cancers*,
© Springer Science+Business Media, LLC 2009

have had other family members (i.e., his father and sister) who were afflicted as well (Antommarchi 1825; Kubba and Young 1999; Lugli et al. 2007). In the Swedish Family Cancer Database, the largest published database of familial gastric cancer to date, standardized incidence rate (SIR) was 1.7 (95% confidence interval: 1.08–1.92) when a patient presented with gastric adenocarcinoma. When a parent presented with gastric carcinoma, offspring showed an increased risk of the concordant carcinoma (SIR 1.59) only at ages older than 50 years. The increased risk from sibling gastric carcinoma probands (SIR of 5.75) was noted for those diagnosed before age 50 years. Taken together, these findings suggest that some of the familial risk factors are likely to be environmental, siblings being at a higher risk than offspring–parent pairs, consistent with the transmission patterns of *Helicobacter pylori* infection (Hemminki and Jiang 2002).

Case-control studies have observed consistent (up to threefold) increases in risk for gastric cancer among relatives of gastric cancer patients (Videbaek and Mosbech 1954; Zanghieri et al. 1990). A population-based control study found an increased risk of developing gastric cancer among first-degree relatives of affected patients [odds ratio (OR) = 1.7 with an affected parent; OR = 2.6 with an affected sibling], with the risk increasing (OR up to 8.5) if more than one first-degree relative was affected (Palli et al. 1994). Interestingly, a higher risk was noted in individuals with an affected mother versus an affected father. Studies have shown a slight trend toward increased concordance of gastric cancers in monozygotic twins compared with dizygotic twins (Gorer 1938; Lee 1971). A genomic analysis of 170 affected sib-pairs from 142 Japanese families with gastric cancer yielded several chromosomal regions, with the strongest linkage at 2q33-35, harboring potential susceptibility genes (Aoki et al. 2005).

Several genetic susceptibility traits with an inherited predisposition to gastric cancer development, some of which are well-characterized clinically and are beginning to be unveiled genetically, are described below.

Hereditary Diffuse Gastric Cancer

E-Cadherin Mutations

Large families with an obvious autosomal dominant, highly penetrant inherited predisposition to the development of gastric cancer, having sufficient power with which to perform productive linkage studies, are rare. A large Maori kindred manifesting early-onset diffuse gastric cancers was investigated for this type of analysis, revealing linkage to the *E-cadherin/CDH1* locus on 16q and associated with mutations in this gene (Guilford et al. 1998). Since then, more than 14 truncating germline *E-cadherin* mutations have been reported, scattered across 8 of the 16 exons this gene encompasses (Caldas et al. 1999; Gayther et al. 1998; Guilford et al. 1998; Shinmura et al. 1999; Stone et al. 1999; Yoon et al. 1999). The ages of onset and diagnosis of diffuse gastric cancer in subjects harboring germline E-cadherin mutations ranged from 14 to 69 years. The incomplete penetrance of germline E-cadherin mutations was seen in obligate carriers, who remained unaffected even in their eighth and ninth decades of life. Whether this incomplete penetrance is attributable to the stochastic nature of the second allele loss or perhaps to the presence of phenotype-altering alleles at other genetic loci, remains to be determined.

One large study of 10 kindreds manifesting diffuse gastric cancers identified three families with germline E-cadherin mutations (Gayther et al. 1998). A study of 25 "sporadic" diffuse gastric adenocarcinomas identified one case with a germline E-cadherin mutation and none in 14 intestinal-type gastric cancers (Ascano et al. 2001). No germline mutations of this gene were detected in apparent "sporadic" diffuse gastric cancer cases with a mean patient age of 62 years in Great Britain (Stone et al. 1999). Based on results of *E-cadherin* mutational screening of probands from 42 new families with diffuse gastric cancers, Brooke-Wilson et al. determined that 40% of families with multiple gastric cancer cases and at least one diffuse gastric cancer case diagnosed younger than age 50 had a pathologic germline *E-cadherin* mutation; thus, they recommended using this molecular alteration as a screening criterion (Lynch et al. 2005). Genetic counseling for kindreds manifesting a strong predisposition toward development of diffuse gastric cancers is imperative to ensure appropriate medical management (see Figure 10.1). Diffuse gastric cancer families are one kindred subgroup to whom genetic testing can now be offered (Blair et al. 2006).

Idiopathic cases

It is noteworthy that two-thirds of hereditary diffuse gastric cancer families reported to date have proven negative for *E-cadherin* gene mutation (Lynch et al. 2005). This finding suggests that other molecular alterations yet to be discovered underlie the majority of these cases.

Hereditary Nonpolyposis Colorectal Cancer

A now well-characterized inherited predisposition syndrome that may include gastric cancer development is hereditary nonpolyposis colorectal cancer (HNPCC) (Lynch et al. 1993). Germline genetic abnormalities of DNA mismatch repair (MMR) genes underlying this disease entity have been discovered, and HNPCC is characterized by tumor development in a variety of tissue types (Kinzler and Vogelstein 1996). Gastric carcinomas occurring in this setting were diagnosed at a mean age of 56 years, predominantly intestinal-type, lacking *H. pylori* infection, and exhibiting microsatellite instability (MSI) in a Finnish HNPCC registry study (Aarnio et al. 1997).

Interestingly, the incidence of gastric cancers associated with HNPCC has decreased, similar to the recent general decrease in incidence of gastric cancer in

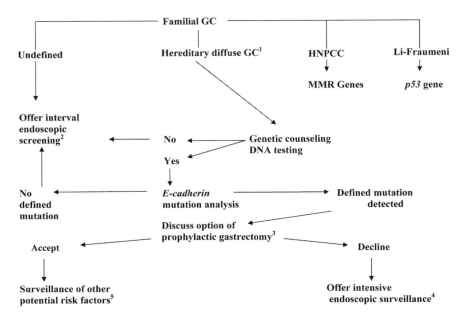

Fig. 10.1 Algorithm for guidance in managing familial gastric cancer (GC) kindreds. Those identified to meet clinical criteria for hereditary nonpolyposis colon cancer syndrome (HNPCC) and the Li-Fraumeni syndromes are counseled for consideration of genetic testing. Microsatellite instability testing can aid in diagnosing HNPCC. [1]This algorithm was initially formulated at the consensus symposia where clinical criteria were generated to define hereditary diffuse gastric cancer: two or more cases in first/second degree relatives, at least one diagnosed before the age of 50; or three or more cases of documented diffuse gastric cancer in first/second degree relatives. [2]Consider individual age-dependent familial expression to determine the initiation, interval, and intensity of screening examinations. Prophylactic gastrectomies have been performed in kindreds expressing highly penetrant phenotypes without prior knowledge of causative mutations. [3]Consider age-dependent familial expression in determining management. [4]Endoscopic ultrasound and chromographic (methylene blue or indigo carmine staining) endoscopy can be applied in attempt to increase the sensitivity of detecting early lesions in the stomach. [5]One should have a high index of suspicion for the potential development of other cancers such as that of the breast, colon, and endometrium. MMR: mismatch repair. (Reproduced from Powell et al. 2002, with permission from Lippincott Williams & Wilkins.)

developed countries (Lynch and Krush 1971). Renkonen-Sinisalo et al. studied gastric histopathology comparing 73 mutation-positive and 32 mutation-negative families for differences in *H. pylori*, atrophy, inflammation, IM, and dysplasia. They identified only a single case of duodenal cancer among mutation-positive individuals, but there was no evidence of gastric neoplastic lesions in either group (Renkonen-Sinisalo et al. 2002). The isolation and characterization of predisposing gene alterations should permit better definition of the proportion of gastric cancers resulting from this trait. Testing for *MSH2* and *MLH1* gene alterations is currently readily available and can be used to identify gastric cancer cases that are a manifestation of this specific cancer-predisposing genotype.

Li-Fraumeni and Peutz-Jeghers Syndromes

Li-Fraumeni syndrome is a rare inherited cancer syndrome defined by a clustering of malignancies including sarcoma, breast cancer, brain tumor, leukemia, and adrenocortical carcinoma attributable to germline mutation of the tumor suppressor gene *p53* (Li and Fraumeni 1969). Gastric cancers have also been included in this syndrome. An extended family affected by this syndrome demonstrated strong evidence of linkage to *p53*, and three of four gastric tumors analyzed showed loss of the wild-type allele (Varley et al. 1995). However, gastric carcinomas account for less than 4% of all neoplasms in this rare syndrome (Kleihues et al. 1997). The Peutz-Jeghers syndrome (PJS) is a rare autosomal dominant disorder characterized by germline mutations of STK11, hamartomatous polyposis, and pigmentation of the lips, buccal mucosa, and digits that carries an increased risk of many types of cancer, including those of the gastrointestinal tract (Hofgartner et al. 1999; Lindor and Greene 1998). Overall, gastric carcinoma is still rare in this setting, and the exact contribution of the polyposis and underlying germline alterations of *LKB1/STK11* (PJS) to gastric cancer development is unclear. Only one missense change of uncertain functional significance in *LKB1/STK11* was noted among 28 sporadic gastric carcinomas (Park et al. 1998).

Adenomatous Polyposis Syndromes

Gastric cancers have been noted to occur in patients with gastrointestinal polyposis disease entities such as familial adenomatous polyposis (FAP). Germline mutations of the *APC* gene are responsible for this disease (Groden et al. 1991; Hofgartner et al. 1999; Lindor and Greene 1998). Interestingly, an increased risk of gastric cancer associated with FAP has been reported in high-risk regions such as Asia (Utsunomiya J. 1990), whereas no increased risk was exhibited in other populations (Offerhaus et al. 1992). Overall, gastric carcinoma is rare in this setting, and the exact contribution of the polyposis and underlying germline alterations of *APC* is again unclear. In the attenuated FAP syndrome, subjects who do not possess germline *APC* gene mutations at the ends of the gene may have inherited allelic defects of the *MYH* base excision repair gene, which predispose to adenomas and cancers of the colon (Jones et al. 2002; Kim et al. 2004; Zhang et al. 2006). Recent studies have also linked this gene defect to 2.1% of sporadic cases of gastric cancer, and even to as many as 9.9% of patients with a familial pattern of gastric cancer (Kim et al. 2004; Zhang et al. 2006).

Other Syndromes

Gastric cancers have been noted to occur in patients with other gastrointestinal polyposis disease entities (such as juvenile polyposis), who harbor germline mutations in

SMAD4 or *BMPR1A*. Up to 24% of those with generalized juvenile gastrointestinal polyposis developed gastric cancer, which was similar to a subset of patients who presented with predominantly gastric polyposis and was found to have neoplastic tissue arising in 25% of resected gastric specimens (Hizawa et al. 1997).

Rare kindreds exhibiting site-specific gastric cancer predilection have been reported, occasionally associated with other inherited abnormalities (Maimon and Zinninger 1953; Woolf and Isaacson 1961). In a 14-year old girl, a constitutional deletion of 18p inherited from her mother, with somatic loss of the remaining long arm of this chromosome, was observed in a gastric carcinoma; the girl had associated mental and cardiac abnormalities, suggesting a predisposing condition (Dellavecchia et al. 1999). A kindred manifesting the autosomal dominant inheritance of familial gastric hyperplastic polyposis and gastric cancer development was demonstrated not to have a germline E-cadherin mutation, and on linkage analysis was not linked to 16q, the locus of E-cadherin (Gayther et al. 1998; Seruca et al. 1993). Thus, it seems that there are other loci that, when altered, may predispose an individual to gastric cancer development. Indeed, patients with stomach cancer have been rarely found to harbor germline mutations in the *ATM* (ataxia-telangiectasia mutated) gene and the protooncogene *MET* (Lee et al. 2000; Massad et al. 1990).

Host and Environment Observations

Host Genotypes Associated with Gastric Adenocarcinoma

The relative importance of bacterial virulence, environmental, and host factors (e.g., age of acquisition, immune responses, acid secretion changes) to the clinical outcome of this infection are currently pressing issues. Several allelic variants of polymorphisms in proinflammatory cytokines such as interleukin (IL)-1β, a potent inhibitor of acid secretion, have been associated with gastric cancers, suggesting the host response may be an important determinant in the clinical outcome of infection (El-Omar et al. 2000, 2003). A polymorphism of tumor necrosis factor alpha and even the toll-like receptor 4 are associated with gastric cancer development as well (Hold et al. 2007; Machado et al. 2003).

A population association study recently demonstrated more frequent occurrence (54%) of the human leukocyte antigen (HLA) DQB1*0301 allele in Caucasian patients with gastric cancer than in a control noncancerous group (27%) (Lee et al. 1996). If confirmed not to represent ethnic heterogeneity, this association may imply that this locus itself directly influences susceptibility to gastric cancer development or is a marker of linkage disequilibrium with a nearby cancer-predisposing locus. In another study, the frequency of allele DQB1*0401 was significantly higher in those infected with *H. pylori* who developed atrophic gastritis than those who were infected and did not develop atrophic gastritis or those not infected (Sakai et al. 1999). The potential role of the HLA locus in gastric tumorigenesis has implications regarding the importance of escape from immune surveillance as a causative factor for this disease. The fact that most people with these alleles do not develop gastric cancer illustrates the complexity of this multifactorial disease.

Of note, the blood group A phenotype is associated with gastric cancers in the 1950s (Aird et al. 1953; Buckwalter et al. 1957). Interestingly, *H. pylori* was shown to adhere to the Lewis[b] blood group antigen, indicating a potentially important host factor that could facilitate this chronic infection and the subsequent risk of gastric cancer (Boren et al. 1993). Additionally, small variant alleles of a mucin gene, *Muc1*, were found to be associated with gastric cancer patients compared with a blood donor control population (Carvalho et al. 1997). Data continue to accumulate that reinforce these earlier findings. A recent study demonstrated that among *H. pylori*–infected patients, the risk of patients with a Lewis (a+b−) phenotype having gastric cancer was 3.15-fold higher than the risk of patients with a Lewis (a−b+) phenotype (p = 0.02, 95% confidence interval: 1.26–7.87) (Sheu et al. 2006).

Several studies have noted variant gene and environmental interactions that potentially impact susceptibility to gastric cancer. Park et al. observed that interactions between CYP2E1 gene polymorphism and smoking have the potential to alter susceptibility to cancer development in the stomach (Park et al. 2003). Furthermore, polymorphisms of ADPRT and XRCC1 that reduced their interaction seemed to confer increased host susceptibility to gastric cancer, particularly in smokers (Miao et al. 2006).

Molecular Alterations

Most molecular analyses of this cancer have involved studies of sporadic tumors for critical, acquired alterations. A detailed, clear working model of gastric tumorigenesis has yet to be formulated. Multiple somatic alterations have been described, but the significance of these changes in gastric tumorigenesis remains to be established in most instances. Molecular studies promise to provide new avenues for lowering the dismal gastric cancer mortality rate.

Cytogenetic Studies

Cytogenetic studies of gastric adenocarcinomas are few in number and have failed to identify any consistent or noteworthy chromosomal abnormalities. A variable number of numerical or structural aberrations have been reported in gastric cancer cells, such as those involving chromosomes 3 (rearrangements), 6 (deletion distal to 6q21), 8 (trisomy), 11 (11p13-p15 aberrations), and 13 (monosomy and translocations) (Panani et al. 1995; Seruca et al. 1993). In a recent study of younger gastric cancer patients, DNA copy number changes in 22 primary gastric cancers and 3 gastric cancer cell lines from patients 45 years or younger were analyzed. Analysis of DNA copy number changes revealed frequent DNA copy number increases at chromosomes 17q (52%), 19q (68%), and 20q (64%), with Southern blotting probes mapping to the 19q12-13.2 region, suggesting cyclin E as one of the candidate target genes (Varis et al. 2003).

Chromosomal Instability

The majority of gastric cancers exhibit significant gross chromosomal aneuploidy. One study found that 72% of differentiated tumors and 43% of undifferentiated gastric tumors were aneuploid (Sasaki et al. 1999). Variability in the classification of instability or histopathologic subtype and in the number of loci examined account for some of variation in aneuploidy, with a trend toward more frequent occurrence in intestinal-type cancers at more advanced stages.

Comparative genomic hybridization (CGH) analyses of xenografted and primary gastric and gastroesophageal junctional adenocarcinomas have revealed several regions of consensus change in DNA copy number indicating the possible location of candidate gastric oncogenes and tumor suppressor genes involved in gastric tumorigenesis (El-Rifai et al. 1998; Moskaluk et al. 1998). A knowledge of these critical alterations is important, because gastric cancers of different histopathologic features have been shown to be associated with distinct patterns of genetic alterations, supporting the notion that these cancers evolve via distinct genetic pathways. Chromosomal arms 4q, 5q, 9p, 17p, and 18q exhibited frequent decreases in DNA copy number. However, chromosomes 8q, 17q, and 20q showed frequent increases in DNA copy number of cases analyzed in this manner (Kimura et al. 2004). These patterns may harbor prognostic information as well. CGH analysis showed an increased frequency of 20q gains and 18q losses in tumors that metastasized to lymph nodes (Hidaka et al. 2003). As described below, several known or candidate tumor suppressor genes have been isolated within some of these frequently lost regions, but the precise targets of genetic loss providing gastric neoplastic cells with survival or growth advantages for clonal expansion remain to be clarified for many of these loci.

Comprehensive loss of heterozygosity (LOH) analysis in xenografted adenocarcinomas has identified 3p, 4p, 4q, 5p, 8q, 13p, 17p, and 18q to be frequently lost well above background (Yustein et al. 1999). LOH analysis has identified several arms and regions of chromosomes that contain tumor suppressor genes important in gastric tumorigenesis, such as 17p (more than 60% at *p53's* locus) (Sano et al. 1991), 18q (more than 60% at the *SMAD4* and *DCC* loci) (Uchino et al. 1992), and 5q (30%–40% at or near the *APC* locus) (Rhyu et al. 1994; Sano et al. 1991). In LOH analysis of more than 100 archived stomach cancers, allelic loss was most frequently noted on chromosome 3p (Schneider et al. 1995). Moreover, three distinct regions of chromosome 4q were found to be frequently lost in gastroesophageal junctional adenocarcinomas, indicating the potential for multiple tumor suppressor genes to be located on this chromosomal arm (Rumpel et al. 1999). Allelic losses of both 17p and 18q in proximal or gastroesophageal junctional tumors were associated with a poorer survival than in those cases without or with only one allelic loss at these loci. The actual targets of genetic loss that provide gastric neoplastic cells with survival or growth advantages for clonal expansion remain to be clarified for many of these loci.

Microsatellite Instability

MSI has been found in 13%–44% of sporadic gastric carcinomas (Iacopetta et al. 1999). The degree of genome-wide instability also varies, with more significant instability [e.g., with MSI-high (MSI-H) tumors exhibiting instability in >33% loci tested] occurring in only 10%–16% of gastric cancers (dos Santos et al. 1996).

Alterations responsible for producing the MSI-H phenotype in a subset of sporadic gastric cancers have been elucidated. Abnormal loss of protein expression of either MLH1 or MSH2 was demonstrated in all cases exhibiting MSI-H (Halling et al. 1999). Altered expression of MLH1 was associated with increased methylation of the promoter region of MLH1 in MSI-H cases suggesting a silencing role of hypermethylation (Fleisher et al. 1999; Leung et al. 1999). Distinct methylation of the promoter of hMLH1 was noted in five of eight MSI-H cases, whereas none of 43 MSI-low (MSI-L) or microsatellite stable (MSS) cases exhibited this methylation (Suzuki et al. 1999). Within the MSI phenotype, some gastric cancers have various defects in HLA class I antigen subunits and APM (antigen processing machinery) and is associated with frequent HLA-A inactivation and frameshift mutations of the $\beta 2m$ and *APM* genes (Hirata et al. 2007).

MSI-H gastric tumors exhibit distinct clinicopathologic characteristics. Consistent associations of the MSI-H phenotype with intestinal subtype, distal location (e.g., antral), and more favorable prognosis have been observed (dos Santos et al. 1996; Oliveira et al. 1998; Ottini et al. 1997; Strickler et al. 1994; Wu et al. 1998; Yamamoto et al. 1999b). Additionally, some but not all studies have noted associations between the MSI-H phenotype and less-frequent lymph node metastasis (dos Santos et al. 1996; Oliveira et al. 1998; Wu et al. 1998), greater depth of invasion (Wu et al. 1998), near-diploid DNA content (dos Santos et al. 1996), and tumoral lymphoid infiltration (dos Santos et al. 1996; Oliveira et al. 1998; Wu et al. 1998). A possible explanation for the unique clinicopathologic phenotype observed in MSI-H gastric tumors may be the occurrence of mutations in a distinct set of cancer-related genes (see below) differing from those in tumors with no or low-level MSI. Because MSI-H gastric carcinomas appear to be clinicopathologically distinct, it may ultimately prove valuable to have markers that identify this subgroup of gastric cancers, such as BAT-26 (Halling et al. 1999). Several tumor suppressor genes have been shown to be critical targets of defective MMR in MSI-H tumors (see below). Moreover, a proapoptotic gene and additional MMR genes have been demonstrated to be altered in MSI gastric cancer cases. These same genes are observed to be infrequently mutated or altered in MSI-L or MSS tumors (Oliveira et al. 1998; Ottini et al. 1998; Yamamoto et al. 1999a).

At least one important target of MSI seems to be the transforming growth factor (TGF)β type II receptor (*TGFβR2*) at a polyadenine tract within its gene (Myeroff et al. 1995). Altered *TGFβR2* could additionally be found in gastric cancers not displaying MSI. Several gastric cancer cell lines resistant to the growth inhibitory and apoptotic effects of TGFβ were shown to have abnormal *TGFβR2* genes and/or transcripts (Park et al. 1994). Moreover, some gastric cancer cell lines and 5 of 40 primary gastric cancers (12.5%) exhibited hypermethylation of the promoter region of TGFβ type 1 receptor gene and decreased mRNA expression (Kang et al. 1999). Another gene involved in this signaling pathway, *ACVR2*, was found to be similarly mutated, even in a biallelic manner, in gastric tumors exhibiting MSI (Mori et al. 2002). Thus, alteration of TGFβ receptors and other members of this signaling path seems to be a critical event in the development of at least a subset of gastric cancers, allowing escape from the growth control signal of TGFβ.

Furthermore, somatic nonframeshift mutations have been reported in *BAX* (Yamamoto et al. 1997). Moreover, the relatively frequent missense mutations at codon 169 of *BAX* was shown to impair its proapoptotic activity (Gil et al. 1999). Additional genes with simple tandem repeat sequences within their coding regions found to be specifically altered in gastric cancers displaying MSI include *IGFRII*, *hMSH3*, *hMSH6*, and *E2F-4*, which are known to be involved in regulation of cell-cycle progression and apoptotic signaling (Souza et al. 1996, 1997; Yamamoto et al. 1997; Yin et al. 1997).

Molecular Profiling of Human Gastric Adenocarcinomas

High-throughput assays such as microarrays that comprehensively determine gene expression or DNA copy number patterns are being explored. Potential biomarkers such as phospholipase A2 or alternatively signature profiles of gene expression or copy number identified with these analyses may provide some prognostic as well as diagnostic information about gastric cancer (Boussioutas et al. 2003; Guan et al. 2000; Hippo et al. 2002; Leung et al. 2002; Weiss et al. 2004).

Comprehensive serial analysis of gene expression has also identified novel genetic alterations including overexpression of calcium-binding proteins (El-Rifai et al. 2002a). A proteomic analysis identified 9 proteins with increased expression and 13 proteins with decreased expression in gastric cancers, which included those involved in mitotic checkpoints such as MAD1L1 and EB1 as well as others such as HSP27, CYR61, and CLPP (Boussioutas et al. 2003).

Specific Somatic Alterations

Trefoil Factor Family 1 Loss

Loss of the trefoil peptide TFF1 (pS2), a stable three-loop molecule synthesized in mucus-secreting cells, has been described in approximately 50% of gastric carcinomas

(Henry et al. 1991; Luqmani et al. 1989; Muller and Borchard 1993; Wu et al. 1998). The biologic significance of this loss was reported in a knockout mouse model of gastric antral neoplasia (Lefebvre et al. 1996). Additionally, expression of TFF1 was observed to be lower in some gastric intestinal metaplasia and gastric adenomatous lesions compared with adjacent normal or hyperplastic mucosa (Machado et al. 1996; Nogueira et al. 1999). *TFF1* resides on chromosome 21q22, a region noted to be deleted in some gastric cancers in LOH studies (Nishizuka et al. 1998; Sakata et al. 1997). Moreover, overexpression of TFF1 in the gastric cancer cell line, AGS, inhibited its growth (Calnan et al. 1999). Furthermore, the overwhelming majority of gastric cancers studied had absent to minimal transcript levels compared with normal gastric mucosa (Beckler et al. 2003). Moreover, C/EBP-β was found to be overexpressed in the majority of gastric cancers in a corresponding manner and to bind to the promoter of TFF1, suggesting a regulatory factor role (Sankpal et al. 2005, 2006). Cytokine signaling through the coreceptor gp130 with increased STAT-3 expression has also been observed to decrease TFF1 and lead to gastric lesions in mice (Howlett et al. 2005; Tebbutt et al. 2002).

E-cadherin Mutations and Loss

Several sporadic gastric cancers have displayed altered E-cadherin, mainly diffuse types. E-cadherin is a transmembrane, calcium ion-dependent adhesion molecule important in epithelial cell homotypic interactions which, when decreased in expression, is associated with invasive properties (Birchmeier and Behrens 1994). Reduced E-cadherin expression determined by immunohistochemical analysis was noted in the majority of gastric carcinomas (92% of 60 cases) compared with their adjacent normal tissue (Mayer et al. 1993). Absent expression of E-cadherin was observed to be significantly associated with undifferentiated, diffuse-type cancers (n = 30) compared with intestinal-type cancers (n = 30). Genetic abnormalities of the *E-cadherin* gene (located on chromosome 16q22.1) and transcripts were demonstrated in half of 26 diffuse gastric cancers on reverse transcriptase–polymerase chain reaction (PCR) analysis (Becker et al. 1994). Moreover, a study of 10 gastric cancer cell lines displaying loose intracellular adhesion found absent E-cadherin transcripts in four lines, and insertions or deletions in two other lines (Oda et al. 1994). *E-cadherin* splice site alterations producing exon deletion and skipping, large deletions including allelic loss, and point mutations mostly of the missense nature have been demonstrated in diffuse-type cancers, some even exhibiting alterations in both alleles (Berx et al. 1998). Seven of 10 diffuse-type gastric carcinomas were found to contain somatic *E-cadherin* alterations including specifically the diffuse component in five of six mixed-type tumors (Machado et al. 1999). Methylation of the *E-cadherin* promoter region was found in 16 (26%) of 61 gastric cancers studied (Suzuki et al. 1999). Additionally, α-catenin, which binds to the intracellular domain of E-cadherin and links it to actin-based cytoskeletal elements, was noted to have reduced immunohistochemical expression in 70% of 60 gastric carcinomas and correlated with infiltrated growth and poor differentiation

(Matsui et al. 1994). Finally, two gene variants of the potential E-cadherin regulator IQGAP1 were found in two diffuse gastric cancers, but not in normal controls (Morris et al. 2005).

p53 Mutations

The *p53* gene has consistently been demonstrated to be significantly altered in gastric adenocarcinomas. Allelic loss occurs in more than 60% of cases and mutations are identified in approximately 30%–50% of cases depending on the mutational screening method used and variable sample sizes (Hollstein et al. 1996). Some mutations of *p53* have even been identified in early dysplastic and apparent intestinal metaplasia gastric lesions; however, most alterations occurred in the advanced stages of neoplasia. The spectrum of mutations in this gene within gastric tumors is not unusual with a predominance of base transitions, especially at CpG dinucleotides. Many studies have used immunohistochemical analysis of tumors in an effort to detect excessive expression of p53 as an indirect means to identify mutations of this gene, but this assay does not seem to have consistent prognostic value in patients with gastric cancers (Gabbert et al. 1995; Hurlimann and Saraga 1994).

Kinases/Phosphatases

Activation of protein kinases by tyrosine kinases and phosphatases has been shown to affect many cellular activities including cell growth, differentiation, and survival (Gschwind et al. 2004). Phosphatidylinositol 3-kinases (PI3K) are lipid kinases that regulate pathways for neoplastic proliferation, adhesion, survival, and motility (Vivanco and Sawyers 2002). Mutations in *PIC3A,* which encodes a catalytic subunit of PI3K, have been characterized in several cancers including glioblastoma, colon, and up to 25% of gastric cancers analyzed (Samuels et al. 2004). In further attempts to identify mutations in *PIC3A*, researchers have reported a lower prevalence of 4.3% in gastric cancers; however, they also observed increased expression of this gene product in gastric tumors and an expression profile associated with this upregulation in these tumors (Li et al. 2005). Furthermore, a kinase involved in chromosome segregation and stability, STK15 (BTAK, Aurora2), was found to be amplified and overexpressed in gastric cancers (Sakakura et al. 2001).

Wang et al. identified 83 somatic mutations in human protein tyrosine phosphatases (PTPs), the majority affecting colon cancer, but also found in 17% of gastric cancers analyzed. The most frequently altered PTP, protein-tyrosinase phosphatase-receptor type (PTPRT) was found to decrease its activity, suggesting a role as a tumor suppressor gene (Wang et al. 2004). However, an additional PCR-based study suggests a much smaller role for alterations in PTPRT in gastric tumorigenesis, which was found in only 1% of cases (Lee et al. 2007).

FHIT Mutations

Evidence of tumor suppressor loci on chromosome 3p has accumulated from a variety of studies including allelic loss in primary gastric tumors (46%) and homozygous deletion in a gastric cancer cell line (KATO III) as well as xenografted tumors (Kastury et al. 1996). The *FHIT* gene was recently isolated from the common fragile site FRA3B region at 3p14.2 and found to have abnormal transcripts with deleted exons in five of nine gastric cancers (Ohta et al. 1996). Furthermore, loss of FHIT protein expression was demonstrated immunohistochemically in the majority of gastric carcinomas (Baffa et al. 1998). One somatic missense mutation was identified in exon 6 of the *FHIT* gene during a coding region analysis of 40 gastric carcinomas (Gemma et al. 1997). Additional studies are needed to identify the critically altered targets on this chromosome, clarify the role *FHIT* has in gastric tumorigenesis, and determine the role breakpoints in this region of 3p has in gastric cancer development.

c-MET Overexpression

Overexpression of the *MET* gene, which encodes a tyrosine kinase receptor for the hepatocyte growth factor, has been reported to have prognostic value to indicate poorer survival in multivariate analysis (Nakajima et al. 1999). There have been numerous reports in the literature indicating that the *MET* gene is amplified in approximately 15% and its expression elevated in up to 50% of gastric tumors (Hara et al. 1998; Kuniyasu et al. 1992, 1993; Taniguchi et al. 1998; Tsugawa et al. 1998). Interestingly, *MET* amplification in colon carcinomas rarely occurs. This evidence, along with the infrequent mutations in the *ras* protooncogene found in gastric cancer (Koshiba et al. 1993; Lee et al. 1995), provides two very good indications that the mechanisms involved in the development of colon and gastric cancers are distinct.

Overexpression of the *MET* gene by itself may not be sufficient for cancer progression because interaction of the precancerous cells with stromal fibroblast-derived growth factors, including hepatocyte growth factor (HGF), may promote the progression of disease. A hypothesis put forth by Tahara proposes that stromal fibroblasts that have been stimulated by IL-1 or TGFβ secrete HGF which can then bind to the MET protein on cancer cells and promote the morphogenesis and progression of the cancer (Tahara 1995). In the absence of (or at low levels of) E-cadherin and catenins, HGF may cause the scattering of cancer cells and the development of the poorly differentiated phenotype. Thus, interactions between the stromal and epithelial layers of the stomach may have profound effects on the development of disease. Inoue et al. demonstrated that TGFβ and HGF produced by a culture of gastric fibroblasts could stimulate the invasiveness of a human scirrhous-type gastric cancer cell line *in vitro* (Inoue et al. 1997). It is tempting to speculate that this may be one of the mechanisms through which *H. pylori* could act to promote the development of gastric cancers. Infection with *H. pylori* has in fact been shown to result in an increase in gastric mucosal HGF levels and an increase in HGF gene expression (Kondo et al. 1995; Taha et al. 1996; Yasunaga et al. 1996).

Methylation Silencing Alterations

No inactivating somatic mutations of *p16^{INK4}* were detected in more than 70 cases of gastric carcinoma screened by PCR–single strand conformational polymorphism analysis (Igaki et al. 1995). However, *p16^{INK4}* somatic mutations were noted in several esophageal adenocarcinomas with LOH and others with loss of p16 expression were found to have abnormal hypermethylation of the *p16^{INK4}* promoter, suggesting this epigenetic gene expression silencing may have a role in esophageal tumorigenesis (Barrett et al. 1996; Klump et al. 1998; Wong et al. 1997). A significant number (41%) of gastric cancers exhibited CpG island methylation in a study of p16's promoter region (Suzuki et al. 1999). Many of these cases with hypermethylation of promoter regions displayed the MSI-H phenotype and multiple sites of methylation including the *MLH1* promoter region (Toyota et al. 1999).

RUNX3, a tumor suppressor gene, apparently suppresses gastric epithelial cell growth by inducing p21 (WAF1/Cip1) expression in cooperation with TGFβ-activated SMAD (Chi et al. 2005). RUNX3 was found to be altered in 82% of gastric cancers through either gene silencing or protein mislocalization to the cytoplasm (Ito et al. 2005). Significant downregulation of RUNX3 through methylation on the promoter region was observed in primary tumors (75%) as well as in all clinical peritoneal metastases of gastric cancers (100%) compared with normal gastric mucosa. Silencing of RUNX3 affects expression of important genes such as *p21* and others involved in aspects of metastasis including cell adhesion, proliferation, and apoptosis. They potentially promote the peritoneal metastasis of gastric cancer (Sakakura et al. 2005).

Multiple tumor suppressor genes and candidates including CDKN2A, GSTP1, APC, MGMT, DAKP, XAF1, THBS-1, RUNX1,CDH1 have been shown to be methylated in gastric cancers and may provide potential molecular markers for these cancers (Tamura 2004; Zou et al. 2006).

Wnt Signaling

Several members of the Wnt signaling pathway, *APC* and *β-Catenin*, have had several somatic alterations noted in a few cases of gastric cancer. β-Catenin participates with the Tcf-Lef family of transcription factors in the transfer of cellular proliferation signals to the nucleus. Thus, inactivating mutations of APC or mutations in specific serine and threonine residues in the NH$_2$ terminus of β-catenin would be expected to lead to the stabilization of β-catenin and the loss of cellular growth control as demonstrated in colon cancers (Morin 1999).

Somatic *APC* mutations, mostly missense in nature and of relatively low frequency, have been reported in Japanese patients' gastric adenocarcinomas and adenomas on partial screening (Nagase and Nakamura 1993). However, several other reports including Japanese patients have not identified significant *APC* mutations in gastric carcinomas on similar partial screening analysis of the frequently mutated region including direct nucleotide sequencing (Maesawa et al. 1995; Ogasawara et al. 1994; Powell et al. 1996). Significant allelic loss at *APC's* loci suggests the existence of a tumor suppressor gene important in gastric tumorigenesis nearby. Indeed, the *interferon regulatory factor-1* gene has been mapped to this frequently deleted region in gastric cancers (Tamura et al. 1996). Thus, defining the critical alteration(s) on chromosome 5q, with frequent loss remains to be delineated.

Missense somatic mutations in *β-Catenin*, which also has connections with the cell adhesion complex involving E-cadherin, were identified in a few cases of intestinal-type gastric cancer (Park et al. 1999). However, the functional consequences of these changes remain to be determined as the four frequently mutated serines or threonines that are known to be involved in regulatory phosphorylation of this signaling pathway were not affected. No *β-Catenin* mutations were observed in diffuse-type gastric cancers (Candidus et al. 1996). The gastric cancer cell line HSC-39, established from a patient with a signet-ring cell carcinoma contains a 321 base pair (bp) in-frame deletion in the NH_2 terminal region including the GSK3-β phosphorylation and α-catenin binding site (Kawanishi et al. 1995). Decreased normal membranous expression of β-catenin was found to be associated with poor differentiation and shorter patient survival time (Ramesh et al. 1999).

Growth Factor Alterations

Tumor and stromal cell interactions have been suggested for several other growth factor signaling systems. Approximately 15% of gastric cancers have been observed to express both epidermal growth factor (EGF) and EGF receptor (EGFR), suggesting the presence of an autocrine mechanism of growth stimulation in these cases (Tokunaga et al. 1995). Some studies suggest a worse prognosis for tumors with positive expression of EGF, EGFR, or both. Involvement of several other growth factor signaling paths have been implicated in gastric tumorigenesis including TGFα, IL-1, criptor, amphiregulin, platelet-derived growth factor, and K-sam (Tahara et al. 1996). Expression of bFGF and FGFR have been found in 70% and 60% of gastric cancers, respectively, and noted to be frequently more undifferentiated and in a more advanced invasive stage (Ueki et al. 1995). Rare amplification of *MYC* and *KSAM* has also been reported in gastric cancers (Hara et al. 1998). Overexpression of ERBB2 (HER-2/neu), a transmembrane tyrosine kinase receptor protooncogene, has been implicated as a potential marker of poor prognosis (Mizutani et al. 1993; Uchino et al. 1993; Yonemura et al. 1991). For ERBB2, polysomy was a more frequent mechanism than amplification in both esophageal (32.5% vs. 7.5%) and gastric (15% vs. 5%) cancers (Bizari et al. 2006). ERBB2 has even been detected immunohistochemically in the serum of gastric cancer patients (Chariyalertsak et al. 1994). In 58 gastric adenocarcinomas with lymph node metastasis, one ERBB2 mutation was detected in the lymph node metastasis, but not in the primary tumor of the same patient suggesting a role in metastasis (Bizari et al. 2006). Other prior studies, however, have not found any prognostic value of ERBB expression in these cancers (Friess et al. 1999; Tateishi et al. 1992).

Apoptosis Alterations

Alterations in the mechanism leading to cell death or apoptosis can allow aberrant cell growth and immortality, having a crucial role in tumorigenesis. Several mutations in genes usually promoting or regulating apoptosis have been identified. Mutations of the *BAX* gene, believed to be a promoter of apoptosis, have been identified in up to 33% of gastric cancers as well as colorectal and endometrial cancers (Ouyang et al. 1998). Another study supporting the role of the *BAX* gene in tumor suppression linked low expression levels of the protein Bax-interacting factor-1 (Bif-1) in gastric cancers, suggesting another breakdown in the pathway of regulating apoptosis (Lee et al. 2006). The bcl-2 homolog BAK, another promoter of apoptosis, was found to have missense mutations in 12.5% of gastric cancers analyzed (Kondo et al. 2000). Defects in cell-surface receptors of the apoptosis pathway have also been observed in gastric tumors such as in the death receptors DR4 and DR5. DR4 encodes a tumor necrosis factor apoptosis receptor gene and prior studies had identified a single nucleotide polymorphism present in nongastric malignancies; however, this polymorphism failed to show increased risk for developing gastric cancer (Kuraoka et al. 2005). Mutations in DR5 seem to have a strong role in the development of gastric cancers because up to 7% of analyzed gastric cancers were found to harbor missense mutations of this gene with transfection studies even showing inhibition of apoptotic cell death in all these mutants (Park et al. 2001).

Angiogenesis Alterations

Thymidine phosphorylase has been noted to correlate with angiogenesis in gastric cancers (Saito et al. 1999). Abnormalities of vascular endothelial growth factor (VEGF) have been associated with gastric cancers, and may have a role in promoting lymphangiogenesis and predisposition to metastasis. Reverse transcription PCR was used to demonstrate VEGF expression in 27.8% in gastric cancers and was strongly associated with tumors with high lymphatic vessel density (Hachisuka et al. 2005). Several of the VEGF gene polymorphisms were found to be an independent prognostic marker for patients with gastric cancer and the analysis of VEGF gene polymorphisms may help identify patients at greatest risk for poor outcomes (Kim et al. 2007).

Other Alterations

Overexpression of DARP32 and a novel truncated isoform was found in the majority of gastric cancers (El-Rifai et al. 2002b). Moreover, an antiapoptotic effect of this overexpression of DARP32 and t-DARP was observed *in vitro* (Belkhiri et al. 2005). Only one of 35 gastric cancers contained an intragenic mutation of *SMAD4* along with allelic loss suggesting this gene is infrequently altered in gastric tumorigenesis (Chi et al. 2005; Powell et al. 1997). Additionally, only one missense change of uncertain functional significance in the *LKB1* (STK11) gene was noted in a study of 28 sporadic gastric carcinomas (Park et al. 1998).

Evidence that cox-2 participates in the proliferation of gastric cancer cells (MKN45) was suggested by Tsuji et al. who also later showed that a role for cox-2 in cancer progression is clearly not limited to tumors of the gastric epithelium (Tsuji et al. 1996). *In vitro* studies suggest that hedgehog signaling contributes to gastric cancer cell growth. Elevated expressions of hedgehog target genes human patched gene 1 (PTCH1) or Glil occurs in 63 of 99 primary gastric cancers. Activation of the hedgehog pathway is associated with poorly differentiated and more-aggressive tumors. Treatment of gastric cancer cells with KAAD-cyclopamine, a hedgehog signaling inhibitor, decreases expressions of Glil and PTCH1, resulting in cell-growth inhibition and apoptosis (Ma et al. 2005).

Additionally, increased plasminogen activation has been reported in several gastric tumors (Ito et al. 1996). A somatic mitochondrial deletion of 50 bp was even demonstrated in four gastric adenocarcinomas (Burgart et al. 1995). Telomerase activity, which seems necessary for cell immortality, has been detected by a PCR-based assay frequently in the late stages of gastric tumors and is associated with a poor prognosis (Hiyama et al. 1995). Furthermore, Tahara et al. have identified telomerase activity in the vast majority of primary gastric tumor tissues whereas no activity was observed in the corresponding normal gastric mucosa (Tahara 1995). Gastric lavage fluid might provide a window to early diagnosis, because in one study, telomerase was detected in 20 patients (80%) with gastric cancer, 7 patients (28%) with peptic ulcer, and none in normal subjects (p < 0.001) (Wong et al. 2006). Specific alterations and the true prevalence of significant changes in these genes, gene products, or phenotypes in human gastric tumors remain to be characterized. Table 10.1 lists alterations that appear consistently in human gastric carcinomas; others will undoubtedly be added as studies accumulate.

Table 10.1 Somatic alterations in human gastric adenocarcinomas

Frequent	Specifically in MSI-H cases	Infrequent
p53 inactivation	Loss of hMLH1 or hMSH2 protein	*APC*
TFF1 loss and C/EBP-β increase	*TGFβIIR*	*Smad4*
FHIT loss	*Bax*	*LKB1*
DARP32/t-DARP/c-MET increase	*IGFIIR*	
HER-2/neu increase	*E2F-4*	
EGFR/K-sam increase	*hMSH3*	
E-cadherin loss (diffuse-type)	*hMSH6*	
P16/p27/ RUNX3/PAI-1 loss		
VEGF/MMP-9/Cox-2 increase		

Potential Biomarkers

Prognostic Markers

Elevated tissue levels of certain proteins are associated with increased invasiveness of tumor cells, and some are even associated with poorer survival, including the following: urokinase-type plasminogen activator, VEGF, MET, MYC, tie-1 protein tyrosine kinase, CD44v6, PDGF-A, TGFβ-1, and cyclin D2 (Han et al. 1999a; Ito et al. 1996; Kuwahara et al. 1999; Lin et al. 1999; Maeda et al. 1999; Takano et al. 1999; Yamamichi et al. 1998). However, decreased levels of other proteins have been noted to be associated with more invasiveness and poorer survival including: loss of p27 (Kip1), p21 (CIP1), plasminogen activator inhibitor type-1 (PAI-1), and tissue-type plasminogen activator (Ganesh et al. 1996; Ito et al. 1996; Jang et al. 1999; Kwonet al. 1999; Xiangming et al. 1999). In a recent study by Zafirellis et al., the expression of p53 and bcl-2 proteins had no significant impact on the outcome of patients with gastric cancer (Zafirellis et al. 2005). The combined analysis of p53 and p27 (kip1) was of added prognostic value, but alone these findings were not compelling as a specific marker in gastric carcinomas (Liu et al. 2001). Furthermore, patients with a high KLK6 expression had a significantly poorer survival rate than those with a low KLK6 expression ($p = 0.03$) (Nagahara et al. 2005). Nishigaki et al. identified 9 proteins with increased expression and 13 proteins with decreased expression in gastric carcinomas, notably including mitotic checkpoint (MAD1L1 and EB1) and mitochondrial functions (CLPP, COX5A, and ECH1) (Nishigaki et al. 2005). Thus, some of these molecules could potentially provide diagnostic and/or prognostic markers that can be evaluated in gastric cancer tissues for clinical utilities.

Chemosensitizati\on Markers

Measuring molecular markers in tumor cells may help in predicting how they will respond to specific treatment agents. The feasibility of this kind of testing was demonstrated in a small study of 30 patients, whereby overexpression of p53 by immunohistochemistry of locally advanced gastric cancers was found to indicate a lower response rate to neoadjuvant cytotoxic therapy (Cascinu et al. 1998). Furthermore, the expression of thymidine phosphorylase or even thymidine synthase in gastric tumors may help predict their response to fluorouracil agents (Saito et al. 1999). In another small study of 23 patients with gastric cancer, those tumors with positive staining for BAX and Bcl-2 were more chemoresistant and had worse prognosis than those negative for Bcl-2 staining (Muguruma et al. 1998; Nakata et al. 1998b). Moreover, an MTT chemosensitivity assay of gastric tumors predicted responses to certain agents and prolonged survival when used (Kurihara et al. 1999).

Serum Markers

Additional diagnostic and prognostic markers for gastric cancer patients have been sought in bone marrow and peripheral blood samples. Concentrations of matrix metalloproteinase 9 were higher in the plasma of gastric cancer patients compared with normal controls (Torii et al. 1997). High concentrations of tissue inhibitor of metalloproteinase 1, IL-10, HGF, soluble receptor for IL-2, and soluble fragment of E-cadherin in the serum or plasma of gastric carcinoma patients are associated with more invasiveness and poorer survival (De Vita et al. 1999; Gofuku et al. 1998; Han et al. 1999b; Saito et al. 1999; Yoshikawa et al. 1999). Furthermore, detection of cytokeratin 20, CD44 variants, CA-125, CEA, CA 19-9, CA 72-4, and anti-p53 antibodies in the peripheral blood of gastric cancer patients tend to indicate disseminated disease (Chausovsky et al. 1999; Marrelli et al. 1999; Nakata et al. 1998a; Pituch-Noworolska et al. 1998; Wu et al. 1999). Elevated plasma levels of osteopontin were an independent risk factor of poor survival (Wu et al. 2007). A proteomic evaluation of serum from gastric cancer patients before and after surgical treatment and compared with controls suggested a panel of five protein markers had diagnostic potential with a sensitivity of 83% and specificity of 95% (Poon et al. 2006).

Potential Targets for Treatment and Intervention

Advancement in our molecular understanding of tumorigenesis can result in specific targeted chemotherapy with dramatic effects in controlling tumor growth and survival. Indeed, the use of imatinib mesylate to inhibit the overexpressed cKIT tyrosin kinase in a relatively specific manner in GIST tumors is a prototype example of this kind of specific targeted treatment with significant results (DeMatteo 2002; Demetri 2001). For gastric carcinomas, some potential candidate molecular targets found to alter gastric cellular growth when altered *in vitro* have been identified. A novel KGF-R2 phosphorylation inhibitor, Ki23057, decreased proliferation of two scirrhous gastric cancer cell lines with K-sam amplification (Nakamura et al. 2006). Additionally, cyclopamine, a hedgehog inhibitor, decreased the growth of gastric cancer cell lines that expressed high levels of SMO (Fukaya et al. 2006). Thus, the prognosis of diagnosed disseminated gastric cancer may soon be changed with new avenues of treatment.

References

Aarnio, M., Salovaara, R., Aaltonen, L. A., Mecklin, J. P., and Jarvinen, H. J. 1997. Features of gastric cancer in hereditary non-polyposis colorectal cancer syndrome. Int J Cancer, 74(5): 551–5.

Aird, I., Bentall, H. H., and Roberts, J. A. 1953. A relationship between cancer of stomach and the ABO blood groups. Br Med J, 1(4814): 799–801.

Antommarchi F. 1825. Les derniers moments de napoleon, en compement du memorial de sainte-helene. 1st ed. Brussels: H. Tarlier.

Aoki, M., Yamamoto, K., Noshiro, H., Sakai, K., Yokota, J., Kohno, T., Tokino, T., Ishida, S., Ohyama, S., Ninomiya, I., Uesaka, K., Kitajima, M., Shimada, S., Matsuno, S., Yano, M., Hiratsuka, M., Sugimura, H., Itoh, F., Minamoto, T., Maehara, Y., Takenoshita, S., Aikou, T., Katai, H., Yoshimura, K., Takahashi, T., Akagi, K., Sairenji, M., Yamamura, Y., and Sasazuki, T. 2005. A full genome scan for gastric cancer. J Med Genet, 42(1): 83–7.

Ascano, J. J., Frierson, H., Jr., Moskaluk, C. A., Harper, J. C., Roviello, F., Jackson, C. E., El-Rifai, W., Vindigni, C., Tosi, P., and Powell, S. M. 2001. Inactivation of the E-cadherin gene in sporadic diffuse-type gastric cancer. Mod Pathol, 14(10): 942–9.

Baffa, R., Veronese, M. L., Santoro, R., Mandes, B., Palazzo, J. P., Rugge, M., Santoro, E., Croce, C. M., and Huebner, K. 1998. Loss of FHIT expression in gastric carcinoma. Cancer Res, 58(20): 4708–14.

Barrett, M. T., Sanchez, C. A., Galipeau, P. C., Neshat, K., Emond, M., and Reid, B. J. 1996. Allelic loss of 9p21 and mutation of the CDKN2/p16 gene develop as early lesions during neoplastic progression in Barrett's esophagus. Oncogene, 13(9): 1867–73.

Becker, K. F., Atkinson, M. J., Reich, U., Becker, I., Nekarda, H., Siewert, J. R., and Hofler, H. 1994. E-cadherin gene mutations provide clues to diffuse type gastric carcinomas. Cancer Res, 54(14): 3845–52.

Beckler, A. D., Roche, J. K., Harper, J. C., Petroni, G., Frierson, H. F., Jr., Moskaluk, C. A., El-Rifai, W., and Powell, S. M. 2003. Decreased abundance of trefoil factor 1 transcript in the majority of gastric carcinomas. Cancer, 98(10): 2184–91.

Belkhiri, A., Zaika, A., Pidkovka, N., Knuutila, S., Moskaluk, C., and El-Rifai, W. 2005. Darpp-32: a novel antiapoptotic gene in upper gastrointestinal carcinomas. Cancer Res, 65(15): 6583–92.

Berx, G., Becker, K. F., Hofler, H., and van Roy, F. 1998. Mutations of the human E-cadherin (CDH1) gene. Hum Mutat, 12(4): 226–37.

Birchmeier, W., and Behrens, J. 1994. Cadherin expression in carcinomas: role in the formation of cell junctions and the prevention of invasiveness. Biochim Biophys Acta, 1198(1): 11–26.

Bizari, L., Borim, A. A., Leite, K. R., Goncalves Fde, T., Cury, P. M., Tajara, E. H., and Silva, A. E. 2006. Alterations of the CCND1 and HER-2/neu (ERBB2) proteins in esophageal and gastric cancers. Cancer Genet Cytogenet, 165(1): 41–50.

Blair, V., Martin, I., Shaw, D., Winship, I., Kerr, D., Arnold, J., Harawira, P., McLeod, M., Parry, S., Charlton, A., Findlay, M., Cox, B., Humar, B., More, H., and Guilford, P. 2006. Hereditary diffuse gastric cancer: Diagnosis and management. Clin Gastroenterol Hepatol, 4(3): 262–75.

Boren, T., Falk, P., Roth, K. A., Larson, G., and Normark, S. 1993. Attachment of helicobacter pylori to human gastric epithelium mediated by blood group antigens. Science, 262(5141): 1892–5.

Boussioutas, A., Li, H., Liu, J., Waring, P., Lade, S., Holloway, A. J., Taupin, D., Gorringe, K., Haviv, I., Desmond, P. V., and Bowtell, D. D. 2003. Distinctive patterns of gene expression in premalignant gastric mucosa and gastric cancer. Cancer Res, 63(10): 2569–77.

Buckwalter, J. A., Wohlwend, C. B., Colter, D. C., Tidrick, R. T., and Knowler, L. A. 1957. The association of the ABO blood groups to gastric carcinoma. Surg Gynecol Obstet, 104(2): 176–9.

Burgart, L. J., Zheng, J., Shu, Q., Strickler, J. G., and Shibata, D. 1995. Somatic mitochondrial mutation in gastric cancer. Am J Pathol, 147(4): 1105–11.

Caldas, C., Carneiro, F., Lynch, H. T., Yokota, J., Wiesner, G. L., Powell, S. M., Lewis, F. R., Huntsman, D. G., Pharoah, P. D., Jankowski, J. A., MacLeod, P., Vogelsang, H., Keller, G., Park, K. G., Richards, F. M., Maher, E. R., Gayther, S. A., Oliveira, C., Grehan, N., Wight, D., Seruca, R., Roviello, F., Ponder, B. A., and Jackson, C. E. 1999. Familial gastric cancer: overview and guidelines for management. J Med Genet, 36(12): 873–80.

Calnan, D. P., Westley, B. R., May, F. E., Floyd, D. N., Marchbank, T., and Playford, R. J. 1999. The trefoil peptide TFF1 inhibits the growth of the human gastric adenocarcinoma cell line AGS. J Pathol, 188(3): 312–7.

Candidus, S., Bischoff, P., Becker, K. F., and Hofler, H. 1996. No evidence for mutations in the alpha- and beta-catenin genes in human gastric and breast carcinomas. Cancer Res, 56(1): 49–52.

Carvalho, F., Seruca, R., David, L., Amorim, A., Seixas, M., Bennett, E., Clausen, H., and Sobrinho-Simoes, M. 1997. MUC1 gene polymorphism and gastric cancer: an epidemiological study. Glycoconjug J, 14(1): 107–11.

Cascinu, S., Graziano, F., Del Ferro, E., Staccioli, M. P., Ligi, M., Carnevali, A., Muretto, P., and Catalano, G. 1998. Expression of p53 protein and resistance to preoperative chemotherapy in locally advanced gastric carcinoma. Cancer, 83(9): 1917–22.

Chariyalertsak, S., Sugano, K., Ohkura, H., and Mori, Y. 1994. Comparison of c-erbB-2 oncoprotein expression in tissue and serum of patients with stomach cancer. Tumour Biol, 15(5): 294–303.

Chausovsky, G., Luchansky, M., Figer, A., Shapira, J., Gottfried, M., Novis, B., Bogelman, G., Zemer, R., Zimlichman, S., and Klein, A. 1999. Expression of cytokeratin 20 in the blood of patients with disseminated carcinoma of the pancreas, colon, stomach, and lung. Cancer, 86(11): 2398–405.

Chi, X. Z., Yang, J. O., Lee, K. Y., Ito, K., Sakakura, C., Li, Q. L., Kim, H. R., Cha, E. J., Lee, Y. H., Kaneda, A., Ushijima, T., Kim, W. J., Ito, Y., and Bae, S. C. 2005. RUNX3 suppresses gastric epithelial cell growth by inducing p21(WAF1/Cip1) expression in cooperation with transforming growth factor β-activated SMAD. Mol Cell Biol, 25(18): 8097–107.

Dellavecchia, C., Guala, A., Olivieri, C., Haintink, O., Cadario, F., Luinetti, O., Fiocca, R., Minelli, A., Danesino, C., and Bona, G. 1999. Early onset of gastric carcinoma and constitutional deletion of 18p. Cancer Genet Cytogenet, 113(1): 96–9.

DeMatteo, R. P. 2002. The GIST of targeted cancer therapy: A tumor (gastrointestinal stromal tumor), a mutated gene (c-kit), and a molecular inhibitor (STI571). Ann Surg Oncol, 9(9): 831–9.

Demetri, G. D. 2001. Targeting c-kit mutations in solid tumors: scientific rationale and novel therapeutic options. Semin Oncol, 28(5 Suppl 17): 19–26.

De Vita, F., Orditura, M., Galizia, G., Romano, C., Infusino, S., Auriemma, A., Lieto, E., and Catalano, G. 1999. Serum interleukin-10 levels in patients with advanced gastrointestinal malignancies. Cancer, 86(10): 1936–43.

dos Santos, N. R., Seruca, R., Constancia, M., Seixas, M., and Sobrinho-Simoes, M. 1996. Microsatellite instability at multiple loci in gastric carcinoma: clinicopathologic implications and prognosis. Gastroenterology, 110(1): 38–44.

El-Omar, E. M., Carrington, M., Chow, W. H., McColl, K. E., Bream, J. H., Young, H. A., Herrera, J., Lissowska, J., Yuan, C. C., Rothman, N., Lanyon, G., Martin, M., Fraumeni, J. F., Jr., and Rabkin, C. S. 2000. Interleukin-1 polymorphisms associated with increased risk of gastric cancer. Nature, 404(6776): 398–402.

El-Omar, E. M., Rabkin, C. S., Gammon, M. D., Vaughan, T. L., Risch, H. A., Schoenberg, J. B., Stanford, J. L., Mayne, S. T., Goedert, J., Blot, W. J., Fraumeni, J. F., Jr., and Chow, W. H. 2003. Increased risk of noncardia gastric cancer associated with proinflammatory cytokine gene polymorphisms. Gastroenterology, 124(5): 1193–201.

El-Rifai, W., Harper, J. C., Cummings, O. W., Hyytinen, E. R., Frierson, H. F., Jr., Knuutila, S., and Powell, S. M. 1998. Consistent genetic alterations in xenografts of proximal stomach and gastro-esophageal junction adenocarcinomas. Cancer Res, 58(1): 34–7.

El-Rifai, W., Moskaluk, C. A., Abdrabbo, M. K., Harper, J., Yoshida, C., Riggins, G. J., Frierson, H. F., Jr., and Powell, S. M. 2002a. Gastric cancers overexpress S100A calcium-binding proteins. Cancer Res, 62(23): 6823–6.

El-Rifai, W., Smith, M. F., Jr., Li, G., Beckler, A., Carl, V. S., Montgomery, E., Knuutila, S., Moskaluk, C. A., Frierson, H. F., Jr., and Powell, S. M. 2002b. Gastric cancers overexpress DARPP-32 and a novel isoform, t-DARPP. Cancer Res, 62(14): 4061–4.

Fleisher, A. S., Esteller, M., Wang, S., Tamura, G., Suzuki, H., Yin, J., Zou, T. T., Abraham, J. M., Kong, D., Smolinski, K. N., Shi, Y. Q., Rhyu, M. G., Powell, S. M., James, S. P., Wilson, K. T., Herman, J. G., and Meltzer, S. J. 1999. Hypermethylation of the hMLH1 gene promoter in human gastric cancers with microsatellite instability. Cancer Res, 59(5): 1090–5.

Friess, H., Fukuda, A., Tang, W. H., Eichenberger, A., Furlan, N., Zimmermann, A., Korc, M., and Buchler, M. W. 1999. Concomitant analysis of the epidermal growth factor receptor family in esophageal cancer: overexpression of epidermal growth factor receptor mRNA but not of c-erbB-2 and c-erbB-3. World J Surg, 23(10): 1010–8.

Fukaya, M., Isohata, N., Ohta, H., Aoyagi, K., Ochiya, T., Saeki, N., Yanagihara, K., Nakanishi, Y., Taniguchi, H., Sakamoto, H., Shimoda, T., Nimura, Y., Yoshida, T., and Sasaki, H. 2006. Hedgehog signal activation in gastric pit cell and in diffuse-type gastric cancer. Gastroenterology, 131(1): 14–29.

Gabbert, H. E., Muller, W., Schneiders, A., Meier, S., and Hommel, G. 1995. The relationship of p53 expression to the prognosis of 418 patients with gastric carcinoma. Cancer, 76(5): 720–6.

Ganesh, S., Sier, C. F., Heerding, M. M., van Krieken, J. H., Griffioen, G., Welvaart, K., van de Velde, C. J., Verheijen, J. H., Lamers, C. B., and Verspaget, H. W. 1996. Prognostic value of the plasminogen activation system in patients with gastric carcinoma. Cancer, 77(6): 1035–43.

Gayther, S. A., Gorringe, K. L., Ramus, S. J., Huntsman, D., Roviello, F., Grehan, N., Machado, J. C., Pinto, E., Seruca, R., Halling, K., MacLeod, P., Powell, S. M., Jackson, C. E., Ponder, B. A., and Caldas, C. 1998. Identification of germ-line E-cadherin mutations in gastric cancer families of European origin. Cancer Res, 58(18): 4086–9.

Gemma, A., Hagiwara, K., Ke, Y., Burke, L. M., Khan, M. A., Nagashima, M., Bennett, W. P., and Harris, C. C. 1997. FHIT mutations in human primary gastric cancer. Cancer Res, 57(8): 1435–7.

Gil, J., Yamamoto, H., Zapata, J. M., Reed, J. C., and Perucho, M. 1999. Impairment of the proapoptotic activity of bax by missense mutations found in gastrointestinal cancers. Cancer Res, 59(9): 2034–7.

Gofuku, J., Shiozaki, H., Doki, Y., Inoue, M., Hirao, M., Fukuchi, N., and Monden, M. 1998. Characterization of soluble E-cadherin as a disease marker in gastric cancer patients. Br J Cancer, 78(8): 1095–101.

Goldgar, D. E., Easton, D. F., Cannon-Albright, L. A., and Skolnick, M. H. 1994. Systematic population-based assessment of cancer risk in first-degree relatives of cancer probands. J Natl Cancer Inst, 86(21): 1600–8.

Gorer, P. A. 1938. Genetic interpretation of studies on cancer in twins. Ann Eugen, 8: 219.

Groden, J., Thliveris, A., Samowitz, W., Carlson, M., Gelbert, L., Albertsen, H., Joslyn, G., Stevens, J., Spirio, L., and Robertson, M. 1991. Identification and characterization of the familial adenomatous polyposis coli gene. Cell, 66(3): 589–600.

Gschwind, A., Fischer, O. M., and Ullrich, A. 2004. The discovery of receptor tyrosine kinases: targets for cancer therapy. Nature reviews. Cancer, 4(5): 361–70.

Guan, X. Y., Fu, S. B., Xia, J. C., Fang, Y., Sham, J. S., Du, B. D., Zhou, H., Lu, S., Wang, B. Q., Lin, Y. Z., Liang, Q., Li, X. M., Du, B., Ning, X. M., Du, J. R., Li, P., and Trent, J. M. 2000. Recurrent chromosome changes in 62 primary gastric carcinomas detected by comparative genomic hybridization. Cancer Genet Cytogenet, 123(1): 27–34.

Guilford, P., Hopkins, J., Harraway, J., McLeod, M., McLeod, N., Harawira, P., Taite, H., Scoular, R., Miller, A., and Reeve, A. E. 1998. E-cadherin germline mutations in familial gastric cancer. Nature, 392(6674): 402–5.

Hachisuka, T., Narikiyo, M., Yamada, Y., Ishikawa, H., Ueno, M., Uchida, H., Yoriki, R., Ohigashi, Y., Miki, K., Tamaki, H., Mizuno, T., and Nakajima, Y. 2005. High lymphatic vessel density correlates with overexpression of VEGF-C in gastric cancer. Oncol Rep, 13(4): 733–7.

Halling, K. C., Harper, J., Moskaluk, C. A., Thibodeau, S. N., Petroni, G. R., Yustein, A. S., Tosi, P., Minacci, C., Roviello, F., Piva, P., Hamilton, S. R., Jackson, C. E., and Powell, S. M. 1999. Origin of microsatellite instability in gastric cancer. Am J Pathol, 155(1): 205–11.

Han, S., Kim, H. Y., Park, K., Cho, H. J., Lee, M. S., Kim, H. J., and Kim, Y. D. 1999a. c-myc expression is related with cell proliferation and associated with poor clinical outcome in human gastric cancer. J Korean Med Sci, 14(5): 526–30.

Han, S. U., Lee, J. H., Kim, W. H., Cho, Y. K., and Kim, M. W. 1999b. Significant correlation between serum level of hepatocyte growth factor and progression of gastric carcinoma. World J Surg, 23(11): 1176–80.

Hara, T., Ooi, A., Kobayashi, M., Mai, M., Yanagihara, K., and Nakanishi, I. 1998. Amplification of c-myc, K-sam, and c-met in gastric cancers: Detection by fluorescence in situ hybridization. Lab Invest, 78(9): 1143–53.

Hemminki, K., and Jiang, Y. 2002. Familial and second gastric carcinomas: a nationwide epidemiologic study from Sweden. Cancer, 94(4): 1157–65.

Henry, J. A., Bennett, M. K., Piggott, N. H., Levett, D. L., May, F. E., and Westley, B. R. 1991. Expression of the pNR-2/pS2 protein in diverse human epithelial tumours. Br J Cancer, 64(4): 677–82.

Hidaka, S., Yasutake, T., Kondo, M., Takeshita, H., Yano, H., Haseba, M., Tsuji, T., Sawai, T., Nakagoe, T., and Tagawa, Y. 2003. Frequent gains of 20q and losses of 18q are associated with lymph node metastasis in intestinal-type gastric cancer. Anticancer Res, 23(4): 3353–7.

Hippo, Y., Taniguchi, H., Tsutsumi, S., Machida, N., Chong, J. M., Fukayama, M., Kodama, T., and Aburatani, H. 2002. Global gene expression analysis of gastric cancer by oligonucleotide microarrays. Cancer Res, 62(1): 233–40.

Hirata, T., Yamamoto, H., Taniguchi, H., Horiuchi, S., Oki, M., Adachi, Y., Imai, K., and Shinomura, Y. 2007. Characterization of the immune escape phenotype of human gastric cancers with and without high-frequency microsatellite instability. J Pathol, 211(5): 516–23.

Hiyama, E., Yokoyama, T., Tatsumoto, N., Hiyama, K., Imamura, Y., Murakami, Y., Kodama, T., Piatyszek, M. A., Shay, J. W., and Matsuura, Y. 1995. Telomerase activity in gastric cancer. Cancer Res, 55(15): 3258–62.

Hizawa, K., Iida, M., Yao, T., Aoyagi, K., and Fujishima, M. 1997. Juvenile polyposis of the stomach: clinicopathological features and its malignant potential. J Clin Pathol, 50(9): 771–4.

Hofgartner, W. T., Thorp, M., Ramus, M. W., Delorefice, G., Chey, W. Y., Ryan, C. K., Takahashi, G. W., and Lobitz, J. R. 1999. Gastric adenocarcinoma associated with fundic gland polyps in a patient with attenuated familial adenomatous polyposis. Am J Gastroenterol, 94(8): 2275–81.

Hold, G. L., Rabkin, C. S., Chow, W. H., Smith, M. G., Gammon, M. D., Risch, H. A., Vaughan, T. L., McColl, K. E., Lissowska, J., Zatonski, W., Schoenberg, J. B., Blot, W. J., Mowat, N. A., Fraumeni, J. F., Jr., and El-Omar, E. M. 2007. A functional polymorphism of toll-like receptor 4 gene increases risk of gastric carcinoma and its precursors. Gastroenterology, 132(3): 905–12.

Hollstein, M., Shomer, B., Greenblatt, M., Soussi, T., Hovig, E., Montesano, R., and Harris, C. C. 1996. Somatic point mutations in the p53 gene of human tumors and cell lines: updated compilation. Nucl Acids Res, 24(1): 141–6.

Howlett, M., Judd, L. M., Jenkins, B., La Gruta, N. L., Grail, D., Ernst, M., and Giraud, A. S. 2005. Differential regulation of gastric tumor growth by cytokines that signal exclusively through the coreceptor gp130. Gastroenterology, 129(3): 1005–18.

Hurlimann, J., and Saraga, E. P. 1994. Expression of p53 protein in gastric carcinomas: association with histologic type and prognosis. Am J Surg Pathol, 18(12): 1247–53.

Iacopetta, B. J., Soong, R., House, A. K., and Hamelin, R. 1999. Gastric carcinomas with micros-atellite instability: clinical features and mutations to the TGF-beta type II receptor, IGFII receptor, and BAX genes. J Pathol, 187(4): 428–32.

Igaki, H., Sasaki, H., Tachimori, Y., Kato, H., Watanabe, H., Kimura, T., Harada, Y., Sugimura, T., and Terada, M. 1995. Mutation frequency of the p16/CDKN2 gene in primary cancers in the upper digestive tract. Cancer Res, 55(15): 3421–3.

Inoue, T., Chung, Y. S., Yashiro, M., Nishimura, S., Hasuma, T., Otani, S., and Sowa, M. 1997. Transforming growth factor-beta and hepatocyte growth factor produced by gastric fibroblasts stimulate the invasiveness of scirrhous gastric cancer cells. Jpn J Cancer Res, 88(2): 152–9.

Ito, K., Liu, Q., Salto-Tellez, M., Yano, T., Tada, K., Ida, H., Huang, C., Shah, N., Inoue, M., Rajnakova, A., Hiong, K. C., Peh, B. K., Han, H. C., Ito, T., Teh, M., Yeoh, K. G., and Ito, Y. 2005. RUNX3, a novel tumor suppressor, is frequently inactivated in gastric cancer by protein mislocalization. Cancer Res, 65(17): 7743–50.

Ito, H., Yonemura, Y., Fujita, H., Tsuchihara, K., Kawamura, T., Nojima, N., Fujimura, T., Nose, H., Endo, Y., and Sasaki, T. 1996. Prognostic relevance of urokinase-type plasminogen activator (uPA) and plasminogen activator inhibitors PAI-1 and PAI-2 in gastric cancer. Virchows Arch, 427(5): 487–96.

Jang, S. J., Park, Y. W., Park, M. H., Lee, J. D., Lee, Y. Y., Jung, T. J., Kim, I. S., Choi, I. Y., Ki, M., Choi, B. Y., and Ahn, M. J. 1999. Expression of cell-cycle regulators, cyclin E and p21WAF1/CIP1, potential prognostic markers for gastric cancer. Eur J Surg Oncol, 25(2): 157–63.

Jones, S., Emmerson, P., Maynard, J., Best, J. M., Jordan, S., Williams, G. T., Sampson, J. R., and Cheadle, J. P. 2002. Biallelic germline mutations in MYH predispose to multiple colorectal adenoma and somatic G:C→T:A mutations. Hum Mol Genet, 11(23): 2961–7.

Kang, S. H., Bang, Y. J., Im, Y. H., Yang, H. K., Lee, D. A., Lee, H. Y., Lee, H. S., Kim, N. K., and Kim, S. J. 1999. Transcriptional repression of the transforming growth factor-beta type I receptor gene by DNA methylation results in the development of TGF-beta resistance in human gastric cancer. Oncogene, 18(51): 7280–6.

Kastury, K., Baffa, R., Druck, T., Ohta, M., Cotticelli, M. G., Inoue, H., Negrini, M., Rugge, M., Huang, D., Croce, C. M., Palazzo, J., and Huebner, K. 1996. Potential gastrointestinal tumor suppressor locus at the 3p14.2 FRA3B site identified by homozygous deletions in tumor cell lines. Cancer Res, 56(5): 978–83.

Kawanishi, J., Kato, J., Sasaki, K., Fujii, S., Watanabe, N., and Niitsu, Y. 1995. Loss of E-cadherin-dependent cell-cell adhesion due to mutation of the beta-catenin gene in a human cancer cell line, HSC-39. Mol Cell Biol, 15(3): 1175–81.

Kim, C. J., Cho, Y. G., Park, C. H., Kim, S. Y., Nam, S. W., Lee, S. H., Yoo, N. J., Lee, J. Y., and Park, W. S. 2004. Genetic alterations of the MYH gene in gastric cancer. Oncogene, 23(40): 6820–2.

Kim, J., Sohn, S., Chae, Y., Cho, Y., Bae, H. I., Yan, G., Park, J., Lee, M. H., Chung, H., and Yu, W. 2007. Vascular endothelial growth factor gene polymorphisms associated with prognosis for patients with gastric cancer. Ann Oncol, 18(6): 1030–6.

Kimura, Y., Noguchi, T., Kawahara, K., Kashima, K., Daa, T., and Yokoyama, S. 2004. Genetic alterations in 102 primary gastric cancers by comparative genomic hybridization: gain of 20q and loss of 18q are associated with tumor progression. Mod Pathol 17(11): 1328–37.

Kinzler, K. W., and Vogelstein, B. 1996. Lessons from hereditary colorectal cancer. Cell, 87(2): 159–70.

Kleihues, P., Schauble, B., zur Hausen, A., Esteve, J., and Ohgaki, H. 1997. Tumors associated with p53 germline mutations: a synopsis of 91 families. Am J Pathol, 150(1): 1–13.

Klump, B., Hsieh, C. J., Holzmann, K., Gregor, M., and Porschen, R. 1998. Hypermethylation of the CDKN2/p16 promoter during neoplastic progression in Barrett's esophagus. Gastroenterology, 115(6): 1381–6.

Kondo, S., Shinomura, Y., Kanayama, S., Higashimoto, Y., Kiyohara, T., Yasunaga, Y., Kitamura, S., Ueyama, H., Imamura, I., and Fukui, H. 1995. Helicobacter pylori increases gene expression of hepatocyte growth factor in human gastric mucosa. Biochem Biophys Res Commun, 210(3): 960–5.

Kondo, S., Shinomura, Y., Miyazaki, Y., Kiyohara, T., Tsutsui, S., Kitamura, S., Nagasawa, Y., Nakahara, M., Kanayama, S., and Matsuzawa, Y. 2000. Mutations of the bak gene in human gastric and colorectal cancers. Cancer Res, 60(16): 4328–30.

Koshiba, M., Ogawa, O., Habuchi, T., Hamazaki, S., Shimada, T., Takahashi, R., and Sugiyama, T. 1993. Infrequent ras mutation in human stomach cancers. Jpn J Cancer Res, 84(2): 163–7.

Kubba, A. K., and Young, M. 1999. The Napoleonic cancer gene? J Med Biogr, 7(3): 175–81.

Kuniyasu, H., Yasui, W., Kitadai, Y., Yokozaki, H., Ito, H., and Tahara, E. 1992. Frequent amplification of the c-met gene in scirrhous type stomach cancer. Bioch Biophys Res Commun, 189(1): 227–32.

Kuniyasu, H., Yasui, W., Yokozaki, H., Kitadai, Y., and Tahara, E. 1993. Aberrant expression of c-met mRNA in human gastric carcinomas. J Int Cancer, 55(1): 72–5.

Kuraoka, K., Matsumura, S., Sanada, Y., Nakachi, K., Imai, K., Eguchi, H., Matsusaki, K., Oue, N., Nakayama, H., and Yasui, W. 2005. A single nucleotide polymorphism in the extracellular domain of TRAIL receptor DR4 at nucleotide 626 in gastric cancer patients in Japan. Oncol Rep, 14(2): 465–70.

Kurihara, N., Kubota, T., Furukawa, T., Watanabe, M., Otani, Y., Kumai, K., and Kitajima, M. 1999. Chemosensitivity testing of primary tumor cells from gastric cancer patients with liver metastasis can identify effective antitumor drugs. Anticancer Res, 19(6B): 5155–8.

Kuwahara, A., Katano, M., Nakamura, M., Fujimoto, K., Miyazaki, K., Mori, M., and Morisaki, T. 1999. New therapeutic strategy for gastric carcinoma: a two-step evaluation of malignant potential from its molecular biologic and pathologic characteristics. J Surg Oncol, 72(3): 142–9.

Kwon, O. J., Kang, H. S., Suh, J. S., Chang, M. S., Jang, J. J., and Chung, J. K. 1999. The loss of p27 protein has an independent prognostic significance in gastric cancer. Anticancer Res, 19(5B): 4215–20.

La Vecchia, C., Negri, E., Franceschi, S., and Gentile, A. 1992. Family history and the risk of stomach and colorectal cancer. Cancer, 70(1): 50–5.

Lee, F. I. 1971. Carcinoma of the gastric antrum in identical twins. Postgrad Med J, 47(551): 622–4.

Lee, J. H., Han, S. U., Cho, H., Jennings, B., Gerrard, B., Dean, M., Schmidt, L., Zbar, B., and Vande Woude, G. F. 2000. A novel germ line juxtamembrane met mutation in human gastric cancer. Oncogene, 19(43): 4947–53.

Lee, J. W., Jeong, E. G., Lee, S. H., Nam, S. W., Kim, S. H., Lee, J. Y., Yoo, N. J., and Lee, S. H. 2007. Mutational analysis of PTPRT phosphatase domains in common human cancers. Acta Pathol Microbiol Immunol Scand, 115(1): 47–51.

Lee, J. W., Jeong, E. G., Soung, Y. H., Nam, S. W., Lee, J. Y., Yoo, N. J., and Lee, S. H. 2006. Decreased expression of tumour suppressor bax-interacting factor-1 (bif-1), a bax activator, in gastric carcinomas. Pathology, 38(4): 312–5.

Lee, K. H., Lee, J. S., Suh, C., Kim, S. W., Kim, S. B., Lee, J. H., Lee, M. S., Park, M. Y., Sun, H. S., and Kim, S. H. 1995. Clinicopathologic significance of the K-ras gene codon 12 point mutation in stomach cancer: an analysis of 140 cases. Cancer, 75(12): 2794–801.

Lee, J. E., Lowy, A. M., Thompson, W. A., Lu, M., Loflin, P. T., Skibber, J. M., Evans, D. B., Curley, S. A., Mansfield, P. F., and Reveille, J. D. 1996. Association of gastric adenocarcinoma with the HLA class II gene DQB10301. Gastroenterology, 111(2): 426–32.

Lefebvre, O., Chenard, M. P., Masson, R., Linares, J., Dierich, A., LeMeur, M., Wendling, C., Tomasetto, C., Chambon, P., and Rio, M. C. 1996. Gastric mucosa abnormalities and tumorigenesis in mice lacking the pS2 trefoil protein. Science, 274(5285): 259–62.

Leung, S. Y., Chen, X., Chu, K. M., Yuen, S. T., Mathy, J., Ji, J., Chan, A. S., Li, R., Law, S., Troyanskaya, O. G., Tu, I. P., Wong, J., So, S., Botstein, D., and Brown, P. O. 2002. Phospholipase A2 group IIA expression in gastric adenocarcinoma is associated with prolonged survival and less frequent metastasis. Proc Natl Acad Sci USA, 99(25): 16203–8.

Leung, S. Y., Yuen, S. T., Chung, L. P., Chu, K. M., Chan, A. S., and Ho, J. C. 1999. hMLH1 promoter methylation and lack of hMLH1 expression in sporadic gastric carcinomas with high-frequency microsatellite instability. Cancer Res, 59(1): 159–64.

Li, F. P., and Fraumeni, J. F., Jr. 1969. Soft-tissue sarcomas, breast cancer, and other neoplasms. A familial syndrome? Ann Intern Med, 71(4): 747–52.

Li, V. S., Wong, C. W., Chan, T. L., Chan, A. S., Zhao, W., Chu, K. M., So, S., Chen, X., Yuen, S. T., and Leung, S. Y. 2005. Mutations of PIK3CA in gastric adenocarcinoma. BMC Cancer, 5: 29.

Lin, W. C., Li, A. F., Chi, C. W., Chung, W. W., Huang, C. L., Lui, W. Y., Kung, H. J., and Wu, C. W. 1999. Tie-1 protein tyrosine kinase: a novel independent prognostic marker for gastric cancer. Clin Cancer Res, 5(7): 1745–51.

Lindor, N. M., and Greene, M. H. 1998. The concise handbook of family cancer syndromes: Mayo familial cancer program. J Natl Cancer Inst, 90(14): 1039–71.

Liu, X. P., Kawauchi, S., Oga, A., Suehiro, Y., Tsushimi, K., Tsushimi, M., and Sasaki, K. 2001. Combined examination of p27(Kip1), p21(Waf1/Cip1) and p53 expression allows precise estimation of prognosis in patients with gastric carcinoma. Histopathology, 39(6): 603–10.

Lugli, A., Zlobec, I., Singer, G., Kopp Lugli, A., Terracciano, L. M., and Genta, R. M. 2007. Napoleon Bonaparte's gastric cancer: a clinicopathologic approach to staging, pathogenesis, and etiology. Nature clinical practice. Gastroenterol Hepatol, 4(1): 52–7.

Luqmani, Y., Bennett, C., Paterson, I., Corbishley, C. M., Rio, M. C., Chambon, P., and Ryall, G. 1989. Expression of the pS2 gene in normal, benign and neoplastic human stomach. J Int Cancer, 44(5): 806–12.

Lynch, H. T., Grady, W., Suriano, G., and Huntsman, D. 2005. Gastric cancer: new genetic developments. J Surg Oncol, 90(3): 114–33; discussion 133.

Lynch, H. T., and Krush, A. J. 1971. Cancer family "G" revisited: 1895–1970. Cancer, 27(6): 1505–11.

Lynch, H. T., Smyrk, T. C., Watson, P., Lanspa, S. J., Lynch, J. F., Lynch, P. M., Cavalieri, R. J., and Boland, C. R. 1993. Genetics, natural history, tumor spectrum, and pathology of hereditary nonpolyposis colorectal cancer: an updated review. Gastroenterology, 104(5): 1535–49.

Ma, X., Chen, K., Huang, S., Zhang, X., Adegboyega, P. A., Evers, B. M., Zhang, H., and Xie, J. 2005. Frequent activation of the hedgehog pathway in advanced gastric adenocarcinomas. Carcinogenesis, 26(10): 1698–705.

Machado, J. C., Carneiro, F., Blin, N., and Sobrinho-Simoes, M. 1996. Pattern of pS2 protein expression in premalignant and malignant lesions of gastric mucosa. Eur J Cancer Prev, 5(3): 169–79.

Machado, J. C., Figueiredo, C., Canedo, P., Pharoah, P., Carvalho, R., Nabais, S., Castro Alves, C., Campos, M. L., Van Doorn, L. J., Caldas, C., Seruca, R., Carneiro, F., and Sobrinho-Simoes, M. 2003. A proinflammatory genetic profile increases the risk for chronic atrophic gastritis and gastric carcinoma. Gastroenterology, 125(2): 364–71.

Machado, J. C., Soares, P., Carneiro, F., Rocha, A., Beck, S., Blin, N., Berx, G., and Sobrinho-Simoes, M. 1999. E-cadherin gene mutations provide a genetic basis for the phenotypic divergence of mixed gastric carcinomas. Lab Invest, 79(4): 459–65.

Maeda, K., Kang, S. M., Onoda, N., Ogawa, M., Kato, Y., Sawada, T., and Chung, K. H. 1999. Vascular endothelial growth factor expression in preoperative biopsy specimens correlates with disease recurrence in patients with early gastric carcinoma. Cancer, 86(4): 566–71.

Maesawa, C., Tamura, G., Suzuki, Y., Ogasawara, S., Sakata, K., Kashiwaba, M., and Satodate, R. 1995. The sequential accumulation of genetic alterations characteristic of the colorectal adenoma-carcinoma sequence does not occur between gastric adenoma and adenocarcinoma. J Pathol, 176(3): 249–58.

Maimon, S. N., and Zinninger, M. M. 1953. Familial gastric cancer. Gastroenterology, 25(2): 139–52; discussion, 153–5.

Marrelli, D., Roviello, F., De Stefano, A., Farnetani, M., Garosi, L., Messano, A., and Pinto, E. 1999. Prognostic significance of CEA, CA 19-9 and CA 72-4 preoperative serum levels in gastric carcinoma. Oncology, 57(1): 55–62.

Massad, M., Uthman, S., Obeid, S., and Majjar, F. 1990. Ataxia-telangiectasia and stomach cancer. Am J Gastroenterol, 85(5): 630–1.

Matsui, S., Shiozaki, H., Inoue, M., Tamura, S., Doki, Y., Kadowaki, T., Iwazawa, T., Shimaya, K., Nagafuchi, A., and Tsukita, S. 1994. Immunohistochemical evaluation of alpha-catenin expression in human gastric cancer. Virchows Arch, 424(4): 375–81.

Mayer, B., Johnson, J. P., Leitl, F., Jauch, K. W., Heiss, M. M., Schildberg, F. W., Birchmeier, W., and Funke, I. 1993. E-cadherin expression in primary and metastatic gastric cancer: down-regulation correlates with cellular dedifferentiation and glandular disintegration. Cancer Res, 53(7): 1690–5.

Miao, X., Zhang, X., Zhang, L., Guo, Y., Hao, B., Tan, W., He, F., and Lin, D. 2006. Adenosine diphosphate ribosyl transferase and x-ray repair cross-complementing 1 polymorphisms in gastric cardia cancer. Gastroenterology, 131(2): 420–7.

Mizutani, T., Onda, M., Tokunaga, A., Yamanaka, N., and Sugisaki, Y. 1993. Relationship of C-erbB-2 protein expression and gene amplification to invasion and metastasis in human gastric cancer. Cancer, 72(7): 2083–8.

Mori, Y., Sato, F., Selaru, F. M., Olaru, A., Perry, K., Kimos, M. C., Tamura, G., Matsubara, N., Wang, S., Xu, Y., Yin, J., Zou, T. T., Leggett, B., Young, J., Nukiwa, T., Stine, O. C., Abraham, J. M., Shibata, D., and Meltzer, S. J. 2002. Instabilotyping reveals unique mutational spectra in microsatellite-unstable gastric cancers. Cancer Res, 62(13): 3641–5.

Morin, P. J. 1999. Beta-catenin signaling and cancer. Bioessays, 21(12): 1021–30.

Morris, L. E., Bloom, G. S., Frierson, H. F., Jr., and Powell, S. M. 2005. Nucleotide variants within the IQGAP1 gene in diffuse-type gastric cancers. Genes Chromosomes Cancer, 42(3): 280–6.

Moskaluk, C. A., Hu, J., and Perlman, E. J. 1998. Comparative genomic hybridization of esophageal and gastroesophageal adenocarcinomas shows consensus areas of DNA gain and loss. Genes Chromosomes Cancer, 22(4): 305–11.

Muguruma, K., Nakata, B., Hirakawa, K., Yamashita, Y., Onoda, N., Inoue, T., Matsuoka, T., Kato, Y., and Sowa, M. 1998. p53 and bax protein expression as predictor of chemotherapeutic effect in gastric carcinoma. Cancer Chemother, 25(Suppl 3): 400–3.

Muller, W., and Borchard, F. 1993. pS2 protein in gastric carcinoma and normal gastric mucosa: association with clinicopathological parameters and patient survival. J Pathol, 171(4): 263–9.

Myeroff, L. L., Parsons, R., Kim, S. J., Hedrick, L., Cho, K. R., Orth, K., Mathis, M., Kinzler, K. W., Lutterbaugh, J., and Park, K. 1995. A transforming growth factor beta receptor type II gene mutation common in colon and gastric but rare in endometrial cancers with microsatellite instability. Cancer Res, 55(23): 5545–7.

Nagahara, H., Mimori, K., Utsunomiya, T., Barnard, G. F., Ohira, M., Hirakawa, K., and Mori, M. 2005. Clinicopathologic and biological significance of kallikrein 6 overexpression in human gastric cancer. Clin Cancer Res, 11(19 Pt 1): 6800–6.

Nagase, H., and Nakamura, Y. 1993. Mutations of the APC (adenomatous polyposis coli) gene. Hum Mutat, 2(6): 425–34.

Nakajima, M., Sawada, H., Yamada, Y., Watanabe, A., Tatsumi, M., Yamashita, J., Matsuda, M., Sakaguchi, T., Hirao, T., and Nakano, H. 1999. The prognostic significance of amplification and overexpression of c-met and c-erb B-2 in human gastric carcinomas. Cancer, 85(9): 1894–902.

Nakamura, K., Yashiro, M., Matsuoka, T., Tendo, M., Shimizu, T., Miwa, A., and Hirakawa, K. 2006. A novel molecular targeting compound as K-samII/FGF-R2 phosphorylation inhibitor, Ki23057, for scirrhous gastric cancer. Gastroenterology, 131(5): 1530–41.

Nakata, B., Hirakawa-YS Chung, K., Kato, Y., Yamashita, Y., Maeda, K., Onoda, N., Sawada, T., and Sowa, M. 1998a. Serum CA 125 level as a predictor of peritoneal dissemination in patients with gastric carcinoma. Cancer, 83(12): 2488–92.

Nakata, B., Muguruma, K., Hirakawa, K., Chung, Y. S., Yamashita, Y., Inoue, T., Matsuoka, T., Onoda, N., Kato, Y., and Sowa, M. 1998b. Predictive value of bcl-2 and bax protein expression for chemotherapeutic effect in gastric cancer. A pilot study. Oncology, 55(6): 543–7.

Nishigaki, R., Osaki, M., Hiratsuka, M., Toda, T., Murakami, K., Jeang, K. T., Ito, H., Inoue, T., and Oshimura, M. 2005. Proteomic identification of differentially-expressed genes in human gastric carcinomas. Proteomics, 5(12): 3205–13.

Nishizuka, S., Tamura, G., Terashima, M., and Satodate, R. 1998. Loss of heterozygosity during the development and progression of differentiated adenocarcinoma of the stomach. J Pathol, 185(1): 38–43.

Nogueira, A. M., Machado, J. C., Carneiro, F., Reis, C. A., Gott, P., and Sobrinho-Simoes, M. 1999. Patterns of expression of trefoil peptides and mucins in gastric polyps with and without malignant transformation. J Pathol, 187(5): 541–8.

Oda, T., Kanai, Y., Oyama, T., Yoshiura, K., Shimoyama, Y., Birchmeier, W., Sugimura, T., and Hirohashi, S. 1994. E-cadherin gene mutations in human gastric carcinoma cell lines. Proc Natl Acad Sci USA, 91(5): 1858–62.

Offerhaus, G. J., Giardiello, F. M., Krush, A. J., Booker, S. V., Tersmette, A. C., Kelley, N. C., and Hamilton, S. R. 1992. The risk of upper gastrointestinal cancer in familial adenomatous polyposis. Gastroenterology, 102(6): 1980–2.

Ogasawara, S., Maesawa, C., Tamura, G., and Satodate, R. 1994. Lack of mutations of the adenomatous polyposis coli gene in oesophageal and gastric carcinomas. Virchows Arch, 424(6): 607–11.

Ohta, M., Inoue, H., Cotticelli, M. G., Kastury, K., Baffa, R., Palazzo, J., Siprashvili, Z., Mori, M., McCue, P., Druck, T., Croce, C. M., and Huebner, K. 1996. The FHIT gene, spanning the chromosome 3p14.2 fragile site and renal carcinoma-associated t(3;8) breakpoint, is abnormal in digestive tract cancers. Cell, 84(4): 587–97.

Oliveira, C., Seruca, R., Seixas, M., and Sobrinho-Simoes, M. 1998. The clinicopathological features of gastric carcinomas with microsatellite instability may be mediated by mutations of different "target genes": a study of the TGFbeta RII, IGFII R, and BAX genes. Am J Pathol, 153(4): 1211–9.

Ottini, L., Falchetti, M., D'Amico, C., Amorosi, A., Saieva, C., Masala, G., Frati, L., Cama, A., Palli, D., and Mariani-Costantini, R. 1998. Mutations at coding mononucleotide repeats in gastric cancer with the microsatellite mutator phenotype. Oncogene, 16(21): 2767–72.

Ottini, L., Palli, D., Falchetti, M., D'Amico, C., Amorosi, A., Saieva, C., Calzolari, A., Cimoli, F., Tatarelli, C., De Marchis, L., Masala, G., Mariani-Costantini, R., and Cama, A. 1997. Microsatellite instability in gastric cancer is associated with tumor location and family history in a high-risk population from Tuscany. Cancer Res, 57(20): 4523–9.

Ouyang, H., Furukawa, T., Abe, T., Kato, Y., and Horii, A. 1998. The BAX gene, the promoter of apoptosis, is mutated in genetically unstable cancers of the colorectum, stomach, and endometrium. Clin Cancer Res, 4(4): 1071–4.

Palli, D., Galli, M., Caporaso, N. E., Cipriani, F., Decarli, A., Saieva, C., Fraumeni, J. F., Jr., and Buiatti, E. 1994. Family history and risk of stomach cancer in Italy. Cancer Epidemiol Biomarkers Prev, 3(1): 15–8.

Panani, A. D., Ferti, A., Malliaros, S., and Raptis, S. 1995. Cytogenetic study of 11 gastric adenocarcinomas. Cancer Genet Cytogenet, 81(2): 169–72.

Park, K., Kim, S. J., Bang, Y. J., Park, J. G., Kim, N. K., Roberts, A. B., and Sporn, M. B. 1994. Genetic changes in the transforming growth factor beta (TGF-beta) type II receptor gene in human gastric cancer cells: correlation with sensitivity to growth inhibition by TGF-beta. Proc Natl Acad Sci USA, 91(19): 8772–6.

Park, G. T., Lee, O. Y., Kwon, S. J., Lee, C. G., Yoon, B. C., Hahm, J. S., Lee, M. H., Hoo Lee, D., Kee, C. S., and Sun, H. S. 2003. Analysis of CYP2E1 polymorphism for the determination of genetic susceptibility to gastric cancer in Koreans. J Gastroenterol Hepatol, 18(11): 1257–63.

Park, W. S., Lee, J. H., Shin, M. S., Park, J. Y., Kim, H. S., Kim, Y. S., Park, C. H., Lee, S. K., Lee, S. H., Lee, S. N., Kim, H., Yoo, N. J., and Lee, J. Y. 2001. Inactivating mutations of KILLER/DR5 gene in gastric cancers. Gastroenterology, 121(5): 1219–25.

Park, W. S., Moon, Y. W., Yang, Y. M., Kim, Y. S., Kim, Y. D., Fuller, B. G., Vortmeyer, A. O., Fogt, F., Lubensky, I. A., and Zhuang, Z. 1998. Mutations of the STK11 gene in sporadic gastric carcinoma. Int J Oncol, 13(3): 601–4.

Park, W. S., Oh, R. R., Park, J. Y., Lee, S. H., Shin, M. S., Kim, Y. S., Kim, S. Y., Lee, H. K., Kim, P. J., Oh, S. T., Yoo, N. J., and Lee, J. Y. 1999. Frequent somatic mutations of the beta-catenin gene in intestinal-type gastric cancer. Cancer Res, 59(17): 4257–60.

Pituch-Noworolska, A., Wieckiewicz, J., Krzeszowiak, A., Stachura, J., Ruggiero, I., Gawlicka, M., Szczepanik, A., Karcz, D., Popiela, T., and Zembala, M. 1998. Evaluation of circulating tumour cells expressing CD44 variants in the blood of gastric cancer patients by flow cytometry. Anticancer Res, 18(5B): 3747–52.

Poon, T. C., Sung, J. J., Chow, S. M., Ng, E. K., Yu, A. C., Chu, E. S., Hui, A. M., and Leung, W. K. 2006. Diagnosis of gastric cancer by serum proteomic fingerprinting. Gastroenterology, 130(6): 1858–64.

Powell, S. M., Cummings, O. W., Mullen, J. A., Asghar, A., Fuga, G., Piva, P., Minacci, C., Megha, T., Tosi, P., and Jackson, C. E. 1996. Characterization of the APC gene in sporadic gastric adenocarcinomas. Oncogene, 12(9): 1953–9.

Powell, S. M., Harper, J. C., Hamilton, S. R., Robinson, C. R., and Cummings, O. W. 1997. Inactivation of Smad4 in gastric carcinomas. Cancer Res, 57(19): 4221–4.

Powell, S. M., and Smith, M. F., Jr. 2002. Gastric Cancer: molecular biology. In: Kelsen, D.P., editor. Gastrointestinal oncology: principle and practice. Philadelphia: Lippincott Williams & Wilkins; pp. 325–40.

Ramesh, S., Nash, J., and McCulloch, P. G. 1999. Reduction in membranous expression of beta-catenin and increased cytoplasmic E-cadherin expression predict poor survival in gastric cancer. Br J Cancer, 81(8): 1392–7.

Renkonen-Sinisalo, L., Sipponen, P., Aarnio, M., Julkunen, R., Aaltonen, L. A., Sarna, S., Jarvinen, H. J., and Mecklin, J. P. 2002. No support for endoscopic surveillance for gastric cancer in hereditary non-polyposis colorectal cancer. Scand J Gastroenterol, 37(5): 574–7.

Rhyu, M. G., Park, W. S., Jung, Y. J., Choi, S. W., and Meltzer, S. J. 1994. Allelic deletions of MCC/APC and p53 are frequent late events in human gastric carcinogenesis. Gastroenterology, 106(6): 1584–8.

Rumpel, C. A., Powell, S. M., and Moskaluk, C. A. 1999. Mapping of genetic deletions on the long arm of chromosome 4 in human esophageal adenocarcinomas. Am J Pathol, 154(5): 1329–34.

Saito, H., Tsujitani, S., Ikeguchi, M., Maeta, M., and Kaibara, N. 1999. Serum level of a soluble receptor for interleukin-2 as a prognostic factor in patients with gastric cancer. Oncology, 56(3): 253–8.

Saito, H., Tsujitani, S., Oka, S., Kondo, A., Ikeguchi, M., Maeta, M., and Kaibara, N. 1999. The expression of thymidine phosphorylase correlates with angiogenesis and the efficacy of chemotherapy using fluorouracil derivatives in advanced gastric carcinoma. Br J Cancer, 81(3): 484–9.

Sakai, T., Aoyama, N., Satonaka, K., Shigeta, S., Yoshida, H., Shinoda, Y., Shirasaka, D., Miyamoto, M., Nose, Y., and Kasuga, M. 1999. HLA-DQB1 locus and the development of atrophic gastritis with Helicobacter pylori infection. J Gastroenterol, 34(Suppl 11): 24–7.

Sakakura, C., Hagiwara, A., Yasuoka, R., Fujita, Y., Nakanishi, M., Masuda, K., Shimomura, K., Nakamura, Y., Inazawa, J., Abe, T., and Yamagishi, H. 2001. Tumour-amplified kinase BTAK is amplified and overexpressed in gastric cancers with possible involvement in aneuploid formation. Br J Cancer, 84(6): 824–31.

Sakakura, C., Hasegawa, K., Miyagawa, K., Nakashima, S., Yoshikawa, T., Kin, S., Nakase, Y., Yazumi, S., Yamagishi, H., Okanoue, T., Chiba, T., and Hagiwara, A. 2005. Possible involvement of RUNX3 silencing in the peritoneal metastases of gastric cancers. Clin Cancer Res, 11(18): 6479–88.

Sakata, K., Tamura, G., Nishizuka, S., Maesawa, C., Suzuki, Y., Iwaya, T., Terashima, M., Saito, K., and Satodate, R. 1997. Commonly deleted regions on the long arm of chromosome 21 in differentiated adenocarcinoma of the stomach. Genes Chromosomes Cancer, 18(4): 318–21.

Samuels, Y., Wang, Z., Bardelli, A., Silliman, N., Ptak, J., Szabo, S., Yan, H., Gazdar, A., Powell, S. M., Riggins, G. J., Willson, J. K., Markowitz, S., Kinzler, K. W., Vogelstein, B., and Velculescu, V. E. 2004. High frequency of mutations of the PIK3CA gene in human cancers. Science, 304(5670): 554.

Sankpal, N. V., Mayo, M. W., and Powell, S. M. 2005. Transcriptional repression of TFF1 in gastric epithelial cells by CCAAT/enhancer binding protein-beta. Biochim Biophys Acta, 1728(1–2): 1–10.

Sankpal, N. V., Moskaluk, C. A., Hampton, G. M., and Powell, S. M. 2006. Overexpression of CEBPbeta correlates with decreased TFF1 in gastric cancer. Oncogene, 25(4): 643–9.

Sano, T., Tsujino, T., Yoshida, K., Nakayama, H., Haruma, K., Ito, H., Nakamura, Y., Kajiyama, G., and Tahara, E. 1991. Frequent loss of heterozygosity on chromosomes 1q, 5q, and 17p in human gastric carcinomas. Cancer Res, 51(11): 2926–31.

Sasaki, O., Soejima, K., Korenaga, D., and Haraguchi, Y. 1999. Comparison of the intratumor DNA ploidy distribution pattern between differentiated and undifferentiated gastric carcinoma. Anal Quant Cytol Histol, 21(2): 161–5.

Schneider, B. G., Pulitzer, D. R., Brown, R. D., Prihoda, T. J., Bostwick, D. G., Saldivar, V., Rodriguez-Martinez, H. A., Gutierrez-Diaz, M. E., and O'Connell, P. 1995. Allelic imbalance in gastric cancer: an affected site on chromosome arm 3p. Genes Chromosomes Cancer, 13(4): 263–71.

Seruca, R., Castedo, S., Correia, C., Gomes, P., Carneiro, F., Soares, P., de Jong, B., and Sobrinho-Simoes, M. 1993. Cytogenetic findings in eleven gastric carcinomas. Cancer Genet Cytogenet, 68(1): 42–8.

Sheu, M. J., Yang, H. B., Sheu, B. S., Cheng, H. C., Lin, C. Y., and Wu, J. J. 2006. Erythrocyte Lewis (A+B−) host phenotype is a factor with familial clustering for increased risk of Helicobacter pylori-related non-cardiac gastric cancer. J Gastroenterol Hepatol, 21(6): 1054–8.

Shinmura, K., Kohno, T., Takahashi, M., Sasaki, A., Ochiai, A., Guilford, P., Hunter, A., Reeve, A. E., Sugimura, H., Yamaguchi, N., and Yokota, J. 1999. Familial gastric cancer: clinico-pathological characteristics, RER phenotype and germline p53 and E-cadherin mutations. Carcinogenesis, 20(6): 1127–31.

Souza, R. F., Appel, R., Yin, J., Wang, S., Smolinski, K. N., Abraham, J. M., Zou, T. T., Shi, Y. Q., Lei, J., Cottrell, J., Cymes, K., Biden, K., Simms, L., Leggett, B., Lynch, P. M., Frazier, M., Powell, S. M., Harpaz, N., Sugimura, H., Young, J., and Meltzer, S. J. 1996. Microsatellite instability in the insulin-like growth factor II receptor gene in gastrointestinal tumours. Nat Genet, 14(3): 255–7.

Souza, R. F., Yin, J., Smolinski, K. N., Zou, T. T., Wang, S., Shi, Y. Q., Rhyu, M. G., Cottrell, J., Abraham, J. M., Biden, K., Simms, L., Leggett, B., Bova, G. S., Frank, T., Powell, S. M., Sugimura, H., Young, J., Harpaz, N., Shimizu, K., Matsubara, N., and Meltzer, S. J. 1997. Frequent mutation of the E2F-4 cell cycle gene in primary human gastrointestinal tumors. Cancer Res, 57(12): 2350–3.

Stone, J., Bevan, S., Cunningham, D., Hill, A., Rahman, N., Peto, J., Marossy, A., and Houlston, R. S. 1999. Low frequency of germline E-cadherin mutations in familial and nonfamilial gastric cancer. Br J Cancer, 79(11–12): 1935–7.

Strickler, J. G., Zheng, J., Shu, Q., Burgart, L. J., Alberts, S. R., and Shibata, D. 1994. P53 mutations and microsatellite instability in sporadic gastric cancer: when guardians fail. Cancer Res, 54(17): 4750–5.

Suzuki, H., Itoh, F., Toyota, M., Kikuchi, T., Kakiuchi, H., Hinoda, Y., and Imai, K. 1999. Distinct methylation pattern and microsatellite instability in sporadic gastric cancer. J Int Cancer, 83(3): 309–13.

Taha, A. S., Curry, G. W., Morton, R., Park, R. H., and Beattie, A. D. 1996. Gastric mucosal hepatocyte growth factor in Helicobacter pylori gastritis and peptic ulcer disease. Am J Gastroenterol, 91(7): 1407–9.

Tahara, E. 1995. Molecular biology of gastric cancer. World J Surg, 19(4): 484–8; discussion 489–90.

Tahara, E., Semba, S., and Tahara, H. 1996. Molecular biological observations in gastric cancer. Semin Oncol, 23(3): 307–15.

Takano, Y., Kato, Y., Masuda, M., Ohshima, Y., and Okayasu, I. 1999. Cyclin D2, but not cyclin D1, overexpression closely correlates with gastric cancer progression and prognosis. J Pathol, 189(2): 194–200.

Tamura, G. 2004. Promoter methylation status of tumor suppressor and tumor-related genes in neoplastic and non-neoplastic gastric epithelia. Histol Histopathol, 19(1): 221–8.

Tamura, G., Ogasawara, S., Nishizuka, S., Sakata, K., Maesawa, C., Suzuki, Y., Terashima, M., Saito, K., and Satodate, R. 1996. Two distinct regions of deletion on the long arm of chromosome 5 in differentiated adenocarcinomas of the stomach. Cancer Res, 56(3): 612–5.

Taniguchi, K., Yonemura, Y., Nojima, N., Hirono, Y., Fushida, S., Fujimura, T., Miwa, K., Endo, Y., Yamamoto, H., and Watanabe, H. 1998. The relation between the growth patterns of gastric carcinoma and the expression of hepatocyte growth factor receptor (c-met), autocrine motility factor receptor, and urokinase-type plasminogen activator receptor. Cancer, 82(11): 2112–22.

Tateishi, M., Toda, T., Minamisono, Y., and Nagasaki, S. 1992. Clinicopathological significance of c-erbB-2 protein expression in human gastric carcinoma. J Surg Oncol, 49(4): 209–12.

Tebbutt, N. C., Giraud, A. S., Inglese, M., Jenkins, B., Waring, P., Clay, F. J., Malki, S., Alderman, B. M., Grail, D., Hollande, F., Heath, J. K., and Ernst, M. 2002. Reciprocal regulation of gastrointestinal homeostasis by SHP2 and STAT-mediated trefoil gene activation in gp130 mutant mice. Nat Med, 8(10): 1089–97.

Tokunaga, A., Onda, M., Okuda, T., Teramoto, T., Fujita, I., Mizutani, T., Kiyama, T., Yoshiyuki, T., Nishi, K., and Matsukura, N. 1995. Clinical significance of epidermal growth factor (EGF), EGF receptor, and c-erbB-2 in human gastric cancer. Cancer, 75(6 Suppl): 1418–25.

Torii, A., Kodera, Y., Uesaka, K., Hirai, T., Yasui, K., Morimoto, T., Yamamura, Y., Kato, T., Hayakawa, T., Fujimoto, N., and Kito, T. 1997. Plasma concentration of matrix metalloproteinase 9 in gastric cancer. Br J Surg, 84(1): 133–6.

Toyota, M., Ahuja, N., Suzuki, H., Itoh, F., Ohe-Toyota, M., Imai, K., Baylin, S. B., and Issa, J. P. 1999. Aberrant methylation in gastric cancer associated with the CpG island methylator phenotype. Cancer Res, 59(21): 5438–42.

Tsugawa, K., Yonemura, Y., Hirono, Y., Fushida, S., Kaji, M., Miwa, K., Miyazaki, I., and Yamamoto, H. 1998. Amplification of the c-met, c-erbB-2 and epidermal growth factor receptor gene in human gastric cancers: correlation to clinical features. Oncology, 55(5): 475–81.

Tsuji, S., Kawano, S., Sawaoka, H., Takei, Y., Kobayashi, I., Nagano, K., Fusamoto, H., and Kamada, T. 1996. Evidences for involvement of cyclooxygenase-2 in proliferation of two gastrointestinal cancer cell lines. Prostaglandins Leukot Essent Fatty Acids, 55(3): 179–83.

Uchino, S., Tsuda, H., Maruyama, K., Kinoshita, T., Sasako, M., Saito, T., Kobayashi, M., and Hirohashi, S. 1993. Overexpression of c-erbB-2 protein in gastric cancer. its correlation with long-term survival of patients. Cancer, 72(11): 3179–84.

Uchino, S., Tsuda, H., Noguchi, M., Yokota, J., Terada, M., Saito, T., Kobayashi, M., Sugimura, T., and Hirohashi, S. 1992. Frequent loss of heterozygosity at the DCC locus in gastric cancer. Cancer Res, 52(11): 3099–102.

Ueki, T., Koji, T., Tamiya, S., Nakane, P. K., and Tsuneyoshi, M. 1995. Expression of basic fibroblast growth factor and fibroblast growth factor receptor in advanced gastric carcinoma. J Pathol, 177(4): 353–61.

Utsunomiya J. 1990. The concept of hereditary colorectal cancer and the implications of its study. Hereditary Colorectal Cancer, 3–16.

Varis, A., van Rees, B., Weterman, M., Ristimaki, A., Offerhaus, J., and Knuutila, S. 2003. DNA copy number changes in young gastric cancer patients with special reference to chromosome 19. Br J Cancer, 88(12): 1914–9.

Varley, J. M., McGown, G., Thorncroft, M., Tricker, K. J., Teare, M. D., Santibanez-Koref, M. F., Martin, J., Birch, J. M., and Evans, D. G. 1995. An extended Li-Fraumeni kindred with gastric carcinoma and a codon 175 mutation in TP53. J Med Genet, 32(12): 942–5.

Videbaek, A., and Mosbech, J. 1954. The aetiology of gastric carcinoma elucidated by a study of 302 pedigrees. Acta Med Scand, 149(2): 137–59.

Vivanco, I., and Sawyers, C. L. 2002. The phosphatidylinositol 3-kinase AKT pathway in human cancer. Nature reviews. Cancer, 2(7): 489–501.

Wang, Z., Shen, D., Parsons, D. W., Bardelli, A., Sager, J., Szabo, S., Ptak, J., Silliman, N., Peters, B. A., van der Heijden, M. S., Parmigiani, G., Yan, H., Wang, T. L., Riggins, G., Powell, S. M., Willson, J. K., Markowitz, S., Kinzler, K. W., Vogelstein, B., and Velculescu, V. E. 2004. Mutational analysis of the tyrosine phosphatome in colorectal cancers. Science, 304(5674): 1164–6.

Weiss, M. M., Kuipers, E. J., Postma, C., Snijders, A. M., Pinkel, D., Meuwissen, S. G., Albertson, D., and Meijer, G. A. 2004. Genomic alterations in primary gastric adenocarcinomas correlate with clinicopathological characteristics and survival. Cell Oncol, 26(5–6): 307–17.

Wong, D. J., Barrett, M. T., Stoger, R., Emond, M. J., and Reid, B. J. 1997. p16INK4a promoter is hypermethylated at a high frequency in esophageal adenocarcinomas. Cancer Res, 57(13): 2619–22.

Wong, S. C., Yu, H., and So, J. B. 2006. Detection of telomerase activity in gastric lavage fluid: a novel method to detect gastric cancer. J Surg Res, 131(2): 252–5.

Woolf, C. M., and Isaacson, E. A. 1961. An analysis of 5 "stomach cancer families" in the state of Utah. Cancer, 14: 1005–16.

Wu, C. W., Lin, Y. Y., Chen, G. D., Chi, C. W., Carbone, D. P., and Chen, J. Y. 1999. Serum anti-p53 antibodies in gastric adenocarcinoma patients are associated with poor prognosis, lymph node metastasis and poorly differentiated nuclear grade. Br J Cancer, 80(3–4): 483–8.

Wu, M. S., Lee, C. W., Shun, C. T., Wang, H. P., Lee, W. J., Sheu, J. C., and Lin, J. T. 1998. Clinicopathological significance of altered loci of replication error and microsatellite instability-associated mutations in gastric cancer. Cancer Res, 58(7): 1494–7.

Wu, M. S., Shun, C. T., Wang, H. P., Lee, W. J., Wang, T. H., and Lin, J. T. 1998. Loss of pS2 protein expression is an early event of intestinal-type gastric cancer. Jpn J Cancer Res 89(3): 278–82.

Wu, C. Y., Wu, M. S., Chiang, E. P., Wu, C. C., Chen, Y. J., Chen, C. J., Chi, N. H., Chen, G. H., and Lin, J. T. 2007. Elevated plasma osteopontin associated with gastric cancer development, invasion and survival. Gut, 56(6): 782–9.

Xiangming, C., Hokita, S., Natsugoe, S., Tanabe, G., Baba, M., Takao, S., Kuroshima, K., and Aikou, T. 1999. Cooccurrence of reduced expression of alpha-catenin and overexpression of p53 is a predictor of lymph node metastasis in early gastric cancer. Oncology, 57(2): 131–7.

Yamamichi, K., Uehara, Y., Kitamura, N., Nakane, Y., and Hioki, K. 1998. Increased expression of CD44v6 mRNA significantly correlates with distant metastasis and poor prognosis in gastric cancer. Int J Cancer, 79(3): 256–62.

Yamamoto, H., Itoh, F., Fukushima, H., Adachi, Y., Itoh, H., Hinoda, Y., and Imai, K. 1999a. Frequent bax frameshift mutations in gastric cancer with high but not low microsatellite instability. J Exp Clin Cancer Res, 18(1): 103–6.

Yamamoto, H., Perez-Piteira, J., Yoshida, T., Terada, M., Itoh, F., Imai, K., and Perucho, M. 1999b. Gastric cancers of the microsatellite mutator phenotype display characteristic genetic and clinical features. Gastroenterology, 116(6): 1348–57.

Yamamoto, H., Sawai, H., and Perucho, M. 1997. Frameshift somatic mutations in gastrointestinal cancer of the microsatellite mutator phenotype. Cancer Res, 57(19): 4420–6.

Yasunaga, Y., Shinomura, Y., Kanayama, S., Higashimoto, Y., Yabu, M., Miyazaki, Y., Kondo, S., Murayama, Y., Nishibayashi, H., Kitamura, S., and Matsuzawa, Y. 1996. Increased production of interleukin 1 beta and hepatocyte growth factor may contribute to foveolar hyperplasia in enlarged fold gastritis. Gut, 39(96): 787–94.

Yin, J., Kong, D., Wang, S., Zou, T. T., Souza, R. F., Smolinski, K. N., Lynch, P. M., Hamilton, S. R., Sugimura, H., Powell, S. M., Young, J., Abraham, J. M., and Meltzer, S. J. 1997. Mutation of hMSH3 and hMSH6 mismatch repair genes in genetically unstable human colorectal and gastric carcinomas. Human Mutat, 10(6): 474–8.

Yonemura, Y., Ninomiya, I., Yamaguchi, A., Fushida, S., Kimura, H., Ohoyama, S., Miyazaki, I., Endou, Y., Tanaka, M., and Sasaki, T. 1991. Evaluation of immunoreactivity for erbB-2 protein as a marker of poor short term prognosis in gastric cancer. Cancer Res, 51(3): 1034–8.

Yoon, K. A., Ku, J. L., Yang, H. K., Kim, W. H., Park, S. Y., and Park, J. G. 1999. Germline mutations of E-cadherin gene in Korean familial gastric cancer patients. J Hum Genet, 44(3): 177–80.

Yoshikawa, T., Saitoh, M., Tsuburaya, A., Kobayashi, O., Sairenji, M., Motohashi, H., Yanoma, S., and Noguchi, Y. 1999. Tissue inhibitor of matrix metalloproteinase-1 in the plasma of patients with gastric carcinoma. A possible marker for serosal invasion and metastasis. Cancer, 86(10): 1929–35.

Yustein, A. S., Harper, J. C., Petroni, G. R., Cummings, O. W., Moskaluk, C. A., and Powell, S. M. 1999. Allelotype of gastric adenocarcinoma. Cancer Res, 59(7): 1437–41.

Zafirellis, K., Karameris, A., Milingos, N., and Androulakis, G. 2005. Molecular markers in gastric cancer: can p53 and bcl-2 protein expressions be used as prognostic factors? Anticancer Res, 25(5): 3629–36.

Zanghieri, G., Di Gregorio, C., Sacchetti, C., Fante, R., Sassatelli, R., Cannizzo, G., Carriero, A., and Ponz de Leon, M. 1990. Familial occurrence of gastric cancer in the 2-year experience of a population-based registry. Cancer, 66(9): 2047–51.

Zhang, Y., Liu, X., Fan, Y., Ding, J., Xu, A., Zhou, X., Hu, X., Zhu, M., Zhang, X., Li, S., Wu, J., Cao, H., Li, J., and Wang, Y. 2006. Germline mutations and polymorphic variants in MMR, E-cadherin and MYH genes associated with familial gastric cancer in Jiangsu of China. Int J Cancer, 119(11): 2592–6.

Zou, B., Chim, C. S., Zeng, H., Leung, S. Y., Yang, Y., Tu, S. P., Lin, M. C., Wang, J., He, H., Jiang, S. H., Sun, Y. W., Yu, L. F., Yuen, S. T., Kung, H. F., and Wong, B. C. 2006. Correlation between the single-site CpG methylation and expression silencing of the XAF1 gene in human gastric and colon cancers. Gastroenterology, 131(6): 1835–43.

Chapter 11
Genomic and Proteomic Advances in Gastric Cancer

Alex Boussioutas and Patrick Tan

Introduction

The "omic" revolution has affected every discipline in medicine and the life sciences. Gastric cancer is no exception to this phenomenon, with significant publications in genomics, proteomics, transcriptomics, and metabolomics appearing in the literature over recent years. In this chapter, we focus on some of the major advances in gastric cancer research uncovered by the availability of new technologies in the areas of genomics and proteomics.

We are cognizant that there is no easier way to date a written piece of work than to write about novel technologies. Indeed, the pace of technology at this time makes it inevitable that by the time of publication the technology has moved forward. Our objective in this chapter is to highlight the principles of existing technologies and concentrate on how these have been used to advance our understanding or management of gastric cancer. Given the disparate technologies that are discussed, we have divided the chapter into two broad sections. The first concentrates on genomics and high-throughput nucleic acid-based technologies, whereas the second deals with proteomics and the large-scale measurements of proteins or peptides.

Although these high-throughput technologies have significant advantages over traditional discovery platforms that have historically adopted a single gene or protein approach, a common issue faced by both platforms is how best to interpret the large amounts of data generated by these studies. This has led to the evolution of a pivotal specialist in this area termed the bioinformatician. This term has different meanings according to the context of use and will be discussed further in the section "Bioinformatic Analysis of Genomic Data."

The molecular biology revolution of the 1970s and beyond has led to a cell-centric view of cancer in which oncogenes and tumor suppressor genes operating in a single cell are thought to be deterministic of cancer. Another facet of malignancy is the reemerging concept of cancer that was postulated in the nineteenth century by Paget which also included the dynamic interplay between a malignant epithelial cell and the tumor microenvironment (Paget 1889). The advent of high-throughput technologies has led to an evolving holistic view of cancer by allowing the simultaneous evaluation of thousands of products (genes or proteins) derived from multicellular

T.C. Wang et al. (eds.) *The Biology of Gastric Cancers*,
© Springer Science+Business Media, LLC 2009

tissues. The complexity of this relationship has become apparent given the multitude of gene or protein expression changes observed in cancer tissue. The challenge has been to determine which cells are expressing the molecules of interest to enable downstream experiments. These concepts are expanded in subsequent sections of the chapter and hopefully will enable an understanding of the current promises and limitations of many of these technologies with respect to gastric cancer.

Genomics

A Summary of Available Genomic Technologies

Genomics, for the purposes of this discussion, refers to the comprehensive analysis of either genetic locus quantification or gene expression measurements by assessing relative amounts of DNA or RNA, respectively.

The ability to comprehensively study the entire human genome has been accelerated because of the enormous advances led by the large-scale sequencing effort of the Human Genome Project (Venter et al. 2001). It is important to clarify that, for the purposes of this chapter, the term "genetic data" refers to either relative amounts of genomic DNA, usually measured by a form of comparative genomic hybridization, or somatic mutations acquired by cancer cells during tumorigenesis. Although the sequencing of the Human Genome has been completed, the number of genetic loci that it encodes is still speculative (Peters et al. 2007); however, it is generally acknowledged that there are a finite number of gene loci, in the order of approximately 20,000 coding regions (Lander et al. 2001; Venter et al. 2001). Gene expression data refers to the transcriptome, which harbors the collective mRNA that can be generated from a cell. The transcriptome is possibly orders of magnitude more complex than the human genome given alternative splicing and other transcriptional control mechanisms. Gene expression by microarrays is another way of measuring these mRNA transcripts. A topic that is not discussed in any detail in this chapter is the evolving area of whole genome sequencing technology, which has been used to study various human diseases (Dahl et al. 2007; Greenman et al. 2007; Sjoblom et al. 2006). This is very new technology and to our knowledge has not yet been specifically applied to gastric cancer research.

All genomic technologies use a modification of the basic first principle of nucleic acid hybridization by complementary base pairing first utilized for DNA/DNA hybridization by Edwin Southern (Southern blotting). The hybridization is coupled with a method to quantify the amount of nucleic acid that hybridizes. The relative amount of hybridization from a given biologic sample can be measured either in absolute terms, or relative to a reference tissue allowing us to obtain normalized quantification of genes (See Figure 11.1, cDNA vs. oligo hybridization). The genomic technologies have progressed from single reactions such as Southern blots and allowed the creation of several hundred thousand reactions on a very small surface area.

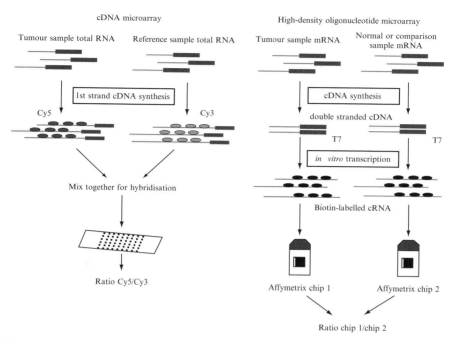

Fig. 11.1 Comparison of cDNA and oligonucleotide microarray platforms. **(A)** Oligonucleotide array using photolithographic techniques to create 25-mer oligo probes onto a chip at very high density. Diagram showing steps involved in hybridization of target to probes and then measurement and comparison of the signal. **(B)** cDNA microarray showing typical design of glass slide with ultraviolet cross-linked cDNA probes then hybridized with relative amounts of target mRNA from two different tissues to be compared

Early high-throughput platforms utilized low-density membrane arrays that had DNA probes spotted onto the membrane. Target cDNAs from biologic samples were hybridized to the membrane and the relative level of hybridization was measured by radiography or fluorescence. Advances in the 1990s, principally from Stanford University and led by Patrick Brown, enabled researchers to robotically attach tens of thousands of polymerase chain reaction (PCR)-based DNA probes to glass microscope slides and then use mixtures of RNA-derived cDNA to hybridize to these arrays. The amount of hybridized target cDNA, and hence level of expression of the gene to which the sequence was homologous, was measured by relative florescence using a confocal scanning laser (see Figure 11.1) (DeRisi et al. 1996). In this version of the DNA microarray, the cDNA technology required two channels: traditionally a Cy3 fluorophor and a Cy5 fluorophor, which are measured relative to each other.

The technology since those halcyon days has progressed with the creation of a whole industry with multiple different platforms that can be used to perform high-throughput genomic experiments. Affymetrix™ was an early leader in the field and adopted a method of photolithographically fixing 25-mer probes to small chips of

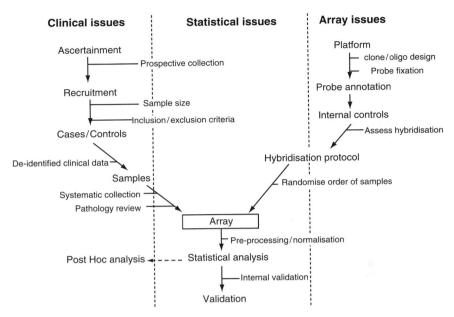

Fig. 11.2 Flowchart of concepts that should be considered when designing clinical-based array studies. The three areas (clinical issues, statistical issues, and array issues) overlap throughout the design process as each affects the other. Sample size, for instance, overlaps with clinical issues to determine the appropriate stratification of important subclasses and statistical issues to allow adequate power to make relevant findings

extremely high density (Lipshutz et al. 1999). This has allowed whole genome arrays to be created and has also opened the way toward tiling arrays, which are more comprehensive still. A notable difference between Affymetrix™ and cDNA platforms is shown in Figure 11.1 where the former does not require a reference RNA to compare the expression given its use of single-channel measures, whereas the latter requires a reference RNA as a comparator to calculate the relative amounts of RNA in the test tissue.

Design of Genomic Experiments

The design of a particular genomic experiment is highly dependent on the nature of the question asked (Fig. 11.2). The specific aim will guide all the aspects of the study design, including the type of tissue to collect and how many samples will be required. Following an appropriate study design is crucial, as the large scale of data that arises from these genomic studies makes them subject to background noise (inherent random or chance changes in gene expression), arising from the sum of all aspects of the study from processing of tissues (consistent collection/handling of tissues to minimize degradation of tissue and RNA/DNA quality) to hybridizing the RNA to a microarray (includes randomizing the order of hybridization of arrays to minimize systematic bias and performing all steps in the process consistently, ideally with the same operator). Thus, a critical aspect of the design that should be mentioned at the outset is the importance of good bioinformatics support in outlining the design of the study and overseeing the appropriate analysis of the data.

Many genomic studies have focused on questions of class discovery or comparison. These aims seek to gain insights about new classes of cancer with a specific gene-expression signature, or to compare existing pathologic classes to seek gene-expression differences, respectively. In clinical-based studies in which investigators hope to use human material, it is important to decide whether to use whole or microdissected tissue. This applies to both DNA-based (studying gene dosage in the genome) and RNA-based (measuring relative gene expression) studies. Historically, whole tissue was used in view of technical limitations of the micro arrays that were used. Improvements in quality of RNA extraction and amplification techniques have allowed microdissection of tissues where specific subtypes of tissues can be dissected by laser-capture devices or by simple microscopy and gross dissection (Bonner et al. 1997; Ohyama et al. 2000; Sgroi et al. 1999). This has enabled more homogeneous populations of cells to be profiled using genomic techniques. An important caveat to the use of amplified material is that it becomes more difficult to compare to unamplified material because of introduction of background signal (noise) into the system. Therefore, the first lesson for investigators is to choose one method (whole tissue or microdissection) and to use it consistently for any one particular study.

Another important aspect of study design for genomic experiments involving primary tissue samples is to ensure that all samples have appropriate pathology and clinical review. This is particularly important when clinical endpoints such as survival are being measured. For instance, the degree of stromal contamination that is included in the sample to be profiled and characteristics of the tumors are important to review because they may lead to heterogeneity of observed results. Other clinical aspects that need to be acknowledged include: whether tissue is taken from surgically resected specimens or endoscopy specimens because the processing of the tissues will be different; whether the patient had neoadjuvant chemotherapy before collection of the tissue because such treatments are likely to influence gene expression; and in the specific case of gastric cancer, the presence or absence of active *Helicobacter pylori* infection might also influence the gene expression profiles obtained. These factors are not exclusion criteria but they create more noise in an already noisy environment and mean that the sample size will need to increase to compensate for these recognizable confounders.

The type of platform used usually depends on budget and experience of the investigator. Most commercial platforms are now robust (with good reproducibility) and have good quality control measures to determine whether RNA/DNA quality is affecting results.

A final aspect of microarray experiments is the validation of the results. This takes a number of forms. First, there is technical validation whereby investigators strive to prove that the findings of gene expression changes reported by the micro-array are real. This usually involves another technology, such as real-time PCR, which can give information regarding the upregulation or downregulation of genes relative to a suitable reference RNA. Anatomic validation is an alternative, whereby investigators strive to determine which cell type expresses the gene of interest. This can be performed by immunohistochemistry (IHC) or *in situ* hybridization.

The other type of validation is biologic: that is, are the findings made in the microarray analysis real and generalizable to other populations? This is usually achieved by an independent analysis of samples, or by testing the initial result on an independent dataset. The key to this form of validation is having access to inde-pendent data. Another way of evaluating the validity of a test result is to artificially subgroup the dataset and randomly permute classes in an iterative process called cross-validation (Simon 2005; Simon et al. 2003).

Bioinformatic Analysis of Genomic Data

An important aspect of large-scale microarray experiments is bioinformatic analy-sis, which is critical to enable reporting of valid results. There are many statistically valid methods of analyzing large-scale datasets but they need to have been incor-porated in the initial design of the study to be effective. There are many bioinfor-matic reviews that address issues of design (Simon et al. 2002; Yang and Speed 2002) or analysis (Dobbin and Simon 2005) as well as the common pitfalls of genomic experiments that need to be factored when performing these types of studies. Herein, we summarize some of these issues.

The type of analysis that is required depends on the question being asked. One of the first questions of such complex data series is to ask whether there is order in the data, or put another way, Is the data divisible into discrete partitions based on gene expression? There are many ways to subdivide data visually and perhaps one of the most common informatic tools used is hierarchical clustering. This method-ology has been used for plotting phylogenetic trees and was first exploited for the analysis of microarrays by Eisen et al. (1998). Hierarchical clustering is a purely unsupervised algorithm, in which there are no *a priori* assumptions of the data. Its principal utility was to visually scale data and enable the understanding of patterns within gene expression information, and to identify associations between particular genes that were previously unexpected.

Since these early days of informatic analysis there have been a multitude of analytic programs that exploit traditional statistical algorithms such as Student t test or analysis of variance (ANOVA) to analyze these data. The principal difference with these types of data and more traditional biostatistics is the extraordinarily large number of experimental observations compared with the relatively small sample sizes of these datasets. For instance, it is not unusual to have an experiment whereby 10 samples are profiled with a microarray comprising 54,000 elements (probes representing genes). When analyzing these data, the rate of false positives in the analysis is very high and needs to have appropriate measures of false discovery rate (Reiner et al. 2003). There are several excellent reviews that discuss the pitfalls of microarray studies, many of which can be overcome by appropriate experimental design (Dobbin and Simon 2005; Simon et al. 2003).

A summary of important aspects of design for a microarray experiment should include:

1. Formulating an appropriate aim that can be answered from microarray data
2. Ensuring excellent collection of tissues that is consistent throughout the life of the project to avoid bias. An essential element is the pathologic review of these tissues to ensure accurate annotation of samples.
3. Deciding on a sample size that would enable Point 1. In scenarios in which scientists are looking for new subgroups of cancer, this is admittedly difficult to predict because the prevalence and recognition of the subgroup is unknown and hence a power calculation would only be based on assumption. However, where endpoints are clear, such as prediction of response to chemotherapy, the sample size would be based on the measurement of that endpoint, which may enable sample-size estimation.
4. Ensuring consistent technical proficiency to avoid systematic bias. This would include ensuring that a single microarray platform is used and that internal variation such as batch to batch variation of microarrays is kept to a minimum.
5. Independent validation of results. In class prediction analyses, such as predicting outcome of a particular class of tumor (see next section), there should be inclusion of independent analysis of the gene predictor on independent cases. Another form of validation would be to perform targeted *in vitro* experiments as predicted by the microarray results.

Limitations of DNA Microarray Experiments

Many of the experimental studies utilizing microarrays have been discovery based and hypothesis generating more than hypothesis testing. The design of these studies in gastric cancer has generally been descriptive but is no different from other tumor types. One of the areas in which microarrays have had the most promise is the correlation of clinical and molecular phenotype, but there are many caveats that not all researchers have heeded.

When conducting a microarray experiment, there needs to be appropriate statistical rigor to ensure that the results obtained are real and not based on chance. The biggest issue is that of multiple hypothesis testing. This results because, when 50,000 transcripts are tested for gene expression level, there are bound to be findings by chance alone. If we were to use a traditional measure of significance such as p value <0.05, then 2,500 false positives will arise. These experiments need to have the number of variables factored into the analysis by multiple testing corrections. Another issue that arises when performing class prediction analysis is overfitting. This occurs when one tries to predict an outcome (e.g., survival) in one dataset and validate on an independent dataset, only to find that the class prediction fails. This is because the number of cases (array experiments) is much lower than the number of variables (numbers of genes/transcripts on the array) tested. The result is predictive algorithms tend to overfit the data to suit a prediction outcome. If the gene set is then used to test an independent cohort, it invariably performs worse because the algorithm has been optimized on the original dataset. Overfitting occurs because small random changes in the data are used to optimize the classifier or prediction algorithm which are not found or not the same in the independent cohort. The way to control for this effect is to reduce the bias in the system by incorporating good estimates of the error rate of finding chance events. One way of testing this is using permutation analysis and leave-one-out cross-validation. The fundamental rule of microarray experiments is to try to validate the result wherever possible. This can be done in a permutation analysis by cross-validation whereby one sample is left out of the analysis and the prediction rule tested and then permuted until all samples are left out. A more stringent validation is to test the result on independent data. There are several excellent reviews of the analysis and problems associated with microarray experiments (Simon et al. 2003) and clinical correlations (Tinker et al. 2006).

Genomic Advancements in Clinical and Molecular Correlations in Gastric Cancer

Researchers have applied this technology to clinical samples obtained either by endoscopic biopsy or from surgical resection specimens. The first question researchers wanted to answer was: What is different about gastric cancer compared with normal gastric epithelium?

Distinguishing Cancer from Noncancer

The most frequently reported observation from gene expression studies has concerned the genes that are differentially expressed between gastric tumors and their adjacent nonneoplastic tissue. There are usually considerable numbers of genes that are listed as either upregulated in tumors compared with normal tissue or downregulated in tumors.

A number of microarray studies have examined the issue of tumor-specific gene expression (Boussioutas et al. 2003; Chen et al. 2003; Hasegawa et al. 2002; Hippo et al. 2002; Kim et al. 2003; Tay et al. 2003). Each study has generated lists of genes that were found to be differentially expressed by gastric cancers compared with adjacent "normal" tissue. The "normal" tissue is usually obtained from adjacent gastric mucosa that has no microscopic evidence of malignancy although often has preneoplastic changes such as intestinal metaplasia or chronic gastritis (Boussioutas et al. 2003).

Genes consistently found associated with gastric cancer include those involved with cell proliferation, cell adhesion, extracellular matrix remodeling, DNA repair and apoptosis (Boussioutas et al. 2003; Chen et al. 2003; Hasegawa et al. 2002; Hippo et al. 2002; Kim et al. 2003; Tay et al. 2003). Many of these gene findings are not surprising given the nature of cancer; however, every list serves as a partial validation of others and the functional groupings listed above are overrepresented when considering all the studies. Interestingly, the gene lists that distinguish cancer from noncancer tissue generated by different research groups are only partially overlapping. There are many genes that are not shared on the lists and this can result from many factors, principally, 1. the use of different array platforms by different laboratories, which means different genes on the arrays; 2. the type of analysis used in each laboratory; and 3. methodologic differences in tissue and array processing. However, when using thematic approaches to the functional annotation of the gene lists, there are remarkable similarities and significant conservation of gene ontology in assigned groupings (Subramanian et al. 2005).

Examples of genes that are consistently reported in the microarray literature as upregulated in gastric cancer are Secreted Protein Acidic and Rich in Cysteine (*SPARC* or *Osteonectin*) and *Fibronectin* (*FN*). *SPARC* is thought to modulate cellular interactions with the extracellular matrix and has been reported to have promigratory effects (Brekken and Sage 2001) and to be linked to angiogenesis (Iruela-Arispe et al. 1991; Jendraschak and Sage 1996; Lau et al. 2006). Studies in gastric cancer have suggested SPARC is a negative prognostic factor (Wang et al. 2004), and studies of prostate cancer have linked *SPARC* expression to metastasis (Chen et al. 2007). Interestingly, there has been an association of *SPARC* and *FN*—these genes are often observed as coexpressed in microarray experiments in gastric cancer (Boussioutas et al. 2003; Jinawath et al. 2004).

There have also been consistently downregulated genes in gastric cancer that are in concordance with published literature showing they could act as tumor suppressor genes. Perhaps the best example of this is the consistent finding that trefoil factor 1 (TFF1) is found to be downregulated in many datasets (Boussioutas et al. 2003; Chen et al. 2003; Jinawath et al. 2004). *Tff1* knockout mice were shown to develop gastric tumors (Lefebvre et al. 1996), which is consistent with the microarray findings of reduced expression specifically in cancers.

The majority of gastric cancer is postulated to arise from premalignant lesions that were noted to be associated with cancer in epidemiologic studies first described by Correa (Correa et al. 1975). For example, *H. pylori* colonization of susceptible hosts is thought to incite mucosal changes resulting in atrophic gastritis and intestinal metaplasia (Correa 1992; Craanen et al. 1992; Honda et al. 1998). These are thought to be precursor lesions to the majority of sporadic gastric cancers.

The study by Boussioutas et al. (2003) was one of the first comprehensive examinations of nonneoplastic lesions associated with the Correa hypothesis. This study examined gene expression signatures that were specific to gastritis and intestinal metaplasia, exhibiting well recognized markers of gastric inflammation and intestinal differentiation, respectively. The latter was characterized by genes that are specifically expressed in intestine rather than gastric mucosa (*CDX1*, *VILL1*, and *VIPR*) and this was independently corroborated by another expression study that had sufficient numbers of nonneoplastic tissues (Chen et al. 2003). Indeed, the study by Chen et al. examined the gene expression profiles of nonneoplastic stomach and compared these with normal small intestinal and colonic (large intestinal) mucosa. The authors found discrete subgroups of intestinal metaplasia of the stomach with similar gene expression to small intestinal mucosa, and another subgroup that was characterized by genes involved in intestinal differentiation that are expressed in small and large intestine, namely, *CDX1*, *CDX2*, and *HNF4A*.

Distinguishing Classes of Gastric Cancer

Microarray results have been very useful in the molecular classification of gastric cancers. It is well recognized that gastric cancer has several clinical classifications that are related to macroscopic morphology [e.g., Borrmann (Borrmann, 1926)] or microscopic morphology [e.g., Lauren classification (Lauren, 1965)]. Several studies investigated the putative molecular differences of the Lauren subgroups using gene expression profiling and found that diffuse-type gastric cancers are more likely to have extracellular components expressed—not surprising, given the close cellular association with the extracellular matrix (Boussioutas et al. 2003; Chen et al. 2003; Jinawath et al. 2004).

An interesting finding by Boussioutas et al. was the predominance of cellular proliferation, by overrepresentation of cell-cycle genes, in intestinal-type compared with diffuse-type gastric cancer. This was validated using IHC to measure expression of Ki67 protein, a marker of proliferation, on gastric tumors. This suggests that the biology of these two subgroups might have fundamental differences that belie their phenotypic differences. One explanation of the observed difference in rate of proliferation between intestinal and diffuse types may be attributable to a dilutional effect by less-proliferative stromal cells contaminating the diffuse-type cancers. Although this could contribute to some difference, it is unlikely to be the only explanation, because the gastric subtypes were matched for tumor content and, in another study that used laser capture microdissection of tumor cells, the same observations were made (Jinawath et al. 2004). One hypothetical explanation of these findings is that diffuse-type cells are more migratory and less proliferative, whereas intestinal-type cancers are less mobile but focus on cellular renewal. This hypothesis remains to be tested.

Although many studies reported molecular differences in existing pathologically identifiable subgroups, microarrays offered the ability to classify groups based solely on molecular differences. Chen et al. reported novel molecular subgroups of gastric cancer based on the molecular signature derived from microarrays (Chen et al. 2003). The intestinal differentiation signature genes described by these researchers were found in certain gastric cancers and these were termed "intestinal-like." Based on an expression signature comprising approximately 41 genes, the authors divided the 90 gastric cancers into "intestinal-like" and "gastric-like" tumors. This result suggests a different cellular derivation of the tumors in the two groups, which may explain the gene expression differences observed.

Making Clinical and Molecular Associations in Gastric Cancer

The promise of microarrays was to value-add to the existing pathologic and clinical definition of gastric cancer using molecular profiles. Several studies have reported clinical/molecular correlations in gastric cancer but these have not reached mainstream medicine with the same impact as in breast cancer, where there are algorithms utilizing molecular signatures to assign chemotherapy to particular cancer cohorts (van de Vijver et al. 2002; van't Veer et al. 2002). Many of the microarray studies in gastric cancer have focused on prognostication and correlation of gene expression to the stage of disease.

Leung et al. (2002) reported that *PLA2G2A* expression was inversely correlated with prognosis, whereby persons with high-expressing tumors had lower stage of disease and better outcomes compared with those who were "low expressors" of this gene. *PLA2G2A* expression was inversely proportional to stage of disease. This finding has intriguing implications in view of the putative relationship of *PLA2G2A* and arachidonic acid metabolism and the finding that PLA2G2A ameliorates the intestinal neoplasia phenotype in the *APC^(Min/+)* mouse model of carcinogenesis (Cormier et al. 1997).

In an extension of their study, Leung et al. performed Cox regression analysis on their microarray data and found that *CCL18* expression was correlated with poor 5-year survival. High expression of *CCL18* was a good prognostic factor for overall and disease-free survival, which was validated on an independent series of gastric cancers remaining significant after multivariate analysis (Leung et al. 2004).

In another study, investigators using a combination of cDNA microarrays and reverse transcriptase (RT)-PCR reported a model comprising three genes (*CD36, SLAM, PIM-1*) and based on logistic regression that could predict survival with a specificity of 73% and positive predictive value of 75% (Chen et al. 2005). Although these results were encouraging, the study was too small to allow definite conclusions. These results are awaiting validation in a prospective clinical trial before advocating real-time use of microarrays in predicting gastric pathology.

The utility of prognostication models, such as those reported above, would be for stratification of individuals who would benefit most from adjuvant therapy. Another manner of investigating the role of adjuvant chemotherapy is to identify genes that are involved in drug resistance. Kang et al. (2004) reported the use of a panel of cell lines resistant to the drugs 5-fluorouracil, doxorubicin, or cisplatin, and compared them with four drug-sensitive parental lines. The authors reported a number of differentially expressed genes in each of the drug groups with few overlapping genes. Eight genes were differentially expressed in more than one drug group. Most of the genes were involved in cell-cycle regulation, although *Survivin*, an antiapoptotic gene (Blanc-Brude et al. 2002), was found to be upregulated. Although this was a study conducted *in vitro* on transformed cell lines, it points toward certain candidate molecules that can potentially be used to molecularly target therapy in gastric cancer. Definitive *in vivo* studies are required because the elements of drug resistance in the clinical setting involve more than the innate sensitivity of the epithelial cell in isolation. The stroma, tumor vascularity, and immunologic reactions need to be considered when extrapolating these studies to the *in vivo* scenario. All of these factors contribute significantly to the efficacy of adjuvant therapy.

Clinical correlations of molecular profiles to the stage of disease are also frequently reported. This has been examined for invasion through the gastric wall (T stage) and lymphatic invasion (N stage). As discussed above, *PLA2G2A* expression levels are inversely correlated with overall stage of disease (Leung et al. 2002).

The finding of lymph node metastasis is a poor prognostic sign for gastric cancer, and is one of the clinical measures of the burden of disease that has been studied by a number of groups. Researchers have used DNA microarrays to investigate genes differentially expressed between node-positive gastric cancer and node-negative disease (Hasegawa et al. 2002; Hippo et al. 2002; Terashima et al. 2005).

In a different type of analysis, comparative genomic hybridization was used to predict lymph node status and survival (Weiss et al. 2003). BAC clones were affixed to glass slides, and hybridization of genomic DNA from samples was used to assess relative copy-number changes. The technique usually compares the copy-number changes to reference DNA, which is derived from circulating lymphocytes of healthy human males. The arrays used in the study were relatively low resolution compared with current standards but were used to assess chromosomal aberrations. The authors used ANOVA and leave-one-out cross-validation to assess variation between the tumors and found three predominant patterns of genomic aberration. The only significant association with these groups was nodal status. The research is weakened by the lack of correlation to the expression of genes that underlie the BAC clones used, which limits the conclusions one can make about the biologic importance of the regions of gain and loss reported.

Terashima et al. (2005) investigated 21 gastric cancer tissues and 6 normal gastric epithelia in a strategy designed to reduce the number of candidate genes. Differentially expressed genes between normal and tumor tissue were then used to determine specific expression in node-positive tumors. The number of genes that were differentially expressed between node-positive and node-negative disease was reduced to three. The researchers then validated *Maspin* expression by RT-PCR and IHC on gastric cancers and metastases in lymph nodes. There was little discussion of the mechanism of *Maspin* in metastasis, but it was recognized that it may act as a tumor suppressor gene in other cancers such as breast, colon, and prostate. One of the concerns of the analysis, which may diminish the reliability of the results, is the absence of correction for multiple testing that should be considered for microarray studies.

Advancements of Gastric Cancer Biology Using Genomics

When examining the role of genomics in the biology of gastric cancer, the experimental design requires the use of cell lines and *in vitro* techniques to determine mechanisms of disease. This section discusses some of the investigations of the biologic properties of gastric cancer, and aspects of the host–pathogen (specifically *H. pylori*) interaction that is paramount in the initiation of gastric cancer.

Investigating Metastasis Using Genomics

An early study used gastric cell lines to investigate differences in gene expression between highly metastatic cell lines and the parental line (Hippo et al. 2002). The researchers found 63 genes upregulated fourfold by a metastatic line characterized by peritoneal spread, and 23 genes upregulated fourfold by a metastatic cell line that metastasized to lymph nodes in an orthotopic transplant model. Thirty-two genes were differentially expressed by the peritoneal metastasizing cell line compared with 10 genes by the lymph node metastasizing cell line. Interestingly, the functional groupings of genes that were putatively involved in metastasis in these models included genes involved in immune response, cell growth, adhesion, and apoptosis. It is notable that some genes reported in this experiment that showed upregulation in more metastatic lines are inconsistent with much of the human gastric cancer literature. One example is the finding that *TFF1* was upregulated in more malignant cell lines, whereas *TFF1* has been suggested to be a tumor suppressor in gastric carcinogenesis (Boussioutas et al. 2003; Park et al. 2000; Tebbutt et al. 2002). This may be attributable to *innate differences* of *in vitro* monoculture experiments, which do not adequately model the influence of stromal components on gene expression of epithelial cells, compared with primary human cancers. This is one of the reasons that researchers study primary human cancers in preference to cell lines. It is important to note the contribution of *in vitro* studies in understanding gene function, but the limitations of such experimentation also need to be recognized.

Sakakura et al. (2002) also conducted a study of gastric cell lines and mechanisms of metastasis. They used RNA obtained from cell lines that were known to metastasize to peritoneum and compared RNA from SNU-1, which is a cell line derived from a primary gastric cancer that does not metastasize. They reported numerous differentially expressed genes (defined as twofold upregulated) and validated these by quantitative mRNA analysis. Specific functional involvement was not assessed. In agreement with previous studies, *CD44* was found to be overexpressed, as were novel genes such as *Desmoplakin, Occludin*, and *Caveolin-3*.

A surprising outcome in these studies was the relatively small gene lists generated by analysis of more than 20,000 probes on the microarray. This is partly attributable to the stringent expression cut-off used in the study by Hippo et al., but even moreso to the microarray platform that was used by Sakakura et al. The latter study used a spotted cDNA microarray that requires two channels to calculate relative gene expression (see the above section, "A Summary of Available Genomic Technologies"). Many two-channel microarray experiments use a reference of RNA derived from a pool of cancer cell lines because of the abundant source of material. The introduction of a cell line reference could be the reason that so few genes were differentially expressed in the study by Sakakura et al. The best way to determine differential expression between two samples is to compare the samples directly. This will result in the lowest variance within each group being compared. When the groups are compared via a common reference, the variance of each group comparison increases because it is being compared indirectly and thus increasing the background noise in the analysis. For a better explanation of relative merits of direct versus indirect comparison in microarray experiments, see the article by Yang and Speed (2002). Sakakura et al. detected fewer differentially expressed genes because there were fewer genes detectable above the noise, inherent in the design of the microarray experiment.

Another interesting aspect of microarray experiments was illustrated by the study by Hippo et al. (2002) They reported an upregulation of MHC class II molecules from cell lines that were considered more metastatic to lymph nodes, which was difficult to reconcile at the time. A feature of genomic research is the broad biology that can be presented in experiments. It is important to establish what "real" differential expression is, and what may be artifact caused by the comparison of two different tissues. Hippo et al. reported that MHC class II protein was not detected on the cell surface of these cells by flow cytometry, suggesting a spurious result from the microarray experiments.

Using Genomics to Understand Pathogenesis of *Helicobacter pylori*

It is widely accepted that *H. pylori* is a class I carcinogen. This was first reported by the International Agency for Research on Cancer in 1994 (International Agency for Research on Cancer 1994). The etiopathogenesis of this host–pathogen interaction is still a subject of considerable research and is thought to require specific host and bacterial genotypes. To determine pathogen-related genes in this interaction, genomic experiments have been designed to identify bacterial genes in *H. pylori* that are involved in human infection (Graham et al. 2002). Graham et al. isolated *H. pylori* from colonized human stomachs and compared the RNA extracted from these strains with that extracted from matched strains of *H. pylori* grown *in vitro*. Despite the methodologic limitations of the study, there were a number of genes differentially expressed, suggesting a gene signature characteristic of *H. pylori* infection. A list of 14 open reading frames (ORFs) were consistently found, with most encoding hypothetical proteins. Two ORFs were validated by RT-PCR in other humans infected with *H. pylori*: one encoded a hypothetical protein and the other was an ORF encoding sulfate permease. This is another example of the utility of genomics, in this case of a pathologic bacterium, to determine the basis of the pathogen–host interaction that is known to be important in gastric oncogenesis.

Using Genomics to Understand Host Pathogenesis from *Helicobacter pylori* Colonization

A very important aspect of research in this area is to determine host responses to *H. pylori* infection that would particularly lead to transformation of the mucosa to gastric cancer. DNA microarrays were useful in delineating the complexity of this host–pathogen interaction. A simple model of infection was used *in vitro* to try to elucidate this host response. AGS cells were infected with a variety of G27 strains of *H. pylori*, which differed by the presence of the key virulence factors *vacA*, *cagA*, and *PAI* [pathogenicity island which encodes a type IV secretion system (TFSS)] expression (Guillemin et al. 2002). Temporal expression differences in AGS cells cocultured with various strains of *H. pylori* were detected. The effect of coculture with wild-type strains was compared with strains deficient in putative pathogenic bacterial loci. The wild-type strain of *H. pylori* induced cytoskeletal changes in AGS cells ("hummingbird phenotype") as well as significant proinflammatory responses. Thus, regulators of actin cytoskeleton, such as Cdc42 effector protein 2, were reported as induced, and genes involved in neutrophil chemoattraction such as *GRO1* and *IL8* were upregulated, as well as genes involved in NF-κB signaling. Interestingly, expression changes of the ΔPAI strain were similar to mock infection, suggesting that *cag PAI* is important for pathogenesis. Another membrane array-based study also confirmed the importance of CagA (Bach et al. 2002). This well-conducted study examined the role of acute infection of *H. pylori* in an *in vitro* system. It addressed the relative pathogenicity of various *H. pylori* genotypes; however, long-term colonization may have different effects on intact human epithelial-stromal components. Importantly, the study showed acute temporal changes in gene expression that were *cagA* and *cagE* dependent which confirms the significance of the TFSS (Odenbreit et al. 2000) in *H. pylori* pathogenesis, because CagE is an important structural component of the TFSS.

Although *in vitro* studies have limitations (discussed above), they offer the possibility of examining cell-specific changes in gene expression in response to specific perturbations. A good example of this is the use of the NCI-60 cell lines to create molecular signatures after treatment with specific drugs. This information is available as a compendium of cell-specific (albeit in transformed lines) genetic responses to drugs that researchers can use as a reference, and may also enable interrogation of cancers using these signatures to predict response to drug treatment (Ross et al. 2000; Scherf et al. 2000). The human immune system obviously has an important role in the pathogenesis of *H. pylori* infection and possibly epithelial cell transformation. Dissecting the relative role of cellular components of tissues in this process is an ongoing research goal and requires appropriate model systems to study the complex interplay of epithelium, stroma, and the immune system.

Introducing Genomics-Based Systems Biology to Gastric Cancer

In contradistinction to the unicellular and unidimensional view of cancer provided by *in vitro* systems, there is an increasing interest in using complex mathematics to study gene interactions in cancer. Recent work has attempted to use mathematical models to deconvolute the interacting pathways in cancer using a systems biology approach (Stuart et al. 2003). This technique takes a step back from individual gene analysis to use pathway signatures and view how they interact to make up gastric cancer. Aggarwal et al. (2006) reported on a human gastrome (an attempt to describe the transcriptional program of gastric cancer), which is a series of interrelated molecular pathways seen consistently across independent cohorts of microarray experiments. This study is the first metaanalysis of gastric cancer transcription. The authors observed gene networks conserved across the different datasets and subnetworks that had strong biologic coherence in view of similar gene functional annotations within those networks. When the interaction was examined at an individual gene level, *PLA2G2A*, a secreted phospholipase, was found to have significant coexpression with *β-catenin* and *EphB2*, which are associated with the Wnt signaling pathway. The results suggest that PLA2G2A might have a role in modulating Wnt signaling in gastric cancer.

Conclusion

This section was a brief foray into the aspects of gastric cancer biology that have been influenced by microarray technology. Microarrays have offered a new perspective on cancer, showing the complexity of the cellular networks involved in distinguishing a malignant cell from its normal counterpart, as well as the complex interactions of the cancer cell in a tumor microenvironment. There are many challenges still facing biologists in the interpretation of microarray experiments and we have reviewed different types of analysis and outlined some of their strengths and weaknesses. The future offers exciting opportunities in applying new technologies and the ability to use smaller tissue samples to effectively study cancer. There is also advancement in the area of systems biology in which we hope to understand how genes work as part of biologic networks to influence phenotype.

Proteomics

Proteomics: A Brief Overview

The word "proteome" was first coined by Wilkins and colleagues in 1994. Originally used to refer to the total repertoire of proteins expressed in a cell or tissue under a set of conditions, the science of "proteomics" has expanded on this basic definition to encompass the study of cellular proteins at multiple levels, ranging from identifying individual constituents (single proteins), to how these constituents are altered (differential expression and posttranslational modifications), and how they interact with one another (protein–protein interactions). In recent years, there has been an explosion of interest in the application of proteomic methods to address a wide variety of problems in human health and disease, as evidenced by the growing numbers of primary reports and review articles on this subject. This interest has been fueled, at least in part, by the recognition that of all the biologic species (DNA, RNA, protein), proteins represent the best potential drug targets and are key determinants of biologic phenotype. Indeed, the majority of drugs in current clinical use are targeted toward proteins, and current drug-discovery processes in pharmaceutical companies are largely optimized toward identifying and developing lead compounds against "drugable" proteins such as kinases and other enzymes. This is particularly true in the area of molecularly targeted therapeutics—traztuzamab, a humanized antibody that targets the amplified ERBB2 tyrosine kinase receptor in cancers, and gefitinib, a small molecular inhibitor of the EGFR growth factor receptor, are classical examples. Beyond therapeutics, a second reason for the burgeoning interest in proteomics is that in the foreseeable future, proteins are likely to remain the preferred choice for developing diagnostic reagents and biomarker assays. Specifically, IHC has long been a mainstay of clinical pathology, and protocols for generating high-affinity antibodies to distinct protein epitopes are well established in the field. Once generated, such high-affinity antibodies can often be readily adapted for other diagnostic applications such as ELISA (enzyme-linked immunosorbent assay), which can be applied to plasma and serum. From a bench research perspective, proteins function as the key effectors of cellular function, and thus the systematic characterization of proteins associated with these processes is likely to provide significant insights into how key biologic processes are mediated and regulated. Although mRNA-based gene expression technologies can address some of these questions, there is a less-than-perfect correlation between gene and protein expression, particularly in eukaryotic cells, and particularly for noncell-structure–related proteins (Nishizuka et al. 2003). Thus, an expressed gene detected by an expression microarray platform may not effectively translate into a bona fide protein product, thereby necessitating the use of techniques that directly measure proteins.

Although the goal of proteomics and genomics is similar—achieving a comprehensive molecular characterization of all possible expressed molecules (in this case proteins)—there are significant differences between the two, particularly in the maturity of technology platforms. In general, genomic technologies, such as DNA sequencing and RNA-based expression profiling, are now considered relatively stable technologies. The main driver of this research is the development of assays that offer higher throughput and decreased cost. However, in proteomics, basic issues of technical reproducibility, proteome coverage, and protein identification remain important challenges in the field. The following sections highlight some of the complexities in proteomics.

Technical Challenges in Proteomics

Lack of High-Throughput Detection

As mentioned above, genomic technologies have relied on the high fidelity of DNA base pairing, allowing sequence-specific nucleic acid probes to unambiguously detect their cognate partners. This enables specific detection and quantitation of individual target sequences. In contrast, proteomic technologies do not have a platform that unambiguously identifies distinct proteins or peptides in a similar high-throughput manner.

Mass spectroscopy (MS) platforms typically rely on *de novo* protein identification, either through peptide mass fingerprinting (PMF) or peptide sequencing (see below), necessitating analysis of each protein and/or peptide independently of all other proteins/peptides. In the case of antibody arrays, although antibodies are clearly capable of binding to protein epitopes with high sensitivity and specificity, two major challenges in antibody creation include: 1. an initial requirement for the target protein or epitope to be expressed and purified as an immunogen, and 2. a downstream requirement of screening, testing, and optimization before an antibody is deemed suitable for use. Thus, it is currently not possible to produce, on a mass scale, collections of high-quality antibodies systematically designed to target large populations of distinct proteins. The use of recombinant antibodies has been proposed as one possible means to address this challenge (see below), although this technology is still under development.

Lack of Amplification Techniques

The realization that human tissue is limiting for many of these high-throughput experiments has led to a concerted effort to use less tissue for the various assays discussed. In genomics, amplification of the starting material is widely accepted practice. Examples of amplification techniques include the PCR to amplify specific regions (e.g., exons) in a genomic template for subsequent sequencing, or the use of *in vitro* transcription to amplify a starting mRNA population before microarray hybridization.

In proteomics, no such technique currently exists for amplifying either a complete proteome or a subset of individual proteins in a protein population. Thus, biologic samples in proteomics projects typically require relatively large amounts of protein, which can prove challenging when analyzing clinical and/or primary samples. A significant portion of proteomics research today is thus focused on developing techniques to analyze low-volume samples. One good example is the surface-enhanced laser desorption ionization–time-of-flight (SELDI-TOF) platform, which requires very little sample for analysis, but disadvantageously is unable to identify specific proteins (see § 1.3. (Engwegen et al. 2006)).

Increased Population Complexity and Dynamic Range

Beyond these technical challenges, perhaps the most striking aspect of proteomics lies in the complexity of the proteome in comparison to the genome. Specifically, whereas there are an estimated 20,000–30,000 gene loci in the human genome, the number of distinct proteins has been estimated to be more than 1,000,000. The reasons for this additional complexity are many, ranging from the presence of different gene splice variants to a variety of different types of posttranslational modifications including protein cleavage events, glycosylation, phosphorylation, and methylation. Proteome complexity is also further increased when one considers that proteins primarily function as multisubunit complexes, in which the individual protein partners are dependent on cellular state, cell compartment, and time.

To add to the complexity, proteins in a typical cellular population are expressed at extremely wide dynamic ranges. For example, in human serum, concentrations of individual proteins span more than 9 orders of magnitude, from 10 mg/mL to < 1 pg/mL (this is 10 orders of magnitude). Moreover, more than 90% of the total protein mass in serum is accounted for by few abundant proteins such as albumin, immunoglobulins, transferrin and α1-antitrypsin. This wide dynamic range poses significant challenges for the accurate detection and measurement of low-abundance proteins. Indeed, it has been estimated that only a small fraction of the entire proteome is typically surveyed in a regular proteomics experiment (Marko-Varga 2004). Therefore, to reduce the noise in the system, the design of proteomics experiments often involves some form of fractionation or separation to target the assay to biologically relevant components of the tissue to be studied. In the case of solid tumor tissue, which is composed of a wide variety of different cellular populations, including tumor cells, immune cells, and stromal cells (such as endothelial cells and fibroblasts), researchers have used microdissection to reduce the heterogeneity of the starting material and target particular cellular types. Other alternatives include immunodepletion strategies whereby multiple-antibody columns can be used to remove abundant proteins from a protein population, thereby raising the chances that low-abundance entities can be successfully detected (Pieper et al. 2003). Another strategy is to perform targeted protein preparations that focus on particular cell compartments of interest, such as nuclear and/or membrane fractions, or organelle fractions such as mitochondria (Mootha et al. 2003).

Proteomic Platforms

Two-Dimensional Polyacrylamide Gel Electrophoresis/Mass Spectroscopy

The vast majority of proteomic projects to date have used MS approaches to determine protein identity. Because MS identification is usually applied to single proteins or peptides, most MS platforms require the prior application of an independent separation step to resolve and distinguish individual proteins and peptide entities from one another, before the actual MS analysis. Over the past decade, the major workhorse of proteomics has been the two-dimensional polyacrylamide gel electrophoresis (2DE-PAGE)/MS platform, otherwise known at 2DE/MS. In 2DE/MS, protein lysates are first separated and resolved using 2DE-PAGE. In the first dimension, proteins are separated on the basis of their isoelectric (pI) charge using immobilized pH gradient gel strips, and in the second dimension, proteins are first separated according to their molecular weight and then a relative amount is quantified by protein stains such as Coomassie or silver stain. The final pattern of protein spot migration is then considered the "proteomic profile" of that sample. Depending on the selection of gel running conditions (e.g., percentage of acrylamide, pH conditions), different repertoires of proteins can thus be targeted and resolved. The generic protein stains mentioned are inexpensive but are limited by low sensitivity or low dynamic range. There has been significant research focused on developing newer protein stains for detecting low-abundance proteins and for more accurate protein quantitation (Wijte et al. 2006).

By comparing the protein migration patterns from different protein extracts (e.g., from normal and tumors), one is then able to identify protein spots that are differentially expressed. To determine the amino acid identity of these protein spots of interest, the relevant spots are excised, digested with trypsin, and subjected to MS analysis such as matrix-assisted laser desorption (MALDI)/TOF. During the MS procedure, the digested peptides are ionized and detected by the MS reader, resulting in a characteristic PMF comprising a pattern of peptides with distinct mass/charge (m/z) ratios that is unique to a particular protein. Using proteomic search programs such as MASCOT (Perkins et al. 1999), the PMFs of the different protein spots are then compared with computer databases of proteins that have been digested *in silico*. Frequently, one is able to match the PMF to a database entry and thus identify the protein.

Although 2DE/MS is a standard in the field, there are several limitations. These include issues of reproducibility—achieving robust results often requires several replicate 2D gels to be run at the same time. Preparing and running 2DE gels, particularly large ones, are also typically laborious and time-consuming tasks. Perhaps the most significant challenge of 2DE/MS is its low sensitivity compared with other, newer proteomics platforms. Most medium- to low-abundance proteins are difficult to detect using this platform and hydrophobic proteins are often not analyzed because of the difficulty of separating them by PAGE (Engwegen et al. 2006). Furthermore, the presence of posttranslational modifications can sometimes prevent a PMF from being correctly identified, although this can be resolved by newer MS/MS approaches (see below). Nevertheless, it must be remembered that, at present, only the 2DE platform is capable of visualizing different protein isoforms, because these typically appear as distinct spots on a gel. Thus, although this platform is being eclipsed by other platforms, it is likely that 2DE will still have an important role in proteomics research for the immediate future. Indeed, some of the newer technologies are analyzing the data in a manner similar to 2DE-PAGE analysis.

Liquid Chromatography/Tandem Mass Spectrometry

Surpassing 2DE/MS in sensitivity and comprehensiveness, liquid chromatography (LC)/MS/MS is fast becoming the technology standard in proteomics research. Conceptually similar to 2DE/MS, LC/MS/MS is also a two-step methodology whereby protein separation is performed before protein identification. The major difference, however, between the two platforms is that in LC/MS/MS, protein lysates are subjected to a predigestion step before undergoing separation, resulting in a highly complex starting population of distinct peptides derived from different proteins. Reflecting this pool of peptides, LC/MS/MS is sometimes referred to as "shotgun proteomics." Peptide separation is usually accomplished by high-performance liquid chromatography, whereby a variety of different biochemical variables can be used to either select or enrich for particular peptides of interest. Typical criteria are peptide charge, size, and hydrophobicity. Once separated, individual peptides are analyzed by MS. However, unlike 2DE/MS, for which multiple peptides corresponding to the same protein can be combined to generate a peptide-mass fingerprint, in LC/MS/MS, peptides constitute the basic unit of identity, and individual peptides are analyzed in isolation from other peptides. Peptide identification is usually achieved through two sequential MS steps. In the first step, the peptides are ionized (first MS) and peptides of a specific m/z ratio are selected. These selected peptides are then subjected to a second MS round where they are further fragmented. By studying this secondary fragmentation pattern, it is possible to infer the primary amino acid sequence of the peptide being analyzed.

Comparisons between 2DE/MS and LC/MS/MS have shown that the latter is superior in identifying overall numbers of proteins, and more importantly, proteins that are expressed in the low/moderate abundance range (Bae et al. 2003). However, one limitation of LC-MS/MS approaches is that, because proteins are digested into peptides before separation by LC, a significant amount of information regarding protein isoforms (e.g., truncated or cleaved proteins, or proteins with posttranslational modifications) is typically lost during the digestion procedure. Another significant concern in LC/MS/MS analysis is that, because protein identification is based on single peptides, some protein assignments are likely to be questionable unless independent peptides belonging to the same protein are also observed in the starting pool. More recently, LC/MS/MS approaches to identify differentially expressed proteins have been developed, the most popular of which is ICAT (isotope coded affinity tags) (Han et al. 2001). In ICAT, proteins from two cellular populations (e.g., drug treated vs. control) are chemical labeled with different atomic weights, digested and mixed together, and analyzed in the same LC/MS/MS run. Differentially expressed proteins can be detected by examining the mass spectra data for protein peaks of differing intensity (i.e., differential expression) that are separated by a characteristic size range associated with the different chemical labels.

Surface-Enhanced Laser Desorption Ionization–Time-of-Flight Platform

Another proteomics platform that has recently gained much attention is SELDI-TOF, a methodology fundamentally different from either 2DE/MS or LC/MS/MS. Protein lysates in SELDI-TOF are applied to "chips" whose surfaces have been chemically treated to demonstrate different biochemical or biologic affinities, including hydrophilic, hydrophobic, immobilized metal affinities, or biologic antigen binding. Once the sample has been applied, the different surfaces thus bind to and capture different repertoires of proteins based on the intrinsic affinities of the applied proteins. Nonbinding and/or weakly binding proteins are removed by washing, and the remaining bound proteins are laser desorbed and ionized for MS analysis. Because different mixtures of proteins are likely to bind to the chip, depending on the specific sample (and affinity chip) used, the SELDI analysis generates a characteristic pattern of masses comprising a unique m/z signature or "sample fingerprint" for the sample. Thus, in SELDI-TOF, the patterns of mass spectral peaks, rather than actual protein identifications, are used to differentiate patient samples from one another, such as cancer or normal. Excitement in this approach was primarily motivated by a high-profile study showing that this technique could identify protein peaks that are differentially expressed in serum from patients with ovarian cancer compared with healthy individuals (Petricoin et al. 2002), and the fact that SELDI-TOF analysis requires very low sample volumes (0.2 mL of serum). However, subsequent statistical analyses have raised significant doubts about the biologic robustness of these original reports and whether their findings might have been attributable to technical artifacts such as systematic differences between sera collected from cancer patients and controls (Baggerly et al. 2004). The major limitations of SELDI-TOF are thus its lack of reproducibility across different centers and the inability to identify the proteins generating a particular mass spectral peak of interest. SELDI-TOF has also been criticized for its poor sensitivity in detecting low-abundance peptides (Diamandis 2004), at concentrations less than 1 μg/mL (Diamandis 2003). New-generation instruments that couple the SELDI-TOF technology with tandem mass spectrometers may address this former problem, although such instruments are still not commonplace.

Antibody Arrays

Perhaps the closest proteomic technology to the microarray platforms often used in genomics is the antibody array. Here, antibodies to specific antigens are arrayed on a solid substrate, and protein lysates are incubated with the arrays. During the incubation period, antigen capture occurs, after which unbound proteins are washed away. Specific protein binding is then visualized by fluorescently labeling the protein lysates, and detection is achieved by a regular microarray scanner. Chief considerations in antibody arrays include issues with correctly immobilizing the antibodies to achieve efficient binding, and the problem of accessing large repertoires of antibodies that are both highly specific and offer similar on-chip performance characteristics. In this regard, the use of recombinant technologies may obviate this problem (see below).

Proteomic Applications in Gastric Cancer

Protein Alterations Associated with *Helicobacter pylori* Infection

Infections by *H. pylori* are a major risk factor for the development of gastric cancer, and *H. pylori* has been classified as a group 1 carcinogen (International Agency for Research on Cancer 1994). However, significantly more individuals are infected by

H. pylori than those that develop gastric pathology. Certain strains of *H. pylori* seem to harbor specific virulence factors, such as CagA, that are likely to be important for *H. pylori* pathogenesis (Censini et al. 1996) by establishing a chronic proinflammatory state in the gastric environment through upregulation of genes such as *COX-2* (Segal et al. 1999). Despite this, the underlying molecular mechanisms relating *H. pylori* infection to gastric tumorigenesis are still unclear, and there is thus considerable interest in achieving a better scientific understanding of this complex host–pathogen interaction.

Researchers in several studies have attempted to characterize the protein alterations occurring in gastric epithelia after infection by *H. pylori in vitro*. Chan et al. (2006) infected AGS gastric cancer cells with *H. pylori* and compared the cytosolic protein fraction of noninfected and *H. pylori*–infected AGS cells. The cytosolic fraction was deliberately used because the membrane fraction needed to be discarded before analysis to avoid contaminating the host proteome with *H. pylori* proteins. In this study, the authors used a 2DE separation technique called fluorescence 2D difference gel electrophoresis (DIGE). In DIGE, two protein samples are independently labeled with different fluorescent dyes (Cy3 and Cy5), and run on the same 2DE-PAGE gel, often with a third internal mixed standard labeled with Cy2. The 2D gel is scanned using a fluorescent gel reader, and differentially expressed protein spots can be identified by their altered Cy3/Cy5 fluorescent ratios. Using this approach, the authors identified 28 significantly and reproducibly altered protein spots associated with *H. pylori* infection. After spot excision and in-gel digestion, the spots were identified using nano-LC/MS/MS spectrometry and found to associate with a variety of cellular processes including cytoskeleton proteins, metabolic enzymes, and gene transcription. Interestingly, although the authors had confirmed the upregulation of COX-2 by Western blotting in infected cells, COX-2 was not identified as a differentially regulated protein in the proteomic analysis. The authors then selected eight proteins that were differentially upregulated upon *H. pylori* infection (laminin γ-1, VCP, HSP70, MMP-P1, 14-3-3-β, FKBP4, TCP1α, enolase A) and found that, by both immunoblotting and IHC, several of these proteins were upregulated in gastric cancers compared with normal tissues. This study raises the possibility that some proteins that are directly upregulated in gastric tissues upon *H. pylori* infection may have a role in gastric carcinogenesis.

In a related study, Ellmark et al. (2006) used antibody microarrays to better understand the host immune response engendered upon infection of stomach tissues with *H. pylori*. From a technology standpoint, this study is notable because the antibodies used were single framework recombinant antibody fragments that can be produced in large quantities and share similar levels of on-chip performance of antigen binding and specificity. In contrast to the previous study, the authors did not use an *in vitro* system but compared primary surgical samples from three populations: *H. pylori*–positive gastric adenocarcinoma samples; *H. pylori*–positive normal gastric samples, obtained as strips of tumor-free tissue 5 cm from the tumor; and *H. pylori*–negative normal gastric tissue samples obtained from patients undergoing pancreatic cancer surgery. Protein lysates from these primary samples were prepared, biotinylated, and incubated with the antibody microarrays containing 127 different antibodies targeted against 60 different immunoregulatory antigens (cytokines, chemokines, and complement factors). Here, one to four antibodies were used to detect each antigen, to avoid overdependence on any one single epitope. After washing the arrays to remove unbound proteins, the bound proteins were fluorescently visualized by incubating the array in Alexa Fluor 647 conjugated streptavidin. By comparing the *H. pylori* +ve normal tissue to the normal *H. pylori* samples, the authors were able to discern an "infection signature" of 14 significantly differentially expressed antigens that were upregulated in infected tissue, including typical TH2 cytokines such as interleukin (IL)-5, IL-6 and IL-13, and the TH1 cytokines interferon-γ and IL-2. This finding is consistent with previous reports demonstrating an enrichment of regulatory T cells in *H. pylori*–infected stomach tissue (Lundgren et al. 2005).

A more dramatic immune response was observed, however, by comparing *H. pylori* + gastric cancers to *H. pylori* normal tissues. The authors identified 30 differentially expressed antigens that were largely upregulated in tumors, including many components of the previously described "infection signature" plus additional immunomodulatory proteins including the chemokines MCP1, MCP3, MCP4, and RANTES.

These results suggest that even in comparison to *H. pylori*–positive normal tissue, *H. pylori*–positive gastric tumors may be undergoing more pronounced inflammation. Thus, these results have shown that *H. pylori*–infected normal gastric mucosa and tumors are associated with distinct immuno-proteomic profiles, and they also identified potential biomarkers for both *H. pylori* infection and gastric adenocarcinoma. A potential criticism of this report is that the authors did not control for inflammation in the stomach by including samples of inflamed gastric mucosa that were *H. pylori* negative, raising the possibility that some of their proteomic profiles may have been the result of gastritis (which could be mediated by chemical ingestion, etc.) rather than *H. pylori* infection.

Protein Alterations Between Nonmalignant and Malignant Gastric Tissues

An obvious application of proteomic technologies to gastric cancer is comparing the proteomes of nonmalignant and malignant tissues. Identifying the repertoire of proteins that are aberrantly expressed in gastric tumors might lead to identifying particular cellular pathways that are dysregulated during carcinogenesis, and also identifying potential biomarkers for early diagnosis and potential drug targets.

In an early report, He et al. (2004) used a 2DE/MS approach to compare the proteomes of 10 gastric tumors to adjacent-matched nonmalignant gastric mucosae. By comparing protein spot migration patterns on silver-stained gels, differentially expressed proteins were identified and characterized by MALDI-TOF MS peptide mass fingerprinting. The authors detected upregulation in tumors of several cytoskeletal components (cytokeratin 8, tropomyosin isoform), glycolytic enzymes, and alterations in several stress and acute-phase related proteins. Interestingly, they found that expression of the acute-phase protein α-1 antitrypsin was strongly suppressed in gastric cancers, and confirmed this by subsequent 2D Western blotting. The identification of α1-AT as a downregulated protein in gastric cancers is particularly intriguing because of the parallel studies, described further in this chapter, addressing the expression of this protein in gastric juice and serum (see below). Using a very similar 2DE/MS approach, Nishigaki et al. (2005) compared the proteomes of 14 gastric tumor/normal matched tissues. Here, the investigators gave special attention to sampling the tumor cores of the cancers in an attempt to avoid contaminating noncancerous tissue, and to limit their sampling of normal tissues to surface epithelium for normal tissues. Possibly necrotic tumor cores, and whether the cells were truly free of stromal contaminants, particularly in cases of diffuse-type gastric cancer, were potential concerns. They found only 22 protein spots that were consistently altered between gastric tumors and normal tissue (nine upregulated and 22 downregulated in tumors) (Figure 11.3). Supporting the reliability of the proteomic analysis, several of these proteins were independently reported as differentially regulated proteins in gastric cancer, including CA11/GSK1 and HSP27 (Kapranos et al. 2002; Shiozaki et al. 2001). The authors subsequently showed that the mitotic checkpoint protein MAD1L1 was indeed significantly downregulated in

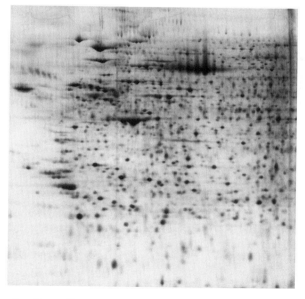

Fig. 11.3 Example of two-dimensional–polyacrylamide gel electrophoresis (2D-PAGE) gels showing differentially expressed spots between human gastric carcinoma and adjacent normal tissues. Proteins were separated by isoelectric (p*I*) (*x*-axis) and molecular mass (*y*-axis) and visualized by silver staining. Proteins identified by matrix-assisted laser desorption–time-of-flight mass spectroscopy (MALDI-TOF MS) as differentially expressed between cancer and normal tissues are shown as symbols on the gels. (Adapted from Nishigaki et al. 2005, with permission from the publisher.)

gastric cancers by IHC, suggesting that the loss of this protein may contribute to the chromosomal instability often associated with gastric cancers. Interestingly, despite their similar experimental designs, there was little overlap in the repertoire of proteins identified in these two studies, showcasing the difficulties in technical reproducibility of these platforms between different centers, which may be attributable to differences in distinct patient populations, sampling and technical protocols.

Contrary to these studies that have focused on the analysis of whole tumor and normal specimens, other researchers have attempted to perform proteomic analysis on a more homogeneous cellular population. Ebert et al. (2005) performed an antibody-based enrichment procedure to select for epithelial cells. Gastric cancers and corresponding normal mucosa were harvested immediately after surgery, homogenized, and incubated with magnetic beads linked to an epithelial-specific BerEP4 antibody which binds to the epithelial cell surface antigen EpCAM. The purified epithelial cell populations from both tumor and normal mucosae were then separated and analyzed by 2DE/MS across two different pH ranges (3–6 and 5–8). The authors identified approximately 200 differentially expressed protein spots between tumors and normals. Not surprisingly, there was once again little overlap in the identification of differentially regulated proteins between this and previous studies, with the exception of CA11 (GSK1) as a downregulated protein in gastric cancer. Nevertheless, this study is notable for the identification of cathepsin B as an overexpressed protein in gastric cancer, which was further confirmed by immunostaining of tumor and normal sections. The authors subsequently performed ELISA on serum samples from gastric cancer patients and found that levels of circulating cathepsin B were significantly elevated in patients with gastric cancer compared with normal controls. Furthermore, within cancer patients, elevated cathepsin B serum levels were associated with higher tumor stage and the presence of distant metastasis. Not surprisingly, given these associations, gastric cancer patients with higher levels of circulating cathepsin B ($>129 \, \text{pmol/L}$) experienced shorter survival times than gastric cancer patients with decreased cathepsin B levels ($<129 \, \text{pmol/L}$). This study exemplifies the utility of an initial proteomic analysis of primary tissues for the purposes of biomarker discovery and application of their findings on serum samples to test its use as a clinical marker of disease.

A major challenge of 2DE separation technologies in proteomic analysis is the amount of biologic material required for effective separation and subsequent downstream analysis (typically $100 \, \mu g$). If a precise demarcation of epithelial and nonepithelial components is desired, this can best be achieved by techniques such as laser capture micro dissection (LCM). However, LCM typically yields insufficient amounts of tissue (3000–5000 cells) for 2DE analysis. To address this problem, Melle et al. (2005) combined the use of LCM-dissected gastric epithelial cells with SELDI-TOF MS to identify differentially expressed proteins between tumors and normal tissues. Using Ciphergen® arrays, the authors identified a specific 24.5-kDa peak that was downregulated in microdissected gastric tumors compared with normal gastric mucosa. This discriminating protein was then identified, by correlating its molecular weight to standard 2DE-separated whole primary samples, to be pepsinogen C, which was previously investigated as a potential diagnostic biomarker for atrophic gastritis and gastric cancer. The reidentification of pepsinogen C in the SELDI-TOF–based study provides reassurance that this technique can indeed identify differentially expressed protein in gastric cancer. However, it remains to be seen if this technique can be universally applied to identify novel, as opposed to previously identified, biomarker proteins.

Comparisons Between Metastatic and Nonmetastatic Gastric Cancers

A major reason for the high mortality rate of gastric cancer is metastasis to several organs, including the peritoneum and lymph nodes. At present, the molecular mechanisms underlying gastric cancer metastasis are unclear, but are likely to involve a complex series of events ranging from loss of tumor cell adhesion in the primary tumor, intravasation and extravasation via the circulation or lymphatics to distinct sites, and finally invasion and proliferation at distant organs (Gupta and Massague 2006).

Several studies have used proteomic approaches to gain insight into the molecular mechanisms of gastric cancer metastasis. For example, Chen et al. (2004) reported a proteomic analysis of normal gastric tissues, gastric cancers, and metastasized tumors in a Wistar rat model. Gastric cancers were chemically induced in these rats by oral administration of the carcinogen N-methyl N-nitro-N-nitrosoguanidine, which causes intestinal metaplasia and eventual adenocarcinoma in these animals. The authors analyzed these tissues using a 2DE/MS approach, and identified 12 and 8 proteins that were, respectively, up- or downregulated in metastatic tumors compared with gastric tumors. However, one such protein (HSP27), identified in this rat model as being downregulated in metastatic tumors, was subsequently revealed by IHC analysis to be expressed at similar levels in both primary gastric cancers and lymph node metastasis from humans. This result, which contrasts with the animal model findings, could be attributed to multiple reasons, such as differences in the specific site of metastases studied in the animal model (lung and liver) and humans (lymph node), differences in the sensitivity and specificity of the antibody used for HSP27 IHC, or that the process of metastasis in rats may differ from humans. Interestingly, another notable gene that was downregulated in the rat gastric metastasis was *PTEN*, which was identified as a negative regulator of the mTOR pathway (Faivre et al. 2006). This finding suggests that one role of mTOR signaling in gastric cancer may be to mediate metastases, which warrants further investigation.

Another strategy to identify proteins involved in gastric cancer metastasis was reported by Chen et al. (2006). The authors elected to perform a comprehensive proteomic analysis of the noninvasive gastric cancer cell line SC-M1, and to compare the SC-M1 proteome against the proteome of the metastatic gastric cancer cell line TMC-1. Instead of separating the proteins using traditional 2DE gel electrophoresis, the authors used LC/MS/MS to identify proteins that were differentially expressed between SC-M1 and TMC-1 cells. In comparison to the 2DE-gel studies above, the throughput of the LC/MS/MS platform was substantially higher, allowing the authors to identify a total of 926 and 909 differentially expressed proteins in SC-M1 and TMC-1 cells, respectively. With this reasonably large number of proteins, the authors were able to perform robust pathway analysis and found that most of the proteins were categorized into functional ontologies such as protein metabolism and modification, nucleic acid metabolism, signal transduction, and cell structure and mobility. In parallel to the shotgun profiling, the authors also performed a cleavable isotope coded affinity tag (cICAT) analysis to identify differentially expressed proteins between SC-M1 and TMC-1 cells. They identified 240 proteins displaying a more than 1.3-fold difference. More than 100 and 140 proteins were either up- or downregulated in TMC-1 cells compared with SC-M1 cells, respectively. The greater number of differentially expressed proteins identified using this approach is consistent with other reports stating that LC/MS/MS is more sensitive, quantitative, and associated with a wider dynamic range than conventional 2DE gel-based technologies (Engwegen et al. 2006).

A prominent set of differentially expressed proteins were those involved in signal transduction, including downregulation of the Rho-related GTP binding protein RhoG, transforming protein RhoA, and transforming protein RhoC in TMC-1 cells. Another notable protein was the cell adhesion and signaling molecule β-catenin, which was upregulated in TMC-1 cells. β-Catenin is a core member of the Wnt signaling pathway. Its upregulation in TMC-1 cells suggests that the Wnt pathway may contribute to gastric cancer metastasis. Other pathways that were upregulated in TMC-1 included those related to cell motility (S100 calcium binding protein family members), cell cycle and proliferation (CDC42 and CDCK6), and protein degradation (cathepsin B and C). Many of these proteins have already been implicated in metastasis in other cancer types (e.g., S100 family members, cathepsin B, galectin 1). Taken collectively, this study identified a host of potential proteins involved in gastric cancer metastasis. The challenge ahead is to determine the generalizability of this result to other cell lines and more importantly in human gastric cancer.

Proteomic Differences Between Nonmalignant and Malignant Gastric Juice

The limitations of using human tissue in view of cellular heterogeneity have already been introduced, and efforts to compensate for this have been discussed. Another strategy that may partially address this issue is the use of gastric-related body fluids, namely, gastric juice and serum. Interest in analyzing these fluids has largely derived from their potential use as early markers of gastric pathology, because obtaining and testing these fluids in apparently healthy individuals is far more feasible than an analysis of gastric tissue or mucosa. Two proteomic studies have been reported analyzing gastric juice in patients with normal stomachs and with gastric pathologies. In normal individuals, gastric juice comprises a complex, multicomponent mixture, including hydrochloric acid, digestive enzymes (lipases, pepsin), intrinsic factor for vitamin B12, mucins, constituents from salivary compartments obtained by swallowing, and gastroduodenal refluxate (which includes bile). Importantly, there have been previous reports, using protein-specific assays, demonstrating increased levels of certain proteins such as leukotrienes in the gastric juice of *H. pylori*–positive patients (Kasirga et al. 1999), thus providing proof-of-concept that proteins in gastric fluids may be fundamentally different between healthy and diseased individuals. To extend these findings, Lee et al. (2004) performed a 2DE gel electrophoresis analysis of gastric juices obtained from 5 healthy subjects and 99 patients with various gastric diseases, 30 of whom had gastric cancer. The team found that, overall, gastric fluids obtained from gastric cancer patients had higher total concentrations of protein and, not surprisingly, had higher pH compared with normal gastric juice. Besides reidentifying expected gastric juice–associated proteins such as pepsins and lipase, the investigators found that 60% of gastric cancer patients had elevated levels of α1-antitrypsin. To determine whether upregulation of α1-AT was specific to cancer, they performed a similar analysis on 69 gastric juice samples from patients with various gastric noncancer diseases, and discovered α1-AT upregulation in 3 of 56 patients with atrophic gastritis, 2 of 7 tubular adenoma cases, 1 of 4 intestinal metaplasia cases, and in one case with a hyperplastic polyp. Because some of these conditions are thought to predispose to gastric cancer, the authors concluded that α1-AT expression in gastric juice may serve as a potential prognostic marker for the future development of gastric cancer. At present, the origin of the elevated α1-AT is unclear, particularly considering the other proteomic studies, cited above, showing that α1-AT expression is strongly *suppressed* in gastric cancers.

In a more recent study, Su et al. (2007) reported a comparative analysis of gastric juices from 120 healthy subjects, 39 gastric ulcer patients, 38 duodenal ulcer patients, and 31 patients with gastric cancer. This team also report that gastric fluids from cancer patients were associated with higher protein concentrations and higher pH levels. They then performed a 2D-PAGE analysis of the gastric fluids, followed by subsequent protein identification of prominent protein spots by MALDI-TOF, MS/MS, and MASCOT amino acid sequencing search engines. Of these expressed proteins, α1-AT was once again identified as being significantly differentially expressed. Expression was as follows: 6%, 42%, 6%, and 93% of healthy, gastric ulcer, duodenal ulcer, and gastric cancer patients, respectively. Notably, in the cancer patients, high expression of α1-AT was also observed in early cancer patients. Further analysis of these patients led these authors to independently conclude that although α1-AT expression was found in gastric cancers, this was not cancer specific because 42% of gastric ulcer patients expressed abundant α1-AT. Instead, they suggested that α1-AT may originate from inflammatory cells and leak into the stomach through cancerous and/or ulcerative lesions, protected from proteolysis by hypoacidity. What is lacking from these studies is anatomic location of the cell expressing the protein, which would help determine the functional significance. Nevertheless, it is still possible that α1-AT expression in gastric juice can be a biomarker for gastric cancer and ulcer. These findings await further large-scale validation studies.

Comparing Serum from Cancer and Noncancer Patients

In addition to gastric juice, there have been several reports that have explored proteomic patterns in serum and whether they can be used to identify potential biomarkers for gastric cancer (Juan et al. 2004; Liang et al. 2006; Poon et al. 2006; Ren et al. 2006; Su et al. 2007). At present, there are no specific serum biomarkers for stomach cancer; conventional markers such as cancer antigen 19-9 and 72-4 have low sensitivities and specificities (20%–30%) (Ishigami et al. 2001; Marrelli et al. 1999, 2001). Furthermore, although the use of *H. pylori* serology and serum pepsinogen levels may be of use in identifying subgroups of individuals at higher risk for cancer development, these cannot be used to distinguish patients with cancer from normal controls. Because of the technical difficulties in analyzing the serum proteome as a result of its high dynamic range, many of these studies have elected to use SELDI-TOF to analyze serum samples, which requires minimal amounts of sample. However, as described above, two major limitations of the SELDI-TOF platform are its reproducibility and the inability to identify the specific proteins contributing to a discriminating spectral peak.

Liang et al. (2006) conducted the first study of SELDI-TOF analysis of gastric cancer patient serum samples. They analyzed samples from 33 gastric cancer patients and 31 normal healthy controls. Of 45 spectral peaks identified, the intensity of 3 peaks was significantly different between the two populations. Using these peaks, the authors then attempted to classify a blinded test set of 15 cancer and 10 healthy sera, and reported that 14 of 15 gastric cancer patients were correctly classified, and 100% of healthy sera were classified as healthy. Interestingly, using these peaks, the authors were also able to exploit the SELDI-TOF spectra to distinguish cancer patients from patients with gastritis.

Taken collectively, these results and similar studies suggest the potential existence of distinct protein serum biomarkers that distinguish gastric cancer patients and patients with other noncancer gastric pathologies. A more extensive study was reported by Poon et al. (2006), who utilized a series of additional analytic steps to confirm the specificity of potential gastric cancer serum biomarkers. Using SELDI-TOF, the authors analyzed sera from 34 cancer and 29 normal controls to identify an initial set of 35 peaks that were significantly different between cancer and normal patients. The authors then investigated whether there was a reduction in the level of the 32 upregulated peaks after complete resection of the primary cancer (Figure 11.4).

This filter reduced the number to five independent peaks, which were then used to create a linear regression model for cancer diagnosis and applied to an independent validation set of 40 cancer and 20 normal controls. In an ROC analysis, the model delivered a 95% specificity and sensitivity/accuracy of 83%/87%, respectively. Interestingly, the sensitivity of the model was correlated with tumor stage— late-stage tumors delivered 95% sensitivity whereas stage I cancers delivered only 20% sensitivity. Of these five peaks, two were subsequently identified as serum amyloid A isoforms (SAA), which are upregulated in various cancers. However, because SAA is an acute-phase protein that is upregulated in many conditions, its use alone may not be ideal to diagnose cancer, hence the necessity to combine it with other markers. Unfortunately, there has not been a rigorous comparison of the overlap in the peaks in these different studies, thus it is difficult to assess their reproducibility, which may be affected by the use of different sampling strategies, different SELDI-chips, and patient populations. The added difficulty of peak identification will make subsequent validation of SELDI-TOF results a challenging undertaking.

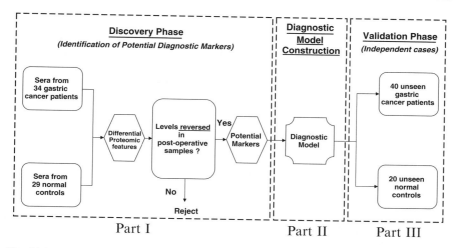

Fig. 11.4 Surface-enhanced laser desorption ionization–time-of-flight (SELDI-TOF) experimental design. The study was divided into three parts: discovery phase, model construction, and validation. (Adapted from Poon et al. 2006, with permission from Elsevier.)

Concluding Remarks

Through this brief survey, we have hoped to provide the reader with a reasonably coherent overview of the state of the art in gastric cancer proteomics. Obviously, proteomics approaches can be applied to a number of key biologic questions in gastric cancer, and in some cases (e.g., serum and gastric juice profiling) may represent quite promising platforms for identifying biomarkers for early cancer diagnosis. However, despite the diversity of these studies, they share several common aspects. First, most are very much discovery-based, and subsequent validation studies are undoubtedly required to assess the robustness of these initial studies. Second, most reported gastric cancer proteomic studies have used 2DE/MS, and it will be quite exciting to see more comprehensive platforms such as LC/MS/MS applied to similar questions. Such studies are undoubtedly ongoing, and promise to revolutionize our understanding of the gastric cancer proteome.

Conclusion

Throughout this chapter, we have introduced key concepts and discussed some limitations of high-throughput technologies in genomics and proteomics rather than specific details of the technologies that future developments may render obsolete. The sections concerning technologies were meant to be brief descriptions and discussions of practical measures to overcome some of the limitations of the technology currently in use to analyze genomic and proteomic data.

The remainder of the chapter was a discussion of specific applications of these technologies to study gastric cancer from the level of *in vitro* analysis of cell lines through to *in vivo* applications that are almost ready for translation to the clinic.

References

Aggarwal, A., Li Guo, D., Hoshida, Y., Tsan Yuen, S., Chu, K. M., So, S., Boussioutas, A., Chen, X., Bowtell, D., Aburatani, H., et al. (2006). Topological and functional discovery in a gene coexpression meta-network of gastric cancer. Cancer Res 66:232–41.

Bach, S., Makristathis, A., Rotter, M., and Hirschl, A. M. (2002). Gene expression profiling in AGS cells stimulated with Helicobacter pylori isogenic strains (cagA positive or cagA negative). Infect Immun 70:988–92.

Bae, S. H., Harris, A. G., Hains, P. G., Chen, H., Garfin, D. E., Hazell, S. L., Paik, Y. K., Walsh, B. J., and Cordwell, S. J. (2003). Strategies for the enrichment and identification of basic proteins in proteome projects. Proteomics 3:569–79.

Baggerly, K. A., Morris, J. S., and Coombes, K. R. (2004). Reproducibility of SELDI-TOF protein patterns in serum: comparing datasets from different experiments. Bioinformatics 20:777–85.

Blanc-Brude, O. P., Yu, J., Simosa, H., Conte, M. S., Sessa, W. C., and Altieri, D. C. (2002). Inhibitor of apoptosis protein survivin regulates vascular injury. Nat Med 8:987–94.

Bonner, R. F., Emmert-Buck, M., Cole, K., Pohida, T., Chuaqui, R., Goldstein, S., and Liotta, L. A. (1997). Laser capture microdissection: molecular analysis of tissue. Science 278:1481–3.

Borrmann, R. (1926). Gescwulste des Magens und Duodenums. In: Henke F, and Lubarch O., editors. Handbuch der speziellen pathologischen Anatomie und Histologie. New York: Springer; pp. 865–79.

Boussioutas, A., Li, H., Liu, J., Waring, P., Lade, S., Holloway, A. J., Taupin, D., Gorringe, K., Haviv, I., Desmond, P. V., and Bowtell, D. D. (2003). Distinctive patterns of gene expression in premalignant gastric mucosa and gastric cancer. Cancer Res 63:2569–77.

Brekken, R. A., and Sage, E. H. (2001). SPARC, a matricellular protein: at the crossroads of cell-matrix communication. Matrix Biol 19:816–27.

Censini, S., Lange, C., Xiang, Z., Crabtree, J. E., Ghiara, P., Borodovsky, M., Rappuoli, R., and Covacci, A. (1996). cag, a pathogenicity island of Helicobacter pylori, encodes type I-specific and disease-associated virulence factors. Proc Natl Acad Sci USA 93:14648–53.

Chen, Y. R., Juan, H. F., Huang, H. C., Huang, H. H., Lee, Y. J., Liao, M. Y., Tseng, C. W., Lin, L. L., Chen, J. Y., Wang, M. J., et al. (2006). Quantitative proteomic and genomic profiling reveals metastasis-related protein expression patterns in gastric cancer cells. J Proteome Res 5:2727–42.

Chen, J., Kahne, T., Rocken, C., Gotze, T., Yu, J., Sung, J. J., Chen, M., Hu, P., Malfertheiner, P., and Ebert, M. P. (2004). Proteome analysis of gastric cancer metastasis by two-dimensional gel electrophoresis and matrix assisted laser desorption/ionization-mass spectrometry for identification of metastasis-related proteins. J Proteome Res 3:1009–16.

Chan, C. H., Ko, C. C., Chang, J. G., Chen, S. F., Wu, M. S., Lin, J. T., and Chow, L. P. (2006). Subcellular and functional proteomic analysis of the cellular responses induced by Helicobacter pylori. Mol Cell Proteomics 5:702–13.

Chen, X., Leung, S. Y., Yuen, S. T., Chu, K. M., Ji, J., Li, R., Chan, A. S., Law, S., Troyanskaya, O. G., Wong, J., et al. (2003). Variation in gene expression patterns in human gastric cancers. Mol Biol Cell 14:3208–15.

Chen, C. N., Lin, J. J., Chen, J. J., Lee, P. H., Yang, C. Y., Kuo, M. L., Chang, K. J., and Hsieh, F. J. (2005). Gene expression profile predicts patient survival of gastric cancer after surgical resection. J Clin Oncol 23:7286–95.

Chen, N., Ye, X. C., Chu, K., Navone, N. M., Sage, E. H., Yu-Lee, L. Y., Logothetis, C. J., and Lin, S. H. (2007). A secreted isoform of ErbB3 promotes osteonectin expression in bone and enhances the invasiveness of prostate cancer cells. Cancer Res 67:6544–8.

Cormier, R. T., Hong, K. H., Halberg, R. B., Hawkins, T. L., Richardson, P., Mulherkar, R., Dove, W. F., and Lander, E. S. (1997). Secretory phospholipase Pla2g2a confers resistance to intestinal tumorigenesis. Nat Genet 17:88–91.

Correa, P. (1992). Human gastric carcinogenesis: a multistep and multifactorial processa—First American Cancer Society Award Lecture on Cancer Epidemiology and Prevention. Cancer Res 52:6735–40.

Correa, P., Haenszel, W., Cuello, C., Tannenbaum, S., and Archer, M. (1975). A model for gastric cancer epidemiology. Lancet 2:58–60.

Craanen, M. E., Blok, P., Dekker, W., Ferwerda, J., and Tytgat, G. N. (1992). Subtypes of intestinal metaplasia and Helicobacter pylori. Gut 33:597–600.

Dahl, F., Stenberg, J., Fredriksson, S., Welch, K., Zhang, M., Nilsson, M., Bicknell, D., Bodmer, W. F., Davis, R. W., and Ji, H. (2007). Multigene amplification and massively parallel sequencing for cancer mutation discovery. Proc Natl Acad Sci USA 104:9387–92.

DeRisi, J., Penland, L., Brown, P. O., Bittner, M. L., Meltzer, P. S., Ray, M., Chen, Y., Su, Y. A., and Trent, J. M. (1996). Use of a cDNA microarray to analyse gene expression patterns in human cancer [see comments]. Nat Genet 14:457–60.

Diamandis, E. P. (2003). Point: proteomic patterns in biological fluids—do they represent the future of cancer diagnostics? Clin Chem 49:1272–5.

Diamandis, E. P. (2004). Analysis of serum proteomic patterns for early cancer diagnosis: drawing attention to potential problems. J Natl Cancer Inst 96:353–6.

Dobbin, K., and Simon, R. (2005). Sample size determination in microarray experiments for class comparison and prognostic classification. Biostatistics 6:27–38.

Ebert, M. P., Kruger, S., Fogeron, M. L., Lamer, S., Chen, J., Pross, M., Schulz, H. U., Lage, H., Heim, S., Roessner, A., et al. (2005). Overexpression of cathepsin B in gastric cancer identified by proteome analysis. Proteomics 5:1693–704.

Eisen, M. B., Spellman, P. T., Brown, P. O., and Botstein, D. (1998). Cluster analysis and display of genome-wide expression patterns. Proc Natl Acad Sci USA 95:14863–8.

Ellmark, P., Ingvarsson, J., Carlsson, A., Lundin, B. S., Wingren, C., and Borrebaeck, C. A. (2006). Identification of protein expression signatures associated with Helicobacter pylori infection and gastric adenocarcinoma using recombinant antibody microarrays. Mol Cell Proteomics 5:1638–46.

Engwegen, J. Y., Gast, M. C., Schellens, J. H., and Beijnen, J. H. (2006). Clinical proteomics: searching for better tumour markers with SELDI-TOF mass spectrometry. Trends Pharmacol Sci 27:251–9.

Faivre, S., Kroemer, G., and Raymond, E. (2006). Current development of mTOR inhibitors as anticancer agents. Nat Rev Drug Discov 5:671–88.

Graham, J. E., Peek, R. M., Jr., Krishna, U., and Cover, T. L. (2002). Global analysis of Helicobacter pylori gene expression in human gastric mucosa. Gastroenterology 123:1637–48.

Greenman, C., Stephens, P., Smith, R., Dalgliesh, G. L., Hunter, C., Bignell, G., Davies, H., Teague, J., Butler, A., Stevens, C., et al. (2007). Patterns of somatic mutation in human cancer genomes. Nature 446:153–8.

Guillemin, K., Salama, N. R., Tompkins, L. S., and Falkow, S. (2002). Cag pathogenicity island-specific responses of gastric epithelial cells to Helicobacter pylori infection. Proc Natl Acad Sci USA 99:15136–41.

Gupta, G. P., and Massague, J. (2006). Cancer metastasis: building a framework. Cell 127:679–95.

Han, D. K., Eng, J., Zhou, H., and Aebersold, R. (2001). Quantitative profiling of differentiation-induced microsomal proteins using isotope-coded affinity tags and mass spectrometry. Nat Biotechnol 19:946–51.

Hasegawa, S., Furukawa, Y., Li, M., Satoh, S., Kato, T., Watanabe, T., Katagiri, T., Tsunoda, T., Yamaoka, Y., and Nakamura, Y. (2002). Genome-wide analysis of gene expression in intestinal-type gastric cancers using a complementary DNA microarray representing 23,040 genes. Cancer Res 62:7012–7.

He, Q. Y., Cheung, Y. H., Leung, S. Y., Yuen, S. T., Chu, K. M., and Chiu, J. F. (2004). Diverse proteomic alterations in gastric adenocarcinoma. Proteomics 4:3276–87.

Hippo, Y., Taniguchi, H., Tsutsumi, S., Machida, N., Chong, J. M., Fukayama, M., Kodama, T., and Aburatani, H. (2002). Global gene expression analysis of gastric cancer by oligonucleotide microarrays. Cancer Res 62:233–40.

Honda, S., Fujioka, T., Tokieda, M., Gotoh, T., Nishizono, A., and Nasu, M. (1998). Gastric ulcer, atrophic gastritis, and intestinal metaplasia caused by Helicobacter pylori infection in Mongolian gerbils. Scand J Gastroenterol 33:454–60.

International Agency for Research on Cancer (1994). Schistosomes, liver flukes and Helicobacter pylori. In: IARC monographs on the evaluation of carcinogenic risks to humans. Lyons: IARC.

Iruela-Arispe, M. L., Hasselaar, P., and Sage, H. (1991). Differential expression of extracellular proteins is correlated with angiogenesis in vitro. Lab Invest 64:174–86.

Ishigami, S., Natsugoe, S., Hokita, S., Che, X., Tokuda, K., Nakajo, A., Iwashige, H., Tokushige, M., Watanabe, T., Takao, S., and Aikou, T. (2001). Clinical importance of preoperative carcinoembryonic antigen and carbohydrate antigen 19-9 levels in gastric cancer. J Clin Gastroenterol 32:41–4.

Jendraschak, E., and Sage, E. H. (1996). Regulation of angiogenesis by SPARC and angiostatin: implications for tumor cell biology. Semin Cancer Biol 7:139–46.

Jinawath, N., Furukawa, Y., Hasegawa, S., Li, M., Tsunoda, T., Satoh, S., Yamaguchi, T., Imamura, H., Inoue, M., Shiozaki, H., and Nakamura, Y. (2004). Comparison of gene-expression profiles between diffuse- and intestinal-type gastric cancers using a genome-wide cDNA microarray. Oncogene 23:6830–44.

Juan, H. F., Chen, J. H., Hsu, W. T., Huang, S. C., Chen, S. T., Yi-Chung Lin, J., Chang, Y. W., Chiang, C. Y., Wen, L. L., Chan, D. C., et al. (2004). Identification of tumor-associated plasma biomarkers using proteomic techniques: from mouse to human. Proteomics 4:2766–75.

Kang, H. C., Kim, I. J., Park, J. H., Shin, Y., Ku, J. L., Jung, M. S., Yoo, B. C., Kim, H. K., and Park, J. G. (2004). Identification of genes with differential expression in acquired drug-resistant gastric cancer cells using high-density oligonucleotide microarrays. Clin Cancer Res 10:272–84.

Kapranos, N., Kominea, A., Konstantinopoulos, P. A., Savva, S., Artelaris, S., Vandoros, G., Sotiropoulou-Bonikou, G., and Papavassiliou, A. G. (2002). Expression of the 27-kDa heat shock protein (HSP27) in gastric carcinomas and adjacent normal, metaplastic, and dysplastic gastric mucosa, and its prognostic significance. J Cancer Res Clin Oncol 128:426–32.

Kasirga, E., Coker, I., Aydogdu, S., Yagci, R. V., Taneli, B., and Gousseinov, A. (1999). Increased gastric juice leukotriene B4, C4 and E4 concentrations in children with Helicobacter pylori colonization. Turk J Pediatr 41:335–9.

Kim, B., Bang, S., Lee, S., Kim, S., Jung, Y., Lee, C., Choi, K., Lee, S. G., Lee, K., Lee, Y., et al. (2003). Expression profiling and subtype-specific expression of stomach cancer. Cancer Res 63:8248–55.

Lander, E. S., Linton, L. M., Birren, B., Nusbaum, C., Zody, M. C., Baldwin, J., Devon, K., Dewar, K., Doyle, M., FitzHugh, W., et al. (2001). Initial sequencing and analysis of the human genome. Nature 409:860–921.

Lau, C. P., Poon, R. T., Cheung, S. T., Yu, W. C., and Fan, S. T. (2006). SPARC and Hevin expression correlate with tumour angiogenesis in hepatocellular carcinoma. J Pathol 210:459–68.

Lauren, P. (1965). The histological main types of gastric carcinoma: diffuse and so-called intestinal-type carcinoma. Acta Path Microbiol Scand 64:31–49.

Lee, K., Kye, M., Jang, J. S., Lee, O. J., Kim, T., and Lim, D. (2004). Proteomic analysis revealed a strong association of a high level of alpha1-antitrypsin in gastric juice with gastric cancer. Proteomics 4:3343–52.

Lefebvre, O., Chenard, M. P., Masson, R., Linares, J., Dierich, A., LeMeur, M., Wendling, C., Tomasetto, C., Chambon, P., and Rio, M. C. (1996). Gastric mucosa abnormalities and tumorigenesis in mice lacking the pS2 trefoil protein. Science 274:259–62.

Leung, S. Y., Chen, X., Chu, K. M., Yuen, S. T., Mathy, J., Ji, J., Chan, A. S., Li, R., Law, S., Troyanskaya, O. G., et al. (2002). Phospholipase A2 group IIA expression in gastric adenocarcinoma is associated with prolonged survival and less frequent metastasis. Proc Natl Acad Sci USA 99:16203–8.

Leung, S. Y., Yuen, S. T., Chu, K. M., Mathy, J. A., Li, R., Chan, A. S., Law, S., Wong, J., Chen, X., and So, S. (2004). Expression profiling identifies chemokine (C-C motif) ligand 18 as an independent prognostic indicator in gastric cancer. Gastroenterology 127:457–69.

Liang, Y., Fang, M., Li, J., Liu, C. B., Rudd, J. A., Kung, H. F., and Yew, D. T. (2006). Serum proteomic patterns for gastric lesions as revealed by SELDI mass spectrometry. Exp Mol Pathol 81:176–80.

Lipshutz, R. J., Fodor, S. P., Gingeras, T. R., and Lockhart, D. J. (1999). High density synthetic oligonucleotide arrays. Nat Genet 21:20–4.

Lundgren, A., Stromberg, E., Sjoling, A., Lindholm, C., Enarsson, K., Edebo, A., Johnsson, E., Suri-Payer, E., Larsson, P., Rudin, A., et al. (2005). Mucosal FOXP3-expressing CD4+ CD25high regulatory T cells in Helicobacter pylori-infected patients. Infect Immun 73:523–31.

Marko-Varga, G. (2004). Proteomics principles and challenges. Pure Appl Chem 76:829–37.

Marrelli, D., Pinto, E., De Stefano, A., Farnetani, M., Garosi, L., and Roviello, F. (2001). Clinical utility of CEA, CA 19-9, and CA 72-4 in the follow-up of patients with resectable gastric cancer. Am J Surg 181:16–9.

Marrelli, D., Roviello, F., De Stefano, A., Farnetani, M., Garosi, L., Messano, A., and Pinto, E. (1999). Prognostic significance of CEA, CA 19-9 and CA 72-4 preoperative serum levels in gastric carcinoma. Oncology 57:55–62.

Melle, C., Ernst, G., Schimmel, B., Bleul, A., Kaufmann, R., Hommann, M., Richter, K. K., Daffner, W., Settmacher, U., Claussen, U., and von Eggeling, F. (2005). Characterization of pepsinogen C as a potential biomarker for gastric cancer using a histo-proteomic approach. J Proteome Res 4:1799–804.

Mootha, V. K., Bunkenborg, J., Olsen, J. V., Hjerrild, M., Wisniewski, J. R., Stahl, E., Bolouri, M. S., Ray, H. N., Sihag, S., Kamal, M., et al. (2003). Integrated analysis of protein composition, tissue diversity, and gene regulation in mouse mitochondria. Cell 115:629–40.

Nishizuka, S., Charboneau, L., Young, L., Major, S., Reinhold, W. C., Waltham, M., Kouros-Mehr, H., Bussey, K. J., Lee, J. K., Espina, V., et al. (2003). Proteomic profiling of the NCI-60 cancer cell lines using new high-density reverse-phase lysate microarrays. Proc Natl Acad Sci USA 100:14229–34.

Nishigaki, R., Osaki, M., Hiratsuka, M., Toda, T., Murakami, K., Jeang, K. T., Ito, H., Inoue, T., and Oshimura, M. (2005). Proteomic identification of differentially-expressed genes in human gastric carcinomas. Proteomics 5:3205–13.

Odenbreit, S., Puls, J., Sedlmaier, B., Gerland, E., Fischer, W., and Haas, R. (2000). Translocation of Helicobacter pylori CagA into gastric epithelial cells by type IV secretion. Science 287:1497–500.

Ohyama, H., Zhang, X., Kohno, Y., Alevizos, I., Posner, M., Wong, D. T., and Todd, R. (2000). Laser capture microdissection-generated target sample for high-density oligonucleotide array hybridization. Biotechniques 29:530–6.

Paget, S. (1889). The distribution of secondary growths in cancer of the breast. Lancet 1:571–3.

Park, W. S., Oh, R. R., Park, J. Y., Lee, J. H., Shin, M. S., Kim, H. S., Lee, H. K., Kim, Y. S., Kim, S. Y., Lee, S. H., et al. (2000). Somatic mutations of the trefoil factor family 1 gene in gastric cancer. Gastroenterology 119:691–8.

Perkins, D. N., Pappin, D. J., Creasy, D. M., and Cottrell, J. S. (1999). Probability-based protein identification by searching sequence databases using mass spectrometry data. Electrophoresis 20:3551–67.

Peters, B. A., St. Croix, B., Sjoblom, T., Cummins, J. M., Silliman, N., Ptak, J., Saha, S., Kinzler, K. W., Hatzis, C., and Velculescu, V. E. (2007). Large-scale identification of novel transcripts in the human genome. Genome Res 17:287–92.

Petricoin, E. F., Ardekani, A. M., Hitt, B. A., Levine, P. J., Fusaro, V. A., Steinberg, S. M., Mills, G. B., Simone, C., Fishman, D. A., Kohn, E. C., and Liotta, L. A. (2002). Use of proteomic patterns in serum to identify ovarian cancer. Lancet 359:572–7.

Pieper, R., Su, Q., Gatlin, C. L., Huang, S. T., Anderson, N. L., and Steiner, S. (2003). Multi-component immunoaffinity subtraction chromatography: an innovative step towards a comprehensive survey of the human plasma proteome. Proteomics 3:422–32.

Poon, T. C., Sung, J. J., Chow, S. M., Ng, E. K., Yu, A. C., Chu, E. S., Hui, A. M., and Leung, W. K. (2006). Diagnosis of gastric cancer by serum proteomic fingerprinting. Gastroenterology 130:1858–64.

Reiner, A., Yekutieli, D., and Benjamini, Y. (2003). Identifying differentially expressed genes using false discovery rate controlling procedures. Bioinformatics 19:368–75.

Ren, H., Du, N., Liu, G., Hu, H. T., Tian, W., Deng, Z. P., and Shi, J. S. (2006). Analysis of variabilities of serum proteomic spectra in patients with gastric cancer before and after operation. World J Gastroenterol 12:2789–92.

Ross, D. T., Scherf, U., Eisen, M. B., Perou, C. M., Rees, C., Spellman, P., Iyer, V., Jeffrey, S. S., Van de Rijn, M., Waltham, M., et al. (2000). Systematic variation in gene expression patterns in human cancer cell lines. Nat Genet 24:227–35.

Sakakura, C., Hagiwara, A., Nakanishi, M., Shimomura, K., Takagi, T., Yasuoka, R., Fujita, Y., Abe, T., Ichikawa, Y., Takahashi, S., et al. (2002). Differential gene expression profiles of gastric cancer cells established from primary tumour and malignant ascites. Br J Cancer 87:1153–61.

Scherf, U., Ross, D. T., Waltham, M., Smith, L. H., Lee, J. K., Tanabe, L., Kohn, K. W., Reinhold, W. C., Myers, T. G., Andrews, D. T., et al. (2000). A gene expression database for the molecular pharmacology of cancer. Nat Genet 24:236–44.

Segal, E. D., Cha, J., Lo, J., Falkow, S., and Tompkins, L. S. (1999). Altered states: involvement of phosphorylated CagA in the induction of host cellular growth changes by Helicobacter pylori. Proc Natl Acad Sci USA 96:14559–64.

Sgroi, D. C., Teng, S., Robinson, G., LeVangie, R., Hudson, J. R., Jr., and Elkahloun, A. G. (1999). In vivo gene expression profile analysis of human breast cancer progression. Cancer Res 59:5656–61.

Shiozaki, K., Nakamori, S., Tsujie, M., Okami, J., Yamamoto, H., Nagano, H., Dono, K., Umeshita, K., Sakon, M., Furukawa, H., et al. (2001). Human stomach-specific gene, CA11, is down-regulated in gastric cancer. Int J Oncol 19:701–7.

Simon, R. (2005). Development and validation of therapeutically relevant multi-gene biomarker classifiers. J Natl Cancer Inst 97:866–7.

Simon, R., Radmacher, M. D., and Dobbin, K. (2002). Design of studies using DNA microarrays. Genet Epidemiol 23:21–36.

Simon, R., Radmacher, M. D., Dobbin, K., and McShane, L. M. (2003). Pitfalls in the use of DNA microarray data for diagnostic and prognostic classification. J Natl Cancer Inst 95:14–8.

Sjoblom, T., Jones, S., Wood, L. D., Parsons, D. W., Lin, J., Barber, T. D., Mandelker, D., Leary, R. J., Ptak, J., Silliman, N., et al. (2006). The consensus coding sequences of human breast and colorectal cancers. Science 314:268–74.

Stuart, J. M., Segal, E., Koller, D., and Kim, S. K. (2003). A gene-coexpression network for global discovery of conserved genetic modules. Science 302:249–55.

Su, Y., Shen, J., Qian, H., Ma, H., Ji, J., Ma, L., Zhang, W., Meng, L., Li, Z., Wu, J., et al. (2007). Diagnosis of gastric cancer using decision tree classification of mass spectral data. Cancer Sci 98:37–43.

Subramanian, A., Tamayo, P., Mootha, V. K., Mukherjee, S., Ebert, B. L., Gillette, M. A., Paulovich, A., Pomeroy, S. L., Golub, T. R., Lander, E. S., and Mesirov, J. P. (2005). Gene set enrichment analysis: a knowledge-based approach for interpreting genome-wide expression profiles. Proc Natl Acad Sci USA 102:15545–50.

Tay, S. T., Leong, S. H., Yu, K., Aggarwal, A., Tan, S. Y., Lee, C. H., Wong, K., Visvanathan, J., Lim, D., Wong, W. K., et al. (2003). A combined comparative genomic hybridization and expression microarray analysis of gastric cancer reveals novel molecular subtypes. Cancer Res 63:3309–16.

Tebbutt, N. C., Giraud, A. S., Inglese, M., Jenkins, B., Waring, P., Clay, F. J., Malki, S., Alderman, B. M., Grail, D., Hollande, F., et al. (2002). Reciprocal regulation of gastrointestinal homeostasis by SHP2 and STAT-mediated trefoil gene activation in gp130 mutant mice. Nat Med 8:1089–97.

Terashima, M., Maesawa, C., Oyama, K., Ohtani, S., Akiyama, Y., Ogasawara, S., Takagane, A., Saito, K., Masuda, T., Kanzaki, N., et al. (2005). Gene expression profiles in human gastric cancer: expression of maspin correlates with lymph node metastasis. Br J Cancer 92:1130–6.

Tinker, A. V., Boussioutas, A., and Bowtell, D. D. (2006). The challenges of gene expression microarrays for the study of human cancer. Cancer Cell 9:333–9.

van de Vijver, M. J., He, Y. D., van't Veer, L. J., Dai, H., Hart, A. A., Voskuil, D. W., Schreiber, G. J., Peterse, J. L., Roberts, C., Marton, M. J., et al. (2002). A gene-expression signature as a predictor of survival in breast cancer. N Engl J Med 347:1999–2009.

van't Veer, L. J., Dai, H., van de Vijver, M. J., He, Y. D., Hart, A. A., Mao, M., Peterse, H. L., van der Kooy, K., Marton, M. J., Witteveen, A. T., et al. (2002). Gene expression profiling predicts clinical outcome of breast cancer. Nature 415:530–6.

Venter, J. C., Adams, M. D., Myers, E. W., Li, P. W., Mural, R. J., Sutton, G. G., Smith, H. O., Yandell, M., Evans, C. A., Holt, R. A., et al. (2001). The sequence of the human genome. Science 291:1304–51.

Wang, C. S., Lin, K. H., Chen, S. L., Chan, Y. F., and Hsueh, S. (2004). Overexpression of SPARC gene in human gastric carcinoma and its clinic-pathologic significance. Br J Cancer 91:1924–30.

Weiss, M. M., Kuipers, E. J., Postma, C., Snijders, A. M., Siccama, I., Pinkel, D., Westerga, J., Meuwissen, S. G., Albertson, D. G., and Meijer, G. A. (2003). Genomic profiling of gastric cancer predicts lymph node status and survival. Oncogene 22:1872–9.

Wijte, D., de Jong, A. L., Mol, M. A., van Baar, B. L., and Heck, A. J. (2006). ProteomIQ blue, a potent post-stain for the visualization and subsequent mass spectrometry based identification of fluorescent stained proteins on 2D-gels. J Proteome Res 5:2033–8.

Yang, Y. H., and Speed, T. (2002). Design issues for cDNA microarray experiments. Nat Rev Genet 3:579–88.

Chapter 12
Animal Models of Gastric Carcinoma

Arlin B. Rogers and James G. Fox

Introduction

Animal models are an indispensable resource for investigating the causes of gastric cancer, and for identifying new preventative and therapeutic measures. In the late 1980s, experimental *Helicobacter mustelae* infection of domestic ferrets provided the first proof-of-principle that *Helicobacter* colonization was sufficient to incite chronic gastritis, helping to dispel skepticism over the pathogenic role of *H. pylori* in human gastrointestinal disease (Fox et al. 1990). Today, there is a choice of well-characterized and emerging animal models of gastric *Helicobacter* infection. In addition to infectious agents, animal models serve as sentinels for chemical and environmental gastric carcinogens, and are used to identify toxic principles within complex chemical compounds. Increasingly, animal models are being used to investigate tumorigenic synergism between multiple risk factors, both environmental and infectious. Rodent models in general, and mice in particular, have emerged as the model of choice for most laboratories, and therefore will receive the most attention in this chapter. Properly designed animal studies address basic questions about the molecular pathogenesis of gastric cancer that cannot be addressed by other methods, and provide a translational bridge from benchtop to bedside.

Animal Model Basics: Comparative Medicine and Genetics

Animal models are selected in part for their ability to reproduce the pathophysiology of human gastric carcinogenesis. Nevertheless, interspecies differences exist, and these must be understood in order to place experimental observations into proper context. Familiarity with the comparative anatomy, physiology, pathology, and genetics of rodent and nonrodent species relative to humans is necessary for any meaningful interpretation of animal-derived laboratory data (Rogers and Fox 2004). This section covers basic similarities and differences between the human stomach and that of model species, with a focus on the mouse.

T.C. Wang et al. (eds.) *The Biology of Gastric Cancers*,
© Springer Science+Business Media, LLC 2009

Rodent Stomach Morphology

Because of their small size, rapid reproduction kinetics, and a broad availability of laboratory reagents, rodents are the most popular species for biomedical research. Rodents frequently used in laboratory investigation include mice, rats, gerbils, hamsters, and guinea pigs. Of these, mice, rats, and Mongolian gerbils are most widely used in gastric cancer research. The rodent and human stomach share many common features, but there are important differences. The proximal third of the rodent stomach, including the anatomic equivalent of the human fundus, is lined by squamous rather than glandular epithelium. Therefore, rodents do not have "fundic glands," although the term appears frequently (and erroneously) in the scientific literature. The interface between the forestomach and glandular stomach is known by various names including the squamocolumnar junction and the forestomach/zymogenic junction (Syder et al. 2003). The band of glandular mucosa immediately distal to this transition parallels in morphology the human cardia but is very narrow, often comprising only 2–3 glandular units. Parietal and chief cells comprising the zymogenic or oxyntic glands that define the beginning of the corpus appear quickly thereafter. In mice, oxyntic glands may ring the corpus discontinuously. Therefore, in some histologic sections, glands with an antral morphology will extend the entire length of glandular stomach, from squamocolumnar junction to pylorus (Figure 12.1). This is normal and must not be confused with oxyntic atrophy or pseudopyloric metaplasia (the replacement of oxyntic glands by antral-type cells; Figure 12.1) (Kang et al. 2005). All of the nonrodent species discussed in this chapter have a glandular stomach that resembles that of the human.

Genetics and Gender

Rodent genetics represent both a powerful tool and potential stumbling block in the design and interpretation of gastric cancer studies. No meaningful conclusions can be drawn from studies in rodents in general, and mice in particular, without knowledge

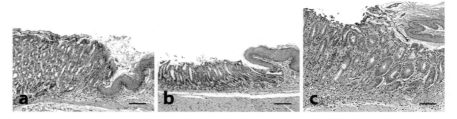

Fig. 12.1 Mouse stomach histopathology. (**a**) Normal stomach demonstrating squamous forestomach at right, squamocolumnar junction and glandular stomach at left. (**b**) Normal stomach with antral-type glands extending to the squamocolumnar junction highlighting the incompletely circumferential development of zymogenic glands in the corpus. (**c**) Pseudopyloric metaplasia characterized by loss of oxyntic mucosa and replacement by poorly differentiated glandular units with a more antral phenotype; note association with inflammation in this *Helicobacter pylori*–infected mouse. All panels bar = 160 μm

of how strain genotype (genetic makeup) influences phenotype (physical characteristics and disease expression) (Brayton et al. 2001). Results may vary when the same protocol is applied to different rodent strains, and care must be taken not to assume that experimental outcomes from a single mouse or rat model will necessarily apply to humans. Inbred mice and rats provide the distinct advantage of genetic homogeneity between individuals, and therefore reproducibility of results within and between laboratories. One of the most common mouse strains in biomedical research is the C57BL/6 (B6; "Black 6"). Genetically engineered mice (GEM) often are backcrossed onto a B6 background for comparisons with the well-characterized parental wild-type (WT) strain. However, B6 mice have peculiar characteristics, including a highly polarized Th1 immune response, that must be borne in mind when interpreting results. Transgenic mice sometimes are left on the FVB strain background in which pronuclear injections are most often performed (Babinet et al. 1989). In contrast, most gene knockout studies utilize embryonic stem cells derived from strain 129 mice (Bolon 2004). Because a backcross onto the C57BL/6 strain is not considered complete until 20 generations of breeding have been performed (or proof of >99.9% genetic homogeneity through molecular means such as speed congenics) (Hickman-Davis 2001; Wakeland et al. 1997), knockout studies sometimes are performed using hybrid B6129 offspring carrying the mutation of interest. Information on rodent strains and their associated phenotypes are available from texts of laboratory animal medicine and from commercial vendors. There are many excellent online sources for mouse genomics and phenomics information (Bolon 2006), including two maintained by The Jackson Laboratory (Bar Harbor, ME): Mouse Genome Informatics (www.informatics.jax. org) and the Mouse Phenome Database (phenome.jax.org).

Gender is another critical determinant in the design and interpretation of rodent studies of gastric carcinoma. In humans, older men have twice the risk of women for the most common form of gastric carcinoma, the intestinal type; however, in younger adults with the less common diffuse-type carcinoma, this gender profile is reversed (Ming and Hirota 1998). Gender-dimorphic cancer risk has not been reported in most animal models of gastric carcinoma. A notable exception is male-predominant disease expression in hypergastrinemic INS-GAS mice on the FVB strain background (Fox, Rogers et al. 2003; Fox, Wang et al. 2003). Whereas gastritis has been reported to be more severe in female *Helicobacter*-infected WT mice on a pure or mixed C57BL/6 background (Court et al. 2003; Ihrig et al. 2005), this finding has not been reproduced in all settings (Rogers et al. 2005). Unfortunately, many studies of gastric carcinogenesis in mice use only one sex, making it impossible to evaluate gender as a risk determinant. Unless gender phenotype for a given model is well established, it is more informative to design animal studies that include both sexes.

Helicobacter-Induced Gastric Carcinogenesis

Pathogenic gastric *Helicobacter* spp. in animals incite chronic gastritis, and in some instances carcinoma, just as *H. pylori* does in humans. The ferret was the first non-human species identified with chronic gastritis associated with *Campylobacter*-like organisms (later identified as *H. mustelae*) (Fox et al. 1986). Until rodent models were developed, this was the most popular animal for studying gastric *Helicobacter* infection (Fox et al. 1990; Gottfried et al. 1990) and it remains in use today (Croinin et al. 2007). Gastric *Helicobacter* spp. have been recovered from rats, dogs, domestic and nondomestic cats, pigs, nonhuman primates, and many wild species including birds and aquatic mammals (Fox et al. 2006; Harper, Feng et al. 2002; Harper et al. 2000, 2003; Harper, Xu et al. 2002; Seymour et al. 1994; Van den Bulck et al. 2005; Whary and Fox 2004). However, the natural history of disease caused by gastric *Helicobacter*s in most species remains unknown. For example, captive cheetahs frequently develop a chronic gastritis that is an important cause of mortality (Munson et al. 2005). Both wild and captive cheetahs harbor gastric *Helicobacter* spp., but only captive populations develop severe gastritis, disease association with infection is weak, and triple therapy for bacterial eradication was shown to be ineffective in reducing morbidity (Citino and Munson 2005; Terio et al. 2005). The two most frequently used animal species for experimental infectious gastric carcinogenesis studies are the mouse and the Mongolian gerbil. In this section, we focus on these well-characterized rodent models, followed by a brief discussion of gastric *Helicobacter* spp. infections in ferrets, domestic cats, and nonhuman primates.

Pathology of Helicobacter-*Induced Gastric Carcinoma in Rodents*

In susceptible rodents, as in humans, pathogenic gastric *Helicobacter* spp. produce chronic progressive gastritis that predisposes to carcinoma. Also as with humans, *Helicobacter* spp. are the only proven infectious cause of gastric cancer in animals, although secondary opportunists may colonize the achlorhydric stomach and contribute to later disease stages (Acheson and Luccioli 2004). Gastric *Helicobacter* spp. colonize glands in the border zone between oxyntic and nonoxyntic regions. Accordingly, organisms are demonstrated most readily at the junction of the corpus with the cardia proximally, and at the corpus–antrum interface antrum distally (Syder et al. 2003). However, although bacteria colonize antral glands early in the course of infection, in mice, inflammatory and dysplastic lesions usually arise in the cardia (squamocolumnar junction) and progress distally in conjunction with oxyntic gland atrophy. Occasionally, neoplasms arise first at the pylorus (Fox, Rogers et al. 2003; Fox et al. 2007; Fox, Wang et al. 2003). In contrast to the anterior distribution of lesions in the mouse, Mongolian gerbils infected with *H. pylori* exhibit disease reflecting the antral predominance found in humans. Neoplasms tend to be more ductular and invasive than those in the mouse, which can contain a significant papillary component. Regardless of anatomic location, the chronologic progression of *Helicobacter*-associated lesions in both mice and Mongolian gerbils closely mimics the classic sequence described by Correa: acute progressing to chronic gastritis, atrophic gastritis, surface epithelial proliferation, metaplasia, dysplasia, and carcinoma (Fox and Wang 2007). Therefore, whereas there are pathologic features unique to each model, both the mouse and gerbil have great value for investigating the molecular events that correspond with tumor promotion, and for evaluating novel therapies aimed at halting or reversing neoplastic progression.

Helicobacter-*Induced Gastric Carcinogenesis in Wild-Type Mice*

Helicobacter felis **in Wild-Type Mice**

Mice can be persistently colonized by a number of gastric *Helicobacter* spp., but the two most often used to study gastritis and cancer are *H. felis* and *H. pylori*. *H. felis*, originally isolated from cats, was the first bacterium shown to induce chronic gastritis in experimentally infected mice (Lee et al. 1990). *H. felis* is a much larger organism than *H. pylori*, with important differentiating ultrastructural and genetic features (De Bock et al. 2006). One limitation of the *H. felis* model is that it lacks the cag pathogenicity-associated island (PAI) found in *H. pylori* (Gasbarrini et al. 2003). Nevertheless, the ability to induce severe gastritis and carcinoma in C57BL/6 and other WT mice makes *H. felis* a highly valuable model for studying inflammation-associated gastric carcinogenesis. Indeed, murine gastritis caused by *H. felis* is more severe than that induced by *H. pylori*, and only the former produces carcinoma in WT C57BL/6 mice (Fox et al. 2002). As with most mouse models of *Helicobacter* infection, phenotypic outcomes are strain specific. Whereas C57BL/6 mice infected with *H. felis* develop severe gastritis progressing to cancer, outbred Swiss Webster mice acquire only a moderate self-limiting gastritis (Fox, Blanco et al. 1993; Lee et al. 1988, 1990). Two different laboratories inoculated different inbred strains of mice with *H. felis* and tabulated the intensity of inflammation: in BALB/c mice, inflammation was minimal; in C3H/He, moderate; and was most severe in C57BL/6 when examined 2–11 weeks postinoculation (PI) (Mohammadi et al. 1996; Sakagami et al. 1994). In a study using congenic strains on the BALB/c and C57BL/6 background, both MHC and non-MHC genes were found to contribute to disease phenotype induced by *H. felis* (Mohammadi et al. 1996).

Because of their susceptibility to severe gastritis and carcinoma, C57BL/6 have been used most widely for *H. felis* studies. For example, the *H. felis* model was used to demonstrate bone marrow-derived stem cell recruitment and carcinogenic transformation in a setting of chronic gastritis, a discovery with significant biologic and therapeutic implications (Houghton et al. 2004). Studies using *H. felis* showed that coinfection with a T_h2-provoking helminth parasite ameliorated gastritis, offering a potential explanation for the "African enigma," a continent with a high *H. pylori* prevalence but relatively low gastric cancer burden, especially in tropical countries with frequent gastrointestinal parasitism (Fox et al. 2000). Conversely, prior infection with the Th1-provoking protozoan *Toxoplasma gondii* confers susceptibility to chronic active gastritis and dysplastic lesions after *H. felis* infection in normally resistant BALB/c mice (Stoicov et al. 2004). These complementary studies highlight the importance of heterologous immunity in host responses to gastric *Helicobacter* infection, and provide insights into the highly variable epidemiologic patterns of disease induced by *H. pylori* in different geographic regions.

Helicobacter pylori in Wild-Type Mice

H. pylori has a limited natural host range, to date having been recovered only from humans, nonhuman primates, and domestic cats from a commercial breeding source (Fox and Lee 1993). Experimentally, *H. pylori* will infect mice, Mongolian gerbils, cats, dogs, and pigs (Blanchard et al. 1996; Fox and Lee 1993; Karita et al. 1994; Kleanthous et al. 1995; Lee et al. 1997; Marchetti et al. 1995; McColm et al. 1995; Poutahidis et al. 2001). Studies of *H. pylori* in rodents have been hampered by poor species adaptation, with low colonization kinetics and a failure to induce disease. A breakthrough came when Lee and colleagues in Australia successfully adapted a strain of *H. pylori* to the mouse stomach through serial inoculation; the Sydney strain (SS1) isolate of *H. pylori* is now frequently used in experimental mouse studies (Lee et al. 1997). Other human strains including B128 have also shown virulence in mouse models (Fox, Wang et al. 2003). At present, there are four published mouse models of gastric carcinogenesis induced by *H. pylori*. Three of these involve genetically engineered mice and are discussed later in this chapter. At present, the only WT mice demonstrated to develop tumors after infection with *H. pylori* are C57BL/6 × 129S6/SvEv (B6129) mice. These mice developed gastric intraepithelial neoplasia as early as 15 months of age when infected with *H. pylori* Sydney strain-1 (Rogers et al. 2005). No difference in tumor susceptibility was observed between B6129F1 first-generation offspring and multigenerationally intercrossed B6;129S hybrids. Identification of a WT mouse model of *H. pylori* carcinogenesis is important because it permits studies of experimental disease pathogenesis in the absence of a specific genetic modification promoting cancer. It is not known at this time whether WT 129S or other inbred strains of mice are susceptible to gastric carcinoma following *H. pylori* infection. However, even in the absence of cancer induction, mouse models of *H. pylori* offer important molecular insights into cancer determinants. For example, *H. pylori* infection was shown to induce oxidative DNA mutations in C57/B6 Big Blue mice carrying a lambda phage mutation biomarker (Touati et al. 2003).

Helicobacter-Induced Gastric Carcinogenesis in Genetically Engineered Mice

Because most strains of WT mice do not develop gastric carcinoma when infected with *H. pylori*, mice with specific genetic modifications have been developed to identify determinants of cancer risk. In some instances, hypotheses about the role of a specific gene in disease progression have been validated whereas in others outcomes were unexpected. Herein, we present a brief introduction to the methods used to generate GEM, followed by a review of studies using such mice to investigate the roles of specific genes in *H. pylori*–associated gastric carcinoma.

Helicobacter-Induced Gastric Carcinogenesis in INS-GAS Transgenic Mice

The INS-GAS model was the first published report of *H. pylori*–induced gastric carcinoma in mice (Fox, Rogers et al. 2003; Fox, Wang et al. 2003). Hypergastrinemic INS-GAS mice that constitutively express humanized gastrin under the rat insulin promoter acquire spontaneous gastric tumors within 2 years, and develop severe gastritis and carcinoma within months when infected with *H. felis* or *H. pylori* (Figure 12.2) (Fox, Wang et al. 2003; Wang et al. 2000). Because of the rapid disease course, putative carcinogenic microbial factors are amenable to study in this system. For example, it was shown that deletion of *H. pylori* cagE (picB) delayed but did not prevent progression to cancer (Fox, Wang et al. 2003). More recently, this model was used to demonstrate that exposure to the Swedish smokeless tobacco snus promoted gastric carcinogenesis (Stenstrom et al. 2007). A unique and important feature of the INS-GAS model is male-predominant disease expression that mirrors human gender dimorphic cancer risk (Fox, Rogers et al. 2003). Studies with INS-GAS mice showed that estrogen has a protective effect against cancer in both males and females infected with *H. pylori* (Ohtani et al. 2007). Importantly, as with other GEM, the strain background on

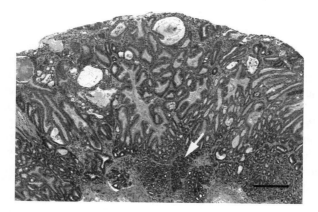

Fig. 12.2 Gastric intraepithelial neoplasia, equivalent to human intramucosal carcinoma, with herniation into the muscularis mucosae (arrow), in a constitutively hypergastrinemic INS-GAS mouse infected with *Helicobacter pylori* for 7 months. Bar = 400 μm

which the INS-GAS transgene is expressed influences phenotypic outcome. Whereas INS-GAS mice on the FVB strain background rapidly progress to cancer when infected with *Helicobacter* spp., C57BL/6 mice expressing the same transgene exhibit an attenuated phenotype and require significantly more time for tumor formation (S. Takaishi, personal communication).

Helicobacter-Induced Gastric Carcinogenesis in p27(kip1)$^{-/-}$ Mice

In contrast to p53+/− and Apc-haploinsufficient mice, deletion of the cell-cycle kinase inhibitor p27(kip1) in B6 mice conferred susceptibility to gastric carcinoma following *H. pylori* infection (Kuzushita et al. 2005). This was the first demonstration of *H. pylori*–induced gastric carcinoma in a knockout (as opposed to the transgenic INS-GAS) mouse. Deletion of this tumor-suppressor gene increased susceptibility to cancer at 60 and 75 weeks postinfection versus WT controls, with increased epithelial proliferation and decreased apoptosis in the p27$^{-/-}$ cohort (Kuzushita et al. 2005). This should prove to be a useful model for the study of *H. pylori* carcinogenesis, and may help spotlight new molecular targets for the interruption of stomach cancer in humans.

Helicobacter-Induced Gastric Carcinogenesis in Tff2$^{-/-}$ Mice

Trefoil factor 2 (TFF2), also known as spasmolytic polypeptide, is normally expressed in mucous neck cells of the corpus and in basally located epithelial cells of the antrum (Goldenring and Nomura 2006). In humans, expansion of TFF2$^+$ cells associated with *H. pylori*–induced antralization of the oxyntic mucosa is called spasmolytic polypeptide-enhancing metaplasia, or SPEM (Goldenring and Nomura 2006). The same process is observed in *Helicobacter*-infected mice with pseudopyloric metaplasia (Kang et al. 2005; Nomura et al. 2004). In mice, pseudopyloric metaplasia is a preneoplastic lesion associated with chronic *Helicobacter* infection. In contrast, mucous metaplasia may occur in the absence of *Helicobacter* infection, and in the absence of additional lesions it is not considered a precursor to neoplasia (Rogers et al. 2005). Because TFF2 is expressed both by cells in mucous metaplasia and pseudopyloric metaplasia, a diagnosis of SPEM in mice must be qualified (Fox et al. 2007). Moreover, whereas SPEM in humans is useful as a biomarker of gastric preneoplasia, recent data from mouse models show that TFF2 has a protective role against inflammation-associated carcinogenesis. TFF2 exhibits an immunoregulatory function in vitro, and mice lacking this gene demonstrate more severe inflammation of the stomach and colon than WT mice challenged with proinflammatory chemicals or *Helicobacter* spp. (Kurt-Jones et al. 2007). Moreover, TFF2$^{-/-}$ mice develop mucous metaplasia and pseudopyloric metaplasia with the same frequency as WT controls infected with *H. pylori*, and the knockouts have a significantly increased risk for tumors of the pyloric antrum (Fox et al. 2007). Taken together, results from mouse models suggest that spasmolytic polypeptide has an antitumorigenic rather than protumorigenic function in gastric carcinogenesis, and that it is dispensable for the development of *Helicobacter*-associated antralization of the gastric corpus. Nevertheless, SPEM remains a promising biomarker for assessing cancer risk in humans infected with *H. pylori* (Halldorsdottir et al. 2003).

Paradoxical Tumor Protection in p53 Haploinsufficient Mice

The p53 tumor suppressor gene is a housekeeping gene that is responsive to DNA damage and its repair helps regulate the cell cycle and has a role in transcription (Harris 1996). Humans heterozygous for p53 have Li-Fraumeni syndrome and are at risk for a variety of tumors. Befitting its role as a central regulator of DNA integrity and cell cycling, double knockout of p53 in mice results in a significantly shortened lifespan (Fox, Li, Cahill et al. 1996). However, p53$^{+/-}$ heterozygotes are viable and susceptible to a variety of spontaneous tumors, but none have been recorded in the glandular stomach (Donehower et al. 1992). A study of chronic *H. felis* infection in p53 hemizygous mice found that after 1 year both WT and p53 hemizygous mice showed severe adenomatous and cystic hyperplasia of the gastric surface foveolar epithelium. The proliferation markers BrdU and PCNA were markedly increased in both groups of infected mice. Not unexpectedly, p53 hemizygous mice had a higher proliferative index than WT mice, although this fell short of statistical significance (Fox, Li, Yan et al. 1996). Despite increased cell turnover, in a separate study, p53 haploinsufficiency resulted in paradoxic protection against tumor development that was attributed to depressed Th1 immune responses in the p53$^{+/-}$ cohort (Fox et al. 2002). The development of frank neoplasia in this mouse model may require either additional mutations, genetic events, or a longer time period of infection (Fox, Li, Cahill et al. 1996). The p53 heterozygous *H. felis*–infected mouse may also be useful for studies on cocarcinogenesis, given the increased sensitivity that p53$^{+/-}$ mice have to tumor induction by a variety of chemicals (Harvey et al. 1993).

Mixed Results in Apc Haploinsufficient Mice

Whereas mutation of the adenomatosis polyposis coli (*APC*) gene is indisputably associated with familial adenomatosis polyposis and increased risk of colon cancer in humans, the syndrome thus far has been linked to increased risk of gastric cancer only in Japan and not in Western countries (Tomita et al. 2007). This may reflect different population exposures to gastric carcinogens such environmental toxins and *H. pylori*. There are two well-described mouse models of Apc deficiency. The first, and most widely adopted, is the Apc$^{Min/+}$ mouse. This mouse was produced from ethylnitrosourea-induced random mutagenesis. The phenotype of widely disseminated intestinal adenomas was discovered after unexpectedly early mortality of mice caused by intestinal hemorrhage (Moser et al. 1995). Whereas early reports identified proliferative lesions limited to the intestinal tract, a more recent study documented spontaneous gastric tumors in Apc$^{Min/+}$ mice after 20 weeks of age (Tomita et al. 2007). Additional work is needed to reconcile conflicting results in Apc$^{Min/+}$ mice. A second mouse model of Apc deficiency, developed through targeted gene deletion, was designated Apc1638 (Yang et al. 1997). The phenotype of Apc1638 mice is similar to that of Apc$^{Min/+}$ mice, characterized by numerous gastroentestinal adenomas developing over the first few months of life, predominantly in the small intestine. *Apc*1638 mice were infected with *H. felis* to determine whether the mutation increased risk of gastric cancer following infection with a tumor-promoting bacterium. *Apc*1638 mice regardless of infection status by 7.5 months of age exhibited atypical proliferation in the mucosa of the antrum and at the pyloric junction (Fox, Dangler, Whary et al. 1997), in agreement with the distal gastric location of spontaneous tumors observed in Apc$^{Min/+}$ mice (Tomita et al. 2007). Interestingly, infected *Apc*1638 mice had less proliferation and inflammation in the corpus than WT mice, lower anti-*H. felis* serum titers, and higher bacterial colonization and gastric urease production compared with WT mice (Fox, Dangler, Whary et al. 1997). Thus, the *Apc*1638 truncating mutation led to spontaneous gastric dysplasia and polyposis of the antrum and pylorus but did not promote gastric cancer within the 7.5-month study period. Similar to p53, *Apc* haploinsufficiency seems to downmodulate immune responses, demonstrating the complex interactions between microbes and different host signaling pathways in determining the outcome of chronic *Helicobacter* spp. infection (Fox, Dangler, Whary et al. 1997).

Helicobacter-*Induced Gastric Carcinogenesis in Mongolian Gerbils*

Mongolian gerbils can be chronically infected with human isolates of *H. pylori*, and may develop gastroduodenitis, ulcers, and antral cancer closely resembling the human disease (Hirayama et al. 1996; Peek and Blaser 2002; Sawada et al. 1999). Gerbils develop both a marked submucosal follicular response combined with moderate to severe diffuse infiltration of the lamina propria with granulocytes and mononuclear cells (Figure 12.3). Dysplastic glands invade the submucosa early in the disease course, and can sometimes markedly expand this compartment with compression of the subjacent tunica muscularis (Figure 12.3). Recently, a rapid model of tumorigenesis was identified in gerbils infected with an *H. pylori* B128 variant (strain 7.13) that transactivated β-catenin signaling through a cagA-dependent mechanism (Franco et al. 2005). This organism produced histologic lesions consistent with invasive carcinoma as early as 4 weeks PI, and cancerous lesions in more than half of infected gerbils by 8–16 weeks PI, making this a very exciting model for studying tumor progression and interventional therapies in a compressed timeframe (Franco et al. 2005). In addition to their value in cancer research, *H. pylori*–infected Mongolian gerbils also are

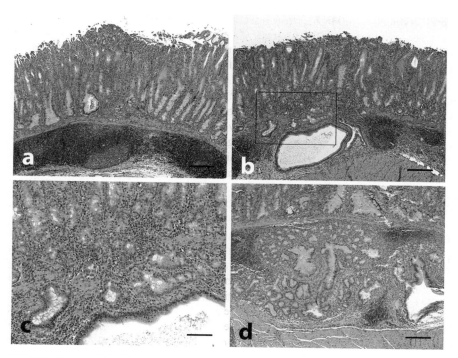

Fig. 12.3 Histopathology of *H. pylori* infection in the Mongolian gerbil. (**a**) Prominent submucosal lymphoid follicles and diffuse moderate-to-severe mixed inflammation in the lamina propria with associated epithelial degeneration in the antrum. (**b**) Low-power magnification of glandular invasion into the submucosa. (**c**) Higher magnification view of area bounded by box in panel b, demonstrating disruption of the muscularis mucosae by downward-migrating mildly dysplastic glands. (**d**) Gastric adenocarcinoma characterized by submucosal invasion of highly dysplastic glands and early extension into the subjacent tunica muscularis. Bar = 400 µm (a, b, d); 160 µm (c)

susceptible to peptic ulcer disease (Lee 2000). However, because *H. pylori* has not reliably produced gastric carcinoma in Mongolian gerbils in all laboratories, the model has come under some scrutiny (Chen et al. 2007; Tatematsu et al. 2005). In some gerbil studies *H. pylori* has produced gastritis and glandular invasion without significant cellular atypia, and the investigators referred to these structures as "heterotopic proliferative glands" (Tatematsu et al. 2005). However, in other settings, gastric epithelial structures are both invasive and highly dysplastic, fulfilling histopathologic criteria of adenocarcinoma (Figure 12.3) (Franco et al. 2005). In some instances, dysplastic glands may extend into the tunica muscularis and dissect between smooth muscle myofibers, resembling human infiltrative gastric carcinoma in the Ming classification system (Ming and Hirota 1998). Interlaboratory differences in disease outcome and interpretation of *H. pylori* infection in Mongolian gerbils are likely the result of genetic, environmental, and microbial circumstances unique to each setting. Unlike inbred strains of mice, Mongolian

gerbils typically used in *H. pylori* infection studies are outbred and exhibit significant genetic diversity (Bergin et al. 2003). Furthermore, resident microbiota in
different animal facilities may greatly influence responses to *H. pylori* infection.
Heterologous immunity, both proinflammatory and antiinflammatory, has been
documented in mouse models of coinfection with enteric protozoa and helminths
(Fox et al. 2000; Stoicov et al. 2004). Once more is learned about the genetic and
environmental factors that influence disease outcomes, the Mongolian gerbil
should gain wider acceptance as an important and highly representational model
of *H. pylori* gastric carcinogenesis.

Helicobacter mustelae *in Ferrets*

H. mustelae, a natural gastric pathogen of ferrets, has many of the same biochemical, molecular, and disease-inducing characteristics as *H. pylori*. It is urease positive, motile, and binds to similar adhesin receptors (Fox et al. 1988, 1989). Similar
to *H. pylori* in humans, *H. mustelae* persistently infects the inflamed mucosa of
ferrets; colonization occurs usually shortly after weaning (Fox et al. 1988). Oral
inoculation of *H. mustelae* into naive ferrets induces a chronic, persistent gastritis
identical to that observed in naturally infected animals, and the organism can be
recovered from feces of both, thus fulfilling Koch's postulates. The ferret stomach
closely resembles that of the human stomach in anatomy, histology, and physiology (Fox 1988). Ferrets secrete gastric acids and proteolytic enzymes under basal
conditions, and similar to humans infected with *H. pylori*, *H. mustelae*–infected
ferrets have hypergastrinemia (Perkins et al. 1996). Naturally occurring pyloric
adenocarcinoma has been linked to *H. mustelae* in ferrets, which is not surprising
because infection is associated with significantly upregulated cell replication as
shown by proliferating cell nuclear antigen immunohistochemistry (Fox, Dangler,
Sager et al. 1997; Yu et al. 1995). The histopathologic changes observed closely
coincide in topography with the presence of *H. mustelae*. Similar to humans and
gerbils infected with *H. pylori*, gastritis in *H. mustelae*–infected ferrets centers on
the distal antrum; the lesion is analogous to human diffuse antral gastritis (Correa
1992; Fox et al. 1990). However, some animals develop patchy lesions at the
corpus–antrum transitional zone, equivalent to human multifocal atrophic gastritis
(Correa 1992; Fox et al. 1990). Superficial gastritis targets the corpus near the
transitional zone, with oxyntic atrophy and metaplasia as found in other species.
Adherent *H. mustelae* organisms are found at the gastric surface, the superficial
portion of glands, and deep within pits. Thus, naturally occurring and experimental *H. mustelae* infection of ferrets represents a suitable model to study the pathogenesis and epidemiology of *H. pylori*–associated chronic gastritis and gastric
cancer (Fox et al. 1990; Fox and Lee 1993). Drawbacks to the ferret model include
the relatively high cost to purchase and maintain animals compared with rodents,
long disease course, and relatively few commercial reagents specifically targeted
to this species. Nevertheless, the ferret model remains highly useful for cocarcinogenesis studies (Figure 12.4) and may prove especially suitable for the evaluation

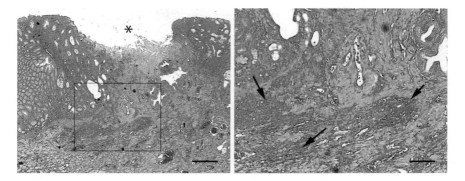

Fig. 12.4 Gastric carcinoma in a *Helicobacter mustelae*–infected ferret exposed to N-methyl-N′-nitro-N-nitrosoguanidine. (**a**) Note surface ulceration denoted by asterisk (*) with subjacent gland atrophy and fibrosis. (**b**) Higher magnification of area bounded by box in previous panel demonstrating deep invasion of tongues and islands *of dysplastic glands (arrows) in a desmoplastic stroma. Bar = 400 μm (a); 160 μm (b)

of therapeutic interventions because of the coadapted nature of the host and gastric pathogen.

Gastric Helicobacter spp. *in Domestic Cats*

Domestic cats may be infected naturally and experimentally with a number of potentially pathogenic gastric *Helicobacter* spp. (Fox 1995; Lecoindre et al. 2000). Furthermore, some *Helicobacter* spp. colonize in both the human and feline stomach, raising the possibility of interspecies transmission (Handt et al. 1994; Newell et al. 1989; Otto et al. 1994). The real-world implications of this were made clear during an outbreak of *H. pylori*–associated gastritis among cats in a commercial cat breeding facility attributed to human-to-animal (anthropozoonotic) transmission (Handt et al. 1995). Experimental *H. pylori* inoculation results in pangastric colonization of the feline stomach with an antral-predominant inflammatory response resembling the human disease both in histologic presentation and cytokine profile (Fox et al. 1995; Perkins et al. 1998; Simpson et al. 2001; Straubinger et al. 2003). Chronic *H. pylori* infection of cats results in prominent follicular gastritis and/or mucosal hyperplasia and dysplasia, although carcinomas have not yet been ascribed to infection in this species (Esteves et al. 2000). In addition to *H. pylori*, cats were identified as the natural source of *H. felis*, which is used widely today in murine experimental models (Paster et al. 1991). Similar to *H. pylori*, *H. felis* infection of cats is characterized by antral-predominant gastritis with expansile submucosal lymphoid follicles and a lesser intramucosal component (Paster et al. 1991; Simpson et al. 2000). Some investigators hypothesize that chronic *Helicobacter* spp. infection is one of several contributing factors to the high incidence of gastrointestinal lymphoid neoplasms in the cat, although Koch's postulates have not yet been fulfilled (Bridgeford in press).

Gastric Helicobacter spp. *in Nonhuman Primates*

In contrast to the extensive study of gastric *Helicobacter* infection in rodent models, reports of experimental infection in nonhuman primates are relatively few in number. This may be partially attributed to the high prevalence of endogenous *Helicobacter* spp. infections including *H. pylori* in animals such as rhesus macaques (Solnick et al. 2003, 2006). However, specific pathogen-free (SPF) macaque colonies with no *H. pylori* have been created through hand-rearing of newborn monkeys (Solnick et al. 1999). As a result of this, controlled experiments of the timing, infectious dose, and chronologic disease progression of *H. pylori* in macaques are now possible (Boonjakuakul et al. 2005; Solnick et al. 2001). Because of the closely related genetic profiles between macaques and humans, documentation of transcriptional profiles following gastric *H. pylori* infection in macaques demonstrate high utility for modeling equivalent human responses (Huff et al. 2004). Further work is needed to determine whether experimental gastric carcinoma can be induced by *H. pylori* in rhesus macaques and other nonhuman primates.

Chemoprophylaxis of Helicobacter-*Induced Gastric Carcinogenesis in Rodent Models*

Whereas *H. pylori* eradication has clinically proven benefit for the treatment of peptic ulcer disease and gastric MALT (mucosa-associated lymphoid tissue) lymphoma (Moss and Malfertheiner 2007), it is less clear whether elimination of bacteria slows or prevents carcinogenic progression in patients with preneoplastic mucosal lesions (de Vries et al. 2007). Mixed results have been reported on tumor prevention in *H. pylori*–infected individuals with histologically confirmed atrophic gastritis, intestinal metaplasia, and/or premalignant dysplasia. Mouse models have been used to help define the "point of no return," the disease stage when carcinogenesis progresses independently of infection. *H. felis* produces atrophic gastritis within 6 months PI (MPI) in C57BL/6 mice, significant dysplasia by 12 MPI, and invasive tumors by 15–18 MPI. Using a timed eradication study design, it was demonstrated that eradication at either 2 or 6 MPI resulted in complete recovery, although the process took longer in the latter group (Cai et al. 2005). In contrast, postponement of eradication until 12 MPI failed to reverse dysplasia. Nevertheless, dysplastic lesions did not appreciably progress in severity after *H. felis* eradication, and tumor incidence was significantly lower in this group at 22 MPI versus persistently infected controls (Cai et al. 2005). A similar time-dependent outcome was observed in Mongolian gerbils infected with *H. pylori*; eradication at 15 weeks PI (WPI) significantly prevented tumors, whereas eradication at 35 WPI was less effective, and eradication at 55 WPI showed no benefit in preventing cancer (Nozaki et al. 2003) Results from these models suggest that *H. pylori* eradication even in later stages of preneoplasia may slow or halt tumorigenic progression, but will not reverse dysplasia or carcinoma. This is consistent with speculation that divergent conclusions regarding late *H. pylori* eradication for cancer prevention may reflect sampling bias at the time of initial biopsy (de Vries et al. 2007). In addition to bacterial eradication, antiinflammatory therapy such as cyclooxygenase (COX)-2 inhibition has demonstrated benefit for preventing gastric cancer in *Helicobacter*-infected mice (Hahm et al. 2003; Nam et al. 2004) and Mongolian gerbils (Futagami et al. 2006; Futagami et al. 2007). This pathway also is implicated by the observation that transgenic mice overexpressing COX-2 and microsomal prostaglandin E synthase have increased susceptibility to *Helicobacter*-induced gastric cancer, and that bacterial eradication suppressed tumorigenesis (Oshima et al. 2004). Nevertheless, in humans, the use of COX-2 inhibitors for the prevention of gastric carcinoma and other cancers has lost significant appeal because of the risk of adverse cardiovascular events (Futagami et al. 2007).

Chemical Gastric Carcinogenesis

Because of a longer record of experimentation and more direct-acting effects, much more is known about mechanisms of chemical than infectious gastric carcinogenesis. Throughout much of the twentieth century, the rat was the most widely utilized laboratory animal in carcinogenesis studies. To this day, rats remain a staple of the National Toxicology Program and other laboratories for long-term chemical carcinogenesis bioassays. The earliest published reports of experimental gastric cancer were induced by polycyclic aromatic hydrocarbons in rats (Stewart and Hare 1950). However, early rodent gastric carcinogenesis studies were hampered by the lack of standardized criteria of malignancy. Attempts to reproduce isolated reports of gastric carcinogenesis often failed because of inadequate descriptions of methodology and differing diagnostic criteria used to assess the cancerous process. Factors required of an in vivo experimental model for gastric carcinogenesis include a suitable animal species, a characterized chemical and/or biologic agent, and an appropriate vehicle or solvent (Bralow 1972; Klein and Palmer 1941; Sugimura and Kawachi 1973). Today, the National Toxicology Program and National Center for Toxicogenomics, along with other laboratories, have well-defined criteria for biomarkers and endpoints in rodent carcinogenesis bioassays, and are developing standardized guidelines for the interpretation of findings as they relate to human health (Fostel et al. 2007; Maronpot et al. 2004; Tennant 2002; Waters et al. 2003).

N-Nitroso Compounds

Whereas a wide variety of chemicals will induce gastric cancer in animal models—for example N,N'2,7-fluorenylenebisacetamide (2,7-FAA) in Buffalo rats

(Morris et al. 1961)—the models that have been used most extensively are based on N-nitroso compounds. Whereas these substances are no longer abundant in the environment, they remain a human health concern because of potential ingestion through charred meat and reduction from nitrites (Haorah et al. 2001; Mensinga et al. 2003). These substances contain reactive nitrogen species capable of interacting with a wide variety of molecular side groups; in the cell nucleus N-nitroso compounds act as powerful DNA alkylating agents (Mirvish 1995). Three of these compounds have been used extensively in rodent and nonrodent models of gastric carcinogenesis: N-methyl-N'-nitro-N-nitrosoguanidine (MNNG), N-ethyl-N'-nitro-N-nitrosoguanidine (ENNG), and N-methyl-N-nitrosourea (MNU) (IARC 1977).

MNNG has been recognized as a powerful gastric carcinogen in rats for more than 40 years (Bralow 1972; Sugimura and Fujimura 1967, 1973). Similar to other alkylnitrosamines, MNNG is converted to the active carcinogen, an unstable alkyldiazo derivative. In acidic conditions, MNNG is converted to N-methyl-N'-nitroguanidine by releasing nitrous acid. In alkaline conditions, it is degraded into diazomethane and nitrocyanamide. The ultimate carcinogen formed under acidic conditions is capable of alkylating the purine bases of DNA, RNA, and some amino acids leading to subsequent mutation. MNNG is administered orally either in drinking water or by gavage (Sugimura and Kawachi 1973). MNNG has emerged as the agent of choice in experimental gastric carcinogenesis studies in various species of laboratory animals including not only the rat, but mice, hamsters, dogs, and nonhuman primates (Sugimura and Kawachi 1973). In solution, MNNG is heat- and light-labile. Because MNNG is relatively insoluble in water, other vehicles are often used including ethanol, dimethyl sulfoxide (DMSO), corn oil, and olive oil. Choice of vehicle can have a significant impact on experimental outcomes, especially in cocarcinogenesis studies.

N-Methyl-N'-nitro-N-nitrosoguanidine in Rats

The majority of studies using MNNG as a gastric carcinogen have been applied to the rat model. Typical drinking water concentrations for the rat range from 30 to 170 µg/mL for 10–12 months with total doses ranging from 400 to 1700 mg. MNNG administered to rats in drinking water at a concentration of 83 µg/mL produced gastric cancer in 70% of the rats within 12 months. MNNG administered by gavage induces predominantly forestomach tumors, whereas administration through drinking water incites predominantly glandular stomach tumors. Because of the difficulties in quantitatively measuring dosage and the safety issues involved with human exposure to MNNG in drinking water studies, investigators have also administered MNNG by gavage. Intragastric gavage of MNNG tends to produce solitary tumors whereas supplementation in drinking water induces tumors at multiple sites. Rats have been given single doses of MNNG ranging from 50 to 300 mg/kg. The susceptibility to MNNG-induced carcinogenesis varies among strains of rats with random-bred Wistar rats most susceptible and inbred Buffalo strain most resistant (Sugimura and Kawachi 1973). This resistance in part is attributable to the proliferative responses of gastric epithelia (Ohgaki et al. 1988). Moreover, resident microbiota seem to influence the phenotype. MNNG-induced gastric neoplasms were threefold higher in Wistar rats raised under conventional housing conditions compared with those raised germ free (Sumi and Miyakawa 1981). The authors suggested that microflora in the stomach of rats, attributed to coprophagism, may have been involved in tumor promotion. MNNG is cytocidal to the gastric mucosa of rats, leading to necrosis and ulceration, which may predispose the animals to tumor-promoting bacterial colonization. The naïve immune status of germ-free rats also may protect against chemical carcinogenesis (Sumi and Miyakawa 1981).

The histopathology of gastric adenocarcinomas produced by MNNG in rats (and hamsters) ranges from well-differentiated neoplasms to highly anaplastic and signet-ring cell carcinomas (Sugimura and Kawachi 1973). MNNG has shown value for examining the sequential histopathologic changes leading to carcinoma of the glandular stomach in rats (Bralow 1972; Sugimura and Kawachi 1973). MNNG has been shown to penetrate and induce cell proliferation more efficiently in the antrum than the corpus as determined by 3H-MNNG and BrdU labeling studies in Wistar rats (Sorbye et al. 1993). This agrees with the predominant location of tumors near the pyloric region. In the stomach, MNNG also can produce squamous cell carcinoma, fibrosarcoma, leiomyoma, and hemangioma, and in the small intestine, adenocarcinoma, hemangioendothelial sarcoma, and fibrosarcoma. Liver and lymph node metastases have been reported.

N-Methyl-N′-nitro-N-nitrosoguanidine in Mice

The mouse is relatively resistant to carcinogenesis of the glandular stomach induced by MNNG. Mice administered MNNG in drinking water over the life of the animal developed adenomatous hyperplasia but not adenocarcinoma (Sugimura and Kawachi 1973). In a study using Swiss outbred mice, administering MNNG in the drinking water with or without a salt-rich diet failed to induce glandular stomach tumors (Kodama et al. 1992). However, in one study, 5 of 69 Swiss albino mice developed gastric adenocarcinoma within 54–68 weeks of receiving 100 μg/mL MNNG in the drinking water (Sigaran and Con-Wong 1979). In another study, the administration of MNNG to five different strains of mice failed to induce gastric tumors (Sugimura and Kawachi 1973). Taken together, these results show that mice are more resistant than rats to carcinomas of the glandular stomach following MNNG exposure, highlighting the importance of host genetics in cancer susceptibility.

N-Methyl-N-nitrosourea in Mice

The chemical action of MNU is similar to that of MNNG, resulting in methylation of purine residues in DNA and RNA, and amino acid residues of proteins. Six-week-old male, BALB/c mice gavaged with 0.5 mg MNU once a week for 10 weeks developed tumors of the forestomach, glandular stomach, and duodenum (Tatematsu et al. 1992). Adenomatous hyperplasia was present in 75% of the mice at 20 weeks and 100% at 40 weeks. Glandular stomach carcinomas were classified into well-differentiated, poorly differentiated and signet-ring cell carcinomas. Importantly, 2 of 27 mice showed regional lymph node involvement, suggesting a utility for this model in investigating mechanisms of metastasis. However, a draw-back to the model was a high incidence of forestomach squamous cell carcinomas, necessitating resection in order to allow mice enough time to develop glandular tumors (Tatematsu et al. 1992). To overcome this problem, in another study, MNU was administered to male C3H/He mice in drinking water at a lower continuous dose for 30 weeks (Tatematsu et al. 1993). Mice administered the highest dose, 120 ppm, developed adenocarcinomas and signet cell carcinomas. Whereas only one mouse acquired a forestomach squamous cell carcinoma, others developed splenic hemangioendothelial sarcomas resulting in significant morbidity (Tatematsu et al. 1993). Taken as a whole, with the caveat of off-target tumors as a potential confounding factor, MNU-induced gastric carcinogenesis in the mouse shows promise as a model of human cancer, especially for studying mechanisms of tumor metastasis.

N-Methyl-N′-nitro-N-nitrosoguanidine/N-Methyl-N-nitrosourea in Mongolian Gerbils and Hamsters

MNNG and MNU orally administered to Mongolian gerbils reliably produces gas-tric carcinomas with features similar to those of the rat and mouse, respectively (Sugiyama et al. 1998; Tatematsu et al. 1998). This makes the Mongolian gerbil an especially valuable rodent model of human stomach cancer, especially because gerbils also develop significant gastritis ± tumors following infection with *H. pylori* (Fujioka et al. 2002). In contrast to Mongolian gerbils, administration of MNNG in the drinking water to hamsters produced fibrosarcomas rather than adenocarcino-mas of the stomach (Fujimura et al. 1970; Kogure et al. 1974).

N-Methyl-N′-nitro-N-nitrosoguanidine in Ferrets

In a study to define the ferret as a model of chemical gastric carcinogenesis, a single dose of MNNG at 50–100 mg/kg body weight (BW) or vehicle only was delivered by gavage to 6-month-old female ferrets (Fox, Wishnok et al. 1993). At 29–55 months after dosing, 9 of 10 ferrets dosed with MNNG had gastric adenocarcinoma whereas none of the control ferrets developed gastric cancer by 63 months. Histologically, the stomachs of MNNG-treated ferrets demonstrate invasive antral adenocarcinomas, including a depressed form correlating with the frequently diag-nosed type IIc carcinoma in humans (Figure 12.4) (Ming and Hirota 1998). In one case, metastasis to a regional lymph node was noted. Because all of the ferrets in the study were infected with *H. mustelae*, it could not be determined whether the

MNNG-initiated tumors were promoted by gastric Helicobacteriosis. However, MNNG at a dose of 100 mg/kg BW did not produce gastric cancer in *H. mustelae*–free ferrets within 48 months of inoculation (J. G. Fox, unpublished data), in agreement with studies in the Mongolian gerbil showing tumorigenic synergism between MNNG and *H. pylori* (Shimizu, Inada, Tsukamoto et al. 1999; Sugiyama et al. 1998).

N-Methyl-N′-nitro-N-nitrosoguanidine and N-Ethyl-N′-nitro-N-nitrosoguanidine in Dogs

Dogs that received MNNG in drinking water at similar concentrations to those administered to rats developed gastric cancer between 24 and 36 months; moreover, the morphology of the canine neoplasms closely resembled human gastric carcinoma (Koyama et al. 1976; Kurihara et al. 1974; Matsukura et al. 1981; Sugimura and Kawachi 1973, 1978; Sugimura et al. 1971). A disadvantage of the canine model was the high frequency of small bowel sarcomas sometimes resulting in premature death (Koyama et al. 1976). Therefore, in recent years, the less mutagenic ENNG has been administered to dogs. ENNG-associated gastric tumors have been described as poorly to well-differentiated adenocarcinomas, as well as signet-ring cell carcinomas. A significant amount of cell necrosis has been observed in ENNG-induced tumors, possibly contributing to the long prepatent periods before clinical disease emergence (O'Gara and Adamson 1972; Sugihara et al. 1985; Szentirmay et al. 1990). Metastasis to regional lymph nodes and livers has been observed (Matsukura et al. 1981). Given the high prevalence of gastric *Helicobacter* sp. in dogs, it is possible that these organisms have a cocarcinogenic role, although studies to address this issue in dogs have not been performed.

N-Methyl-N′-nitro-N-nitrosoguanidine and N-Ethyl-N′-nitro-N-nitrosoguanidine in Nonhuman Primates

Nonhuman primates exhibit remarkable anatomic and physiologic similarities to humans. However, because of high costs and limited availability, these species are rarely used to investigate gastric carcinogenesis. Nevertheless, macaques were used as an early model for both naturally occurring and experimentally induced *H. pylori* gastritis, helping to cement the role of the bacterium as an important human pathogen (Dubois et al. 1994). In limited studies, researchers have documented that both rhesus (*Macaca mulatta*) and cynomolgus (*M. fasicularis*) macaques are susceptible to ENNG-induced gastric carcinoma (Ohgaki et al. 1986). Two male rhesus and three female cynomolgus macaques administered 200 or 300 µg/mL ENNG in drinking water for 11–26 months, but not those receiving lower chemical doses, developed antral carcinomas between 11 and 38 months (Ohgaki et al. 1986). The authors asserted that the neoplasms bore a closer resemblance to human tumors than carcinomas induced by MNNG in rats. However, the tumors were slow growing, as a follow-up publication noted that after 2 years, one of the biopsy-confirmed intramucosal signet-ring cell carcinomas still had not invaded the submucosa (Szentirmay et al. 1990). MNNG delivered via stomach tube resulted in gastric carcinoma in 2 of 4 male monkeys at 256 weeks. However, statistical evaluation was impeded by acute toxicity, morbidity, and mortality in the study group associated with this dosing regimen (Sharashidze et al. 1989).

Duodenogastric Reflux

Gastroenterostomy in humans has been shown to increase risk of stomach carcinoma at the site of anastomosis, although the rate of this complication is highly variable between studies (Sinning et al. 2007). Reflux of alkaline pancreaticoduodenal and or bile juices is believed to be the primary mechanism of tumor promotion. Rat models have been used to investigate the cause and progression of this syndrome. Depending on the type of experimental surgery performed, gastroenterostomy in rats results in relatively rapid development of premalignant and malignant lesions at the gastroenteric anastomosis site (Kondo et al. 1995). Surgery plus MNNG accelerates tumor progression in Wistar rats compared with MNNG only (Kondo 2002). In both human and rat models, studies have produced divergent results on the relative importance of pancreaticoduodenal versus bile juices toward tumor promotion. However, most studies agree that the presence of both mixtures is more tumorigenic than either alone. Oral administration of bile acids to rats promoted gastric tumors, alone or in combination with MNNG (Kondo et al. 1995). Gastric stump tumors associated with anastomotic operations may be of either the diffuse or intestinal type, with one or the other predominating under different clinical conditions. Concurrent *H. pylori* infection has been implicated as a copromoter of gastric stump carcinoma in some studies but not others (Seoane et al. 2005; Sinning et al. 2007). In the Mongolian gerbil, *H. pylori* was found to be a stronger tumor copromoter in MNNG-treated animals than was duodenal reflux, although the latter also increased tumor burden versus MNNG alone (Tanaka et al. 2004). Interestingly, *H. pylori*–associated lesions were attenuated by concurrent duodenal reflux in that system.

Cigarette Smoke and Tobacco Products

Epidemiologically there is strong evidence of an association between smoking and increased risk of stomach and esophageal cancers (Freedman et al. 2007; Ohtani et al. 2007). However, few studies have been performed in animal models on the ability of inhaled smoke products to promote gastric cancer. In one of the few reported inhalation studies, cigarette smoke did not cause gastric cancer in a Syrian Golden hamster model, nor did it promote tumors initiated by dimethylbenzanthracene (DMBA) (Dontenwill et al. 1973). In contrast, injected nicotine was shown to promote tumor growth of transplanted AGS stomach cancer cells in nude mice, and N′-nitrosonornicotine promoted cancer of the squamous but not glandular rat stomach (Shin et al. 2004; Singer and Taylor 1976). The food flavoring additive hickory smoke condensate did not promote stomach tumors in MNNG-initiated rats (Nishikawa et al. 1993). Catechol, a phenolic compound present in cigarette smoke, wood smoke, coffee, crude beet sugar, and onions, as well as hair dyes and photographic developers, produced gastric carcinoma in F344 rats, but only adenomas of the glandular stomach in B6C3F1 mice (Hirose et al. 1993). As described above, the Swedish moist snuff "snus" copromoted gastric carcinoma with *H. pylori* in INS-GAS mice (Stenstrom et al. 2007). Lycopene (a nutrient found in tomatoes and other foods) was shown to improve p53 and apoptosis profiles in ferrets exposed to cigarette smoke; however, cancer was not an endpoint in the untreated control group, nor was tumor copromotion by concurrent *H. mustelae* infection evaluated (Liu et al. 2006). Compared with other risk factors such as *Helicobacter* infection and N-nitroso compounds, the putative role of cigarette smoke and other tobacco products in gastric carcinogenesis has not received widespread attention in animal models. Additional work in this area is clearly indicated.

High Salt Intake and Gastric Cancer: Risk Factor or Red Herring?

Whether or not high salt ingestion promotes gastric cancer has been debated for decades (Geboers et al. 1985; Tuyns 1988). Advocates of the protumorigenic role of salt cite evidence of the deleterious effects of high salt from epidemiologic studies, biochemical analyses, and *in vivo* experiments (Joossens and Geboers 1981; Takahashi and Hasegawa 1986). Human epidemiologic surveys from different parts of the world have reported a relative risk for gastric cancer between 1.5 and 2.0 among cohorts with a self-reported preference for salt as a food seasoning (Buiatti et al. 1989; Correa et al. 1985; Joossens and Geboers 1981). Furthermore, those with a self-reported preference for inherently salty foods (such as miso soup), or foods cured in salt for preservation, were found to be at increased risk for stomach cancer (Kelley and Duggan 2003; Tajima and Tominaga 1985; Tuyns 1988; You et al. 1988). A prospective study in Japan reported an association between self-reported salt ingestion and stomach cancer risk; however, significance was only found among men, and correlations weakened when geographically stratified (Tsugane et al. 2004). Direct measurements of salt ingestion, or indirect measures through urinary sodium concentration, have shown a positive correlation with gastric cancer risk in some populations but not others (Correa et al. 1985; LaVecchia et al. 1987; Tsugane et al. 1991, 2004). In contrast, a prospective study in the United States found no association between self-reported salt preference and subsequent risk of stomach cancer, whereas high salt intake was significantly correlated with hypertension (Friedman and Parsonnet 1992). Nevertheless, based on the composite of current evidence, a Joint WHO/FAO Expert Consultation group has determined that high salt intake "probably increase[s] the risk of stomach cancer" (WHO 2003). Because data from humans remain inconclusive, animal models have been used to more directly evaluate the potential of salt to promote gastric carcinoma.

High Salt Intake Alone Does Not Induce Gastric Cancer

Few animal studies have been performed in which high salt intake was evaluated as a sole risk factor for gastric cancer. The normal proportion of sodium chloride in rodent chow is <0.5%. Mice fed a rice diet containing high-salt food (salted codfish and yakuri sold for food consumption in Japan) developed acute gastric mucosal damage but there was no report of tumors (Sato et al. 1959). In later studies, Swiss/ICR mice fed salted (10% w/w NaCl) rice diets for 3–12 months developed hypertrophy of the squamous forestomach coupled with oxyntic atrophy of the glandular stomach, but not tumors (Kodama et al. 1984). Most experiments that have included rodents receiving high salt alone have been part of a larger study examining the role of salt as a cocarcinogen. In virtually all species, high salt has been associated with gastric mucosal irritation, alterations in mucin production, variable oxyntic gland atrophy, and increased surface epithelial proliferation (Charnley and Tannenbaum 1985; Furihata et al. 1984; Ohgaki et al. 1989; Sorbye et al. 1988). However, no animal study to date has reported that high salt alone is sufficient to induce gastric carcinoma.

High-salt Variably Promotes Chemical Gastric Carcinogenesis

In rats initiated with MNNG, NaCl was shown to have a promotional effect on induction of gastric adenocarcinoma when given weekly by gavage or as a dietary additive in combination with MNNG in the drinking water (Newberne et al. 1987; Takahashi et al. 1983; Tatematsu et al. 1975). Interestingly, most of these animals developed adenocarcinomas of the antrum but not corpus (Takahashi et al. 1983). ACI rats given a single initiation dose of MNNG (0.25 mL/10 g BW) by intubation and fed a 10% NaCl diet had significantly increased tumors of both the forestomach and glandular stomach after 1 year compared with rats fed a standard diet (Watanabe et al. 1992). MNNG and NaCl ingestion in rats induced up to 100-fold increases of ornithine decarboxylase activity and about a 10-fold increase in DNA synthesis in the pyloric region of the stomach (Furihata et al. 1984). In a study using Wistar male rats administered a single intragastric dose of 250 mg/kg of MNNG, with or without a hypertonic salt solution (4.5 M) by intragastric intubation, coadministration of salt increased the risk of tumors in the squamous forestomach but not glandular stomach (Sorbye, Kvinnsland, and Svanes 1994). The quality and quantity of gastric surface mucus seems to influence any tumor-promoting effects of high salt. For example, in the above study using Wistar rats, the authors used DMSO as vehicle. DMSO is a gastric irritant that results in mucosal coverage by a thick mucus layer. The authors speculated that this may have protected the animals by impeding salt absorption, as well as improving circulation through vasodilation (Sorbye et al. 1993; Sorbye, Maaartmann-Moe, and Svanes 1994). Others have shown that the cocarcinogenic effect of salt on MNNG-induced cancer in rats was abolished when mucin was added to the diet (Tatematsu et al. 1976). Thus, a variety of host and environmental factors contribute to the cocarcinogenic effect of salt in gastric cancer.

High Salt Does Not Promote *Helicobacter*-Induced Gastric Carcinogenesis

Because of the significant geographic overlap among *H. pylori*, high-salt diet, and gastric cancer in places such as the Far East, some investigators have postulated that high salt intake may cooperate with *H. pylori* to promote tumors (Tsugane 2007; Tsugane and Sasazuki 2007). However, animal studies performed to date have not supported this hypothesis. No tumor synergism among high salt, constitutive hypergastrinemia, and *H. pylori* infection was found in the INS-GAS mouse model; indeed, male mice on the high-salt diet had less severe gastric lesions than those on the standard chow diet (Fox, Rogers et al. 2003). Similarly, in WT B6129 mice infected with *H. pylori*, a 7.5% high-salt diet (versus 0.25% NaCl in standard chow) was associated with oxyntic atrophy, foveolar hyperplasia, and mucous metaplasia defined as the repopulation of zymogenic glands with mucous neck cell-phenotype cells, but not dysplasia, neoplasia, or tumor copromotion (Rogers et al. 2005). In all mouse models, high salt has been associated with a shift in anti-*Helicobacter* humoral immunity from a Th1- to a Th2-predominant pattern, which may explain in part the absence of tumor promotion (Fox et al. 1999; Rogers et al. 2005). In a study of Mongolian gerbils, a 2.5% high-salt diet was associated with oxyntic atrophy and foveolar cell hyperplasia; however, there was no dysplasia or evidence of cotumor promotion with *H. pylori* (Bergin et al. 2003). A synergistic promoting effect of *H. pylori* and high salt on MNU-initiated gastric cancer was identified in gerbils (Nozaki et al. 2002). However, chemical initiation was required to induce cancer, and no tumors developed in gerbils untreated with MNU regardless of *H. pylori* or high salt exposure. Thus, to date, animal models have not supported the assertion that high salt promotes gastric tumorigenesis in the absence of chemical initiation. The complex interactions that determine whether or not high salt has any effect on gastric carcinogenesis in animals agree with disparate results from human epidemiologic studies. Whereas high salt clearly has negative effects on human health parameters such as hypertension, its culpability in the promotion of gastric cancer remains circumspect.

Combined Chemical and Infectious Gastric Carcinogenesis

Exposure to N-nitroso and nitrosamine compounds is a significant risk factor for human gastric carcinoma, as is *H. pylori* infection. However, interactions between these chemical and infectious agents as cocarcinogens have only recently become the subject of scientific inquiry. Despite the presence of high nitrites in drinking water, and ingestion of nitrosated foods in some populations, the inconsistency of epidemiologic data suggests that high dietary nitrites/nitrates alone are insufficient to incite gastric cancer in many individuals. For many years, bacteria have been incriminated as cocarcinogens in experimental animal models of chemical carcinogenesis (Schreiber et al. 1972; Sumi and Miyakawa 1981). Many studies have shown that conventional and SPF mice can be infected with protumorigenic *H. felis* and *H. pylori*. However, studies with MNNG and MNU in the mouse have been limited, and the dosage of the carcinogens needed in this species is largely undetermined. In a study involving BALB/c mice, MNU-treated animals infected with *H. felis* exhibited significantly greater epithelial hyperplasia than mice treated with MNU alone (p <0.05) (Shimizu et al. 1998). Because of the limited information on chemical carcinogenesis in mice, further work is needed to more fully develop mouse models of combined chemical and infectious gastric carcinoma. The same may be said of rats, but for the opposite reason. Whereas the rat is an historically important model of chemical carcinogenesis, few studies have been conducted to characterize gastric *Helicobacter* infection in this species. One study demonstrated that germ-free rats can be colonized with *H. felis* (Fox et al. 1991), but there is a paucity of data on other gastric *Helicobacter* spp., including *H. pylori*. Given the historical importance of rats in chemical carcinogenesis studies, characterization of infectious gastritis and tumor promotion in this species is a significant unmet need. Fortunately, the Mongolian gerbil model has been well characterized with regard to combined chemical and infectious gastric carcinogenesis. Numerous studies have shown that gerbils initiated either with MNNG or MNU and infected with *H. pylori* develop significantly more tumors of the glandular stomach than do gerbils treated with the chemical carcinogen alone (Shimizu, Inada, Nakanishi et al. 1999; Shimizu, Inada, Tsukamoto et al. 1999; Sugiyama et al. 1998). Therefore, animal models are helping to crystallize the fact that *Helicobacter* infection can synergistically promote the action of chemical mutagens, resulting in accelerated gastric carcinogenesis. This is highly relevant to human populations in areas endemic for both of these risk factors, including sub-Saharan Africa and portions of the Far East.

Naturally Occurring Gastric Carcinoma in Animals

Naturally Occurring Gastric Carcinoma in Rodents

Gastric cancer is an uncommon naturally occurring neoplasm of most laboratory animal species (Barker and Dreumel 1985; Jones and Hunt 1983; McClure et al. 1978; Squire et al. 1978). Spontaneous forestomach squamous cell carcinomas occur rarely, but experimentally they have been associated with parasitic infection, dietary deficiencies, and many chemical agents (Bralow 1972; Klein and Palmer

1941; Sugimura and Kawachi 1973). Because forestomach tumors are more analogous to human esophageal cancer, they will not be discussed further here. Spontaneous glandular stomach tumors in rodents are rare, but have been reported to occur more frequently in the Syrian hamster and Strain I mice. Gastric carcinoids (tumors of intramucosal endocrine cells) occur frequently in multi-mammate mice (*Praomys [Mastomys] natalensis*) which are used as a model of this uncommon tumor, but carcinoids bear little relationship to gastric carcinoma and have no known association with *Helicobacter* infection. Idiopathic hyperplasia, dysplasia, and carcinoma of the glandular stomach were reported in a colony of cotton rats (*Sigmodon hispidus*) (Kawase and Ishikura 1995). In this instance, the authors did not determine whether or not the rats were infected with gastric *Helicobacter*. Nevertheless, the emergence of carcinoma in a background of adenomatous hyperplasia suggests that the cotton rat model may be useful for studying multistep tumor progression (Kawase and Ishikura 1995).

Naturally Occurring Gastric Carcinoma in Dogs and Cats

Of the species often used for biomedical research, spontaneous gastric adenocarcinoma is most frequently reported in the dog although it accounts for ≤1% of all neoplasms in this species (Swann and Holt 2002). As is the case in humans, canine gastric adenocarcinoma usually originates along the lesser curvature (Campbell and Lauder 1952; Murray et al. 1972). All of the common histologic variants that affect humans have also been reported in the dog (Lingeman et al. 1971). Surgical interventions including partial gastrectomy, Billroth I gastroduodenostomy, and gastro-jejunostomy all have been performed in canine patients, with variable success (Swann and Holt 2002). Unfortunately, long-term survival rates are poor (Swann and Holt 2002). In cats, gastric lymphoma is much more common than carcinoma (Gualtieri et al. 1999). Nevertheless, carcinomas do occur, as shown in a report of gastritis-associated adenocarcinomas in two cats from the same household. The cause of gastritis was not determined, although surface-associated spiral bacteria (as well as nematodes) were detected by histopathology (Dennis et al. 2006). Further work is needed to determine whether naturally acquired *Helicobacter* infection increases the risk of gastric carcinoma in dogs and cats.

Naturally Occurring Gastric Carcinoma in Nonhuman Primates

Spontaneous gastric tumors in monkeys and apes, although reported, are rare (Kent 1960; Kimbrough 1966; McCarrison 1915; O'Gara and Adamson 1972; Ohgaki et al. 1984; Ruch 1959). In a survey from a German nonhuman primate center, two gastric adenocarcinomas were diagnosed in cotton-top tamarins (*Saguinus oedipus*) who also had colitis-associated colorectal carcinomas (Brack 1998). Monkeys parasitized with the nematode *Nochtia nochti*, however, are known to develop gastric tumors (Bonne and Sandground 1939). One monkey developed a gastric tumor after long-term cholesterol feeding (Lapin and Iakovleva 1959). Because gastric *Helicobacters* are commonly found in these species and may be acquired at an early age, further studies are needed to characterize the natural history of gastric carcinogenesis in nonhuman primates (Solnick et al. 2003, 2006).

Conclusions

Molecular interactions and specific gene functions can be inferred from cell-based and other *in vitro* experiments, but only in animal models can the complex interactions among host, infectious agents, and environment be fully recapitulated. Animal models are particularly vital to the study of gastric carcinogenesis, where all three of these elements individually and combinatorially significantly influence disease outcome. Animal models first demonstrated the carcinogenic potential of specific environmental hazards such as N-nitroso compounds, and later validated the initially provocative hypothesis that *H. pylori* infection is a major risk determinant for human intestinal-type gastric adenocarcinoma (Forman 1991; Mirvish 1971, 1983). Rodent models in particular have been used to demonstrate specific pathways in gastric carcinogenesis including those associated with cell proliferation, inflammation, metaplasia, and endocrine regulation. With the exploding number of genetically engineered mouse models, additional molecular pathways, including currently unsuspected ones, undoubtedly will be revealed. Nevertheless, cutting-edge technology will never replace insightful observation. Arguably, the most important discovery in gastroenterology was made by Warren and Marshall using only a common silver stain and a light microscope (Marshall and Warren 1984; Warren 1984).

References

Acheson, D. W., and Luccioli, S. 2004. Microbial-gut interactions in health and disease. Mucosal immune responses. Best Pract Res Clin Gastroenterol 18:387–404.

Babinet, C., Morello, D., and Renard, J. P. 1989. Transgenic mice. Genome 31:938–949.

Barker, I., and Dreumel, A. 1985. The alimentary system. In: Jubb, K., Kennedy, P., and Palmer, N., editors. Pathology of domestic animals. New York: Academic Press; pp. 1–237.

Bergin, I. L., Sheppard, B. J., and Fox, J. G. 2003. Helicobacter pylori infection and high dietary salt independently induce atrophic gastritis and intestinal metaplasia in commercially available outbred Mongolian gerbils. Dig Dis Sci 48:475–485.

Blanchard, T., Nedrud, J., and Czinn, S. 1996. Development of a rat model of H. pylori infection and disease to study the role of H. pylori in gastric cancer incidence. Gut 39:A76.

Bolon, B. 2004. Genetically engineered animals in drug discovery and development: a maturing resource for toxicologic research. Basic Clin Pharmacol Toxicol 95:154–161.

Bolon, B. 2006. Internet resources for phenotyping engineered rodents. Ilar J 47:163–171.

Bonne, C., and Sandground, J. 1939. On the production of gastric tumors, bordering on malignancy in Javenese monkeys through the agency of Nochtia nochti, a parasitic nematoda. Am J Cancer 37:173–185.

Boonjakuakul, J. K., Canfield, D. R., and Solnick, J. V. 2005. Comparison of Helicobacter pylori virulence gene expression in vitro and in the Rhesus macaque. Infect Immun 73:4895–4904.

Brack, M. 1998. Gastrointestinal tumors observed in nonhuman primates at the German primate center. J Med Primatol 27:319–324.

Bralow, S. P. 1972. Experimental gastric carcinogenesis. Digestion 5:290–310.

Brayton, C., Justice, M., and Montgomery, C. A. 2001. Evaluating mutant mice: anatomic pathology. Vet Pathol 38:1–19.

Buiatti, E., Palli, D., Decarli, A., Amadori, D., Avellini, C., Bianchi, S., Biserni, R., Cipriani, F., Cocco, P., Giacosa, A., et al. 1989. A case-control study of gastric cancer and diet in Italy. Int J Cancer 44:611–616.

Cai, X., Carlson, J., Stoicov, C., Li, H., Wang, T. C., and Houghton, J. 2005. Helicobacter felis eradication restores normal architecture and inhibits gastric cancer progression in C57BL/6 mice. Gastroenterology 128:1937–1952.

Campbell, R. S., and Lauder, I. M. 1952. Gastric carcinoma in the dog. J Comp Pathol 62:275–278.

Charnley, G., and Tannenbaum, S. R. 1985. Flow cytometric analysis of the effect of sodium chloride on gastric cancer risk in the rat. Cancer Res 45:5608–5616.

Chen, D., Stenstrom, B., Zhao, C. M., and Wadstrom, T. 2007. Does Helicobacter pylori infection per se cause gastric cancer or duodenal ulcer? Inadequate evidence in Mongolian gerbils and inbred mice. FEMS Immunol Med Microbiol 50:184–189.

Citino, S. B., and Munson, L. 2005. Efficacy and long-term outcome of gastritis therapy in cheetahs (Acinonyx jubatus). J Zoo Wildl Med 36:401–416.

Correa, P. 1992. Human gastric carcinogenesis: a multistep and multifactorial process—first American Cancer Society Award Lecture on Cancer Epidemiology and Prevention. Cancer Res 52:6735–6740.

Correa, P., Fontham, E., Pickle, L. W., Chen, V., Lin, Y. P., and Haenszel, W. 1985. Dietary determinants of gastric cancer in south Louisiana inhabitants. J Natl Cancer Inst 75:645–654.

Court, M., Robinson, P. A., Dixon, M. F., Jeremy, A. H., and Crabtree, J. E. 2003. The effect of gender on Helicobacter felis-mediated gastritis, epithelial cell proliferation, and apoptosis in the mouse model. J Pathol 201:303–311.

Croinin, T. O., McCormack, A., van Vliet, A. H., Kusters, J. G., and Bourke, B. 2007. Random mutagenesis to identify novel Helicobacter mustelae virulence factors. FEMS Immunol Med Microbiol 50:257–263.

De Bock, M., D'Herde, K., Duchateau, L., Hellemans, A., Decostere, A., Haesebrouck, F., and Ducatelle, R. 2006. The effect of Helicobacter felis and Helicobacter bizzozeronii on the gastric mucosa in Mongolian gerbils: a sequential pathological study. J Comp Pathol 135:226–236.

Dennis, M. M., Bennett, N., and Ehrhart, E. J. 2006. Gastric adenocarcinoma and chronic gastritis in two related Persian cats. Vet Pathol 43:358–362.

de Vries, A. C., Haringsma, J., and Kuipers, E. J. 2007. The detection, surveillance and treatment of premalignant gastric lesions related to Helicobacter pylori infection. Helicobacter 12:1–15.

Donehower, L. A., Harvey, M., Slagle, B. L., McArthur, M. J., Montgomery, C. A., Jr., Butel, J. S., and Bradley, A. 1992. Mice deficient for p53 are developmentally normal but susceptible to spontaneous tumours. Nature 356:215–221.

Dontenwill, W., Chevalier, H. J., Harke, H. P., Lafrenz, U., Reckzeh, G., and Schneider, B. 1973. Investigations on the effects of chronic cigarette-smoke inhalation in Syrian golden hamsters. J Natl Cancer Inst 51:1781–1832.

Dubois, A., Fiala, N., Heman-Ackah, L. M., Drazek, E. S., Tarnawski, A., Fishbein, W. N., Perez-Perez, G. I., and Blaser, M. J. 1994. Natural gastric infection with Helicobacter pylori in monkeys: a model for spiral bacteria infection in humans. Gastroenterology 106:1405–1417.

Esteves, M. I., Schrenzel, M. D., Marini, R. P., Taylor, N. S., Xu, S., Hagen, S., Feng, Y., Shen, Z., and Fox, J. G. 2000. Helicobacter pylori gastritis in cats with long-term natural infection as a model of human disease. Am J Pathol 156:709–721.

Forman, D. 1991. Helicobacter pylori infection: a novel risk factor in the etiology of gastric cancer. J Natl Cancer Inst 83:1702–1703.

Fostel, J. M., Burgoon, L., Zwickl, C., Lord, P., Corton, J. C., Bushel, P. R., Cunningham, M., Fan, L., Edwards, S. W., Hester, S., Stevens, J., Tong, W., Waters, M., Yang, C., and Tennant, R. 2007. Toward a checklist for exchange and interpretation of data from a toxicology study. Toxicol Sci 99:26–34.

Fox, J. G. 1995. Non-human reservoirs of Helicobacter pylori. Aliment Pharmacol Ther 9(Suppl 2): 93–103.

Fox, J. G. 1998. Biology and diseases of the ferret. Baltimore: Williams and Wilkins.

Fox, J. G., Batchelder, M., Marini, R., Yan, L., Handt, L., Li, X., Shames, B., Hayward, A., Campbell, J., and Murphy, J. C. 1995. Helicobacter pylori-induced gastritis in the domestic cat. Infect Immun 63:2674–2681.

Fox, J. G., Beck, P., Dangler, C. A., Whary, M. T., Wang, T. C., Shi, H. N., and Nagler-Anderson, C. 2000. Concurrent enteric helminth infection modulates inflammation and gastric immune responses and reduces Helicobacter-induced gastric atrophy. Nat Med 6:536–542.

Fox, J. G., Blanco, M., Murphy, J. C., Taylor, N. S., Lee, A., Kabok, Z., and Pappo, J. 1993. Local and systemic immune responses in murine Helicobacter felis active chronic gastritis. Infect Immun 61:2309–2315.

Fox, J. G., Cabot, E. B., Taylor, N. S., and Laraway, R. 1988. Gastric colonization by Campylobacter pylori subsp. mustelae in ferrets. Infect Immun 56:2994–2996.

Fox, J. G., Chilvers, T., and Goodwin, C. 1989. Campylobacter mustelae, a new species resulting from the elevation of Campylobacter pylori subsp. mustelae to species status. Int J Sys Bacteriol 39:301–303.

Fox, J. G., Correa, P., Taylor, N. S., Lee, A., Otto, G., Murphy, J. C., and Rose, R. 1990. Helicobacter mustelae-associated gastritis in ferrets. An animal model of Helicobacter pylori gastritis in humans. Gastroenterology 99:352–361.

Fox, J. G., Dangler, C. A., Sager, W., Borkowski, R., and Gliatto, J. M. 1997. Helicobacter mustelae-associated gastric adenocarcinoma in ferrets (Mustela putorius furo). Vet Pathol 34:225–229.

Fox, J. G., Dangler, C. A., Taylor, N. S., King, A., Koh, T. J., and Wang, T. C. 1999. High-salt diet induces gastric epithelial hyperplasia and parietal cell loss, and enhances Helicobacter pylori colonization in C57BL/6 mice. Cancer Res 59:4823–4828.

Fox, J. G., Dangler, C. A., Whary, M. T., Edelman, W., Kucherlapati, R., and Wang, T. C. 1997. Mice carrying a truncated Apc gene have diminished gastric epithelial proliferation, gastric inflammation, and humoral immunity in response to Helicobacter felis infection. Cancer Res 57:3972–3978.

Fox, J. G., Edrise, B. M., Cabot, E. B., Beaucage, C., Murphy, J. C., and Prostak, K. S. 1986. Campylobacter-like organisms isolated from gastric mucosa of ferrets. Am J Vet Res 47:236–239.

Fox, J. G., and Lee, A. 1993. Gastric Helicobacter infection in animals: natural and experimental infections. In: Goodwin, C., and Worsley, B., editors. Biology and clinical practice. Boca Raton, FL: CRC Press; pp. 407–430.

Fox, J. G., Lee, A., Otto, G., Taylor, N. S., and Murphy, J. C. 1991. Helicobacter felis gastritis in gnotobiotic rats: an animal model of Helicobacter pylori gastritis. Infect Immun 59:785–791.

Fox, J. G., Li, X., Cahill, R. J., Andrutis, K., Rustgi, A. K., Odze, R., and Wang, T. C. 1996. Hypertrophic gastropathy in Helicobacter felis-infected wild-type C57BL/6 mice and p53 hemizygous transgenic mice. Gastroenterology 110:155–166.

Fox, J. G., Li, X., Yan, L., Cahill, R. J., Hurley, R., Lewis, R., and Murphy, J. C. 1996. Chronic proliferative hepatitis in A/JCr mice associated with persistent Helicobacter hepaticus infection: a model of Helicobacter-induced carcinogenesis. Infect Immun 64:1548–1558.

Fox, J. G., Rogers, A. B., Ihrig, M., Taylor, N. S., Whary, M. T., Dockray, G., Varro, A., and Wang, T. C. 2003. Helicobacter pylori-associated gastric cancer in INS-GAS mice is gender specific. Cancer Res 63:942–950.

Fox, J. G., Rogers, A. B., Whary, M. T., Ge, Z., Ohtani, M., Jones, E. K., and Wang, T. C. 2007. Accelerated progression of gastritis to dysplasia in the pyloric antrum of TFF2/C57BL6 x Sv129 Helicobacter pylori-infected mice. Am J Pathol 171:1520–1528.

Fox, J. G., Sheppard, B. J., Dangler, C. A., Whary, M. T., Ihrig, M., and Wang, T. C. 2002. Germ-line p53-targeted disruption inhibits Helicobacter-induced premalignant lesions and invasive gastric carcinoma through down-regulation of Th1 proinflammatory responses. Cancer Res 62:696–702.

Fox, J. G., Taylor, N. S., Howe, S., Tidd, M., Xu, S., Paster, B. J., and Dewhirst, F. E. 2006. Helicobacter anseris sp. nov. and Helicobacter brantae sp. nov., isolated from feces of resident Canada geese in the greater Boston area. Appl Environ Microbiol 72:4633–4637.

Fox, J. G., and Wang, T. C. 2007. Inflammation, atrophy, and gastric cancer. J Clin Invest 117:60–69.

Fox, J. G., Wang, T. C., Rogers, A. B., Poutahidis, T., Ge, Z., Taylor, N., Dangler, C. A., Israel, D. A., Krishna, U., Gaus, K., and Peek, R. M., Jr. 2003. Host and microbial constituents influence Helicobacter pylori-induced cancer in a murine model of hypergastrinemia. Gastroenterology 124:1879–1890.

Fox, J. G., Wishnok, J. S., Murphy, J. C., Tannenbaum, S. R., and Correa, P. 1993. MNNG-induced gastric carcinoma in ferrets infected with Helicobacter mustelae. Carcinogenesis 14:1957–1961.

Franco, A. T., Israel, D. A., Washington, M. K., Krishna, U., Fox, J. G., Rogers, A. B., Neish, A. S., Collier-Hyams, L., Perez-Perez, G. I., Hatakeyama, M., Whitehead, R., Gaus, K., O'Brien, D. P., Romero-Gallo, J., and Peek, R. M., Jr. 2005. Activation of beta-catenin by carcinogenic Helicobacter pylori. Proc Natl Acad Sci USA 102:10646–10651.

Freedman, N. D., Abnet, C. C., Leitzmann, M. F., Mouw, T., Subar, A. F., Hollenbeck, A. R., and Schatzkin, A. 2007. A prospective study of tobacco, alcohol, and the risk of esophageal and gastric cancer subtypes. Am J Epidemiol 165:1424–1433.

Friedman, G. D., and Parsonnet, J. 1992. Salt intake and stomach cancer: some contrary evidence. Cancer Epidemiol Biomarkers Prev 1:607–608.

Fujimura, S., Kogure, K., Oboshi, S., and Sugimura, T. 1970. Production of tumors in glandular stomach of hamsters by N-methyl-N'-nitro-N-nitrosoguanidine. Cancer Res 30:1444–1448.

Fujioka, T., Murakami, K., Kodama, M., Kagawa, J., Okimoto, T., and Sato, R. 2002. Helicobacter pylori and gastric carcinoma—from the view point of animal model. Keio J Med 51(Suppl 2): 69–73.

Furihata, C., Sato, Y., Hosaka, M., Matsushima, T., Furukawa, F., and Takahashi, M. 1984. NaCl induced ornithine decarboxylase and DNA synthesis in rat stomach mucosa. Biochem Biophys Res Commun 121:1027–1032.

Futagami, S., Suzuki, K., Hiratsuka, T., Shindo, T., Hamamoto, T., Tatsuguchi, A., Ueki, N., Shinji, Y., Kusunoki, M., Wada, K., Miyake, K., Gudis, K., Tsukui, T., and Sakamoto, C. 2006. Celecoxib inhibits Cdx2 expression and prevents gastric cancer in Helicobacter pylori-infected Mongolian gerbils. Digestion 74:187–198.

Futagami, S., Suzuki, K., Hiratsuka, T., Shindo, T., Hamamoto, T., Ueki, N., Kusunoki, M., Miyake, K., Gudis, K., Tsukui, T., and Sakamoto, C. 2007. Chemopreventive effect of celecoxib in gastric cancer. Inflammopharmacology 15:1–4.

Gasbarrini, A., Carloni, E., Gasbarrini, G., and Menard, A. 2003. Helicobacter pylori and extragastric diseases—other Helicobacters. Helicobacter 8(Suppl 1):68–76.

Geboers, J., Joosens, J., and Kesteloot, H. 1985. Epidemiology of stomach cancer. In: Joossens, J., Hill, M., and Geboers, J., editors. Diet and human carcinogenesis. Amsterdam: Elsevier; pp. 81–95.

Goldenring, J. R., and Nomura, S. 2006. Differentiation of the gastric mucosa. III. Animal models of oxyntic atrophy and metaplasia. Am J Physiol Gastrointest Liver Physiol 291:G999–1004.

Gottfried, M. R., Washington, K., and Harrell, L. J. 1990. Helicobacter pylori-like microorganisms and chronic active gastritis in ferrets. Am J Gastroenterol 85:813–818.

Gualtieri, M., Monzeglio, M. G., and Scanziani, E. 1999. Gastric neoplasia. Vet Clin North Am Small Anim Pract 29:415–440.

Hahm, K. B., Song, Y. J., Oh, T. Y., Lee, J. S., Surh, Y. J., Kim, Y. B., Yoo, B. M., Kim, J. H., Han, S. U., Nahm, K. T., Kim, M. W., Kim, D. Y., and Cho, S. W. 2003. Chemoprevention of

Helicobacter pylori-associated gastric carcinogenesis in a mouse model: is it possible? J Biochem Mol Biol 36:82–94.

Halldorsdottir, A. M., Sigurdardottrir, M., Jonasson, J. G., Oddsdottir, M., Magnusson, J., Lee, J. R., and Goldenring, J. R. 2003. Spasmolytic polypeptide-expressing metaplasia (SPEM) associated with gastric cancer in Iceland. Dig Dis Sci 48:431–441.

Handt, L. K., Fox, J. G., Dewhirst, F. E., Fraser, G. J., Paster, B. J., Yan, L. L., Rozmiarek, H., Rufo, R., and Stalis, I. H. 1994. Helicobacter pylori isolated from the domestic cat: public health implications. Infect Immun 62:2367–2374.

Handt, L. K., Fox, J. G., Stalis, I. H., Rufo, R., Lee, G., Linn, J., Li, X., and Kleanthous, H. 1995. Characterization of feline Helicobacter pylori strains and associated gastritis in a colony of domestic cats. J Clin Microbiol 33:2280–2289.

Haorah, J., Zhou, L., Wang, X., Xu, G., and Mirvish, S. S. 2001. Determination of total N-nitroso compounds and their precursors in frankfurters, fresh meat, dried salted fish, sauces, tobacco, and tobacco smoke particulates. J Agric Food Chem 49:6068–6078.

Harper, C. M., Dangler, C. A., Xu, S., Feng, Y., Shen, Z., Sheppard, B., Stamper, A., Dewhirst, F. E., Paster, B. J., and Fox, J. G. 2000. Isolation and characterization of a Helicobacter sp. from the gastric mucosa of dolphins, Lagenorhynchus acutus and Delphinus delphis. Appl Environ Microbiol 66:4751–4757.

Harper, C. G., Feng, Y., Xu, S., Taylor, N. S., Kinsel, M., Dewhirst, F. E., Paster, B. J., Greenwell, M., Levine, G., Rogers, A., and Fox, J. G. 2002. Helicobacter cetorum sp. nov., a urease-positive Helicobacter species isolated from dolphins and whales. J Clin Microbiol 40:4536–4543.

Harper, C. M., Xu, S., Feng, Y., Dunn, J. L., Taylor, N. S., Dewhirst, F. E., and Fox, J. G. 2002. Identification of novel Helicobacter spp. from a beluga whale. Appl Environ Microbiol 68:2040–2043.

Harper, C. G., Xu, S., Rogers, A. B., Feng, Y., Shen, Z., Taylor, N. S., Dewhirst, F. E., Paster, B. J., Miller, M., Hurley, J., and Fox, J. G. 2003. Isolation and characterization of novel Helicobacter spp. from the gastric mucosa of harp seals Phoca groenlandica. Dis Aquat Organ 57:1–9.

Harris, C. C. 1996. p53 tumor suppressor gene: from the basic research laboratory to the clinic— an abridged historical perspective. Carcinogenesis 17:1187–1198.

Harvey, M., McArthur, M. J., Montgomery, C. A., Jr., Butel, J. S., Bradley, A., and Donehower, L. A. 1993. Spontaneous and carcinogen-induced tumorigenesis in p53-deficient mice. Nat Genet 5:225–229.

Hickman-Davis, J. M. 2001. Implications of mouse genotype for phenotype. News Physiol Sci 16:19–22.

Hirayama, F., Takagi, S., Kusuhara, H., Iwao, E., Yokoyama, Y., and Ikeda, Y. 1996. Induction of gastric ulcer and intestinal metaplasia in mongolian gerbils infected with Helicobacter pylori. J Gastroenterol 31:755–757.

Hirose, M., Fukushima, S., Tanaka, H., Asakawa, E., Takahashi, S., and Ito, N. 1993. Carcinogenicity of catechol in F344 rats and B6C3F1 mice. Carcinogenesis 14:525–529.

Houghton, J., Stoicov, C., Nomura, S., Rogers, A. B., Carlson, J., Li, H., Cai, X., Fox, J. G., Goldenring, J. R., and Wang, T. C. 2004. Gastric cancer originating from bone marrow-derived cells. Science 306:1568–1571.

Huff, J. L., Hansen, L. M., and Solnick, J. V. 2004. Gastric transcription profile of Helicobacter pylori infection in the rhesus macaque. Infect Immun 72:5216–5226.

IARC. 1977. N-nitroso-N-methylurea. In: Monographs on the evaluation of carcinogenic risk of chemicals to man. Lyon: IARC; pp. 227–255.

Ihrig, M., Whary, M. T., Dangler, C. A., and Fox, J. G. 2005. Gastric Helicobacter infection induces a Th2 phenotype but does not elevate serum cholesterol in mice lacking inducible nitric oxide synthase. Infect Immun 73:1664–1670.

Jones, T., and Hunt, R. 1983. Veterinary pathology. Philadelphia: Lea & Febiger.

Joossens, J. V., and Geboers, J. 1981. Nutrition and gastric cancer. Proc Nutr Soc 40:37–46.

Kang, W., Rathinavelu, S., Samuelson, L. C., and Merchant, J. L. 2005. Interferon gamma induction of gastric mucous neck cell hypertrophy. Lab Invest 85:702–715.

Karita, M., Li, Q., Cantero, D., and Okita, K. 1994. Establishment of a small animal model for human Helicobacter pylori infection using germ-free mouse. Am J Gastroenterol 89:208–213.

Kawase, S., and Ishikura, H. 1995. Female-predominant occurrence of spontaneous gastric adeno-carcinoma in cotton rats. Lab Anim Sci 45:244–248.

Kelley, J. R., and Duggan, J. M. 2003. Gastric cancer epidemiology and risk factors. J Clin Epidemiol 56:1–9.

Kent, S. P. 1960. Spontaneous and induced malignant neoplasms in monkeys. Ann NY Acad Sci 85:819–827.

Kimbrough, R. 1966. Spontaneous malignant gastric tumor in a Rhesus monkey (Macaca mulatta). Arch Pathol 81:343–351.

Kleanthous, H., Tibbitts, T., Bakios, T., et al. 1995. In vivo selection of a highly adapted H. pylori isolate and the development of an H. pylori mouse model for studying vaccine efficacy and attenuating lesions. Gut 37:A94.

Klein, A., and Palmer, W. 1941. Experimental gastric carcinoma: a critical review with comments on the criteria of induced malignancy. J Natl Cancer Inst 1:559–584.

Kodama, M., Kodama, T., Fukami, H., Ogiu, T., and Kodama, M. 1992. Comparative genetics of host response to N-methyl 1-N'-nitro-N-nitrosoguanidine. I. A lack of tumor production in the glandular stomach of Swiss mouse. Anticancer Res 12:441–449.

Kodama, M., Kodama, T., Suzuki, H., and Kondo, K. 1984. Effect of rice and salty rice diets on the structure of mouse stomach. Nutr Cancer 6:135–147.

Kogure, K., Sasadaira, H., Kawachi, T., Shimosato, Y., and Tokunaga, A. 1974. Further studies on induction of stomach cancer in hamsters by N-methyl-N'-nitro-N-nitrosoguanidine. Br J Cancer 29:132–142.

Kondo, K. 2002. Duodenogastric reflux and gastric stump carcinoma. Gastric Cancer 5:16–22.

Kondo, K., Kojima, H., Akiyama, S., Ito, K., and Takagi, H. 1995. Pathogenesis of adenocarci-noma induced by gastrojejunostomy in Wistar rats: role of duodenogastric reflux. Carcinogenesis 16:1747–1751.

Koyama, Y., Omori, K., Hirota, T., Sano, R., and Ishihara, K. 1976. Leiomyosarcomas of the small intestine induced in dogs by N-methul-N'-nitro-N-nitrosoguanidine. Gann 67:241–251.

Kurihara, M., Shirakabe, H., Murakami, T., Yasui, A., and Izumi, T. 1974. A new method for producing adenocarcinomas in the stomach of dogs with N-ethyl-N'-nitro-N-nitrosoguanidine. Gann 65:163–177.

Kurt-Jones, E. A., Cao, L., Sandor, F., Rogers, A. B., Whary, M. T., Nambiar, P. R., Cerny, A., Bowen, G., Yan, J., Takaishi, S., Chi, A. L., Reed, G., Houghton, J., Fox, J. G., and Wang, T. C. 2007. Trefoil family factor 2 is expressed in murine gastric and immune cells and controls both gastrointestinal inflammation and systemic immune responses. Infect Immun 75:471–480.

Kuzushita, N., Rogers, A. B., Monti, N. A., Whary, M. T., Park, M. J., Aswad, B. I., Shirin, H., Koff, A., Eguchi, H., and Moss, S. F. 2005. p27kip1 deficiency confers susceptibility to gastric carcinogenesis in Helicobacter pylori-infected mice. Gastroenterology 129:1544–1556.

Lapin, B. A., and Iakovleva, L. A. 1959. [Tumors of the gastrointestinal tract in monkeys follow-ing prolonged intake of cholesterol.]. Arkh Patol 21(10):25–30.

LaVecchia, C., Negri, E., Decarli, A., et al. 1987. A case-control study of diet and gastric cancer in northern Italy. Int J Cancer 40:198–205.

Lecoindre, P., Chevallier, M., Peyrol, S., Boude, M., Ferrero, R. L., and Labigne, A. 2000. Gastric Helicobacters in cats. J Feline Med Surg 2:19–27.

Lee, A. 2000. Animal models of gastroduodenal ulcer disease. Baillieres Best Pract Res Clin Gastroenterol 14:75–96.

Lee, A., Fox, J. G., Otto, G., and Murphy, J. 1990. A small animal model of human Helicobacter pylori active chronic gastritis. Gastroenterology 99:1315–1323.

Lee, A., Hazell, S. L., O'Rourke, J., and Kouprach, S. 1988. Isolation of a spiral-shaped bacterium from the cat stomach. Infect Immun 56:2843–2850.

Lee, A., O'Rourke, J., De Ungria, M. C., Robertson, B., Daskalopoulos, G., and Dixon, M. F. 1997. A standardized mouse model of Helicobacter pylori infection: introducing the Sydney strain. Gastroenterology 112:1386–1397.

Lingeman, C. H., Garner, F. M., and Taylor, D. O. 1971. Spontaneous gastric adenocarcinomas of dogs: a review. J Natl Cancer Inst 47:137–153.

Liu, C., Russell, R. M., and Wang, X. D. 2006. Lycopene supplementation prevents smoke-induced changes in p53, p53 phosphorylation, cell proliferation, and apoptosis in the gastric mucosa of ferrets. J Nutr 136:106–111.

Marchetti, M., Arico, B., Burroni, D., Figura, N., Rappuoli, R., and Ghiara, P. 1995. Development of a mouse model of Helicobacter pylori infection that mimics human disease. Science 267:1655–1658.

Maronpot, R. R., Flake, G., and Huff, J. 2004. Relevance of animal carcinogenesis findings to human cancer predictions and prevention. Toxicol Pathol 32(Suppl 1):40–48.

Marshall, B. J., and Warren, J. R. 1984. Unidentified curved bacilli in the stomach of patients with gastritis and peptic ulceration. Lancet 1:1311–1315.

Matsukura, N., Morino, K., Ohgaki, H., et al. 1981. Canine gastric carcinoma as an animal model of human gastric carcinoma. Stomach Intest 16:715–722.

McCarrison, R. 1915. The pathogenesis of deficiency disease on the occurrence of recently developed cancer of the stomach in a monkey fed on food deficient in vitamins. Indian J Med Res 7:342–345.

McClure, H., Chapman, W., Hooper, B., et al. 1978. The digestive system. In: Benirschke, K., Garner, F., and Jones, T., editors. Pathology of laboratory animals. New York: Springer-Verlag; pp. 125–317.

McColm, A., Bagshaw, J., O'Malley, C., et al. 1995. Development of a mouse model of gastric colonisation with Helicobacter pylori. Gut 37:A50.

Mensinga, T. T., Speijers, G. J., and Meulenbelt, J. 2003. Health implications of exposure to environmental nitrogenous compounds. Toxicol Rev 22:41–51.

Ming, S.-C., and Hirota, T. 1998. Malignant epithelial tumors of the stomach. In: Ming, S.-C., and Goldman, H., editors. Pathology of the gastrointestinal tract. Baltimore: Williams & Wilkins; pp. 607–650.

Mirvish, S. S. 1971. Kinetics of nitrosamide formation from alkylureas, N-alkylurethans, and alkylguanidines: possible implications for the etiology of human gastric cancer. J Natl Cancer Inst 46:1183–1193.

Mirvish, S. S. 1983. The etiology of gastric cancer. Intragastric nitrosamide formation and other theories. J Natl Cancer Inst 71:629–647.

Mirvish, S. S. 1995. Role of N-nitroso compounds (NOC) and N-nitrosation in etiology of gastric, esophageal, nasopharyngeal and bladder cancer and contribution to cancer of known exposures to NOC. Cancer Lett 93:17–48.

Mohammadi, M., Redline, R., Nedrud, J., and Czinn, S. 1996. Role of the host in pathogenesis of Helicobacter-associated gastritis: H. felis infection of inbred and congenic mouse strains. Infect Immun 64:238–245.

Morris, H. P., Wagner, B. P., Ray, F. E., Snell, K. C., and Stewart, H. L. 1961. Comparative study of cancer and other lesions of rats fed N,N'-2,7-fluorenylene-bisacetamide or N-2-fluorenylacetamide. Natl Cancer Inst Monogr 5:1–53.

Moser, A. R., Luongo, C., Gould, K. A., McNeley, M. K., Shoemaker, A. R., and Dove, W. F. 1995. ApcMin: a mouse model for intestinal and mammary tumorigenesis. Eur J Cancer 31A:1061–1064.

Moss, S. F., and Malfertheiner, P. 2007. Helicobacter and gastric malignancies. Helicobacter 12(Suppl 1):23–30.

Munson, L., Terio, K. A., Worley, M., Jago, M., Bagot-Smith, A., and Marker, L. 2005. Extrinsic factors significantly affect patterns of disease in free-ranging and captive cheetah (Acinonyx jubatus) populations. J Wildl Dis 41:542–548.

Murray, M., Robinson, P. B., McKeating, F. J., Baker, G. J., and Lauder, I. M. 1972. Primary gastric neoplasia in the dog: a clinico-pathological study. Vet Rec 91:474–479.

Nam, K. T., Hahm, K. B., Oh, S. Y., Yeo, M., Han, S. U., Ahn, B., Kim, Y. B., Kang, J. S., Jang, D. D., Yang, K. H., and Kim, D. Y. 2004. The selective cyclooxygenase-2 inhibitor nimesulide prevents Helicobacter pylori-associated gastric cancer development in a mouse model. Clin Cancer Res 10:8105–8113.

Newberne, P. M., Charnley, G., Adams, K., Cantor, M., Suphakarn, V., Roth, D., and Schrager, T. F. 1987. Gastric carcinogenesis: a model for the identification of risk factors. Cancer Lett 38:149–163.

Newell, D. G., Lee, A., Hawtin, P. R., Hudson, M. J., Stacey, A. R., and Fox, J. 1989. Antigenic conservation of the ureases of spiral- and helical-shaped bacteria colonising the stomachs of man and animals. FEMS Microbiol Lett 53:183–186.

Nishikawa, A., Furukawa, F., Imazawa, T., Toyoda, K., Mitsui, M., Hasegawa, T., and Takahashi, M. 1993. Effects of hickory smoke condensate on gastric carcinogenesis in Wistar rats after treatment with N-methyl-N′-nitro-N-nitrosoguanidine and sodium chloride. Food Chem Toxicol 31:25–30.

Nomura, S., Baxter, T., Yamaguchi, H., Leys, C., Vartapetian, A. B., Fox, J. G., Lee, J. R., Wang, T. C., and Goldenring, J. R. 2004. Spasmolytic polypeptide expressing metaplasia to preneoplasia in H. felis-infected mice. Gastroenterology 127:582–594.

Nozaki, K., Shimizu, N., Ikehara, Y., Inoue, M., Tsukamoto, T., Inada, K., Tanaka, H., Kumagai, T., Kaminishi, M., and Tatematsu, M. 2003. Effect of early eradication on Helicobacter pylori-related gastric carcinogenesis in Mongolian gerbils. Cancer Sci 94:235–239.

Nozaki, K., Shimizu, N., Inada, K., Tsukamoto, T., Inoue, M., Kumagai, T., Sugiyama, A., Mizoshita, T., Kaminishi, M., and Tatematsu, M. 2002. Synergistic promoting effects of Helicobacter pylori infection and high-salt diet on gastric carcinogenesis in Mongolian gerbils. Jpn J Cancer Res 93:1083–1089.

O'Gara, R., and Adamson, R. 1972. Spontaneous and induced neoplasms in nonhuman primates. In: Fiennes, R., editor. Pathology of Simian primates. Part I. Basel: Karger; pp. 190–238.

Ohgaki, H., Hasegawa, H., Kusama, K., Morino, K., Matsukura, N., Sato, S., Maruyama, K., and Sugimura, T. 1986. Induction of gastric carcinomas in nonhuman primates by N-ethyl-N′-nitro-N-nitrosoguanidine. J Natl Cancer Inst 77:179–186.

Ohgaki, H., Hasegawa, H., Kusama, K., Sato, S., Kawachi, T., Masui, M., Tanabe, K., Kawasaki, I., Hiramatsu, H., and Saito, K. 1984. Squamous cell carcinoma found in the cardiac region of the stomach of an aged orangoutang. Gann 75:415–417.

Ohgaki, H., Szentirmay, Z., Take, M., and Sugimura, T. 1989. Effects of 4-week treatment with gastric carcinogens and enhancing agents on proliferation of gastric mucosa cells in rats. Cancer Lett 46:117–122.

Ohgaki, H., Tomihari, M., Sato, S., Kleihues, P., and Sugimura, T. 1988. Differential proliferative response of gastric mucosa during carcinogenesis induced by N-methyl-N′-nitro-N-nitrosoguanidine in susceptible ACI rats, resistant Buffalo rats, and their hybrid F1 cross. Cancer Res 48:5275–5279.

Ohtani, M., Garcia, A., Rogers, A. B., Ge, Z., Taylor, N. S., Xu, S., Watanabe, K., Marini, R. P., Whary, M. T., Wang, T. C., and Fox, J. G. 2007. Protective role of 17β-estradiol against the development of Helicobacter pylori-induced gastric cancer in INS-GAS mice. Carcinogenesis 28(12):2597–2604.

Oshima, H., Oshima, M., Inaba, K., and Taketo, M. M. 2004. Hyperplastic gastric tumors induced by activated macrophages in COX-2/mPGES-1 transgenic mice. Embo J 23:1669–1678.

Otto, G., Hazell, S. H., Fox, J. G., Howlett, C. R., Murphy, J. C., O'Rourke, J. L., and Lee, A. 1994. Animal and public health implications of gastric colonization of cats by Helicobacter-like organisms. J Clin Microbiol 32:1043–1049.

Paster, B. J., Lee, A., Fox, J. G., Dewhirst, F. E., Tordoff, L. A., Fraser, G. J., O'Rourke, J. L., Taylor, N. S., and Ferrero, R. 1991. Phylogeny of Helicobacter felis sp. nov., Helicobacter mustelae, and related bacteria. Int J Syst Bacteriol 41:31–38.

Peek, R. M., Jr., and Blaser, M. J. 2002. Helicobacter pylori and gastrointestinal tract adenocarcinomas. Nat Rev Cancer 2:28–37.

Perkins, S. E., Fox, J. G., Marini, R. P., Shen, Z., Dangler, C. A., and Ge, Z. 1998. Experimental infection in cats with a cagA+ human isolate of Helicobacter pylori. Helicobacter 3:225–235.

Perkins, S. E., Fox, J. G., and Walsh, J. H. 1996. Helicobacter mustelae-associated hypergastrinemia in ferrets (Mustela putorius furo). Am J Vet Res 57:147–150.

Poutahidis, T., Tsangaris, T., Kanakoudis, G., Vlemmas, I., Iliadis, N., and Sofianou, D. 2001. Helicobacter pylori-induced gastritis in experimentally infected conventional piglets. Vet Pathol 38:667–678.

Rogers, A. B., and Fox, J. G. 2004. Inflammation and Cancer. I. Rodent models of infectious gastrointestinal and liver cancer. Am J Physiol Gastrointest Liver Physiol 286:G361–366.

Rogers, A. B., Taylor, N. S., Whary, M. T., Stefanich, E. D., Wang, T. C., and Fox, J. G. 2005. Helicobacter pylori but not high salt induces gastric intraepithelial neoplasia in B6129 mice. Cancer Res 65:10709–10715.

Ruch, T. 1959. Neoplasia. In: Ruch, T., editor. Diseases of laboratory primates. Philadelphia: Saunders; pp. 529–567.

Sakagami, T., Shimoyama, T., O'Rourke, J., et al. 1994. Back to the host: severity of inflammation induced by Helicobacter felis in different strains of mice [abstract]. Am J Gastroenterol 89:1345.

Sato, T., Fukuyama, T., Urata, F., et al. 1959. Studies of the causation of gastric cancer. I. Bleeding in the glandular stomach of mice by feeding with highly salted foods and comment on salted foods in Japan. Jap Med J 1835:25.

Sawada, Y., Yamamoto, N., Sakagami, T., Fukuda, Y., Shimoyama, T., Nishigami, T., Uematsu, K., and Nakagawa, K. 1999. Comparison of pathologic changes in Helicobacter pylori-infected Mongolian gerbils and humans. J Gastroenterol 34(Suppl 11):55–60.

Schreiber, H., Nettesheim, P., Lijinsky, W., Richter, C. B., and Walburg, H. E., Jr. 1972. Induction of lung cancer in germfree, specific-pathogen-free, and infected rats by N-nitrosoheptamethyleneimine: enhancement by respiratory infection. J Natl Cancer Inst 49:1107–1114.

Seoane, A., Bessa, X., Alameda, F., Munne, A., Gallen, M., Navarro, S., O'Callaghan, E., Panades, A., Andreu, M., and Bory, F. 2005. Role of Helicobacter pylori in stomach cancer after partial gastrectomy for benign ulcer disease. Rev Esp Enferm Dig 97:778–785.

Seymour, C., Lewis, R. G., Kim, M., Gagnon, D. F., Fox, J. G., Dewhirst, F. E., and Paster, B. J. 1994. Isolation of Helicobacter strains from wild bird and swine feces. Appl Environ Microbiol 60:1025–1028.

Sharashidze, L. K., Beniashvili, D., Sherenesheva, N. I., and Turkia, N. G. 1989. Induction of gastric cancer in monkeys by N-methyl-N-nitro-N-nitrosoguanidine (MNNG). Neoplasma 36:129–133.

Shimizu, N., Inada, K., Nakanishi, H., Tsukamoto, T., Ikehara, Y., Kaminishi, M., Kuramoto, S., Sugiyama, A., Katsuyama, T., and Tatematsu, M. 1999. Helicobacter pylori infection enhances glandular stomach carcinogenesis in Mongolian gerbils treated with chemical carcinogens. Carcinogenesis 20:669–676.

Shimizu, N., Inada, K. I., Tsukamoto, T., Nakanishi, H., Ikehara, Y., Yoshikawa, A., Kaminishi, M., Kuramoto, S., and Tatematsu, M. 1999. New animal model of glandular stomach carcinogenesis in Mongolian gerbils infected with Helicobacter pylori and treated with a chemical carcinogen. J Gastroenterol 34(Suppl 11):61–66.

Shimizu, N., Kaminishi, M., Tatematsu, M., Tsuji, E., Yoshikawa, A., Yamaguchi, H., Aoki, F., and Oohara, T. 1998. Helicobacter pylori promotes development of pepsinogen-altered pyloric glands, a preneoplastic lesion of glandular stomach of BALB/c mice pretreated with N-methyl-N-nitrosourea. Cancer Lett 123:63–69.

Shin, V. Y., Wu, W. K., Ye, Y. N., So, W. H., Koo, M. W., Liu, E. S., Luo, J. C., and Cho, C. H. 2004. Nicotine promotes gastric tumor growth and neovascularization by activating extracellular signal-regulated kinase and cyclooxygenase-2. Carcinogenesis 25:2487–2495.

Sigaran, M. F., and Con-Wong, R. 1979. Production of proliferation lesions in gastric mucosa of albino mice by oral administration of N-methyl-N'-nitro-N-nitrosoguanidine. Gann 70:343–352.

Simpson, K. W., Strauss-Ayali, D., Scanziani, E., Straubinger, R. K., McDonough, P. L., Straubinger, A. F., Chang, Y. F., Domeneghini, C., Arebi, N., and Calam, J. 2000. Helicobacter felis infection is associated with lymphoid follicular hyperplasia and mild gastritis but normal gastric secretory function in cats. Infect Immun 68:779–790.

Simpson, K. W., Strauss-Ayali, D., Straubinger, R. K., Scanziani, E., McDonough, P. L., Straubinger, A. F., Chang, Y. F., Esteves, M. I., Fox, J. G., Domeneghini, C., Arebi, N., and

Calam, J. 2001. Helicobacter pylori infection in the cat: evaluation of gastric colonization, inflammation and function. Helicobacter 6:1–14.

Singer, G. M., and Taylor, H. W. 1976. Carcinogenicity of N′-nitrosonornicotine in Sprague-Dawley rats. J Natl Cancer Inst 57:1275–1276.

Sinning, C., Schaefer, N., Standop, J., Hirner, A., and Wolff, M. 2007. Gastric stump carcinoma: epidemiology and current concepts in pathogenesis and treatment. Eur J Surg Oncol 33:133–139.

Solnick, J. V., Canfield, D. R., Yang, S., and Parsonnet, J. 1999. Rhesus monkey (Macaca mulatta) model of Helicobacter pylori: noninvasive detection and derivation of specific-pathogen-free monkeys. Lab Anim Sci 49:197–201.

Solnick, J. V., Chang, K., Canfield, D. R., and Parsonnet, J. 2003. Natural acquisition of Helicobacter pylori infection in newborn rhesus macaques. J Clin Microbiol 41:5511–5516.

Solnick, J. V., Fong, J., Hansen, L. M., Chang, K., Canfield, D. R., and Parsonnet, J. 2006. Acquisition of Helicobacter pylori infection in rhesus macaques is most consistent with oral-oral transmission. J Clin Microbiol 44:3799–3803.

Solnick, J. V., Hansen, L. M., Canfield, D. R., and Parsonnet, J. 2001. Determination of the infectious dose of Helicobacter pylori during primary and secondary infection in rhesus monkeys (Macaca mulatta). Infect Immun 69:6887–6892.

Sorbye, H., Guttu, K., Gislason, H., Grong, K., and Svanes, K. 1993. Gastric mucosal injury and associated changes in mucosal blood flow and gastric fluid secretion caused by dimethyl sulfoxide (DMSO) in rats. Dig Dis Sci 38:1243–1250.

Sorbye, H., Kvinnsland, S., and Svanes, K. 1994. Effect of salt-induced mucosal damage and healing on penetration of N-methyl-N′-nitro-N-nitrosoguanidine to proliferative cells in the gastric mucosa of rats. Carcinogenesis 15:673–679.

Sorbye, H., Maaartmann-Moe, H., and Svanes, K. 1994. Gastric carcinogenesis in rats given hypertonic salt at different times before a single dose of N-methyl-N′-nitro-N-nitrosoguanidine. J Cancer Res Clin Oncol 120:159–163.

Sorbye, H., Svanes, C., Stangeland, L., Kvinnsland, S., and Svanes, K. 1988. Epithelial restitution and cellular proliferation after gastric mucosal damage caused by hypertonic NaCl in rats. Virchows Arch A Pathol Anat Histopathol 413:445–455.

Squire, R., Goodman, D., Valerio, M., et al. 1978. Tumors. In: Benirschke, K., Garner, F., and Jones, T., editors. Pathology of laboratory animals. New York: Springer-Verlag; pp. 1051–1283.

Stenstrom, B., Zhao, C. M., Rogers, A. B., Nilsson, H. O., Sturegard, E., Lundgren, S., Fox, J. G., Wang, T. C., Wadstrom, T., and Chen, D. 2007. Swedish moist snuff accelerates gastric cancer development in Helicobacter pylori-infected wild-type and gastrin transgenic mice. Carcinogenesis 28(9):2041–2046.

Stewart, H. L., and Hare, W. V. 1950. Variation in susceptibility of the fundic and pyloric portions of the glandular stomach of the rat to induction of neoplasia by 20-methylcholanthrene. Acta Unio Int Contra Cancrum 7:176–177.

Stoicov, C., Whary, M., Rogers, A. B., Lee, F. S., Klucevsek, K., Li, H., Cai, X., Saffari, R., Ge, Z., Khan, I. A., Combe, C., Luster, A., Fox, J. G., and Houghton, J. 2004. Coinfection modulates inflammatory responses and clinical outcome of Helicobacter felis and Toxoplasma gondii infections. J Immunol 173:3329–3336.

Straubinger, R. K., Greiter, A., McDonough, S. P., Gerold, A., Scanziani, E., Soldati, S., Dailidiene, D., Dailide, G., Berg, D. E., and Simpson, K. W. 2003. Quantitative evaluation of inflammatory and immune responses in the early stages of chronic Helicobacter pylori infection. Infect Immun 71:2693–2703.

Sugihara, H., Tsuchihashi, Y., Hattori, T., Fukuda, M., and Fujita, S. 1985. Cell proliferation and cell loss in intramucosal signet ring cell carcinoma of canine stomachs induced by N-ethyl-N′-nitro-N-nitrosoguanidine. J Cancer Res Clin Oncol 110:87–94.

Sugimura, T., and Fujimura, S. 1967. Tumour production in glandular stomach of rat by N-methyl-N′-nitro-N-nitrosoguanidine. Nature 216:943–944.

Sugimura, T., and Kawachi, T. 1973. Experimental stomach cancer. Methods Cancer Res 7:245–308.

Sugimura, T., and Kawachi, T. 1978. Experimental stomach carcinogenesis In: Lipkin, M., and Good R., editors. Gastrointestinal tract cancer. New York: Plenum Press; pp. 327–341.

Sugimura, T., Tanaka, N., Kawachi, T., Kogure, K., and Fujimura, S. 1971. Production of stomach cancer in dogs by N-methyl-N'-nitro-N-nitrosoguanidine. Gann 62:67.

Sugiyama, A., Maruta, F., Ikeno, T., Ishida, K., Kawasaki, S., Katsuyama, T., Shimizu, N., and Tatematsu, M. 1998. Helicobacter pylori infection enhances N-methyl-N-nitrosourea-induced stomach carcinogenesis in the Mongolian gerbil. Cancer Res 58:2067–2069.

Sumi, Y., and Miyakawa, M. 1981. Comparative studies on the production of stomach tumors following the intubation of several doses of N-methyl-N'-nitro-N-nitrosoguanidine in germ-free and conventional newborn rats. Gann 72:700–704.

Swann, H. M., and Holt, D. E. 2002. Canine gastric adenocarcinoma and leiomyosarcoma: a retrospective study of 21 cases (1986–1999) and literature review. J Am Anim Hosp Assoc 38:157–164.

Syder, A. J., Oh, J. D., Guruge, J. L., O'Donnell, D., Karlsson, M., Mills, J. C., Bjorkholm, B. M., and Gordon, J. I. 2003. The impact of parietal cells on Helicobacter pylori tropism and host pathology: an analysis using gnotobiotic normal and transgenic mice. Proc Natl Acad Sci USA 100:3467–3472.

Szentirmay, Z., Ohgaki, H., Maruyama, K., Esumi, H., Takayama, S., and Sugimura, T. 1990. Early gastric cancer induced by N-ethyl-N'-nitro-N-nitrosoguanidine in a cynomolgus monkey six years after initial diagnosis of the lesion. Jpn J Cancer Res 81:6–9.

Tajima, K., and Tominaga, S. 1985. Dietary habits and gastro-intestinal cancers: a comparative case-control study of stomach and large intestinal cancers in Nagoya, Japan. Jpn J Cancer Res 76:705–716.

Takahashi, M., and Hasegawa, H. 1986. Enhancing effects of dietary salt on both initiation and promotion stages of rat gastric carcinogenesis. In: Hayashi, Y., Nagao, M., Sugimura, T., Takayama, S., Tomatis, T., et al., editors. Diet, nutrition and cancer. Utrecht: Japan Sci Sco Press; pp. 169–182.

Takahashi, M., Kokubo, T., Furukawa, F., Kurokawa, Y., Tatematsu, M., and Hayashi, Y. 1983. Effect of high salt diet on rat gastric carcinogenesis induced by N-methyl-N'-nitro-N-nitrosoguanidine. Gann 74:28–34.

Tanaka, Y., Osugi, H., Morimura, K., Takemura, M., Ueno, M., Kaneko, M., Fukushima, S., and Kinoshita, H. 2004. Effect of duodenogastric reflux on N-methyl-N'-nitro-N-nitrosoguanidine-induced glandular stomach tumorigenesis in Helicobacter pylori-infected Mongolian gerbils. Oncol Rep 11:965–971.

Tatematsu, M., Ogawa, K., Hoshiya, T., Shichino, Y., Kato, T., Imaida, K., and Ito, N. 1992. Induction of adenocarcinomas in the glandular stomach of BALB/c mice treated with N-methyl-N-nitrosourea. Jpn J Cancer Res 83:915–918.

Tatematsu, M., Takahashi, M., Fukushima, S., Hananouchi, M., and Shirai, T. 1975. Effects in rats of sodium chloride on experimental gastric cancers induced by N-methyl-N-nitro-N-nitrosoguanidine or 4-nitroquinoline-1-oxide. J Natl Cancer Inst 55:101–106.

Tatematsu, M., Takahashi, M., Hananouchi, M., Shirai, T., and Hirose, M. 1976. Protective effect of mucin on experimental gastric cancer induced by N-methyl-N'-nitro-N-nitrosoguanidine plus sodium chloride in rats. Gann 67:223–229.

Tatematsu, M., Tsukamoto, T., and Mizoshita, T. 2005. Role of Helicobacter pylori in gastric carcinogenesis: the origin of gastric cancers and heterotopic proliferative glands in Mongolian gerbils. Helicobacter 10:97–106.

Tatematsu, M., Yamamoto, M., Iwata, H., Fukami, H., Yuasa, H., Tezuka, N., Masui, T., and Nakanishi, H. 1993. Induction of glandular stomach cancers in C3H mice treated with N-methyl-N-nitrosourea in the drinking water. Jpn J Cancer Res 84:1258–1264.

Tatematsu, M., Yamamoto, M., Shimizu, N., Yoshikawa, A., Fukami, H., Kaminishi, M., Oohara, T., Sugiyama, A., and Ikeno, T. 1998. Induction of glandular stomach cancers in Helicobacter pylori-sensitive Mongolian gerbils treated with N-methyl-N-nitrosourea and N-methyl-N'-nitro-N-nitrosoguanidine in drinking water. Jpn J Cancer Res 89:97–104.

Tennant, R. W. 2002. The National Center for Toxicogenomics: using new technologies to inform mechanistic toxicology. Environ Health Perspect 110:A8–10.

Terio, K. A., Munson, L., Marker, L., Aldridge, B. M., and Solnick, J. V. 2005. Comparison of Helicobacter spp. in Cheetahs (Acinonyx jubatus) with and without gastritis. J Clin Microbiol 43:229–234.

Tomita, H., Yamada, Y., Oyama, T., Hata, K., Hirose, Y., Hara, A., Kunisada, T., Sugiyama, Y., Adachi, Y., Linhart, H., and Mori, H. 2007. Development of gastric tumors in Apc(Min/+) mice by the activation of the beta-catenin/Tcf signaling pathway. Cancer Res 67:4079–4087.

Touati, E., Michel, V., Thiberge, J. M., Wuscher, N., Huerre, M., and Labigne, A. 2003. Chronic Helicobacter pylori infections induce gastric mutations in mice. Gastroenterology 124:1408–1419.

Tsugane, S. 2007. Dietary factors in gastrointestinal tract cancers: an Asian perspective. Acta Oncol 46:405–406.

Tsugane, S., Akabane, M., Inami, T., Matsushima, S., Ishibashi, T., Ichinowatari, Y., Miyajima, Y., and Watanabe, S. 1991. Urinary salt excretion and stomach cancer mortality among four Japanese populations. Cancer Causes Control 2:165–168.

Tsugane, S., and Sasazuki, S. 2007. Diet and the risk of gastric cancer: review of epidemiological evidence. Gastric Cancer 10:75–83.

Tsugane, S., Sasazuki, S., Kobayashi, M., and Sasaki, S. 2004. Salt and salted food intake and subsequent risk of gastric cancer among middle-aged Japanese men and women. Br J Cancer 90:128–134.

Tuyns, A. J. 1988. Salt and gastrointestinal cancer. Nutr Cancer 11:229–232.

Van den Bulck, K., Decostere, A., Baele, M., Driessen, A., Debongnie, J. C., Burette, A., Stolte, M., Ducatelle, R., and Haesebrouck, F. 2005. Identification of non-Helicobacter pylori spiral organisms in gastric samples from humans, dogs, and cats. J Clin Microbiol 43:2256–2260.

Wakeland, E., Morel, L., Achey, K., Yui, M., and Longmate, J. 1997. Speed congenics: a classic technique in the fast lane (relatively speaking). Immunol Today 18:472–477.

Wang, T. C., Dangler, C. A., Chen, D., Goldenring, J. R., Koh, T., Raychowdhury, R., Coffey, R. J., Ito, S., Varro, A., Dockray, G. J., and Fox, J. G. 2000. Synergistic interaction between hyper-gastrinemia and Helicobacter infection in a mouse model of gastric cancer. Gastroenterology 118:36–47.

Warren, J. R. 1984. Spiral bacteria of the gastric antrum. Med J Aust 141:477–478.

Watanabe, H., Takahashi, T., Okamoto, T., Ogundigie, P. O., and Ito, A. 1992. Effects of sodium chloride and ethanol on stomach tumorigenesis in ACI rats treated with N-methyl-N'-nitro-N-nitrosoguanidine: a quantitative morphometric approach. Jpn J Cancer Res 83:588–593.

Waters, M. D., Olden, K., and Tennant, R. W. 2003. Toxicogenomic approach for assessing toxicant-related disease. Mutat Res 544:415–424.

Whary, M. T., and Fox, J. G. 2004. Natural and experimental Helicobacter infections. Comp Med 54:128–158.

WHO. 2003. Diet, nutrition and the prevention of chronic diseases. World Health Organ Tech Rep Ser 916:i–viii, 1–149.

Yang, K., Edelmann, W., Fan, K., Lau, K., Kolli, V. R., Fodde, R., Khan, P. M., Kucherlapati, R., and Lipkin, M. 1997. A mouse model of human familial adenomatous polyposis. J Exp Zool 277:245–254.

You, W. C., Blot, W. J., Chang, Y. S., Ershow, A. G., Yang, Z. T., An, Q., Henderson, B., Xu, G. W., Fraumeni, J. F., Jr., and Wang, T. G. 1988. Diet and high risk of stomach cancer in Shandong, China. Cancer Res 48:3518–3523.

Yu, J., Fox, J. G., Blanco, M. C., Yan, L., Correa, P., and Russell, R. M. 1995. Long-term supplementation of canthaxanthin does not inhibit gastric epithelial cell proliferation in Helicobacter mustelae-infected ferrets. J Nutr 125:2493–2500.

Chapter 13
Insights into the Development of Preneoplastic Metaplasia: Spasmolytic Polypeptide-Expressing Metaplasia and Oxyntic Atrophy

James R. Goldenring and Sachiyo Nomura

Introduction

Deaths from gastric adenocarcinoma are the third largest cause of cancer-related mortality in the world (Pisani et al. 1999). Early recognition and resection of gastric cancers remains the mainstay of gastric cancer therapy, and adjuvant treatment regimens provide only minimal benefits beyond surgery (Wainess et al. 2003). Although aggressive endoscopic screening procedures in Asian countries, such as Japan and South Korea, have led to earlier discovery and surgical removal of gastric cancers (Rembacken et al. 2001), little is known of the cellular etiology of gastric neoplasms. Although the concept of preneoplasia does spur screening endoscopy, there is a general lack of knowledge of the sequence of events leading first to metaplasia and later to preneoplastic lesions and gastric cancer. This uncertainty in the analysis of precancerous events hobbles efforts to develop effective screening methods for patient populations at risk for gastric cancer. Thus, detailed analyses of the pathways to carcinogenesis in humans and in animal models of gastric carcinogenesis are critical for elucidation of the precancerous process.

Human Gastric Cancer Pathogenesis

Inherent to a broader understanding of gastric cancer pathogenesis is the critical concept that cancer arises in a field of altered mucosal lineages, either focally or globally. Studies over the past 15 years have demonstrated that the major proximate cause of gastric cancer in humans is chronic infection with particular strains of the bacterium *Helicobacter pylori* (Blaser and Parsonnet 1991, 1994; Peura 1997). The World Health Organization has designated *H. pylori* as a class I carcinogen (Peura 1997). Two important factors contribute to the evolution of gastric cancer in the presence of chronic *H. pylori* infection: first, the infection elicits a prominent inflammatory response throughout the gastric mucosa (Blaser 1992; Blaser and Parsonnet 1994); second, chronic infection leads to loss of glandular lineages in the gastric fundus, especially acid-secreting parietal cells and pepsin-secreting chief cells (Ormand et al. 1991). The overall picture of the stomach mucosa in patients with chronic *Helicobacter* infection is one of dynamic mucosal lineage changes and prominent inflammatory response. Oxyntic atrophy (the loss of parietal cells) either focal or global is a prerequisite for the development of gastric cancer (El-Zimaity et al. 2002). Whereas an association of human gastric cancer with gastric atrophy and inflammation are now well accepted, the intervening cellular events that mediate the progression from atrophy to neoplasia remain controversial.

Studies over the last decade have increasingly emphasized the association of mucous cell metaplasias as precursors for upper gastrointestinal cancers in the esophagus, pancreas, and stomach. Development of esophageal adenocarcinoma is closely linked with Barrett's epithelial metaplasia and pancreatic adenocarcinoma arises from discrete mucous cell metaplasias (Biankin et al. 2003; Cameron et al. 1985, 1995). In the case of pancreatic adenocarcinoma, sequential preneoplastic PANIN lesions evolve from mucous cell metaplasia of pancreatic acinar cells (Biankin et al. 2003; Means et al. 2005). Whereas the association of intestinal-type cancers with chronic *H. pylori* infection and oxyntic atrophy is well accepted (Figure 13.1A), the connections between discrete metaplasias and cancer are less clear. Most Western authorities have considered goblet cell intestinal metaplasia (Figure 13.1A) as the leading candidate for origination of gastric cancer (Correa 1988; Filipe et al. 1994). Goblet cells are not found in the normal stomach, so the presence of cells with goblet cell morphology represents a clear metaplastic process with intestinal phenotype cells. Nevertheless, little evidence exists linking directly intestinal metaplastic cells with dysplastic transformation (Brito et al. 1995; Hattori 1986; Hattori and Fujita 1979; Takizawa and Koike 1998). Interestingly, recent investigations have found that intestinal metaplasias uniformly express the transcription factor Pdx1, which is expressed in the antrum and duodenum (distal foregut) (Leys et al. 2006). Thus, although many have thought of the presence of goblet cells in the stomach as a colonic-type metaplasia, it now seems more likely that intestinal metaplasia in the stomach actually represents a duodenal metaplasia, mimicking cells from the contiguous gut segment.

A number of recent studies have revealed that intestinal metaplasia is not the only possible metaplastic precursor of cancer. A number of investigators, especially in Asia, have reported on metaplastic glands in the fundus with a general phenotype similar to that of the antral or pyloric glands (Hattori 1986; Hattori and Fujita 1979; Hattori et al. 1982; Xia et al. 2000). This antralization of the fundus, also known as pseudopyloric metaplasia, is frequently associated with intestinal-type adenocarcinoma (Figure 13.1A). We have described a similar metaplastic process as spasmolytic polypeptide-expressing metaplasia or SPEM (Schmidt et al. 1999), which is characterized by the presence of trefoil factor 2 (TFF2 or spasmolytic polypeptide) immunoreactive cells in the gastric fundus with morphologic characteristics similar to those of deep antral gland cells or Brunner's gland cells. We have observed SPEM in association with greater than 90% of resected gastric cancers in three studies in the United States, Japan, and Iceland (Halldorsdottir et al. 2003; Schmidt et al. 1999; Yamaguchi et al. 2001). SPEM is observed as TFF2 immunostaining cells emerging from the bases of fundic glands, often associated

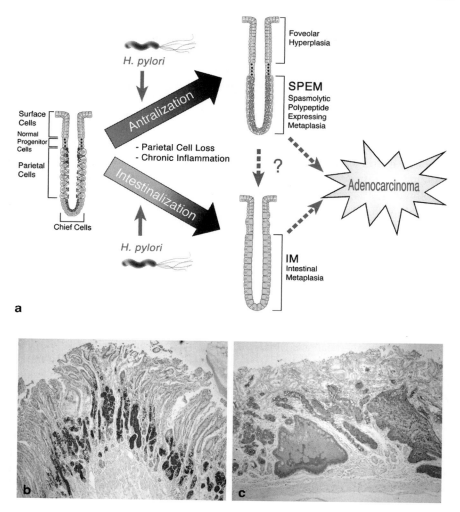

Fig. 13.1 Gastric metaplasia in humans. (**A**) Scheme for the development of gastric metaplasias in humans and their relationship to gastric cancer. (Adapted from Goldenring and Nomura 2006.) (**B, C**) Trefoil factor 2 (TFF2) staining in specimens from patients resected for gastric cancer. (**B**) Spasmolytic polypeptide-expressing metaplasia (SPEM) stained with TFF2 antibodies arises from the base of the glands, and foveolar hyperplasia is observed luminal to the SPEM cells. (**C**) TFF2 staining shows strong labeling of gastritis cystica profunda, a preneoplastic pathology, surrounded by further SPEM glands

with foveolar hyperplasia in the same glands (Figure 13.1B). TFF2 immunostaining is also observed in gastritis cystica profunda, which is associated with the development of intramucosal dysplasia (Figure 13.1C). In addition, similar findings were recently reported for patients from Japan, where expression of TFF2 correlated with metastasis (Dhar et al. 2003). In all of these studies, SPEM was present as often or more often in association with cancer than goblet cell intestinal metaplasia. Whereas TFF2 immunoreactivity was less prominent in advanced cancers, in the Iceland study, TFF2 immunoreactivity was observed in greater than 50% of early gastric cancers (Halldorsdottir et al. 2003). It is important to note at this point that TFF2 is expressed in the normal fundic mucosa and the expression that they are referring to here is an overexpression in an aberrant cell lineage. All of these investigations have suggested that SPEM and intestinal metaplasia share equal importance as putative preneoplastic lesions in the stomach.

Several pathologic processes lead to oxyntic atrophy in humans. First, antibodies generated against parietal cell proteins, especially the H/K-ATPase, in patients with autoimmune gastritis lead to profound oxyntic atrophy (Marshall et al. 2005). Autoimmune gastritis is associated with carcinoid tumors in the stomach, but is less associated with gastric adenocarcinoma (Borch et al. 1985). Second, chronic infection with *H. pylori* leads to loss of parietal cells and other attendant changes in the gastric mucosa (Peura 1997). Chronic *H. pylori* infection represents the most prominent worldwide cause for gastric adenocarcinoma. Third, in Ménétrier's disease, patients demonstrate loss of parietal cells along with massive foveolar hyperplasia often leading to loss of serum proteins through the leaky gastric mucosa (Coffey et al. 1987; Wolfsen et al. 1993). Although Ménétrier's disease has been linked with gastric cancer in the past, it seems likely that this association is more appropriately assigned to the lymphocytic gastritis subtype which is associated with concurrent *H. pylori* infection. In patients without *H. pylori* infection and significant inflammatory infiltrate, the pathology seems to stem from vast overproduction of transforming growth factor (TGF)α in the gastric mucosa (Dempsey et al. 1992). Administration of antibodies, which block binding to the epidermal growth factor (EGF) receptor, can radically ameliorate the disease and reverse oxyntic atrophy (Burdick et al. 2000). Importantly, although the former two pathways to oxyntic atrophy implicate degenerative influences on parietal cells, recent investigations suggest that the influence of TGFα leads to the induction of a true antral phenotype with expression of both antral mucosal lineage related genes, such as *Pdx1*, and the presence of gastrin cells in the affected fundic mucosa (Nomura et al. 2005).

Mouse Models of Oxyntic Atrophy and Metaplasia

Over the past decade, a number of mouse models have been devised that have led to insights into the ramifications of oxyntic atrophy. These studies can be divided into three general categories: 1. studies of chronic *Helicobacter* sp. infection, 2. studies of genetic manipulations that lead to oxyntic atrophy, and 3. models of toxicity against parietal cells. Critical to these studies is the analysis of gastric mucosal lineages using a number of histologic and immunohistochemical markers. As we have noted previously, classification of mouse lineages is best accomplished through specific immunostains. Intestinal metaplasia in humans is defined by goblet cells stained with Alcian blue. Human intestinal metaplasia also stains with markers of intestinal goblet cells, including TFF3 and Muc2. Only a few Alcian blue–staining mucous neck cells are present in the normal human stomach. However, in mice, essentially all of the deep gland cells of the antrum are stained with Alcian blue (Goldenring and Nomura 2006). These deep antral cells express TFF2 and MUC6, rather than TFF3 and MUC2 as in goblet cells. Thus, assignment of pathologies as "intestinal metaplasia" in mice should only be made when TFF3 or MUC2 staining have been established. Similarly, SPEM assignment should require staining with TFF2 or MUC6. Nevertheless, TFF2 and MUC6 are also present in normal fundic mucous neck cells in both humans and rodents. Thus, assignment of lineages as SPEM requires immunostaining as well as morphologic correlation. More recently, we have also identified novel markers, such as HE-4, which are not present in the normal stomach, but are present in both SPEM and intestinal metaplasia (Nozaki et al. 2008). Thus, a battery of immunostains can be utilized to assign more precisely the origins of metaplasia. Caution is also recommended in the use of periodic acid–Schiff (PAS) to assign surface cell lineages. SPEM is usually also PAS positive, although the color intensity is usually less in SPEM and, indeed, the contrast between deep carmine staining of surface cells versus a more pink staining of SPEM can be useful in rapid histologic analysis of foveolar hyperplasia and SPEM.

Oxyntic Atrophy and Metaplasia After Chronic Helicobacter felis *Infection*

C57BL/6 mice infected with *Helicobacter* sp., particularly *H. felis*, demonstrate profound loss of parietal cells and the replacement of fundic mucosa with a mucous cell metaplasia, which is highly proliferative and expresses TFF2 (Figure 13.2A) (Fox et al. 1996, 2003; Wang et al. 1998). These mice demonstrate prominent SPEM by 6 months of infection. SPEM is also observed in other models of *Helicobacter* infection using *H. pylori* strains in both mice and gerbils (Fox et al. 1996, 2003; Kirchner et al. 2001; Wang et al. 1998). Importantly, in these *Helicobacter* infection models, SPEM represents the only observed metaplasia and no goblet cell intestinal metaplasia is present (Figure 13.2A). Also importantly, inflammation is critical to the development of atrophy and metaplasia. Thus, immunodeficient mice do not develop atrophy and metaplasia after *H. felis* infection (Fox et al. 1993). Similarly, TNFα-deficient mice do not develop SPEM after chronic *H. pylori* infection (Oshima et al. 2005).

Just as in humans, a number of studies have indicated that, in *Helicobacter*-infected mice, SPEM can progress to dysplasia and intramucosal cancer. This process is accelerated in infected insulin-gastrin mice, where gastritis cystica profunda and dysplasia develop along a more rapid time course (Wang et al. 2000). The importance of SPEM as a precursor of gastric dysplasia was recently highlighted in studies on the engraftment of bone marrow cells in the stomachs of *H. felis*–infected C57BL/6 mice (Houghton et al. 2004). The questions of bone marrow engraftment

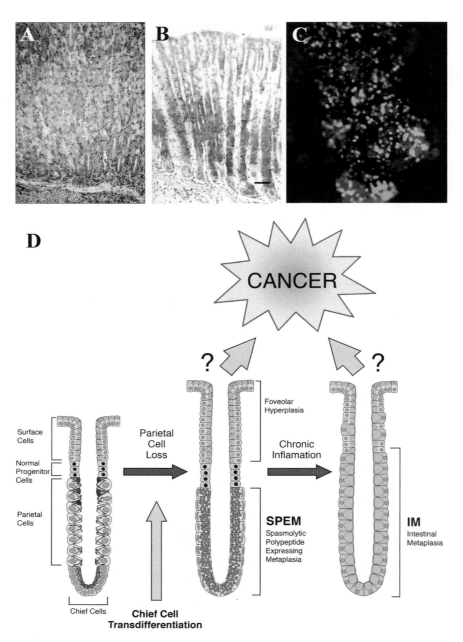

Fig. 13.2 Spasmolytic polypeptide-expressing metaplasia (SPEM) in mice. (**A**) Trefoil factor 2 (TFF2) staining reveals near complete replacement of fundic glands with SPEM in a C57BL/6 mouse infected with *Helicobacter felis* for 9 months. (**B**) TFF2 staining shows prominent SPEM in a mouse treated with DMP-777 for 14 days. (**C**) Dual immunofluorescence staining for intrinsic factor (green) and TFF2 (red) demonstrates dual-staining SPEM cells at the bases of glands from a mouse treated with DMP-777 for 14 days. Note that the intrinsic factor and TFF2 are present in separate granule populations in the SPEM cells. (**D**) A unified hypothesis for the origin of gastric metaplasias. Our studies suggest that SPEM arises from transdifferentiation of chief cells after parietal cell loss. In addition, we hypothesize that, in the presence of chronic inflammation in humans, intestinal metaplasia emerges from further differentiation of SPEM. (Adapted from Goldenring and Nomura 2006.) (*See Color Plates*)

are addressed in detail elsewhere in this volume. However, it is important to note that these studies demonstrated that bone marrow–derived cells engrafted into the SPEM cell lineage in the fundus. Moreover, over time, markers for bone marrow–derived cell origin also were found in gastritis cystica profunda and dysplasia, suggesting that dysplastic pathologies develop from SPEM. Whether similar engraftment can be observed in mice with goblet cell intestinal metaplasia is unclear because mice infected with *Helicobacter* sp. do not develop intestinal metaplasia. Nevertheless, the bone marrow transplantation studies demonstrate a clear connection between SPEM and the progression to dysplasia in mice, and implicate the SPEM lineage in the origin of gastric neoplasia in mice.

Oxyntic Atrophy, Hyperplasia, and Metaplasia After Genetic Manipulation

A number of mouse models of transgenic and targeted deletion have led to oxyntic atrophy phenotypes. Most of these manipulations develop either hyperplastic or metaplastic phenotypes. It should be noted that not all of these changes may truly represent metaplasias, because the mucosal phenotypes are often established during development because of a change in global lineage patterning. Several models are illustrative of mucosal changes associated with oxyntic atrophy. The metallothionein (MT)-TGFα mice demonstrate foveolar hyperplasia and oxyntic atrophy similar to that seen in patients with Ménétrier's disease (Dempsey et al. 1992; Sharp et al. 1995). These mice, as in Ménétrier's disease patients, demonstrate upregulation in the gastric fundus of the distal foregut transcription factor Pdx-1 (Nomura et al. 2005). The presence of neoplasia in these mice is controversial. Although the mice do appear to have marked changes in their mucosa, most of the pathologies are more consistent with benign cystic changes than dysplasia. Indeed, although Ménétrier's disease is considered preneoplastic in humans, it now seems probable that patients without inflammatory infiltrates are likely not precancerous and experience disease based on overexpression of TGFα. More recently, others have studied the effects of the knockout of the H_2-histamine receptor in mice and also found a Ménétrier's disease–like phenotype (Ogawa et al. 2003). H_2-receptor knockout mice develop profound oxyntic atrophy and foveolar hyperplasia. Interestingly, these mice also show prominent elevations in gastric TGFα levels and demonstrate some the serum albumin losses seen in a subset of Ménétrier's disease patients. It is not clear whether these mice develop truly dysplastic changes in the gastric mucosa.

Recent investigations have noted that KLF4-deficient mice develop oxyntic atrophy and SPEM phenotype throughout the fundus (Katz et al. 2005). As in the *H. felis* mice, the antrum is spared of changes. KLF4-deficient mice develop extensive TFF2-expressing metaplasia throughout the gastric fundus. It is notable that these mice do not seem to develop any significant inflammatory response in the mucosa and no dysplastic changes have been reported to date.

Cdx-2 is an intestinal transcription factor that is expressed throughout the small and large intestines. Although Cdx-2 is not expressed in the normal stomach, forced expressing of Cdx-2 in the stomach using a short H/K-ATPase promoter leads to intestinalization of the gastric fundus (Mutoh et al. 2002; Silberg et al. 2002). H/K-Cdx-2 mice demonstrate profound oxyntic atrophy with expression of intestinal goblet cells throughout the fundus of the stomach. Recent investigations have noted that dysplasia develops in the stomachs of older H/K-Cdx-2 mice. Thus, although *Helicobacter*-infected mice do not develop intestinal goblet cell metaplasia, the presence of intestinal metaplasia in the stomach in mice does represent a potentially preneoplastic scenario with analogy to humans. The metaplasia profile after *Helicobacter* infection seems to be species dependent, because *H. pylori*–infected Mongolian gerbils do develop intestinal metaplasia. Importantly, recently we have noted that goblet cell intestinal metaplasia in gerbils develops from SPEM (Yoshizawa et al. 2007).

Somewhat paradoxically, a number of models have reported oxyntic atrophy after long-term induction of acid hypersecretion. Although all three of these models lead to the loss of parietal cells at greater than six months of age, the atrophic phenotypes are different. Expression in parietal cells of a point mutant of the H/K-ATPase, which cannot be endocytosed efficiently after delivery to the apical lumen, leads to eventual atrophic gastritis with cystic changes in older animals (Courtois-Coutry et al. 1997). The phenotype in these mice seems to be primarily attributable to foveolar hyperplasia although no formal analysis of metaplasias has been performed. Insulin-gastrin transgenic mice demonstrate elevated serum gastrin levels with acid hypersecretion early in life followed by oxyntic atrophy in older animals (Wang et al. 2000). These older animals develop SPEM and gastritis cystica profunda. Furthermore, infection with *H. felis* leads to accelerated development of SPEM and dysplastic cystic changes. Most recently, Samuelson and colleagues have studied the phenotype of transgenic mice with targeted expression of cholera toxin in parietal cells(Lopez-Diaz et al. 2006). These mice show a progression of oxyntic atrophy after 6 months of age with initial mucous neck cell hyperplasia followed later by development of fundic glands fully replaced with cells showing a SPEM-like morphology. This phenotype could reflect progressive expansion of mucous neck cells (mucous neck cell hyperplasia) combined with eventual SPEM development. Notably, the full manifestation of this phenotype correlates with the detection of anti-H/K-ATPase antibodies. At present it is not clear whether other models of acid hypersecretion followed by oxyntic atrophy also might accrue from antiparietal cell antibodies. Interestingly, human patients with pernicious anemia associated with antiparietal cell antibodies do not exhibit SPEM as an associated metaplasia (our unpublished results).

Two models of direct genetic parietal cell ablation have been reported in mice. H/K-diphtheria toxin mice demonstrate parietal cell–specific expression of tetanus toxin, leading to the rapid demise of parietal cells as they begin to express the proton pump (Li et al. 1996). This genetic ablation model leads to an expansion of preparietal cells in the midportion of gastric glands. At ages older than 1 year, these mice develop cystic changes and alterations consistent with dysplasia. No analysis of metaplastic lineages was performed in these mice, so it is presently unclear whether dysplastic lesions may arise from SPEM or some other metaplastic process. One other genetic ablation model has been reported in H/K-thymidine kinase mice (Canfield et al. 1996). These mice demonstrated parietal cell–specific expression of thymidine kinase, but treatment with ganciclovir resulted in complete loss of the glandular fundic mucosa likely because of the exchange of toxic adducts through the extensive system of gap junctions among the mucosal cells. Thus, this model could not address issues of reactive metaplasia. Nevertheless, it is notable that after cessation of ganciclovir treatment, animals could reassemble the normal pattern of mucosal lineages in the reconstituted fundic mucosa.

Insights into the Origin of Spasmolytic Polypeptide-Expressing Metaplasia After Acute Oxyntic Atrophy

Although the above discussion has detailed a number of scenarios leading to the observation of SPEM, they have not provided particular insights into the origin of metaplasia. One difficulty with these models has been that induction of SPEM was a chronic process. Thus, we have turned to a model of acute oxyntic atrophy to provide information on the emergence of SPEM after the loss of parietal cells. The orally active, cell permeant neutrophil elastase inhibitor, DMP-777, at high doses (>200 mg/kg/day) induces acute oxyntic atrophy in all mammals tested without inducing a significant inflammatory infiltrate. This has allowed investigations of the influences of parietal cell loss in the absence of the intramucosal inflammation seen in *Helicobacter* infection models. In mice or rats, administration of high doses of oral DMP-777 leads to a rapid loss of parietal cells within 3 days of treatment (Goldenring et al. 2000; Nomura et al. 2005). The acute loss of parietal cells leads within 1 day of treatment to rapid increases in gastrin levels and prominent foveolar hyperplasia. After this initial reactive surface mucous cell hyperplasia, SPEM then develops in the gastric fundus between 7 to 10 days of treatment (Figure 13.2B). The entire process is the result of parietal cell loss caused by the action of DMP-777 as a parietal cell secretory membrane protonophore. Parietal cell necrosis presumably follows acid reflux back into the cell through protonophore channels in the apical membrane. Indeed, pretreatment of animals with the proton pump inhibitor omeprazole blocks the loss of parietal cells after DMP-777 treatment (Ogawa et al. 2006a).

This acute model of oxyntic atrophy has proven amenable to detailed study, providing important knowledge into the mucosal responses to parietal cell loss. In this model, gastrin is the major driving force for foveolar hyperplasia, because gastrin knockout mice do not develop surface cell hyperplasia in response to DMP-777 treatment (Nomura et al. 2004). These results confirmed previous results in other rodent models (Konda et al. 1999). Nevertheless, the absence of gastrin seems to promote the development of SPEM, with rapid induction of metaplasia after only 1 day of DMP-777 treatment. Gastrin-deficient mice develop SPEM after only one dose of DMP-777, compared with the 7–10 days required in wild-type C57BL/6

(Nomura et al. 2004). This rapid development of SPEM in gastrin-deficient mice is too rapid for metaplasia to arise from the normal progenitor zone located in the neck region. Although we had previously suggested that SPEM might develop from cryptic progenitor cells located at the bases of fundic glands, more recent studies suggest that SPEM develops through transdifferentiation of chief cells. Indeed, we have found that the presence of cells at the base of glands expressing both intrinsic factor (a chief cell marker in mice) and TFF2 in separate granules is the best reflection of SPEM induction (Figure 13.2C). Gastrin-deficient mice treated with DMP-777 have a rapid increase in these dual-expressing cells after only 1 day of treatment. Electron micrographs of SPEM cells demonstrate separate populations of granules with characteristics of either zymogen or mucous granules (Nozaki et al. 2008). In addition to dual-expressing cells, we also observed the presence of BrdU labeling S phase cells at the bases of fundic glands, distinct from the normal progenitor zone located near the lumen (Goldenring et al. 2000; Nomura et al. 2004). Although the proliferative rate observed in DMP-777–treated mice is considerably lower than that in SPEM in *H. felis*–infected mice, the rapid induction basally located proliferating cells suggest that some transdifferentiating cells can reenter the cell cycle. Thus, SPEM cells may eventually become self-renewing or be influenced by inflammatory regulators toward metaplastic expansion or dysplastic transformation.

The reduction in EGF-receptor signaling in *waved-2* mice carrying a hypomorphic mutation, which reduces EGF-receptor tyrosine kinase activity, also causes acceleration of SPEM development after DMP-777 treatment (Ogawa et al. 2006b). More recent investigations have demonstrated that loss of amphiregulin signaling seems to account for most of the effects observed in *waved-2* mice (Nam et al. 2007). In these studies, specific loss of amphiregulin accelerated the development of SPEM similar to the findings observed in *waved-2* mice. However, TGFα-deficient mice developed SPEM along a time course similar to that observed in wild-type mice. Thus, specific EGF-receptor ligands seem to have differing influences on mucosal lineage differentiation. In addition, it is notable that the amphiregulin-deficient mice seemed to have altered somatostatin dynamics (Nam et al. 2007). These studies demonstrate how alteration of an important intramucosal factor may have widespread effects on the dynamic regulation of the mucosal milieu.

It is clear that intrinsic paracrine and endocrine regulators modulate the emergence of metaplasia after the loss of parietal cells. Although the DMP-777 treatment model allows rapid induction of metaplasia, it should be noted that even after prolonged administration of drug for up to a year, no dysplastic lesions are ever observed in mice or rats despite the profound oxyntic atrophy and SPEM. These results seem to accrue from the absence of significant inflammatory infiltrate in DMP-777–treated animals. The lack of infiltrate likely is a result from the major action of this drug as a cell-permeant inhibitor of neutrophil elastase. It is also notable that no bone marrow–derived cell engraftment was observed in DMP-777–treated mice, even though they showed extensive SPEM (Houghton et al. 2004). Thus, SPEM can develop solely in response to the loss of parietal cells, especially in the absence of inflammatory infiltrate. Moreover, these investigations point out that the presence of chronic inflammation is a requirement for development of dysplasia from metaplasia.

Toward a Unified Hypothesis for the Origin of Gastric Metaplasias

Studies analyzing the phenotypes of the growing number of genetic mouse models in induced-atrophy scenarios in mice have demonstrated how complicated the influences are within the gastric mucosa. Chronic overexpression or knockout of key regulators may lead to a watershed of alterations in a number of cytokines and growth factors that normally regulate the dynamics of mucosal homeostasis. Thus, a loss of gastrin could lead to decreases in sonic hedgehog expression, which could in turn lead to alterations in growth factor expression. Similarly, loss of an EGF-receptor ligand such as amphiregulin may lead to augmentation of the influence of other EGF-receptor ligands such as HB-EGF. Thus, emergence and persistence of metaplastic lesions are likely regulated by the balance of an array of intramucosal factors. Alterations in the balance of these factors likely lead to a number of aberrant mucosal lineage phenotypes, from foveolar hyperplasia and SPEM to dysplasia.

Studies in rodents have demonstrated that SPEM develops in the setting of oxyntic atrophy from activation of cryptic progenitor cells at the bases of gastric glands distinct from the normal progenitor cells in the gland neck (Goldenring et al. 2000; Wang et al. 1998; Yamaguchi et al. 2002). As noted above, studies in gastrin- and amphiregulin-deficient mice support the origin of SPEM from transdifferentiation of chief cells (Figure 13.2D). The origin of intestinal metaplasia remains elusive (Hattori and Fujita, 1979), in part because goblet cell intestinal metaplasia is not observed in mouse models of *Helicobacter* infection (Fox et al. 1996). Thus, intestinal metaplasia could arise separately from SPEM or could represent a further differentiation of a metaplastic lineage from SPEM (Figure 13.2D). The evolution of one mucous cell metaplasia from another has been observed in the setting of injury and repair associated with Crohn's disease (Wright et al. 1990). No investigation has sought to evaluate systematically the relative contribution of discrete metaplasias to the development of gastric dysplasia and gastric cancer in humans. One is therefore left with a series of hypothetical constructs for the development of cancer from precedent metaplasias. Either of the observed metaplasias, SPEM or intestinal metaplasia, could be paracancerous, whereas the other is truly preneoplastic. Alternatively, as noted above, intestinal metaplasia could evolve from SPEM, either as a precancerous transition or a paracancerous transition. There presently is no evidence for the evolution of SPEM from intestinal metaplasia, because antralization and SPEM seem to develop earlier in the process of oxyntic atrophy (Takizawa and Koike 1998). Finally, it is possible that each metaplasia gives rise to a distinct type of cancer; for example, intestinal metaplasia could evolve into intestinal-type cancers while SPEM evolves into gastric-type cancers.

One should also note that there is even a possibility that both SPEM and intestinal metaplasia are paracancerous, but investigations in mice, at least, suggest that SPEM can lead to cancer (Wang et al. 1998, 2000).

The predominance of studies in mice now indicates that SPEM is the proximate precursor of dysplasia and cancer. But as noted above, there is no evidence for intestinal metaplasia in most models of murine gastric cancer after *Helicobacter* infection. Thus, it remains uncertain how SPEM is related to intestinal metaplasia. Recent studies in mongolian gerbils, where SPEM precedes the development of intestinal metaplasia, have indicated that intestinal metaplasia develops from SPEM glands (Yoshizawa et al. 2007). Indeed, we have observed a number of examples of intestinal metaplasia emanating from basal SPEM in human resection specimens. These studies now lead to a unified hypothesis: Loss of parietal cells during the initial stages of infection with *H. pylori* leads to the induction of SPEM through transdifferentiation of chief cells. In the course of chronic and sustained infection, the SPEM lineage may undergo further differentiation into intestinal metaplasia (Figure 13.2D). It remains to be determined whether either or both of these metaplasias can progress to dysplasia or neoplasia. Given the results in mice, intestinal metaplasia may reflect a further benign attempt by the mucosa to increase repair in the face of chronic infection and inflammation. Further studies of human metaplastic lineages are required to determine relationships of individual metaplasias to observed neoplastic progression.

References

Biankin, A.V., J.G. Kench, F.P. Dijkman, S.A. Biankin, and S.M. Henshall. 2003. Molecular pathogenesis of precursor lesions of pancreatic ductal adenocarcinoma. Pathology. 35:14–24.
Blaser, M. 1992. Hypotheses on the pathogenesis and natural history of Helicobacter pylori-induced inflammation. Gastroenterology. 102:720–727.
Blaser, M., and J. Parsonnet. 1994. Parasitism by the 'slow' bacterium Helicobacter pylori leads to altered gastric homeostasis and neoplasia. J Clin Invest. 94:4–8.
Borch, K., H. Renvall, and G. Liedberg. 1985. Gastric endocrine cell hyperplasia and carcinoid tumors in pernicious anemia. Gastroenterology. 88:638–648.
Brito, M.J., M.I. Filipe, J. Linehan, and J. Jankowski. 1995. Association of transforming growth factor alpha (TGFA) and its precursors with malignant change in Barrett's epithelium: biological and clinical variables. Int J Cancer. 60:27–32.
Burdick, J.S., E. Chung, G. Tanner, M. Sun, J.E. Paciga, J.Q. Cheng, K. Washington, J.R. Goldenring, and R.J. Coffey. 2000. Treatment of Menetrier's disease with a monoclonal antibody against the epidermal growth factor receptor. N Engl J Med. 343:1697–1701.
Cameron, A.J., C.T. Lomboy, M. Pera, and H.A. Carpenter. 1995. Adenocarcinoma of the esophagogastric junction and Barrett's esophagus. Gastroenterology. 109:1541–1546.
Cameron, A.J., B.J. Ott, and W.S. Payne. 1985. The incidence of adenocarcinoma in columnar-lined (Barrett's) esophagus. New Engl J Med. 313:857–859.
Canfield, V.A., A.B. West, J.R. Goldenring, and R. Levenson. 1996. Targeted ablation of parietal cells in transgenic mice. Proc Natl Acad Sci USA. 93(6):2431–2435.
Coffey, R.J., R. Derynck, J.N. Wilcox, T.S. Bringman, A.S. Goustin, H.L. Moses, and M.R. Pittelkow. 1987. Production and auto-induction of TGF alpha in human keratinocytes. Nature. 328:817–820.

Correa, P. 1988. A human model of gastric carcinogenesis. Cancer Res. 48:3554–3560.

Courtois-Coutry, N., D. Rousch, V. Rajendran, J.B. McCarthy, J. Geibel, M. Kashgarian, and M.J. Caplan. 1997. A tyrosine-based signal targets H/K-ATPase to a regulated compartment and is required for the cessation of gastric acid secretion. Cell. 90:501–510.

Dempsey, P.J., J.R. Goldenring, C.J. Soroka, I.M. Modlin, R.W. McClure, C.D. Lind, D.A. Ahlquist, M.R. Pittlekow, D.C. Lee, E.P. Sandgren, D.L. Page, and R.J. Coffey. 1992. Possible role of TGFa in the pathogenesis of Menetrier's disease: supportive evidence from humans and transgenic mice. Gastroenterology. 103:1950–1963.

Dhar, D.K., T.C. Wang, R. Maruyama, J. Udagawa, H. Kubota, T. Fuji, M. Tachibana, T. Ono, H. Otani, and N. Nagasue. 2003. Expression of cytoplasmic TFF2 is a marker of tumor metastasis and negative prognostic factor in gastric cancer. Lab Invest. 83:1343–1352.

El-Zimaity, H.M.T., H. Ota, D.Y. Graham, T. Akamatsu, and T. Katsuyama. 2002. Patterns of gastric atrophy in intestinal type gastric carcinoma. Cancer. 94:1428–1436.

Filipe, M.I., N. Munoz, I. Matko, I. Kato, V. Pompe-Kirn, A. Juersek, S. Teuchmann, M. Benz, and T. Prijon. 1994. Intestinal metaplasia types and the risk of gastric cancer: a cohort study in Slovenia. Int J Cancer. 57:324–329.

Fox, J.G., M. Blanco, J.C. Murphy, N.S. Taylor, A. Lee, Z. Kabok, and J. Pappo. 1993. Local and systemic immune responses in murine Helicobacter felis active chronic gastritis. Infect Immun. 61:2309–2315.

Fox, J.G., X. Li, R.J. Cahill, K. Andrutis, A.K. Rustgi, R. Odze, and T.C. Wang. 1996. Hypertrophic gastropathy in Helicobacter felis-infected wild type C57BL/6 mice and p53 hemizygous transgenic mice. Gastroenterology. 110:155–166.

Fox, J.G., T.C. Wang, A.B. Rogers, T. Poutahidis, Z. Ge, N. Taylor, C.A. Dangler, D.A. Israel, U. Krishna, K. Gaus, and R.M. Peek, Jr. 2003. Host and microbial constituents influence Helicobacter pylori-induced cancer in a murine model of hypergastrinemia. Gastroenterology. 124:1879–1890.

Goldenring, J.R., and S. Nomura. 2006. Differentiation of the gastric mucosa III. Animal models of oxyntic atrophy and metaplasia. Am J Physiol Gastrointest Liver Physiol. 291:G999–1004.

Goldenring, J.R., G.S. Ray, R.J. Coffey, P.C. Meunier, P.J. Haley, T.B. Barnes, and B.D. Car. 2000. Reversible drug-induced oxyntic atrophy in rats. Gastroenterology. 118:1080–1093.

Halldorsdottir, A.M., M. Sigurdardottir, J.G. Jonasson, M. Oddsdottir, J. Magnusson, J.R. Lee, and J.R. Goldenring. 2003. Spasmolytic polypeptide expressing metaplasia (SPEM) associated with gastric cancer in Iceland. Dig Dis Sci. 48:431–441.

Hattori, T. 1986. Development of adenocarcinomas in the stomach. Cancer. 57:1528–1534.

Hattori, T., and S. Fujita. 1979. Tritiated thymidine autoradiographic study on histogenesis and spreading of intestinal metaplasia in human stomach. Pathol Res Pract. 164:224–237.

Hattori, T., B. Helpap, and P. Gedigk. 1982. The morphology and cell kinetics of pseudopyloric glands. Virchows Arch B Cell Pathol Incl Mol Pathol. 39:31–40.

Houghton, J., C. Stoicov, S. Nomura, J. Carlson, H. Li, A.B. Rogers, J.G. Fox, J.R. Goldenring, and T.C. Wang. 2004. Gastric cancer originating from bone marrow derived cells. Science. 306:1568–1571.

Katz, J.P., N. Perreault, B.G. Goldstein, L. Actman, S.R. McNally, D.G. Silberg, E.E. Furth, and K.H. Kaestner. 2005. Loss of Klf4 in mice causes altered proliferation and differentiation and precancerous changes in the adult stomach. Gastroenterology. 128:935–945.

Kirchner, T., S. Muller, T. Hattori, K. Mukaisyo, T. Papadopoulos, T. Brabletz, and A. Jung. 2001. Metaplasia, intraepithelial neoplasia and early cancer of the stomach are related to dedifferentiated epithelial cells defined by cytokeratin-7 expression in gastritis. Virchows Arch. 439:512–522.

Konda, Y., H. Kamimura, H. Yokota, N. Hayashi, K. Sugano, and T. Takeuchi. 1999. Gastrin stimulates the growth of gastric pit with less-differentiated features. Am J Physiol. 277: G773–784.

Leys, C.M., S. Nomura, E. Rudzinski, M. Kaminishi, E. Montgomery, M.K. Washington, and J.R. Goldenring. 2006. Expression of Pdx-1 in human gastric metaplasia and gastric adenocarcinoma. Hum Pathol. 37:1162–1168.

Li, Q., S.M. Karam, and J.I. Gordon. 1996. Diphtheria toxin-mediated ablation of parietal cells in the stomach of transgenic mice. J Biol Chem. 271:3671–3676.
Lopez-Diaz, L., K.L. Hinkle, R.N. Jain, Y. Zavros, C.S. Brunkan, T. Keeley, K.A. Eaton, J.L. Merchant, C.S. Chew, and L.C. Samuelson. 2006. Parietal cell hyperstimulation and autoimmune gastritis in cholera toxin transgenic mice. Am J Physiol Gastrointest Liver Physiol. 290: G970–979.
Marshall, A.C., F. Alderuccio, K. Murphy, and B.H. Toh. 2005. Mechanisms of gastric mucosal cell loss in autoimmune gastritis. Int Rev Immunol. 24:123–134.
Means, A.L., I.M. Meszoely, K. Suzuki, Y. Miyamoto, A.K. Rustgi, R.J. Coffey, Jr., C.V. Wright, D.A. Stoffers, and S.D. Leach. 2005. Pancreatic epithelial plasticity mediated by acinar cell transdifferentiation and generation of nestin-positive intermediates. Development. 132:3767–3776.
Mutoh, H., Y. Hakamata, K. Sato, A. Eda, I. Yanaka, S. Honda, H. Osawa, Y. Kaneko, and K. Sugano. 2002. Conversion of gastric mucosa to intestinal metaplasia in Cdx2-expressing transgenic mice. Biochem Biophys Res Commun. 294:470–479.
Nam, K.T., A. Varro, R.J. Coffey, and J.R. Goldenring. 2007. Potentiation of oxyntic atrophy-induced gastric metaplasia in amphiregulin-deficient mice. Gastroenterology. 132:1804–1819.
Nomura, S., S.H. Settle, C. Leys, A.L. Means, R.M. Peek, Jr., S.D. Leach, C.V.E. Wright, R.J. Coffey, and J.R. Goldenring. 2005. Evidence for repatterning of the gastric fundic epithelium associated with Menetrier's disease and TGFa overexpression. Gastroenterology. 128:1292–1305.
Nomura, S., H. Yamaguchi, T.C. Wang, J.R. Lee, and J.R. Goldenring. 2004. Alterations in gastric mucosal lineages induced by acute oxyntic atrophy in wild type and gastrin deficient mice. Am J Physiol. 288:G362–G375.
Nozaki, K., M. Ogawa, J.A. Williams, B.J. LaFleur, V. Ng, R.I. Drapkin, J.C. Mills, S.F. Konieczny, S. Nomura, and J.R. Goldenring. 2008. A molecular signature of gastric metaplasia arising in response to acute parietal cell loss. Gastroenterology. 134(2):511–522.
Ogawa, T., K. Maeda, S. Tonai, T. Kobayashi, T. Watanabe, and S. Okabe. 2003. Utilization of knockout mice to examine the potential role of gastric histamine H2-receptors in Menetrier's disease. J Pharmacol Sci. 91:61–70.
Ogawa, M., S. Nomura, B.D. Car, and J.R. Goldenring. 2006a. Omeprazole treatment ameliorates oxyntic atrophy induced by DMP-777. Dig Dis Sci. 51:431–439.
Ogawa, M., S. Nomura, A. Varro, T.C. Wang, and J.R. Goldenring. 2006b. Altered metaplastic response of waved-2 EGF receptor mutant mice to acute oxyntic atrophy. Am J Physiol Gastrointest Liver Physiol. 290:G793–804.
Ormand, J.E., N.J. Talley, R.G. Shorter, C.R. Conley, H.A. Carpenter, A. Fich, W.R. Wilson, and S.F. Phillips. 1991. Prevalence of Helicobacter pylori in specific forms of gastritis: further evidence supporting a pathogenic role for H. pylori in chronic nonspecific gastritis. Dig Dis Sci. 36:142–145.
Oshima, M., H. Oshima, A. Matsunaga, and M.M. Taketo. 2005. Hyperplastic gastric tumors with spasmolytic polypeptide-expressing metaplasia caused by tumor necrosis factor-alpha-dependent inflammation in cyclooxygenase-2/microsomal prostaglandin E synthase-1 transgenic mice. Cancer Res. 65:9147–9151.
Parsonnet, J., G.D. Friedman, D.P. Vandersteen, Y. Chang, J.H. Vogelman, N. Orentreich, and R.K. Sibley. 1991. Helicobacter pylori infection and the risk of gastric cancer. New Engl J Med. 325:1127–1131.
Peura, D.A. 1997. The report of the Digestive Health Initiative international update conference on Helicobacter pylori. Gastroenterology. 113:S4–S8.
Pisani, P., D.M. Parkin, F. Bray, and J. Ferlay. 1999. Estimates of the worldwide mortality from 25 cancers in 1990. Int J Cancer. 83:18–29.
Rembacken, B., T. Fujii, and H. Kondo. 2001. The recognition and endoscopic treatment of early gastric and colonic cancer. Best Pract Res Clin Gastroenterol. 15:317–336.
Schmidt, P.H., J.R. Lee, V. Joshi, R.J. Playford, R. Poulsom, N.A. Wright, and J.R. Goldenring. 1999. Identification of a metaplastic cell lineage associated with human gastric adenocarcinoma. Lab Invest. 79:639–646.

Sharp, R., M.W. Babyatsky, H. Takagi, S. Tagerud, T.C. Wang, D.E. Bockman, S.E. Brand, and G. Merlino. 1995. Transforming growth factor alpha disrupts the normal program of cellular differentiation in the gastric mucosa of transgenic mice. Development. 121:149–161.

Silberg, D.G., J. Sullivan, E. Kang, G.P. Swain, J. Moffett, N.J. Sund, S.D. Sackett, and K.H. Kaestner. 2002. Cdx2 ectopic expression induces gastric intestinal metaplasia in transgenic mice. Gastroenterology. 122:689–696.

Takizawa, T., and M. Koike. 1998. Minute gastri carcinoma from pathomorphological aspect—reconsideration concerning histogenesis of gastric carcinomas. Stomach Intestine. 23:791–800.

Wainess, R.M., J.B. Dimick, G.R. Upchurch, Jr., J.A. Cowan, and M.W. Mulholland. 2003. Epidemiology of surgically treated gastric cancer in the United States, 1988–2000. J Gastrointest Surg. 7:879–883.

Wang, T.C., C.A. Dangler, D. Chen, J.R. Goldenring, T. Koh, R. Raychowdhury, R.J. Coffey, S. Ito, A. Varro, G.J. Dockray, and J.G. Fox. 2000. Synergistic interaction between hypergastrinemia and Helicobacter infection in a mouse model of gastric cancer. Gastroenterology. 118:36–47.

Wang, T.C., J.R. Goldenring, C. Dangler, S. Ito, A. Mueller, W.K. Jeon, T.J. Koh, and J.G. Fox. 1998. Mice lacking secretory phospholipase A2 show altered apoptosis and differentiation with Helicobacter felis infection. Gastroenterology. 114:675–689.

Wolfsen, H.C., H.A. Carpenter, and N.J. Talley. 1993. Menetrier's disease: a form of hypertrophic gastropathy or gastritis? Gastroenterology. 104:1310–1319.

Wright, N.A., C. Pike, and G. Elia. 1990. Induction of a novel epidermal growth factor-secreting cell lineage by mucosal ulceration in human gastrointestinal stem cells. Nature. 343:82–85.

Xia, H.H., J.S. Kalantar, N.J. Talley, J.M. Wyatt, S. Adams, K. Cheung, and H.M. Mitchell. 2000. Antral-type mucosa in the gastric incisura, body and fundus (antralization): a link between Helicobacter pylori infection and intestinal metaplasia. Am J Gastroenterol. 95:114–121.

Yamaguchi, H., J.R. Goldenring, M. Kaminishi, and J.R. Lee. 2001. Identification of spasmolytic polypeptide expressing metaplasia (SPEM) in remnant gastric cancer and surveillance post-gastrectomy biopsies. Dig Dis Sci. 47:573–578.

Yamaguchi, H., J.R. Goldenring, M. Kaminishi, and J.R. Lee. 2002. Association of spasmolytic polypeptide expressing metaplasia (SPEM) with carcinogen administration and oxyntic atrophy in rats. Lab Invest. 82:1045–1052.

Yoshizawa, N., Y. Takenaka, H. Yamaguchi, T. Tetsuya, H. Tanaka, M. Tatematsu, S. Nomura, J.R. Goldenring, and M. Kaminishi. 2007. Emergence of spasmolytic polypeptide-expressing metaplasia in Mongolian gerbils infected with Helicobacter pylori. Lab Invest. 87:1265–1276.

Chapter 14
Deregulation of E-Cadherin in Precancerous Lesions and Gastric Cancer

Annie On On Chan and Benjamin Chun-Yu Wong

Introduction

Gastric cancer remains the second major cause of cancer-related deaths in the world. China and Hong Kong are among the high incidence areas in the world, showing no trend of reduction in gastric cancer incidence over the past 10 years and no improvement in the survival rate. *Helicobacter pylori* infection is an important etiologic risk factor in gastric cancer, and has been classified as a class I carcinogen by the World Health Organization. However, only a minority of the infected population eventually develop gastric cancer. Molecular studies of the host may give definite proof of the etiologic role of *H. pylori* in inducing gastric carcinoma.

E-cadherin is a transmembrane glycoprotein expressed on epithelial cells, and is responsible for homotypic cell adhesion. Defective cell adhesion may contribute to loss of contact inhibition of growth, which is an early step in carcinogenesis. The critical role of E-cadherin is underlined by the observation that familial gastric cancer is related to germline mutation of the E-cadherin gene. Somatic mutations of E-cadherin were found in approximately 50% of gastric carcinomas of the diffuse histologic type. CpG island methylation was found to be the second "genetic hit" in abrogating E-cadherin expression. These results show that E-cadherin is an important putative tumor suppressor gene involved in gastric carcinogenesis. In this chapter, we review the role of E-cadherin in gastric carcinogenesis.

E-Cadherin and Cancer: Not Only the Adhesive Glue

Cadherin is a superfamily of calcium-mediated membrane glycoproteins, with a molecular mass of 120 kDa. It belongs to one of the four classes of adhesion molecules (Gumbiner 2000; Takeichi 1995; Yagi and Takeichi 2000). Some common cadherins expressed by epithelial cells are E-, N-, and P-cadherin. The intracellular domains of classical cadherins interact with β-catenin, γ-catenin (also called plakoglobin), and p120ctn to assemble the cytoplasmic cell adhesion complex (CCC) that is critical for the formation of extracellular cell–cell adhesion. β-Catenin and

T.C. Wang et al. (eds.) *The Biology of Gastric Cancers*,
© Springer Science+Business Media, LLC 2009

γ-catenin bind directly to α-catenin, which links the CCC to the actin cytoskeleton (Grunwald et al. 1993; Takeichi et al. 1990). The cadherins are responsible for the homotypic cell–cell adhesion. However, knowledge gained in the recent few decades showed that the role of E-cadherin is more than just an adhesive glue.

The Role of E-Cadherin in Metastasis

E-cadherin is the prototype of the cadherin class. It is expressed in all epithelial cell types. Underexpression of E-cadherin is found in gastric, hepatocellular, esophageal, breast, prostatic, bladder, and gynecologic carcinomas and correlates with infiltrative and metastatic ability (Takeichi et al. 1993). Loss of E-cadherin-mediated cell–cell adhesion is a prerequisite for tumor cell invasion and metastasis formation (Birchmeier and Behrens 1994). Reestablishing the functional cadherin complex, e.g., by forced expression of E-cadherin, results in a reversion from an invasive, mesenchymal, to a benign, epithelial phenotype of cultured tumor cells (Birchmeier and Behrens 1994; Vleminckx et al. 1991). Hence, the E-cadherin gene is also called an invasion suppressor gene.

Certain human cancers expressing an abundance of cadherins can metastasize, posing the question of how they leave the primary tumor. One possible mechanism for such a process would be a transient and local loss of cadherins caused by down-regulation of the expression or by selective proteolysis. Another possible mechanism is pertubation of the cadherin cell adhesion system without loss of cadherin. Pertubation of the cadherin adhesion system may also occur as a result of biochemical modification of catenins. Phosphorylation of catenins might interfere with cadherin action bringing about unstable cell–cell adhesion. It has been shown that epidermal growth factor receptor, c-erb-2, hepatocyte growth receptor c-met, and the oncoprotein pp60^{vsrc} all phosphorylate β-catenin.

Unstable or reduced expression of E-cadherin has been postulated to account for the invasive ability or metastatic potential of gastric adenocarcinomas. Decreased expression of E-cadherin has been observed in gastric cancer ranging from 17% (Shimoyama et al. 1991) to 92% (Mayer et al. 1993), depending on the method and the definition used (please see the section below).

The Role of E-Cadherin in Carcinogenesis

Recently, it has been postulated that the role of E-cadherin in carcinogenesis is not only limited to metastasis and invasion. It is now increasingly recognized that there is also a role for E-cadherin in modulating intracellular signaling, and thus promoting tumor growth. There are several lines of evidence. Cadherin-mediated cell–cell adhesion can affect the Wnt-signaling pathway (Bienz and Clevers 2000; Polakis 2000). β-Catenin (and γ-catenin) is usually sequestered by cadherins in the cadherin–catenin

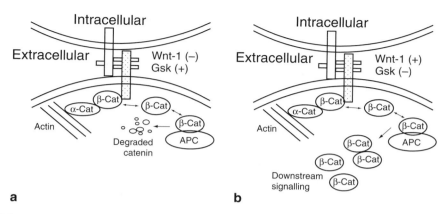

Fig. 14.1 Illustrated interaction between the cadherin–catenin complex and the APC protein. (**a**) In the absence of Wnt-1 and the presence of glycogen synthase kinase (GSK), β-catenin (β-cat) is stabilized and bound to cadherin or APC protein. Cadherin acts as a negative regulator of β-catenin by regulating the amount of free β-catenin. The free cytoplasmic β-catenin is degraded. (**b**) In the presence of Wnt-1, GSK is antagonized and mutant APC and tyrosine phosphorylated β-catenin cannot bind to each other, cytosolic free β-catenin concentrations increase, which leads to downstream cell signaling and may be involved in carcinogenesis. (Reproduced from Chan et al. 1999.)

complex. Upon loss of E-cadherin function, nonsequestered, free β-catenin is usually phosphorylated by glycogen synthase kinase 3β (GSK-3β) in the adenomatous polyposis coli (APC)-axin-GSK-3β complex and subsequently degraded by the ubiquitin-proteasome pathway. In many cancer cells, loss of function of the tumor suppressor APC, mutations in β-catenin, or inhibition of GSK-3β by the activated Wnt-signaling pathway leads to the stabilization of β-catenin in the cytoplasm. Subsequently, it translocates to the nucleus, where it binds to members of the Tcf/Lef-1 family of transcription factors and modulates expression of Tcf/Lef-1-target genes, including the protooncogene c-Myc and cyclin D1 (Figure 14.1). In addition, recent studies on familial gastric cancer indicate that E-cadherin can also act at a much earlier stage during tumor development. Mutations of the E-cadherin gene were found in three familial gastric cancer kindreds from New Zealand (Guilford et al. 1998) and this observation was confirmed in kindreds of European origin (Gayther et al. 1998). These results demonstrate that loss of function of E-cadherin may have a role in susceptibility to initial tumor development in addition to its role as an inhibitor of tumor invasion.

Mechanism of Inactivation of E-Cadherin

Genetic Inactivation

The E-cadherin gene can be genetically inactivated by a number of mechanisms. The first hints of a role for E-cadherin in tumor development, particularly in suppression of invasion, came from studies of loss of heterozygosity on chromosome

16. Subsequently, mutations were reported in tumor samples in gynecologic cancers (Risinger et al. 1994), diffuse-type gastric carcinomas (Becker et al. 1994), and infiltrative lobular breast cancer (Berx et al. 1995; Kanai et al. 1994). In gastric cancer, E-cadherin mutations are common in diffuse-type carcinomas, but are not seen in the intestinal type (Becker et al. 1994). Mutations of E-cadherin in other types of cancer are rarely observed, although deletion of one E-cadherin allele, as detected by loss of heterozygosity, is more widely observed. The specificity of types of cancer affected by mutations of the E-cadherin gene, despite the prevalence of reduced E-cadherin expression in many cancer types, suggests that E-cadherin mutations may be of particular importance in the development of these tumors. In addition, mutations in both diffuse gastric (Muta et al. 1996) and lobular breast cancer (Vos et al. 1997) have been detected early in tumor development, suggesting a role in tumor suppression, as opposed to invasion suppression. Further evidence for this comes from the observations that mutations of E-cadherin have also been observed in several kindreds exhibiting familial gastric cancer (Gayther et al. 1998; Guilford et al. 1998). Furthermore, at least one kindred exhibited both diffuse gastric cancer and early-onset breast cancer (Guilford et al. 1999).

Inactivation by Hypermethylation

In recent years it has become increasingly apparent that increased methylation within the promoter regions of genes has a key role in the inactivation of many important genes during the development of cancer (Costello and Plass 2001). Subsequently, numerous reports of E-cadherin promoter methylation, associated with reduced E-cadherin expression, have been published. Methylation is frequently associated with disease progression and metastasis (Nass et al. 2000; Tamura et al. 2000). Unlike mutational inactivation, which is frequent in only two specific tumor types, hypermethylation of E-cadherin is seen in a wide range of human tumors. In addition, hypermethylation has also been shown to have a role in familial gastric cancer, acting as the second hit in inactivation of E-cadherin (Grady et al. 2000).

Transcriptional Regulation

Loss of E-cadherin function during tumor progression can be caused by transcriptional repression binding to the CDH1-E box elements, e.g., by the repressors Snail (Batlle et al. 2000; Cano et al. 2000), Sip-1 (Comijn et al. 2001), Slug (Bolos et al. 2003), E12/E47 (Perez-Moreno et al. 2001), ZEB1 (Grooteclaes and Frisch, 2000), and ZEB2 (Comijn et al. 2001), which have been found to be upregulated. These transcription factors, the most well-characterized of which are members of the Snail/Slug family, have the compound effect of repressing certain epithelial genes such as E-cadherin and cytokeratin-8 while increasing the expression of mesenchymal genes.

Posttranslational Modification

Posttranslational mechanisms that have been shown to regulate cell–cell adhesion include tyrosine phosphorylation and relocalization of E-cadherin away from the plasma membrane. Tyrosine phosphorylation is implicated in the regulation of cadherin function: RTKs, such as EGFR, c-Met, and FGFR, and the nonreceptor tyrosine kinase, c-Src, phosphorylate E-cadherin, N-cadherin, β-catenin, γ-catenin, and p120ctn, resulting in the disassembly of the cytoplasmic adhesion complex and a disruption of cadherin-mediated cell adhesion and cell scattering (Behrens et al. 1993; Fujita et al. 2002). An E3 ubiquitin ligase, Hakai, has recently been identified that binds to E-cadherin when E-cadherin is tyrosine phosphorylated, ubiquitinating it and enhancing endocytosis as a result (Fujita et al. 2002). Hence, regulation of E-cadherin ubiquitination by Hakai in response to growth factor stimulation provides another mechanism for growth factors to regulate the amount of plasma membrane–associated E-cadherin. Downregulation of E-cadherin by endocytosis could potentially be the "immediate early response" event leading to loss of cell–cell adhesion.

Changes of E-Cadherin in Gastric Cancer and Precursors

Expression of E-Cadherin

The expression of E-cadherin has been widely studied by immunohistochemical method. Decreased expression has been observed in gastric cancer, ranging from 17% (Shimoyama and Hirohashi 1991) to 92% (Mayer et al. 1993), depending on the method and the definition used by the investigators. The decreased expression of E-cadherin was mainly observed in the diffuse-type and less in the intestinal-type of gastric cancer. Direct correlation between E-cadherin and the grade of tumor differentiation has been observed in all these studies. In addition, it was shown in a study of 413 gastric cancers by Gabbert et al. (1996) that patients with E-cadherin–positive tumors had significantly better 3- and 5-year survival rates than patients with E-cadherin–negative tumors. However, methods to evaluate, qualitatively or quantitatively, protein expression in biopsies from human tumors may have serious limitations because of sampling from heterogeneous tissue, non-stoichiometric labeling, and subjective evaluation. We have also shown that expression of E-cadherin was decreased early in intestinal metaplasia and a progressive decrease was observed along the Correa's cascade (Chan et al. 2003).

Mutation of E-Cadherin Gene

Germline E-cadherin mutations was first reported in a familial diffuse type of gastric cancers from New Zealand (Guilford 1998) and subsequently in a number of other kindreds (Gayther et al. 1998; Richards et al. 1999; Salahshor et al. 2001). These families are characterized by a highly penetrant susceptibility to diffuse gastric cancer with an autosomal dominant pattern of inheritance, predominantly in young persons. Familial aggregation of gastric cancer occurs in about 1% of gastric cancer patients. Although the genetic factors resulting in this aggregation have been unclear, the study indicates that germline mutations of the E-cadherin gene do contribute to such a clustering. The analysis of all the reported genetic abnormalities in E-cadherin found in familial diffuse gastric cancer reveals that the majority are inactivating mutations (splice-site, frameshift, and nonsense) rather then missense. Furthermore, the E-cadherin germline mutations are evenly distributed along the E-cadherin gene, in contrast to the clustering in exons 7–9 observed in sporadic diffuse gastric cancer (Berx et al. 1998). These germline mutations affect one allele, leaving a wild-type E-cadherin allele. The "second hit" is normally deletion of the whole gene, or silencing of the gene by promoter methylation. Aberrant CDH1 promoter methylation has been demonstrated in three of six familial diffuse gastric cancer with negative E-cadherin expression (Grady et al. 2000).

Mutations of the E-cadherin gene has a high penetrance and confers a lifetime risk of gastric cancer of 75%–80% for carriers (Berx et al. 1998; Giarelli 2002; Hunstsman et al. 2001; Oliveira et al. 2002); the average age of gastric cancer patients is 37 years (Chun et al. 2001; Humar et al. 2002; Lewis et al. 2001). These characteristics have led to the consideration of prophylactic total gastrectomy in family members with CDH1 mutations.

In sporadic gastric cancer, whereas nearly half of the undifferentiated-scattered (diffuse) type gastric carcinomas contain E-cadherin mutations (Becker 1994; Tamura 1996), such mutations are rare in early undifferentiated carcinomas (Muta 1996; Tamura 1996), and are only detected in the undifferentiated component of mixed differentiated/undifferentiated carcinomas (Machado 1999). This suggests that E-cadherin mutations are involved in the dedifferentiation of such tumors. In contrast, E-cadherin methylation, which is associated with decreased E-cadherin expression, is observed in >50% of early-stage undifferentiated carcinomas (Tamura 2000, 2001), and is also observed in surrounding noncancerous gastric epithelia (Suzuki 1999; Waki 2002) (see the section below).

CpG Island Methylation of E-Cadherin Gene

E-cadherin methylation was reported to be present in 49%–75% of gastric cancers surveyed (Tamura et al. 2000, 2001). Interestingly, E-cadherin methylation was absent in nonneoplastic gastric mucosa from a Western population (Herman et al. 1996; Tamura et al. 2000), but present in gastric mucosa of Asian patients from

Japan, Hong Kong, and Korea (Chan et al. 2003; Kang et al. 2001; Waki et al. 2002). Toyota et al. (1999) reported that methylation could be age related (Type A), or cancer related (Type C). However, it is also well recognized that methylation could be the result of underlying chronic inflammation (Chan et al. 2003). In gastric cancer, the interplay of age, *H. pylori*, and the subsequent chronic gastritis complicates the development of E-cadherin methylation. The prevalence of *H. pylori* increases with age. In addition, the presence of *H. pylori* is almost invariably associated with gastritis. When a person is infected with *H. pylori*, a superficial gastritis results within hours or days and can progress to chronic gastritis. The prevalence of *H. pylori* is higher in China, Japan, and Korea than in Western countries. This fact might explain why E-cadherin methylation was absent in nonneoplastic gastric mucosa when tested in Western patients, but present in studies performed in Asian patients (Chan et al. 2003; Kang et al. 2001; Waki et al. 2002).

In our initial study on E-cadherin expression in the Correa cascade leading to gastric cancer, we observed an early decrease in expression of E-cadherin at intestinal metaplasia (Chan et al. 2003). The decrease in immunohistochemical staining for E-cadherin was highly concordant with E-cadherin methylation; methylation was present in 57% of intestinal metaplasias in patients with gastric cancer (Chan et al. 2003). More importantly, our study showed that, by multivariate analysis, methylation of E-cadherin was associated with *H. pylori* infection in normal gastric mucosa in patients with dyspepsia (Chan et al. 2003). We believe that the association of *H. pylori* infection and methylation of E-cadherin could be explained by the fact that patients with interleukin-1β polymorphisms and infected with *H. pylori* will have upregulation of interleukin-1β, which will lead to the production of nitric oxide and the subsequent activation of DNA methyltransferase, hence inducing gene methylation (Chan et al. 2003). To further confirm the role of *H. pylori* in inducing E-cadherin methylation, we performed a randomized study and observed that those patients from whom *H. pylori* had been eradicated showed reduced E-cadherin methylation (Chan et al. 2006). Moreover, the E-cadherin methylation status was stable after long-term follow-up and was not related to age (Chan et al. 2006). To further support our hypothesis of the role of interleukin-1β in E-cadherin methylation, we have demonstrated that patients with *H. pylori* infection and with the interleukin-1β T/T genotype have the strongest association with methylation of multiple CpG islands, including E-cadherin, in gastric cancer. Thus, our results further suggest an association exists among E-cadherin gene methylation, *H. pylori*, and interleukin-1β, which is a novel observation.

Soluble E-Cadherin

E-cadherin has a cleavage site near the transmembrane domain and artificially produces a soluble 80-kDa amino-terminal fragment in the culture medium upon trypsin digestion in the presence of calcium (Damsky et al. 1983). This soluble E-cadherin fragment is considered to be a degradation product of the 120-kDa intact E-cadherin generated by a calcium-dependent proteolytic action, and can be detected in the protein extract of tissue samples from peripheral blood and urine (Katayama et al. 1994). Serum soluble E-cadherin is reported to be increased in dermatologic disorders (bullous pemphigoid, pemphigus vulgaris, psoriasis vulgaris), multiorgan failure, and various tumors such as bladder cancer, prostate cancer, lung cancer, and gastric cancer (Gofuko et al. 1998; Griffiths et al. 1996; Katayama et al. 1994; Matsuyoshi et al. 1995; Pittard et al. 1996). The role of soluble E-cadherin and its biologic significance is still unclear. Recent studies indicate that, in inflammatory conditions, serum soluble E-cadherin is induced by inflammatory mediators and cytokines (Perry et al. 1999), whereas in cancerous diseases, soluble E-cadherin is increased by cleavage of tissue E-cadherin because of overexpressed proteases (Noe et al. 2001). The potential of soluble E-cadherin to be a prognostic marker in gastric cancer has been shown (Juhasz et al. 2003).

Marker for Prognosis and Evaluation

We have examined the potential clinical role of E-cadherin in gastric cancer. We first studied the correlation between E-cadherin immunostaining expression and the concentration in sera (Chan et al. 2003). We found that normal strong membranous staining, such as in normal gastric epithelium, was associated with low levels of serum soluble E-cadherin, but a partially reduced, cytoplasmic staining, such as in intestinal type of gastric cancer, was associated with high levels, whereas complete absence of staining, such as in diffuse type of gastric cancer, was associated with low levels of soluble E-cadherin. We also found that soluble E-cadherin may be a potentially useful prognostic marker. High levels of soluble E-cadherin correlated with the depth of tumor invasion, as well as inoperability (Chan et al. 2001). More importantly, levels greater than 10,000 ng/mL predicts survival less than 3 years in more than 90% of patients (Chan et al. 2003). In addition, we have also shown that the increase in postoperative soluble E-cadherin levels predicts tumor recurrence in patients who received curative surgery for gastric cancer (Chan et al. 2005).

Conclusion

E-cadherin has an important role in invasion and metastasis in almost all kinds of epithelial malignancies, hence its designation as an invasion suppressor gene. However, recent studies have shown that E-cadherin actually has an early and important role in carcinogenesis and acts as a tumor suppressor gene, particularly in gastric cancer. Further thorough understanding of the role of E-cadherin and its association with the extracellular environment and intracellular functions will be extremely important in prevention of gastric carcinogenesis.

References

Batlle E, Sancho E, Franci C, Dominguez D, Monfar M, Baulida J, Garcia De Herreros A. The transcription factor snail is a repressor of E-cadherin gene expression in epithelial tumor cells. Nat Cell Biol 2000; 2: 84–89.

Becker KF, Atkinson MJ, Reich U, Becker I, Nekarda H, Siewert JR, Hofler H. E-cadherin gene mutations provide clues to diffuse type gastric carcinomas. Cancer Res 1994; 54: 3845–3852.

Behrens J, Mareel MM, Van Roy FM, et al. Dissecting tumor cell invasion: epithelial cells acquire invasive properties after the loss of uvomorulin-mediated cell-cell adhesion. J Cell Biol 1989; 108: 2435–2447.

Behrens J, Vakaet L, Friis R, Winterhager E, Van Roy F, Mareel MM, Birchmeier W. Loss of epithelial differentiation and gain of invasiveness correlates with tyrosine phosphorylation of the E-cadherin/beta-catenin complex in cells transformed with a temperature-sensitive v-SRC gene. J Cell Biol 1993; 120: 757–766.

Berx G, Becker KF, Hofler H, van Roy F. Mutations of the human E-cadherin (CDH1) gene. Hum Mutat 1998; 12: 226–237.

Berx G, Cleton-Jansen AM, Nollet F, de Leeuw WJ, van de Vijver M, Cornelisse C, van Roy F. E-cadherin is a invasion/invasion suppressor gene mutated in human lobular breast cancers. EMBO J 1995; 14: 6107–6115.

Bienz M, Clevers H. Linking colorectal cancer to Wnt signaling. Cell 2000; 103: 311–320.

Birchmeier W, Behrens J. Cadherin expression in carcinomas: role in the formation of cell junctions and the prevention of invasiveness. Biochem Biophys Acta 1994; 1198: 11–26.

Bolós V, Peinado H, Pérez-Moreno MA, Fraga MF, Esteller M, Cano AJ. The transcription factor Slug represses E-cadherin expression and induces epithelial to mesenchymal transitions: a comparison with Snail and E47 repressors. J Cell Sci 2003; 116: 499–511.

Cano A, Perez-Moreno MA, Rodrigo I, Locascio A, Blanco MJ, del Barrio MG, Portillo F, Nieto MA. The transcription factor snail controls epithelial–mesenchymal transitions by repressing E-cadherin expression. Nat Cell Biol 2000; 2: 76–83.

Chan AO, Chu KM, Huang C, Lam KF, Leung SY, Sun YW, Ko S, Xia HH, Cho CH, Hui WM, Lam SK, Rashid A. Association between Helicobacter pylori infection and interleukin 1beta polymorphism predispose to CpG island methylation in gastric cancer. Gut 2007; 56: 595–597.

Chan AO, Chu KM, Lam SK, Cheung KL, Law S, Kwok KF, Wong WM, Yuen MF, Wong BC. Early prediction of tumor recurrence after curative resection of gastric carcinoma by measuring soluble E-cadherin. Cancer 2005; 104: 740–746.

Chan AO, Chu KM, Lam SK, Wong BC, Kwok KF, Law S, Ko S, Hui WM, Yueng YH, Wong J. Soluble E-cadherin is an independent pretherapeutic factor for long-term survival in gastric cancer. J Clin Oncol 2003; 21: 2288–2293.

Chan AO, Huang C, Hui WM, Cho CH, Yuen MF, Lam SK, Rashid A, Wong BC. Stability of E-cadherin methylation status in gastric mucosa associated with histology changes. Aliment Pharmacol Ther 2006; 24: 831–836.

Chan AO, Lam SK, Chu KM, Lam CM, Kwok E, Leung SY, Yuen ST, Law SY, Hui WM, Lai KC, Wong CY, Hu HC, Lai CL, Wong J. Soluble E-cadherin is a valid prognostic marker in gastric carcinoma. Gut 2001; 48: 808–811.

Chan AO, Lam SK, Wong BC, Kwong YL, Rashid A. Gene methylation in non-neoplastic mucosa of gastric cancer: age or Helicobacter pylori related? Am J Pathol 2003; 163: 370–371.

Chan AO, Lam SK, Wong BC, Wong WM, Yuen MF, Yeung YH, Hui WM, Rashid A, Kwong YL. Promoter methylation of E-cadherin gene in gastric mucosa associated with Helicobacter pylori infection and in gastric cancer. Gut 2003; 52: 502–506.

Chan AO, Luk JM, Hui WM, Lam SK. Molecular biology of gastric carcinoma: from laboratory to bedside. J Gastroenterol Hepatol 1999; 14(12): 1150–1160.

Chan AO, Peng JZ, Lam SK, Lai KC, Yuen MF, Cheung HK, Kwong YL, Rashid A, Chan CK, Wong BC. Eradication of Helicobacter pylori infection reverses E-cadherin promoter hypermethylation. Gut 2006; 55: 463–468.

Chan AO, Wong BC, Lan HY, Loke SL, Chan WK, Hui WM, Yuen YH, Ng I, Hou L, Wong WM, Yuen MF, Luk JM, Lam SK. Deregulation of E-cadherin-catenin complex in precancerous lesions of gastric adenocarcinoma. J Gastroenterol Hepatol 2003; 18: 534–539.

Chun YS, Lindor NM, Smyrk TC, Petersen BT, Burgart LJ, Guilford PJ, Donohue JH. Germline E-cadherin gene mutations: is prophylactic total gastrectomy indicated? Cancer 2001; 92: 181–187.

Comijn J, Berx G, Vermassen P, Verschueren K, van Grunsven L, Bruyneel E, Mareel M, Huylebroeck D, van Roy F. The two-handed E box binding zinc finger protein SIP1 downregulates E-cadherin and induces invasion. Mol Cell 2001; 7: 1267–1278.

Costello JF, Plass C. Methylation matters. J Med Genet 2001; 38: 285–303.

Damsky CH, Richa J, Solter D, Knudsen K, Buck CA. Identification and purification of a cell surface glycoprotein mediating intercellular adhesion in embryonic and adult tissue. Cell 1983; 34: 455–466.

Fujita Y, Krause G, Scheffner M, Zechner D, Leddy HE, Behrens J, Sommer T, Birchmeier W. Hakai, a c-Cbl-like protein, ubiquitinates and induces endocytosis of the E-cadherin complex. Nat Cell Biol 2002; 4: 222–231.

Fujita K, Ohuchi N, Yao T, et al. Frequent overexpression, but not activation by point mutation, of ras genes in primary human gastric cancers. Gastroenterology 1987; 93: 1339–1345.

Gabbert HE, Mueller W, Schneiders A, Meier S, Moll R, Birchmeier W, Hommel G. Prognostic value of E-cadherin expression in 413 gastric carcinomas. Int J Cancer 1996; 69: 184–189.

Gayther SA, Gorringe KL, Ramus SJ, et al. Identification of germ-line E-cadherin mutations in gastric cancer families of European origin. Cancer Res 1998; 58: 4086–4089.

Giarelli E. Prophylactic gastrectomy for CDH1 mutation carriers. Clin J Oncol Nurs 2002; 6: 161–162.

Gofuku J, Shiozaki H, Doki Y, et al. Characterization of soluble E-cadherin as a disease marker in gastric cancer patients. Br J Cancer 1998; 78: 1095–1101.

Grady WM, Willis J, Guilford PJ, et al. Methylation of the CDH1 promoter as the second genetic hit in hereditary diffuse gastric cancer. Nat Genet 2000; 26: 16–17.

Griffiths TR, Brotherick I, Bishop RI, et al. Cell adhesion molecules in bladder cancer: soluble serum E-cadherin correlates with predictors of recurrence. Br J Cancer 1996; 74: 579–584.

Grooteclaes ML, Frisch SM. Evidence for a function of CtBP in epithelial gene regulation and anoikis. Oncogene 2000; 19: 3823–3828.

Grunwald G. The structural and functional analysis of cadherin calcium-dependent cell adhesion molecules. Curr Opin Cell Biol 1993; 5: 797–805.

Guilford PJ, Hopkins JBW, Grady WM, Markowitz SD, Willis J, Lynch H, Rajput A, Wiesner GL, Lindor NM, Burgart LJ, Toro TT, Lee D, Limacher JM, Shaw DW, Findlay MPN, Reeve AE. E-cadherin germline mutations define an inherited cancer syndrome dominated by diffuse gastric cancer. Hum Mutat 1999; 14: 249–255.

Guilford P, Hopkins J, Harraway J, McLeod M, McLeod N, Harawira P, Taite H, Scoular R, Miller A, Reeve AE. E-cadherin germline mutations in familial gastric cancer. Nature 1998; 392: 402–405.

Gumbiner BM. Regulation of cadherin adhesive activity. J Cell Biol 2000; 148: 399–404.

Herman JG, Graff JR, Myohanen S, Nelkin BD, Baylin SB. Methylation-specific PCR: a novel PCR assay for methylation status of CpG islands. Proc Natl Acad Sci USA 1996; 93: 9821–9826.

Humar B, Toro T, Graziano F, Muller H, Dobbie Z, Kwang-Yang H, Eng C, Hampel H, Gilbert D, Winship I, Parry S, Ward R, Findlay M, Christian A, Tucker M, Tucker K, Merriman T, Guilford P. Novel germline CDH1 mutations in hereditary diffuse gastric cancer families. Hum Mutat 2002; 19: 518–525.

Huntsman DG, Carneiro F, Lewis FR, MacLeod PM, Hayashi A, Monaghan KG, Maung R, Seruca R, Jackson CE, Caldas C. Early gastric cancer in young, asymptomatic carriers of germ-line E-cadherin mutations. N Engl J Med 2001; 344: 1904–1909.

Juhasz M, Ebert MP, Schulz HU, Rocken C, Molnar B, Tulassay Z, Malfertheiner P. Dual role of serum soluble E-cadherin as a biological marker of metastatic development in gastric cancer. Scand J Gastroenterol 2003; 38: 850–855.

Kanai Y, Oda T, Tsuda H, Ochiai A, Hirohashi S. Point mutation of the E-cadherin gene in invasive lobular carcinoma of the breast. Jpn J Cancer Res 1994; 85: 1035–1039.

Kang GH, Shim YH, Jung HY, Kim WH, Ro JY, Rhyu MG. CpG island methylation in premalignant stages of gastric carcinoma. Cancer Res 2001; 61: 2847–2851.

Katayama M, Hirai S, Kamihagi K, et al. Soluble E-cadherin fragments increased in circulation of cancer patients. Br J Cancer 1994; 69: 580–585.

Lewis FR, Mellinger JD, Hayashi A, Lorelli D, Monaghan KG, Carneiro F, Huntsman DG, Jackson CE, Caldas C. Prophylactic total gastrectomy for familial gastric cancer. Surgery 2001; 130: 612–617.

Machado JC, Soares P, Carneiro F, Rocha A, Beck S, Blin N, Berx G, Sobrinho-Simoes M. E-cadherin gene mutations provide a genetic basis for the phenotypic divergence of mixed gastric carcinomas. Lab Invest 1999; 79: 459–465.

Matsuyoshi N, Tanaka T, Toda K, Okamoto H, Furukawa F, Imamura S. Soluble E-cadherin: a novel cutaneous disease marker. Br J Dermatol 1995; 132: 745–749.

Mayer B, Johnson JP, Leitl F, Jauch KW, Heiss MM, Schildberg FW, Birchmeier W, Funke J. E-cadherin expression in primary and metastatic gastric cancer: down-regulation correlates with cellular dedifferentiation and glandular disintegration. Cancer Res 1993; 53: 1690–1695.

Muta H, Noguchi M, Kanai Y, Ochiai A, Nawata H, Hirohashi S. E-cadherin gene mutation in signet ring cell carcinoma of the stomach. Jpn J Cancer Res 1996; 87: 843–848.

Nass SJ, Herman JG, Gabrielson E, Iversen PW, Parl EF, Davidson NE, Graff JR. Aberrant methylation of the estrogen receptor and E-cadherin 5' CpG islands increases with malignant progression in human breast cancer. Cancer Res 2000; 60: 4346–4348.

Noe V, Fingleton B, Jacobs K, Crawford HC, Vermeulen S, Steelant W, Bruyneel E, Matrisian LM, Mareel M. Release of an invasion promoter E-cadherin fragment by matrilysin and stromelysin-1. J Cell Sci 2001; 114: 111–118.

Oliveira C, Bordin MC, Grehan N, Huntsman D, Suriano G, Machado JC, Kiviluoto T, Aaltonen L, Jackson CE, Seruca R, Caldas C. Screening E-cadherin in gastric cancer families reveals germline mutations only in hereditary diffuse gastric cancer kindred. Hum Mutat 2002; 19: 510–517.

Perez-Moreno MA, Locascio A, Rodrigo I, Dhondt G, Portillo F, Nieto MA, Cano A. A new role for E12/E47 in the repression of E-cadherin expression and epithelial-mesenchymal transitions. J Biol Chem 2001;276:27424–27431.

Perry I, Tselepis C, Hoyland J, Iqbal TH, Scott D, Sanders SA, Cooper BT, Jankowski JA. Reduced cadherin/catenin complex expression in celiac disease can be reproduced in vitro by cytokine stimulation. Lab Invest 1999; 79: 1489–1499.

Pittard AJ, Banks RE, Galley HF, et al. Soluble E-cadherin concentrations in patients with systemic inflammatory response syndrome and multiorgan dysfunction syndrome. Br J Anaesth 1996; 76: 629–631.

Polakis P. Wnt signaling and cancer. Genes Dev 2000; 14: 1837–1851.

Richards FM, McKee SA, Rajpar MH, et al. Germline E-cadherin gene (CDH1) mutations predispose to familial gastric cancer and colorectal cancer. Hum Mol Genet 1999; 8: 607–610.

Risinger JI, Berchuck A, Kohler MF, Boyd J. Mutations of the E-cadherin gene in gynecologic cancers. Nat Genet 1994; 7: 98–102.

Salahshor S, Hou H, Diep CB, et al. A germline E-cadherin mutation in a family with gastric and colon cancer. Int J Mol Med 2001; 8: 439–443.

Shimoyama Y, Hirohashi S. Expression of E- and P-cadherin in gastric carcinomas. Cancer Res 1991; 51: 2185–2192.

Suzuki H, Itoh F, Toyota M, Kikuchi T, Kakiuchi H, Hinoda Y, Imai K. Distinct methylation pattern and microsatellite instability in sporadic gastric cancer. Int J Cancer 1999; 83: 309–313.

Takeichi M. Cadherins: a molecular family important in selective cell-cell adhesion. Annu Rev Biochem 1990; 59: 237–252.

Takeichi M. Morphogenetic roles of classic cadherins. Curr Opin Cell Biol 1995; 7: 619–627.

Tamura G, Sakata K, Nishizuka S, Maesawa C, Suzuki Y, Iwaya T, Terashima M, Saito K, Satodate R. Inactivation of the E-cadherin gene in primary gastric carcinomas and gastric carcinoma cell lines. Jpn J Cancer Res 1996; 87: 1153–1159.

Tamura G, Sato K, Akiyama S, Tsuchiya T, Endoh Y, Usuba O, Kimura W, Nishizuka S, Motoyama T. Molecular characterization of undifferentiated-type gastric carcinoma. Lab Invest 2001; 81: 593–598.

Tamura G, Yin J, Wang S, Fleisher AS, Zou T, Abraham JM, Kong D, Smolinski KN, Wilson KT, James SP, Silverberg SG, Nishizuka S, Terashima M, Motoyama T, Meltzer SJ. E-Cadherin gene promoter hypermethylation in primary human gastric carcinomas. J Natl Cancer Inst 2000; 92: 569–573.

Toyota M, Ahuja N, Ohe-Toyota M, Herman JG, Baylin SB, Issa JP. CpG island methylator phenotype in colorectal cancer. Proc Natl Acad Sci USA 1999; 96: 8681–8686.

Vleminckx K, Vakaet L Jr, Mareel M, et al. Genetic manipulation of E-cadherin expression by epithelial tumor cells reveals an invasion suppressor role. Cell 1991; 66: 107–119.

Vos CBJ, Cleton-Jansen AM, Berx G, De Leeuw WJF, Ter Haar NT, Van Roy F, Cornelisse CJ, Peters JL, Van de Vijver M.J. E-cadherin inactivation in lobular carcinoma in situ of the breast: an early event in tumorigenesis. Br J Cancer 1997; 76: 1131–1133.

Waki T, Tamura G, Tsuchiya T, Sato K, Nishizuka S, Motoyama T. Promoter methylation status of E-cadherin, hMLH1, and p16 genes in nonneoplastic gastric epithelia. Am J Pathol 2002; 161: 399–403.

Wijnhoven BP, Dinjens WN, Pignatelli M. E-cadherin-catenin cell-cell adhesion complex and human cancer. Br J Surg 2000; 87: 992–1005.

Yagi T, Takeichi M. Cadherin superfamily genes: functions, genomic organization and neurologic diversity. Genes Dev 2000; 14: 1169–1180.

Chapter 15
Role of CagA in *Helicobacter pylori* Infection and Pathology

Takeshi Azuma

Introduction

Gastric cancer is the second leading cause of cancer-related deaths worldwide (Parkinet al. 2001). Various epidemiologic studies identified a role for *Helicobacter pylori* in gastric carcinoma in humans (The Eurogast Study Group 1991; Kikuchi et al. 1995; Parsonnet et al. 1991). In 1994, the World Health Organization International Agency for Research on Cancer classified *H. pylori* as a group I carcinogen in humans (IARC 1994). CagA protein, encoded by the *cagA* gene, is one of the most studied virulence factors of *H. pylori*, and is a highly immunogenic protein. The *cagA* gene is one of many genes located within the pathogenicity island (PAI) known as the *cag* PAI. The presence of *cagA* is considered a marker for the presence of *cag* PAI (Covacci et al. 1993). The *cag* PAI is a 40-kb locus in the chromosomal glutamate racemase gene. Its G + C content (35%) differs from the G + C content of the remainder of the genome (39%), suggesting that it was acquired from another organism by horizontal transfer (Censini et al. 1996; Covacci et al. 1999; Tomb et al. 1997). At some point during evolution, IS605, a mobile sequence encoding two transposases, entered the *H. pylori* genome, and in some strains these transposases interrupted, mutated, or deleted parts of the PAI (Censini et al. 1996). The severity of *H. pylori*–related disease is correlated with the presence of *cag* PAI. Infection with *cag* PAI–positive *H. pylori* is statistically associated with gastric cancer (Blaser et al. 1995; Censini et al. 1996; Covacci et al. 1993). Recent studies have provided a molecular basis for the pathologic actions of CagA on gastric epithelial cells. In this chapter, recent molecular analysis of *cagA*, and the relationship between CagA protein diversity and gastric cancer in Asia, are summarized.

Type IV Secretion System of *Helicobacter pylori*

H. pylori attaches specifically and tightly to gastric epithelial cells. The adherence of *H. pylori* to the gastric epithelial cells is an important determinant of pathogenesis. Bacterial attachment causes microvilli effacement, actin rearrangement, pedestal formation, and induces interleukin (IL)-8 synthesis and secretion. Segal et al. first

T.C. Wang et al. (eds.) *The Biology of Gastric Cancers*,
© Springer Science+Business Media, LLC 2009

reported that attachment of *H. pylori* to cultured gastric epithelial cells, such as AGS cells, can induce tyrosine phosphorylation of a 145-kDa host protein and accumulation of F-actin beneath the bacterium; furthermore, attachment of *H. pylori* leads to activation of nuclear factor-κB and release of IL-8 (Segal et al. 1996). The ability of bacteria to induce protein tyrosine phosphorylation, IL-8 production, and the rearrangement of the actin cytoskeleton is closely correlated with the presence of the *cag* PAI. The *cag* PAI contains 31 genes, 6 of which are thought to encode a putative type IV secretion system, which specializes in the transfer of a variety of multimolecular complexes across the bacterial membrane to the extracellular space or into other cells (Covacci et al. 1999). For example, the *cag* homologs of VirB4 (CagE), VirB7 (CagT), VirB9 (*cag*ORF528), VirB10 (*cag*ORF527), VirB11 (*cag*ORF525), and VirD4 (*cag*ORF524) of *Agrobacterium tumefaciens* have been shown to be assembled as a complex, and form the type IV transport machinery. The potential of *H. pylori* to deliver bacterial effector molecules through the putative type IV secretion system into the attached host cells has been suggested, thus enabling the bacteria to alter host cell signaling, such as that required for protein tyrosine phosphorylation, stimulation of IL-8 release, and induction of actin dynamics (Covacci et al. 1999).

Helicobacter pylori CagA Is Translocated From the Bacteria to Gastric Epithelial Cells and Receives Tyrosine Phosphorylation

Adherence of *H. pylori* to gastric epithelial cells can induce host cellular responses, including the reorganization of the actin cytoskeleton, the tyrosine phosphorylation of a 145-kDa protein, and release of IL-8. The 145-kDa phosphorylated protein induced in gastric epithelial cells infected with *H. pylori* strain 87A300 was originally reported by Segal and colleagues, who proposed that the 145-kDa protein was a host cellular component (Segal et al. 1996). Further investigation by Asahi et al. (2000) into the origin of the 145-kDa protein identified the phosphorylated protein as *H. pylori*–derived CagA. After attachment of *cagA*-positive *H. pylori* to gastric epithelial cells, CagA is directly injected from the bacteria into the cells via the bacterial type IV secretion system and undergoes tyrosine phosphorylation in the host cells. Several other investigators confirmed these important findings regarding CagA (Odenbreit et al. 2000; Segal et al. 1999; Stein et al. 2000). In addition, it has been reported that translocated CagA localizes to the inner surface of the host plasma membrane, where it undergoes tyrosine phosphorylation by Src family kinases (SFKs) (Selbach et al. 2002; Stein et al. 2002). Protein tyrosine phosphorylation is a major signaling mechanism by which proliferation, survival, differentiation, and migration of mammalian cells is regulated. Dysregulation of kinases and/or phosphatases that control the level of tyrosine phosphorylation is fundamentally associated with cellular transformation. This fact raises the possibility that upon tyrosine phosphorylation the bacterial protein dysregulates intracellular signaling, which directly or indirectly contributes to gastric carcinogenesis.

Intracellular Host Cell Targets of *Helicobacter pylori* CagA Protein

After attachment of *cagA*-positive *H. pylori* to gastric epithelial cells, CagA is directly injected from the bacteria into the cells via the bacterial type IV secretion system and undergoes tyrosine phosphorylation by host cell SFKs. The *cagA*-positive *H. pylori*–host cell interaction triggers morphologic changes *in vitro* where cells become elongated and "scattered" in a manner reminiscent of cells treated by hepatocyte growth factor (HGF). First described as the *hummingbird* phenotype by Segal et al. (1996), the changes in cell morphology arise from cytoskeletal rearrangements that disrupt cell–cell adhesion. The loss of epithelial barrier function by rearrangement is hypothesized to be one of the initial steps in the etiology of *H. pylori*–associated gastric diseases *in vivo*. Further dissection of Segal's model of HGF-dependent scattering of cells, has identified that SHP-2, an SH2 containing cytoplasmic tyrosine-phosphatase of mammalian cells, has as major role in inducing the hummingbird phenotype (Kodama et al. 2000). SHP-2 regulates signal transduction events from a variety of activated receptor tyrosine kinases including the mitogen-activated protein (MAP) kinase signaling pathway (Ahmad et al. 1993; Feng et al. 1993; Freeman et al. 1992), by both Ras-dependent and independent mechanisms (Neel et al. 2003). Because SH2 domains are phosphotyrosine-binding modules, we investigated the capacity of CagA to bind SHP-2. In lysates from AGS cells transfected with the CagA expression vector, CagA coimmunoprecipitated endogenous SHP-2 and vice versa. In contrast, the phosphorylation-resistant CagA, which was mutated at the phosphorylation sites, did not coimmunoprecipitate with SHP-2 thus demonstrating that CagA binds SHP-2 in AGS cells in a tyrosine phosphorylation-dependent manner resulting in increased activation of the downstream mitogenic MAP kinase pathway (Higashi et al. 2002b). Furthermore, we also demonstrated the binding of SHP-2 to CagA *in vivo* in human gastric mucosa (Yamazaki et al. 2003). Interestingly, Mimuro et al. (2002) reported that the activation of MAP kinase can also arise as a result of CagA's interaction with Grb2, an interaction that is reported to be independent of CagA's phosphorylation status. Further studies by Churin et al. (2003) demonstrated that CagA can also target the c-Met receptor and activate MAP kinase signaling, which results in increased mitosis of the affected cells. Consistently, the presence of CagA in gastric epithelial cells results in prolonged activation of Erk MAP kinase activity (Tsutsumi et al. 2006). Sustained Erk activation has been suggested to have an important role in the progression of G1 to S phase (Roovers and Assoian 2000). CagA may therefore predispose gastric epithelial cells to unscheduled proliferation, and disruption of cell–cell adhesion, at least partly through sustained MAP kinase signaling. A recent human study by Jackson et al. gives credence to this hypothesis. These authors demonstrated that *in vivo H. pylori* CagA+ strains induce a stronger MAP kinase response than their CagA− counterparts in preneoplastic lesions (Jackson et al. 2007).

Yamazaki et al. demonstrated that CagA protein and tyrosine phosphorylated CagA were not present in the gastric mucosa of patients with intestinal metaplasia

or cancer, although they were detected in the noncancer mucosal tissue from
H. pylori–positive early gastric cancer patients (Yamazaki et al. 2003). Correa's
model of gastric carcinogenesis suggests that the process may be a continuous progression whereby lesions of increasing severity develop over 2 or 3 decades: acute
gastritis progresses to chronic gastritis, then to chronic atrophic gastritis with the
development of intestinal metaplasia, and finally to frank gastric carcinoma (Correa
1988). It has been estimated that 10% of patients with chronic atrophic gastritis
develop gastric cancer over a 15-year period. Intestinal metaplasia with goblet cells
has been associated with gastric cancer in up to 90% of cases and is held to be a
good histologic marker of premalignancy. Chronic gastritis induced by *H. pylori*
infection usually progresses to atrophic gastritis. The risk for gastric cancer
increases with the degree and extent of atrophic gastritis. Severe glandular atrophy
develops, and intestinal metaplasia occurs, accompanied by the disappearance of *H.
pylori* colonization. It is therefore suspected that *H. pylori* infection has a causative
role at a relatively early phase in gastric carcinogenesis. Dysregulation of SHP-2 by
CagA may have a role in the acquisition of a cellular transformed phenotype at a
relatively early stage of multistep gastric carcinogenesis.

Biologic Activities of CagA

Current studies indicate that CagA can induce changes in cell morphology *in vitro*
through its interaction with many host cell proteins associated with cytoskeletal
remodeling. To date, host cell protein targets of CagA involved with cell morphology and/or cell–cell adhesion include E-Cadherin (Murata-Kamiya et al. 2007),
Grb2 (Mimuro et al. 2002), c-Met (HGF receptor) (Churin et al. 2003), ZO-1 and
JAM (Amieva 2003), Crk (Suzuki et al. 2005), and SHP-2 (Higashi et al. 2002b).
It was recently found that CagA-activated SHP-2 directly dephosphorylates focal
adhesion kinase (FAK) at the activating tyrosine phosphorylation sites in gastric
epithelial cells. Coexpression of constitutively active FAK with CagA inhibits
induction of the hummingbird phenotype, whereas expression of dominant-negative
FAK elicits an elongated cell shape characteristic of the hummingbird phenotype.
Thus, inhibition of FAK kinase activity by CagA-activated SHP-2 has a crucial role
in the morphogenetic activity of CagA (Tsutsumi et al. 2006). Besides its ability to
deregulate intracellular signaling, translocated CagA perturbs the apical junctions
of epithelial cells that regulate cell–cell adhesion and maintain the integrity of the
cell barrier. In polarized epithelial cells, CagA disrupts the tight junction and causes
loss of apical-basolateral polarity (Amieva et al. 2003; Bagnoli et al. 2005).
Recently, it has been reported that CagA specifically interacts with partitioning-
defective 1 (PAR-1)/microtubule affinity-regulating kinase, which has an essential
role in epithelial cell polarity. Association of CagA inhibits PAR-1 kinase activity
and prevents aPKC-mediated PAR-1 phosphorylation that dissociates PAR-1 from
the membrane, collectively causing junctional and polarity defects. Induction of the
hummingbird phenotype by CagA-activated SHP-2 requires simultaneous inhibition

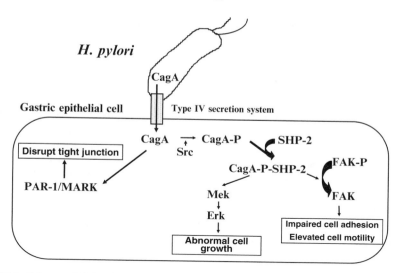

Fig. 15.1 Scheme of *Helicobacter pylori*–cell interaction

of PAR-1 kinase activity by CagA. Thus, the CagA–PAR-1 interaction not only elicits the junctional and polarity defects but also promotes the morphogenetic activity of CagA (Saadat et al. 2007). This recent report indicated that PAR-1 is a key target of CagA in the disorganization of gastric epithelial architecture underlying mucosal damage, inflammation, and carcinogenesis (Figure 15.1).

Diversity of CagA

H. pylori is genetically more diverse than most bacterial species. Strain-specific genetic diversity has been proposed to be involved in the organism's ability to cause different diseases. There are also indications of significant geographic differences among strains. Only half to two-thirds of Western isolates carry *cagA*. In contrast, nearly all East Asian strains carry *cagA* (Ito et al. 1997; van Doorn et al. 1999). The CagA protein is highly immunogenic with a molecular weight of 128–140 kDa. Variation in the size of the protein has been correlated with the presence of a variable number of repeat sequences located in the 3' region of the gene (Covacci et al. 1993; Yamaoka et al. 1998). The phosphorylation sites are located in the repeat region of CagA (Backert et al. 2001). The sequence variation raises an intriguing possibility that the biologic activity of CagA can vary from one strain to the next, which may influence the pathogenicities of different *cagA*-positive *H. pylori* strains. The alignment of the deduced amino acid sequence in the 3' region of the *cagA* gene (CagA repeat domain) among strains 26695 and F32, which are typical Western and East Asian CagA, respectively, is shown in

Figure 15.2A. We previously reported that the Glu-Pro-Ile-Tyr-Ala (EPIYA) motifs are potential targets of tyrosine phosphorylation by SFKs. These EPIYA motifs are involved in the interaction of CagA with SHP-2. The first and the second EPIYA motifs (which we designated EPIYA-A and EPIYA-B, respectively) are present in almost all CagA proteins, whereas the remaining three EPIYA motifs (which we designated EPIYA-C) were made by duplication of an EPIYA-containing 34-amino-acid sequence. Because the sequence exists in various numbers ranging from 1 to 3 in most CagA proteins from *H. pylori* isolated in Western countries, we designated it the *Western CagA-specific, SHP-2 binding sequence* (*WSS*) (Higashi et al. 2002a).The WSS contains D1, D2, and D3 motifs as defined by Covacci et al. (1993) or R1 and WSR regions as defined by Yamaoka et al. (1999). The strain 26695 had a single WSS and was thus classified as the "A-B-C" type. The amino acid sequence of CagA from *H. pylori* isolated in East Asian countries is quite different from that of Western CagA. Predominant East Asia CagA proteins do not have the WSS, but instead possess a distinct sequence that we designated *East Asian CagA-specific, SHP-2-binding sequence* (*ESS*) in the corresponding region

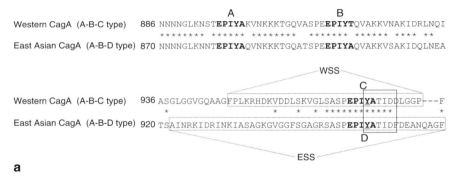

a

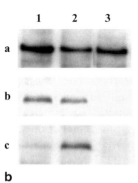

Fig. 15.2 (**A**) Amino acid sequence alignment of the EPIYA regions (CagA repeat domains) for Western (strain 26695) and East Asian CagA (strain F32) proteins. The Western CagA-specific sequence (WSS) and the East Asian CagA-specific sequence (ESS) are indicated in the figure. Numbers at the head of sequences represent the positions of amino acid residues of both strains. Stars denote identity. A hyphen indicates the absence of an amino acid residue. The tyrosine residue of the phosphorylation site is underlined. The binding site of SHP-2 is boxed. (**B**) Immunoblot analysis of AGS cells infected with Western CagA-positive strain (A-B-C type) (lane 1), East Asian CagA-positive strain (A-B-D type) (lane 2), or A-B type CagA-positive stain (lane 3) by anti-CagA (a), anti-phosphotyrosine (b), and anti–SHP-2 antibodies (c). CagA phosphorylation and CagA–SHP-2 binding were not detected in the infection with A-B type CagA-positive strain. East Asian type CagA protein conferred stronger SHP-2 binding to the Western type CagA protein. (Modified from Azuma et al. 2004.)

(Higashi et al. 2002a). ESS contains a JSR sequence previously defined by Yamaoka et al. (1999) and also possesses an EPIYA motif, denoted EPIYA-D. F32 had a single ESS and was thus classified as the "A-B-D" type. The tyrosine residue in EPIYA-C or -D motifs is the major phosphorylation site. Upon tyrosine phosphorylation, the East Asian–specific sequence confers stronger SHP-2 binding and transforming activities to Western CagA (Figure 15.2B) (Higashi et al 2002a). The CagA–SHP-2 interaction requires the SH2 domains of SHP-2. De Souza et al. reported that the two SH2 domains from SHP-2 bind to highly related sequences, and the consensus ligand-binding motif for the N- and C-SH2 domains of SHP-2 is pY-(S/T/A/V/I)-X-(V/I/L)-X-(W/F) (De Souza et al. 2002). Intriguingly, the consensus motif perfectly matches the SHP-2-binding site of East Asian CagA, pY-A-T-I-D-F. Furthermore, replacement of the pY + 5 position from W/F with any other amino acids, such as aspartic acid in the case of WSS in Western CagA, reduces the binding affinity to SHP-2. Hence, differential SHP-2 binding activities observed between WSS and ESS of CagA proteins are caused by the difference in a single amino acid at the pY + 5 position. The potential of CagA to disturb host cell functions as a virulence factor could be determined by the degree of SHP-2 binding activity. The diversity of the CagA phosphorylation site, which collectively determines binding affinity of CagA to SHP-2, may be an important variable in determining the clinical outcome of infection with different *H. pylori* strains.

The Distribution of CagA Protein Diversity and Association Between the CagA Protein Diversity and Gastric Cancer

The incidence and mortality rate attributable to gastric cancer in Japan is high compared with that in other developed countries. However, large intracountry differences in the mortality rates of gastric cancer have been reported (Ito et al. 1996). Fukui is a typical rural prefecture located on the central Japanese mainland (Honshu), whereas Okinawa consists of islands in the southwestern part of Japan and has a history and food culture different from those of other parts of Japan. The two areas are separated by more than 1,300 km. The prevalence of atrophic gastritis, a precursor lesion of gastric cancer, is more frequent in Fukui, and the mortality rate from gastric cancer is more than 2.4 times higher in Fukui (43.7/100,000 in 1999) than in Okinawa (18.2/100,000 in 1999). We previously investigated the diversity of CagA phosphorylation sites in isolates from two different areas in Japan (Fukui and Okinawa) where gastric cancer risk is different in order to examine the association between diversity and gastric cancer. We demonstrated that the prevalence of *cagA*-positive *H. pylori* was significantly different between Fukui and Okinawa. All isolates examined from Fukui (64 strains) were *cagA*-positive strains. In contrast, 12.0% (6/50) isolates from Okinawa were *cagA*-negative strains. All *cagA*-negative strains were isolated from patients with chronic gastritis. In addition, the distribution of CagA protein diversity was different in Fukui and Okinawa. Almost all strains isolated from

Table 15.1 Distribution of the diversity of the CagA protein

	Fukui			Okinawa		
	cagA (−)	East Asian	Western	cagA (−)	East Asian	Western
Chronic gastritis	0	35	0	6	28	8
Gastric cancer	0	29	0	0	8	0
Total	0	64	0	6	36	8*

Source: Modified from Azuma et al. 2004.
*The prevalence of Western CagA-positive strain was significantly higher in Okinawa than in Fukui (p = 0.001).

Fukui were East Asian CagA-positive strains containing the ESS sequence. Predominant Fukui strains had a single ESS region and were classified "A-B-D." In contrast, 16.0% (8/50) of Okinawa strains were Western CagA-positive strains containing a WSS sequence. The prevalence of the Western CagA-positive strain was significantly higher in Okinawa than in Fukui. All gastric cancer strains (29 Fukui and 8 Okinawa strains) were East Asian CagA-positive strains in both Fukui and Okinawa (Table 15.1) (Azuma et al. 2004).

A high CagA diversity in the strains isolated in Okinawa was observed, although CagA was homogenous in other parts of Japan. Therefore, Okinawa is the best place to investigate the association between the diversity of *H. pylori* and clinical outcome in Japan. A phylogenetic tree constructed on the basis of the full-length *cagA* sequences showed that all gastric cancer strains with the East Asian–type of CagA were in the East Asian cluster, and that most duodenal ulcer strains were in the Western cluster (Figure 15.3). Two gastric cancer strains were in the Western cluster. These strains had the repeats of "C" in CagA: AB‴CC (OK310) and AB´CCC (OK308) (Satomi et al. 2006). These findings indicate that the origins of *H. pylori* isolates are different between gastric cancer strains and duodenal ulcer strains, and that East Asian CagA-positive *H. pylori* infection is associated with gastric cancer. We have reported that the number of "C" sites is directly correlated with levels of tyrosine phosphorylation and SHP-2 binding activity among Western CagA species (Segal et al. 1996). Yamaoka et al. also reported that the number of "C" sites was associated with the severity of atrophic gastritis and gastric cancer in patients infected with Western CagA-positive strains (Yamaoka et al. 1999). These findings suggest that deregulation of SHP-2 by CagA may have a key role in gastric carcinogenesis.

The Distribution of CagA Protein Diversity in the World

Huang et al. conducted a metaanalysis of the relationship between CagA seropositivity and gastric cancer and demonstrated that infection with *cagA*-positive strains of *H. pylori* increases the risk for gastric cancer over the risk associated with *H. pylori* infection alone. The analysis, based on 7 studies of 1,707 gastric cancer

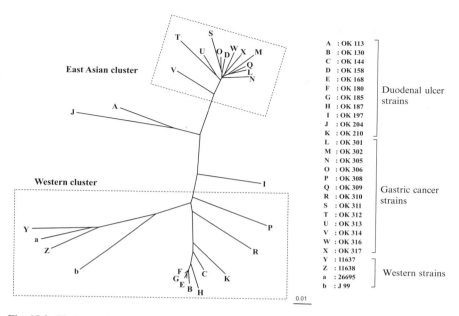

Fig. 15.3 Phylogenetic tree constructed on the basis of the full-length *cagA* sequences. All except three strains were divided into two major groups, a Western and an East Asian cluster, corresponding to the diversity of the SHP-2 binding site of the CagA protein. All gastric cancer strains with the East Asian–type CagA were in the East Asian cluster, and most duodenal ulcer strains were in the Western cluster. Two gastric cancer strains were in the Western cluster. These strains had the repeats of "C" in CagA: AB‴CC (OK310) and AB´CCC (OK308). (Modified from Satomi et al. 2006.)

patients and 2,124 matched controls, showed that infection with *cagA*-positive strains of *H. pylori* increases the risk for gastric cancer by 2.87-fold, irrespective of *H. pylori* status. In contrast, overall risk for gastric cancer associated with *H. pylori* infection was 2.28 (Huang et al. 2003).The greater magnitude of risk observed with CagA seropositivity over *H. pylori* infection alone suggests that patients harboring *cagA*-positive strains of *H. pylori* are at higher risk for developing gastric cancer than those infected with *cagA*-negative strains. We investigated the distribution of CagA protein diversity in the world using our data and data deposited in the GenBank (Table 15.2). The prevalence of East Asian CagA-positive strains seemed to be associated with the rate of gastric cancer mortality worldwide. Endemic circulation of *H. pylori* populations with more virulent East Asian CagA proteins may affect the prevalence of gastric cancer in East Asian countries. Patients harboring East Asian CagA-positive *H. pylori* are at a higher risk for developing gastric cancer than those infected with Western CagA-positive strains (Azuma et al. 2004). Searching for CagA diversity rather than just *H. pylori* infection may confer an additional benefit in identifying populations at greater risk for gastric cancer.

Table 15.2 The distribution of CagA protein diversity worldwide

Country	CagA type		Mortality rate of gastric cancer (per 100,000 males in 2000)
	East Asian	Western	
The West			
Ireland	0	3	13.07
Austria	0	1	14.12
Italy	0	1	27.84
England	0	1	17.66
United States	0	6	6.15
Australia	0	3	8.64
East Asia			
Japan	123	12	58.39
Korea	5	0	36.71
China	20	0	24.56
Asia			
Vietnam	10	0	12.83
Thailand	13	9	3.31
India	0	3	3.83

Conclusions

Strain-specific diversity has been proposed to be involved in *H. pylori*'s ability to cause different diseases. The *cagA* gene product, CagA, was demonstrated to be injected into the host cytoplasm through a type IV secretion system, and phosphorylated by the host cellular kinases (Asahi et al. 2000; Covacci et al. 1999; Odenbreit et al. 2000). In addition, CagA forms a physical complex with SHP-2, which is known to have an important positive role in mitogenic signal transduction, and stimulates phosphatase activity (Higashi et al. 2002b). In Japan, nearly 100% of the strains possess functional *cag* PAI (Azuma et al. 2002; Ito et al. 1997), and the incidence of atrophic gastritis and gastric cancer is quite high compared with Western countries. There are two major CagA subtypes: the East Asian and the Western type. We recently discovered that East Asian CagA has a distinct sequence at the SHP-2 binding site, and that the East Asian–specific sequence confers stronger SHP-2 binding and transforming activities than Western CagA (Higashi et al. 2002a). East Asian CagA-positive *H. pylori* infection is associated with atrophic gastritis and gastric cancer, and persistent active inflammation induced by the East Asian CagA-positive strain may have a role in its pathogenesis (Azuma et al. 2004). Endemic circulation of *H. pylori* populations carrying biologically more active CagA proteins in East Asian countries, where the mortality rate of gastric cancer is among the highest in the world, may be involved in increasing the risk of gastric cancer in these populations.

   H. pylori CagA–host cell interaction is introducing a new paradigm for "bacterial carcinogenesis." The bacterial protein targets the tyrosine phosphorylation system in mammalian cells, the disturbance of which has a crucial role in the development of cancer. Further elucidation of host cell signaling targeted by CagA should give insights into general understanding of inflammation-mediated cancer.

References

Ahmad, S., Banville, D., Zhao, Z., et al. 1993. A widely expressed human protein-tyrosine phosphatase containing src homology 2 domains. Proc Natl Acad Sci USA 90:2197–2201.

Amieva, M.R., Vogelmann, R., Covacci, A., et al. 2003. Disruption of the epithelial apical-junctional complex by Helicobacter pylori CagA. Science 300:1430–1434.

Asahi, M., Azuma, T., Ito, S., et al. 2000. Helicobacter pylori CagA protein can be tyrosine phosphorylated in gastric epithelial cells. J Exp Med 191:593–602.

Azuma, T., Yamakawa, A., Yamazaki, S., et al. 2002. Correlation between variation of the 3'region of the cagA gene in Helicobacter pylori and disease outcome in Japan. J Infect Dis 186:1621–1630.

Azuma, T., Yamazaki, S., Yamakawa, A., et al. 2004. Variation in the SHP-2 binding site of Helicobacter pylori CagA protein is associated with gastric atrophy and cancer. J Infect Dis 189:820–827.

Backert, S., Moese, S., Selbach, M., et al. 2001. Phosphorylation of tyrosine 972 of the Helicobacter pylori CagA protein is essential for induction of a scattering phenotype in gastric epithelial cells. Mol Microbiol 42:631–644.

Bagnoli, F., Buti, L., Tompkins, L., et al. 2005. Helicobacter pylori CagA induces a transition from polarized to invasive phenotypes in MDCK cells. Proc Natl Acad Sci USA 102:16339–16344.

Blaser, M.J., Perez-Perez, G.I., Kleanthous, H., et al. 1995. Infection with Helicobacter pylori strains possessing cagA is associated with an increased risk of developing adenocarcinoma of the stomach. Cancer Res 55:2111–2115.

Censini, S., Lange, C., Xiang, Z., et al. 1996. Cag, a pathogenicity island of Helicobacter pylori, encodes type I-specific and disease-associated virulence factors. Proc Natl Acad Sci USA 93:14648–14653.

Churin, Y., Al-Ghoul, L., Kepp, O., et al. 2003. Helicobacter pylori CagA protein targets the c-Met receptor and enhances the motogenic response. J Cell Biol 161:249–255.

Correa, P. 1988. A human model of gastric carcinogenesis. Cancer Res 48:3554–3560.

Covacci, A., Censini, S., Bugnoli, M., et al. 1993. Molecular characterization of the 128-kDa immunodominant antigen of Helicobacter pylori associated with cytotoxicity and duodenal ulcer. Proc Natl Acad Sci USA 90:5791–5795.

Covacci, A., Telford, J.L., Del Giudice, G., et al. 1999. Helicobacter pylori virulence and genetic geography. Science 284:1328–1333.

De Souza, D., Fabri, L.J., Nash, A., et al. 2002. SH2 domains from suppressor of cytokine signaling-3 and protein tyrosine phosphatase SHP-2 have similar binding specificities. Biochemistry 41:9229–9236.

The Eurogast Study Group. 1993. An association between Helicobacter pylori infection and gastric cancer. Lancet 341:1359–1362.

Feng, G.S., Hui, C.C., Pawson, T. 1993. SH2-containing phosphotyrosine phosphatase as a target of protein-tyrosine kinases. Science 259:1607–1611.

Freeman, R.M., Jr., Plutzky, J., Neel, B.G. 1992. Identification of a human src homology 2-containing protein-tyrosine-phosphatase: a putative homolog of Drosophila corkscrew. Proc Natl Acad Sci USA 89:11239–11243.

Higashi, H., Tsutsumi, R., Fujita, A., et al. 2002a. Biological activity of the Helicobacter pylori virulence factor CagA is determined by variation in the tyrosine phosphorylation sites. Proc Natl Acad Sci USA 99:14428–14433.

Higashi, H., Tsutsumi, R., Muto, S., et al. 2002b. SHP-2 tyrosine phosphatase as an intracellular target of Helicobacter pylori CagA protein. Science 295:683–686.

Huang, J.Q., Zheng, G.F., Sumanac, K., et al. 2003. Meta-analysis of the relationship between cagA seropositivity and gastric cancer. Gastroenterology 125:1636–1644.

IARC. 1994. Schistosomes, liver flukes and Helicobacter pylori. Monographs on the evaluation of carcinogenic risks to humans. IARC Sci Publ 61:1–241.

Ito, Y., Azuma, T., Ito, S., et al. 1997. Analysis and typing of the vacA gene from cagA-positive strains of Helicobacter pylori isolated in Japan. J Clin Microbiol 35:1710–1714.

Ito, S., Azuma, T., Murakita, H., et al. 1996. Profile of Helicobacter pylori cytotoxin derived from two areas of Japan with different prevalence of atrophic gastritis. Gut 39:800–806.

Jackson, C.B., Judd, L.M., Menheniott, T.R., et al. 2007. Augmented gp130-mediated cytokine signalling accompanies human gastric cancer progression. J Pathol 213:140–151.

Kikuchi, S., Wada, O., Nakajima, T., et al. 1995. Serum anti-Helicobacter pylori antibody and gastric carcinoma among young adults. Research group on prevention of gastric carcinoma among young adults. Cancer 75:2789–2793.

Kodama, A., Matozaki, T., Fukuhara, A., et al. 2000. Involvement of an SHP-2-Rho small G protein pathway in hepatocyte growth factor/scatter factor-induced cell scattering. Mol Biol Cell 11:2565–2575.

Mimuro, H., Suzuki, T., Tanaka, J., et al. 2002. Grb2 is a key mediator of Helicobacter pylori CagA protein activities. Mol Cell 10:745–755.

Murata-Kamiya, N., Kurashima, Y., Teishikata, Y., et al. 2007. Helicobacter pylori CagA interacts with E-cadherin and deregulates the beta-catenin signal that promotes intestinal transdifferentiation in gastric epithelial cells. Oncogene 26:4617–4626.

Neel, B.G., Gu, H., Pao, L. 2003. The "Shp"ing news: SH2 domain-containing tyrosine phosphatases in cell signaling. Trends Biochem Sci 28:284–293.

Odenbreit, S., Puls, J., Sedlmaier, B., et al. 2000. Translocation of Helicobacter pylori CagA into gastric epithelial cells by type IV secretion. Science 287:1497–1500.

Parkin, D.M., Bray, F.I., Devesa, S.S. 2001. Cancer burden in the year 2000. The global picture. Eur J Cancer 37:S4–S66.

Parsonnet, J., Friedman, G.D., Vandersteen, D.P., et al. 1991. Helicobacter pylori infection and the risk of gastric carcinoma. N Engl J Med 325:11127–11131.

Roovers, K., Assoian, R.K. 2000. Integrating the MAP kinase signal into the G1 phase cell cycle machinery. Bioessays 22:818–826.

Saadat, I., Higashi, H., Obuse, C., et al. Helicobacter pylori CagA targets PAR1/MARK kinase to disrupt epithelial cell polarity. Nature 447:330–333.

Satomi, S., Yamakawa, A., Matsunaga, S., et al. 2006. Relationship between the diversity of the cagA gene of Helicobacter pylori and gastric cancer in Okinawa, Japan. J Gastroenterol 41:668–673.

Segal, E.D., Cha, J., Lo, J., et al. 1999. Altered states: involvement of phosphorylated CagA in the induction of host cellular growth changes by Helicobacter pylori. Proc Natl Acad Sci USA 96:14559–14564.

Segal, E.D., Falkow, S., Tompkins, L.S. 1996. Helicobacter pylori attachment to gastric cells induces cytoskeletal rearrangements and tyrosine phosphorylation of host cell protein. Proc Natl Acad Sci USA 93:1259–1264.

Selbach, M., Moese, S., Hauck, C.R., et al. 2002. Src is the kinase of the Helicobacter pylori CagA protein in vitro and in vivo. J Biol Chem 277:6775–6778.

Stein, M., Bagnoli, F., Halenbeck, R., et al. 2002. c-Src/Lyn kinase activate Helicobacter pylori CagA through tyrosine phosphorylation of the EPIYA motifs. Mol Microbiol 43:971–980.

Stein, M., Rappuoli, R., Covacci, A. 2000. Tyrosine phosphorylation of the Helicobacter pylori CagA antigen after cag-driven host cell translocation. Proc Natl Acad Sci USA 97:1263–1268.

Suzuki, M., Mimuro, H., Suzuki, T., et al. 2005. Interaction of CagA with Crk plays an important role in Helicobacter pylori-induced loss of gastric epithelial cell adhesion. J Exp Med 202:1235–1247.

Tomb, J.F., White, O., Kerlavage, A.R., et al. 1997. The complete genome sequence of the gastric pathogen Helicobacter pylori. Nature 388:539–547.

Tsutsumi, R., Takahashi, A., Azuma, T., et al. 2006. FAK is a substrate and downstream effector of SHP-2 complexed with Helicobacter pylori CagA. Mol Cell Biol 26:261–276.

van Doorn, L.J., Figueiredo, C., Megraud, F., et al. 1999. Geographic distribution of vacA allelic types of Helicobacter pylori. Gastroenterology 116:823–830.

Yamaoka, Y., El-Zimaity, H.M., Gutierrez, O., et al. 1999. Relationship between the cagA 3' repeat region of Helicobacter pylori, gastric histology, and susceptibility to low pH. Gastroenterology 117:342–349.

Yamaoka, Y., Kodama, T., Kashima, K., et al. 1998. Variants of the 3' region of the cagA gene in Helicobacter pylori isolates from patients with different H. pylori-associated diseases. J Clin Microbiol 36:2258–2263.

Yamazaki, S., Yamakawa, A., Ito, Y., et al. 2003. The CagA protein of Helicobacter pylori is translocated into epithelial cells and binds to SHP-2 in human gastric mucosa. J Infect Dis 187:334–337.

Chapter 16
The Role of *Helicobacter pylori* Virulence Factors in Rodent and Primate Models of Disease

Dawn A. Israel and Richard M. Peek, Jr.

Introduction

Helicobacter pylori colonizes the stomach of at least half of the world's population, and this process usually persists for the lifetime of the host. Virtually all persons infected with *H. pylori* develop gastric inflammation, which confers an increased risk for developing gastric cancer. However, only a fraction of infected persons ever develop these clinical sequelae. Identification of bacterial biomarkers associated with increased disease risk has profound ramifications because such findings will not only provide mechanistic insights into inflammatory carcinogenesis, but may also identify a subpopulation of *H. pylori*–infected individuals who can then be targeted for intervention.

Animal models provide a unique opportunity to study mechanisms by which *H. pylori* colonizes its host and to identify bacterial factors related to virulence and the development of disease (Ferraro and Fox 2001; O'Rourke and Lee 2003). Rodents, including mice and gerbils, as well as monkeys represent the principal models that have been utilized in these studies. Each model has its own distinct advantages and disadvantages, and as such, can be viewed as complementary systems. Mice are typically inbred, permitting host variables to be carefully controlled, although most strains of mice develop only mild inflammation and not cancer following challenge with *H. pylori*. Mongolian gerbils are outbred and are not as useful for the detailed study of host factors, but gerbils have been shown to develop cancer when colonized with certain strains of *H. pylori*. Monkeys are the most closely related of these models to the human host. However, studies and manipulations in monkeys cannot be conducted on the same scale as rodents, and large, long-term studies are impractical because of costs. This chapter reviews *H. pylori* virulence factors that have been identified using these models and discusses their importance within the context of host defense mechanisms that the bacterium must overcome to establish persistent infection and induce gastric injury.

T.C. Wang et al. (eds.) *The Biology of Gastric Cancers*,
© Springer Science+Business Media, LLC 2009

Gastric Peristalsis

Motility

One of the primary obstacles *H. pylori* must overcome to successfully colonize its host is gastric peristalsis. Indeed, *H. pylori* has evolved several mechanisms to elude this host defense including motility, chemotaxis, and the ability to adhere to host gastric epithelium. Studies utilizing Mongolian gerbils have shown that *H. pylori* resides deep within the gastric mucus gel layer near the epithelia, similar in topography to where the bacteria are found in human gastric specimens (Schreiber et al. 2004). Localization to these sites is dependent on bacterial motility and is frequently accompanied by adherence to gastric epithelial cells. *H. pylori* possesses polar flagella (Figure 16.1A) consisting of two major subunits, FlaA and FlaB (O'Toole et al. 2000; Table 16.1), and deletion of *flaA* results in flagellar truncation and decreased motility (Josenhans et al. 1995). Using signature-tagged mutagenesis, Kavermann et al. identified *flaA* as an essential gene for the colonization of gerbils, along with several other genes involved either in flagellar assembly or that encode structural components of flagella (Kavermann et al. 2003). Using a different mining technique (Microarray tracking of transposon mutants), another study identified genes involved in flagellar biosynthesis as being required for colonization of mice (Baldwin et al. 2007). A role for FlaA and FlaB in persistent colonization was also detected in a gnotobiotic piglet model (Eaton et al. 1996). However, these studies did not directly address the question of whether flagella *per se* are required for prolonged infection or if the motility phenotype is necessary for colonization. Therefore, additional studies have utilized *motB* mutants of *H. pylori* that are deficient in the motor protein MotB (required for motility) but that retain flagellar structures (Ottemann and Lowenthal 2002). It was determined that MotB-deficient bacteria could colonize mice; however, the 50% infectious dose (ID_{50}) for this mutant was at least 4 logs higher than that of the wild-type strain. Furthermore, nonmotile strains were found to compete poorly with wild-type *H. pylori* in a suckling mouse model over a 24-hour period of infection. Also, when mice were cochallenged with the *motB* mutant and wild-type parental bacteria, the mutant could not be recovered after 2 days (Guo and Mekalanos 2002; Ottemann and Lowenthal 2002). Taken together, these data strongly support a role for motility in colonization.

Related studies have examined the role of chemotaxis in *H. pylori* colonization of rodents. *H. pylori* possesses four methyl-accepting chemotaxis proteins (MCPs) that sense external stimuli: TlpA, TlpB, TlpC, and TlpD/HlyB. Homologs of the *Escerichia coli* receptor-kinase adaptor CheW, histidine kinase CheA, and response regulator CheY are predicted within the *H. pylori* genome (Alm and Trust 1999; Tomb et al. 1997) (Figure 16.1B). As predicted, mutants defective in CheW, CheA, or CheY are nonchemotactic. They are, however, able to infect mice, albeit at lower densities than wild-type *H. pylori* (Terry et al. 2005). In coinfection experiments, wild-type *H. pylori* were able to out-compete *cheW* and *cheY* mutants (Guo and Mekalanos 2002; Terry et al. 2005). The *cheW* mutant seems to be attenuated in the

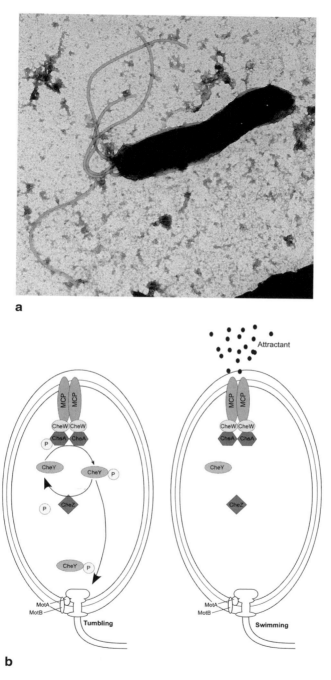

Fig. 16.1 *Helicobacter pylori* motility. (**A**) Electron micrograph of *H. pylori* with polar flagella. (Courtesy of A. Franco, unpublished data.) (**B**) Model of *H. pylori* regulation of chemotaxis and motility

Table 16.1 Virulence factors identified or confirmed using animal models of *Helicobacter pylori* pathogenesis

Protein	Predicted function/activity	Role in virulence
FlaA, FlaB	Major components of flagella	Motility
MotB	Flagellar motor protein	Motility
TlpA, TlpB, TlpC, TlpD	Methyl-accepting chemotaxis proteins	Motility
CheW, CheA, CheY	Response regulator system	Motility
BabA	OMP, Le^b adhesin	Adherence
SabA	OMP, sialyl-Le^X adhesin	Adherence
OipA (HopH)	OMP	Adherence
UreA, UreB	Components of urease	Acid resistance
AhpC	Alkyl hydroperoxide reductase	Oxidative stress resistance
Tpx	Thioperoxidase	Oxidative stress resistance
BCP	Thioperoxidase	Oxidative stress resistance
SodB	Peroxidase	Oxidative stress resistance
KatA	Catalase	Oxidative stress resistance
KapA	Resistance to H_2O_2	Oxidative stress resistance
MdaB	NADPH quinone reductase	Oxidative stress resistance
NapA	Neutrophil recruitment	Oxidative stress resistance
Msr	Methionine sulfoxide reductase	Oxidative stress resistance
Endo III	DNA endonuclease III	Oxidative stress resistance
MutY	Adenine glycosylase	Oxidative stress resistance
MutS	Component of DNA methyl-directed mismatch repair system	Oxidative stress resistance
RuvC	DNA endonuclease required for homologous recombination	Oxidative stress resistance
LPS	Lipopolysaccharide	Immune evasion
Le^x, Le^y	Lewis antigens	Immune evasion
cag PAI proteins	Numerous proteins including a T4SS and CagA effector protein	Delivery of effector proteins to host cells
VacA	Vacuolating cytotoxin	Potential role in colonization

Abbreviations: NADPH: nicotinamide adenine dinucleotide phosphate; OMP: outer membrane protein; LPS: lipopolysaccharide; PAI: pathogenicity island.

initial steps of colonization because the ID_{50} was 10- to 100-fold greater than that of the wild-type strain. However, once established, the *cheW* mutant persisted at levels comparable to wild type out to 6 months postinoculation. Interestingly, histologic examination revealed that wild-type bacteria were found in both the corpus and antrum of infected mice, but the *cheW* mutant was only found in the antrum (Terry et al. 2005).

Three of the *H. pylori* MCPs are predicted to possess canonical membrane spanning domains similar to those of *E. coli* MCPs, suggesting that they may act as environmental sensors to regulate chemotaxis. *In vitro* studies have demonstrated that TlpA is a receptor for arginine and sodium bicarbonate (Cerda et al. 2003) whereas TlpB is required for pH taxis (Croxen et al. 2006). Disruption of *tlpA* or *tlpC* did not alter motility *in vitro* or the ability to colonize mice

(Andermann et al. 2002; Croxen et al. 2006). In competition assays, however, the wild-type strain had a survival advantage over either mutant *in vivo* (Andermann et al. 2002). Disruption of *tlpB* leads to differing effects on colonization, depending on the particular study. In one study, TlpB mutants were unable to colonize interleukin (IL)-12 knockout mice (Croxen et al. 2006). However, in a different study, a TlpB mutant was able to colonize both mice and gerbils (McGee et al. 2005; Williams et al. 2007). Interestingly, loss of TlpA, TlpB, CheY, or CheA was associated with a decreased inflammatory response in mice, and bacteria lacking any of these proteins were not juxtaposed as closely to gastric epithelial cells *in vivo* as wild type, *tlpC*- or *hylB*- *H. pylori* (Williams et al. 2007). Collectively, these studies indicate that motility and chemotaxis have critical roles in establishing colonization and maintaining persistence in the gastric mucosal niche.

Adherence

Although the majority of *H. pylori* are found free-living within the gastric mucus layer, adherence has an important role in the establishment of colonization and induction of pathologic sequelae, and several *H. pylori* adhesins have been studied in animal models. One well-studied adhesin is the outer membrane protein (OMP) BabA, which binds the Lewis[b] (Le[b]) blood-group antigen on gastric epithelial cells. Not all *H. pylori* strains express this adhesin; however, BabA-expressing strains are associated with an increased risk for gastric adenocarcinoma (Gerhard et al. 1999). Expression of the gene encoding BabA, *babA2*, is regulated by at least two distinct mechanisms. First, slipped strand mispairing during replication of a CT repeat region can render this transcript in- or out-of-frame for translation, dictating whether the full-length protein is transcribed (Ilver et al. 1998; Solnick et al. 2004). In addition, studies in which Rhesus macaques were experimentally infected with *H. pylori* demonstrated that *babA2* could be replaced with the highly related gene *babB* via recombination (Solnick et al. 2004). As expected, recovered isolates that did not express BabA were deficient in their ability to bind Le[b] when tested *in vitro*. These results indicate that *H. pylori* can utilize antigenic variation to regulate its interaction with host cells.

The *H. pylori* adhesin SabA binds sialylated glycans and was initially identified by a contact-dependent retagging technique as the specific adhesin for sialyl-Le[X] (Mahdavi et al. 2002). Using gastric biopsy tissue from a Rhesus monkey, the investigators further demonstrated that, *in situ*, an *H. pylori* BabA mutant bound to gastric epithelium in a pattern that reflected expression of sialyl-Le[X]. Similarly, when *H. pylori* were pretreated with sialyl-Le[X], the ability of the BabA mutant to bind gastric tissue from Le[b]-, sialyl-Le[X]-expressing mice was reduced >90% compared with the wild-type strain. Experimental infection of a monkey with the Le[b]-binding, sialyl-Le[X]-binding *H. pylori* strain J166 revealed that *H. pylori* induces increased expression of sialyl-Le[X] in the gastric epithelium (Mahdavi et al. 2002).

Another OMP important for *H. pylori* adherence is OipA/HopH. Similar to BabA, expression of OipA can be regulated by slipped strand mispairing that is dependent on the number of CT dinucleotide repeats within the 5' region of the gene (Yamaoka et al. 2000). Expression of OipA has been associated with increased secretion of the proinflammatory cytokine IL-8 *in vitro* and infection of mice with *H. pylori oipA* mutants has revealed a role for this OMP in inflammation (Yamaoka et al. 2002). Additionally, there are recent data indicating a role for OipA for *H. pylori*–induced cancer in Mongolian gerbils (Franco et al. 2008). Results of numerous *in vitro* studies of OMPs and cell signaling as well as epidemiologic studies of OMPs and gastric disease suggest that *in vivo* studies using animal models for the study of these proteins will continue to be a fertile area of *H. pylori* research.

Acid Resistance

The harsh, acidic environment of the stomach is another defense that *H. pylori* must overcome to successfully colonize its host. It has been well documented that *H. pylori* grows optimally *in vitro* at a pH near neutral and fails to grow at pH levels less than 4 (Bijlsma et al. 1998; McGowan et al. 1994, 1996). This implies that *H. pylori* must have a mechanism for surviving the low pH conditions within the gastric lumen (frequently <2.0) before invading the mucous layer.

Urease is an enzyme produced by a number of diverse bacteria which catalyzes the hydrolysis of urea to ultimately yield ammonia and carbonic acid. Generation of ammonia has been proposed as a strategy used by *H. pylori* to alter the pH of the local gastric environment to make it more habitable. An early study in which a structural subunit of this enzyme, UreB, was disrupted, demonstrated that the *ureB* mutant was unable to colonize nude mice (Tsuda et al. 1994). This finding was later confirmed in a study using C57BL/6 mice, which further demonstrated that reintroduction of UreB expression restored the ability of the bacteria to colonize the mouse stomach (Eaton et al. 2002). *H. pylori* urease activity is also required for colonization of Rhesus macaques (Hansen and Solnick 2001). In that study, a human *H. pylori* strain (J166) that had previously been shown to readily colonize Rhesus monkeys was used for inoculation, along with two additional human strains. While characterizing the input strains, the investigators found that the J166 strain exhibited low levels of urease activity compared with the other two strains; nevertheless, J166 was the only strain recovered from the animals. When assayed after animal passage, J166 isolates were found to possess high levels of urease activity. The authors determined that the low urease activity of the input J166 strain was attributable to a mutation within the *ureA* gene. However, a small subpopulation of J166 with wild-type *ureA* and high urease activity existed within the input pool, and this population became dominant during *in vivo* infection, emphasizing the critical role of urease activity for persistent colonization.

Resistance to Oxidative Stress

Prevention of Damage

Oxidative stress limits the survival of virtually all living organisms. Reactive oxygen species (ROS) include superoxide, hydrogen peroxide, and hydroxy radicals, and each of these constituents exerts a broad range of toxic effects on DNA, proteins, and membranes and heightens the risk for neoplastic transformation (Imlay 2003). Reactive nitrogen intermediates, such as nitric oxide, comprise another class of oxidant molecules that can exert damaging consequences to bacteria. *H. pylori* induces a strong inflammatory response during the course of its colonization, resulting in the production of seemingly harmful oxidants. However, *H. pylori* possesses multiple mechanisms to evade this stress, aiding in its persistence (Figure 16.2).

One group of enzymes with a role in resistance to oxidative stress is the peroxiredoxins (Prx) family. *H. pylori* possesses three homologs of this enzyme family, one of which is alkyl hydroperoxide reductase (AhpC). *In vitro*, AhpC mutants fail to grow under normal conditions, but can be recovered under conditions of low oxygen tension (Olczak et al. 2002). *In vivo* experiments in mice have demonstrated that disruption of *ahpC* results in a deficiency in colonization, indicating an essential role for this antioxidant activity within inflamed mucosa (Olczak et al. 2003). Another Prx family member expressed by *H. pylori* is the thiolperoxidase Tpx, which is important for resistance to peroxide and superoxide (Comtois et al. 2003). Tpx mutants colonize only 5% of challenged mice, compared with a colonization

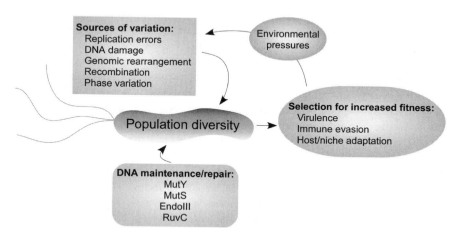

Fig. 16.2 Genetic diversity of *Helicobacter pylori* contributes to fitness. *H. pylori* exhibits diversity both among hosts and within an individual host. Sources of diversity include replication infidelity, DNA damage, genomic rearrangement, recombination including horizontal gene transfer, and phase variation. Several enzymes including MutY, MutS, EndoIII, and RuvC contribute to the maintenance of DNA integrity. The diversity of the *H. pylori* population is important for colonization, persistence, and continued adaptation to environmental changes

frequency of 63% for the wild-type parental strain (Olczak et al. 2003). The third member of this family is bacterioferritin comigratory protein (BCP), which also possesses thiolperoxidase activity. *In vitro* studies have demonstrated that the function of BCP overlaps with that of AhpC, but that BCP is present at much lower levels (Wang et al. 2005b). When mice were inoculated with a BCP-deficient mutant, only 30% were colonized at 3 weeks compared with 91% of mice challenged with the wild-type strain (Wang et al. 2005b). In conjunction with data regarding AhpC and Tpx, *in vivo* rodent experiments utilizing a mutant deficient in BCP emphasize that, although these three proteins are all members of the Prx family, they each have individual roles in colonization.

The most well-studied enzyme for resisting oxidative stress, superoxide dismutase (SOD), is virtually ubiquitous among organisms that encounter ROS in vertebrate organisms. SOD catalyzes the conversion of superoxide to H_2O_2, which is then further degraded by peroxidase or catalase. In *H. pylori*, this enzyme is encoded by the gene *sodB*. *H. pylori* mutants deficient in SodB grow poorly at O_2 levels more than 5% and viability is severely impaired (Seyler et al. 2001). When used to challenge mice, such mutants colonized at a rate of only 4%, compared with wild-type colonization rates of 88% of animals challenged (Seyler et al. 2001).

As mentioned above, catalase is an enzyme important for the conversion of H_2O_2 into innocuous oxygen and water. *H. pylori* catalase is encoded by *katA* and mutations of this gene result in heightened sensitivity to H_2O_2 as expected, although these mutants have no growth defects *in vitro* (Harris et al. 2002; Manos et al. 1998; Odenbreit et al. 1996). *In vivo*, *katA* mutants colonize mice with virtually the same frequency as wild type (90% vs. 100%) after 1 week of infection; however, colonization efficiency at 24 weeks decreases to 50% in mice challenged with KatA-deficient mutants, suggesting a defect in persistence rather than initial colonization (Harris et al. 2003). Distal to *katA* in the *H. pylori* genome is the gene *kapA*, which encodes the KatA-associated protein (KapA). Disruption of *kapA* does not affect catalase activity *per se* (Harris et al. 2002; Odenbreit et al. 1996), but it does increase sensitivity to H_2O_2 (Harris et al. 2002). Similar to inactivation of catalase, experiments with a KapA-deficient mutant demonstrate that this gene is not required for initial colonization, but the number of mice that remained colonized at 24 weeks was only 20%, indicating a similar important role in persistence (Harris et al. 2002). An additional *H. pylori* enzyme involved in resistance to H_2O_2 is the nicotinamide adenine dinucleotide phosphate (NADPH) quinone reductase encoded by *mdaB* (Wang and Maier 2004). Mutations in this gene also lead to increased sensitivity to organic hydroperoxides and oxygen (Wang and Maier 2004). *In vivo* studies demonstrated that MdaB has a role in colonization efficiency and density, because only 25% of mice challenged with the *mdaB* mutant harbored *H. pylori* after 3 weeks and the bacterial density in animals colonized with the mutant was significantly lower than for those challenged with the wild-type strain (Wang and Maier 2004).

H. pylori NapA was first identified for its role in neutrophil recruitment and represents a major antigen that stimulates the human immune response (Evans et al. 1995). Also, *napA* is one of the most highly expressed genes in Rhesus macaques experimentally infected with *H. pylori* (Boonjakuakul et al. 2005). NapA also has homology to members of the Dps family of proteins, which function in iron binding and protection of DNA (Andrews et al. 2003; Grant et al. 1998). *In vitro* experiments have demonstrated that disruption of *napA* leads to an increased sensitivity to oxidative stress and DNA damage compared with wild-type *H. pylori* (Cooksley et al. 2003; Wang et al. 2006). When assessed in *H. pylori* mutants deficient in AhpC, KatA, or SodB, levels of NapA are increased, suggesting a compensatory role for this protein in protection against oxidative damage (Olczak et al. 2002, 2005). Surprisingly, challenge of mice with a *napA* mutant did not lead to alterations in colonization efficiency compared with wild-type *H. pylori*. However, when mice were challenged first with nonviable wild-type *H. pylori* (to stimulate production of reactive oxygen intermediates) and subsequently inoculated with the *napA* mutant, only one-third of the mice were colonized, and the density of colonization was significantly lower than for those challenged with the wild-type strain (Wang et al. 2006). These data suggest a dual function for NapA in which it induces the production of oxidative stress by the host, yet can also provide protection from this stress.

H. pylori also encodes the arginase RocF, which catalyzes the hydrolysis of arginine to ornithine and urea (McGee et al. 1999; Mendz and Hazell 1996). Studies have demonstrated that RocF-deficient *H. pylori* are 1,000-fold more sensitive to acid exposure compared with the wild-type parental strain (McGee et al. 1999). Additional *in vitro* experiments have demonstrated that RocF can also compete with macrophages for the iNOS substrate arginine, downregulating host iNOS production and allowing the bacterium to evade the eukaryotic NO defense mechanism (Chaturvedi et al. 2007; Gobert et al. 2001). Furthermore, urea generated as a product of arginine catabolism is hydrolyzed by *H. pylori* urease to yield ammonia, which may contribute to acid resistance by neutralizing the bacterial microenvironment (McGee et al. 1999). Interestingly, disruption of *rocF* affects colonization efficiency only slightly, if at all (McGee et al. 1999).

Repair of Damage

One consequence of oxidative stress is protein damage caused by the oxidation of susceptible amino acids. Methionine is an amino acid that is particularly sensitive to oxidation and such modification results in the formation of methionine sulfoxide, which may lead to structural changes or inactivation of proteins (Ezraty et al. 2005; Vogt 1995). Similar to most bacteria, *H. pylori* possesses a defense mechanism to combat oxidation of methionine in the form of methionine sulfoxide reductase (Msr), an enzyme responsible for converting oxidized methionine residues back to their reduced state. As expected, *H. pylori* strains that are Msr deficient are more susceptible to oxidative stress (Alamuri and Maier 2006), and when used in mouse colonization experiments, Msr-deficient bacteria were only recovered at the earliest time point tested (7 days vs. 14 or 21 days postchallenge) (Alamuri and Maier 2006). Furthermore, only 4 of 12 mice (33%) were colonized with the mutant at the

7-day time point and colonization density was significantly lower than for mice harboring wild-type *H. pylori* (Alamuri and Maier 2006).

As mentioned above, one of the potential consequences of oxidative stress is DNA damage, which includes generation of base derivatives, sugar damage, strand breakage, DNA/protein crosslinking, and abasic site formation (Wang et al. 1998). Organisms must, therefore, possess mechanisms that ensure the integrity of their DNA, thereby avoiding potentially lethal consequences. Endonuclease III (Endo III) is encoded by *nth* in *H. pylori* and is critical for the excision of oxidized pyrimidines such as thymine glycol in many bacteria (Cooke et al. 2003; Evans et al. 2004). *H. pylori nth* mutants exhibit a fourfold increase in spontaneous mutation rate and a fivefold increase in the rate of macrophage-induced mutations (O'Rourke et al. 2003). In mice, colonization densities of *nth* mutants were 15-fold lower than those of the wild-type strain after 15 days and the bacterial load recovered from the *H. pylori nth⁻*-infected mice continued to decrease until the terminal time point of 60 days, whereas it remained high for mice challenged with the wild-type strain (O'Rourke et al. 2003). Furthermore, the wild-type strain was able to out-compete the mutant *in vivo*, regardless of which strain predominated at the time of initial inoculation (O'Rourke et al. 2003).

MutY is an adenine glycosylase that catalyzes the removal of adenine from A:(8-oxo-G) mismatches that can result from oxidation of a guanine residue which, if not repaired, leads to a GC to TA transversion (Wang et al. 1998). *H. pylori* MutY mutants have a >300-fold higher spontaneous mutation rate than wild-type strains under conditions of 5% oxygen. This difference increases to >700-fold by exposure to atmospheric oxygen for 4 hours (Eutsey et al. 2007). Experimental infections in mice have demonstrated that *mutY* mutants colonized 30% less densely than the wild-type parental strain, suggesting a role for this enzyme in protection against oxidative stress *in vivo* (Eutsey et al. 2007).

MutS is another protein involved in the repair of DNA damage. In *E. coli*, MutS is a member of a DNA methyl-directed mismatch repair (MMR) system, which includes two additional proteins, MutL and MutH. *H. pylori* only encodes a MutS homolog, suggesting that it does not possess an intact MMR system for DNA repair (Alm et al. 1999; Tomb et al. 1997). Consistent with this hypothesis, *H. pylori* exhibits a mutation rate higher than that of many other bacteria and disruption of *mutS* does not alter this rate (Bjorkholm et al. 2001b; Pinto et al. 2005). These data suggest that *H. pylori* MutS is not part of a classical MMR mechanism, but may serve another function. In support of this, Pinto and colleagues demonstrated a role for MutS in regulating the rate of homologous recombination in *H. pylori* (Pinto et al. 2005). MutS-deficient mutants exhibited a 5- to 25-fold increase in recombination rates, as determined by transformation requiring incorporation of DNA into the chromosome for recovery (Pinto et al. 2005). It was further demonstrated that MutS mutants have an approximately 10-fold increase in mutagenesis in response to oxidative stress (Wang et al. 2005a). The role of MutS has been examined *in vivo* by challenging mice with wild-type or *mutS⁻ H. pylori*. After 3 weeks, 90% of mice challenged with the wild-type strain were colonized at high densities. In contrast, only 57% of mice challenged with the MutS mutant were colonized and the density of *H. pylori* was much lower (Wang et al. 2005a).

Another important component of the DNA recombination machinery is the endonuclease RuvC, which is required for resolution of Holliday Junction branched intermediates (Shinagawa and Iwasaki 1996). As expected, disruption of *ruvC* in *H. pylori* decreases the frequency of homologous recombination by approximately 20-fold (Loughlin et al. 2003). Furthermore, *ruvC* mutants were more sensitive to DNA damage from mitomycin C and ultraviolet irradiation as well as the antimicrobials metronidazole and levofloxicin, and RuvC-deficient mutants displayed an increased sensitivity to oxidative stress. Mice challenged with this mutant were colonized significantly less efficiently (50%) and at a much lower density than mice inoculated with the wild-type strain 21 days postchallenge. (Loughlin et al. 2003). When examined over time, significantly fewer *ruvC⁻* bacteria were recovered at 7 days than wild-type *H. pylori* and, whereas colonization density remained constant for the wild-type strain, the number of recoverable mutants continued to decline until day 36, at which time no RuvC-deficient *H. pylori* were present (Loughlin et al. 2003). The results of these mouse challenge experiments with mutants lacking EndoIII, MutS, MutY, and RuvC indicate the collective importance of maintaining the integrity of the *H. pylori* genome through a variety of repair systems during infection. In total, these data suggest that *H. pylori* resists self-inflicted oxidative damage by numerous mechanisms, which, based on *in vivo* experiments, are required for efficient and persistent colonization. A consequence of such microbial-induced oxidative stress, however, is the development of gastric cancer.

Immune Evasion/Modulation

One of the most formidable obstacles microbial agents encounter in the establishment of infection is the host immune system. Emerging data have indicated that *H. pylori* has multiple mechanisms to both evade and modify the immune response of its host. Gram-negative bacteria, including *H. pylori*, possess lipopolysaccharide (LPS) as a component of the cell wall. Typically, LPS elicits a strong inflammatory response; however, the LPS of *H. pylori* is relatively anergic, having as little as 10^3 less endotoxin activity than LPS from other Gram-negative bacteria (Birkholz et al. 1993; Nielsen et al. 1994). Interestingly, mice immunized with *H. pylori* LPS-containing sonicate elicited a strong Th1-type immune response, whereas with sonicates depleted of LPS induced a Th2 response (Nielsen et al. 1994; Taylor et al. 2007). These data indicate that *H. pylori* LPS is not completely nonimmunogenic, but may have an important role in modulation of the host immune response.

Another role of *H. pylori* LPS in pathogenesis is related to glycosylation of the O-antigen component, resulting in molecular mimicry of Lewis antigens that are expressed on human red blood cells and epithelial cells. Data using *H. pylori* clinical isolates indicate that approximately 85% of strains express Le^x and Le^y (Heneghan et al. 2000; Simoons-Smit et al. 1996), and that both antigens can be expressed by individual strains, although one type frequently predominates (Wirth et al. 1999). It has been well documented *in vitro* that Lewis epitopes undergo phase variation, that is, switching of phenotypes at a higher frequency than would otherwise be expected (van der Woude and Baumler 2004). In an initial study to examine Lewis antigen phase variation *in vivo*, Mongolian gerbils and multiple strains of mice were challenged with a single well-characterized *H. pylori* strain and, at time points ranging from 2 weeks to 5 months, bacteria were recovered for analysis. The results demonstrated that there was little evidence of phase variation during the course of infection (Wirth et al. 1999). However, although this study produced important data, the *in vivo* experiments were performed over a relatively short time period. More recently, it has been demonstrated that gastric epithelial cells of Rhesus monkeys, like humans, express Le^x and Le^y (Linden et al. 2004). Of interest, *H. pylori* infection in this animal model over a longer time course (40 weeks) demonstrated that the Lewis antigen profile of the bacteria changed based on the expression pattern of the host (Wirth et al. 2006). That is, *H. pylori* recovered from Le^x-expressing monkeys expressed predominantly Le^x, whereas *H. pylori* isolated from monkeys expressing Le^y expressed predominantly Le^y, indicating adaptation of the bacteria to its individual host.

Bacterial Diversity/Recombination

H. pylori is one of the most genetically diverse bacterial species known and exhibits variation attributable to point mutations, allelic variation, recombination, and phase variation (Figure 16.2). Additionally, the *H. pylori* genome consists of both core (invariant) and flexible (strain-specific) components. Numerous studies have examined the extent of this diversity using *in vitro* assays (Israel et al. 2001; Kersulyte et al. 2000; Kuipers et al. 2000; Lundin et al. 2005; Salama et al. 2000; van Doorn et al. 1998) to identify mechanisms of variation (Aras et al. 2003; Salaun et al. 2004). Others have extrapolated from these data to consider *H. pylori* diversity from an evolutionary perspective (Gressmann et al. 2005). Data using experimentally infected animals strongly suggest that *H. pylori* adapts to its host at a genomic level. For example, the adhesin BabA can undergo genetic variation during *in vivo* colonization as a result of slipped-strand mispairing as well as intergenomic recombination (Ilver et al. 1998; Solnick et al. 2004). Slipped-strand mispairing also regulates expression of the fucosyl transferases that alter LPS glycosylation (Nilsson et al. 2006; Wang et al. 2000) and results of animal experiments suggest that these changes are a direct response to the individual host environment (Wirth et al. 2006). Infection of monkeys with an *H. pylori* strain expressing low urease activity led to the recovery of a derivative with high urease activity attributed to a 1-bp insertion in the gene for UreA (Hansen and Solnick 2001). Similarly, infection of mice with a clinical *H. pylori* strain led to the recovery of a derivative isolate with a 102-bp deletion in the gene encoding polyphosphate kinase leading to increased enzyme activity and increased bacterial growth in the stationary phase (Ayraud et al. 2003). The importance of this enzyme for efficient colonization was later confirmed using gene disruption techniques (Ayraud et al. 2005; Tan et al. 2005). These data support a significant role for genetic diversity in regulation of DNA repair/recombination as well as immune evasion.

Another study focused on *H. pylori* genetic variability compared differences in colonization efficiency of two single-colony isolates harvested from a single individual. RFLP and whole genome microarray analysis demonstrated that these isolates were closely related derivatives of a single strain, differing only slightly in gene content. One of the significant differences was absence of the *cag* pathogenicity island (PAI) in one of the two isolates. The *cag* PAI⁻ isolate was unable to colonize Lewis[b] transgenic mice, whereas the *cag* PAI⁺ derivative colonized 74% of challenged mice (Bjorkholm et al. 2001a). Another study addressed the role of diversity in mouse colonization by coinfecting several strains of mice with two unrelated *H. pylori* isolates. One strain of *H. pylori* predominated in the antrum, whereas the second strain predominated in the corpus (Akada et al. 2003). When monkeys were used for coinfection experiments, similar findings ensued in that one of the challenge strains typically predominated and out-competed all other strains, suggesting that certain *H. pylori* strains are better suited for the colonization of a specific host (Dubois et al. 1999, 1996; Solnick et al. 2001).

The occurrence of genomic changes *in vivo* has been additionally assessed by examining the on/off status of 31 predicted phase-variable genes after mouse colonization in each of three *H. pylori* strains (Salaun et al. 2005). Changes that would lead to altered expression were identified in 10 genes in at least one of the *H. pylori* strains examined at the longest time point (150 or 360 days). Proteins encoded by the identified genes include members of several different functional groups, such as those required for LPS biosynthesis, OMPs, and hypothetical proteins. Furthermore, when analyzed at time points between inoculation and the termination of the experiment, it was determined that the timing of these changes varied. That is, some changes were found as early as 3 days postinoculation, whereas others were not identified until the end of the experiment. Of the 10 genes identified, four exhibited the same status among the three strains (all "on" or all "off"); however, each strain had an overall unique pattern of gene variation at the end of the experiment, again suggesting specific adaptation of each strain to its host (Salaun et al. 2005). These experiments underscore an important role for both preexisting diversity in the initial inoculum as well as the ability to undergo changes *in vivo* to permit successful colonization and persistent infection.

Virulence Factors Associated with Carcinogenesis

CagA

One of the earliest identified virulence factors of *H. pylori*, as well as one of the most well studied, is the product of the cytotoxin-associated gene, *cagA*. CagA is encoded by the terminal gene of the *cag* PAI and is translocated into host cells by a type IV secretion system, which is also encoded by genes of the PAI. Numerous *in vitro* studies have demonstrated roles for CagA in the alteration of epithelial cell signaling and morphologic changes (Higashi et al. 2002; Odenbreit et al. 2000; Selbach et al. 2002). *In vivo* studies, however, have not readily recapitulated these observations and infection of mice with *cag* PAI⁺ strains frequently leads to PAI deletions (Crabtree et al. 2002; Philpott et al. 2002; Sozzi et al. 2001). One study tested the role of several *cag* PAI genes including CagA in colonization by challenging CD1 mice with *H. pylori* strains deficient in the targeted genes (Marchetti and Rappuoli 2002). Ten days postinfection, mice challenged with mutants of *cagX*, *cagY*, or *cagα* (encoding a VirB11 homolog) were colonized less densely than those inoculated with wild-type *H. pylori*, suggesting a role for these genes in the establishment of infection. Another study utilized the insulin–gastrin hyper-gastrinemic mouse model of gastric cancer and demonstrated that inactivation of *cagE*, which encodes a component of the secretion apparatus, temporally delayed, but did not prevent, the development of cancer in *H. pylori*–infected male mice (Fox et al. 2003).

In contrast to mice, Mongolian gerbils have been very useful for investigations focused on the *cag* PAI primarily because of the ability of these rodents to sustain infection with *cag⁺ H. pylori* strains. When gerbils were infected with wild-type or CagE-deficient *H. pylori* for up to 1 year, those colonized with the mutant had no evidence of gastric ulcer or metaplasia and had significantly less severe gastritis than those infected with the wild-type strain (Ogura et al. 2000; Saito et al. 2005). Furthermore, gerbils infected with a *cagE* mutant had significantly lower anti-*H. pylori* antibody titers and lower levels of neutrophil infiltration within the gastric mucosa over the course of the study (Aspholm et al. 2006).

Rieder et al. challenged gerbils with a *cagA* mutant, a *cagY* mutant (secretion system deficient), or wild-type *H. pylori* and analyzed results at 32 weeks of infection (Rieder et al. 2005). The wild-type strain was recovered at similar levels from both the antrum and corpus. However, although the bacterial density of the mutants was high in the antrum, the level of recovery of mutants from the corpus was at least 10-fold lower, suggesting that CagA and the *cag* secretion system are important for robust colonization of the corpus, a histologic precursor for hypochlorhydria and gastric cancer. In support of this, gerbils infected with the wild-type strain had higher levels of inflammation and a higher frequency of premalignant lesions, whereas those infected with *cag* mutants had less-severe gastritis and histologic changes associated with increased cancer risk in the corpus. Our laboratory has utilized the Mongolian gerbil animal model along with *H. pylori* strain 7.13 which reproducibly induces gastric cancer in this model as early as 24 weeks postinfection (Franco et al. 2005) and is dependent on CagA (Franco et al. 2008). Using this system, we have demonstrated that strain 7.13 increases the nuclear accumulation of the host effector β-catenin in gastric epithelial cells. *In vitro* experiments revealed that nuclear translocation of β-catenin is CagA dependent. Concordant with these data, persons infected with *cag⁺* vs. *cag⁻ H. pylori* were found to have increased nuclear accumulation of β-catenin (Franco et al. 2005).

Another study examined expression of virulence factors, including the *cag* PAI, in Rhesus monkeys, by quantitative real-time reverse-transcriptase polymerase chain reaction (Boonjakuakul et al. 2005). These data indicated that *cagA* was expressed at relatively high levels during the entire time course of infection examined. Interestingly, some *cag* PAI genes, such as *cagY*, were more highly expressed at 1 week postinfection compared with later time points, whereas expression of others, such as *cagC*, increased between 2 and 3 months and then decreased by the 4- to 6-month time point. Taken together, these data indicate an important role for CagA and other products of the *cag* PAI in the development of disease, particularly gastric cancer.

VacA

Another bacterial factor that has been associated with *H. pylori*–induced cancer is the vacuolating cytotoxin VacA. All *H. pylori* strains have the gene encoding VacA (*vacA*); however, some *H. pylori* possess allelic variants of *vacA* (s1/m1, s1/m2) that result in high levels of vacuolating activity such as cancer. Furthermore, these strains are associated with more severe disease outcome. *In vitro* studies have identified many effects of VacA including vacuolation, induction of apoptosis, inhibition of T-cell proliferation, and membrane channel formation. Despite the wealth of *in vitro* data regarding VacA, much less is known about the function of VacA related to pathogenesis *in vivo*. Although multiple studies in various animal models were unable to detect a significant difference in wild-type *H. pylori* and a *vacA* mutant (Eaton et al. 1997; Guo and Mekalanos 2002; Wirth et al. 1998), one study demonstrated that a *vacA* mutant was less efficient than wild-type *H. pylori* in colonization of mice with an $ID_{50} > 300$-fold higher than that of wild type (Salama et al. 2001). However, mice that were colonized by the mutant were colonized to the same density as those colonized by wild type. Furthermore, coinfection experiments demonstrated a competitive disadvantage for the VacA-deficient bacteria, because the colonizing population was always overtaken by wild-type *H. pylori*, as early as 2 days postchallenge.

Conclusion

Animal models have provided an invaluable resource with which to study *H. pylori* pathogenesis and carcinogenesis. Although each model has its own advantages and disadvantages, there can be little doubt of their significant contributions to our current understanding of *H. pylori*–related diseases in humans (Table 16.1). As additional candidate virulence factors are identified by genomic, epidemiologic, or *in vitro* studies, these models will continue to facilitate translational studies that can elucidate mechanisms through which *H. pylori* elicits pathologic responses in its host.

References

Akada J. K., Ogura K., Dailidiene D., Dailide G., Cheverud J. M., Berg D. E. 2003. Helicobacter pylori tissue tropism: mouse-colonizing strains can target different gastric niches. Microbiology 149:1901–1909.

Alamuri P., Maier R. J. 2006. Methionine sulfoxide reductase in Helicobacter pylori: interaction with methionine-rich proteins and stress-induced expression. J. Bacteriol. 188:5839–5850.

Alm R. A., Ling L. S., Moir D. T., King B. L., Brown E. D., Doig P. C., Smith D. R., Noonan B., Guild B. C., deJonge B. L., Carmel G., Tummino P. J., Caruso A., Uria-Nickelsen M., MillsD. M., Ives C., Gibson R., Merberg D., Mills S. D., Jiang Q., Taylor D. E., Vovis G. F., Trust T. J. 1999. Genomic-sequence comparison of two unrelated isolates of the human gastric pathogen Helicobacter pylori. Nature 397:176–180.

Alm R. A., Trust T. J. 1999. Analysis of the genetic diversity of Helicobacter pylori: the tale of two genomes. J. Mol. Med. 77:834–846.

Andermann T. M., Chen Y. T., Ottemann K. M. 2002. Two predicted chemoreceptors of Helicobacter pylori promote stomach infection. Infect. Immun. 70:5877–5881.

Andrews S. C., Robinson A. K., Rodriguez-Quinones F. 2003. Bacterial iron homeostasis. FEMS Microbiol. Rev. 27:215–237.

Aras R. A., Kang J., Tschumi A. I., Harasaki Y., Blaser M. J. 2003. Extensive repetitive DNA facilitates prokaryotic genome plasticity. Proc. Natl. Acad. Sci. USA 100:13579–13584.

Aspholm M., Olfat F. O., Norden J., Sonden B., Lundberg C., Sjostrom R., Altraja S., Odenbreit S., Haas R., Wadstrom T., Engstrand L., Semino-Mora C., Liu H., Dubois A., Teneberg S., Arnqvist A., Boren T. 2006. SabA is the H. pylori hemagglutinin and is polymorphic in binding to sialylated glycans. PLoS Pathog. 2:e110.

Ayraud S., Janvier B., Labigne A., Ecobichon C., Burucoa C., Fauchere J. L. 2005. Polyphosphate kinase: a new colonization factor of Helicobacter pylori. FEMS Microbiol. Lett. 243:45–50.

Ayraud S., Janvier B., Salaun L., Fauchere J. L. 2003. Modification in the ppk gene of Helicobacter pylori during single and multiple experimental murine infections. Infect. Immun. 71:1733–1739.

Baldwin D. N., Shepherd B., Kraemer P., Hall M. K., Sycuro L. K., Pinto-Santini D. M., Salama N. R. 2007. Identification of Helicobacter pylori genes that contribute to stomach colonization. Infect. Immun. 75:1005–1016.

Bijlsma J. J., Gerrits M. M., Imamdi R., Vandenbroucke-Grauls C. M., Kusters J. G. 1998. Urease-positive, acid-sensitive mutants of Helicobacter pylori: urease-independent acid resistance involved in growth at low pH. FEMS Microbiol. Lett. 167:309–313.

Birkholz S., Knipp U., Nietzki C., Adamek R. J., Opferkuch W. 1993. Immunological activity of lipopolysaccharide of Helicobacter pylori on human peripheral mononuclear blood cells in comparison to lipopolysaccharides of other intestinal bacteria. FEMS Immunol. Med. Microbiol. 6:317–324.

Bjorkholm B., Lundin A., Sillen A., Guillemin K., Salama N., Rubio C., Gordon J. I., Falk P., Engstrand L. 2001a. Comparison of genetic divergence and fitness between two subclones of Helicobacter pylori. Infect. Immun. 69:7832–7838.

Bjorkholm B., Sjolund M., Falk P. G., Berg O. G., Engstrand L., Andersson D. I. 2001b. Mutation frequency and biological cost of antibiotic resistance in Helicobacter pylori. Proc. Natl. Acad. Sci. USA 98:14607–14612.

Boonjakuakul J. K., Canfield D. R., Solnick J. V. 2005. Comparison of Helicobacter pylori virulence gene expression in vitro and in the Rhesus macaque. Infect. Immun. 73:4895–4904.

Cerda O., Rivas A., Toledo H. 2003. Helicobacter pylori strain ATCC700392 encodes a methyl-accepting chemotaxis receptor protein (MCP) for arginine and sodium bicarbonate. FEMS Microbiol. Lett. 224:1751–1781.

Chaturvedi R., Asim M., Lewis N. D., Algood H. M., Cover T. L., Kim P. Y., Wilson K. T. 2007. L-arginine availability regulates inducible nitric oxide synthase-dependent host defense against Helicobacter pylori. Infect. Immun. 75:4305–4315.

Comtois S. L., Gidley M. D., Kelly D. J. 2003. Role of the thioredoxin system and the thiol-peroxidases Tpx and Bcp in mediating resistance to oxidative and nitrosative stress in Helicobacter pylori. Microbiology 149:121–129.

Cooke M. S., Evans M. D., Dizdaroglu M., Lunec J. 2003. Oxidative DNA damage: mechanisms, mutation, and disease. FASEB J. 17:1195–11214.

Cooksley C., Jenks P. J., Green A., Cockayne A., Logan R. P., Hardie K. R. 2003. NapA protects Helicobacter pylori from oxidative stress damage, and its production is influenced by the ferric uptake regulator. J. Med. Microbiol. 52:461–469.

Crabtree J. E., Ferrero R. L., Kusters J. G. 2002. The mouse colonizing Helicobacter pylori strain SS1 may lack a functional cag pathogenicity island. Helicobacter 7:139–140.

Croxen M. A., Sisson G., Melano R., Hoffman P. S. 2006. The Helicobacter pylori chemotaxis receptor TlpB (HP0103) is required for pH taxis and for colonization of the gastric mucosa. J. Bacteriol. 188:2656–2665.

Dubois A., Berg D. E., Incecik E. T., Fiala N., Heman-Ackah L. M., Del Valle J., Yang M., Wirth H. P., Perez-Perez G. I., Blaser M. J. 1999. Host specificity of Helicobacter pylori strains and host responses in experimentally challenged nonhuman primates. Gastroenterology 116:90–96.

Dubois A., Berg D. E., Incecik E. T., Fiala N., Heman-Ackah L. M., Perez-Perez G. I., Blaser M. J. 1996. Transient and persistent experimental infection of nonhuman primates with Helicobacter pylori: implications for human disease. Infect. Immun. 64:2885–2891.

Eaton K. A., Cover T. L., Tummuru M. K., Blaser M. J., Krakowka S. 1997. Role of vacuolating cytotoxin in gastritis due to Helicobacter pylori in gnotobiotic piglets. Infect. Immun. 65:3462–3464.

Eaton K. A., Gilbert J. V., Joyce E. A., Wanken A. E., Thevenot T., Baker P., Plaut A., Wright A. 2002. In vivo complementation of ureB restores the ability of Helicobacter pylori to colonize. Infect. Immun. 70:771–778.

Eaton K. A., Suerbaum S., Josenhans C., Krakowka S. 1996. Colonization of gnotobiotic piglets by Helicobacter pylori deficient in two flagellin genes. Infect. Immun. 64:2445–2448.

Eutsey R., Wang G., Maier R. J. 2007. Role of a MutY DNA glycosylase in combating oxidative DNA damage in Helicobacter pylori. DNA Repair (Amst) 6:19–26.

Evans D. J., Jr., Evans D. G., Takemura T., Nakano H., Lampert H. C., Graham D. Y., Granger D. N., Kvietys P. R. 1995. Characterization of a Helicobacter pylori neutrophil-activating protein. Infect. Immun. 63:2213–2220.

Evans M. D., Dizdaroglu M., Cooke M. S. 2004. Oxidative DNA damage and disease: induction, repair and significance. Mutat. Res. 567:1–61.

Ezraty B., Aussel L., Barras F. 2005. Methionine sulfoxide reductases in prokaryotes. Biochim. Biophys. Acta 1703:221–229.

Ferraro R. L., Fox J. G. 2001. In vivo modeling of Helicobacter-associated gastrointestinal diseases. In: Mobley H. L. T., Mendz G. L., Hazell S. L., editors. Helicobacter pylori: physiology and genetics. Washington, D.C.: ASM Press; pp. 565–582.

Fox J. G., Wang T. C., Rogers A. B., Poutahidis T., Ge Z., Taylor N., Dangler C. A., Israel D. A., Krishna U., Gaus K., Peek R. M., Jr. 2003. Host and microbial constituents influence Helicobacter pylori-induced cancer in a murine model of hypergastrinemia. Gastroenterology 124:1879–1890.

Franco A. T., Israel D. A., Washington M. K., Krishna U., Fox J. G., Rogers A. B., Neish A. S., Collier-Hyams L., Perez-Perez G. I., Hatakeyama M., Whitehead R., Gaus K., O'Brien D. P., Romero-Gallo J., Peek R. M., Jr. 2005. Activation of beta-catenin by carcinogenic Helicobacter pylori. Proc. Natl. Acad. Sci. USA 102:10646–10651.

Franco A. T., Johnston E., Krishna U., Yamaoka Y., Israel D. A., Nagy T. A., Wroblewski L. E., Piazuelo M. B., Correa P., Peek R. M., Jr. 2008. Regulation of gastric carcinogenesis by Helicobacter pylori virulence factors. Cancer Res. 68(2):379–387.

Gerhard M., Lehn N., Neumayer N., Boren T., Rad R., Schepp W., Miehlke S., Classen M., Prinz C. 1999. Clinical relevance of the Helicobacter pylori gene for blood-group antigen-binding adhesin. Proc. Natl. Acad. Sci. USA 96:12778–12783.

Gobert A. P., McGee D. J., Akhtar M., Mendz G. L., Newton J. C., Cheng Y., Mobley H. L., Wilson K. T. 2001. Helicobacter pylori arginase inhibits nitric oxide production by eukaryotic cells: a strategy for bacterial survival. Proc. Natl. Acad. Sci. USA 98:13844–13849.

Grant R. A., Filman D. J., Finkel S. E., Kolter R., Hogle J. M. 1998. The crystal structure of Dps, a ferritin homolog that binds and protects DNA. Nat. Struct. Biol. 5:294–303.

Gressmann H., Linz B., Ghai R., Pleissner K. P., Schlapbach R., Yamaoka Y., Kraft C., Suerbaum S., Meyer T. F., Achtman M. 2005. Gain and loss of multiple genes during the evolution of Helicobacter pylori. PLoS Genet. 1:e43.

Guo B. P., Mekalanos J. J. 2002. Rapid genetic analysis of Helicobacter pylori gastric mucosal colonization in suckling mice. Proc. Natl. Acad. Sci. USA 99:8354–8359.

Hansen L. M., Solnick J. V. 2001. Selection for urease activity during Helicobacter pylori infection of rhesus macaques (Macaca mulatta). Infect. Immun. 69:3519–3522.

Harris A. G., Hinds F. E., Beckhouse A. G., Kolesnikow T., Hazell S. L. 2002. Resistance to hydrogen peroxide in Helicobacter pylori:role of catalase (KatA) and Fur, and functional analysis of a novel gene product designated 'KatA-associated protein,' KapA (HP0874). Microbiology 148:3813–3825.

Harris A. G., Wilson J. E., Danon S. J., Dixon M. F., Donegan K., Hazell S. L. 2003. Catalase (KatA) and KatA-associated protein (KapA) are essential to persistent colonization in the Helicobacter pylori SS1 mouse model. Microbiology 149:665–672.

Heneghan M. A., McCarthy C. F., Moran A. P. 2000. Relationship of blood group determinants on Helicobacter pylori lipopolysaccharide with host lewis phenotype and inflammatory response. Infect. Immun. 68:937–941.

Higashi H., Tsutsumi R., Muto S., Sugiyama T., Azuma T., Asaka M., Hatakeyama M. 2002. SHP-2 tyrosine phosphatase as an intracellular target of Helicobacter pylori CagA protein. Science 295:683–686.

Ilver D., Arnqvist A., Ogren J., Frick I. M., Kersulyte D., Incecik E. T., Berg D. E., Covacci A., Engstrand L., Boren T. 1998. Helicobacter pylori adhesin binding fucosylated histo-blood group antigens revealed by retagging. Science 279:373–377.

Imlay J. A. 2003. Pathways of oxidative damage. Annu. Rev. Microbiol. 57:395–418.

Israel D. A., Salama N., Krishna U., Rieger U. M., Atherton J. C., Falkow S., Peek R. M., Jr. 2001. Helicobacter pylori genetic diversity within the gastric niche of a single human host. Proc. Natl. Acad. Sci. USA 98:14625–14630.

Josenhans C., Labigne A., Suerbaum S. 1995. Comparative ultrastructural and functional studies of Helicobacter pylori and Helicobacter mustelae flagellin mutants: both flagellin subunits, FlaA and FlaB, are necessary for full motility in Helicobacter species. J. Bacteriol. 177:3010–3020.

Kavermann H., Burns B. P., Angermuller K., Odenbreit S., Fischer W., Melchers K., Haas R. 2003. Identification and characterization of Helicobacter pylori genes essential for gastric colonization. J. Exp. Med. 197:813–822.

Kersulyte D., Mukhopadhyay A. K., Velapatino B., Su W., Pan Z., Garcia C., Hernandez V., Valdez Y., Mistry R. S., Gilman R. H., Yuan Y., Gao H., Alarcon T., Lopez-Brea M., Balakrish Nair G., Chowdhury A., Datta S., Shirai M., Nakazawa T., Ally R., Segal I., Wong B. C., Lam S. K., Olfat F. O., Boren T., Engstrand L., Torres O., Schneider R., Thomas J. E., Czinn S., Berg D. E. 2000. Differences in genotypes of Helicobacter pylori from different human populations. J. Bacteriol. 182:3210–3218.

Kuipers E. J., Israel D. A., Kusters J. G., Gerrits M. M., Weel J., van Der Ende A., van Der Hulst R. W., Wirth H. P., Hook-Nikanne J., Thompson S. A., Blaser M. J. 2000. Quasispecies development of Helicobacter pylori observed in paired isolates obtained years apart from the same host. J. Infect. Dis. 181:273–282.

Linden S., Boren T., Dubois A., Carlstedt I. 2004. Rhesus monkey gastric mucins: oligomeric structure, glycoforms and Helicobacter pylori binding. Biochem. J. 379:765–775.

Loughlin M. F., Barnard F. M., Jenkins D., Sharples G. J., Jenks P. J. 2003. Helicobacter pylori mutants defective in RuvC Holliday junction resolvase display reduced macrophage survival and spontaneous clearance from the murine gastric mucosa. Infect. Immun. 71:2022–2031.

Lundin A., Bjorkholm B., Kupershmidt I., Unemo M., Nilsson P., Andersson D. I., Engstrand L. 2005. Slow genetic divergence of Helicobacter pylori strains during long-term colonization. Infect. Immun. 73:4818–4822.

Mahdavi J., Sonden B., Hurtig M., Olfat F. O., Forsberg L., Roche N., Angstrom J., Larsson T., Teneberg S., Karlsson K. A., Altraja S., Wadstrom T., Kersulyte D., Berg D. E., Dubois A., Petersson C., Magnusson K. E., Norberg T., Lindh F., Lundskog B. B., Arnqvist A., Hammarstrom L., Boren T. 2002. Helicobacter pylori SabA adhesin in persistent infection and chronic inflammation. Science 297:573–578.

Manos J., Kolesnikow T., Hazell S. L. 1998. An investigation of the molecular basis of the spontaneous occurrence of a catalase-negative phenotype in Helicobacter pylori. Helicobacter 3:28–38.

Marchetti M., Rappuoli R. 2002. Isogenic mutants of the cag pathogenicity island of Helicobacter pylori in the mouse model of infection: effects on colonization efficiency. Microbiology 148:1447–1456.

McGee D. J., Langford M. L., Watson E. L., Carter J. E., Chen Y. T., Ottemann K. M. 2005. Colonization and inflammation deficiencies in Mongolian gerbils infected by Helicobacter pylori chemotaxis mutants. Infect. Immun. 73:1820–1827.

McGee D. J., Radcliff F. J., Mendz G. L., Ferrero R. L., Mobley H. L. 1999. Helicobacter pylori rocF is required for arginase activity and acid protection in vitro but is not essential for colonization of mice or for urease activity. J. Bacteriol. 181:7314–7322.

McGowan C. C., Cover T. L., Blaser M. J. 1994. The proton pump inhibitor omeprazole inhibits acid survival of Helicobacter pylori by a urease-independent mechanism. Gastroenterology 107:1573–1578.

McGowan C. C., Cover T. L., Blaser M. J. 1996. Helicobacter pylori and gastric acid: biological and therapeutic implications. Gastroenterology 110:926–938.

Mendz G. L., Hazell S. L. 1996. The urea cycle of Helicobacter pylori. Microbiology 142:2959–2967.

Nielsen H., Birkholz S., Andersen L. P., Moran A. P. 1994. Neutrophil activation by Helicobacter pylori lipopolysaccharides. J. Infect. Dis. 170:135–139.

Nilsson C., Skoglund A., Moran A. P., Annuk H., Engstrand L., Normark S. 2006. An enzymatic ruler modulates Lewis antigen glycosylation of Helicobacter pylori LPS during persistent infection. Proc. Natl. Acad. Sci. USA 103:2863–2868.

O'Rourke E. J., Chevalier C., Pinto A. V., Thiberge J. M., Ielpi L., Labigne A., Radicella J. P. 2003. Pathogen DNA as target for host-generated oxidative stress: role for repair of bacterial DNA damage in Helicobacter pylori colonization. Proc. Natl. Acad. Sci. USA 100:2789–2794.

O'Rourke J. L., Lee A. 2003. Animal models of Helicobacter pylori infection and disease. Microbes Infect. 5:741–748.

O'Toole P. W., Lane M. C., Porwollik S. 2000. Helicobacter pylori motility. Microbes Infect. 2:1207–1214.

Odenbreit S., Puls J., Sedlmaier B., Gerland E., Fischer W., Haas R. 2000. Translocation of Helicobacter pylori CagA into gastric epithelial cells by type IV secretion. Science 287:1497–1500.

Odenbreit S., Wieland B., Haas R. 1996. Cloning and genetic characterization of Helicobacter pylori catalase and construction of a catalase-deficient mutant strain. J. Bacteriol. 178:6960–6967.

Ogura K., Maeda S., Nakao M., Watanabe T., Tada M., Kyotoku T., Yoshida H., Shiratori Y., Omata M. 2000. Virulence factors of Helicobacter pylori responsible for gastric diseases in Mongolian gerbil. J. Exp. Med. 192:1601–1610.

Olczak A. A., Olson J. W., Maier R. J. 2002. Oxidative-stress resistance mutants of Helicobacter pylori. J. Bacteriol. 184:3186–3193.

Olczak A. A., Seyler R. W., Jr., Olson J. W., Maier R. J. 2003. Association of Helicobacter pylori antioxidant activities with host colonization proficiency. Infect. Immun. 71:580–583.

Olczak A. A., Wang G., Maier R. J. 2005. Up-expression of NapA and other oxidative stress proteins is a compensatory response to loss of major Helicobacter pylori stress resistance factors. Free Radic. Res. 39:1173–1182.

Ottemann K. M., Lowenthal A. C. 2002. Helicobacter pylori uses motility for initial colonization and to attain robust infection. Infect. Immun. 70:1984–1990.

Philpott D. J., Belaid D., Troubadour P., Thiberge J. M., Tankovic J., Labigne A., Ferrero R. L. 2002. Reduced activation of inflammatory responses in host cells by mouse-adapted Helicobacter pylori isolates. Cell Microbiol. 4:285–296.

Pinto A. V., Mathieu A., Marsin S., Veaute X., Ielpi L., Labigne A., Radicella J. P. 2005. Suppression of homologous and homeologous recombination by the bacterial MutS2 protein. Mol. Cell 17:113–120.

Rieder G., Merchant J. L., Haas R. 2005. Helicobacter pylori cag-type IV secretion system facilitates corpus colonization to induce precancerous conditions in Mongolian gerbils. Gastroenterology 128:1229–1242.

Saito H., Yamaoka Y., Ishizone S., Maruta F., Sugiyama A., Graham D. Y., Yamauchi K., Ota H., Miyagawa S. 2005. Roles of virD4 and cagG genes in the cag pathogenicity island of Helicobacter pylori using a Mongolian gerbil model. Gut 54:584–590.

Salama N., Guillemin K., McDaniel T. K., Sherlock G., Tompkins L., Falkow S. 2000. A wholegenome microarray reveals genetic diversity among Helicobacter pylori strains. Proc. Natl. Acad. Sci. USA 97:14668–14673.

Salama N. R., Otto G., Tompkins L., Falkow S. 2001. Vacuolating cytotoxin of Helicobacter pylori plays a role during colonization in a mouse model of infection. Infect. Immun. 69:730–736.

Salaun L., Ayraud S., Saunders N. J. 2005. Phase variation mediated niche adaptation during prolonged experimental murine infection with Helicobacter pylori. Microbiology 151:917–923.

Salaun L., Linz B., Suerbaum S., Saunders N. J. 2004. The diversity within an expanded and redefined repertoire of phase-variable genes in Helicobacter pylori. Microbiology 150:817–830.

Schreiber S., Konradt M., Groll C., Scheid P., Hanauer G., Werling H. O., Josenhans C., Suerbaum S. 2004. The spatial orientation of Helicobacter pylori in the gastric mucus. Proc. Natl. Acad. Sci. USA 101:5024–5029.

Selbach M., Moese S., Hauck C. R., Meyer T. F., Backert S. 2002. Src is the kinase of the Helicobacter pylori CagA protein in vitro and in vivo. J. Biol. Chem. 277:6775–6778.

Seyler R. W., Jr., Olson J. W., Maier R. J. 2001. Superoxide dismutase-deficient mutants of Helicobacter pylori are hypersensitive to oxidative stress and defective in host colonization. Infect. Immun. 69:4034–4040.

Shinagawa H., Iwasaki H. 1996. Processing the Holliday junction in homologous recombination. Trends Biochem. Sci. 21:107–111.

Simoons-Smit I. M., Appelmelk B. J., Verboom T., Negrini R., Penner J. L., Aspinall G. O., Moran A. P., Fei S. F., Shi B. S., Rudnica W., Savio A., de Graaff J. 1996. Typing of Helicobacter pylori with monoclonal antibodies against Lewis antigens in lipopolysaccharide. J. Clin. Microbiol. 34:2196–2200.

Solnick J. V., Hansen L. M., Canfield D. R., Parsonnet J. 2001. Determination of the infectious dose of Helicobacter pylori during primary and secondary infection in rhesus monkeys (Macaca mulatta). Infect. Immun. 69:6887–6892.

Solnick J. V., Hansen L. M., Salama N. R., Boonjakuakul J. K., Syvanen M. 2004. Modification of Helicobacter pylori outer membrane protein expression during experimental infection of rhesus macaques. Proc. Natl. Acad. Sci. USA 101:2106–2111.

Sozzi M., Crosatti M., Kim S. K., Romero J., Blaser M. J. 2001. Heterogeneity of Helicobacter pylori cag genotypes in experimentally infected mice. FEMS Microbiol. Lett. 203:109–114.

Tan S., Fraley C. D., Zhang M., Dailidiene D., Kornberg A., Berg D. E. 2005. Diverse phenotypes resulting from polyphosphate kinase gene (ppk1) inactivation in different strains of Helicobacter pylori. J. Bacteriol. 187:7687–7695.

Taylor J. M., Ziman M. E., Canfield D. R., Vajdy M., Solnick J. V. 2007. Effects of a Th1- versus a Th2-biased immune response in protection against Helicobacter pylori challenge in mice. Microb. Pathog. 44(1):20–27.

Terry K., Williams S. M., Connolly L., Ottemann K. M. 2005. Chemotaxis plays multiple roles during Helicobacter pylori animal infection. Infect. Immun. 73:803–811.

Tomb J. F., White O., Kerlavage A. R., Clayton R. A., Sutton G. G., Fleischmann R. D., Ketchum K. A., Klenk H. P., Gill S., Dougherty B. A., Nelson K., Quackenbush J., Zhou L., Kirkness E. F., Peterson S., Loftus B., Richardson D., Dodson R., Khalak H. G., Glodek A., McKenney K., Fitzegerald L. M., Lee N., Adams M. D., Hickey E. K., Berg D. E., Gocayne J. D., Utterback T. R., Peterson J. D., Kelley J. M., Cotton M. D., Weidman J. M., Fujii C., Bowman C., Watthey L., Wallin E., Hayes W. S., Borodovsky M., Karp P. D., Smith H. O., Fraser C. M., Venter J. C. 1997. The complete genome sequence of the gastric pathogen Helicobacter pylori. Nature 388:539–547.

Tsuda M., Karita M., Morshed M. G., Okita K., Nakazawa T. 1994. A urease-negative mutant of Helicobacter pylori constructed by allelic exchange mutagenesis lacks the ability to colonize the nude mouse stomach. Infect. Immun. 62:3586–3589.

van der Woude M. W., Baumler A. J. 2004. Phase and antigenic variation in bacteria. Clin. Microbiol. Rev. 17:581–611.

van Doorn N. E., Namavar F., Kusters J. G., van Rees E. P., Kuipers E. J., de Graaff J. 1998. Genomic DNA fingerprinting of clinical isolates of Helicobacter pylori by REP-PCR and restriction fragment end-labeling. FEMS Microbiol. Lett. 160:145–150.

Vogt W. 1995. Oxidation of methionyl residues in proteins: tools, targets, and reversal. Free Radic. Biol. Med. 18:93–105.

Wang D., Kreutzer D. A., Essigmann J. M. 1998. Mutagenicity and repair of oxidative DNA damage: insights from studies using defined lesions. Mutat. Res. 400:99–115.

Wang G., Alamuri P., Humayun M. Z., Taylor D. E., Maier R. J. 2005a. The Helicobacter pylori MutS protein confers protection from oxidative DNA damage. Mol. Microbiol. 58:166–176.

Wang G., Ge Z., Rasko D. A., Taylor D. E. 2000. Lewis antigens in Helicobacter pylori: biosynthesis and phase variation. Mol. Microbiol. 36:1187–1196.

Wang G., Hong Y., Olczak A., Maier S. E., Maier R. J. 2006. Dual roles of Helicobacter pylori NapA in inducing and combating oxidative stress. Infect. Immun. 74:6839–6846.

Wang G., Maier R. J. 2004. An NADPH quinone reductase of Helicobacter pylori plays an important role in oxidative stress resistance and host colonization. Infect. Immun. 72:1391–1396.

Wang G., Olczak A. A., Walton J. P., Maier R. J. 2005b. Contribution of the Helicobacter pylori thiol peroxidase bacterioferritin comigratory protein to oxidative stress resistance and host colonization. Infect. Immun. 73:378–384.

Williams S. M., Chen Y. T., Andermann T., Carter J. E., McGee D. J., Ottemann K. M. 2007. Helicobacter pylori chemotaxis modulates inflammation and gastric-epithelium interactions in infected mice. Infect. Immun. 75:3747–3757.

Wirth H. P., Beins M. H., Yang M., Tham K. T., Blaser M. J. 1998. Experimental infection of Mongolian gerbils with wild-type and mutant Helicobacter pylori strains. Infect. Immun. 66:4856–4866.

Wirth H. P., Yang M., Peek R. M., Jr., Hook-Nikanne J., Fried M., Blaser M. J. 1999. Phenotypic diversity in Lewis expression of Helicobacter pylori isolates from the same host. J. Lab. Clin. Med. 133:488–500.

Wirth H. P., Yang M., Sanabria-Valentin E., Berg D. E., Dubois A., Blaser M. J. 2006. Host Lewis phenotype-dependent Helicobacter pylori Lewis antigen expression in rhesus monkeys. FASEB J. 20:1534–1536.

Yamaoka Y., Kita M., Kodama T., Imamura S., Ohno T., Sawai N., Ishimaru A., Imanishi J., Graham D. Y. 2002. Helicobacter pylori infection in mice: role of outer membrane proteins in colonization and inflammation. Gastroenterology 123:1992–2004.

Yamaoka Y., Kwon D. H., Graham D. Y. 2000. A M(r) 34,000 proinflammatory outer membrane protein (oipA) of Helicobacter pylori. Proc. Natl. Acad. Sci. USA 97:7533–7538.

Chapter 17
Host Immunity in the Development of Gastric Preneoplasia

Peter B. Ernst, Mohammad S. Alam, Asima Bhattacharyya, and Sheila E. Crowe

Introduction

Correa proposed a sequence of events that were associated with the development of gastric cancer that included gastritis as one of the earliest phases of a multistep process (Correa and Houghton 2007) (Figure 17.1). With the identification of *Helicobacter pylori*, it quickly became apparent that infection caused the gastritis that subsequently led to gastric preneoplasia and the development of gastric cancer in a subset of infected subjects. Several important issues have emanated from these observations including the role of the host response in the persistence of *H. pylori* infection and, given the chronic nature of the infection and inflammation, the role of immune and inflammatory factors in the pathogenesis of gastric cancer itself. This chapter reviews the initial immunologic responses that lead to gastritis and explains the persistence of infection. The consequence of chronic inflammation on epithelial cell damage that may lead to gastric neoplasia is discussed in the context of animal models and translational studies that have shed light on our understanding of the pathogenesis of diseases associated with *H. pylori* infection.

Initiation of the Host Response to *Helicobacter pylori*

As reviewed in more detail elsewhere (Ernst et al. 2006), the host response to any infection begins with the host recognizing the infection and discriminating between a potential pathogen and commensal organisms. In the case of *H. pylori*, the organism has adapted to the harsh gastric environment with its ability to minimize the effect of gastric acid and to propel itself to the safety of the epithelial surface underlying the mucus blanket. Class II major histocompatibility complex (MHC) molecules were the first epithelial cell receptor for *H. pylori* demonstrated to directly affect signaling in host cells (Fan et al. 1998), and binding of urease to epithelial cells via class II MHC is sufficient to induce apoptosis (Fan et al. 2000). Furthermore, these molecules are expressed *in situ* by gastric epithelial cells at increased levels during infection, and their expression is increased in response to

T.C. Wang et al. (eds.) *The Biology of Gastric Cancers*,
© Springer Science+Business Media, LLC 2009

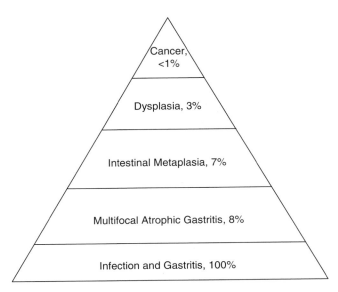

Fig. 17.1 Gastric cancer: the tip of the iceberg. Dr. P. Correa is credited with proposing the progression of gastritis to gastric cancer, which became even more relevant when *Helicobacter pylori* was identified as a major cause of gastritis. The model has been modified from what has been described elsewhere (Correa and Houghton 2007). Because not everyone infected with *H. pylori* develops gastric cancer, several modifiers have been identified. Currently, the perception is that gastric cancer arises from multiple "hits" that include oxidative stress and environmental toxins, which increase mutation rates. Diet, bacterial factors, and genes regulating the host response likely affect the degree of oxidative stress and DNA damage. Once key genes are mutated, the enhanced epithelial growth associated with infection drives the tumor proliferation. The percentages represent lifetime frequency of progression estimates from various studies

interferon (IFN)-γ (a cytokine that predominates in the gastric mucosa of infected individuals). Because the ability of *H. pylori* to induce cytokine production in the epithelium is also well known, Reyes and colleagues have extended these studies to demonstrate that both cytokine production and apoptosis are mediated through class II molecules (Beswick et al. 2005). This suggests that the class II MHC complex has an important role in binding and signaling some of the events that initiate the host response to infection.

The pathogen-associated molecular pattern (PAMP) recognition receptors have also been examined to determine if they have a role in binding of *H. pylori* to the host epithelial cells in order to initiate a host response. The toll-like receptors (TLRs) belong to the family of PAMPs of which 13 members have been identified thus far and are joined by two nucleotide oligomerization domain (NOD) proteins. Each of these seems to have a different specificity for various bacterial molecules and trigger host responses (Shibolet and Podolsky 2007). Cytokines and *H. pylori* particles increase the expression of TLR4 (Su et al. 2003) but other reports suggest that *H. pylori* lipopolysaccharide (LPS) stimulates monocytes (Bliss et al. 1998) and gastric epithelial cell responses in guinea pigs (Kawahara et al. 2001), presumably via TLR4. Other PAMPs, including TLR2, are also activated by highly purified *H. pylori* LPS (Mandell et al. 2004; Smith et al. 2003). TLR5 binds bacterial flagellins and, similarly to TLR2, induces a signaling response that would trigger acute inflammation. *H. pylori* produces flagellin that binds TLR5 and activates a response *in vitro* (Smith et al. 2003). It should be remembered that gastric epithelial cell lines that do not express TLRs still respond well to *H. pylori*, as demonstrated by interleukin (IL)-8 production. One possible mechanism to explain this is the NOD proteins. NOD1 has been described as recognizing peptidoglycan that is translocated into gastric epithelial cells by means of the *H. pylori* type IV secretion system (Viala et al. 2004). Together, these receptors may bind bacterial products, and thereby enhance both bacterial binding and cell signaling (Su et al. 2003) that lead to gastritis and associated changes in epithelial cell biology.

Several other binding mechanisms have been explored. For example, many strains of *H. pylori* express a vacuolating cytotoxin (VacA), which attaches to epithelial cells via an interaction with a protein tyrosine phosphatase (Yahiro et al. 2003). Several studies have examined the structure and function of VacA and its association with disease (Wilson and Crabtree 2007). For example, specific VacA alleles (s1 and m1) are associated with disease (Atherton et al. 1995) and the induction of epithelial cell apoptosis (Cover et al. 2003). The interaction between VacA and its receptor therefore seems to be important in the pathogenesis of gastroduodenal disease. Another important interaction between *H. pylori* and the host is the type IV secretion system (Blaser 2005). Interestingly, the epithelial integrins have now been identified as the essential receptors for facilitating the binding of the secretion system to the epithelium and the subsequent downstream effects (Kwok et al. 2007).

The gastric trefoil protein TFF1 was demonstrated to serve as a receptor for *H. pylori* (Clyne et al. 2004). This molecule is predominantly expressed in the gastric mucosa and found in association with gastric mucous, where it may account for some of the gastric tropism. Interestingly, when the expression of TFF1 is decreased in genetically engineered mice, a spontaneous, antral adenoma develops, suggesting that this molecule provides an element of control over gastric epithelial cell growth (Lefebvre et al. 1996; Tebbutt et al. 2002). The attachment of *H. pylori* to TFF1 in humans could impair its function (mimicking the decreased expression in the animal models), thus favoring the proliferation of epithelial cells and the subsequent development of cancer in some infected individuals.

One major consequence of *H. pylori*'s contact with epithelial cells is the release of chemokines that recruit and activate neutrophils and antigen-presenting cells. The latter are particularly important to effect the transition from an innate response to an adaptive immune response. There is evidence that T cells can be recruited to the stomach from the peripheral blood after being stimulated in the Peyer's patches (Nagai et al. 2007). However, it is less clear if gastric antigen-presenting cells (APCs) in humans contribute to the initiation of T-cell response versus the perpetuation of their activation. Histologic studies have demonstrated that molecules associated with APCs are indeed found in the gastric mucosa (Sarsfield et al. 1996). These studies have identified cells resembling both macrophages and dendritic cells although heterogeneity of these lineages may not have been discerned completely at this point. This is quite important as mucosal APCs vary significantly from their counterparts in the blood or other systemic tissues (Smith et al. 2005). Interestingly, the epithelium also expresses many of the markers associated with APCs (Beswick et al. 2005; Sakai et al. 1987; Ye et al. 1997) suggesting that they may be able to modulate, if not initiate, gastric T-cell responses.

Although the most relevant work on the role of the APC in responding to *H. pylori* would be obtained using cells isolated from the gastric mucosa, most studies to date have relied on cell lines or cells from systemic tissue such as the peripheral blood. Using these approaches, it is evident that APCs can bind *H. pylori* or its components and trigger substantial cytokine responses (Ferrero 2005; Guiney et al. 2003; Kranzer et al. 2004; Nishi et al. 2003; Rad et al. 2007). However, in limited studies, the potential for *H. pylori* to stimulate responses in primary mucosal APCs differs substantially (Harris et al. 1998). Reyes and colleagues have shown that gastric epithelial cells favor the development of Th cells which may help to limit gastritis (Beswick et al. 2007). The implications of these cells will be discussed below, but suffice it to say that research related to APC function in the stomach remains to be addressed completely but will inevitably aid in our understanding of the events that trigger chronic inflammation during persistent infection with *H. pylori*.

Activation of the Adaptive Response

In addition to the robust innate response, gastritis associated with *H. pylori* infection elicits the expansion of T and B cell responses. It is generally accepted that the Th1 response predominates in the gastric mucosa (Ernst et al. 2006). However, most of these studies have been based on a limited panel of cytokine and as Th cell heterogeneity becomes more apparent, additional definition to the pathogenic Th cell response will continue to evolve. For example, the Th1 responses were largely based on the presence of IFN-γ and absence of IL-4 (Bamford et al. 1998; D'Elios et al. 1997; Karttunen et al. 1995). More recently, so-called Th17 cells have emerged as important contributors in the proinflammatory response in Th cell–mediated disease (Figure 17.2). These cells produce IL-17, which can stimulate the release of chemokines, including IL-8 from gastric epithelial cells (Sebkova et al. 2004), and enhance inflammation. It should be noted that *H. pylori*–induced gastritis was the first model of inflammation in which IL-17 was described as an important component (Luzza et al. 2000) and, in fact, this work precedes the modern enthusiasm for Th17 cells by several years. As this field continues to unfold, it seems that both IFN-γ and IL-17 are often made by the same Th cell necessitating a reassessment of the model for Th cell–induced gastritis. The likely scenario is that both Th1 and Th17 cells are involved although the relative contribution of each is unknown. This nuance may be relevant as we seek to understand how inflammation is triggered and controlled, which could have implications for vaccine development

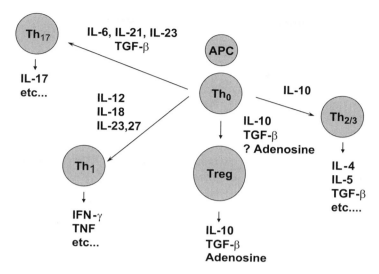

Fig. 17.2 Heterogeneity of helper T-cell subsets. Helper T cells emerge from an undifferentiated precursor (Th$_0$) and assume various phenotypes based on the exposure to several factors such as cytokines. Th$_1$ and Th$_2$ have been studied the most. Th$_1$ and Th$_{17}$ cells are likely present together in gastritis whereas Th$_2$ are the most infrequent in the stomach. Both Th$_{17}$ and Treg emerge in response to transforming growth factor (TGF)-β; however, interleukin (IL)-6 can depress Treg while it enhances Th17 differentiation along with IL-21 and IL-23. Adenosine is now recognized as a factor that favors IL-10 and TGF-β accumulation, thus has the potential to stimulate Treg. Adenosine is also a mediator of suppression by Treg because adenosine receptor expression by immune and inflammatory cells is required for optimal Treg function. APC: antigen-presenting cell; IFN: interferon; TNF: tumor necrosis factor

and the control of gastritis during the persistent infection with *H. pylori*. Regardless, both of these cell types can contribute to the gastritis and likely have a role in the milieu that can culminate in the development of neoplasia.

Normally, antibodies in the gastrointestinal tract are of the immunoglobulin (Ig)A isotype, which are highly adapted for mucosal protection. Moreover, IgA confers protective immunity without activating complement efficiently and stimulating deleterious amounts of inflammation. During infection with *H. pylori*, the number of IgA producing cells increases. IgG and IgM are also detected (Ernst et al. 2006) along with activated complement (Berstad et al. 1997). As reviewed previously, several groups have suggested that the level of autoantibodies in humans and animal models correlates with the severity of gastritis (Ernst et al. 2006). If these observations are correct, one can conclude that local immune complexes contribute to gastroduodenal inflammation and tissue damage during infection.

Direct evidence for autoantibody production in response to *H. pylori* requires the identification of the self-antigens that are recognized by these antibodies. Monoclonal antibodies that recognize *H. pylori* cross-react with both human and murine gastric epithelial cells (Appelmelk et al. 1996; Negrini et al. 1991). Adoptive transfer of monoclonal antibodies recognizing *H. pylori* to recipient mice induces gastritis (Negrini et al. 1991), as does the transfer of B cells that recognize heat shock proteins from individuals with MALToma (Yokota et al. 1999). *H. pylori* induced antibodies that recognize Le$^{x/y}$ and are also capable of inducing gastritis in mice (Appelmelk et al. 1996). The cellular target for these antibodies in mice seems to be the gastric H$^+$, K$^+$-ATPase localized in the parietal cell canaliculi (Claeys et al. 1998). Anti-Le antibodies have been described in humans and occur independently of the Le phenotype of the host (Heneghan et al. 2001). However, they do not seem to be autoreactive. Thus, one has to surmise that autoantibodies induced in mice may recognize different targets within the gastric mucosa even though they may cross-react with human gastric tissue whereas autoantibodies induced in humans may have a completely different specificity.

Th cells That Control Gastrointestinal Inflammation

Whereas many antigenic challenges induce a response that eventually decreases the antigenic burden (or even creates sterile immunity), some microbial triggers, such as *H. pylori* in the stomach, become persistent infections. As these bacteria can induce pathogenic responses, perhaps because of a failed attempt to clear the infection, the host must attenuate its reactivity in order to accommodate the persistent infection. Therefore, an acute stimulation would be necessarily followed by one or more responses that attenuate the inflammatory response. Tolerance itself is attributed to clonal deletion, anergy, or a state of immune regulation that suppresses adverse host responses. This redundancy reflects the fact that tolerance is so important that nature has evolved several complementary mechanisms (Bach and Francois 2003). In the digestive tract, most of the discoveries to date suggest that T cells mediate much of the restraint that protects against excessive responses.

   In tissues such as the gut, the host can "tolerate" chronic exposure to ever-changing luminal antigens in part, through the antiinflammatory effects conferred by Th cells. For example, the local administration of antigen induces oral tolerance that is mediated by Th3 cells producing IL-10 and transforming growth factor (TGF)-β (Weiner 2001). Luminal flora also stimulate Th cells that function as Treg (Cong et al. 2002; Elson and Cong 2002; Kullberg et al. 2002). Whereas it has been appreciated for a long time that intestinal T cells prevent excessive responses to luminal antigen by the induction of oral tolerance (Weiner 2001), other Th cell subsets with "regulatory" function exist. For example, Th cell subsets that expand to protect from antigens encountered after birth can be described as "adaptive" or "acquired" regulatory T cells (Treg) (Bluestone and Abbas 2003).

   Excessive exuberance on the part of Th1 or Th2 cells can stimulate disease but Treg attenuate diseases caused by cell-mediated (Th1) immunity (Elson and Cong 2002) or Th2 responses (Akbari et al. 2002). These observations favor the postulate that normal intestinal Th cell responses include cells that produce increased amounts of antiinflammatory cytokines (such as IL-10 or TGF-β) (Jump and Levine 2002; Maloy et al. 2003) and prevent excessive inflammatory responses to luminal flora (Cong et al. 2002; Duchmann et al. 1995; Zhou et al. 2004). This model is supported by data showing that T cells in normal mice acquire a "regulatory" T-cell phenotype to potentially pathogenic bacteria, such as *H. hepaticus* (Kullberg et al. 2002) whereas mice deficient in the ability to generate Treg or lacking effector proteins (such as IL-10) develop colitis that is exacerbated by the same bacteria (Kullberg et al. 2002). Thus, Th cells that are induced in response to acquired antigens, such as those associated with *H. hepaticus*, can develop a phenotype that prevents chronic inflammation. However, little is known about the events that directly control the differentiation of Th cell function in the stomach after infection with *H. pylori*.

The complete picture describing the factors favoring the emergence of adaptive Treg remains unclear but includes factors that affect T cells directly, such as TGF-β (Chen et al. 2003; Davidson et al. 2007) or indirectly via antigen-presenting cells (APCs). APCs manipulated such that costimulation is impaired, lead to the emergence of Th cells resembling Treg (Bour-Jordan et al. 2004). Within the gut, macrophages seem to select for Treg more easily than dendritic cells (DCs) (Denning et al. 2007). However, culturing DCs *in vitro* with IL-10 yields an APC that selects for Treg capable of preventing colitis (Wakkach et al. 2003). Others have shown that NF-κB function in DCs is essential for activating effector Th cells (Teff) (Yoshimura et al. 2001) and that blocking NF-κB impairs their maturation and leaves an APC that selects for Th cells with an antiinflammatory phenotype (Cong et al. 2005). Adenosine shares the ability to impair gene expression in APCs in response to LPS (Hasko and Cronstein 2004; Lukashev et al. 2004) suggesting that it may also shape the phenotype of the APCs and their ability to select for adaptive/acquired Treg.

The Role of Adenosine in Th Cell Function

Adenosine is a signaling molecule released from inflamed or hypoxic tissues that mediates many physiologic responses. Initially during T-cell activation, adenosine deaminase binds to CD26 in the immunologic synapse where it enhances T-cell response (Pacheco et al. 2005), for example, by degrading endogenous adenosine. Subsequently, extracellular enzymes that synthesize adenosine are induced, including CD39 and CD73 (Sitkovsky et al. 2004), which increase the local adenosine concentration.

The effects of adenosine are controlled by four G protein-coupled receptors (A1, A_{2A}, A_{2B}, and A3), which are variably expressed on immune cells depending on cell type and species (Sitkovsky et al. 2004). The $A_{2A}AR$ are found on bone marrow–derived cells including neutrophils, monocytes, macrophages, dendritic cells, and lymphocytes. The $A_{2A}AR$ are of particular interest because they are believed to be the major adenosine receptor on T cells (Huang et al. 1997) and their activation on immune cells produces a series of responses that can be categorized as antiinflammatory (Linden 2001).

Adenosine limits *collateral damage* that occurs as a result of severe inflammation (Hasko and Cronstein 2004) including chemical-induced colitis (Mabley et al. 2003) and ileitis (Odashima et al. 2005). The activation of $A_{2A}AR$ protects animals from sepsis and ischemia/reperfusion injury in the heart, lung, liver, and kidney by reducing neutrophil accumulation, superoxide generation, and endothelial expression of adhesion molecules (Day et al. 2004; Okusa et al. 2000). Furthermore, activation of $A_{2A}AR$ on human monocytes and murine macrophages inhibits the secretion of IL-12 and TNF-α, both proinflammatory cytokines implicated in the pathogenesis of gastritis and associated with Th1 responses (Hasko et al. 2000). The notion that adenosine has a direct role in attenuating the proinflammatory capacity of Th cells is supported by the fact that Th cells express inhibitory $A_{2A}AR$ (Huang et al. 1997) and the observation that $A_{2A}AR$ agonists impair IFN-γ production (Lappas et al. 2005; Naganuma et al. 2006). Moreover, adenosine is implicated in optimal Treg function (Deaglio et al. 2007; Naganuma et al. 2006).

The Selection of Th Cell Response During *Helicobacter pylori* Infection

If Th cells can be pathogenic, antiinflammatory, or indeed, protective, then it would be of substantial interest to understand the factors that favor the differentiation of these T-cell subsets in order to permit therapeutic manipulations. This widely studied field cannot be easily reviewed in this limited space but it is fair to say that the phenotype of the APC, the cytokine milieu and factors produced by the organism itself are often the most important in selecting for a Th-cell phenotype.

H. pylori is something of an enigma in that it is overwhelmingly an extracellular pathogen that is one additional step removed from the host by residing predominantly in the lumen. Normally, Th1 responses are targeted against intracellular infections because under the cloak of the host's own cells, they do not provide an available target for extracellular defense mechanisms. In this regard, *H. pylori* is similar to several nematodes that have a close association with the host epithelium but reside mainly in the lumen. However, these pathogens generally are associated with marked Th2 responses, likely favored by pathogen-associated structures that have the ability to select for Th2 responses (Thomas et al. 2005). Thus, *H. pylori* lacks the efficiency of nematodes at inducing Th2 responses. This may be an adaptive trait that favors persistence because *H. pylori* vaccines that induce Th2 responses induce protective immunity in mice (Mohammadi et al. 1997).

Several reports have described an important role for the phenotype of APCs on the selection of Th-cell subsets. For example, impairment of dendritic cell maturation allows them to favor the induction of Treg versus effector Th cells (Cong et al. 2005). APCs can also influence the phenotype of the Th cell, mainly because of the density of accessory molecule expression or the cytokines they induce. For example, IL-12 promotes Th1 cells; IL-10 promotes Th2 cells (Sher et al. 1991); TGF-β alone can favor Treg (Chen et al. 2003); whereas TGF-β, IL-6, IL-21, and IL-23 can favor the differentiation of Th17 cells (Harrington et al. 2005; Zhou et al. 2007). The effects of these cytokines can be direct on Th cells, or indirect on APCs (Fiorentino et al. 1991). Using cells isolated from the intestine itself, Denning et al. have shown that macrophages and dendritic cells in the gut differ in their ability to select for Th-cell subsets with the former selecting for Treg and the latter Th17 cells (Denning et al. 2007). The approach of studying APCs from gastric tissue directly has never been taken to understand the functional role of these cells in response to any gastric disease.

The Implications of Treg Development and Persistence

Our working model is that, after infection, the initial burst of Th-cell activation is followed by an increase in the ability to generate Treg. For example, the ectoenzymes that form adenosine, CD39, CD73, as well as adenosine receptor expression on APCs and Th cells, are induced within 48–72 hours of activation. Therefore, if the host response does not succeed in clearing the infection and removing the antigenic impetus to drive the inflammatory response, the host must attenuate these responses. As suggested in the IL-10–deficient mouse, marked inflammatory responses can clear *Helicobacter* spp. infection (Ismail et al. 2003). Thus, attenuating these responses may favor persistence.

In fact, several reports have described T cells with markers associated with Treg cells in the gastroduodenal tissue in response to *H. pylori* infection (Enarsson et al. 2006; Lundgren et al. 2005; Rad et al. 2006; Raghavan and Holmgren 2005). These Th cells may produce sufficient levels of IL-10, TGF-β, or adenosine to attenuate responses rather than to prevent gastritis totally. It has also been shown that the bacterial burden is increased in the stomach of animal models when Treg function is impaired (Rad et al. 2006). Their importance is illustrated in animal models whereby the absence of IL-10 is associated with fewer Treg and more profound gastritis as well as colitis (Kullberg et al. 2002). Interestingly, the colitis in these mice is also associated with changes that precede the onset of colon cancer (Berg et al. 1996). In humans, gastric cancer is associated with polymorphisms in IL-10 (El-Omar et al. 2003; Wu et al. 2002). A theoretic consequence of Treg impairing immune effector mechanisms is that they may also interfere with tumor immunity. In fact, increased numbers of Treg can be found in areas of the stomach where gastric cancer is present (Enarsson et al. 2006).

How Does the Gastric Immune/Inflammatory Response Lead to Neoplasia?

With the background on the host response above, it is now possible to consider where these responses lie in the path to neoplasia and how these factors contribute to this outcome. Several lines of evidence support a role for the host response in gastric cancer. Importantly, these responses occur before the development of cancer and in direct response to the known cause of gastric cancer— *H. pylori*. The challenge is that many decades intervene between infection, the development of inflammation, and the onset of cancer. Correa noted the association, but stronger evidence now exists. As discussed elsewhere in this text, population genetic studies have shown that gastric cancer is associated with polymorphisms in genes that encode proteins that regulate the host response (El-Omar et al. 2003). Generally, therefore, if the polymorphism is in a gene encoding a proinflammatory cytokine, such as IL-1, then one would predict that it leads to an overproduction and enhanced inflammation. Similarly, if the genetic link is in a gene encoding an antiinflammatory cytokine, then if the hypothesis is correct, one would predict that it would be decreased, hence favor inflammation by the removal of a control mechanism. This would be the case for IL-10, for example. Indeed, as indicated above, there is evidence that supports this model and the hypothesis would be favored by providing more direct, mechanistic evidence.

Epithelial Responses

With the persistence of infection, there is a chronic host response that gnaws away at the integrity of the epithelium and invokes changes that are believed to contribute to the preneoplasia pathway. The effects of the host response on the gastric epithelium begin with changes in signaling that lead to effects on gene expression and function. Whereas the effects of the bacteria are described in detail elsewhere, it should be clear that the combined effects of infection and cytokines greatly magnify the impact on the epithelium.

One of the obvious changes in gene expression is the induction of class II MHC molecules (Fan et al. 1998). This response, along with an increase in various accessory molecules associated with antigen presentation, can facilitate T-cell activation by the gastric epithelial cell (Beswick et al. 2007; Ye 1997). However, there is another consequence. As described above, the complex of class II MHC and invariant chain can bind *H. pylori*, so increasing their expression on the apical surface of the epithelium can increase the bacterial load (Fan et al. 1998). In fact, the number of bacteria able to bind the class II MHC complex is directly proportional to the expression of these receptors. Moreover, this complex is sufficient to initiate the signaling that leads to epithelial cell apoptosis and associated changes in chemokine expression that lead to the recruitment of phagocytes to engulf and remove the dead cells.

In addition to increasing the expression of class II MHC molecules and accessory molecules that facilitate T-cell activation, infection and the associated cytokine storm boost the expression of both Fas and Fas ligand (FasL) on gastric epithelial cells (Bennett et al. 1999; Rudi et al. 1998; Wagner et al. 1997; Wang et al. 2000). The expression of Fas can be recognized by immune cells, particularly T cells, that express FasL, which leads to the induction of apoptosis in the epithelial cells (Wang et al. 2000). In addition, there are reports that epithelial cells can express both Fas and FasL, which would lead to cell death by suicide or fratricide as receptors and ligands interact and stimulate cell death (Rudi et al. 1998). Thus, there are several bacterial and host responses that can induce epithelial cell death through apoptosis in response to infection. However, it is also possible to contemplate that the induction of FasL on malignant gastric epithelial cells could kill an incoming T cell and thereby block its ability to mediate tumor immunity (Bennett et al. 1999; Kume et al. 1999). This would be another level of immune privilege that may favor gastric cancer development.

Regulation of Cellular Responses by Redox: AP Endonuclease-1/Redox Factor-1

Another consequence of the host inflammatory response to *H. pylori* is the influx of inflammatory cells that contribute to a burst of reactive oxygen species (ROS) and reactive nitrogen species. The induction of ROS at low levels within cells is important for the regulation of ROS-sensitive gene expression. Physiologic levels of ROS are induced by cytokines such as TNF-α, as well as the infection itself. In turn, electrons generated during this reaction are passed from one molecule to the next ending in the reductive activation of transcription factors.

The net response to oxidative stress varies and includes aberrant proliferation, adaptation cytotoxicity, and cell death (Janssen et al. 1993; Ryter et al. 2007). There are multiple levels at which oxidative stress regulates cellular responses including activation of MAP kinases (Karin 1998) and a variety of early response genes induced by the activation of transcription factors that are mainly dependent on redox-dependent processes. For example, the ability of c-Jun and c-Fos to bind to the DNA regulatory elements containing the activator protein-1 (AP-1) binding site depends on reductive activation by AP endonuclease-1/redox factor-1 (APE-1/Ref-1) (Xanthoudakis et al. 1992) (see Figure 17.3). Oxidants and *H. pylori* have been shown to activate NF-κB in gastric epithelial cells (Keates et al. 1997; O'Hara et al. 2006; Sharma et al. 1998) and AP-1 activation also occurs during *H. pylori* infection (Naumann et al. 1999; O'Hara et al. 2006). Of particular interest is the central role that APE-1/Ref-1 may have in orchestrating these responses to oxidative stress and infection.

APE-1/Ref-1 is a multifunctional protein that activates a growing list of transcription factors—c-Jun, c-Fos, C-Myb, CREB, p53, pax5, pax8, and HLF (Evans et al. 2000)—and participates in the base excision DNA repair pathway (Demple and Sung 2005; Izumiet al. 1996; Jayaraman et al. 1997; Tell et al. 2000; Xanthoudakis et al. 1992). Its essential role in cellular homeostasis is evident by the embryonically lethal nature of homozygous deletion of the gene (Xanthoudakis et al. 1996). In cell culture systems, increased APE-1/Ref-1 activity decreases

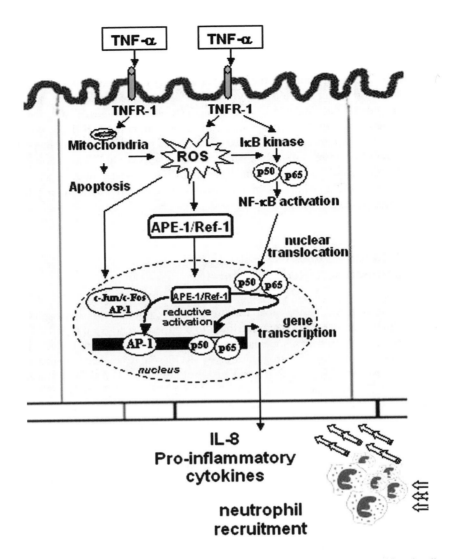

Fig. 17.3 The role of AP endonuclease-1/redox factor-1 (APE-1Ref-1) in redox-sensitive signaling. *Helicobacter pylori* or the associated inflammatory mediators, such as TNF-α, lead to the accumulation of intracellular reactive oxygen species (ROS). ROS damage cellular proteins, lipids, and DNA as well as regulate the activation of transcription factors that induce the expression of redox-sensitive genes encoding DNA repair enzymes, antioxidant proteins, cytokines, and factors regulating cell growth and apoptosis. APE-1, also known as Ref-1, is necessary for the reduction of cysteine residue 252 in the DNA binding portion of c-Jun and c-Fos. *H. pylori* binds gastric epithelial cells, induces ROS, then apoptosis. *H. pylori* also increases the expression of APE-1/Ref-1. The ROS induced by infection activate APE-1/Ref-1 which in turn reductively activates transcription factors, such as p53 or c-Jun. This permits DNA binding of the reduced transcription factor to the binding site upstream of several proapoptotic genes including *fasL* or, in the case of p53, *bax*. As such, the expression of genes regulated by APE-1/Ref-1 is "redox sensitive." Interestingly, many of the genes regulating apoptosis are redox sensitive as are signaling pathways such as JNK/SAPK. By focusing on the effects of *H. pylori* on ROS, APE-1/Ref-1, and specific genes regulated by c-Jun/AP-1 and p53, a representative system for studying redox-sensitive gene expression can be developed and applied to other genes in future studies. TNF: tumor necrosis factor; TNFR: tumor necrosis factor receptor; IL: interleukin

DNA damage induced by oxidants, ionizing radiation, and other agents while antisense APE-1/Ref-1 RNA sensitizes cells to such stresses (Walker et al. 1994). The carboxy-terminal portion of APE-1/Ref-1 is responsible for DNA repair activities, whereas its ability to activate transcription factors including c-Jun, c-Fos, c-Myb, and p53, lies in the N-terminal region of the molecule (Izumi and Mitra 1998). Other studies by Mitra (Ramana et al. 1998), for example, have shown that ROS enhance the expression and activity of APE-1/Ref-1 in fibroblasts, HeLa, and CHO cells. For example, H_2O_2 increases APE-1/Ref-1 expression demonstrating that gastric epithelial cells resemble other cells in that the expression of the APE-1/Ref-1 gene is very sensitive to the oxidative environment (Ding et al. 2004).

Regulation of Epithelial Cell Proliferation and Apoptosis by *Helicobacter pylori* Infection

H. pylori infection increases the rate of epithelial cell proliferation *in vivo* with a return to normal rates after eradication of the infection (Brenes et al. 1993; Lynch et al. 1995). Similar findings have been shown for epithelial apoptosis with an increase in the stomach of *H. pylori*–infected patients and a return to control levels after eradication of infection (Jones et al. 1997; Moss et al. 1996). Infection also induces apoptosis in gastric epithelial cell lines. Studies from our laboratory and others have demonstrated that apoptosis in cultured cells is mediated directly by *H. pylori* and that programmed cell death is augmented by cytokines that are increased in *H. pylori* infection, IFN-γ and TNF-α (Fan et al. 1998; Wagner et al. 1997). Bacterial genotype seems to affect the gastric epithelial cell cycle (Peek et al. 1999), but the full mechanisms by which *H. pylori* regulate host cell growth and death remain to be elucidated.

There is a growing understanding of the role programmed cell death has in normal gastrointestinal growth and development and its relationship to gut disease. Evidence supporting alterations in apoptosis leading to gastric cancer comes from studies of the Bcl-2 family of genes. Some members of the family, including *bcl-2*, *bcl-x_L*, and *mcl-1*, are antiapoptotic and inhibit the action of caspases, which form the final common pathway of events culminating in apoptosis by binding a factor that activates caspase-3 and/or by preventing the release of cytochrome c from mitochondria (Hetts 1998). Other members are proapoptotic, including *bax*, *bak*, and *bad*. The proapoptotic *bak* and *bax* are upregulated in *H. pylori*–associated gastritis (Moss 1998) in association with an increase in apoptosis (Yamasaki et al. 2006) and changes in epithelial barrier function. In gastric cancer, Bcl-x expression is increased whereas Bak is decreased (Krajewska et al. 1996). As discussed above, alterations in balance of Fas and FasL expression in the gut mucosa are involved in apoptosis associated with *H. pylori* infection (Rudi et al. 1998; Wagner et al. 1997) and may have a role in the development of gastric cancer (Bennett et al. 1999). Thus, the regulation of these genes governing cell survival may affect development of cancer.

Redox-Sensitive Pathways That Control Apoptosis

There is good evidence that ROS are involved in programmed cell death because increased formation of ROS and depletion of cellular antioxidants lead to apoptosis whereas antioxidant compounds and overexpression of antioxidant enzymes inhibit apoptosis (Bauer et al. 1998). There are multiple levels in which ROS have a role in apoptotic events. Redox-sensitive mechanisms are involved in p53-mediated apoptosis with p53-related generation of ROS resulting in mitochondrial damage, which leads to the release of proteins that activate caspases, then mediating apoptosis (Polyak et al. 1997). Other studies suggest that ROS are common mediators of ceramide-induced apoptosis (Quillet-Mary et al. 1997) and may be involved in CD95 (APO-1/Fas) mediated apoptosis (Bauer et al. 1998). We have shown that antioxidants inhibit *H. pylori*–induced apoptosis of gastric epithelial cells (Ding et al. 20075). Indirect evidence that signaling through APE-1/Ref-1 can lead to apoptosis includes observations in which induction of APE-1/Ref-1 expression was shown to enhance transcription factor activity of p53 as demonstrated by its effect on *p21, bax, gadd45*, and *cyclin G* promoters. APE-1/Ref-1 levels were previously correlated with the extent of apoptosis induced by p53 (Gaiddon et al. 1999). Because both AP-1 and p53 are involved in the induction of apoptosis (Ishikawa et al. 1997; Singh et al. 1995; Wyllie 1997), APE-1/Ref-1 represents an important common element in the pathways leading to programmed cell death that may be regulated by oxidative stress.

Redox-Dependent Activities of AP Endonuclease -1Redox factor-1

The affinity of c-Jun and other APE-1Ref-1–regulated transcription factors to bind to DNA sequences is dependent on the redox status of specific cysteine residues located near the DNA binding domain of the factor. APE-1/Ref-1 maintains the reduced state of these cysteine residues thereby enhancing DNA binding activity. DNA binding activity of c-Jun is dependent on the reduction of the cysteine residue at position 272 by APE-1Ref-1 (Xanthoudakis et al. 1994). We have shown that both *H. pylori* infection and TNF-α enhance AP-1 and NF-κB binding of nuclear proteins in gastric epithelial cells and this is decreased in cells treated with siRNA for APE-1Ref-1 (O'Hara et al. 2006). A luciferase-linked IL-8 promoter with binding sites for AP-1 and NF-κB demonstrate that enhanced APE-1/Ref-1 increases transcriptional activity (O'Hara et al. 2006). Other recent data indicate that another transcription factor reported to be regulated by APE-1Ref-1, HIF-1α (Lando et al. 2000), is not detected in resting AGS cells but it is increased in a dose-dependent manner by *H. pylori* infection and hypoxia. Interestingly, hypoxia results in intracellular oxidative stress generated at the mitochondrial level and is considered a form of cellular stress associated with altered APE-1Ref-1 levels (Hall et al. 2001).

Direct Effects of the Host Response That Predispose to Neoplasia

How *H. pylori* infection leads to gastric cancer is unknown. Several indirect effects were discussed above, including the impairment of antitumor immunity. However, the most significant direct effects of *H. pylori* may reflect similarities to other chronic inflammatory diseases such as ulcerative colitis in which dysplasia and cancer may develop, presumably as a consequence of chronic inflammation (Correa and Houghton 2007). One likely mechanism is that ROS and reactive nitrogen species released from immune/inflammatory cells may damage DNA and activate signaling pathways to alter patterns of cell growth (Obst et al. 2000; Wiseman and Halliwell 1996). Mutations of p53 leading to a loss of function so that the protective pathways of cell cycle arrest and apoptosis cannot be activated, are among the more common abnormalities found in human cancers and have been reported in intestinal metaplasia, dysplasia, and carcinoma of the stomach (Moss 1998). Mutations of other genes including *apc* and *dcc* have also been reported in gastric cancer but their interaction with *H. pylori* is less clear (Wu et al. 1997).

Evidence in support of this model also includes the increased oxidative DNA damage in the gastric mucosa of *H. pylori*–infected subjects (Baik et al. 1996; Farinati et al. 1998). Activated neutrophils and macrophages are potent sources of ROS including superoxide anion, hydrogen peroxide (H_2O_2), and hydroxyl radical in addition to reactive nitrogen species that can react with ROS to generate peroxynitrite. Given the close proximity of neutrophils to the epithelium in *H. pylori*–infected gastric mucosa, it is likely that epithelial cells are exposed to ROS on a chronic basis. As we and others have shown, increased levels of ROS can be measured in the gastric mucosa of infected subjects (Davies et al. 1994). Bacterial expression of superoxide dismutase and catalase may serve to regulate the local redox status of infected mucosa (Wu et al. 1997). As a result of the typical lifelong infection with *H. pylori*, the persistence of ROS accumulation in gastric epithelial cells may eventually have significant effects. It is possible that host DNA repair mechanisms—including APE-1Ref-1—attempt to correct this DNA damage but should these processes fail, they attempt to cull the damaged cell by regulating apoptosis.

Summary

H. pylori has the ability to stimulate a robust gastritis, which may favor its survival because of the enhanced expression of host receptors for the bacteria and sufficient tissue damage that allows nutrients that leak into the lumen. However, the magnitude of the inflammation seems to be kept in check by several mechanisms that contribute to the persistent infection and a life of exposure to potentially dangerous oxidative stress. Exposure to various cofactors that contribute to cancer, failure to mediate correct DNA repair, and the possible impairment of tumor immunity may all favor the expansion of potentially malignant precursor cells that eventually lead to cancer in a small subset of infected individuals.

References

Akbari, O., G. J. Freeman, E. H. Meyer, E. A. Greenfield, T. T. Chang, A. H. Sharpe, G. Berry, R. H. DeKruyff, and D. T. Umetsu. 2002. Antigen-specific regulatory T cells develop via the ICOS-ICOS-ligand pathway and inhibit allergen-induced airway hyperreactivity. Nat Med 8(9):1024–1032.

Appelmelk, B. J., I. Simoons-Smit, R. Negrini, A. P. Moran, G. O. Aspinall, J. G. Forte, T. De Vries, et al. 1996. Potential role of molecular mimicry between Helicobacter pylori lipopoly-saccharide and host Lewis blood group antigens in autoimmunity. Infect Immun 64:2031–2040.

Atherton, J. C., P. Cao, R. M. Peek, M. K. Tummuru, M. J. Blaser, and T. L. Cover. 1995. Mosaicism in vacuolating cytotoxin alleles of Helicobacter pylori. Association of specific vacA types with cytotoxin production and peptic ulceration. J Biol Chem 270:17771–17777.

Bach, J. F. 2003. Regulatory T cells under scrutiny. Nat Rev Immunol 3(3):189–198.

Baik, S.-C., H.-S. Youn, M.-H. Chung, W.-K. Lee, M.-J. Cho, G.-H. Ko, C.-K. Park, H. Kasai, and K.-H. Rhee. 1996. Increased oxidative DNA damage in Helicobacter pylori-infected human gastric mucosa. Cancer Res 56:1279–1282.

Bamford, K. B., X. J. Fan, S. E. Crowe, J. F. Leary, W. K. Gourley, G. K. Luthra, E. G. Brooks, et al. 1998. Lymphocytes in the human gastric mucosa during Helicobacter pylori have a T helper cell 1 phenotype. Gastroenterology 114:482–492.

Bauer, M. K. A., M. Vogt, M. Los, J. Siegel, S. Weselborg, and K. Schulze-Osthoff. 1998. Role of reactive oxygen intermediates in activation-induced CD95 (APO-1/Fas) ligand expression. J Biol Chem 273:8048–8055.

Bennett, M. W., J. O'Connell, G. C. O'Sullivan, D. Roche, C. Brady, J. K. Collins, and F. Shanahan. 1999. Expression of Fas ligand by human gastric adenocarcinomas: a potential mechanism of immune escape in stomach cancer. Gut 44:156–162.

Berg, D. J., N. Davidson, R. Kuhn, W. Muller, S. Menon, G. Holland, L. Thompson-Snipes, M. W. Leach, and D. Rennick. 1996. Enterocolitis and colon cancer in interleukin-10 deficient mice are associated with aberrant cytokine production and CD4+ Th1-like responses. J Clin Invest 98:1010–1020.

Berstad, A. E., P. Brandtzaeg, R. Stave, and T. S. Halstensen. 1997. Epithelium related deposition of activated complement in Helicobacter pylori associated gastritis. Gut 40:196–203.

Beswick, E. J., S. Das, I. V. Pinchuk, P. Adegboyega, G. Suarez, Y. Yamaoka, and V. E. Reyes. 2005. Helicobacter pylori-induced IL-8 production by gastric epithelial cells up-regulates CD74 expression. J Immunol 175(1):171–176.

Beswick, E. J., I. V. Pinchuk, S. Das, D. W. Powell, and V. E. Reyes. 2007. B7-H1 expression on gastric epithelial cells after Helicobacter pylori exposure promotes the development of CD4+ CD25+ FoxP3+ regulatory T cells. Infect Immun 75(9):4334–4341.

Blaser, M. J. 2005. The biology of cag in the Helicobacter pylori-human interaction. Gastroenterology 128(5):1512–1515.

Bliss, C. M., Jr., D. T. Golenblock, S. Keates, J. K. Linevsky, and C. P. Kelly. 1998. Helicobacter pylori lipopolysaccharide binds to CD14 and stimulates release of interleukin-8, epithelial neutrophil-activating peptide 78, and monocyte chemotactic protein 1 by human monocytes. Infect Immun 66:5357–5363.

Bluestone, J. A. and A. K. Abbas. 2003. Natural versus adaptive regulatory T cells. Nat Rev Immunol 3(3):253–257.

Bour-Jordan, H., B. L. Salomon, H. L. Thompson, G. L. Szot, M. R. Bernhard, and J. A. Bluestone. 2004. Costimulation controls diabetes by altering the balance of pathogenic and regulatory T cells. J Clin Invest 114(7):979–987.

Brenes, F., B. Ruiz, P. Correa, F. Hunter, T. Rhamakrishnan, E. Fontham, and T.-Y. Shi. 1993. Helicobacter pylori causes hyperproliferation of the gastric epithelium: pre- and post-eradication indices of proliferating cell nuclear antigen. Am J Gastroenterol 88:1870–1875.

Chen, W., W. Jin, N. Hardegen, K. J. Lei, L. Li, N. Marinos, G. McGrady, and S. M. Wahl. 2003. Conversion of peripheral CD4+CD25– naive T cells to CD4+CD25+ regulatory T cells by TGF-beta induction of transcription factor Foxp3. J Exp Med 198(12):1875–1886.

Claeys, D., G. Faller, B. J. Appelmelk, R. Negrini, and T. Kirchner. 1998. The gastric H+, K+-ATPase is a major autoantigen in chronic Helicobacter pylori gastritis with body mucosa atrophy. Gastroenterology 115:340–347.

Clyne, M., P. Dillon, S. Daly, R. O'Kennedy, F. E. May, B. R. Westley, and B. Drumm. 2004. Helicobacter pylori interacts with the human single-domain trefoil protein TFF1. Proc Natl Acad Sci USA 101(19):7409–7414.

Cong, Y., A. Konrad, N. Iqbal, R. D. Hatton, C. T. Weaver, and C. O. Elson. 2005. Generation of antigen-specific, Foxp3-expressing CD4+ regulatory T cells by inhibition of APC proteosome function. J Immunol 174(5):2787–2795.

Cong, Y., C. T. Weaver, A. Lazenby, and C. O. Elson. 2002. Bacterial-reactive T regulatory cells inhibit pathogenic immune responses to the enteric flora. J Immunol 169(11):6112–6119.

Correa, P. and J. Houghton. 2007. Carcinogenesis of Helicobacter pylori. Gastroenterology 133(2):659–672.

Cover, T. L., U. S. Krishna, D. A. Israel, and R. M. Peek, Jr. 2003. Induction of gastric epithelial cell apoptosis by Helicobacter pylori vacuolating cytotoxin. Cancer Res 63(5):951–957.

Davidson, T. S., R. J. DiPaolo, J. Andersson, and E. M. Shevach. 2007. Cutting edge: IL-2 is essential for TGF-beta-mediated induction of Foxp3+ T regulatory cells. J Immunol 178(7):4022–4026.

Davies, G. R., N. J. Simmonds, T. R. J. Stevens, M. T. Sheaff, N. Banatvala, I. F. Laurenson, D. R. Blake, and D. S. Rampton. 1994. Helicobacter pylori stimulates antral mucosal reactive oxygen metabolite production in vivo. Gut 35:179–185.

Day, Y. J., M. A. Marshall, L. Huang, M. J. McDuffie, M. D. Okusa, and J. Linden. 2004. Protection from ischemic liver injury by activation of A2A adenosine receptors during reperfusion: inhibition of chemokine induction. Am J Physiol Gastrointest Liver Physiol 286(2): G285–G293.

Deaglio, S., K. M. Dwyer, W. Gao, D. Friedman, A. Usheva, A. Erat, J. F. Chen, et al. 2007. Adenosine generation catalyzed by CD39 and CD73 expressed on regulatory T cells mediates immune suppression. J Exp Med 204(6):1257–1265.

D'Elios, M. M., M. Manghetti, M. De Carli, F. Costa, C. T. Baldari, D. Burroni, J. Telford, S. Romagnani, and G. Del Prete. 1997. T helper 1 effector cells specific for Helicobacter pylori in gastric antrum of patients with peptic ulcer disease. J Immunol 158:962–967.

Demple, B. and J. S. Sung. 2005. Molecular and biological roles of Ape1 protein in mammalian base excision repair. DNA Repair (Amst) 4(12):1442–1449.

Denning, T. L., Y. C. Wang, S. R. Patel, I. R. Williams, and B. Pulendran. 2007. Lamina propria macrophages and dendritic cells differentially induce regulatory and interleukin 17-producing T cell responses. Nat Immunol 8(10):1086–1094.

Ding, S. Z., Y. Minohara, X. J. Fan, J. Wang, V. E. Reyes, J. Patel, B. Dirden-Kramer, et al. 2007. Helicobacter pylori infection induces oxidative stress and programmed cell death in human gastric epithelial cells. Infect Immun 75(8):4030–4039.

Ding, S. Z., A. M. O'Hara, T. L Denning, B. Dirden-Kramer, R. C. Mifflin, V. E. Reyes, K. A. Ryan, et al. 2004. Helicobacter pylori and H_2O_2 increases AP endonuclease-1/redox factor-1 expression in human gastric epithelial cells. Gastroenterology 127:845–858.

Duchmann, R., I. Kaiser, E. Hermann, W. Mayet, K. Ewe, and K. H. Meyer zum Buschenfelde. 1995. Tolerance exists towards resident intestinal flora but is broken in active inflammatory bowel disease (IBD). Clin Exp Immunol 102:445–447.

El-Omar, E. M., C. S. Rabkin, M. D. Gammon, T. L. Vaughan, H. A. Risch, J. B. Schoenberg, J. L. Stanford, et al. 2003. Increased risk of noncardia gastric cancer associated with proinflammatory cytokine gene polymorphisms. Gastroenterology 124(5):1193–1201.

Elson, C. O. and Y. Cong. 2002. Understanding immune-microbial homeostasis in intestine. Immunol Res 26(1–3):87–94.

Enarsson, K., A. Lundgren, B. Kindlund, M. Hermansson, G. Roncador, A. H. Banham, B. S. Lundin, and M. Quiding-Jarbrink. 2006. Function and recruitment of mucosal regulatory T cells in human chronic Helicobacter pylori infection and gastric adenocarcinoma. Clin Immunol 121(3):358–368.

Ernst, P. B., D. A. Peura, and S. E. Crowe. 2006. The translation of Helicobacter pylori basic research to patient care. Gastroenterology 130(1):188–206.

Evans, A. R., M. Limp-Foster, and M. R. Kelley. 2000. Going APE over ref-1. Mutat Res 461(2):83–108.

Fan, X. J., S. E. Crowe, S. Behar, H. Gunasena, G. Ye, H. Haeberle, N. Van Houten, et al. 1998. The effect of class II MHC expression on adherence of Helicobacter pylori and induction of apoptosis in gastric epithelial cells: a mechanism for Th1 cell-mediated damage. J Exp Med 187:1659–1669.

Fan, X., H. Gunasena, Z. Cheng, R. Espejo, S. E. Crowe, P. B. Ernst, and V. E. Reyes. 2000. Helicobacter pylori urease binds to class II MHC on gastric epithelial cells and induces their apoptosis. J Immunol 165:1918–1924.

Farinati, F., R. Cardin, P. Degan, M. Rugge, F. D. Mario, P. Bonvicini, and R. Naccarato. 1998. Oxidative DNA damage accumulation in gastric carcinogenesis. Gut 42:351–356.

Ferrero, R. L. 2005. Innate immune recognition of the extracellular mucosal pathogen, Helicobacter pylori. Mol Immunol 42(8):879–885.

Fiorentino, D. F., A. Zlotnik, P. Vieira, T. R. Mosmann, M. Howard, K. W. Moore, and A. O'Garra. 1991. IL-10 acts on the antigen-presenting cell to inhibit cytokine production by Th1 cells. J Immunol 146:3444–3451.

Gaiddon, C., N. C. Moorthy, and C. Prives. 1999. Ref-1 regulates the transactivation and pro-apoptotic functions of p53 in vivo. EMBO J 18:5609–5621.

Guiney, D. G., P. Hasegawa, and S. P. Cole. 2003. Helicobacter pylori preferentially induces interleukin 12 (IL-12) rather than IL-6 or IL-10 in human dendritic cells. Infect Immun 71(7):4163–4166.

Hall, J. L., X. Wang, Adamson Van, Y. Zhao, and G. H. Gibbons. 2001. Overexpression of Ref-1 inhibits hypoxia and tumor necrosis factor- induced endothelial cell apoptosis through nuclear factor-kappa b- independent and -dependent pathways. Circ Res 88(12):1247–1253.

Harrington, L. E., R. D. Hatton, P. R. Mangan, H. Turner, T. L. Murphy, K. M. Murphy, and C. T. Weaver. 2005. Interleukin 17-producing CD4+ effector T cells develop via a lineage distinct from the T helper type 1 and 2 lineages. Nat Immunol 6(11):1123–1132.

Harris, P. R., P. B. Ernst, S. Kawabata, H. Kiyono, M. F. Graham, and P. D. Smith. 1998. Recombinant Helicobacter pylori urease activates primary mucosal macrophages. J Infect Dis 178(5):1516–1520.

Hasko, G. and B. N. Cronstein. 2004. Adenosine: an endogenous regulator of innate immunity. Trends Immunol 25(1):33–39.

Hasko, G., D. G. Kuhel, J. F. Chen, M. A. Schwarzschild, E. A. Deitch, J. G. Mabley, A. Marton, and C. Szabo. 2000. Adenosine inhibits IL-12 and TNF-[alpha] production via adenosine A2a receptor-dependent and independent mechanisms. FASEB J 14(13):2065–2074.

Heneghan, M. A., C. F. McCarthy, D. Janulaityte, and A. P. Moran. 2001. Relationship of anti-Lewis x and anti-Lewis y antibodies in serum samples from gastric cancer and chronic gastritis patients to Helicobacter pylori-mediated autoimmunity. Infect Immun 69:4774–4781.

Hetts, S. W. 1998. To die or not to die. An overview of apoptosis and its role in disease. JAMA 279:300–307.

Huang, S., S. Apasov, M. Koshiba, and M. Sitkovsky. 1997. Role of A2a extracellular adenosine receptor-mediated signaling in adenosine-mediated inhibition of T-cell activation and expansion. Blood 90(4):1600–1610.

Ishikawa, Y., T. Yokoo, and M. Kiamura. 1997. c-Jun/AP-1, but not NF-kB, is a mediator for oxidant-initiated apoptosis in glomerular mesangial cells. Biochem Biophys Res Commun 240:496–501.

Ismail, H. F., J. Zhang, R. G. Lynch, Y. Wang, and D. J. Berg. 2003. Role for complement in development of Helicobacter-induced gastritis in interleukin-10-deficient mice. Infect Immun 71(12):7140–7148.

Izumi, T., W. D. Henner, and S. Mitra. 1996. Negative regulation of the major human AP-endonuclease, a multifunctional protein. Biochemistry 35:14679–14683.

Izumi, T. and S. Mitra. 1998. Deletion analysis of human AP-endonuclease: minimum sequence required for the endonuclease activity. Carcinogenesis 19:525–527.

Janssen, Y. M. W., B. Van Houten, P. J. A. Borm, and B. T. Mossman. 1993. Biology of Disease. Cell and tissue responses to oxidative damage. Lab Invest 69:261–274.

Jayaraman, L., K. G. K. Murthy, C. Zhu, T. Curran, S. Xanthoudakis, and C. Prives. 1997. Identification of redox/repair protein Ref-1 as a potent activator of p53. Genes Dev 11:558–570.

Jones, N. L., P. T. Shannon, E. Cutz, H. Yeger, and P. M. Sherman. 1997. Increase in proliferation and apoptosis of gastric epithelial cells early in the natural history of Helicobacter pylori infection. Am J Pathol 151:1695–1703.

Jump, R. L. and A. D. Levine. 2002. Murine Peyer's patches favor development of an IL-10-secreting, regulatory T cell population. J Immunol 168(12):6113–6119.

Karin, M. 1998. The regulation of AP-1 activity by mitogen-activated protein kinases. J Biol Chem 270:16483–16486.

Karttunen, R., T. Karttunen, H.-P. T. Ekre, and T. T. MacDonald. 1995. Interferon gamma and interleukin 4 secreting cells in the gastric antrum in Helicobacter pylori positive and negative gastritis. Gut 36:341–345.

Kawahara, T., Y. Kuwano, S. Teshima-Kondo, T. Kawai, T. Nikawa, K. Kishi, and K. Rokutan. 2001. Toll-like receptor 4 regulates gastric pit cell responses to Helicobacter pylori infection. J Med Invest 48(3–4):190–197.

Keates, S., Y. S. Hitti, M. Upton, and C. P. Kelly. 1997. Helicobacter pylori infection activates NF-κB in gastric epithelial cells. Gastroenterology 113:1099–1109.

Krajewska, M., C. M. Fenoglio-Preiser, S. Krajewski, K. Song, J. S. Macdonald, G. Stemmerman, and J. C. Reed. 1996. Immunohistochemical analysis of Bcl-2 family proteins in adenocarcinomas of the stomach. Am J Pathol 149:1449–1457.

Kranzer, K., A. Eckhardt, M. Aigner, G. Knoll, L. Deml, C. Speth, N. Lehn, M. Rehli, and W. Schneider-Brachert. 2004. Induction of maturation and cytokine release of human dendritic cells by Helicobacter pylori. Infect Immun 72(8):4416–4423.

Kullberg, M. C., D. Jankovic, P. L. Gorelick, P. Caspar, J. J. Letterio, A. W. Cheever, and A. Sher. 2002. Bacteria-triggered CD4(+) T regulatory cells suppress Helicobacter hepaticus-induced colitis. J Exp Med 196(4):505–515.

Kume, T., K. Oshima, Y. Yamashita, T. Shirakusa, and M. Kikuchi. 1999. Relationship between Fas-ligand expression on carcinoma cell and cytotoxic T-lymphocyte response in lymphoepithelioma-like cancer of the stomach. Int J Cancer 84:339–343.

Kwok, T., D. Zabler, S. Urman, M. Rohde, R. Hartig, S. Wessler, R. Misselwitz, et al. 2007. Helicobacter exploits integrin for type IV secretion and kinase activation. Nature 449(7164):862–866.

Lando, D., I. Pongratz, L. Poellinger, and M. L. Whitelaw. 2000. A redox mechanism controls differential DNA binding activities of hypoxia-inducible factor (HIF) 1a and the HIF-like factor. J Biol Chem 275:4618–4627.

Lappas, C. M., J. M. Rieger, and J. Linden. 2005. A2A adenosine receptor induction inhibits IFN-gamma production in murine CD4+ T cells. J Immunol 174(2):1073–1080.

Lefebvre, O., M. P. Chenard, R. Masson, J. Linares, A. Dierich, M. LeMeur, C. Wendling, et al. 1996. Gastric mucosa abnormalities and tumorigenesis in mice lacking the pS2 trefoil protein. Science 274(5285):259–262.

Linden, J. 2001. Molecular approach to adenosine receptors: receptor-mediated mechanisms of tissue protection. Annu Rev Pharmacol Toxicol 41:775–787.

Lukashev, D., A. Ohta, S. Apasov, J. F. Chen, and M. Sitkovsky. 2004. Cutting edge: physiologic attenuation of proinflammatory transcription by the Gs protein-coupled A2A adenosine receptor in vivo. J Immunol 173(1):21–24.

Lundgren, A., E. Stromberg, A. Sjoling, C. Lindholm, K. Enarsson, A. Edebo, E. Johnsson, et al. 2005. Mucosal FOXP3-expressing CD4+ CD25high regulatory T cells in Helicobacter pylori-infected patients. Infect Immun 73(1):523–531.

Luzza, F., T. Parrello, G. Monteleone, L. Sebkova, M. Romano, R. Zarrilli, M. Imeneo, and F. Pallone. 2000. Up-regulation of IL-17 is associated with bioactive IL-8 expression in Helicobacter pylori-infected human gastric mucosa. J Immunol 165(9):5332–5337.

Lynch, D. A. F., N. P. Mapstone, A. M. T. Clarke, G. M. Sobala, P. Jackson, L. Morrison, M. F. Dixon, P. Quirke, and A. T. R. Axon. 1995. Cell proliferation in Helicobacter pylori associated gastritis and the effect of eradication therapy. Gut 36:346–350.

Mabley, J. G., P. Pacher, L. Liaudet, F. G. Soriano, G. Hasko, A. Marton, C. Szabo, and A. L. Salzman. 2003. Inosine reduces inflammation and improves survival in a murine model of colitis. Am J Physiol Gastrointest Liver Physiol 284(1):G138–G144.

Maloy, K. J., L. Salaun, R. Cahill, G. Dougan, N. J. Saunders, and F. Powrie. 2003. CD4(+)CD25(+) T(R) cells suppress innate immune pathology through cytokine-dependent mechanisms. J Exp Med 197(1):111–119.

Mandell, L., A. P. Moran, A. Cocchiarella, J. Houghton, N. Taylor, J. G. Fox, T. C. Wang, and E. A. Kurt-Jones. 2004. Intact gram-negative Helicobacter pylori, Helicobacter felis, and Helicobacter hepaticus bacteria activate innate immunity via toll-like receptor 2 but not toll-like receptor 4. Infect Immun 72(11):6446–6454.

Mohammadi, M., J. Nedrud, R. Redline, N. Lycke, and S. J. Czinn. 1997. Murine CD4 T-cell response to Helicobacter infection: TH1 cells enhance gastritis and TH2 cells reduce bacterial load. Gastroenterology 113:1848–1857.

Moss, S. F. 1998. Cellular markers in the gastric precancerous process. Aliment Pharmacol Ther 12(Suppl 1):91–109.

Moss, S. F., J. Calam, B. Agarwal, S. Wang, and P. G. Holt. 1996. Induction of gastric epithelial apoptosis by Helicobacter pylori. Gut 38:498–501.

Nagai, S., H. Mimuro, T. Yamada, Y. Baba, K. Moro, T. Nochi, H. Kiyono, et al. 2007. Role of Peyer's patches in the induction of Helicobacter pylori-induced gastritis. Proc Natl Acad Sci USA 104(21):8971–8976.

Naganuma, M., E. B. Wiznerowicz, C. M. Lappas, J. Linden, M. T. Worthington, and P. B. Ernst. 2006. Cutting Edge: critical role for adenosine A_{2A} receptors in the T cell mediated regulation of colitis. J Immunol 177:2765–2769.

Naumann, M., S. Wessler, C. Bartsch, B. Wieland, A. Covacci, R. Haas, and T. F. Meyer. 1999. Activation of activator protein 1 and stress response kinases in epithelial cells colonized by Helicobacter pylori encoding the cag pathogenicity island. J Biol Chem 274:31655–31662.

Negrini, R., L. Lisato, I. Zanella, L. Cavazzini, S. Gullini, V. Villanacci, C. Poiesi, A. Albertini, and S. Ghielmi. 1991. Helicobacter pylori infection induces antibodies cross-reacting with human gastric mucosa. Gastroenterology 101:437–445.

Nishi, T., K. Okazaki, K. Kawasaki, T. Fukui, H. Tamaki, M. Matsuura, M. Asada, et al. 2003. Involvement of myeloid dendritic cells in the development of gastric secondary lymphoid follicles in Helicobacter pylori-infected neonatally thymectomized BALB/c mice. Infect Immun 71(4):2153–2162.

Obst, B., S. Wagner, K. F. Sewing, and W. Beil. 2000. Helicobacter pylori causes DNA damage in gastric epithelial cells. Carcinogenesis 21:1111–1115.

Odashima, M., G. Bamias, J. Rivera-Nieves, J. Linden, C. C. Nast, C. A. Moskaluk, M. Marini, et al. 2005. Activation of A2A adenosine receptor attenuates intestinal inflammation in animal models of inflammatory bowel disease. Gastroenterology 129(1):26–33.

O'Hara, A. M., A. Bhattacharya, J. Bai, R. C. Mifflin, M. F. Smith, Jr., K. A. Ryan, K. G-E. Scott, et al. 2006. Interleukin-8 induction by Helicobacter pylori in human gastric epithelial cells is dependent on apurinic/apyrimidinic endonuclease-1/redox factor-1. J Immunol 177:7990–7999.

Okusa, M. D., J. Linden, L. Huang, J. M. Rieger, T. L. Macdonald, and L. P. Huynh. 2000. A(2A) adenosine receptor-mediated inhibition of renal injury and neutrophil adhesion. Am J Physiol Renal Physiol 279(5):F809–F818.

Pacheco, R., J. M. Martinez-Navio, M. Lejeune, N. Climent, H. Oliva, J. M. Gatell, T. Gallart, et al. 2005. CD26, adenosine deaminase, and adenosine receptors mediate costimulatory signals in the immunological synapse. Proc Natl Acad Sci USA 102(27):9583–9588.

Peek, R. M., Jr., M. J. Blaser, D. J. Mays, M. H. Forsyth, T. L. Cover, S. Y. Song, U. Krishna, and J. A. Pietenpol. 1999. Helicobacter pylori strain-specific genotypes and modulation of the gastric epithelial cell cycle. Cancer Res 59:6124–6131.

Polyak, K., Y. Xia, J. L. Zweier, K. W. Kinzler, and B. Vogelstein. 1997. A model of p53-induced apoptosis. Nature 389:300–305.

Quillet-Mary, A., J. P. Jaffrezou, V. Mansat, C. Bordier, J. Naval, and G. Laurent. 1997. Implication of mitochondrial hydrogen peroxide generation in ceramide-induced apoptosis. J Biol Chem 272:21388–21395.

Rad, R., L. Brenner, S. Bauer, S. Schwendy, L. Layland, C. P. da Costa, W. Reindl, et al. 2006. CD25+/Foxp3+ T cells regulate gastric inflammation and Helicobacter pylori colonization in vivo. Gastroenterology 131(2):525–537.

Rad, R., L. Brenner, A. Krug, P. Voland, J. Mages, R. Lang, S. Schwendy, et al. 2007. Toll-like receptor-dependent activation of antigen-presenting cells affects adaptive immunity to Helicobacter pylori. Gastroenterology 133(1):150–163.

Raghavan, S. and J. Holmgren. 2005. CD4+CD25+ suppressor T cells regulate pathogen induced inflammation and disease. FEMS Immunol Med Microbiol 44(2):121–127.

Ramana, C. V., I. Boldogh, T. Izumi, and S. Mitra. 1998. Activation of apurinic/apyrimidinic endonuclease in human cells by reactive oxygen species and its correlation with their adaptive response to genotoxicity of free radicals. Proc Natl Acad Sci 95:5061–5065.

Rudi, J., D. Kuck, S. Strand, A. Von Herbay, S. M. Mariani, P. H. Krammer, P. R. Galle, and W. Stremmel. 1998. Involvement of the CD95 (APO-1/Fas) receptor and ligand system in Helicobacter pylori-induced gastric epithelial apoptosis. J Clin Invest 102:1506–1514.

Ryter, S. W., H. P. Kim, A. Hoetzel, J. W. Park, K. Nakahira, X. Wang, and A. M. Choi. 2007. Mechanisms of cell death in oxidative stress. Antioxid Redox Signal 9(1):49–89.

Sakai, K., M. Takiguchi, S. Mori, O. Kobori, Y. Morioka, H. Inoko, M. Sekiguchi, and K. Kano. 1987. Expression and function of class II antigens on gastric carcinoma cells and gastric epithelia: differential expression of DR, DQ, and DP antigens. J Natl Cancer Inst 79:923–932.

Sarsfield, P., D. B. Jones, A. C. Wotherspoon, T. Harvard, and D. H. Wright. 1996. A study of accessory cells in the acquired lymphoid tissue of Helicobacter gastritis. J Pathol 180(1):18–25.

Sebkova, L., A. Pellicano, G. Monteleone, B. Grazioli, G. Guarnieri, M. Imeneo, F. Pallone, and F. Luzza. 2004. Extracellular signal-regulated protein kinase mediates interleukin 17 (IL-17)-induced IL-8 secretion in Helicobacter pylori-infected human gastric epithelial cells. Infect Immun 72(9):5019–5026.

Sharma, S. A., M. K. Tummuru, M. J. Blaser, and L. D. Kerr. 1998. Activation of IL-8 gene expression by Helicobacter pylori is regulated by transcription factor nuclear factor-kappa B in gastric epithelial cells. J Immunol 160:2401–2407.

Sher, A., D. Fiorentino, P. Caspar, E. Pearce, and T. R. Mosmann. 1991. Production of IL-10 by CD4+ T lymphocytes correlates with down-regulation of Th1 cytokine synthesis in helminth infection. J Immunol 147:2713–2716.

Shibolet, O. and D. K. Podolsky. 2007. TLRs in the Gut. IV. Negative regulation of Toll-like receptors and intestinal homeostasis: addition by subtraction. Am J Physiol Gastrointest Liver Physiol 292(6):G1469–G1473.

Singh, N., Y. Sun, K. Nakamura, M. R. Smith, and N. H. Colburn. 1995. C-JUN/AP-1 as possible mediators of tumor necrosis factor-a-induced apoptotic response in mouse JB6 tumor cells. Oncol Res 7:353–362.

Sitkovsky, M. V., D. Lukashev, S. Apasov, H. Kojima, M. Koshiba, C. Caldwell, A. Ohta, and M. Thiel. 2004. Physiological control of immune response and inflammatory tissue damage by hypoxia-inducible factors and adenosine A2A receptors. Annu Rev Immunol 22:657–682.

Smith, M. F., Jr., A. Mitchell, G. Li, S. Ding, A. M. Fitzmaurice, K. Ryan, S. E. Crowe, and J. B. Goldberg. 2003. TLR2 and TLR5, but not TLR4, are required for Helicobacter pylori-induced NF-kappa B activation and chemokine expression by epithelial cells. J Biol Chem 278(35):32552–32560.

Smith, P. D., C. Ochsenbauer-Jambor, and L. E. Smythies. 2005. Intestinal macrophages: unique effector cells of the innate immune system. Immunol Rev 206:149–159.

Su, B., P. J. Ceponis, S. Lebel, H. Huynh, and P. M. Sherman. 2003. Helicobacter pylori activates Toll-like receptor 4 expression in gastrointestinal epithelial cells. Infect Immun 71:3496–3502.

Tebbutt, N. C., A. S. Giraud, M. Inglese, B. Jenkins, P. Waring, F. J. Clay, S. Malki, et al. 2002. Reciprocal regulation of gastrointestinal homeostasis by SHP2 and STAT-mediated trefoil gene activation in gp130 mutant mice. Nat Med 8(10):1089–1097.

Tell, G., A. Zecca, L. Pellizzari, P. Spessotto, A. Colombatti, M. R. Kelley, G. Damante, and C. Pucillo. 2000. An "environment to nucleus" signaling system operates in B lymphocytes: redox status modulates BSAP/Pax-5 activation through Ref-1 nuclear translocation. Nucleic Acids Res 28:1099–1105.

Thomas, P. G., M. R. Carter, A. A. Da'dara, T. M. DeSimone, and D. A. Harn. 2005. A helminth glycan induces APC maturation via alternative NF-kappa B activation independent of I kappa B alpha degradation. J Immunol 175(4):2082–2090.

Viala, J., C. Chaput, I. G. Boneca, A. Cardona, S. E. Girardin, A. P. Moran, R. Athman, et al. 2004. Nod1 responds to peptidoglycan delivered by the Helicobacter pylori cag pathogenicity island. Nat Immunol 5(11):1166–1174.

Wagner, S., W. Beil, J. Westermann, R. P. Logan, C. T. Bock, C. Trautwein, J. S. Bleck, and M. P. Manns. 1997. Regulation of gastric epithelial cell growth by Helicobacter pylori: evidence for a major role of apoptosis. Gastroenterology 113(6):1836–1847.

Wakkach, A., N. Fournier, V. Brun, J. P. Breittmayer, F. Cottrez, and H. Groux. 2003. Characterization of dendritic cells that induce tolerance and T regulatory 1 cell differentiation in vivo. Immunity 18(5):605–617.

Walker, L. J., R. B. Craig, A. L. Harris, and I. D. Hickson. 1994. A role for the human DNA repair enzyme HAP1 in cellular protection against DNA damaging agents and hypoxic stress. Nucleic Acids Res 22:4884–4889.

Wang, J., X. J. Fan, C. Lindholm, M. Bennet, J. O'Connell, F. Shanahan, E. G. Brooks, V. E. Reyes, and P. B. Ernst. 2000. Helicobacter pylori modulates lymphoepithelial cell interactions leading to epithelial cell damage through Fas/Fas Ligand interactions. Infect Immun 68:4303–4311.

Weiner, H. L. 2001. Induction and mechanism of action of transforming growth factor-beta-secreting Th3 regulatory cells. Immunol Rev 182:207–214.

Wilson, K. T. and J. E. Crabtree. 2007. Immunology of Helicobacter pylori: insights into the failure of the immune response and perspectives on vaccine studies. Gastroenterology 133(1):288–308.

Wiseman, H. and B. Halliwell. 1996. Damage to DNA by reactive oxygen and nitrogen species: role in inflammatory disease and progression to cancer. Biochem J 313:17–29.

Wu, M. S., S. P. Huang, Y. T. Chang, C. T. Shun, M. C. Chang, M. T. Lin, H. P. Wang, and J. T. Lin. 2002. Tumor necrosis factor-alpha and interleukin-10 promoter polymorphisms in Epstein-Barr virus-associated gastric carcinoma. J Infect Dis 185(1):106–109.

Wu, M.-S., C.-T. Shun, H.-P. Wang, J.-C. Sheu, W.-J. Lee, T.-H. Wang, and J.-T. Lin. 1997. Genetic alterations in gastric cancer: Relation to histological subtypes, tumor stage, and Helicobacter pylori infection. Gastroenterology 112:1457–1465.

Wyllie, A. H. 1997. Apoptosis and carcinogenesis. Eur J Cell Biol 73:189–197.

Xanthoudakis, S., G. G. Miao, and T. Curran. 1994. The redox and DNA-repair activities of Ref-1 are encoded by nonoverlapping domains. Proc Natl Acad Sci USA 91(1):23–27.

Xanthoudakis, S., G. Miao, F. Wang, Y.-C. E. Pan, and T. Curran. 1992. Redox activation of Fos-Jun DNA binding activity is mediated by a DNA repair enzyme. EMBO J 11:3323–3335.

Xanthoudakis, S., R. J. Smeyne, J. D. Wallace, and T. Curran. 1996. The redox DNA repair protein, Ref-1, is essential for early embryonic development in mice. Proc Natl Acad Sci 93:8919–8923.

Yahiro, K., A. Wada, M. Nakayama, T. Kimura, K. Ogushi, T. Niidome, H. Aoyagi, et al. 2003. Protein-tyrosine phosphatase alpha, RPTP alpha, is a Helicobacter pylori VacA receptor. J Biol Chem 278(21):19183–19189.

Yamasaki, E., A. Wada, A. Kumatori, I. Nakagawa, J. Funao, M. Nakayama, J. Hisatsune, et al. 2006. Helicobacter pylori vacuolating cytotoxin induces activation of the proapoptotic proteins Bax and Bak, leading to cytochrome c release and cell death, independent of vacuolation. J Biol Chem 281(16):11250–11259.

Ye, G., C. Barrera, X. J. Fan, W. K. Gourley, S. E. Crowe, P. B. Ernst, and V. E. Reyes. 1997. Expression of B7-1 and B7-2 costimulatory molecules by human gastric epithelial cells. Potential role in CD4[+] T cell activation during Helicobacter pylori infection. J Clin Invest 99:1628–1636.

Yokota, K., K. Kobayashi, Y. Kawahara, S. Hayashi, Y. Hirai, M. Mizuno, H. Okada, et al. 1999. Gastric ulcers in SCID mice induced by Helicobacter pylori infection after transplanting lymphocytes from patients with gastric lymphoma. Gastroenterology 117:893–899.

Yoshimura, S., J. Bondeson, B. M. Foxwell, F. M. Brennan, and M. Feldmann. 2001. Effective antigen presentation by dendritic cells is NF-kappaB dependent: coordinate regulation of MHC, co-stimulatory molecules and cytokines. Int Immunol 13(5):675–683.

Zhou, P., R. Borojevic, C. Streutker, D. Snider, H. Liang, and K. Croitoru. 2004. Expression of dual TCR on DO11.10 T cells allows for ovalbumin-induced oral tolerance to prevent T cell-mediated colitis directed against unrelated enteric bacterial antigens. J Immunol 172(3):1515–1523.

Zhou, L., I. I. Ivanov, R. Spolski, R. Min, K. Shenderov, T. Egawa, D. E. Levy, W. J. Leonard, and D. R. Littman. 2007. IL-6 programs T(H)-17 cell differentiation by promoting sequential engagement of the IL-21 and IL-23 pathways. Nat Immunol 8(9):967–974.

Chapter 18
Atrophy and Altered Mesenchymal–Epithelial Signaling Preceding Gastric Cancer

Juanita L. Merchant and Yana Zavros

Introduction

Chronic inflammation in the stomach (gastritis) induces gastric atrophy characterized by the loss of the acid-secreting oxyntic glands. Although the initial report of phenotypic changes that precede gastric cancer suggest that these steps occur sequentially (Correa et al. 1975), whether atrophy precedes metaplasia or occurs concurrently is unclear. Nevertheless, the metaplastic cell that begins to repopulate the gastric epithelium under hypochlorhydric conditions is a mucous cell of gastric or intestinal origin (Goldenring and Nomura 2006; Kang et al. 2005; Nomura et al. 2004; Oshima et al. 2006). Although *Helicobacter pylori* infection is the major reason chronic inflammation develops in the stomach, the molecular networks linking chronic inflammation to the atrophic/metaplastic changes are not well understood. This chapter reviews the causes of gastric atrophy and highlights the role that some activated signaling pathways have in committing the mucosa to neoplastic transformation.

Causes of Gastric Atrophy

The chronic lack of gastric acid secretion generally indicates significant loss of parietal cell mass. Coincident with parietal cell atrophy is also the inability to activate acid-dependent proteases, e.g., pepsinogens A and C. In the original Correa model, it was reported that gastric atrophy was one of several changes that occurred with chronic inflammation (Correa et al. 1975). We now understand that the major cause of chronic inflammation in the normal, acid-secreting stomach is *H. pylori*, a spiral organism that has adapted itself to survival in the hostile environment of the stomach (Peek and Crabtree 2006). Chronic infection by *H. pylori* induces an inflammatory response that eventually results in gastric atrophy and low or absent acid secretion (hypo- or achlorhydria). The lack of acid secretion removes *H. pylori*'s unique niche and allows other organisms to compete with it for the gastric mucosal niche. Because the number and type of organisms that eventually colonize the stomach are too numerous to classify, repopulation of the stomach with these generally aerobic, rapidly growing bacteria has simply been described as bacterial overgrowth.

T.C. Wang et al. (eds.) *The Biology of Gastric Cancers*,
© Springer Science + Business Media, LLC 2009

449

The Correa Paradigm

The human stomach is separated into four major regions that include the cardia, fundus, body, and antrum. The fundus and body contain the acid-secreting parietal cells and zymogenic or chief cells, whereas the antrum contains the gastrin-producing G cells (Figure 18.1). Gastric cancer in humans develops in either the proximal (cardia) or distal (antrum) regions of the stomach, with the distal tumors characteristically associated with *H. pylori* infection (El-Zimaity et al. 2002; Shiotani et al. 2005b). However, in the United States, it is cancer of the gastric cardia that is on the increase (Corley and Kubo 2004). There are two distinct histopathologic subtypes of gastric cancer—intestinal-type (well differentiated) and diffuse-type (undifferentiated). Intestinal-type gastric cancer predominates in high-risk populations and is preceded by well-characterized precancerous lesions. Diffuse cancer is frequent in low-risk populations and is not preceded by a series of well-characterized histologic changes (Blair et al. 2006). The precancerous lesions of the intestinal type represent a series of stages that culminate in neoplastic transformation to cells that retain a glandular morphology (Correa 1992) (Figure 18.2). The first step consists of chronic inflammation that is typically triggered by a bacterial infection such as *H. pylori* (Blaser and Parsonnet 1994) or possibly bacterial overgrowth (Stockbruegger et al. 1984; Williams and McColl 2006; Zavros et al.

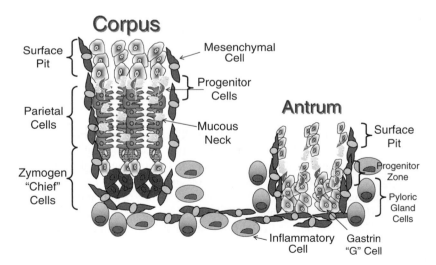

Fig. 18.1 The human stomach is composed of four regions: cardia, fundus, corpus (body), and antrum. The glands of the cardia and fundus are similar to the antrum. However, the mouse stomach is composed of a forestomach comprising squamous epithelium such as the esophagus, corpus, and antrum. The cells comprising the glands of the murine corpus and antrum are the same as in the human. For the sake of simplicity, the murine stomach and antrum are modeled here. (Drawn by Meghna Waghray, University of Michigan.)

2002b). After several years of residual, smoldering inflammation, the intermediate precancerous changes of parietal cell atrophy and intestinal metaplasia emerge.

Although gastric atrophy and intestinal metaplasia are terms that are often used synonymously, they are distinct events. Intestinal metaplasia is typically defined as the replacement of gastric glands with glandular structures lined by cells normally found in the small intestine, e.g., goblet, Paneth, enteroendocrine cells, and enterocytes (Mesquita et al. 2006). In short, a phenotypically normal intestinal cell is present in the wrong place (stomach). Matsukura and coworkers proposed the original definition of intestinal metaplasia in 1980 on the basis of mucin secretion patterns that only distinguished between small and large intestine phenotypes (Matsukura et al. 1980) using immunohistochemical determination of the disaccharides (Kawachi et al. 1974). More recently, Tsukamoto has considered the

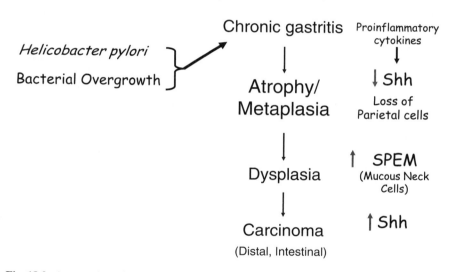

Model of Gastric Neoplasia

Helicobacter pylori

Bacterial Overgrowth

Chronic gastritis   Proinflammatory cytokines

↓

Atrophy/ Metaplasia   ↓ Shh   Loss of Parietal cells

↓

Dysplasia   ↑ SPEM (Mucous Neck Cells)

↓

Carcinoma (Distal, Intestinal)   ↑ Shh

Fig. 18.2 An overview of the paradigm for intestinal gastric cancer. In this model modified from the original report by Correa, *Helicobacter pylori* is the major trigger of chronic gastritis in the acid-secreting stomach (Correa et al. 1975). However, there remains the possibility that *H. pylori* loses its niche in the hypochlorhydric stomach, permitting bacterial overgrowth to occur and promote gastritis. Regardless of the trigger, the stomach responds to foreign bacteria by generating a Th1 proinflammatory cytokine response, which typically includes the production of interferon-γ from T cells and interleukin-1β and tumor necrosis factor-α from tissue or invading macrophages. Hypochlorhydria can occur even in the absence of parietal cell loss (atrophy). Loss of parietal cells in the body (corpus) of the stomach coincides with an increase in metaplastic cells. Characteristically, the metaplastic cell in the corpus is phenotypically an antral cell (mucous or pseudopyloric gland) or intestinal cell (villin-positive enterocyte, mucous goblet cell, lysozyme-positive Paneth cell). Some time after these changes occur, dysplastic cells arise within the atrophic/metaplastic cell environment generally in the distal or antral portion of the stomach. Shh: sonic hedgehog; SPEM: spasmolytic peptide-expressing metaplasia

gastric and intestinal properties to propose a new classification of metaplasia-based phenotypic markers that better indicate cell differentiation status (Tsukamoto et al. 2006). Tsukamoto et al. divide the label "intestinal metaplasia" into two types—a mixed phenotype consisting of cells with gastric and intestinal properties (GI), and exclusively an intestinal phenotype (I). The mixed GI phenotype contains antral-pyloric–type cells that express the class III mucin, Muc6 and pepsinogen (II or C) reactivity. These cells will also be trefoil factor (TFF)2 positive. Tsukamoto, therefore, equates the mixed GI phenotype with what has been described as incomplete metaplasia. "G"-type metaplasia are glands with pyloric cells and no goblet cells and are likely equivalent to what has been called spasmolytic peptide-expressing metaplasia (SPEM) (Schmidt et al. 1999).

El-Zimaity et al. (2002), in their histopathologic analysis, referred to the antral-type metaplasia as pseudopyloric or mucous gland metaplasia. Pseudopyloric gland metaplasia is a gastric mucous cell that is phenotypically of antral origin appearing in the corpus (Wright 1996). Goldenring and coworkers described the presence of a mucous cell lineage or SPEM in *H. pylori*–infected subjects that appears to arise from the base of the corpus glands (Halldorsdottir et al. 2003; Nomura et al. 2004; Schmidt et al. 1999). Spasmolytic peptide is the original term for TFF2. TFF2 and the mucin Muc6 are expressed in the mucous neck lineage, pyloric glands of the antrum, and Brunner's glands of the small intestine. Mucous neck cells are precursors of the chief cell lineage in the corpus (Karam and Leblond 1993b; Ramsey et al. 2007). Therefore, SPEM likely represents the arrested differentiation program of mucous cells transitioning into chief cells and would explain why the appearance of SPEM at the base of the gland tends to correlate with chief cell atrophy.

Classically, chronic atrophic gastritis is grouped into two categories according to the site of the inflammation and etiology, i.e., autoimmune- versus inflammatory-based (Table 18.1) (Whittingham and Mackay 2005). In autoimmune (type A) gastritis, the corpus has atrophied, whereas in "bacterial" (type B) gastritis, corpus atrophy is a late event. Type A atrophic gastritis is attributed to T-cell–mediated destruction of parietal cells, which causes a deficiency of intrinsic factor produced by human parietal cells. The lack of intrinsic factor–mediated vitamin B12 absorption required for proper erythrocyte develop-

Table 18.1 Types of chronic atrophic gastritis

Type	Characteristic
Type A	Atrophy in the fundus and body
	Development of antibodies to parietal cells/intrinsic factor
	Low serum pepsinogen A concentrations
	Hypochlorhydria (low gastric acidity)
	Hypergastrinemia
	Vitamin B12 deficiency associated with pernicious anemia
Type B	Atrophy late and variable; appearing in antrum > corpus
	Loss of parietal and chief cells
	Helicobacter pylori infection
	Gastrin levels variable

ment is the mechanism underlying the development of pernicious anemia. The stomachs of patients with pernicious anemia develop a mononuclear cell infiltrate of T, B, and plasma cells in the submucosa extending into the lamina propria between the gastric glands (Kaye et al. 1983; Toh et al. 1997). The infiltrating plasma cells contain autoantibodies directed toward parietal cell autoantigens H^+, K^+-ATPase and intrinsic factor (Baur et al. 1968). As a result, the pathologic lesion of type A gastritis is restricted to the fundus and body regions of the stomach where the loss of parietal and zymogenic cells is replaced by metaplastic mucus-containing cells (Toh et al. 1997). In addition, the loss of parietal cells results in achlorhydria, low pepsinogen A levels, hypergastrinemia, and hyperplasia.

By contrast, type B gastritis affects the antrum and corpus/body often in association with *H. pylori* infection, which usually begins in the antrum and spreads proximally to the corpus (Bayerdorffer et al. 1992; Rugge et al. 1993). In a histologic mapping study of the entire human stomach, El-Zimaity et al. (2002) described the pattern of gastric atrophy that arises in *H. pylori*–infected subjects. In that study, the entire stomach of 16 patients was fixed and analyzed by immunohistochemistry. Most of the specimens (14/16) exhibited atrophic changes beginning at the lesser curvature and progressing into the corpus with a tendency to spare the greater curvature. The antral pattern of atrophy (2/16 cases) exhibited nearly complete replacement of the antrum with intestinal metaplastic cells but retained a nonatrophic corpus. In all cases, the cancers arose in the atrophic zones, 8/16 in the antrum, 6/16 in the corpus, and 2/16 at the incisura. Whereas the cancer site varied in the specimens with atrophic corpuses, in the two cases with atrophic antrums, the tumors arose at the antral–corpus junction and in the distal antrum. This study reconfirms the generally understood principle that gastric atrophy is the critical step possibly irreversible in the development of intestinal-type gastric cancer because loss of the normal cellular phenotype accompanies metaplastic replacement. Therefore, a more enlightened definition of atrophy is described anatomically by El-Zimaity et al. (2002) and molecularly defined by Tsukamoto et al. (2006). In short, it is the loss of normal gastric glands and gradual replacement with both types of metaplasia, i.e., intestinal and gastric (pseudopyloric), progressing outward from the lesser curvature of the antral–corpus border (incisura) into the antrum, corpus, or both.

Causes of Parietal Cell Atrophy

Although several factors may influence an individual's predisposition to gastric cancer, the reason why chronic inflammation suppresses long-lived parietal and zymogenic cells is unclear. The turnover time for parietal cells is 54 days and 194 for zymogenic cells (Karam 1993; Karam and Leblond 1993b). By contrast, the surface pit cells turn over every 3–5 days and mucous neck cells about every 7–14 days (Karam and Leblond 1993a).

Hypochlorhydria—Cause or Effect?

H. pylori is a gram-negative organism that survives in the deep mucous layer of the stomach and attaches to the epithelial cells (Correa 1995; Hofman et al. 2004). It is estimated that half of the world's population is infected with *H. pylori*, making it perhaps the most common chronic infection in humans. Colonization by *H. pylori* is currently the most recognized cause of inflammatory changes in the human stomach (Blaser and Parsonnet 1994). In addition, *H. pylori* infection has long been associated with altered acid secretion, which is a function of both the severity and location of the gastritis (Blaser and Parsonnet 1994; McColl et al. 1998). *H. pylori* infection causes type B gastritis and is associated with variable gastrin levels. On the one hand, the inflammatory cytokines produced during the infection can increase gastrin production and subsequently induce hyperacidity as a consequence of parietal cell stimulation (McColl 2006; Whittingham and Mackay 2005). On the other hand, longstanding inflammation induces metaplastic changes that in the corpus might result in parietal cell atrophy and hypergastrinemia from the hypochlorhydria. However, if antral G cells have also succumbed to the chronic inflammation and have been replaced by metaplastic cells, hypochlorhydria will result (El-Zimaity et al. 2002; Tsukamoto et al. 2006). In mice, we found that elevated gastrin levels might be attributable to cytokine stimulation of gastrin initially, then suppression of somatostatin, eventually permitting enhanced acid secretion (Zavros et al. 2003). However, chronically, there might be normal or reduced plasma gastrin levels because of feedback activation of somatostatin caused by hyperacidity that subsequently inhibits gastrin.

Whereas it is known that hypochlorhydria and parietal cell atrophy are precursors for the development of gastric cancer, the consequences of long-term drug-induced hypochlorhydria are unclear. The addition of antisecretory drugs to antimicrobial therapy is considered the standard of care for *H. pylori*–infected patients with active or inactive peptic ulcer. In developed countries, eradication therapy is successful in approximately 90% of cases. In the developing world, eradication is more difficult because of the frequency of antibiotic resistance and recurrent infection (Graham 1998). Currently in the United States, triple-drug regimens have yielded the best eradication rates (Chey and Wong 2007). These regimens usually consist of combining a proton pump inhibitor (PPI) (e.g., lansoprazole, omeprazole) with amoxicillin and clarithromycin for 10–14 days. The addition of bismuth subsalicylate for quadruple therapy is another option. More recently, levofloxacin is being substituted for clarithromycin because of increasing antibiotic resistance (Gisbert et al. 2007). However, when gastric acid is inhibited by PPIs, there is the risk of bacterial overgrowth that flourishes because of the relatively alkaline stomach pH probably crowding out and eventually excluding the slower-growing *H. pylori* organisms. In humans, long-term PPI treatment and H2 receptor-blocker treatment have been associated with an increased risk of intestinal metaplasia and is the primary rationale for linking these drugs to an increased risk for gastric cancer (Klinkenberg-Knol et al. 2000). Yet, in the nearly 30 years since these drugs have been available, gastric cancer has never been a confirmed complication of their use (Lamberts et al. 2001;

Rindi et al. 2005). In a mouse model of PPI-induced bacterial overgrowth, the major species detected were *Lactobacillus*, *Enterobacter*, *Staphylococcus*, and *Propionibacterium* (Zavros et al. 2002b). A similar broad spectrum of organisms are also detected in human subjects treated with PPIs (Williams and McColl 2006). Thus, hypochlorhydria alone might not be a prerequisite for gastric cancer but might require synergy with additional risk factors such as high salt diet, inflammation, and bacterial products (N-nitrosamines) (Fox and Wang 2007).

Recent studies have suggested that acidic conditions might also be required for the gene expression of factors that are critical in the growth and differentiation of the gastric mucosa. For example, acidic conditions increase the expression of sonic hedgehog (Shh) in gastric cell lines (Dimmler et al. 2003). Shh, a member of a family of developmental peptides, is critical for patterning and growth in a number of tissues during embryogenesis (Fietz et al. 1994). Nevertheless, it has been shown that Shh is a secreted signal important in the maintenance of oxyntic glands (van den Brink et al. 2001, 2002). Immunohistochemical data show that the lesions that develop in intestinal metaplasia display reduced expression of Shh (Shiotani et al. 2005a). Whereas Shh is expressed in normal gastric epithelium, expression of the intestine-specific caudal-type homeobox transcription factors (Cdx) 1 and 2 is normally absent. However, forced gastric expression of these genes is associated with intestinal metaplasia and cancer (Mutoh et al. 2004a, 2004b; Silberg et al. 2002). *In vitro*, acid conditions suppress the expression of Cdx2 in CaCo2 cells (Faller et al. 2004) suggesting that acidic conditions might also modulate gene expression to prevent the appearance of intestinal metaplasia.

Inflammatory Cytokines

The first step in the Correa model consists of a chronic active inflammatory response that is usually triggered by a bacterial infection such as *H. pylori*. The type of T helper (Th) lymphocytes that are recruited to the site of the infection are characterized by the type of cytokines they secrete. Th type 1 (Th1) cells are CD4-positive lymphocytes that mediate the innate immune response and produce, e.g., interferon (IFN)-γ, interleukin (IL)-2, tumor necrosis factor (TNF)-α cytokines. By contrast, the adaptive immune response is characterized by the presence of CD4-positive Th type 2 (Th2) cells that typically secrete, e.g., IL-4, -5, -6, -10, and -13 to recruit B lymphocytes that differentiate into immunoglobulin-secreting plasma cells.

Infection by *H. pylori* typically induces an innate (Th1) immune response that is characterized by an initial infiltration of neutrophils and production of IFN-γ, TNF-α, and IL-1β (Bamford et al. 1998). In addition, Oshima et al. showed that transgenic mice simultaneously expressing COX-2 and microsomal prostaglandin E synthase-1 in the gastric epithelium also develop metaplasia, hyperplasia, and cancer because of induced macrophage infiltration (Oshima et al. 2004). El-Omar and colleagues have shown that polymorphisms in the IL-1β and TNF-α gene loci are associated with gastric atrophy (El-Omar et al. 2000, 2003; Furuta et al. 2002; Hou et al. 2007; Macarthur et al. 2004). More extensive analysis of the IL-1β base pair change has shown that this promoter polymorphism results in higher cytokine levels. IFN-γ is believed to play a pivotal role in initiating the mucosal damage observed during gastritis. For example, exogenous infusion of only IFN-γ into mice for 7 days is sufficient to induce significant mucous gland metaplasia and hypergastrinemia (Zavros et al. 2003). Collectively, these results suggest that the mucosal changes observed during bacterial infection are attributable in part to Th1 cytokines.

Not all models of gastric cancer develop in the setting of atrophy. A mouse model of particular interest is one in which point mutations in a cytoplasmic phosphorylation domain of the gp130 cytokine receptor (gp130$^{757F/F}$) results in overstimulation of the signal transducer and activator of transcription 3 (STAT3) signaling pathway (Giraud et al. 2007; Howlett et al. 2005; Judd et al. 2004). These mice develop antral gastric cancer within 2 months without evidence of parietal cell atrophy, although they are hypogastrinemic probably from the displacement of the G cells by tumor (Judd et al. 2004, 2006). Subsequent analysis of this mouse model has revealed that elevated STAT3 signaling desensitizes the cytostatic effect of the transforming growth factor (TGF) β pathway by transcriptional induction of the inhibitory SMAD7 protein (Jenkins et al. 2005). Indeed, the crosstalk between the IFN-γ/STAT pathway and the TGFβ/SMAD pathway has been studied in other tissues (Sobral et al. 2007). Inferred primarily from studies performed in liver, TGFβ typically is expressed in the mesenchyme where it stimulates myofibroblasts migration and extracellular matrix production (Bataller and Brenner 2005; Uhal et al. 2007). Disruption of the TGFβ pathway or its downstream targets, e.g., RUNX3, promotes gastric tumor formation (Bae and Choi 2004; Friedrich et al. 2006; Jenkins et al. 2005; Kim et al. 2006; Li et al. 2002; Mishra et al. 2005; Zavros et al. 2005). Thus, because expression of TGFβ and its signaling network are expressed in the mesenchyme (Powell et al. 1999a, 1999b), we deduced that the loss of TGFβ signaling promotes transformation by disturbing the normal mesenchymal-to-epithelial crosstalk (Ahmed and Nawshad 2007; Sheehan et al. 2007).

Mouse Models of Gastric Atrophy

Bacterial Infection

Because *H. pylori* infection is a major cause of human gastric cancer, many investigators have developed animal models of *Helicobacter* infection to recapitulate the human phenotype (Table 18.2). Although the *H. pylori*–infected Mongolian gerbil model has been shown to induce gastric atrophy resulting in carcinoma (Hirayama et al. 1999; Honda et al. 1998; Watanabe et al. 1998), mice chronically infected with *H. pylori* seem to be resistant to neoplastic transformation. By contrast, mice chronically infected with *H. felis* demonstrate significant parietal cell atrophy and replacement of fundic mucosa with mucous cell metaplasia (Fox et al. 1996; Wang et al. 1998). Moreover, mutant mice with elevated gastrin levels develop corpus tumors within 15 months when infected with *H. felis* (Wang et al. 2000).

Table 18.2 Mouse models of gastric atrophy

Mouse model	Gastric phenotype	References
Bacterial infection		
Helicobacter pylori infection	• Inflammation	
	• Parietal cell atrophy	
	• No gastric cancer	
	• Inflammation	
H. felis infection	• Gastric atrophy, intestinal metaplasia, dysplasia, gastric cancer	
Bacterial overgrowth caused by achlorhydria (gastrin-deficient mice)	• Inflammation (increase interferon-γ expression)	(Zavros et al. 2002b, 2005)
	• Parietal cell atrophy	
	• Intestinal metaplasia	
	• Antral dysplasia	
	• Antral carcinoma after 12 months of age	
Genetic manipulation		
H^+, K^+-ATPase β subunit knockout	• Mucosal hypertrophy	(Scarff et al. 1999)
	• Mononuclear cell in lamina propria and submucosa	
	• Loss of parietal and chief cells	
	• Hypochlorhydria	
	• Hypergastrinemia	
NHE2-deficient mice	• Loss of parietal and zymogenic cells	(Schultheis et al. 1998)
	• Hypochlorhydria	
	• Hypergastrinemia	
Ins-GAS	• Hypergastrinemic	(Wang et al. 1995, 2000)
	• Hyperacidity	
	• Parietal cell atrophy	
H^+, K^+-diphtheria toxin	• Expansion of preparietal cells	(Li et al. 1996)
	• Dysplasia at 12 months of age	
	• Hypochlorhydria	
Gastrin-deficient mice	• Inflammation	(Zavros et al. 2007)
	• Decreased sonic hedgehog (shh) protein expression in fundus	
	• Gastric atrophy, metaplasia, dysplasia	
	• Antral carcinoma after 12 months of age	
Toxicity		
DMP-777 treatment	• Absence of inflammatory infiltrate	(Barrett et al. 1996)
	• Gastric atrophy	
	• SPEM induction	

(continued)

Table 18.2 (continued)

Mouse model	Gastric phenotype	References
Autoimmune gastritis		
H⁺, K⁺-ATPase–Cholera toxin mouse	• H⁺, K⁺-ATPase autoantibodies • Mucous neck cell hyperplasia • Parietal cell atrophy • Hyperacidity • Reduced plasma gastrin levels	(Lopez-Diaz et al. 2006)
Thymectomy model	• Submucosal infiltration of mono-nuclear cells into the lamina propria • Parietal cell atrophy • Loss of zymogenic (chief) cells	(Gleeson et al. 1996; Kojima and Prehn 1981; Smith et al. 1992) (Barrett et al. 1996)

Inflammation generated in mutant mice in the absence of *Helicobacter* spp. can induce neoplastic transformation with or without gastric atrophy (Judd et al. 2004; Zavros et al. 2005). In particular, hypochlorhydric mice, because of genetically induced gastrin deficiency, develop severe inflammation from bacterial overgrowth that resolves with antibiotics (Zavros et al. 2002b). Chronically, persistent inflammation in gastrin-deficient mice causes atrophic gastritis, metaplasia, and eventually dysplasia (Zavros et al. 2005), which follows the sequence of events observed in the human model of gastric cancer (Correa et al. 1975).

Genetic Manipulation

There are a number of mouse models that have altered parietal cell function and develop gastric atrophy from genetic manipulation. One example is targeted deletion of the H⁺, K⁺-ATPase β subunit. These hypochlorhydric mice develop inflammation and hypergastrinemia (Scarff et al. 1999). Conditional deletion of the Na⁺, H⁺ isoform 2 exchanger (NHE2) in parietal cells blocks acid secretion and causes hypergastrinemia in the setting of a dramatic loss of parietal and zymogenic cells (Schultheis et al. 1998). Whereas the H⁺, K⁺-ATPase β subunit and NHE2 knockout mice are models of hypochlorhydria, the Ins-Gas mouse, which overexpresses human gastrin from the insulin promoter, initially exhibits chronically elevated gastrin levels and hyperacidity then subsequently gastric atrophy after 1 year (Wang et al. 2000). Parietal cell ablation models such as the tissue-specific expression of diphtheria toxin initially results in expansion of preparietal cells in the gastric glands then dysplasia after 1 year (Li et al. 1996). As in humans, the extended timeframe observed until neoplasia develops in mice suggests the convergence of several signaling pathways that generally couple the secretion of inflammatory mediators with defects in cellular differentiation.

Chemical Ablation of Parietal Cells

Mice treated with the orally active, cell-permanent neutrophil elastase inhibitor DMP-777 represent a toxicity model of gastric atrophy that does not exhibit inflammation (Nomura et al. 2005b). DMP-777–treated mice develop SPEM, which is characterized by the presence of TFF2-immunoreactive cells in the corpus (Schmidt et al. 1999). Such a model supports the concept that mucosal factors other than inflammatory mediators may modulate the development of gastric atrophy and metaplasia. However, chronic inflammation ultimately seems to be required for the development of frank neoplasia.

Autoimmune G astritis

There have been several mouse models that recapitulate autoimmune gastritis. Autoimmune disease including gastritis develops in BALB/c mice after thymectomy or cyclosporine treatment (Barrett et al. 1995; Gleeson et al. 1996; Kojima and Prehn, 1981; Smith et al. 1992). In the absence of the thymus, naïve T cells do not receive the appropriate instructive signals to distinguish between foreign and self-antigens, a function that is mediated by IFN-γ–secreting Th1-type CD4 T cells (Barrett et al. 1996). Thymectomy-induced autoimmune gastritis such as pernicious anemia is characterized by submucosal infiltration of mononuclear cells extending into the lamina propria, parietal and zymogenic cell atrophy (Barrett et al. 1995; Gleeson et al. 1996; Kojima and Prehn 1981; Smith et al. 1992).

More recently, Lopez-Diaz et al. developed a transgenic mouse model that targets cholera toxin expression to the parietal cell (Ctox) using the H^+, K^+-ATPase promoter (Lopez-Diaz et al. 2006). Because of adenosine 5'-diphosphate ribosylation of elongation factors by the toxin, these mice express constitutively high levels of adenosine 3',5'-cyclic monophosphate, resulting in hyperstimulation of the parietal cell to secrete acid. These mice have elevated basal gastric acidity and a compensatory reduction in plasma gastrin. Interestingly, after 6 months of age, the Ctox mice develop parietal cell atrophy and antiparietal cell antibodies that are consistent with the gastric phenotype observed in patients with autoimmune gastritis (Lopez-Diaz et al. 2006). Similar to the Ins-Gas mouse, the hyperstimulated parietal cell atrophies in mice older than 1 year. Thus, it would be interesting to test whether *Helicobacter* infection of Ctox mice also results in accelerated atrophy and tumor formation.

Zymogenic (Chief) Cell Atrophy

In addition to acid-secreting parietal cells, zymogen-secreting chief cells are the other differentiated cell type that disappear with atrophy and intestinal metaplasia (Houghton et al. 2004; Nomura et al. 2005b; Shiotani et al. 2005b). One plausible molecular explanation for the loss of chief cells during gastric atrophy is Mist1 (Bhlhb8), a class B basic helix-loop-helix transcription factor (Ramsey et al. 2007). Typically expressed in specialized secretory cell lineages such as acinar cells of the salivary gland and pancreas (Johnson et al. 2004; Pin et al. 2000), Mist1 regulates zymogenic cell maturation in the stomach (Ramsey et al. 2007). The authors reported that Mist1 null mice transitional cells, which express features of both mucous neck and zymogen cells (Ramsey et al. 2007). Thus, the suppression of Mist1, perhaps from proinflammatory cytokines, prevents the complete transition of mucous neck to chief cells. A complete block in the differentiation pathway from mucous neck to chief cell seems to occur during *Helicobacter* infection and SPEM formation. Additional regulatory factors are likely to be required to affect chief cell atrophy. Nevertheless, this novel study is the first to provide us with a molecular basis to distinguish between parietal and chief cell atrophy.

Signaling Networks Regulated During Gastric Atrophy

As described above, mucosal changes in the stomach are tightly linked to loss of acid secretion, which eventually is attributable to loss of the parietal cells but can also occur in the presence of parietal cells that have ceased secreting acid. Without acid, many rapidly proliferating bacterial species ingested or residing in the oral pharynx are able to colonize the stomach. Thus, *H. pylori* might be displaced from its normally low pH niche. The question has arisen as to whether these other bacteria are able to generate inflammation. Indeed, we have found in a mouse model that bacterial overgrowth species are still recognized by the immune system as foreign invaders and generate a robust immune response (Table 18.2) (Zavros et al. 2002a). The presence of inflammation in turn means that there are soluble mediators, e.g., cytokines and growth factors, especially epidermal growth factor (EGF) receptor ligands available to regulate cell function. Although the epidemiologic steps described by Correa suggest that hypochlorhydria follows the loss of differentiated cells, one might consider that chronic exposure to cytokines and growth factors as part of the inflammatory milieu may hasten cell atrophy through dedifferentiation, transdifferentiation, cell death, or redirection of progenitor differentiation potential. Moreover, the initial steps in this process might be simply to prevent the parietal cell from producing acid.

In rabbit parietal or mouse gastric primary cell cultures, IL-1β and TNF-α each inhibit acid secretion (^{14}C-aminopyrine uptake) (Beales 2000; Beales and Calam 1998; Nompleggi et al. 1994). These studies have fueled the argument that chronic inflammation inhibits parietal cell acid secretion and supports the notion that functional hypochlorhydria might precede parietal cell atrophy. In that situation, the stomach would become relatively hypochlorhydric despite normal or possibly increased quantities of parietal cells. The latter effect might be a feedback mechanism of gastric progenitors to restore gastric acid by increasing parietal cell mass.

Young mice (4 months of age) that are deficient in the hormone gastrin also exhibit increased parietal cell mass despite the ability to produce acid, supporting the argument that reduced acid secretion does not necessarily correlate with reduced parietal cell mass (Friis-Hansen et al. 2006; Zavros et al. 2005). These studies underscore that fact that gastrin is not the only mechanism capable of regulating expansion of the parietal cell compartment and likely includes nonhormone mediators. Nevertheless, gastrin-deficient mice develop parietal cell atrophy and distal gastric tumors by 9–12 months of age, demonstrating that parietal cell mass cannot be sustained without gastrin (Friis-Hansen et al. 2006; Zavros et al. 2005). The prolonged time course suggests that a number of events must converge to result in tumor development. Not surprisingly, these hypochlorhydric mice develop bacterial overgrowth and inflammation, which might be one of the processes that contributes to tumor development (Friis-Hansen et al. 2006; Zavros et al. 2005).

Epidermal Growth Factor Receptor Signaling

Although there are likely several signal transduction pathways activated during gastric atrophy, most of the studies have been directed toward understanding parietal cell, as opposed to chief cell, atrophy. TGFβ and heparin-binding EGF (HB-EGF) are highly expressed in the gastric mucosa and therefore have been implicated in regulation of parietal cell differentiation (Beauchamp et al. 1989; Chen et al. 1993; Ford et al. 1997). Todisco and coworkers have examined the effect of chronic EGF receptor signaling on primary cultures of parietal cells and found differences in EGF receptor activation depending on the length of time exposed to EGF (Stepan et al. 2004). Although brief exposure to the growth factor (7–16 hours) rapidly induces both the Erk and Akt pathways and initially stimulates H^+, K^+-ATPase gene expression, chronic exposure to EGF for 72 h results in a second peak of Erk induction without an increase in Akt. This late peak in Erk activity correlates with inhibition of H^+, K^+-ATPase β subunit gene expression and a morphologic change in the parietal cell from a round to a fusiform shape. In addition, overexpression of Akt kinase blocks the morphologic and biochemical changes, i.e., α subunit gene expression, observed with chronic EGF treatment. Thus, the authors concluded that Akt kinase contributes to maintenance of parietal cell differentiation. Although not directly examined, it is assumed that chronic EGF treatment also blocks acid secretion because expression of the α subunit is inhibited. Targeted deletion of either the α or β subunit prevents acid secretion despite development of both parietal and chief cell lineages (Scarff et al. 1999; Spicer et al. 2000). Therefore, as with the gastrin-deficient mice, achlorhydria may exist despite the presence of parietal cells. Parietal cell differentiation, but not development, is clearly coupled to the presence of the H^+, K^+-ATPase enzyme with the presence of gastric acid serving as an indicator of functional maturity.

The extracellular signals that link H^+, K^+-ATPase gene expression and acid secretion can now be used as tools to further dissect the signaling networks that control gastric cell differentiation. Although the role of EGF receptor activation in regulating parietal cell differentiation and function are not completely understood, it is known that the ectodomains of the EGF receptor ligand family (EGF, HB-EGF, amphiregulin, Epigen, Epiregulin, betacellulin) are released from the cell surface by cytokine-induced matrix metalloproteases (MMPs) or a disintegrin and metalloprotease (ADAMs) (Merlos-Suarez et al. 2001; Sahin and Blobel 2007; Sanderson et al. 2005, 2006). Proteases themselves have been shown to bind and activate protease-activated G-protein coupled receptors, which subsequently transactivate the EGF receptor (Caruso et al. 2006) and are overexpressed in gastric cancer (Coughlin and Camerer 2003; Daub et al. 1996; Pai et al. 2002; Prenzel et al. 1999). Taken together, proinflammatory cytokines through several mechanisms increase growth factor ligands that in turn create an environment ripe for uncontrolled cell growth. For example, the *H. pylori*–infected Ins-Gas mouse develops tumors and expresses high levels of EGF receptor ligands and metalloproteases (MMPs, ADAMs) by 1 year (Takaishi and Wang 2007; Wang et al. 2000). An explanation for tumor development in these mice may be that a G-protein–coupled receptor (gastrin) activates the EGF signaling pathway (Daub et al. 1996; Prenzel et al. 1999). Notably, the acute response might not activate the proliferative signals, whereas chronically a different set of signaling networks might be activated by the same cytokine contributing to enhanced cell growth. In fact, EGF receptor antagonists are under investigation to evaluate their effectiveness in treating gastric cancer (Kishida et al. 2005).

Notch Signaling

Although the EGF ligand family is known to have a critical role in gastric transformation and can clearly be linked to inflammation, other signaling networks with their origins in development have been implicated in gastric cancer. The Notch signaling pathway controls cell fate and has been more extensively analyzed in the small bowel compared with the stomach (Artavanis-Tsakonas et al. 1999; Lai 2004; Yang et al. 2001). Notch proteins are transmembrane receptors that release their intracellular tail when the extracellular domain binds Notch ligands (Jagged1/2 or Delta-like (Dll) 1,3,/4) produced by an adjacent cell. A series of proteolytic cleavages (by the MMP and γ-secretase family of proteases) release the Notch intracellular domain that translocates to the nucleus to activate a helix-loop-helix transcription factor called Hes-1 (Hairy/enhancer of split). Hes is a repressor of another Notch target gene, the helix-loop-helix transcription factor called *Hath1* (human atonal homolog) or *Math1* (mouse atonal homolog). In the small intestine, the action of a progenitor cell to differentiate into a secretory (goblet, Paneth, enteroendocrine) versus an absorptive cell (enterocyte) involves the Notch pathway and specifically expression of the *atonal homolog*. An increase in Notch signaling induces Hes1 (repressor), suppresses Hath/Math and the secretory lineages favoring development of the absorptive cell lineage (Suzuki et al. 2005a; Yang et al. 2001). Suppression of the secretory pathway (elevated Notch signaling) is thought to be oncogenic (Katoh and Katoh 2007). A recent study has shown that increased levels of Hath1 in gastric cell lines stimulate expression of Muc5AC and Muc6, mucins expressed by the surface pit (foveolar) and mucous neck cells in the stomach, respectively (Sekine et al. 2006). This result is consistent with the ability of *atonal* homologs to induce progenitor differentiation to a mucous producing cell such as observed for goblet cells in the small intestine. Thus, one might surmise that elevated Notch signaling should block expression of the secretory lineage, e.g, the mucous population in the stomach. At present, the relationship between inflammation and Notch signaling in the stomach has not been evaluated, but one might speculate that this network will have a central role because of its ability to influence cellular differentiation. Because activation of the Notch pathway requires proteases, these molecules have become therapeutic targets that have spanned the development of pharmaceutical agents already used in clinical trials for Alzheimer's disease (Carlson and Conboy 2007; Lundkvist and Naslund 2007; Pissarnitski 2007).

Wnt Signaling

The role of the Wnt signaling pathway in colon cancer is well described. Nearly 80% of colon cancers have mutations in the APC gene encoding a large, multidomain protein that prevents β-catenin from translocating to the nucleus to activate proliferative genes (Bertario et al. 2004; Davidson 2007). Wnts bind to their receptor Frizzled to release β-catenin from a multiprotein complex composed of APC, Axin, and the kinase GSKβ. Phosphorylation of β-catenin by GSKβ targets the protein for proteasome degradation. Thus, mutations in any of these proteins can modulate targets of this pathway. Yet, most colon cancers seem to inactivate APC through mutations or deletions in the coding sequence with the remainder exhibiting constitutive activating mutations of β-catenin (Jass 2006). By contrast, APC mutations are rare in gastric cancer, despite nuclear accumulation of β-catenin in these tumors (Ogasawara et al. 2006; Rocco et al. 2006). Recently, it has been shown that inactivation of Wnt inhibitors called secreted frizzled-related peptides (sFRPs) may contribute to constitutive activation of Wnt signaling (Nojima et al. 2007). Moreover, methylation is a mechanism by which sFRPs are inactivated in gastric cancers (Kim et al. 2005c; Zhao et al. 2007).

Using serial analysis of gene expression (SAGE), Shivdasani and coworkers identified a homeobox domain transcription factor called Barx1 expressed during mouse embryonic day 12 that specifies gastric epithelial identity by inhibiting Wnt signaling (Kim et al. 2005a). In Barx1 null mouse embryos, the characteristic anterior stomach/corpus phenotype fails to develop, whereas the posterior stomach/antrum is hyperplastic and described as 5–6 cell layers (rather than 2–3 layers) of heaping undifferentiated epithelium. Intestine-specific genes are overexpressed, e.g., Cdx2, Muc2, defensin-2, FABP, and sFRP1, 2, IGFBP4 expression is suppressed. Because sFRPs antagonize the Wnt pathway, the authors examined whether suppression of the Wnt pathway recapitulates the Barx1 expression phenotype. In this experiment, Dickkopf (Dkk1), another Wnt antagonist, that blocks signaling by binding the Wnt coreceptor LRP5/6 was used to prevent ligand binding. Indeed, overexpression of Dkk1 induced reexpression of stomach specific markers (Muc1, gastrin, intrinsic factor) in the E12 mouse epithelium. Thus, suppression of the Wnt pathway seems to be essential to appropriate differentiation of the anterior embryonic gut tube into normal corpus. Barx1, sFRPs, and Wnts are expressed in the gastric mesenchyme, whereas Wnt targets, e.g., β-catenin is expressed in the epithelium. Notably, we have also observed a decrease in sFRPs and Barx1 in antral tumors that developed in 12-month-old gastrin null mice (Table 18.3). Thus, the embryonic Wnt pathway is an example of mesenchymal–epithelial crosstalk required to maintain the gastric phenotype that might be reexpressed during gastric transformation. Considering the role of Wnts and sFRPs in stomach development, one might predict that Hh signals have a critical role in maintaining a differentiated anterior stomach/corpus phenotype, i.e., the acid-secreting oxyntic glands.

Recently, it has been shown that Hh signaling inhibits the Wnt pathway by increasing sFRP gene expression (He et al. 2006). Overexpression of Gli1 in AGS cells induced sFRP1 and decreased β-catenin accumulation, suggesting that Hh signaling modulates the Wnt pathway by direct induction of sFRP1 (Figure 18.4). By contrast, overexpression of Shh in the 10T1/2 mesenchymal cell line suppressed sFRP1 expression in a microarray analysis of Hh target genes (Ingram et al. 2002). Collectively, these studies suggest that the Wnt pathway in the stomach opposes Hh signaling, but that the source of the ligands and responding target cells probably influences the overall phenotype observed.

Hedgehog Signaling Pathways

A prior study showed that gastric "metaplasia" develops in the Shh null mouse, suggesting that loss of this peptide morphogen might contribute to gastric atrophy (Ramalho-Santos et al. 2000). However, reexamination of these mice revealed that they retain normal gastric glands, but that the glands are hyperplastic (Kim et al. 2005b). Because these mice expire soon after birth, the impact of Shh in adult stomach has not been intensively studied. Prior reports of a gastric phenotype in the Shh null mouse (Kim et al. 2005b) suggest that Shh has an important role in gastric epithelial cell differentiation (van den Brink et al. 2001, 2002). However, because of its overexpression in a number of cancers, Shh has emerged as a potential target for cancer chemotherapy (Berman et al. 2003; Fukaya et al. 2006; Ma et al. 2005; Yanai et al. 2007). What is unusual about Shh expression in the adult stomach is that it is highest in the parietal cell compartment, but then is expressed in transformed mucous-like cells of the stomach (Berman et al. 2003; Lee et al. 2007; Ma et al. 2005; van den Brink et al. 2001,

Table 18.3 Three antral tumors from 12-month-old gastrin-deficient mice (Gas–/–) were individually excised and total RNA generated for microarray analysis. Shown are the results of upregulated and downregulated genes

Upregulated (fold change)
- MMP 13 (59×)
- Mmp 10 (17×)
- IL 1R-like (13×)
- Mast cell protease (9×)
- Ceacam1 (9×)
- Epiregulin (7×)
- Vanin (7×)
- IL 1b (7×)
- CXCl1 (7×)
- Follistatin (7×)
- IGFbp3 (6×)
- Cathepsin E (6×)
- IL1rn (4×)
- Nephronectin (4×)
- Claudin 4 and 23 (3×)
- Egr1 (3×)
- Amphiregulin (3×)
- Clusterin (3×)
- Sox 2 (3×)

Jagged1 (2.5×)
Activin b (2.5×)
Run x2 (2.5×)
TNFrsf (2×)
Sox 18 (2×)
CyclinD2 (2×)
EphRb2 (2×)
HB-EGF (1.8×)
Klf5 (1.6×)

Downregulated (fold change)
- Angiogenin1 (–43×)
- H, K-ATPase (–29×)
- Gastrin (–20×)
- Sfrp1 (–6×)
- GABAr (–5×)
- Necdin (–5×)
- Msi2h (–4×)
- FGFbp 1 (–4×)
- TGFbR3 (–3×)
- Frzd 7 (–3×)
- Barx1 (–2.5×)
- Frzd 2 (–2.5×)
- Supervillin (–2)

Source: Kang W. and Merchant J.L., unpublished observations. × = Fold change of three separate tumors compared with age-matched control antrum.

2002) (Figure 18.2). This raises the question of whether Shh expression and activity are regulated differently in the two cell compartments or alter the target cell it modulates to affect the hyperplastic events observed in the neoplastic stomach. Collectively, these studies set the stage for opposing functions of Shh and Hh signaling in the stomach as a function of extracellular cues. The following section focuses on Shh as an important developmental link to the preneoplastic changes occurring in the stomach.

Hedgehog Signaling

Drosophila *Hedgehog* encodes a 471 residue protein that is required to establish cell polarity by directing morphologic patterning in the tissues adjacent to cells expressing the ligand (Nusslein-Volhard and Wieschaus 1980). In mammals, there are three known hedgehog genes of slightly smaller size designated Sonic (Shh), Indian (Ihh), and Desert (Dhh); Shh is the most homologous to the original Drosophila *Hedgehog* gene sequence (Katoh and Katoh 2005; Marigo et al. 1995; Ruiz i Altaba et al. 2002). The extracellular signals regulating Shh gene expression in adult organisms are generally not known. Yet, recently, a study using the gastric cancer cell line 23132 suggests a connection between increased Shh expression and acidic conditions (Dimmler et al. 2003). Moreover, IFN-γ has been shown to regulate Shh gene expression in medulloblastomas during the initiation of neoplastic transformation but not progression (Berman et al. 2002; Lin et al. 2004; Wang et al. 2003, 2004). Therefore, these two sets of studies may provide clues as to how Shh is regulated in the stomach by inflammation.

Shh undergoes a complex series of processing steps that includes the initial generation of an approximately 45-kDa precursor, removal of the first 24 N-terminal amino acid residues comprising the signal peptide, then cleavage of the amino terminus to generate a 19-kDa protein that can be posttranslationally modified by palmitate and cholesterol (Bumcrot et al. 1995; Goetz et al. 2002; Hammerschmidt et al. 1997; Lee et al. 1994; Wendler et al. 2006) (Figure 18.3). Prior studies primarily in Drosophila indicate that there is cholesterol transferase activity within the C-terminal portion of Shh that esterifies cysteine 198 to mediate intramolecular, autocatalytic cleavage of the 45-kDa precursor (Porter et al. 1995, 1996a, 1996b; Roelink et al. 1995). Recent studies have revealed that the extent of lipid modification modulates the range of Shh diffusion away from the cell of origin and is mediated by a protein called Dispatched1 (Disp1) (Kawakami et al. 2002; Tian et al. 2005). Apparently, a shorter range of Shh diffusion correlates with a higher degree of lipid (cholesterol) modification and membrane association (Goetz et al. 2006; Gritli-Linde et al. 2001; Zeng et al. 2001).

Hh signaling involves binding of the ligands to the multitransmembrane receptor Patched (Ptch), which ultimately relieves its inhibition of another transmembrane protein called Smoothened (Smo). In cilia, recent studies show that binding of Shh to Ptch removes the receptor from the plasma membrane surface, allowing Smo to accumulate then signal to cytoplasmic Gli (Rohatgi et al. 2007). Once Smo is activated, it restrains a trimeric complex residing in the cytoplasm that normally prevents the translocation of the Gli transcription factors. If full-length Gli translocates to the nucleus, it can bind and activate Hh target genes. If Smo is inactive, the trimeric complex restrains Gli so that it can be cleaved by a protease called Slimb. The N-terminal fragment that is generated from this processing event can also translocate to the nucleus, but when it does so it represses Hh targets rather than activates (Riobo and Manning 2007) (Figure 18.4).

Shh processing

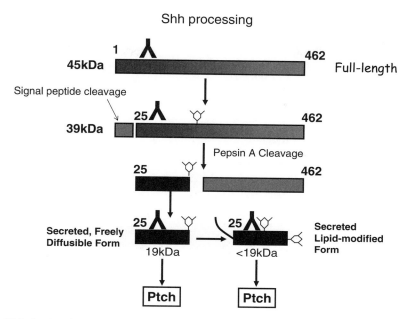

Fig. 18.3 Sonic hedgehog (Shh) processing scheme. There is little information regarding regulators of Shh gene expression in adult tissues. However, it is known that Shh must be processed from the nascent 45-kDa peptide to a 19-kDa form that is lipid modified with cholesterol and at the amino terminus by palmitate. The cholesterol esterification at residue 198 is thought to facilitate autocleavage by protease activity residing with the C terminus. Prior mutational analysis has demonstrated that residue H271 is required for autocatalytic cleavage. The 19-kDa forms bind the receptor Patched (Ptch), whereas the C-terminus fragment does not. However, we have recently found that parietal cells, which produce pepsin A cleave Shh at a consensus site (GGCF^{200}P) that includes residue 200 (Zavros et al. 2007)

Sonic Hedgehog in the Stomach

Transgenic mouse studies have furthered our understanding of Shh in the developing gut. During embryonic development, Shh is expressed throughout the gut and in other forgut organs, e.g., lung, pancreas (Kim and Melton 1998; Litingtung et al. 1998; Shannon and Hyatt 2004). However, in the adult stomach, our understanding of Shh function has generally been limited to descriptive studies because null mouse lines do not survive long after birth (Ramalho-Santos et al. 2000). Shh is downregulated in the epithelial cells of the intestine but remains highly expressed in the gastric epithelium. Recently, it has been suggested that Shh regulates epithelial cell maturation and differentiation in the adult stomach (van den Brink et al. 2001, 2002). Normally, Shh is expressed in the mature acid-secreting glands of the stomach, primarily within parietal cells (Fukaya et al. 2006; Stepan et al. 2005; Van Den Brink et al.

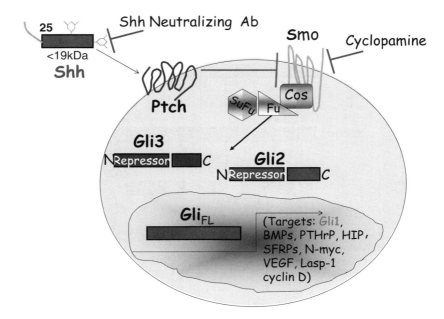

Fig. 18.4 Hedgehog (Hh) signal transduction. When a hedgehog ligand, Sonic, Indian, or Desert (Shh, Ihh, Dhh) binds its receptor Patched (Ptch), repression of the 7-transmembrane receptor Smoothened (Smo) is removed. Smo interacts with a trimolecular complex, which includes Costal (Cos)2, Fused (Fu), and Suppressor of Fused (SuFu). When Smo interacts with Cos2, SuFu is inactive and a full-length Gli2 or Gli3 protein translocates to the nucleus and activates Hh target genes. Several of the Hh target genes are listed. In the absence of a Hh ligand, Cos2 binds Gli2 or 3 and retains them in the nucleus where they are subject to proteolytic cleavage. The N-terminal cleavage product translocates to the nucleus and represses Hh target genes. Gli2 and Gli3 are post-translationally regulated, whereas the Hh signaling pathway transcriptionally regulates Gli1. The pathway can be inhibited using neutralizing Shh antibodies, the alkaloid Smoothened antagonist, cyclopamine, or its natural inhibitor HIP (hedgehog interacting peptide)

2001). As a result of parietal cell atrophy, Shh expression in the corpus is also lost (Shiotani et al. 2005b; Suzuki et al. 2005b). Clinically, this is reflected by the development of hypochlorhydria and reduced serum pepsinogen A levels (Iijima et al. 2005; Kokkola et al. 2005; Nomura et al. 2005a; Shiotani et al. 2005c; Sierra et al. 2006; Sipponen et al. 2002).

Recently, we found that both processing and expression of Shh is linked to acid secretion in the adult stomach (Zavros et al. 2007). The hormone gastrin (infused over 2 weeks) stimulated Shh gene expression in the hypochlorhydric gastrin-deficient mouse model in concert with reestablishing gastric acid secretion. Moreover, posttranslational processing of the Shh precursor to the secreted form was acid-dependent and cleaved by the acid-activated protease pepsin A. Thus, we have discovered that even generation of the biologically active form of Shh in native parietal cells depends on the gastric pH. If gastric acidity is reduced, because of inhibition of acid secretion or loss of the parietal cell (atrophy), then pepsin A is not activated and most of the precursor Shh protein produced in the stomach is not cleaved. Review of the literature revealed little information on the cell origin of pepsin A production, whereas pepsin C is thought to originate from the mucous neck and chief cells (Shao et al. 1998). We found as suggested by Alpers and cow-orkers that the parietal cells express some pepsin A (Shao et al. 1998). This conclusion is consistent with the observation that gastric atrophy (specifically loss of the parietal cell) correlates with reduced pepsinogen A to C ratios observed in atrophic human stomachs (Shiotani et al. 2005c).

Regulation of Sonic Hedgehog by Inflammation and Interferon-γ

Perhaps a result of its essential role in developmental processes, there is surprisingly little information regarding the regulation of Shh by inflammation and specifically proinflammatory cytokines. However, of the eight articles identified in Medline in 2007, half describe changes in Shh expression in the stomach during *H. pylori* infection. Three of the gastric articles focused on how chronic inflammation in the stomach initiated by the organism decreases Shh expression and is reexpressed upon eradication (Nishizawa et al. 2007; Shiotani et al. 2007; Suzuki et al. 2005b). However, IFN-γ induces Shh gene expression by a STAT1 mechanism in cerebellar granule neurons, which ultimately leads to the development of tumors (medulloblastoma) (Wang et al. 2003). IFN-γ overexpressed in the mouse cerebellum of newborn mice under the control of an inducible tissue-specific promoter greatly upregulated Shh expression and initiated medulloblastoma tumor development (Lin et al. 2004). However, progression of the tumor was found to be IFN-γ independent because the tumor progressed despite termination of IFN-γ expression by removing doxycycline (Lin et al. 2004). When the cerebellums of the doxycycline-treated mice were analyzed further, the authors found that elevated levels of IFN-γ induced STATS1, 3, and 5 (Wang et al. 2004). In addition, the 19-kDa processed form of Shh increased with IFN-γ infusion without a corresponding increase in the nascent 45-kDa precursor (Wang et al. 2004). The authors concluded that IFN-γ might be affecting Shh translation, an observation that would support our findings that Shh processing might be regulated in the stomach. These results have clinical significance because of the linkage of medulloblastomas to aberrant Hh signaling pathways (Goodrich et al. 1997; Kimura et al. 2005; Piedimonte et al. 2005). It is possible that a similar mechanism might be occurring in the stomach, forming the basis for investigating the role of cytokines, including IFN-γ in the regulation of Shh gene expression and activity. A specific role for IL-1β and TNF-α in the regulation of Shh gene expression has not been evaluated. However, NFκB, a downstream transcriptional mediator of IL-1β and TNF-α signaling is upregulated in the antrums of human subjects with chronic active gastritis (van Den Brink et al. 2000). Therefore, these cytokines are worth investigating as potential regulators of Shh expression. In fact, chronic gastritis increases Shh expression and

BMP4, a downstream target of Hh signaling (Bleuming et al., 2006) in archived human tissue (documented by *in situ* hybridization) (Nielsen et al. 2004). Consistent with this concept, El-Zaatari et al. (2007) showed that IL-8, another proinflammatory cytokine released during *Helicobacter* infection, increases Shh gene expression in two human gastric adenocarcinoma cell lines, AGS and MGLVA1.

Sonic Hedgehog in Gastric Cancer

Multiple studies clearly link elevated Shh and signaling to gastric cancers (Berman et al. 2003; Fukaya et al. 2006; Ma et al. 2005; Yanai et al. 2007). However, its functional role has not been established in a primary cancer model. Beachy and coworkers showed that Hh signaling in xenografts of human gastric cancer lines is required for tumor growth because progression could be blocked with cyclopamine (Berman et al. 2003). Cyclopamine is a plant alkaloid found in corn lilies that inhibits the Hh signaling pathway by binding Smoothened (Figure 18.4) (Chen et al. 2002; James et al. 2004; Taipale et al. 2000). Nevertheless, examination of primary gastric tumors has led to conflicting information about the source of the ligand and responding cell type (Shiotani et al. 2005b). There are multiple explanations for the variability, e.g., differences in the site of Shh expression and tumor subtype. The normal pattern of Shh expression in the stomach seems to fit the paracrine model initially established in developing fetal tissues. However, application of the paracrine model to cancer development suggests that the Shh ligand originating in the tumor stimulates responding cells in the stroma. Whereas this mechanism should result in diffuse-type cancers, it is unclear how a paracrine mechanism would explain hyperplasia of the epithelium. Rather, an autocrine mechanism or paracrine mechanism in which stromal cells produce most of the Shh would better predict development of intestinal-type cancers. Considering a role for at least one inflammatory cytokine, e.g, IFN-γ, in the activation of Shh gene expression, there is a strong possibility that a source of the Shh modulating tumor growth is from infiltrating leukocytes.

Regional Differences in Hedgehog Signaling and Response

Hh signal transduction involves activation of the *Gli* gene family of transcription factors (Altaba 1997; Kinzler and Vogelstein 1990). Of the three *Gli* family members, only *Gli1* is transcriptionally activated by Hh signals (Bai et al. 2002; Lee et al. 1997) (Figure 18.4). Posttranslational processing regulates the activity of Gli2 and 3 (Motoyama et al. 1998; Pan et al. 2006) (Figure 18.4). For example, Yanai et al. (2007) reported nuclear Gli1 staining of epithelial cells in human tumors. However, examination of those tissues exhibited faint cytoplasmic staining rather than the nuclear staining expected from Hh-activated Gli. Also, in some specimens, staining was observed in the stroma. By contrast, Fukaya et al. (2006) reported that Shh expression in gastric cancer is stromal. Still other studies suggest that the predominant Shh phenotype is loss of expression (Shiotani et al. 2005b). Coste et al. (2007) examined precancerous lesions in a transgenic mouse model of sporadic β-catenin activation for the association of gastric cancer with the loss of Shh. They found that incomplete metaplasia develops where β-catenin is expressed and Shh expression is suppressed. These studies are consistent with the prior notion that cancers recapitulate fetal phenotypes. In this instance, Hh signaling opposes the pro-proliferative signals mediated by the Wnt-β-catenin pathway. However, none of the studies, so far, addressed what role elevated levels of Shh has in actual gastric cancers.

To address this critical issue, it will be important to determine whether or not dysplastic gastric epithelium is responding to Hh signals and, if so, whether the signal is of epithelial origin (autocrine), stromal origin (paracrine), or both. A likely outcome is that Shh may be expressed in either the epithelium or stroma as a function of tumor biology, e.g., diffuse versus well differentiated, slow growing versus rapidly expanding. In particular, expression of Shh in the epithelium and Gli in the stroma of the tumor would suggest a paracrine response in which the tumor might have a "diffuse" cancer phenotype and more aggressive behavior. The results might impact ways to design antagonists of Hh signaling for gastric cancer subjects. Therefore, whether the cellular source of Shh expression and the type of cell responding to the ligand affects tumor behavior will need to be clarified.

Currently, the only functional study of Hh signaling in a gastric cancer model is the use of human cell lines to generate xenografts in nude mice (Berman et al. 2003). In that study, the authors found that xenograft tumor growth was blocked using the Smo inhibitor cyclopamine or Shh neutralizing antibodies. Thus, Shh inhibits ectopic tumors arising from cell lines. But it is still unclear what role Shh and Hh signals have in tumors arising in the milieu of the stomach. The several mouse models described here will be quite useful in examining the role of Shh in the development and progression of gastric cancer because human studies are likely to be primarily correlative until Hh antagonists are evaluated for efficacy in clinical trials. Clearly, Shh expression is lost in the corpus, but then is overexpressed in the mucous-derived cells of the tumor. However, it is not known what function Shh has in these two cell types, requiring further analysis.

With respect to regional differences, it is clear that Shh expression is highest in the corpus and essentially absent in the antrum. If we apply what is understood from fetal patterning, we would propose that Shh produced in the oxyntic gland modulates signals in the epithelium, e.g., Wnts. As proposed by Shivdasani, the Wnt pathway must be suppressed to maintain a normal acid-secreting epithelium (Kim et al. 2005a). Thus in the adult, we propose that epithelial secretion of Shh subsequently activating sFRPs in the mesenchyme is an important connection, perhaps linking events in the corpus to those in the antrum. Nevertheless, if sFRP expression in the mesenchyme is lost because of methylation (Nojima et al. 2007; Zhao et al. 2007), or loss of Shh (He et al. 2006), it may explain how β-catenin–mediated proliferation emerges in the antrum.

In summary, this review has focused on the critical step between inflammation and epithelial cell remodeling into precancerous lesions. Some potential extracellular signals that affect cells of the gastric gland have been reviewed, and how specific growth factor and developmental networks modulate cell fate resulting in parietal and chief cell atrophy have been highlighted. In particular, the Hh signal transduction pathway was further scrutinized, because it might be an essential signal linking corpus atrophy to antral proliferation.

References

Ahmed, S., and A. Nawshad. 2007. Complexity in interpretation of embryonic epithelial-mesenchymal transition in response to transforming growth factor-beta signaling. Cells Tissues Organs. 185:131–45.

Altaba, R.R.I. 1997. Catching a Gli-mpse of Hedgehog. Cell. 90:193–6.

Artavanis-Tsakonas, S., M.D. Rand, and R.J. Lake. 1999. Notch signaling: cell fate control and signal integration in development. Science. 284:770–6.

Bai, C.B., W. Auerbach, J.S. Lee, D. Stephen, and A.L. Joyner. 2002. Gli2, but not Gli1, is required for initial Shh signaling and ectopic activation of the Shh pathway. Development. 129:4753–61.

Bae, S.C., and J.K. Choi. 2004. Tumor suppressor activity of RUNX3. Oncogene. 23:4336–40.

Bamford, K.B., X. Fan, S.E. Crowe, J.F. Leary, W.K. Gourley, G.K. Luthra, E.G. Brooks, D.Y. Graham, V.E. Reyes, and P.B. Ernst. 1998. Lymphocytes in the human gastric mucosa during Helicobacter pylori have a T helper cell 1 phenotype. Gastroenterology. 114:482–92.

Barrett, S.P., P.A. Gleeson, H. de Silva, B.H. Toh, and I.R. van Driel. 1996. Interferon-gamma is required during the initiation of an organ-specific autoimmune disease. Eur J Immunol. 26:1652–5.

Barrett, S.P., B.H. Toh, F. Alderuccio, I.R. van Driel, and P.A. Gleeson. 1995. Organ-specific autoimmunity induced by adult thymectomy and cyclophosphamide-induced lymphopenia. Eur J Immunol. 25:238–44.

Bataller, R., and D.A. Brenner. 2005. Liver fibrosis. J Clin Invest. 115:209–18.

Baur, S., J.M. Fisher, R.G. Strickland, and K.B. Taylor. 1968. Autoantibody-containing cells in the gastric mucosa in pernicious anaemia. Lancet. 2:887–94.

Bayerdorffer, E., N. Lehn, R. Hatz, G.A. Mannes, H. Oertel, T. Sauerbruch, and M. Stolte. 1992. Difference in expression of Helicobacter pylori gastritis in antrum and body. Gastroenterology. 102:1575–82.

Beales, I.L. 2000. Effects of pro-inflammatory cytokines on acid secretion. Dig Dis Sci. 45:289–90.

Beales, I.L., and J. Calam. 1998. Interleukin 1 beta and tumour necrosis factor alpha inhibit acid secretion in cultured rabbit parietal cells by multiple pathways. Gut. 42:227–34.

Beauchamp, R.D., J.A. Barnard, C.M. McCutchen, J.A. Cherner, and R.J. Coffey, Jr. 1989. Localization of transforming growth factor a and its receptor in gastric mucosal cells. J Clin Invest. 84:1017–23.

Berman, D.M., S.S. Karhadkar, A.R. Hallahan, J.I. Pritchard, C.G. Eberhart, D.N. Watkins, J.K. Chen, M.K. Cooper, J. Taipale, J.M. Olson, and P.A. Beachy. 2002. Medulloblastoma growth inhibition by hedgehog pathway blockade. Science. 297:1559–61.

Berman, D.M., S.S. Karhadkar, A. Maitra, R. Montes De Oca, M.R. Gerstenblith, K. Briggs, A.R. Parker, Y. Shimada, J.R. Eshleman, D.N. Watkins, and P.A. Beachy. 2003. Widespread requirement for Hedgehog ligand stimulation in growth of digestive tract tumours. Nature. 425:846–51.

Bertario, L., A. Russo, P. Sala, L. Varesco, R. Crucianelli, M. Frattini, M.A. Pierotti, and P. Radice. 2004. APC genotype is not a prognostic factor in familial adenomatous polyposis patients with colorectal cancer. Dis Colon Rectum. 47:1662–9.

Blair, V., I. Martin, D. Shaw, I. Winship, D. Kerr, J. Arnold, P. Harawira, M. McLeod, S. Parry, A. Charlton, M. Findlay, B. Cox, B. Humar, H. More, and P. Guilford. 2006. Hereditary diffuse gastric cancer: diagnosis and management. Clin Gastroenterol Hepatol. 4:262–75.

Blaser, M.J., and J. Parsonnet. 1994. Parasitism by the "slow" bacterium Helicobacter pylori leads to altered gastric homeostasis and neoplasia. J Clin Invest. 94:4–8.

Bleuming, S.A., L.L. Kodach, M.J. Garcia Leon, D.J. Richel, M.P. Peppelenbosch, P.H. Reitsma, J.C. Hardwick, and G.R. van den Brink. 2006. Altered bone morphogenetic protein signalling in the Helicobacter pylori-infected stomach. J Pathol. 209:190–7.

Bumcrot, D.A., R. Takada, and A.P. McMahon. 1995. Proteolytic processing yields two secreted forms of sonic hedgehog. Mol Cell Biol. 15:2294–303.

Carlson, M.E., and I.M. Conboy. 2007. Regulating the Notch pathway in embryonic, adult and old stem cells. Curr Opin Pharmacol. 7:303–9.

Caruso, R., F. Pallone, E. Fina, V. Gioia, I. Peluso, F. Caprioli, C. Stolfi, A. Perfetti, L.G. Spagnoli, G. Palmieri, T.T. Macdonald, and G. Monteleone. 2006. Protease-activated receptor-2 activation in gastric cancer cells promotes epidermal growth factor receptor trans-activation and proliferation. Am J Pathol. 169:268–78.

Chen, M.C., A.T. Lee, W.E. Karnes, D. Avedian, M. Martin, J.M. Sorvillo, and A.H. Soll. 1993. Paracrine control of gastric epithelial cell growth in culture by transforming growth factor-α. Am J Physiol. 264:G390–6.

Chen, J.K., J. Taipale, K.E. Young, T. Maiti, and P.A. Beachy. 2002. Small molecule modulation of Smoothened activity. Proc Natl Acad Sci USA 99:14071–6.

Chey, W.D., and B.C. Wong. 2007. American College of Gastroenterology guideline on the management of Helicobacter pylori infection. Am J Gastroenterol. 102(8):1808–25.

Corley, D.A., and A. Kubo. 2004. Influence of site classification on cancer incidence rates: an analysis of gastric cardia carcinomas. J Natl Cancer Inst. 96:1383–7.

Correa, P. 1992. Human gastric carcinogenesis: a multistep and multifactorial process—first American Cancer Society award lecture on cancer epidemiology and prevention. Cancer Res. 52:6735–40.

Correa, P. 1995. Helicobacter pylori and gastric carcinogenesis. Am J Surg Pathol. 19(Suppl 1): S37–43.

Correa, P., W. Haenszel, C. Cuello, S. Tannenbaum, and M. Archer. 1975. A model for gastric cancer epidemiology. Lancet. 2:58–60.

Coste, I., J.N. Freund, S. Spaderna, T. Brabletz, and T. Renno. 2007. Precancerous lesions upon sporadic activation of beta-catenin in mice. Gastroenterology. 132:1299–308.

Coughlin, S.R., and E. Camerer. 2003. PARticipation in inflammation. J Clin Invest. 111:25–7.

Daub, H., F.U. Weiss, C. Wallasch, and A. Ullrich. 1996. Role of transactivation of the EGF receptor in signalling by G-protein-coupled receptors. Nature. 379:557–60.

Davidson, N.O. 2007. Genetic testing in colorectal cancer: who, when, how and why. Keio J Med. 56:14–20.

Dimmler, A., T. Brabletz, F. Hlubek, M. Hafner, T. Rau, T. Kirchner, and G. Faller. 2003. Transcription of sonic hedgehog, a potential factor for gastric morphogenesis and gastric mucosa maintenance, is up-regulated in acidic conditions. Lab Invest. 83:1829–37.

El-Omar, E.M., M. Carrington, W.H. Chow, K.E. McColl, J.H. Bream, H.A. Young, J. Herrera, J. Lissowska, C.C. Yuan, N. Rothman, G. Lanyon, M. Martin, J.F. Fraumeni, Jr., and C.S. Rabkin. 2000. Interleukin-1 polymorphisms associated with increased risk of gastric cancer. Nature. 404:398–402.

El-Omar, E.M., C.S. Rabkin, M.D. Gammon, T.L. Vaughan, H.A. Risch, J.B. Schoenberg, J.L. Stanford, S.T. Mayne, J. Goedert, W.J. Blot, J.F. Fraumeni, Jr., and W.H. Chow. 2003. Increased risk of noncardia gastric cancer associated with proinflammatory cytokine gene polymorphisms. Gastroenterology. 124:1193–201.

El-Zaatari, M., A. Tobias, A.M. Grabowska, R. Kumari, P.J. Scotting, P. Kaye, J. Atherton, P.A. Clarke, D.G. Powe, and S.A. Watson. 2007. De-regulation of the sonic hedgehog pathway in the InsGas mouse model of gastric carcinogenesis. Br J Cancer. 96:1855–61.

El-Zimaity, H.M., H. Ota, D.Y. Graham, T. Akamatsu, and T. Katsuyama. 2002. Patterns of gastric atrophy in intestinal type gastric carcinoma. Cancer. 94:1428–36.

Faller, G., A. Dimmler, T. Rau, S. Spaderna, F. Hlubek, A. Jung, T. Kirchner, and T. Brabletz. 2004. Evidence for acid-induced loss of Cdx2 expression in duodenal gastric metaplasia. J Pathol. 203:904–8.

Fietz, M.J., J.P. Concordet, R. Barbosa, R. Johnson, S. Krauss, A.P. McMahon, C. Tabin, and P.W. Ingham. 1994. The hedgehog gene family in Drosophila and vertebrate development. Dev Suppl 43–51.

Ford, M.G., J.D. Valle, C.J. Soroka, and J.L. Merchant. 1997. EGF receptor activation stimulates endogenous gastrin gene expression in canine G cells and human gastric cell cultures. J Clin Invest. 99:2762–71.

Fox, J.G., X. Li, R.J. Cahill, K. Andrutis, A.K. Rustgi, R. Odze, and T.C. Wang. 1996. Hypertrophic gastropathy in Helicobacter felis-infected wild-type C57BL/6 mice and p53 hemizygous transgenic mice. Gastroenterology. 110:155–66.

Fox, J.G., and T.C. Wang. 2007. Inflammation, atrophy, and gastric cancer. J Clin Invest. 117:60–9.

Friedrich, M.J., R. Rad, R. Langer, P. Voland, H. Hoefler, R.M. Schmid, C. Prinz, and M. Gerhard. 2006. Lack of RUNX3 regulation in human gastric cancer. J Pathol. 210:141–6.

Friis-Hansen, L., K. Rieneck, H.O. Nilsson, T. Wadstrom, and J.F. Rehfeld. 2006. Gastric inflammation, metaplasia, and tumor development in gastrin-deficient mice. Gastroenterology. 131:246–58.

Fukaya, M., N. Isohata, H. Ohta, K. Aoyagi, T. Ochiya, N. Saeki, K. Yanagihara, Y. Nakanishi, H. Taniguchi, H. Sakamoto, T. Shimoda, Y. Nimura, T. Yoshida, and H. Sasaki. 2006. Hedgehog signal activation in gastric pit cell and in diffuse-type gastric cancer. Gastroenterology. 131:14–29.

Furuta, T., E.M. El-Omar, F. Xiao, N. Shirai, M. Takashima, H. Sugimura, and H. Sugimurra. 2002. Interleukin 1beta polymorphisms increase risk of hypochlorhydria and atrophic gastritis and reduce risk of duodenal ulcer recurrence in Japan. Gastroenterology. 123:92–105.

Giraud, A.S., C. Jackson, T.R. Menheniott, and L.M. Judd. 2007. Differentiation of the gastric mucosa. IV. Role of trefoil peptides and IL-6 cytokine family signaling in gastric homeostasis. Am J Physiol Gastrointest Liver Physiol. 292:G1–5.

Gisbert, J.P., M. Fernandez-Bermejo, J. Molina-Infante, B. Perez-Gallardo, A.B. Prieto-Bermejo, J.M. Mateos-Rodriguez, P. Robledo-Andres, and G. Gonzalez-Garcia. 2007. First-line triple therapy with levofloxacin for Helicobacter pylori eradication. Aliment Pharmacol Ther. 26:495–500.

Gleeson, P.A., B.H. Toh, and I.R. van Driel. 1996. Organ-specific autoimmunity induced by lymphopenia. Immunol Rev. 149:97–125.

Goetz, J.A., S. Singh, L.M. Suber, F.J. Kull, and D.J. Robbins. 2006. A highly conserved amino-terminal region of sonic hedgehog is required for the formation of its freely diffusible multimeric form. J Biol Chem. 281:4087–93.

Goetz, J.A., L.M. Suber, X. Zeng, and D.J. Robbins. 2002. Sonic hedgehog as a mediator of long-range signaling. Bioessays. 24:157–65.

Goldenring, J.R., and S. Nomura. 2006. Differentiation of the gastric mucosa. III. Animal models of oxyntic atrophy and metaplasia. Am J Physiol Gastrointest Liver Physiol. 291:G999–1004.

Goodrich, L.V., L. Milenkovic, K.M. Higgins, and M.P. Scott. 1997. Altered neural cell fates and medulloblastoma in mouse patched mutants. Science. 277:1109–13.

Graham, D.Y. 1998. Antibiotic resistance in Helicobacter pylori: implications for therapy. Gastroenterology. 115:1272–7.

Gritli-Linde, A., P. Lewis, A.P. McMahon, and A. Linde. 2001. The whereabouts of a morphogen: direct evidence for short- and graded long-range activity of hedgehog signaling peptides. Dev Biol. 236:364–86.

Halldorsdottir, A.M., M. Sigurdardottrir, J.G. Jonasson, M. Oddsdottir, J. Magnusson, J.R. Lee, and J.R. Goldenring. 2003. Spasmolytic polypeptide-expressing metaplasia (SPEM) associated with gastric cancer in Iceland. Dig Dis Sci. 48:431–41.

Hammerschmidt, M., A. Brook, and A.P. McMahon. 1997. The world according to hedgehog. Trends Genet. 13:14–21.

He, J., T. Sheng, A.A. Stelter, C. Li, X. Zhang, M. Sinha, B.A. Luxon, and J. Xie. 2006. Suppressing Wnt signaling by the hedgehog pathway through sFRP-1. J Biol Chem. 281:35598–602.

Hirayama, F., S. Takagi, E. Iwao, Y. Yokoyama, K. Haga, and S. Hanada. 1999. Development of poorly differentiated adenocarcinoma and carcinoid due to long-term Helicobacter pylori colonization in Mongolian gerbils. J Gastroenterol. 34:450–4.

Hofman, P., B. Waidner, V. Hofman, S. Bereswill, P. Brest, and M. Kist. 2004. Pathogenesis of Helicobacter pylori infection. Helicobacter. 9(Suppl 1):15–22.

Honda, S., T. Fujioka, M. Tokieda, R. Satoh, A. Nishizono, and M. Nasu. 1998. Development of Helicobacter pylori-induced gastric carcinoma in Mongolian gerbils. Cancer Res. 58:4255–9.

Hou, L., E.M. El-Omar, J. Chen, P. Grillo, C.S. Rabkin, A. Baccarelli, M. Yeager, S.J. Chanock, W. Zatonski, L.H. Sobin, J. Lissowska, J.F. Fraumeni, Jr., and W.H. Chow. 2007. Polymorphisms in Th1-type cell-mediated response genes and risk of gastric cancer. Carcinogenesis. 28:118–23.

Houghton, J., C. Stoicov, S. Nomura, A.B. Rogers, J. Carlson, H. Li, X. Cai, J.G. Fox, J.R. Goldenring, and T.C. Wang. 2004. Gastric cancer originating from bone marrow-derived cells. Science. 306:1568–71.

Howlett, M., L.M. Judd, B. Jenkins, N.L. La Gruta, D. Grail, M. Ernst, and A.S. Giraud. 2005. Differential regulation of gastric tumor growth by cytokines that signal exclusively through the coreceptor gp130. Gastroenterology. 129:1005–18.

Iijima, K., H. Sekine, T. Koike, A. Imatani, S. Ohara, and T. Shimosegawa. 2005. Serum pepsinogen concentrations as a measure of gastric acid secretion in Helicobacter pylori-negative and -positive Japanese subjects. J Gastroenterol. 40:938–44.

Ingram, W.J., C.A. Wicking, S.M. Grimmond, A.R. Forrest, and B.J. Wainwright. 2002. Novel genes regulated by sonic hedgehog in pluripotent mesenchymal cells. Oncogene. 21:8196–205.

James, L.F., K.E. Panter, W. Gaffield, and R.J. Molyneux. 2004. Biomedical applications of poisonous plant research. J Agric Food Chem. 52:3211–30.

Jass, J.R. 2006. Colorectal cancer: a multipathway disease. Crit Rev Oncog. 12:273–87.

Jenkins, B.J., D. Grail, T. Nheu, M. Najdovska, B. Wang, P. Waring, M. Inglese, R.M. McLoughlin, S.A. Jones, N. Topley, H. Baumann, L.M. Judd, A.S. Giraud, A. Boussioutas, H.J. Zhu, and M. Ernst. 2005. Hyperactivation of Stat3 in gp130 mutant mice promotes gastric hyperproliferation and desensitizes TGF-beta signaling. Nat Med. 11:845–52.

Johnson, C.L., A.S. Kowalik, N. Rajakumar, and C.L. Pin. 2004. Mist1 is necessary for the establishment of granule organization in serous exocrine cells of the gastrointestinal tract. Mech Dev. 121:261–72.

Judd, L.M., B.M. Alderman, M. Howlett, A. Shulkes, C. Dow, J. Moverley, D. Grail, B.J. Jenkins, M. Ernst, and A.S. Giraud. 2004. Gastric cancer development in mice lacking the SHP2 binding site on the IL-6 family co-receptor gp130. Gastroenterology. 126:196–207.

Judd, L.M., K. Bredin, A. Kalantzis, B.J. Jenkins, M. Ernst, and A.S. Giraud. 2006. STAT3 activation regulates growth, inflammation, and vascularization in a mouse model of gastric tumorigenesis. Gastroenterology. 131:1073–85.

Kang, W., S. Rathinavelu, L.C. Samuelson, and J.L. Merchant. 2005. Interferon gamma induction of gastric mucous neck cell hypertrophy. Lab Invest. 85(5):702–15.

Karam, S.M. 1993. Dynamics of epithelial cells in the corpus of the mouse stomach. IV. Bidirectional migration of parietal cells ending in their gradual degeneration and loss. Anat Rec. 236:314–32.

Karam, S.M., and C.P. Leblond. 1993a. Dynamics of epithelial cells in the corpus of the mouse stomach. II. Outward migration of pit cells. Anat Rec. 236:280–96.

Karam, S.M., and C.P. Leblond. 1993b. Dynamics of epithelial cells in the corpus of the mouse stomach. III. Inward migration of neck cells followed by progressive transformation into zymogenic cells. Anat Rec. 236:297–313.

Katoh, Y., and M. Katoh. 2005. Comparative genomics on sonic hedgehog orthologs. Oncol Rep. 14:1087–90.

Katoh, M., and M. Katoh. 2007. Notch signaling in gastrointestinal tract [review]. Int J Oncol. 30:247–51.

Kawachi, T., K. Kogure, N. Tanaka, A. Tokunaga, and T. Sugimura. 1974. Studies of intestinal metaplasia in the gastric mucosa by detection of disaccharidases with "Tes-Tape." J Natl Cancer Inst. 53:19–30.

Kawakami, T., T. Kawcak, Y.J. Li, W. Zhang, Y. Hu, and P.T. Chuang. 2002. Mouse dispatched mutants fail to distribute hedgehog proteins and are defective in hedgehog signaling. Development. 129:5753–65.

Kaye, M.D., P.J. Whorwell, and R. Wright. 1983. Gastric mucosal lymphocyte subpopulations in pernicious anemia and in normal stomach. Clin Immunol Immunopathol. 28:431–40.

Kim, B.M., G. Buchner, I. Miletich, P.T. Sharpe, and R.A. Shivdasani. 2005a. The stomach mesenchymal transcription factor Barx1 specifies gastric epithelial identity through inhibition of transient Wnt signaling. Dev Cell. 8:611–22.

Kim, J.H., Z. Huang, and R. Mo. 2005b. Gli3 null mice display glandular overgrowth of the developing stomach. Dev Dyn. 234:984–91.

Kim, H.C., J.C. Kim, S.A. Roh, C.S. Yu, J.H. Yook, S.T. Oh, B.S. Kim, K.C. Park, and R. Chang. 2005c. Aberrant CpG island methylation in early-onset sporadic gastric carcinoma. J Cancer Res Clin Oncol. 131:733–40.

Kim, S.K., and D.A. Melton. 1998. Pancreas development is promoted by cyclopamine, a hedgehog signaling inhibitor. Proc Natl Acad Sci USA. 95:13036–41.

Kim, S.S., K. Shetty, V. Katuri, K. Kitisin, H.J. Baek, Y. Tang, B. Marshall, L. Johnson, B. Mishra, and L. Mishra. 2006. TGF-beta signaling pathway inactivation and cell cycle deregulation in the development of gastric cancer: role of the beta-spectrin, ELF. Biochem Biophys Res Commun. 344:1216–23.

Kimura, H., D. Stephen, A. Joyner, and T. Curran. 2005. Gli1 is important for medulloblastoma formation in Ptc1+/– mice. Oncogene. 24:4026–36.

Kinzler, K.W., and B. Vogelstein. 1990. The GLI gene encodes a nuclear protein which binds specific sequences in the human genome. Mol Cell Biol. 10:634–42.

Kishida, O., Y. Miyazaki, Y. Murayama, M. Ogasa, T. Miyazaki, T. Yamamoto, K. Watabe, S. Tsutsui, T. Kiyohara, I. Shimomura, and Y. Shinomura. 2005. Gefitinib (Iressa, ZD1839) inhibits SN38-triggered EGF signals and IL-8 production in gastric cancer cells. Cancer Chemother Pharmacol. 55:584–94.

Klinkenberg-Knol, E.C., F. Nelis, J. Dent, P. Snel, B. Mitchell, P. Prichard, D. Lloyd, N. Havu, M.H. Frame, J. Roman, and A. Walan. 2000. Long-term omeprazole treatment in resistant gastroesophageal reflux disease: efficacy, safety, and influence on gastric mucosa. Gastroenterology. 118:661–9.

Kojima, A., and R.T. Prehn. 1981. Genetic susceptibility to post-thymectomy autoimmune diseases in mice. Immunogenetics. 14:15–27.

Kokkola, A., J. Louhimo, P. Puolakkainen, H. Alfthan, C. Haglund, and H. Rautelin. 2005. Helicobacter pylori infection and low serum pepsinogen I level as risk factors for gastric carcinoma. World J Gastroenterol. 11:1032–6.

Lai, E.C. 2004. Notch signaling: control of cell communication and cell fate. Development. 131:965–73.

Lamberts, R., G. Brunner, and E. Solcia. 2001. Effects of very long (up to 10 years) proton pump blockade on human gastric mucosa. Digestion. 64:205–13.

Lee, J.J., S.C. Ekker, D.P. von Kessler, J.A. Porter, B.I. Sun, and P.A. Beachy. 1994. Autoproteolysis in hedgehog protein biogenesis. Science. 266:1528–37.

Lee, S.Y., H.S. Han, K.Y. Lee, T.S. Hwang, J.H. Kim, I.K. Sung, H.S. Park, C.J. Jin, and K.W. Choi. 2007. Sonic hedgehog expression in gastric cancer and gastric adenoma. Oncol Rep. 17:1051–5.

Lee, J., K.A. Platt, P. Censullo, and A. Ruiz i Altaba. 1997. Gli1 is a target of Sonic hedgehog that induces ventral neural tube development. Development. 124:2537–52.

Li, Q.L., K. Ito, C. Sakakura, H. Fukamachi, K. Inoue, X.Z. Chi, K.Y. Lee, S. Nomura, C.W. Lee, S.B. Han, H.M. Kim, W.J. Kim, H. Yamamoto, N. Yamashita, T. Yano, T. Ikeda, S. Itohara, J. Inazawa, T. Abe, A. Hagiwara, H. Yamagishi, A. Ooe, A. Kaneda, T. Sugimura, T. Ushijima, S.C. Bae, and Y. Ito. 2002. Causal relationship between the loss of RUNX3 expression and gastric cancer. Cell. 109:113–24.

Li, Q., S.M. Karam, and J.I. Gordon. 1996. Diphtheria toxin-mediated ablation of parietal cells in the stomach of transgenic mice. J Biol Chem. 271:3671–6.

Lin, W., A. Kemper, K.D. McCarthy, P. Pytel, J.P. Wang, I.L. Campbell, M.F. Utset, and B. Popko. 2004. Interferon-gamma induced medulloblastoma in the developing cerebellum. J Neurosci. 24:10074–83.

Litingtung, Y., L. Lei, H. Westphal, and C. Chiang. 1998. Sonic hedgehog is essential to foregut development. Nat Genet. 20:58–61.

Lopez-Diaz, L., K.L. Hinkle, R.N. Jain, Y. Zavros, C.S. Brunkan, T. Keeley, K.A. Eaton, J.L. Merchant, C.S. Chew, and L.C. Samuelson. 2006. Parietal cell hyperstimulation and autoimmune gastritis in cholera toxin transgenic mice. Am J Physiol Gastrointest Liver Physiol. 290:G970–9.

Lundkvist, J., and J. Naslund. 2007. Gamma-secretase: a complex target for Alzheimer's disease. Curr Opin Pharmacol. 7:112–8.

Ma, X., K. Chen, S. Huang, X. Zhang, P.A. Adegboyega, B.M. Evers, H. Zhang, and J. Xie. 2005. Frequent activation of the hedgehog pathway in advanced gastric adenocarcinomas. Carcinogenesis. 26:1698–705.

Macarthur, M., G.L. Hold, and E.M. El-Omar. 2004. Inflammation and cancer. II. Role of chronic inflammation and cytokine gene polymorphisms in the pathogenesis of gastrointestinal malignancy. Am J Physiol Gastrointest Liver Physiol. 286:G515–20.

Marigo, V., D.J. Roberts, S.M. Lee, O. Tsukurov, T. Levi, J.M. Gastier, D.J. Epstein, D.J. Gilbert, N.G. Copeland, C.E. Seidman, et al. 1995. Cloning, expression, and chromosomal location of SHH and IHH: two human homologues of the Drosophila segment polarity gene hedgehog. Genomics. 28:44–51.

Matsukura, N., K. Suzuki, T. Kawachi, M. Aoyagi, T. Sugimura, H. Kitaoka, H. Numajiri, A. Shirota, M. Itabashi, and T. Hirota. 1980. Distribution of marker enzymes and mucin in intestinal metaplasia in human stomach and relation to complete and incomplete types of intestinal metaplasia to minute gastric carcinomas. J Natl Cancer Inst. 65:231–40.

McColl, K.E. 2006. Cancer of the gastric cardia. Best Pract Res Clin Gastroenterol. 20:687–96.

McColl, K.E., E. el-Omar, and D. Gillen. 1998. Interactions between H. pylori infection, gastric acid secretion and anti-secretory therapy. Br Med Bull. 54:121–38.

Merlos-Suarez, A., S. Ruiz-Paz, J. Baselga, and J. Arribas. 2001. Metalloprotease-dependent pro-transforming growth factor-alpha ectodomain shedding in the absence of tumor necrosis factor-alpha-converting enzyme. J Biol Chem. 276:48510–7.

Mesquita, P., A. Raquel, L. Nuno, C.A. Reis, L.F. Silva, J. Serpa, I. Van Seuningen, H. Barros, and L. David. 2006. Metaplasia—a transdifferentiation process that facilitates cancer development: the model of gastric intestinal metaplasia. Crit Rev Oncog. 12:3–26.

Mishra, L., K. Shetty, Y. Tang, A. Stuart, and S.W. Byers. 2005. The role of TGF-beta and Wnt signaling in gastrointestinal stem cells and cancer. Oncogene. 24:5775–89.

Motoyama, J., J. Liu, R. Mo, Q. Ding, M. Post, and C.C. Hui. 1998. Essential function of Gli2 and Gli3 in the formation of lung, trachea and oesophagus. Nat Genet. 20:54–7.

Mutoh, H., S. Sakurai, K. Satoh, H. Osawa, Y. Hakamata, T. Takeuchi, and K. Sugano. 2004a. Cdx1 induced intestinal metaplasia in the transgenic mouse stomach: comparative study with Cdx2 transgenic mice. Gut. 53:1416–23.

Mutoh, H., S. Sakurai, K. Satoh, K. Tamada, H. Kita, H. Osawa, T. Tomiyama, Y. Sato, H. Yamamoto, N. Isoda, T. Yoshida, K. Ido, and K. Sugano. 2004b. Development of gastric carcinoma from intestinal metaplasia in Cdx2-transgenic mice. Cancer Res. 64:7740–7.

Nielsen, C.M., J. Williams, G.R. van den Brink, G.Y. Lauwers, and D.J. Roberts. 2004. Hh pathway expression in human gut tissues and in inflammatory gut diseases. Lab Invest. 84:1631–42.

Nishizawa, T., H. Suzuki, T. Masaoka, Y. Minegishi, E. Iwasahi, and T. Hibi. 2007. Helicobacter pylori eradication restored sonic hedgehog expression in the stomach. Hepatogastroenterology. 54:697–700.

Nojima, M., H. Suzuki, M. Toyota, Y. Watanabe, R. Maruyama, S. Sasaki, Y. Sasaki, H. Mita, N. Nishikawa, K. Yamaguchi, K. Hirata, F. Itoh, T. Tokino, M. Mori, K. Imai, and Y. Shinomura. 2007. Frequent epigenetic inactivation of SFRP genes and constitutive activation of Wnt signaling in gastric cancer. Oncogene. 26:4699–713.

Nompleggi, D.J., M. Beinborn, A. Roy, and M.M. Wolfe. 1994. The effect of recombinant cytokines on [14C]-aminopyrine accumulation by isolated canine parietal cells. J Pharmacol Exp Ther. 270:440–5.

Nomura, S., T. Baxter, H. Yamaguchi, C. Leys, A.B. Vartapetian, J.G. Fox, J.R. Lee, T.C. Wang, and J.R. Goldenring. 2004. Spasmolytic polypeptide expressing metaplasia to preneoplasia in H. felis-infected mice. Gastroenterology. 127:582–94.

Nomura, A.M., L.N. Kolonel, K. Miki, G.N. Stemmermann, L.R. Wilkens, M.T. Goodman, G.I. Perez-Perez, and M.J. Blaser. 2005a. Helicobacter pylori, pepsinogen, and gastric adenocarcinoma in Hawaii. J Infect Dis. 191:2075–81.

Nomura, S., S.H. Settle, C.M. Leys, A.L. Means, R.M. Peek, Jr., S.D. Leach, C.V. Wright, R.J. Coffey, and J.R. Goldenring. 2005b. Evidence for repatterning of the gastric fundic epithelium associated with Menetrier's disease and TGFalpha overexpression. Gastroenterology. 128:1292–305.

Nusslein-Volhard, C., and E. Wieschaus. 1980. Mutations affecting segment number and polarity in Drosophila. Nature. 287:795–801.

Ogasawara, N., T. Tsukamoto, T. Mizoshita, K. Inada, X. Cao, Y. Takenaka, T. Joh, and M. Tatematsu. 2006. Mutations and nuclear accumulation of beta-catenin correlate with intestinal phenotypic expression in human gastric cancer. Histopathology. 49:612–21.

Oshima, H., A. Matsunaga, T. Fujimura, T. Tsukamoto, M.M. Taketo, and M. Oshima. 2006. Carcinogenesis in mouse stomach by simultaneous activation of the Wnt signaling and prostaglandin E2 pathway. Gastroenterology. 131:1086–95.

Oshima, H., M. Oshima, K. Inaba, and M.M. Taketo. 2004. Hyperplastic gastric tumors induced by activated macrophages in COX-2/mPGES-1 transgenic mice. EMBO J. 23:1669–78.

Pai, R., B. Soreghan, I.L. Szabo, M. Pavelka, D. Baatar, and A.S. Tarnawski. 2002. Prostaglandin E2 transactivates EGF receptor: a novel mechanism for promoting colon cancer growth and gastrointestinal hypertrophy. Nat Med. 8:289–93.

Pan, Y., C.B. Bai, A.L. Joyner, and B. Wang. 2006. Sonic hedgehog signaling regulates Gli2 transcriptional activity by suppressing its processing and degradation. Mol Cell Biol. 26:3365–77.

Peek, R.M., Jr., and J.E. Crabtree. 2006. Helicobacter infection and gastric neoplasia. J Pathol. 208:233–48.

Piedimonte, L.R., I.K. Wailes, and H.L. Weiner. 2005. Medulloblastoma: mouse models and novel targeted therapies based on the Sonic hedgehog pathway. Neurosurg Focus. 19:E8.

Pin, C.L., A.C. Bonvissuto, and S.F. Konieczny. 2000. Mist1 expression is a common link among serous exocrine cells exhibiting regulated exocytosis. Anat Rec. 259:157–67.

Pissarnitski, D. 2007. Advances in gamma-secretase modulation. Curr Opin Drug Discov Dev. 10:392–402.

Porter, J.A., S.C. Ekker, W.J. Park, D.P. von Kessler, K.E. Young, C.H. Chen, Y. Ma, A.S. Woods, R.J. Cotter, E.V. Koonin, and P.A. Beachy. 1996a. Hedgehog patterning activity: role of a lipophilic modification mediated by the carboxy-terminal autoprocessing domain. Cell. 86:21–34.

Porter, J.A., D.P. von Kessler, S.C. Ekker, K.E. Young, J.J. Lee, K. Moses, and P.A. Beachy. 1995. The product of hedgehog autoproteolytic cleavage active in local and long-range signalling. Nature. 374:363–6.

Porter, J.A., K.E. Young, and P.A. Beachy. 1996b. Cholesterol modification of hedgehog signaling proteins in animal development. Science. 274:255–9.

Powell, D.W., R.C. Mifflin, J.D. Valentich, S.E. Crowe, J.I. Saada, and A.B. West. 1999a. Myofibroblasts. I. Paracrine cells important in health and disease. Am J Physiol. 277:C1–9.

Powell, D.W., R.C. Mifflin, J.D. Valentich, S.E. Crowe, J.I. Saada, and A.B. West. 1999b. Myofibroblasts. II. Intestinal subepithelial myofibroblasts. Am J Physiol. 277:C183–201.

Prenzel, N., E. Zwick, H. Daub, M. Leserer, R. Abraham, C. Wallasch, and A. Ullrich. 1999. EGF receptor transactivation by G-protein-coupled receptors requires metalloproteinase cleavage of proHB-EGF. Nature. 402:884–8.

Ramalho-Santos, M., D.A. Melton, and A.P. McMahon. 2000. Hedgehog signals regulate multiple aspects of gastrointestinal development. Development. 127:2763–72.

Ramsey, V.G., J.M. Doherty, C.C. Chen, T.S. Stappenbeck, S.F. Konieczny, and J.C. Mills. 2007. The maturation of mucus-secreting gastric epithelial progenitors into digestive-enzyme secreting zymogenic cells requires Mist1. Development. 134:211–22.

Rindi, G., R. Fiocca, A. Morocutti, A. Jacobs, N. Miller, and B. Thjodleifsson. 2005. Effects of 5 years of treatment with rabeprazole or omeprazole on the gastric mucosa. Eur J Gastroenterol Hepatol. 17:559–66.

Riobo, N.A., and D.R. Manning. 2007. Pathways of signal transduction employed by vertebrate hedgehogs. Biochem J. 403:369–79.

Rocco, A., R. Caruso, S. Toracchio, L. Rigoli, F. Verginelli, T. Catalano, M. Neri, M.C. Curia, L. Ottini, V. Agnese, V. Bazan, A. Russo, G. Pantuso, G. Colucci, R. Mariani-Costantini, and G. Nardone. 2006. Gastric adenomas: relationship between clinicopathological findings, Helicobacter pylori infection, APC mutations and COX-2 expression. Ann Oncol. 17(Suppl 7): vii103–8.

Roelink, H., J.A. Porter, C. Chiang, Y. Tanabe, D.T. Chang, P.A. Beachy, and T.M. Jessell. 1995. Floor plate and motor neuron induction by different concentrations of the amino-terminal cleavage product of sonic hedgehog autoproteolysis. Cell. 81:445–55.

Rohatgi, R., L. Milenkovic, and M.P. Scott. 2007. Patched1 regulates hedgehog signaling at the primary cilium. Science. 317:372–6.

Rugge, M., F. Di Mario, M. Cassaro, R. Baffa, F. Farinati, J. Rubio, Jr., and V. Ninfo. 1993. Pathology of the gastric antrum and body associated with Helicobacter pylori infection in non-ulcerous patients: is the bacterium a promoter of intestinal metaplasia? Histopathology. 22:9–15.

Ruiz i Altaba, A., P. Sanchez, and N. Dahmane. 2002. Gli and hedgehog in cancer: tumours, embryos and stem cells. Nat Rev Cancer. 2:361–72.

Sahin, U., and C.P. Blobel. 2007. Ectodomain shedding of the EGF-receptor ligand epigen is mediated by ADAM17. FEBS Lett. 581:41–4.

Sanderson, M.P., P.J. Dempsey, and A.J. Dunbar. 2006. Control of ErbB signaling through metalloprotease mediated ectodomain shedding of EGF-like factors. Growth Factors. 24:121–36.

Sanderson, M.P., S.N. Erickson, P.J. Gough, K.J. Garton, P.T. Wille, E.W. Raines, A.J. Dunbar, and P.J. Dempsey. 2005. ADAM10 mediates ectodomain shedding of the betacellulin precursor activated by p-aminophenylmercuric acetate and extracellular calcium influx. J Biol Chem. 280:1826–37.

Scarff, K.L., L.M. Judd, B.H. Toh, P.A. Gleeson, and I.R. Van Driel. 1999. Gastric H(+),K(+)-adenosine triphosphatase beta subunit is required for normal function, development, and membrane structure of mouse parietal cells. Gastroenterology. 117:605–18.

Schmidt, P.H., J.R. Lee, V. Joshi, R.J. Playford, R. Poulsom, N.A. Wright, and J.R. Goldenring. 1999. Identification of a metaplastic cell lineage associated with human gastric adenocarcinoma. Lab Invest. 79:639–46.

Schultheis, P.J., L.L. Clarke, P. Meneton, M. Harline, G.P. Boivin, G. Stemmermann, J.J. Duffy, T. Doetschman, M.L. Miller, and G.E. Shull. 1998. Targeted disruption of the murine Na+/H+ exchanger isoform 2 gene causes reduced viability of gastric parietal cells and loss of net acid secretion. J Clin Invest. 101:1243–53.

Sekine, A., Y. Akiyama, K. Yanagihara, and Y. Yuasa. 2006. Hath1 up-regulates gastric mucin gene expression in gastric cells. Biochem Biophys Res Commun. 344:1166–71.

Shannon, J.M., and B.A. Hyatt. 2004. Epithelial-mesenchymal interactions in the developing lung. Annu Rev Physiol. 66:625–45.

Shao, J.S., W. Schepp, and D.H. Alpers. 1998. Expression of intrinsic factor and pepsinogen in the rat stomach identifies a subset of parietal cells. Am J Physiol. 274:G62–70.

Sheehan, K.M., C. Gulmann, G.S. Eichler, J.N. Weinstein, H.L. Barrett, E.W. Kay, R.M. Conroy, L.A. Liotta, and E.F. Petricoin, 3rd. 2007. Signal pathway profiling of epithelial and stromal compartments of colonic carcinoma reveals epithelial-mesenchymal transition. Oncogene. 27(3):323–31.

Shiotani, A., H. Iishi, S. Ishiguro, M. Tatsuta, Y. Nakae, and J.L. Merchant. 2005a. Epithelial cell turnover in relation to ongoing damage of the gastric mucosa in patients with early gastric cancer: increase of cell proliferation in paramalignant lesions. J Gastroenterol. 40:337–44.

Shiotani, A., H. Iishi, N. Uedo, S. Ishiguro, M. Tatsuta, Y. Nakae, M. Kumamoto, and J.L. Merchant. 2005b. Evidence that loss of sonic hedgehog is an indicator of Helicobater pylori-induced atrophic gastritis progressing to gastric cancer. Am J Gastroenterol. 100:581–7.

Shiotani, A., H. Iishi, N. Uedo, M. Kumamoto, Y. Nakae, S. Ishiguro, M. Tatsuta, and D.Y. Graham. 2005c. Histologic and serum risk markers for noncardia early gastric cancer. Int J Cancer. 115:463–9.

Shiotani, A., N. Uedo, H. Iishi, M. Tatsuta, S. Ishiguro, Y. Nakae, T. Kamada, K. Haruma, and J.L. Merchant. 2007. Re-expression of sonic hedgehog and reduction of CDX2 after Helicobacter pylori eradication prior to incomplete intestinal metaplasia. Int J Cancer. 121(6):1182–9.

Sierra, R., C. Une, V. Ramirez, M.I. Gonzalez, J.A. Ramirez, A. de Mascarel, R. Barahona, R. Salas-Aguilar, R. Paez, G. Avendano, A. Avalos, N. Broutet, and F. Megraud. 2006. Association of serum pepsinogen with atrophic body gastritis in Costa Rica. Clin Exp Med. 6:72–8.

Silberg, D.G., J. Sullivan, E. Kang, G.P. Swain, J. Moffett, N.J. Sund, S.D. Sackett, and K.H. Kaestner. 2002. Cdx2 ectopic expression induces gastric intestinal metaplasia in transgenic mice. Gastroenterology. 122:689–96.

Sipponen, P., P. Ranta, T. Helske, I. Kaariainen, T. Maki, A. Linnala, O. Suovaniemi, A. Alanko, and M. Harkonen. 2002. Serum levels of amidated gastrin-17 and pepsinogen I in atrophic gastritis: an observational case-control study. Scand J Gastroenterol. 37:785–91.

Smith, H., Y.H. Lou, P. Lacy, and K.S. Tung. 1992. Tolerance mechanism in experimental ovarian and gastric autoimmune diseases. J Immunol. 149:2212–8.

Sobral, L.M., P.F. Montan, H. Martelli-Junior, E. Graner, and R.D. Coletta. 2007. Opposite effects of TGF-beta1 and IFN-gamma on transdifferentiation of myofibroblast in human gingival cell cultures. J Clin Periodontol. 34:397–406.

Spicer, Z., M.L. Miller, A. Andringa, T.M. Riddle, J.J. Duffy, T. Doetschman, and G.E. Shull. 2000. Stomachs of mice lacking the gastric H,K-ATPase alpha -subunit have achlorhydria, abnormal parietal cells, and ciliated metaplasia. J Biol Chem. 275:21555–65.

Stepan, V., N. Pausawasdi, S. Ramamoorthy, and A. Todisco. 2004. The Akt and MAPK signal-transduction pathways regulate growth factor actions in isolated gastric parietal cells. Gastroenterology. 127:1150–61.

Stepan, V., S. Ramamoorthy, H. Nitsche, Y. Zavros, J.L. Merchant, and A. Todisco. 2005. Regulation and function of the sonic hedgehog signal transduction pathway in isolated gastric parietal cells. J Biol Chem. 280:15700–8.

Stockbruegger, R.W., P.B. Cotton, G.G. Menon, J.O. Beilby, B.A. Bartholomew, M.J. Hill, and C.L. Walters. 1984. Pernicious anaemia, intragastric bacterial overgrowth, and possible consequences. Scand J Gastroenterol. 19:355–64.

Suzuki, K., H. Fukui, T. Kayahara, M. Sawada, H. Seno, H. Hiai, R. Kageyama, H. Okano, and T. Chiba. 2005a. Hes1-deficient mice show precocious differentiation of Paneth cells in the small intestine. Biochem Biophys Res Commun. 328:348–52.

Suzuki, H., Y. Minegishi, Y. Nomoto, T. Ota, T. Masaoka, G.R. van den Brink, and T. Hibi. 2005b. Down-regulation of a morphogen (sonic hedgehog) gradient in the gastric epithelium of Helicobacter pylori-infected Mongolian gerbils. J Pathol. 206:186–97.

Taipale, J., J.K. Chen, M.K. Cooper, B. Wang, R.K. Mann, L. Milenkovic, M.P. Scott, and P.A. Beachy. 2000. Effects of oncogenic mutations in Smoothened and Patched can be reversed by cyclopamine. Nature. 406:1005–9.

Takaishi, S., and T.C. Wang. 2007. Gene expression profiling in a mouse model of Helicobacter-induced gastric cancer. Cancer Sci. 98:284–93.

Tian, H., J. Jeong, B.D. Harfe, C.J. Tabin, and A.P. McMahon. 2005. Mouse Disp1 is required in sonic hedgehog-expressing cells for paracrine activity of the cholesterol-modified ligand. Development. 132:133–42.

Toh, B.H., I.R. van Driel, and P.A. Gleeson. 1997. Pernicious anemia. N Engl J Med. 337:1441–8.

Tsukamoto, T., T. Mizoshita, and M. Tatematsu. 2006. Gastric-and-intestinal mixed-type intestinal metaplasia: aberrant expression of transcription factors and stem cell intestinalization. Gastric Cancer. 9:156–66.

Uhal, B.D., J.K. Kim, X. Li, and M. Molina-Molina. 2007. Angiotensin-TGF-beta 1 crosstalk in human idiopathic pulmonary fibrosis: autocrine mechanisms in myofibroblasts and macrophages. Curr Pharm Des. 13:1247–56.

van den Brink, G.R., J.C. Nielsen, C. Xu, F.J. ten Kate, J. Glickman, S.J. van Deventer, D.J. Roberts, and M.P. Peppelenbosch. 2002. Sonic hedgehog expression correlates with fundic gland differentiation in the adult gastrointestinal tract. Gut. 51:628–33.

Van Den Brink, G.R., J.C. Hardwick, G.N. Tytgat, M.A. Brink, F.J. Ten Kate, S.J. Van Deventer, and M.P. Peppelenbosch. 2001. Sonic hedgehog regulates gastric gland morphogenesis in man and mouse. Gastroenterology. 121:317–28.

van Den Brink, G.R., F.J. ten Kate, C.Y. Ponsioen, M.M. Rive, G.N. Tytgat, S.J. van Deventer, and M.P. Peppelenbosch. 2000. Expression and activation of NF-kappa B in the antrum of the human stomach. J Immunol. 164:3353–9.

Wang, T.C., M.W. Babyatsky, P.S. Oates, Z. Zhang, L. Tillotson, M. Chulak, S.J. Brand, and E.V. Schmidt. 1995. A rat gastrin-human gastrin chimeric transgene directs antral G cell-specific expression in transgenic mice. Am. J. Physiol. 268:G1025–36.

Wang, T.C., C.A. Dangler, D. Chen, J.R. Goldenring, T. Koh, R. Raychowdhury, R.J. Coffey, S. Ito, A. Varro, G.J. Dockray, and J.G. Fox. 2000. Synergistic interaction between hypergastrinemia and Helicobacter infection in a mouse model of gastric cancer. Gastroenterology. 118:36–47.

Wang, T.C., J.R. Goldenring, C. Dangler, S. Ito, A. Mueller, W.K. Jeon, T.J. Koh, and J.G. Fox. 1998. Mice lacking secretory phospholipase A$_2$ show altered apoptosis and differentiation with Helicobacter felis infection. Gastroenterology. 114:675–689.

Wang, J., W. Lin, B. Popko, and I.L. Campbell. 2004. Inducible production of interferon-gamma in the developing brain causes cerebellar dysplasia with activation of the Sonic hedgehog pathway. Mol Cell Neurosci. 27:489–96.

Wang, J., N. Pham-Mitchell, C. Schindler, and I.L. Campbell. 2003. Dysregulated Sonic hedgehog signaling and medulloblastoma consequent to IFN-alpha-stimulated STAT2-independent production of IFN-gamma in the brain. J Clin Invest. 112:535–43.

Watanabe, T., M. Tada, H. Nagai, S. Sasaki, and M. Nakao. 1998. Helicobacter pylori infection induces gastric cancer in mongolian gerbils. Gastroenterology. 115:642–8.

Wendler, F., X. Franch-Marro, and J.P. Vincent. 2006. How does cholesterol affect the way hedgehog works? Development. 133:3055–61.

Whittingham, S., and I.R. Mackay. 2005. Autoimmune gastritis: historical antecedents, outstanding discoveries, and unresolved problems. Int Rev Immunol. 24:1–29.

Williams, C., and K.E. McColl. 2006. Review article: proton pump inhibitors and bacterial overgrowth. Aliment Pharmacol Ther. 23:3–10.

Wright, N.A. 1996. Migration of the ductular elements of gut-associated glands gives clues to the histogenesis of structures associated with responses to acid hypersecretory state: the origins of "gastric metaplasia" in the duodenum of the specialized mucosa of Barrett's esophagus and of pseudopyloric metaplasia. Yale J Biol Med. 69:147–53.

Yanai, K., S. Nagai, J. Wada, N. Yamanaka, M. Nakamura, N. Torata, H. Noshiro, M. Tsuneyoshi, M. Tanaka, and M. Katano. 2007. Hedgehog signaling pathway is a possible therapeutic target for gastric cancer. J Surg Oncol. 95:55–62.

Yang, Q., N.A. Bermingham, M.J. Finegold, and H.Y. Zoghbi. 2001. Requirement of Math1 for secretory cell lineage commitment in the mouse intestine. Science. 294:2155–8.

Zavros, Y., K.A. Eaton, W. Kang, S. Rathinavelu, V. Katukuri, J.Y. Kao, L.C. Samuelson, and J.L. Merchant. 2005. Chronic gastritis in the hypochlorhydric gastrin-deficient mouse progresses to adenocarcinoma. Oncogene. 24:2354–66.

Zavros, Y., S. Rathinavelu, J.Y. Kao, A. Todisco, J. DelValle, J.V. Weinstock, M.J. Low, and J. L. Merchant. 2003. Treatment of Helicobacter gastritis with interleukin-4 requires somatostatin. Proc Natl Acad Sci. 100:12944–9.

Zavros, Y., G. Rieder, A. Ferguson, and J.L. Merchant. 2002a. Gastritis and hypergastrinemia due to Acinetobacter lwoffii in mice. Infect Immun. 70:2630–9.

Zavros, Y., G. Rieder, A. Ferguson, L.C. Samuelson, and J.L. Merchant. 2002b. Genetic or chemical hypochlorhydria is associated with inflammation that modulates parietal and G-cell populations in mice. Gastroenterology. 122:119–33.

Zavros, Y., M. Waghray, A. Tessier, L. Bai, A. Todisco, D.L. Gumucio, L.C. Samuelson, A. Dlugosz, and J.L. Merchant. 2007. Reduced pepsin a processing of sonic hedgehog in parietal cells precedes gastric atrophy and transformation. J Biol Chem. 282(46):33265–74.

Zeng, X., J.A. Goetz, L.M. Suber, W.J. Scott, Jr., C.M. Schreiner, and D.J. Robbins. 2001. A freely diffusible form of Sonic hedgehog mediates long-range signalling. Nature. 411:716–20.

Zhao, C.H., X.M. Bu, and N. Zhang. 2007. Hypermethylation and aberrant expression of Wnt antagonist secreted frizzled-related protein 1 in gastric cancer. World J Gastroenterol. 13:2214–7.

Chapter 19
Genetic Models of Gastric Cancer in the Mouse

Andrew S. Giraud and Louise M. Judd

Introduction

This chapter concentrates on murine genetic models that mimic gastric adenocarcinoma development, the most common stomach cancer of humans, rather than gastric lymphoma, or endocrine (ECL) carcinoids for which only a handful of mouse models have been described. Epithelial tumors are much less common in mice than humans; in the former, extragastric lymphomas and sarcomas predominate (Rangarajan and Weinberg 2003). Therefore, for epithelial tumors of the mouse stomach to be a dominant phenotype, the gene defect is likely to be important, and by homology may also contribute to human gastric cancer pathogenesis.

To date, the vast majority of genetic models recapitulate some or all of the developmental stages associated with intestinal-type gastric adenocarcinoma development rather than the less common sporadic (diffuse) gastric cancer. Thus, stepwise progression beginning with gastritis, atrophy, intestinal metaplasia, spasmolytic polypeptide expressing metaplasia (SPEM), dysplasia, and submucosal invasion over variable timeframes, have been documented in numerous models. It should be noted that although this progression closely mimics that of human gastric adenocarcinoma development, there are several significant differences. First, differentiation of goblet cells associated with intestinal metaplasia is not observed apart from Cdx1 and 2 transgenics (Mutoh et al. 2004). Second, despite the often robust tumor phenotype and frequent observation of submucosal vessel invasion, no genetic models to date show true disseminating metastasis, with the possible exception of the H^+, K^+-ATPase SV40 T antigen transgenic (Li et al. 1995). Metastasis is therefore likely to be polygenic in nature, requiring multiple genetic hits that have yet to be recapitulated in a mouse model, or the mouse is unusually resistant to gastric cancer-induced metaplasia.

Most mouse genetic models have been established on the C57Bl6 or mixed C57Bl6/129SvJ background. In many of the models discussed below, there are differences in tumor penetration and phenotype, which are dependent on genetic background and gender. The gender-based differences are variable, with instances of increased penetrance in both females and males. One possible reason for gender-based differences is a role for estrogen-driven signaling in the gastric mucosa; another

T.C. Wang et al. (eds.) *The Biology of Gastric Cancers*,
© Springer Science+Business Media, LLC 2009

is the sensitivity of the gastric mucosa to gender disparities in immunologic responses. As might be expected for inbred mouse populations, the inherent genetic variability in mouse strains results in different comparative susceptibilities to the development of gastric pathologies and cancers. For instance, it is well documented that development of autoimmunity in the stomach is determined by genetic background with BALB/c being susceptible and C57BL/6 resistant (Baxter et al. 2005). Regions of DNA have been isolated that account for these differing susceptibilities and some of these regions colocalize with genes that determine diabetes susceptibility in NOD mice (Ang et al. 2007). Because many epithelial cancers are dependent on strong inflammatory responses, it is clear that these strain differences help shape autoimmune susceptibility in the stomach, and may also be important in determining gastric tumor susceptibility. As such, it is crucial when utilizing particular genetic models of gastric cancer, that wild-type control mice of an identical genetic background (and preferably litter mates) to mice with pathology are used as comparators.

The epithelium of the mouse stomach can be separated into two distinct anatomic compartments: the proximal fundus (or body) and the distal antrum. These two compartments perform distinct functions. The fundus produces acid and digestive enzyme secretions, and the antrum has an endocrine and mucus-secretory role. As such, gastric tumors develop mainly independently in these two regions and have very different genetic triggers. Most of the models discussed develop tumors in either the fundus or the antrum, but in some cases, both are involved Table 19.1.

Trefoil Factor 1–/– Mutant

Trefoil factor 1 (TFF1 or pS2) is a member of the trefoil domain peptides of which there are two other members: TFF2 and TFF3. These peptides are highly expressed in a regional manner predominantly in the gut; TFF1 and 2 synthesized and secreted by surface/pit mucus and mucus neck/antral gland cells, respectively. Before the engineering of the TFF1–/– mouse, the function of this peptide was thought to be primarily associated with mucosal repair and barrier function; however, TFF1–/– mice exhibit gastric antral/pylorus-specific hyperplasia by 1 week of age (S. Karam, personal communication), and by 20 weeks, approximately one-third of mice develop dysplasia and multifocal intraepithelial carcinomas, some of which penetrate the submucosa (Lefebvre et al. 1996). Unlike many other genetic models of gastric cancer, extensive transmural inflammation has not been described in TFF1–/– stomach pathology. TFF1–/– tumors are fully penetrant, and are restricted to the distal stomach. This, together with subsequent *in vivo* and *in vitro* evidence, supports a role for TFF1 as a stomach-specific tumor-suppressor gene (Tomasetto and Rio 2005).

Secondary phenotypes observed in the TFF1–/– mouse were the increased length of small intestinal villi with associated lymphocytic infiltrate (Lefebvre et al. 1996) and the loss of neutral glycoprotein (mucus) from surface and pit cells of the stomach. This latter observation suggests a role for TFF1 in regulating gastric differentiation pathways. This is supported by recent developmental data, based on a

Table 19.1 Temporal, secretory, and pathologic development in the stomach exemplified by different genetic models of gastric hyperplasia and tumorigenesis in the mouse

Model	Age of onset	Achlorhydria	Hypergastrinemia	Fundus Hyperplasia	Fundus Atrophy	Fundus Inflammation	Fundus Metaplasia	Fundus Dysplasia	Antrum Hyperplasia	Antrum Atrophy	Antrum Inflammation	Antrum Metaplasia	Antrum Dysplasia	Carcinoma in situ	Invasion	Metastasis
TFF1−/−	1 week	N	ND	−	−	−	++	−	++	−	−	++	++	Y	Y	N
gp130 757F757F	3–4 weeks	Y	N	+++	++	+++	+++	++	+++	++	+++	+++	+++	Y	Y	N
Cdx2 transgenic	12 weeks	Y	Y	++	ND	++	++	++	−	−	−	−	−	N	Y	N
INS-GAS	24 months	N	Y	+++	+++	ND	+++	+++	−	−	−	−	−	Y	N	N
ACT-GAS	20 months	N	Y	+++	+++	ND	+++	+++	−	−	−	−	−	Y	N	N
Gastrin−/−	12 months	Y	N	ND	ND	ND	ND	ND	+++	++	+++	+++	+++	Y	N	N
H/K-ATPase α subunit−/−	3 months	Y	Y	+++	+++	+++	+++	+	−	−	−	−	−	N	N	N
H/K-ATPase β subunit−/−	20 months	Y	Y	+++	+++	+++	+++	+	−	−	−	−	−	N	N	
NHE2−/−	3 months	Y	Y	++	+++	ND	ND	−	−	−	−	−	−	N	N	N
NHE4−/−	3 months	Y	Y	++	+++	ND	ND	−	−	−	−	−	−	N	N	N
Kv1qt1−/−	ND	ND	ND	+++	+++	+++	+++	+	−	−	−	−	−	N	N	N
KCNE2−/−	ND	Y	Y	+++	+++	+++	+++	+	−	−	−	−	−	N	N	N
Kcnq1−/−	ND	ND	ND	+++	+++	+++	+++	++	−	−	−	−	−	N	N	N

(continued)

Table 19.1 (continued)

Model	Age of onset	Achlorhydria	Hypergastrinemia	Fundus Hyperplasia	Fundus Atrophy	Fundus Inflammation	Fundus Metaplasia	Fundus Dysplasia	Antrum Hyperplasia	Antrum Atrophy	Antrum Inflammation	Antrum Metaplasia	Antrum Dysplasia	Carcinoma in situ	Invasion	Metastasis
Histamine H2 receptor−/−	17 months	N	Y	+++	−+?	+++	–	–	–	–	–	–	–	N	Herniation	N
IQGAP1−/−	24 months	ND	ND	+++	ND	ND	ND	+	–	–	–	–	–	N	N	N
MTH1−/−	ND	ND	ND	–	–	–	–	–	++	ND	ND	ND	++	Y	ND	ND
K19-C2mE transgenic	5 weeks	ND	ND	+++	ND	++	+++	–	–	–	–	–	–	N	N	N
TSP-1−/−	ND	ND	ND	ND	ND	ND	ND	ND	++	ND	ND	++	ND	ND	ND	ND
TGFα transgenic	4-6 week	Y	N	+++	++	+/−	++	ND	–	–	–	–	–	N	N	N
AhR transgenic	12 months	ND	ND	+++	+++	ND	+++	++	–	–	–	–	–	Y	Y	N
Klf4 conditional−/−	6 months	ND	–	+++	+++	–	+++	–	+++	+++	–	+++	+++	N	N	N
p27^{Kip1}−/−	12 months	ND	ND	+++	ND	ND	+++	+	–	–	–	–	–	N	N	N
β-Catenin transgenic	ND	ND	ND	ND	ND	ND	+	+++	–	–	–	–	–	N	N	N
MHC Class II−/−	6 months	ND	+	+	–	+++	ND	ND	–	–	–	–	–	N	N	N
CA IX−/−	ND	N	N	++	ND	ND	ND	ND	–	–	–	–	–	N	N	N

Model	Age														
CEA SV40 transgenic	5 months	ND	−	−	−	−	−	+++	+++	ND	++	+++	Y	Y	N
H+/K+-ATPase β subunit SV40 transgenic	12 months	ND	+++	+++	ND	ND	+++	−	−	−	−	−	Y	Y	Y
TGFβ1−/−	0.5 month	ND	+++	+++	+++	ND	ND	−	−	−	−	−	N	N	N
SMAD4+/−	18 months	ND	−	−	−	−	−	+++	ND	ND	+++	+++	Y	Y	N
elf+/− SMAD4+/−	ND	ND	−	−	−	−	−	+++	+++	ND	+++	+++	Y	Y	N
Runx3−/−	8 months	ND	+++	+++	+++	++	+	ND	ND	ND	ND	ND	Y	Y	N
Fkh6−/−	0.1 month	ND	++	ND	ND	++	ND	−	−	−	−	−	ND	ND	ND
Shh−/−	18.5-day embryo	ND	++	ND	ND	+++	ND	++	ND	ND	ND	+++	ND	ND	ND
Occludin−/−	10 months	ND	+++	ND	ND	+++	−	−	−	−	−	−	ND	ND	ND
CCR7−/−	12 months	ND	+++	ND	+++	+++	+	−	−	−	−	−	N	Y	N
NF-κB-2−/−	12 months	ND	−	−	−	−	−	+++	+++	++	+	−	ND	ND	ND

Abbreviations: ND: not done or not reported; + to +++ : positive; −: negative; Y: yes; N: no.

panel of lineage-specific markers in which it was found that in the TFF1−/− mouse, pit cells were amplified at the expense of parietal cells, and that TFF1−/− mucus neck cells were dysfunctional because they were depleted of TFF2 (Karam et al. 2004). Moreover, the loss of mucin indicates that TFF1 is required for correct maturation and functionality of surface and pit cells (Tomasetto and Rio 2005).

The utility of the TFF1−/− model is exemplified by investigations of the protumorigenic role of constitutive expression of cyclooxygenase-2 (COX-2) and its metabolic products, as well as the mechanisms of action of the specific COX-2 inhibitor celecoxib in this mouse (Saukkonen et al. 2003). Similar to adenocarcinoma of the stomach in humans, the TFF1−/− mouse has elevated COX-2 expression (stromal cells) specifically associated with distal stomach tumorigenesis, and oral application of celecoxib for 3 months caused adenoma shrinkage because of focal ulceration and was also associated with localized mononuclear cell inflammation (Saukkonen et al. 2003). This outcome broadens our understanding of the role of COX-2 and the use of its inhibitors in inhibiting neoplasia, as well as the critical role of macrophages in tumor pathology, and supports the use of the TFF1−/− mouse as a relevant model of gastric cancer.

Because trefoil peptide receptors are yet to be identified, and recombinant TFF1 is not widely available, limited information has been generated on gene targets for TFF1 action. An alternative approach has been to compare the antral stomach gene expression profile between age-matched TFF1−/− mice and wild-type litter mates. Thus, differential expression analysis by cDNA array identified a number of mainly endoplasmic reticulum genes associated with the unfolded protein response including GRP78, ERp72, p58IPK, CHOP10, and clusterin. This, along with complementary ultrastructural data, suggests that lack of TFF1 results in an accumulation of misfolded proteins in the gastric epithelium (Torres et al. 2002), and that one of the normal functions of TFF1 is to regulate correct protein folding and likely aspects of the secretory process. It is not yet clear how this is accomplished because TFF1 is primarily a secreted protein; however, it must also act intracellularly because it has properties of a tumor-suppressor gene.

Taking advantage of the slow transition to carcinoma *in situ* exemplified by the TFF1−/− mouse, an oligonucleotide microarray approach has also been used to identify genes involved in mediating preneoplastic events, particularly metaplasia and dysplasia. Genes most overexpressed in the TFF1 null stomach were claudin 7 (encoding a tight junction protein), early growth response 1 (encoding a nuclear transcription factor), and epithelial membrane protein 1 (encoding a junctional membrane protein) (Johnson et al. 2005). Claudin 7 upregulation was subsequently confirmed in human preneoplastic stomach and in gastric adenocarcinoma, thus underscoring the utility of the TFF1−/− mouse model in gastric cancer gene discovery.

The gp130^{757F757F} Knockin Mutant

The gp130^{757F757F} or gp130^{FF} mouse was originally generated by knockin mutation of the common SHP2/SOCS3 binding site on the interleukin (IL)-6 family coreceptor gp130, in order to genetically dissect the independent contribution of the two main signaling pathways downstream of this common signal transducing receptor (Tebbutt et al. 2002). The critical tyrosine (Y) residue at position 757 on the intracellular arm of gp130 was replaced by a phenylalanine residue in both alleles (hence 757F757F), thereby preventing SHP2 (and SOCS3) docking after ligand binding with the receptor complex, and blocking signal transduction via the ras/ERK/AP-1 signaling cascade. Inhibition of this cascade not only prevents activation of target genes by the AP-1 transcription factors, but also serves to augment signaling via an alternate second messenger pathway frequently utilized by IL-6 cytokines, which involves recruitment, activation, and binding of the transcription factor STAT3 to *cis* regulatory elements of target genes. Thus, normal feedback inhibition of STAT3 activation by SHP-2/ras/ERK and SOCS3 is lost, resulting in constitutive oncogenic signaling by STAT3 dimers. The phenotypic outcome is splenomegaly and rapid gastric tumorigenesis, with genes normally regulated by IL-6 family cytokine activation of SHP-2/ras/ERK/AP-1 being downregulated, such as the tumor suppressor gene TFF1, whereas genes regulated by STAT3, such as the growth factors like Reg1 (Judd et al. 2004), and antiapoptotic, proangiogenic, and cytostatic genes such as bcl-2, bcl-x, vascular endothelial growth factor, and transforming growth factor (TGF)β (Jenkins et al. 2005) are strongly upregulated in a constitutive manner (Figure 19.1).

An important feature of this model is that it displays many of the temporal phenotypic changes characteristic of the intestinal form of human gastric adenocarcinoma including gastritis, atrophy, intestinal-type mucus metaplasia and SPEM, dysplasia, and submucosal invasion, but without disseminating metastasis (Figure 19.1). This progression occurs independently of *Helicobacter pylori* infection, hypergastrinemia (mice are hypogastrinemic), and constitutive epidermal growth factor receptor activation (Judd et al. 2004) characteristic of many other stomach cancer models, and highlights the importance of IL-6 family cytokine signaling in maintaining gastric homeostasis. Other important features of the model include the following: gastric tumorigenesis is 100% penetrant; timing and site of tumor initiation and dysplastic changes are consistent from mouse to mouse; development is rapid so that initiating antral tumors with associated transmural gastritis are macroscopically evident by 4 weeks of age, then growth progresses rapidly along the lesser curvature of the stomach to encompass the entire secretory mucosa by 20 weeks (Judd et al. 2004); tumor development and degree of penetrance are prevalent in both the C57Bl6/129Sv/J and C57Bl6 (but not BALB/c) background; and tumor growth occurs at the same rate in specific pathogen-free and nonbarrier animal facilities. Interestingly, mice heterozygous for the mutation (gp130[757F+/−]) show variable tumor disposition (frequently antro-fundic junction) and delayed initiation compared with gp130[757FF] mice, as well as greatly attenuated tumor growth with age (Judd et al. 2004).

The role of initiating factors in driving the gastric tumor phenotype has recently been evaluated. Inflammation and attendant proinflammatory cytokine and chemokine expression accompanies tumor growth, which can be delayed by antimicrobial treatment (Judd et al. 2006). The main stomach ligands for the gp130 receptor are the cytokines IL-6 and IL-11 (Howlett et al. 2005;Judd et al. 2004), and both of these are strongly induced in the stomach during tumorigenesis in the gp130[FF] mouse. IL-6 is

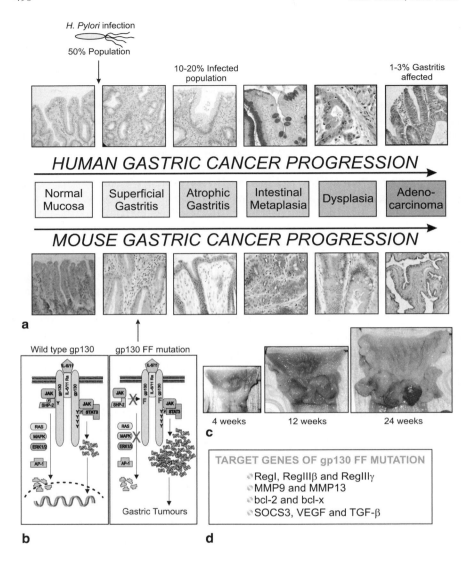

Fig. 19.1 Development of gastric cancer in the gp130[757FF] mouse. (**A**) Comparison among events leading to gastric adenocarcinoma initiated after *Helicobacter pylori* infection in the human stomach, and that of the gp130[757FF] mouse in which pathology occurs after genetic imbalancing of interleukin-6 family cytokine signaling downstream of the signal transducing receptor gp130. (**B**) Diagrammatic representation of imbalanced signaling in mice with a mutation at position 757 on the intracellular arm of gp130, which leads to inhibition of SHP-2 and SOCS3 docking, with consequent ablation of MAPK/AP-1 signaling, and constitutive activation of STAT3-mediated gene transcription. (**C**) Phenotypic progression of tumorigenesis in the gp130[757FF] mouse at 4, 12, and 24 weeks (×1). (**D**) Some target genes affected by the gp130 mutation in gastric tumors. Reg 1, 3β, 3γ growth factors, the matrix metalloproteinases (MMP)-9, -13, antiapoptotic genes bcl-2 and bcl-x, SOCS3, and the angiogenic gene vascular endothelial growth factor (VEGF) are strongly induced, whereas the cytostatic gene transforming growth factor (TGF)β is inhibited in a SMAD7-dependent manner

an established proinflammatory cytokine with elevated expression in gastric cancer; however, genetic polymorphisms in IL-6 have been shown *not* to contribute to *H. pylori*–induced stomach cancer in susceptible humans (Hwang et al. 2003) and IL-6 mRNA expression is negatively correlated with distant tumor metastasis (Garcia de Galdeano et al. 2001; Wang et al. 2002), suggesting that this pleomorphic cytokine may have a protective rather than damaging role in the stomach. To evaluate the role of IL-6 in gp130^{FF} gastric inflammation and tumor growth compound, gp130^{FF} × IL-6^{KO} mice were generated. Unexpectedly, in the absence of IL-6, gastric tumor growth occurred at the same rate and with the same incidence as gp130^{FF} controls (Howlett et al. 2005). In addition, STAT3 activation was undiminished and correlated with tumor progression. A striking finding was that although IL-6 was dispensable for tumor development, in its absence, the incidence and extent of submucosal invasion was increased 10-fold. Comparison of compound mutants with gp130^{FF} mice showed that IL-6 may inhibit submucosal invasion in part by reducing the expression of several key genes previously shown to augment metastatic migration of cancer cells, particularly the matrix metalloproteinases (MMP)-9 and -13 (Howlett et al. 2005).

The increased incidence of submucosal invasion in compound mutants along with reduced inflammation suggests that local tumor surveillance mechanisms might be compromised. This idea is supported by the fact that IL-6–type cytokines potently inhibit T-cell apoptosis in the gut, thereby extending T-cell life (Atreya et al. 2000) and cytotoxic activity by promoting perforin gene expression (Yu et al. 1999). IL-6 also augments B-cell development and immunoglobulin A expression (Ramsay et al. 1994). To test this, gp130^{FF} × perforin^{KO} (unpublished observations) and gp130^{FF} × Rag1^{KO} (Howlett et al. 2005) mice were generated. Subsequent analysis ruled out a role for immunosurveillance mediated by CD8 T-cell killing, as well as other T, B, or NKT cells, because mutant mice lacking these immunocytes or perforin did not have increased submucosal invasion or altered tumor growth (Howlett et al. 2005).

Whereas IL-6 is dispensable for gastric tumor initiation and growth in the gp130^{FF} mouse, it seems that IL-11 signaling is absolutely required for both gastric STAT3 activation and tumorigenesis, because both are completely abrogated in gp130^{FF} × IL-11α receptor null (IL-11αr^{KO}) mice. This strongly suggests that IL-11 acts as a STAT3-dependent growth factor in mouse gastric tumorigenesis. Moreover, gp130^{FF} × STAT3$^{+/-}$ mice, with only one active STAT3 allele, have diminished STAT3 activation, and much smaller tumors than gp130^{FF} litter mates (Jenkins et al. 2005; Judd et al. 2006), underscoring the importance of STAT3-mediated transcription in tumor progression. The cellular source of IL-11 in tumors as well as the upstream regulatory events that allow this antiinflammatory cytokine to be subverted in oncogenic transformation, and its main gene targets, are yet to be determined. Evidence in support of a mitogenic role for IL-11 in the stomach comes from other mouse models, and from human gastric cancer. For instance, it has been shown that the gastric hyperproliferation accompanying autoimmune gastritis in thymectomized BALB-c mice, and hyperplasia in H$^+$, K$^+$-β ATPase null mouse stomach, are strongly associated with increased IL-11 expression (Franic et al. 2005), and analysis of human gastric cancer cDNA microarrays and resected material show increased IL-11 expression in adenocarcinoma (Jackson et al. 2007).

Cdx1 and 2 Transgenic

Cdx1 and 2 are adult intestine/colon-specific transcription factors that also function as caudal-related homeobox genes in development. They are associated with intestinal metaplasia (IM), the intestinalization of the gastric mucosa associated with progression to gastric intestinal-type adenocarcinoma development in both mice and humans. The Cdx2 transgenic mouse, created using the β subunit of the H/K-ATPase promoter on a C57Bl6 background under specific pathogen-free (SPF) conditions, exhibited incomplete IM throughout gastric fundic glands at 12 weeks of age (Mutoh et al. 2002), coincident with hypergastrinemia, achlorhydria, as well as SPEM (Mutoh et al. 2004). By 100 weeks, fundic tumors were observed in 100% transgenic mice (but none of wild-type litter mates). These tumors were hyperproliferative, showed submucosal invasion, and exhibited mutations in both APC and p53 genes (Mutoh et al. 2004). Introduction of the APC$^{min/+}$ mutation, or loss of one p53 allele (p53$^{+/-}$) resulted in earlier tumor development (Mutoh et al. 2004). Similar to the Cdx2 transgenic mouse, Cdx1 transgenic stomachs also exhibit rapidly expanding IM, but this was complete with all differentiated cell types including Paneth cells and a range of hormone-expressing endocrine cells (Mutoh et al. 2004). However, the metaplastic stomachs of Cdx1 overexpressing mice do not develop gastric tumors, and neither transgenic mouse produces metastatic lesions in other organs, as is the case in human adenocarcinoma.

Gastrin Mutants

Gastrin, produced by G cells in the antral mucosa, is a crucial regulatory hormone in the gastric mucosa, able to transcriptionally regulate cell division, invasion, angiogenesis, and antiapoptotic activity (Friis-Hansen 2002; Watson et al. 2006). It functions to regulate acid secretion by the fundic mucosa in response to feeding, and developmentally in maintaining epithelial cell homeostasis in both the fundic and antral mucosa. As such, a failure in the tight regulation of gastrin expression will lead to perturbations of gastric epithelial cell dynamics and potentially gastric cancer. These characteristics of gastrin dysregulation have been utilized in a number of mouse models of stomach cancer.

Insulin-Gastrin Transgenic

In this transgenic mouse model, the expression of the human gastrin transgene is directed by a mouse insulin promoter, and expression of processed forms of gastrin are observed in the pancreas, stomach, and colon (Wang et al. 1993, 1996). In 1-year-old INS-GAS mice, the consequence of overexpression of gastrin in the stomach is evident as moderate to marked thickening of the fundic mucosa, multifocal hyperplasia, and mild clonal expansion of ECL cells (Wang et al. 1996), and by 2 years, a proportion of these mice spontaneously develop atrophy and cancer (Wang et al. 2000). The INS-GAS mouse is now frequently used in conjunction with other agents as a model for gastric cancer development, because these mice have a lower threshold for carcinogenesis. Notably, 7 months postinfection with *H. pylori* or *H. felis*, male INS-GAS mice develop atrophy, intestinal metaplasia, dysplasia, and eventually gastric adenocarcinoma at a much greater frequency than either uninfected INS-GAS mice or infected wild-type mice (Fox et al. 2003; Wang et al. 2000). Cancers that develop are either *in situ* or intramucosal carcinoma (Fox et al. 2003). Cancer development is potentially mediated by reactivation of sonic hedgehog expression (El-Zaatari et al. 2007). Inhibition of the gastrin/CCK2 and histamine H2 receptors are able to synergistically limit the development of gastric cancer in these mice (Takaishi et al. 2005).

Actin-Gastrin Transgenic

The actin promoter has also been used to drive gastrin expression in the Act-Gas transgenic model. In this model, the gastrin transgene was mutated, allowing the processed forms of gastrin to be expressed by nonendocrine cells (Konda et al. 1999). By 16 weeks of age, these mice develop gastric mucosal hypertrophy, consisting mainly of foveolar hyperplasia accompanied by parietal cell atrophy (Kanda et al. 2006). By 80 weeks of age, a proportion develop gastric intramucosal adenocarcinoma, with epithelial atypia, hyperchromia, and loss of cell polarity but no invasion (Kanda et al. 2006). Similar to the TFF1–/– model, treatment of these mice with a selective COX-2 inhibitor reduced cell proliferation and foveolar thickness but did not affect serum gastrin levels, suggesting that COX-2 and prostaglandin E2 (PGE2) might act downstream of gastrin (Kanda et al. 2006).

Gastrin Knockout

Gastrin-deficient mice are hypochlorhydric because of the absence of the gastrin-hydrochloric acid secretory pathway (Friis-Hansen et al. 1998; Koh et al. 1997). In humans, the absence of gastric acid provides permissive conditions for bacterial overgrowth (Friis-Hansen 2006) and the same is true in these hypochlorhydric mice (Zavros et al. 2002). This overgrowth with coincident inflammation was resolved by treatment with antibiotics (Zavros et al. 2002). In young mice (12 weeks), the absence of gastrin had only very minor effects on the gastric mucosa including a slight reduction in the number of parietal cells and reduction in ECL cell activity (Friis-Hansen 2002). By 12 months of age, this hypochlorhydric mouse develops chronic gastritis, atrophy, metaplasia, dysplasia, and eventually intramucosal carcinoma in the antral mucosa (Zavros et al. 2005) that is completely dependent on mucosal infection and inflammation. The metaplasia that develops is not a true intestinal metaplasia (marked by the presence of goblet cells), rather an antralization caused by an expansion of a mucous cell lineage often seen in mouse gastric metaplasia (SPEM), and it seems that development of this lineage is inflammation dependent (Goldenring and Nomura 2006; Kang et al. 2005). Carcinomas developed independent of gastrin, but were associated with large increases in the amount of activated STAT3 and loss of RUNX3 expression (Zavros et al. 2005).

Parietal Cell Mutants

Parietal cells of the fundic stomach synthesize and secrete hydrochloric acid, which serves to sterilize the gastric contents and promotes the activation of stomach enzymes such as pepsin to initiate protein digestion. Physical loss of parietal cells (gastric atrophy), or their acid synthetic function (achlorhydria), promotes opportunistic bacterial colonization of the gastric lumen, and this along with the constitutive inflammatory response of the host likely predisposes to gastric pathology including cancer. It follows therefore that a wide variety of mutations in parietal cell function, particularly those affecting acid secretion, will be contribute to understanding mechanisms predisposing metaplasia and cancer in mouse models.

H⁺, K⁺-ATPase Knockout

As already mentioned, gastric acid has a role in minimizing infection and the subsequent inflammatory response in the stomach. The H^+, K^+-ATPase, expressed by fundic parietal cells, is responsible for acidification of the gastric contents. Knockout of either the α (Spicer et al. 2000) or β (Scarff et al. 1999) subunit of the H^+, K^+-ATPase demonstrates that both are required for full enzymatic function. In young (12 week) H^+, K^+-ATPase α or β subunit knockout mice, there was some hypertrophy, parietal cell atrophy and pathology, gastritis, and achlorhydria induced hypergastrinemia (Scarff et al. 1999; Spicer et al. 2000); however, as with many of these mouse models, the penetrance of the pathologies is dependent on the genetic background of the mice. In H^+, K^+-ATPase β subunit knockout mice, hypertrophy was attributed to hyperproliferation of immature cells and this was completely dependent on the overexpression of gastrin (Franic et al. 2001). The H^+, K^+-ATPase α subunit–deficient mice were aged to 20 months and, at this time point, fundic hypertrophy progressed to hyaline transformation, mucocystic and ciliated metaplasia, and chronic gastritis (particularly in female mice), yet never progressed to frank neoplasia, despite the consistently elevated gastrin levels over this time (Judd et al. 2005). Coincident with the pathology was a decrease in mRNA expression of the gastric tumor suppressor gene TFF1 (Judd et al. 2005).

Na⁺/H⁺ Exchanger Knockout

Na^+/H^+ exchangers (NHE) are a family of proteins expressed in the basolateral membrane of gastric epithelial cells, particularly parietal cells and are purported to have a role in the mechanisms of acid secretion and in maintaining epithelial cell viability. Knockout of either NHE2 (Schultheis et al. 1998) or 4 (Gawenis et al. 2005) results in a fundic atrophy, parietal cell loss, achlorhydria, hypergastrinemia, and resultant glandular hyperplasia. There is also an increased frequency of apoptotic cells in the mucosa of these mice, reinforcing the role that this family of proteins has in maintaining epithelial cell viability (Gawenis et al. 2005; Schultheis et al. 1998).

Potassium Channel Knockouts

Because of the high activity of the H^+, K^+-ATPase in gastric parietal cells, potassium channels have a crucial role in the maintenance of correct ionic balance. This is evidenced by the knockout of Kvlqt1, a voltage-gated potassium channel expressed in parietal cells as well as the developing myocardium and inner ear. The fundic mucosa of these mice morphologically resembles the H^+, K^+-ATPase knockouts, likely because these mice develop achlorhydria, hypergastrinemia, and hyperplasia, although this has not been directly tested (Lee et al. 2000). KCNE2 is an ancillary subunit that coassembles with voltage-gated potassium channels to modulate their function. The fundic mucosa of KCNE2 knockouts again resembles the H^+, K^+-ATPase knockouts and, in this instance, a study confirmed that there was coincident achlorhydria, and hypergastrinemia-induced hyperplasia (Roepke et al. 2006), accounting for the gastric pathology that was observed in these mice.

Kcnq1 is a potassium channel localized to parietal cells and is required for normal acid secretion (Grahammer 2001). Targeted mutation of Kcnq1 in mice resulted in an expanded fundic proliferation zone with severe hyperplasia, achlorhydria, and hypergastrinemia (Lee et al. 2000). More recently, several mouse lines with a defective *Kcnq1* locus (14Gso) were generated in a random mutagenesis program induced by X-ray irradiation of spermatogonia (Elso et al. 2004). The consequences of these mutations were very similar to that previously described for H^+, K^+ATPase α and β null mutant mice. Pathology was evident from 8 days of age and progressively developed with inflammation, achlorhydria, fundic metaplasia, dysplasia, and adenoma independent of microbial infection.

Histamine (H2) Receptor Knockout

The histamine H2 receptor (H2R) is present on the acid-secreting parietal cells of the gastric mucosa and has a role in the stimulation of gastric acid secretion. H2R knockout mice are viable, fertile, and have normal basal gastric acid secretion, maintained by muscarinic receptors (Kobayashi et al. 2000). Despite the normal gastric acid output, 12-week-old knockout mice have hypergastrinemia and coincident fundic mucosal hyperplasia. However, as expected, studies showed that the antral mucosa was unaffected. Fundic hyperplasia occurred as a direct result of progressively increased numbers of parietal and ECL cells to 17 months of age (Kobayashi et al. 2000), after which the pathology worsened to include mucocystic metaplasia, with a proportion of the mice developing herniation of the epithelium which penetrated the muscularis mucosa, producing a phenotype closely mimicking Ménétrier's disease in a humans (Ogawa et al. 2003).

IQGAP1 Knockout

Parietal cells in the gastric mucosa undergo massive subcellular reorganization each time they become activated to secrete acid. This reorganization is thought to be dependent on the precise formation of F-actin structures directed by the Rho family of Ras-related GTPases. IQGAP1 is involved in these processes because it contains an F-actin binding domain and interacts with Rho family proteins (Li et al. 2000). Mice with heterozygous and homozygous mutations in IQGAP1 develop late onset (approximately 12 months) gastric lesions. In a study by Li et al. (2000), the lesions consisted of marked hyperplasia and dysplasia with no evidence of carcinoma *in situ* or submucosal invasion. No further studies on this model have been published, so it is not clear precisely what causes this phenotype; however, IQGAP1 is also bound by the APC protein and may therefore modulate Wnt signaling pathways (Senda et al. 2007).

MTH1 Knockout

MTH1 is the mammalian homolog of the *Escherichia coli mutT* gene which inhibits the incorporation of 8-oxodGTPase residues into DNA. Such incorporation promotes nucleotide oxidation and transversion during DNA synthesis, and is likely to promote carcinogenesis in susceptible tissues. Thus, MTH1 acts as a tumor-suppressor gene. The MTH1 gene has been ablated globally and a mixed cohort of mice were aged for 18 months under SPF conditions along with litter mate controls and then organ pathology was assessed (Tsuzuki et al. 2001). As well as lung and liver tumors, stomach adenomatous polyps and adenocarcinomas were identified in 14% MTH1−/−, but only 4% wild-type controls. Tumors were more prevalent in males than females, but no comprehensive histologic assessment was performed (Tsuzuki et al. 2001). Recently, a second MTH gene (MTH2) has been described with a similar activity profile to MTH1 (Cai et al. 2003), raising the possibility that this gene family may contribute to the prevention of tumorigenesis by mediating local oxidative damage.

K19-C2mE Transgenic and Variants

The K19-C2mE mouse overexpresses COX-2 and the microsomal PGE synthase 1 genes under the control of the cytokeratin 19 gene promoter (Oshima et al. 2004). Its importance as a model of gastric cancer initiation and progression is described in detail in Chapter 20. In summary, this mouse develops macrophage-dependent gastric hyperplasia in a constitutively upregulated PGE2 environment. Although hyperplasia is evident by 12 weeks, maximal tumorigenesis is not apparent until about 50 weeks or more. Loss of the key proinflammatory cytokine IL-1β, or the adaptive immune response (Rag2−/−) in the context of the K19-C2mE transgene has no effect on tumorigenesis; however, depletion of tumor necrosis factor α severely retards inflammation, hyperplasia, and TFF2-associated mucous cell metaplasia (SPEM) development (Oshima et al. 2005), thus emphasizing the importance of this proinflammatory cytokine in gastric metaplasia and oncogenesis.

The importance of the Wnt pathway in gastric tumorigenesis is underscored in the compound K19-C2mE × K19-Wnt1 transgenic mouse, which develops accelerated gastric tumors compared with the K19-C2mE mice alone, and with the former showing severe dysplasia, hyperplasia, inflammation, and submucosal invasion by 20 weeks of age (Oshima et al. 2006).

Thrombospondin 1 Knockout

Thrombospondins are a family of extracellular calcium-binding proteins that have been reported to regulate cellular attachment, migration, differentiation, and proliferation (Lawler et al. 1998). The best-characterized thrombospondin (TSP-1) has also been shown to activate TGFβ, and TGFβ1−/− and TSP-1−/− mice show a similar broad-tissue spectrum of pathology including intestinal metaplasia and increased mitosis and hyperplasia of the gastric epithelium in 17- to 21-day pups (Crawford et al. 1998). Compound TSP-1 × αVβ6 integrin null mice develop stomach hyperplasia (21%) as well as gastric papillomas and squamous cell carcinomas (Ludlow et al. 2005). The utility of TSP-1 mice in studying gastric cancer pathogenesis is tempered by the variable fetal lethality of the mutation, the low penetrance of the stomach phenotype, and the broad spectrum of compromising multiorgan pathology.

Transforming Growth Factor α Transgenic

TGFα transgenic mice in which the transgene is preferentially overexpressed in fundic stomach mucus cells (MT-TGFα), present a phenotype of giant fundic mucosal folds brought about through massive cellular hyperplasia and glandular cystic dilation (Dempsey et al. 1992; Takagi et al. 1992). This phenotypic pattern is reminiscent of the rare human condition Ménétrier's disease, which is also associated with elevated TGFα expression (Dempsey et al. 1992). Adult TGFα transgenic mice develop an expanded surface mucous cell population at the expense of both parietal and chief (zymogenic) cells with the isthmus-located stem cell zone being displaced nearer the base of the glands (Bockman et al. 1995; Goldenring et al. 1996; Sharp et al. 1995), and the mucosa becoming much more fibrotic (Takagi et al. 1997). Thus, the fundic mucosa phenocopies the antralization seen in precancerous metaplasia, including expression of the antral transcription factor Pdx1 (Nomura et al. 2005); however, true invasive gastric tumors with intestinal metaplasia are not observed, so this model is more applicable for understanding Ménétrier's disease than gastric adenocarcinoma progression.

Dioxin/Aryl Hydrocarbon Receptor Transgenic

Activation of the dioxin/aryl hydrocarbon receptor (AhR) by environmental triggers results in transcriptional activation of a range of genes encoding xenobiotic metabolizing enzymes. Ligands for AhR include environmental pollutants such as dioxins and biphenyls (Andersson et al. 2002). Endogenous expression of AhR is predominantly in the lung, although mice transgenic for AhR also express the transgene in the thymus, spleen, liver, skin, and stomach. As early as 3 months of age, AhR transgenic mice developed grossly apparent cysts in the fundic gastric mucosa. By 12 months, at which point transgene expression generally results in lethality, these cystic lesions developed into dysplastic structures that penetrated the muscularis mucosa into the submucosa and subserosa. The penetrating mucosal cells maintained a well-differentiated, even benign appearance; however, their presence seemed to be invasive rather than a result of herniation (Andersson et al. 2002). Similar lesions have also been described in laboratory animals treated with AhR ligands (Andersson et al. 2002). Differential gene expression analysis in these fundic lesions have pointed to decreased expression of osteopontin as a possible cause of lesion development, with localized reexpression where submucosal invasion occurred. Osteopontin is typically involved with physiologic and pathologic tissue remodeling, so its altered expression in the AhR mouse may be responsible for the particularly invasive lesions seen in this mouse (Kuznetsov et al. 2005). In contrast, the H^+, K^+-ATPase α subunit knockout mice, which have very severe but noninvasive fundic lesions, have elevated levels of osteopontin compared with wild types; this elevated osteopontin expression may inhibit invasion of lesions into the gastric muscularis (Judd et al. 2005).

Klf4−/− Knockout

Klf4 is an epithelial-specific, zinc finger transcription factor important in the regulation of cellular proliferation and differentiation. Knockout mice for Klf4 show early lethality (Segre et al. 1999), so mutants with a conditional deletion of Klf4 in the glandular gastric mucosa alone were developed. At 6 months of age, these conditional knockout mice developed epithelial hypertrophy, hyperproliferation, mucous metaplasia, atrophy, and polypoid lesions in the fundus and antrum, but interestingly, no inflammation, hypergastrinemia, dysplasia, or malignancies. (Katz et al. 2005).

p27^{kip1}–/– Knockout

The p27^{Kip1} protein principally inhibits cyclin-dependent kinases thus blocking cell-cycle progression, and it also has roles in cell migration, mitosis, apoptosis, differentiation, and in mediating inflammatory responses. Young p27^{Kip1} knockout mice develop mild epithelial hyperplasia, and at approximately 1 year of age, the mice further develop mucous cell metaplasia and some low-grade dysplasia. This preneoplastic condition makes p27^{Kip1} mice significantly more susceptible to the development of high-grade dysplasia or intramucosal carcinoma after *H. pylori* infection (Kuzushita et al. 2005). Worsening pathology in p27^{Kip1} *H. pylori*–infected mice is probably a result of a combination of increased cell turnover and an exaggerated inflammatory response (Kuzushita et al. 2005).

APC$^{min/+}$ and Wnt Signaling Pathway Mutants

The Wnt signaling pathway has a major role in determining cell fate and morphogenesis during embryogenesis, whereas in the adult, it is important in homeostatic control of rapidly renewing tissues. β-Catenin has an important role in the Wnt pathway because it is stabilized and translocated to the nucleus upon activation of Wnt ligands. Transgenic overexpression of activated β-catenin is linked to tumor progression (Harada et al. 1999; Romagnolo et al. 1999); however, in such models, β-catenin is overexpressed in a large number of cells simultaneously. To more closely mimic human oncogenic development in which cancer is believed to arise from a mutation in single cell, a transgenic model was established in which overexpression of the activated form of β-catenin occurs in sporadic cells. In a very small proportion of these mice, the gastric mucosa alone developed discrete multifocal dysplastic lesions. The lesions have lost expression of gastric specific markers and may be promoted because of a loss of the *Sonic hedgehog* gene (Coste et al. 2007).

Loss of the APC tumor-suppressor gene is clearly associated with colorectal cancer initiation in humans and mice via nuclear accumulation of β-catenin and activation of β-catenin/Tcf target genes. Additionally, APC mutant mice (codon 1638) develop gastric tumors with low frequency as they age (Fox et al. 1997; Yang et al. 1997); however, analysis of gastric cancer in these mice is compromised by the spectrum of nongastric pathologies. Gastric pathology associated with the APC$^{min/+}$/C57Bl6 background may be additionally infrequently observed because of the attenuated lifespan of these mice. However, it has recently been shown that APC[min/+]/C57Bl6 mice maintained under SPF conditions develop gastric adenomas predominantly of the antrum by 20 weeks, suggesting that microbial infection has a role in the development of the some of the pathologies in these mice. The antral adenomas show hyperplasia and nuclear atypia accompanied by increased expression of Myc, cyclin D1, and β-catenin (Tomita et al. 2007).

Major Histocompatibility Complex Class II Knockout

Major histocompatibility complex (MHC) Class II molecules have a major role in regulating the CD4+ arm of the adaptive immune response. Mice deficient in MHC Class II, and are unable to produce a functional CD4+-mediated immune response, also lack CD4+ T cells because MHC Class II is required for the maturation of these cells in the thymus. The fundic stomachs of 6-month-old MHC Class II–deficient mice had gastrin-dependent mild hyperplasia with infiltration of granulocytes and macrophages but no epithelial cell atrophy (Fukui et al. 2006). This study demonstrates that persistent activation of the innate immune system can cause hyperplastic changes in the fundic mucosa through mechanisms that are yet to be established.

Carbonic Anhydrase IX Knockout

Carbonic anhydrases (CAs) are zinc-containing metalloenzymes that are involved in pH regulation by catalyzing the reversible hydration of CO_2. CA IX is unique among the family because of its tumor-related expression, high catalytic activity, and dual function as an adhesion molecule. Its role in tumors may be as a pH regulator in the hypoxic tumor mass. Ca IX knockout mice develop nonprogressive glandular expansion, mostly restricted to a foveolar hyperplasia in the fundus (Gut et al. 2002). There were subtle variations in the degree of pathology when CA IX knockouts were crossed onto a pure genetic background of C57BL/6 or BALB/c, or fed a high salt diet (Leppilampi et al. 2005). The knockout of CA IX did not affect gastric acid secretion or serum pH, electrolytes, or gastrin, suggesting that the hyperplastic phenotype is probably attributable to the role of CA IX as an adhesion molecule (Gut et al. 2002).

p53 Hemizygous Knockout

p53 is a transcription factor that acts as a master regulator of proliferation, apoptosis, and genomic repair. Loss of heterozygosity mutations in p53 have been described in gastric cancer and precancerous lesions. Hemizygous p53 knockout mice have a low level (2%) of spontaneous carcinogenesis in a range of organs. These mice were infected with *H. felis* to determine if they had increased susceptibility to gastric carcinogenesis. The *H. felis*–infected hemizygous p53 knockout mice had an increased proliferative index and significant growth advantage compared with wild types, but no frank neoplasia (Fox et al. 1996). In a subsequent study, the same group demonstrated that wild-type C57BL/6 mice developed early invasive adenocarcinomas at 15 months postinfection, with associated metaplasia and a greater inflammatory response. This suggests that the hemizygosity of p53 results in depressed immune responsiveness, particularly Th1, and subsequently less pathology (Fox et al. 2002).

SV40 T Antigen Transgenics

SV40 T antigen, derived from the simian virus, is a very potent transforming agent and oncogene. Its aberrant expression has been used to generate transformations in a number of different cells lines and tissues.

Carcinoembryonic Antigen SV40 T Antigen Transgenic

Carcinoembryonic antigen (CEA) is predominantly expressed in embryonic development and is widely expressed in a range of tumors, including colorectal, breast, lung, and pancreatic carcinomas. The CEA promoter was used to drive the transgenic expression of SV40 T antigen. Despite detectable levels of SV40 T antigen transgene expression only in the stomach, these animals variably developed carcinomas, lymphomas, and sarcomas. Only one transgenic line reproducibly developed temporally predictable tumors in the same tissue, and this tissue was the antral stomach. The tumors were poorly differentiated adenocarcinomas that had lost gastric mucin expression. Tumors were visible macroscopically from as young as 5 weeks of age, and caused the death of the mice at approximately 20 weeks, by which stage the tumors had penetrated all tissue layers of the stomach and eventually invaded and blocked the duodenum (Thompson et al. 2000). Cell lines have been generated from these gastric adenocarcinomas and are currently the only mouse adenocarcinoma cell line available (Nockel et al. 2006).

H⁺, K⁺-ATPase β Subunit SV40 T Antigen Transgenic

The promoter of the H^+, K^+-ATPase β subunit was used to direct transgenic expression of the SV40 T antigen, producing a massive increase in the normally rare preparietal cells by 12 weeks of age; however, these cells were unable to differentiate into mature parietal cells (Li et al. 1995). The stomachs of these mice were also hyperplastic, with a reduction in the number of mature zymogenic as well as parietal cells (Li et al. 1995). By 40 weeks of age, there was progressive hyperplasia, cystic dilations, and some regions of focal dysplasia, and by 1 year, all mice had developed invasive gastric cancer with evidence of lymphatic-vascular invasion and associated lymph node and hepatic metastasis. Invasive tumor cells were only weakly positive for SV40 T antigen and negative for the H^+, K^+-ATPase β subunit. Additionally, the stomachs no longer exhibited the typical mucous-glandular structures that are evident in adenocarcinomas of the human stomach. A detailed transcriptome analysis of these stomachs revealed that the invasive tumor cells had transdifferentiated into neuroendocrine cells, based on their expression of dopa decarboxylase, chromogranin A, and tryptophan hydroxylase. Several genes typically involved in neuroendocrine differentiation, Sox2, Hey1, and NeuroD1, had increased expression in these invasive tumor cells (Snyder et al. 2004).

The same H^+, K^+-ATPase β subunit promoter was used to direct the transgenic expression of a temperature-sensitive SV40 T antigen, with a view to inducing cultured progenitor cells into mature parietal cells at the nonpermissive temperature for the SV40 T antigen. Again, at 12 weeks of age, transgenic mice developed fundic hypertrophy. In one transgenic line, the hypertrophy resulted because of an expansion of immature stem cells and mucous cells, whereas in another line, there was an expansion of immature stem cells and preparietal cells (Stewart et al. 2002).

TGFβ, TGFβ Receptor, and Signaling Mutants

There are three isoforms of TGFβ: 1, 2, and 3, and all three of them can bind to TGFβ receptor II, although TGFβ1 is most frequently implicated in regulation of tumorigenesis. Upon ligand binding to TGFβ receptor II, heterodimerization with and activation of TGFβ receptor I occurs. Activation of TGFβ receptor I results in activation of the SMAD complexes, which are translocated to the nucleus and can activate TGFβ-responsive genes. TGFβ is typically inhibitory to proliferation of most cell types. Many of the downstream components of the TGFβ signaling pathway are considered to have tumor-suppressor activity, including the TGFβ receptors 1 and 2 and SMAD2 and 4. As such, loss of control in TGFβ signaling can act as a potent stimulator of tumor progression, invasion, and metastasis (Akhurst et al. 2001; Derynck et al. 2001).

TGFβ1 Knockout

TGFβ1 normally suppresses cell growth, and tumors often develop mechanisms to evade such suppression, sometimes by reduced expression or activity of TGFβ receptors and also by alterations in downstream signal transduction pathways. The development of TGFβ1 resistance may represent a significant step in the process of carcinogenesis (Crawford et al. 1998; Shull et al. 1992). TGFβ1 also has roles in control of extracellular matrix protein production and degradation and cellular differentiation. Approximately 20 days after birth, mice homozygous for the TGFβ1 null mutation develop a severe wasting syndrome followed by death, as a result of multifocal, mixed inflammatory cell infiltration in a range of tissues including the stomach. Other pathologies in the stomach include ulceration, hyperplasia, and nodule formation in the mucosa (Crawford et al. 1998; Shull et al. 1992). The early lethality of this mutation makes it impossible to fully assess the effects of the absence of TGFβ1 in gastric carcinogenesis in this mouse.

TGFβ Type II Receptor Dominant-Negative Transgenic

A dominant-negative transgene of the TGFβ type II receptor was generated under control of the TFF1 promoter to direct expression predominantly to the stomach. These mice are no longer able to respond to TGFβ ligands in the stomach; however, they showed no overt gastric abnormalities, possibly because signaling through TGFβ is not completely ablated. However, after infection with *H. pylori*, these transgenic mice had a more severe pathology in the fundus and antrum including greater hyperplasia, inflammation, and dysplasia, as well as intramucosal carcinoma (Hahm et al. 2002).

SMAD4 Hemizygous Knockout

As already mentioned, SMAD4 belongs to a family of molecules involved in the signal transduction pathways emanating from TGFβ, in which SMAD activation results in their nuclear translocation and activation of gene expression. Homozygous SMAD4 knockout mice are embryonic lethal, whereas heterozygotes are fertile and appear normal up to approximately 1 year of age. On closer examination, a small proportion of 1-year-old heterozygous mice developed predominantly antral gastric polyps and by 1.5 years of age, the penetrance was 100%. Microscopically, the polyps were elongated tubular structures with moderate stromal cell expansion, the epithelia of which was predominantly hyperplastic, showing little cellular atypia, although foci with obvious dysplastic signs and nuclear atypia, and with abundant mucinous adenocarcinoma-like cells, were observed (Takaku et al. 1999; Xu et al. 2000). Submucosal invasion was also observed in a limited number of mice (Xu et al. 2000). Polymerase chain reaction and immunohistochemical analysis revealed that loss of heterozygosity of SMAD4 was evident in these polyps (Takaku et al. 1999; Xu et al. 2000), and that loss of a single SMAD4 allele was sufficient for hyperplasia. However, loss of heterozygosity is required for dysplasia to occur (Xu et al. 2000). A mutation in SMAD4 has now been confirmed to be evident in 50% of familial juvenile polyposis cases.

To further expand on this finding, studies were performed in which SMAD4 was conditionally ablated in either T cells or epithelia, including the intestinal epithelium. Mice with a specific SMAD4 deletion in T cells began to display signs of illness by 3 months age, and had a significantly shortened lifespan as a result of pathology initiated in the small and large intestine that developed into invasive and then metastatic epithelial cancer (Kim et al. 2006). SMAD4-deficient T cells activated in culture were skewed toward a Th2 phenotype. However, specific deletion of SMAD4 in epithelial cells was not sufficient to induce these cancers in the gut (Kim et al. 2006). A deficiency of SMAD4 in T cells also leads to an increase in IL-6 receptor α expression in the gastric epithelium (Kim et al. 2006). These experiments demonstrate that, in some epithelial cancers at least, it is the control of TGFβ signaling in the inflammatory cells that is crucial for regulating tumor development.

Elf and SMAD4 Compound Hemizygous Knockouts

Embryonic liver fodrin (elf), a novel β-spectrin, is a membrane-associated cytoskeletal component important in cellular differentiation. Elf acts as an adaptor for the SMAD proteins, and without elf, nuclear translocation of SMAD3 and SMAD4 is disrupted. Normally, elf is strongly expressed in the glandular tissue of the stomach and weakly expressed in the squamous forestomach epithelium (Kim et al. 2006). Mice with homozygous deficiency in elf4 die during embryonic development, and mice with heterozygous deletions for both SMAD4 and elf had a higher incidence of severe gastric lesions than those with mutations in SMAD4 alone (Redman et al. 2005). Further examination revealed that the loss of elf resulted in the accelerated entrance into the cell cycle and resistance to senescent and apoptotic stimuli (Kim et al. 2006).

Runx3 Knockout

Runx proteins act as regulators of gene expression in several important developmental pathways; for example, they form complexes with SMAD2 and SMAD3 to enable transmission of TGFβ/activin signals (Blyth et al. 2005; Levanon et al. 2003; Li et al. 2002). Runx3 knockout mice on a C57Bl/6 genetic background have reduced viability and none survive beyond 10 days (Li et al. 2002). The gastric mucosa of knockout Runx3 mice was thickened, with increased proliferation and decreased apoptosis in the fundic and antral mucosae (Li et al. 2002). Proliferation of epithelial cells from the mucosa of these mice was no longer effectively inhibited by TGFβ (Li et al. 2002). It is not possible to assess whether in the long term these mice develop gastric cancer because of their reduced viability, and it would be very informative to have available a mouse that has a conditional deletion of Runx3 in adult life.

In contrast, another research group has knocked out Runx3 and maintained the line on an ICRxMF1 genetic background. These mice survive for several months and do not develop any early signs of gastric hypertrophy or carcinogenesis (Levanon et al. 2002; 2003), but on either a C57Bl/6 or BALB/c background, the mice do not survive postnatally (Brenner et al. 2004). Runx3 mice on an ICRxMF1 genetic background successfully aged to 8 months, developed marked hyperplasia, glandular atrophy, hyaline degeneration, hyperproliferation, and gastritis in the fundic mucosa. Some penetration of hyperplastic epithelial cells through the muscularis mucosa was noted (Brenner et al. 2004). These researchers fully attribute the late-onset gastric pathology observed to a loss of Runx3 in the gut leucocytes, rather than a direct effect of a loss of Runx3 expression in the gastric epithelium itself (Brenner et al. 2004). It is not clear whether the differences in gastric phenotype are a result of different genetic background or differences in gene-targeting strategy by the investigators, but given that both knockouts display similar defects in neurogenesis and thymopoiesis, it is likely that any differences observed are unique to the stomach phenotype (Bae and Ito 2003; Levanon et al. 2003).

Meanwhile, various *in vitro* experiments have demonstrated that Runx3 is able to mediate apoptosis induced by TGFβ by upregulating transcription of Bim (Yano et al. 2006), in a manner dependent on FoxO3a (Yamamura et al. 2006). Isolated epithelial cell lines from Runx3 mice only weakly attach to each other and do not form glandular structures in culture. They can form glandular structures and indeed tumors when injected into nude mice, whereas cells from wild-type mice are not able to do so. True intestinal metaplasia developed in these tumors, with cells resembling goblet cells that expressed Cdx2 (Fukamachi et al. 2004).

Forkhead Homolog 6 Knockout

Forkhead homolog 6 (Fkh6) is a mesenchymal winged helix transcription factor, expressed in the gastrointestinal tract in the mesenchyme directly adjacent to the endoderm-derived epithelium. Fkh6 knockout mice have a compromised viability, but those that do survive demonstrate progressively worsening gastric pathology evident from 3 days of age. Stomachs displayed epithelial hyperplasia, cyst formation, mucous cell metaplasia, increased cell proliferation, and a diffuse submucosal mesenchyme. There was also a significant reduction in bmp4, a soluble polypeptide implicated in epithelial signaling processes, suggesting that bmp4 may be a downstream target of Fkh6 (Kaestner et al. 1997). The dramatic and rapid onset of the phenotype in this knockout mouse reinforces the important role of mesenchymal–epithelial cell interaction in the growth and differentiation of the gastric mucosa.

Sonic Hedgehog Knockout

Sonic hedgehog (Shh) knockouts die at or shortly after birth. Therefore, to enable analysis of the effects of this important cell-signaling molecule in stomach development, knockout mice were analyzed at 18.5 days of embryonic life. These knockout mice demonstrated a hyperplastic gastric epithelium, yet paradoxically with no increase in cell proliferation, but showed occlusion of the duodenum caused by overgrown villi. The stomachs also showed intestinalization, evidenced by increased expression of intestinal alkaline phosphatase, a marker of the brush border of enterocytes (Ramalho-Santos et al. 2000).

Occludin Knockout

Occludin is a functional component of tight junctions, responsible for cell–cell adhesion and maintaining the integrity of intercellular spaces. Knockout of occludin in mice does not affect their viability but does cause a slight but significant reduction in body weight. The loss of occludin did not seem to affect the morphology, protein content, or functionality of tight junctions in intestinal epithelial cells. However, by 3–6 weeks of age, knockout mice had atrophy of the fundic mucosa. Gastritis progressively developed and by 40 weeks the fundic mucosa was hyperplastic with mucous cell metaplasia (Saitou et al. 2000). No discussion is made of the possible increased susceptibility that these mice might have to infections and the role this might have in contributing to gastric pathology.

CCR7 Knockout

Although not typically associated with gastrointestinal mucosal homeostasis, genetic ablation of the chemokine receptor CCR7 results in a pathology that contributes to understanding the initiation of gastric hyperplasia. CCR7 is a chemokine receptor that regulates the trafficking and retention of leucocytes in secondary lymphoid organs, such as the gut. Loss of expression of CCR7 leads to a number of phenotypes including accumulation of functional lymphoid follicles in the stomachs of 8- to 10-week-old mice, with concomitant development of histopathology of the gastric mucosa, including accumulation of cells in the mucous neck region (Hopken et al. 2007). In 12-month-old mice, profound hyperplasia was observed with cystic dilatation reminiscent of Ménétrier's disease. Therefore, homeostatic regulation of differentiation and proliferation of fundic tissue is influenced by the presence of nonspecific, noninflammatory lymphoid aggregates in the gastric mucosa.

NFκB2 Knockout

Further demonstrating the role of the inflammatory system in regulating the gastric mucosa is a model in which the COOH terminus of NFκB2, an important transcription factor mediating inflammatory signals, has been deleted (Ishikawa et al. 1997). Deletion of this region of NFκB2 results in increased activation of Rel/NFKb transcription factors. In this model, the mice developed a very severe gastric phenotype whereby, at 3 weeks of age, the antral epithelial tissue was so severely hyperplastic that the lumen was occluded, resulting in premature death (Ishikawa et al. 1997). No further work on this model has been published, so it is unclear which genes are affected by the increased activation of NFκB leading to this gastric phenotype.

Conclusion

The mouse models of gastric hyperplasia and tumorigenesis described herein have furthered our understanding of initiating events in the context of chronic inflammation, the role of different metaplasias in pathologic progression, and the primary steps in submucosal invasion by dysplastic gastric epithelial cells. A major deficit is the lack of a tractable model of metastasis, which if developed would likely promote significant advances in development of therapeutics.

References

Akhurst, B., E. J. Croager, et al. (2001). A modified choline-deficient, ethionine-supplemented diet protocol effectively induces oval cells in mouse liver. Hepatology 34(3): 519–22.

Andersson, P., J. McGuire, et al. (2002). A constitutively active dioxin/aryl hydrocarbon receptor induces stomach tumors. Proc Natl Acad Sci USA 99(15): 9990–5.

Ang, D. K., T. C. Brodnicki, et al. (2007). Two genetic loci independently confer susceptibility to autoimmune gastritis. Int Immunol 19(9): 1135–44.

Atreya, R., J. Mudter, et al. (2000). Blockade of interleukin 6 trans signaling suppresses T-cell resistance against apoptosis in chronic intestinal inflammation: evidence in Crohn disease and experimental colitis in vivo. Nat Med 6(5): 583–8.

Bae, S. C. and Y. Ito (2003). Comment on Levanon et al., Runx3 knockouts and stomach cancer, in EMBO reports (June 2003). EMBO Rep 4(6): 538–9.

Baxter, A. G., M. A. Jordan, et al. (2005). Genetic control of susceptibility to autoimmune gastritis. Int Rev Immunol 24(1–2): 55–62.

Blyth, K., E. R. Cameron, et al. (2005). The RUNX genes: gain or loss of function in cancer. Nat Rev Cancer 5(5): 376–87.

Bockman, D. E., R. Sharp, et al. (1995). Regulation of terminal differentiation of zymogenic cells by transforming growth factor alpha in transgenic mice. Gastroenterology 108(2): 447–54.

Brenner, O., D. Levanon, et al. (2004). Loss of Runx3 function in leukocytes is associated with spontaneously developed colitis and gastric mucosal hyperplasia. Proc Natl Acad Sci USA 101(45): 16016–21.

Cai, J. P., T. Ishibashi, et al. (2003). Mouse MTH2 protein which prevents mutations caused by 8-oxoguanine nucleotides. Biochem Biophys Res Commun 305(4): 1073–7.

Coste, I., J. N. Freund, et al. (2007). Precancerous lesions upon sporadic activation of beta-catenin in mice. Gastroenterology 132(4): 1299–308.

Crawford, S. E., V. Stellmach, et al. (1998). Thrombospondin-1 is a major activator of TGF-beta1 in vivo. Cell 93(7): 1159–70.

Dempsey, P. J., J. R. Goldenring, et al. (1992). Possible role of transforming growth factor alpha in the pathogenesis of Menetrier's disease: supportive evidence form humans and transgenic mice. Gastroenterology 103(6): 1950–63.

Derynck, R., R. J. Akhurst, et al. (2001). TGF-beta signaling in tumor suppression and cancer progression. Nat Genet 29(2): 117–29.

Elso, C. M., X. Lu, et al. (2004). Heightened susceptibility to chronic gastritis, hyperplasia and metaplasia in Kcnq1 mutant mice. Hum Mol Genet 13(22): 2813–21.

El-Zaatari, M., A. Tobias, et al. (2007). De-regulation of the sonic hedgehog pathway in the InsGas mouse model of gastric carcinogenesis. Br J Cancer 96(12): 1855–61.

Fox, J. G., C. A. Dangler, et al. (1997). Mice carrying a truncated Apc gene have diminished gastric epithelial proliferation, gastric inflammation, and humoral immunity in response to Helicobacter felis infection. Cancer Res 57(18): 3972–8.

Fox, J. G., X. Li, et al. (1996). Hypertrophic gastropathy in Helicobacter felis-infected wild-type C57BL/6 mice and p53 hemizygous transgenic mice. Gastroenterology 110(1): 155–66.

Fox, J. G., A. B. Rogers, et al. (2003a). Helicobacter pylori-associated gastric cancer in INS-GAS mice is gender specific. Cancer Res 63(5): 942–50.

Fox, J. G., B. J. Sheppard, et al. (2002). Germ-line p53-targeted disruption inhibits Helicobacter-induced premalignant lesions and invasive gastric carcinoma through down-regulation of Th1 proinflammatory responses. Cancer Res 62(3): 696–702.

Fox, J. G., T. C. Wang, et al. (2003b). Host and microbial constituents influence Helicobacter pylori-induced cancer in a murine model of hypergastrinemia. Gastroenterology 124(7): 1879–90.

Franic, T. V., L. M. Judd, et al. (2001). Regulation of gastric epithelial cell development revealed in H(+)/K(+)-ATPase beta-subunit- and gastrin-deficient mice. Am J Physiol Gastrointest Liver Physiol 281(6): G1502–11.

Franic, T. V., I. R. van Driel, et al. (2005). Reciprocal changes in trefoil 1 and 2 expression in stomachs of mice with gastric unit hypertrophy and inflammation. J Pathol 207(1): 43–52.

Friis-Hansen, L. (2002). Gastric functions in gastrin gene knock-out mice. Pharmacol Toxicol 91(6): 363–7.

Friis-Hansen, L. (2006). Achlorhydria is associated with gastric microbial overgrowth and development of cancer: lessons learned from the gastrin knockout mouse. Scand J Clin Lab Invest 66(7): 607–21.

Friis-Hansen, L., F. Sundler, et al. (1998). Impaired gastric acid secretion in gastrin-deficient mice. Am J Physiol 274(3 Pt 1): G561–8.

Fukamachi, H., K. Ito, et al. (2004). Runx3−/− gastric epithelial cells differentiate into intestinal type cells. Biochem Biophys Res Commun 321(1): 58–64.

Fukui, T., A. Nishio, et al. (2006). Gastric mucosal hyperplasia via upregulation of gastrin induced by persistent activation of gastric innate immunity in major histocompatibility complex class II deficient mice. Gut 55(5): 607–15.

Garcia de Galdeano, A., J. C. Cruz-Conde, et al. (2001). Effect of IL-2 and IL-6 on parameters related to metastatic activity in a murine melanoma. Pathobiology 69(4): 230–6.

Gawenis, L. R., J. M. Greeb, et al. (2005). Impaired gastric acid secretion in mice with a targeted disruption of the NHE4 Na+/H+ exchanger. J Biol Chem 280(13): 12781–9.

Goldenring, J. R. and S. Nomura (2006). Differentiation of the gastric mucosa. III. Animal models of oxyntic atrophy and metaplasia. Am J Physiol Gastrointest Liver Physiol 291(6): G999–1004.

Goldenring, J. R., G. S. Ray, et al. (1996). Overexpression of transforming growth factor-alpha alters differentiation of gastric cell lineages. Dig Dis Sci 41(4): 773–84.

Gut, M. O., S. Parkkila, et al. (2002). Gastric hyperplasia in mice with targeted disruption of the carbonic anhydrase gene Car9. Gastroenterology 123(6): 1889–903.

Hahm, K. B., K. M. Lee, et al. (2002). Conditional loss of TGF-beta signalling leads to increased susceptibility to gastrointestinal carcinogenesis in mice. Aliment Pharmacol Ther 16(Suppl 2): 115–27.

Harada, N., Y. Tamai, et al. (1999). Intestinal polyposis in mice with a dominant stable mutation of the beta-catenin gene. EMBO J 18(21): 5931–42.

Hopken, U. E., A. M. Wengner, et al. (2007). CCR7 deficiency causes ectopic lymphoid neogenesis and disturbed mucosal tissue integrity. Blood 109(3): 886–95.

Howlett, M., L. M. Judd, et al. (2005). Differential regulation of gastric tumor growth by cytokines that signal exclusively through the coreceptor gp130. Gastroenterology 129(3): 1005–18.

Hwang, I. R., P. I. Hsu, et al. (2003). Interleukin-6 genetic polymorphisms are not related to Helicobacter pylori-associated gastroduodenal diseases. Helicobacter 8(2): 142–8.

Ishikawa, H., D. Carrasco, et al. (1997). Gastric hyperplasia and increased proliferative responses of lymphocytes in mice lacking the COOH-terminal ankyrin domain of NF-kappaB2. J Exp Med 186(7): 999–1014.

Jackson, C. B., L. M. Judd, et al. (2007). Augmented gp130-mediated cytokine signalling accompanies human gastric cancer progression. J Pathol 213(2): 140–51.

Jenkins, B. J., D. Grail, et al. (2005). Hyperactivation of Stat3 in gp130 mutant mice promotes gastric hyperproliferation and desensitizes TGF-beta signaling. Nat Med 11(8): 845–52.

Johnson, A. H., H. F. Frierson, et al. (2005). Expression of tight-junction protein claudin-7 is an early event in gastric tumorigenesis. Am J Pathol 167(2): 577–84.

Judd, L. M., B. M. Alderman, et al. (2004). Gastric cancer development in mice lacking the SHP2 binding site on the IL-6 family co-receptor gp130. Gastroenterology 126(1): 196–207.

Judd, L. M., A. Andringa, et al. (2005). Gastric achlorhydria in H/K-ATPase-deficient (Atp4a(−/−)) mice causes severe hyperplasia, mucocystic metaplasia and upregulation of growth factors. J Gastroenterol Hepatol 20(8): 1266–78.

Judd, L. M., K. Bredin, et al. (2006). STAT3 activation regulates growth, inflammation, and vascularization in a mouse model of gastric tumorigenesis. Gastroenterology 131(4): 1073–85.

Kaestner, K. H., D. G. Silberg, et al. (1997). The mesenchymal winged helix transcription factor Fkh6 is required for the control of gastrointestinal proliferation and differentiation. Genes Dev 11(12): 1583–95.

Kanda, N., H. Seno, et al. (2006). Involvement of cyclooxygenase-2 in gastric mucosal hypertrophy in gastrin transgenic mice. Am J Physiol Gastrointest Liver Physiol 290(3): G519–27.

Kang, W., S. Rathinavelu, et al. (2005). Interferon gamma induction of gastric mucous neck cell hypertrophy. Lab Invest 85(5): 702–15.

Karam, S. M., C. Tomasetto, et al. (2004). Trefoil factor 1 is required for the commitment programme of mouse oxyntic epithelial progenitors. Gut 53(10): 1408–15.

Katz, J. P., N. Perreault, et al. (2005). Loss of Klf4 in mice causes altered proliferation and differentiation and precancerous changes in the adult stomach. Gastroenterology 128(4): 935–45.

Kim, B. G., C. Li, et al. (2006a). Smad4 signalling in T cells is required for suppression of gastrointestinal cancer. Nature 441(7096): 1015–9.

Kim, S. S., K. Shetty, et al. (2006b). TGF-beta signaling pathway inactivation and cell cycle deregulation in the development of gastric cancer: role of the beta-spectrin, ELF. Biochem Biophys Res Commun 344(4): 1216–23.

Kobayashi, T., S. Tonai, et al. (2000). Abnormal functional and morphological regulation of the gastric mucosa in histamine H2 receptor-deficient mice. J Clin Invest 105(12): 1741–9.

Koh, T. J., J. R. Goldenring, et al. (1997). Gastrin deficiency results in altered gastric differentiation and decreased colonic proliferation in mice. Gastroenterology 113(3): 1015–25.

Konda, Y., H. Kamimura, et al. (1999). Gastrin stimulates the growth of gastric pit with less-differentiated features. Am J Physiol 277(4 Pt 1): G773–84.

Kuznetsov, N. V., P. Andersson, et al. (2005). The dioxin/aryl hydrocarbon receptor mediates downregulation of osteopontin gene expression in a mouse model of gastric tumourigenesis. Oncogene 24(19): 3216–22.

Kuzushita, N., A. B. Rogers, et al. (2005). p27kip1 deficiency confers susceptibility to gastric carcinogenesis in Helicobacter pylori-infected mice. Gastroenterology 129(5): 1544–56.

Lawler, J., M. Sunday, et al. (1998). Thrombospondin-1 is required for normal murine pulmonary homeostasis and its absence causes pneumonia. J Clin Invest 101(5): 982–92.

Lee, M. P., J. D. Ravenel, et al. (2000). Targeted disruption of the Kvlqt1 gene causes deafness and gastric hyperplasia in mice. J Clin Invest 106(12): 1447–55.

Lefebvre, O., M. P. Chenard, et al. (1996). Gastric mucosa abnormalities and tumorigenesis in mice lacking the pS2 trefoil protein. Science 274(5285): 259–62.

Leppilampi, M., T. J. Karttunen, et al. (2005). Gastric pit cell hyperplasia and glandular atrophy in carbonic anhydrase IX knockout mice: studies on two strains C57/BL6 and BALB/C. Transgenic Res 14(5): 655–63.

Levanon, D., D. Bettoun, et al. (2002). The Runx3 transcription factor regulates development and survival of TrkC dorsal root ganglia neurons. EMBO J 21(13): 3454–63.

Levanon, D., O. Brenner, et al. (2003). Runx3 knockouts and stomach cancer. EMBO Rep 4(6): 560–4.

Li, Q., S. M. Karam, et al. (1995). Simian virus 40 T antigen-induced amplification of pre-parietal cells in transgenic mice. Effects on other gastric epithelial cell lineages and evidence for a p53-independent apoptotic mechanism that operates in a committed progenitor. J Biol Chem 270(26): 15777–88.

Li, Q. L., K. Ito, et al. (2002). Causal relationship between the loss of RUNX3 expression and gastric cancer. Cell 109(1): 113–24.

Li, S., Q. Wang, et al. (2000). Gastric hyperplasia in mice lacking the putative Cdc42 effector IQGAP1. Mol Cell Biol 20(2): 697–701.

Ludlow, A., K. O. Yee, et al. (2005). Characterization of integrin beta6 and thrombospondin-1 double-null mice. J Cell Mol Med 9(2): 421–37.

Mutoh, H., Y. Hakamata, et al. (2002). Conversion of gastric mucosa to intestinal metaplasia in Cdx2-expressing transgenic mice. Biochem Biophys Res Commun 294(2): 470–9.

Mutoh, H., S. Sakurai, et al. (2004a). Cdx1 induced intestinal metaplasia in the transgenic mouse stomach: comparative study with Cdx2 transgenic mice. Gut 53(10): 1416–23.

Mutoh, H., S. Sakurai, et al. (2004b). Development of gastric carcinoma from intestinal metaplasia in Cdx2-transgenic mice. Cancer Res 64(21): 7740–7.

Nockel, J., N. K. van den Engel, et al. (2006). Characterization of gastric adenocarcinoma cell lines established from CEA424/SV40 T antigen-transgenic mice with or without a human CEA transgene. BMC Cancer 6: 57.

Nomura, S., S. H. Settle, et al. (2005). Evidence for repatterning of the gastric fundic epithelium associated with Menetrier's disease and TGFalpha overexpression. Gastroenterology 128(5): 1292–305.

Ogawa, T., K. Maeda, et al. (2003). Utilization of knockout mice to examine the potential role of gastric histamine H2-receptors in Menetrier's disease. J Pharmacol Sci 91(1): 61–70.

Oshima, H., A. Matsunaga, et al. (2006). Carcinogenesis in mouse stomach by simultaneous activation of the Wnt signaling and prostaglandin E2 pathway. Gastroenterology 131(4): 1086–95.

Oshima, H., M. Oshima, et al. (2004). Hyperplastic gastric tumors induced by activated macrophages in COX-2/mPGES-1 transgenic mice. EMBO J 23(7): 1669–78.

Oshima, M., H. Oshima, et al. (2005). Hyperplastic gastric tumors with spasmolytic polypeptide-expressing metaplasia caused by tumor necrosis factor-alpha-dependent inflammation in cyclooxygenase-2/microsomal prostaglandin E synthase-1 transgenic mice. Cancer Res 65(20): 9147–51.

Ramalho-Santos, M., D. A. Melton, et al. (2000). Hedgehog signals regulate multiple aspects of gastrointestinal development. Development 127(12): 2763–72.

Ramsay, A. J., A. J. Husband, et al. (1994). The role of interleukin-6 in mucosal IgA antibody responses in vivo. Science 264(5158): 561–3.

Rangarajan, A. and R. A. Weinberg (2003). Opinion: comparative biology of mouse versus human cells—modelling human cancer in mice. Nat Rev Cancer 3(12): 952–9.

Redman, R. S., V. Katuri, et al. (2005). Orofacial and gastrointestinal hyperplasia and neoplasia in smad4+/− and elf+/−/smad4+/− mutant mice. J Oral Pathol Med 34(1): 23–9.

Roepke, T. K., A. Anantharam, et al. (2006). The KCNE2 potassium channel ancillary subunit is essential for gastric acid secretion. J Biol Chem 281(33): 23740–7.

Romagnolo, B., D. Berrebi, et al. (1999). Intestinal dysplasia and adenoma in transgenic mice after overexpression of an activated beta-catenin. Cancer Res 59(16): 3875–9.

Saitou, M., M. Furuse, et al. (2000). Complex phenotype of mice lacking occludin, a component of tight junction strands. Mol Biol Cell 11(12): 4131–42.

Saukkonen, K., C. Tomasetto, et al. (2003). Cyclooxygenase-2 expression and effect of celecoxib in gastric adenomas of trefoil factor 1-deficient mice. Cancer Res 63(12): 3032–6.

Scarff, K. L., L. M. Judd, et al. (1999). Gastric H(+),K(+)-adenosine triphosphatase beta subunit is required for normal function, development, and membrane structure of mouse parietal cells. Gastroenterology 117(3): 605–18.

Schultheis, P. J., L. L. Clarke, et al. (1998). Targeted disruption of the murine Na+/H+ exchanger isoform 2 gene causes reduced viability of gastric parietal cells and loss of net acid secretion. J Clin Invest 101(6): 1243–53.

Segre, J. A., C. Bauer, et al. (1999). Klf4 is a transcription factor required for establishing the barrier function of the skin. Nat Genet 22(4): 356–60.

Senda, T., A. Iizuka-Kogo, et al. (2007). Adenomatous polyposis coli (APC) plays multiple roles in the intestinal and colorectal epithelia. Med Mol Morphol 40(2): 68–81.

Sharp, R., M. W. Babyatsky, et al. (1995). Transforming growth factor alpha disrupts the normal program of cellular differentiation in the gastric mucosa of transgenic mice. Development 121(1): 149–61.

Shull, M. M., I. Ormsby, et al. (1992). Targeted disruption of the mouse transforming growth factor-beta 1 gene results in multifocal inflammatory disease. Nature 359(6397): 693–9.

Snyder, A. J., S. M. Karam, et al. (2004). A transgenic mouse model of metastatic carcinoma involving transdifferentiation of a gastric epithelial lineage progenitor to a neuroendocrine phenotype. Proc Natl Acad Sci USA 101(13): 4471–6.

Spicer, Z., M. L. Miller, et al. (2000). Stomachs of mice lacking the gastric H,K-ATPase alpha-subunit have achlorhydria, abnormal parietal cells, and ciliated metaplasia. J Biol Chem 275(28): 21555–65.

Stewart, L. A., I. R. van Driel, et al. (2002). Perturbation of gastric mucosa in mice expressing the temperature-sensitive mutant of SV40 large T antigen. Potential for establishment of an immortalised parietal cell line. Eur J Cell Biol 81(5): 281–93.

Takagi, H., T. Fukusato, et al. (1997). Histochemical analysis of hyperplastic stomach of TGF-alpha transgenic mice. Dig Dis Sci 42(1): 91–8.

Takagi, H., C. Jhappan, et al. (1992). Hypertrophic gastropathy resembling Menetrier's disease in transgenic mice overexpressing transforming growth factor alpha in the stomach. J Clin Invest 90(3): 1161–7.

Takaishi, S., G. Cui, et al. (2005). Synergistic inhibitory effects of gastrin and histamine receptor antagonists on Helicobacter-induced gastric cancer. Gastroenterology 128(7): 1965–83.

Takaku, K., H. Miyoshi, et al. (1999). Gastric and duodenal polyps in Smad4 (Dpc4) knockout mice. Cancer Res 59(24): 6113–7.

Tebbutt, N. C., A. S. Giraud, et al. (2002). Reciprocal regulation of gastrointestinal homeostasis by SHP2 and STAT-mediated trefoil gene activation in gp130 mutant mice. Nat Med 8(10): 1089–97.

Thompson, J., T. Epting, et al. (2000). A transgenic mouse line that develops early-onset invasive gastric carcinoma provides a model for carcinoembryonic antigen-targeted tumor therapy. Int J Cancer 86(6): 863–9.

Tomasetto, C. and M. C. Rio (2005). Pleiotropic effects of trefoil factor 1 deficiency. Cell Mol Life Sci 62(24): 2916–20.

Tomita, H., Y. Yamada, et al. (2007). Development of gastric tumors in Apc(Min/+) mice by the activation of the beta-catenin/Tcf signaling pathway. Cancer Res 67(9): 4079–87.

Torres, L. F., S. M. Karam, et al. (2002). Trefoil factor 1 (TFF1/pS2) deficiency activates the unfolded protein response. Mol Med 8(5): 273–82.

Tsuzuki, T., A. Egashira, et al. (2001). Spontaneous tumorigenesis in mice defective in the MTH1 gene encoding 8-oxo-dGTPase. Proc Natl Acad Sci USA 98(20): 11456–61.

Wang, T. C., S. Bonner-Weir, et al. (1993). Pancreatic gastrin stimulates islet differentiation of transforming growth factor alpha-induced ductular precursor cells. J Clin Invest 92(3): 1349–56.

Wang, T. C., C. A. Dangler, et al. (2000). Synergistic interaction between hypergastrinemia and Helicobacter infection in a mouse model of gastric cancer. Gastroenterology 118(1): 36–47.

Wang, T. C., T. J. Koh, et al. (1996). Processing and proliferative effects of human progastrin in transgenic mice. J Clin Invest 98(8): 1918–29.

Wang, Y. F., S. Y. Chang, et al. (2002). Clinical significance of interleukin-6 and interleukin-6 receptor expressions in oral squamous cell carcinoma. Head Neck 24(9): 850–8.

Watson, S. A., A. M. Grabowska, et al. (2006). Gastrin—active participant or bystander in gastric carcinogenesis? Nat Rev Cancer 6(12): 936–46.

Xu, X., S. G. Brodie, et al. (2000). Haploid loss of the tumor suppressor Smad4/Dpc4 initiates gastric polyposis and cancer in mice. Oncogene 19(15): 1868–74.

Yamamura, Y., W. L. Lee, et al. (2006). RUNX3 cooperates with FoxO3a to induce apoptosis in gastric cancer cells. J Biol Chem 281(8): 5267–76.

Yang, K., W. Edelmann, et al. (1997). A mouse model of human familial adenomatous polyposis. J Exp Zool 277(3): 245–54.

Yano, T., K. Ito, et al. (2006). The RUNX3 tumor suppressor upregulates Bim in gastric epithelial cells undergoing transforming growth factor beta-induced apoptosis. Mol Cell Biol 26(12): 4474–88.

Yu, C. R., J. R. Ortaldo, et al. (1999). Role of a STAT binding site in the regulation of the human perforin promoter. J Immunol 162(5): 2785–90.

Zavros, Y., K. A. Eaton, et al. (2005). Chronic gastritis in the hypochlorhydric gastrin-deficient mouse progresses to adenocarcinoma. Oncogene 24(14): 2354–66.

Zavros, Y., G. Rieder, et al. (2002). Genetic or chemical hypochlorhydria is associated with inflammation that modulates parietal and G-cell populations in mice. Gastroenterology 122(1): 119–33.

Chapter 20
Prostaglandin and Transforming Growth Factor β Signaling in Gastric Cancer

Masanobu Oshima, Hiroko Oshima, and Makoto Mark Taketo

Introduction

It has been established that induction of cyclooxygenase (COX)-2 expression, activation of Wnt signaling, and suppression of transforming growth factor β (TGF-β) signaling are important for colorectal cancer development. Involvement of these pathways in gastric tumorigenesis has also been suggested by various studies using clinical samples or *in vitro* experiments. Recent genetic studies using mouse models have shown that activation of the COX-2/prostaglandin E_2 (PGE_2) pathway leads to hyperplastic metaplasia in the stomach, and cooperation of PGE_2 and Wnt activation causes gastric cancer development. Moreover, several lines of genetic evidence indicate that inactivation of TGF-β signaling or novel tumor suppressor RUNX3 contributes to gastric tumorigenesis. In this chapter, involvement of COX-2, PGE_2, and TGF-β pathways in gastric tumorigenesis is discussed with reference to *in vitro* as well as *in vivo* evidence.

Cyclooxygenae-2 in Gastric Cancer

Cyclooxygenases and Prostanoids in Tumor Development

Arachidonic Acid Cascade and Cyclooxygenase-2

Epidemiologic studies indicate that regular use of nonsteroidal antiinflammatory drugs (NSAIDs) lowers the mortality rate of gastrointestinal cancer (Thun et al. 1991). The major target of NSAIDs is COX that is the rate-limiting enzyme for prostaglandin biosynthesis (Figure 20.1). COX enzyme catalyze synthesis of prostaglandin (PG)H_2 that is subsequently converted to various prostanoids including PGE_2 by tissue-specific converting enzymes. There are two COX isozymes, COX-1 and COX-2, which share a high degree of structural and enzymatic homology. COX-1 is constitutively expressed in most tissues and considered to be responsible for physiologic levels of PG biosynthesis (Dewitt and Smith 1988). On the other

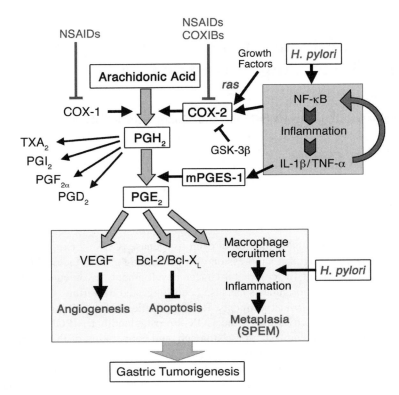

Fig. 20.1 Schematic presentation of arachidonic acid metabolism in the context of gastric tumorigenesis. Expression of cyclooxygenase-2 (COX-2) and microsomal prostaglandin E synthase-1 (mPGES-1) is induced by *Helicobacter pylori*–associated inflammatory response (pink box). Other possible mechanisms for induction of these enzymes are also indicated. Increased PGE$_2$ signaling by simultaneous expression of both COX-2 and mPGES-1 leads to stimulation of angiogenesis, suppression of apoptosis, and development of preneoplastic metaplasia, SPEM (yellow box), all of which can contribute to gastric tumorigenesis

hand, COX-2 expression is induced in inflammatory and tumor tissues by various stimuli including cytokines and growth factors (Figure 20.1) (Fletcher et al. 1992; Hla and Neilson 1992; Xie et al. 1991).

Suppression of Intestinal Polyposis by Inhibition of Cyclooxygenae-2 Pathway

Expression of COX-2 is induced in colorectal cancer tissues, and numerous studies have demonstrated the important role of COX-2 in intestinal tumorigenesis (Gupta and DuBois 2001). Mutation in the tumor suppressor gene *Apc* (*Adenomatous polyposis coli*) causes development of intestinal polyps (Oshima et al. 1995). Disruption of the COX-2 gene in *Apc*-mutant mice resulted in significant suppression of intestinal polyposis, indicating that the COX-2 pathway has an essential role in intestinal tumorigenesis (Chulada et al. 2000; Oshima et al. 1996). Moreover, various animal studies have confirmed that conventional NSAIDs or COX-2 selective inhibitors (COXIBs) suppress formation of chemically induced intestinal tumors and growth of tumor xenografts (Oshima and Taketo 2002).

Expression of Cyclooxygenae-2 and Microsomal Prostaglandin E Synthase-1 in Gastric Cancer

Cyclooxygenae-2 Expression in Gastric Tumorigenesis

Expression of COX-2 is induced also in gastric cancer tissues, whereas COX-1 expression is not elevated (Ristimäki et al. 1997; Saukkonen et al. 2003). Numerous reports using clinical samples have indicated that COX-2 expression is found in 43%–100% (mean 67%) of gastric cancer. This variation in the COX-2–positive frequency is possibly attributable to the sensitivity of antibodies used in these studies. Gastric cancer can be divided into two histologic subtypes: intestinal and diffuse types. The intestinal-type gastric cancer has certain precursor lesions such as chronic gastritis, intestinal metaplasia and dysplasia, although it is still controversial as to whether intestinal metaplasia is a precursor of intestinal-type gastric cancer. It has been reported that COX-2 induction is found predominantly in the intestinal-type gastric cancer (58%), whereas it is detected only in 6% of the diffuse type (Saukkonen et al. 2001). Accordingly, it is possible that development of intestinal-type gastric cancer requires the COX-2 pathway, whereas diffuse-type gastric cancer does not.

COX-2 has been detected in nontumorous gastric hyperplasia that arises in chronic gastritis (Kawada et al. 2003). However, the level of COX-2 expression increases gradually during tumor development from intestinal metaplasia and dysplasia to gastric adenocarcinoma (Lim et al. 2000; Ristimäki et al. 1997; Saukkonen et al. 2001; van Rees et al. 2002; Yamagata et al. 2002). Moreover, the level of COX-2 expression in gastric cancer is correlated with the tumor size, depth of invasion, and lymph-node metastasis (Murata et al. 1999; Ohno et al. 2001; Uefuji et al. 2000). It has been also reported that patients with high levels of COX-2 expression show poor survival, and that a high COX-2 level is an independent prognostic factor (Mrena et al. 2005). These expression analyses suggest that COX-2 has important roles in both initiation and progression stages of gastric tumorigenesis, and that increase in its level is important for malignant progression.

Cyclooxygenae-2 Induction by *Helicobacter pylori* Infection

Helicobacter pylori has been classified as "class I carcinogen" for gastric cancer by the World Health Organization. Infection with *H. pylori* induces a chronic superficial gastritis, which then leads to atrophic gastritis and intestinal metaplasia that seems to be a precursor of intestinal-type gastric cancer (Correa 2003). In the *H. pylori*–infected gastric mucosa, COX-2 expression is induced significantly, which is decreased by successful *H. pylori* eradication (McCarthy et al. 1999). Consistently, it has been shown with *in vitro* experiments that infection with *H. pylori* or the closely related strain *H. felis* induces COX-2 expression and PGE_2 production in normal gastric epithelial cells (Oshima et al. 2004; Shen et al. 2006). Moreover, in *H. pylori*–associated intestinal metaplasia, dysplasia, and gastric adenocarcinoma, levels of COX-2 expression are also increased significantly, and reduced by eradication of *H. pylori* (Chen et al. 2001; Sun et al. 2004; Sung et al. 2000). These results indicate that *H. pylori* infection is responsible for COX-2 induction from an early infection stage to later formation of adenocarcinoma.

It is possible that an activated cytokine network also induces COX-2 expression in the *H. pylori*–infected gastric mucosa. *H. pylori* can also stimulate Toll-like receptors (TLRs), causing activation of the NF-κB pathway, which is one of the transcription signals responsible for COX-2 expression (Chang et al. 2004; Smith et al. 2003). It has also been reported that the TLR2/TLR9 signaling by *H. pylori* activates mitogen-activated protein kinases (MAPK) including p38, resulting in the activation of CRE and AP-1 elements on the COX-2 gene promoter (Chang et al. 2005). These results suggest that infectious stimuli with *H. pylori* induce expression of COX-2 through various pathways (see also the section "Molecular Mechanisms of Cyclooxygenae-2 Induction in Gastric Cancer").

Induction of Microsomal Prostaglandin E Synthase-1 Expression and Production of Prostaglandin E_2 in Gastric Cancer

Microsomal PGE synthase-1 (mPGES-1) is an inducible enzyme that seems to be functionally coupled with COX-2 (Murakami et al. 2000). Simultaneous induction of mPGES-1 and COX-2 was found in colon cancer (Yoshimatsu et al. 2001) as well as in mouse intestinal polyps (Takeda et al. 2004), suggesting that the level of PGE_2 is increased in intestinal tumors. Genetic studies using *Apc* mutant mice ($Apc^{\Delta716}$) have indicated that PGE_2 signaling through its receptor EP2, is responsible for intestinal tumor development through a positive feedback loop of induction of COX-2 and induction of angiogenesis (Seno et al. 2002; Sonoshita et al. 2001).

Expression of mPGES-1 is also found in gastric cancer tissues, suggesting that PGE_2 production is enhanced in gastric cancer (Jang et al. 2004; van Rees et al. 2003). The level of mPGES-1 was decreased after therapeutic eradication of *H. pylori* (Nardone et al. 2004). These results suggest that *H. pylori* infection enhances PGE_2 production through induction of both COX-2 and mPGES-1 (Figure 20.1). Consistently, the PGE_2 level in gastric cancer is increased significantly

(Uefuji et al. 2000), and it is higher in *H. pylori*–associated cases than in those without *H. pylori* (Al-Marhoon et al. 2004).

Suppression of Intestinal-Type Gastric Cancer by Cyclooxygenae-2 Inhibition

Suppression of Gastric Cancer Development by Nonsteroidal Anti-inflammatory Drugs

Epidemiologic studies indicate that regular use of NSAIDs is association with decreased incidence or mortality of gastric cancer (Akre et al. 2001; Langman et al. 2000; Schreinemachers and Everson 1994; Thun et al. 1993; Zaridze et al. 1999). Protective effects of NSAIDs against gastric cancer were observed only in individuals who were *H. pylori* positive. Furthermore, reduction of the gastric cancer risk by NSAIDs was associated with intestinal-type tumors, but was uncertain for diffuse-type cancers (Akre et al. 2001).

It has also been shown that single nucleotide polymorphism in the COX-2 gene promoter ($-1195A/A$) is associated with increased risk of gastric cancer (Liu et al. 2006). Moreover, infection with *H. pylori* significantly stimulates COX-2 gene promoter activity driven by $-1195A$ compared with the $-1195G$, suggesting that the COX-2 gene polymorphism affects the level of COX-2 induction in *H. pylori*–infected lesions and subsequent gastric tumorigenesis.

Suppression of Gastric Cancer by Cyclooxygenae-2 Inhibition in Animal Models

Suppression of gastric tumorigenesis by treatment with COXIBs has been examined in several animal-model experiments. Growth of gastric cancer cell xenografts was inhibited by treatment with COXIBs in immunodeficient mice (Sawaoka et al. 1998; Tang et al. 2004). Moreover, rat gastric cancer induced by N-methyl-N′-nitro-N-nitrosoguanidine treatment and mouse gastric tumors induced by *H. pylori* infection or combination of *H. pylori* with N-methyl-N-nitrosourea were suppressed by treatment with NSAIDs or COXIBs (Hu et al. 2004; Nam et al. 2004; Xiao et al. 2001). A significant decrease in the PGE_2 level and substantial induction of apoptosis were also found in these drug-treated tumor tissues.

H. pylori infection of Mongolian gerbils is an established model to study gastric tumorigenesis by *H. pylori*. Again, treatment of the *H. pylori*–infected Mongolian gerbils with COXIBs suppressed development of gastric cancer as well as intestinal metaplasia (Futagami et al. 2006; Magari et al. 2005). These animal studies indicate that COX-2 pathway is required for gastric tumorigenesis.

Induction Mechanism and Outcome Effects
of Cyclooxygenae-2 Pathway in Gastric Cancer

Molecular Mechanisms of Cyclooxygenae-2 Induction in Gastric Cancer

In the COX-2 gene promoter, there are several *cis* elements such as NF-κB binding element, cyclic AMP response element (CRE), and NF-IL6 (C/EBPβ). NF-κB is a key transcription factor that regulates inflammatory responses. In gastric cancer cells, inhibition of NF-κB resulted in downregulation of COX-2 expression, indicating that NF-κB has a role in COX-2 induction (Lim et al. 2001). In gastric mucosa, NF-κB is activated either by *H. pylori* infection or inflammatory cytokine tumor necrosis factor (TNF)-α. Moreover, TNF-α and interleukin (IL)-1β can induce expression of mPGES-1 (Stichtenoth et al. 2001), suggesting that the inflammatory network increases the PGE$_2$ level through induction of both COX-2 and mPGES-1 by activation of NF-κB (Figure 20.1).

Recently, it has been reported that expression of COX-2 is induced by inhibition of glycogen synthase kinase (GSK)-3β in gastric cancer cells, but not by MAPK (Thiel et al. 2006). GSK-3β is one of the target substrates of the PI3K/Akt pathway, and its phosphorylation causes inhibition of the kinase activity. Accordingly, it is possible that activation of the PI3K/Akt pathway is important for gastric cancer development through inhibition of GSK-3β, resulting in COX-2 induction (Figure 20.1).

Angiogenesis Mediated by the Cyclooxygenae-2 Pathway in Gastric Cancer

Angiogenesis is one of the key mechanisms that supports tumor development. In intestinal tumors, the COX-2 pathway has a key role in angiogenesis through induction of vascular endothelial growth factor (VEGF) and basic fibroblast growth factor (bFGF) (Seno et al. 2002; Sonoshita et al. 2001; Tsujii et al. 1998). In gastric cancer, it has been shown that COX-2 expression is associated with VEGF expression and increased microvessel density (MVD), whereas COX-2–negative gastric tumors show decreased levels of VEGF and MVD (Joo et al. 2003; Shi et al. 2003; Tatsuguchi et al. 2004; Uefuji et al. 2000; Yu et al. 2003). Transfection of the COX-2 gene in gastric cancer cells led to induction of PGE$_2$ production and VEGF expression (Leung et al. 2003), and treatment of gastric cancer cells with αCOXIB suppressed proliferation of cocultured endothelial cells (Fu et al. 2005). Moreover, treatment of the human gastric cancer xenografts with αCOXIB reduced the expression level of VEGF and FGF in tumor tissues together with decreased MVD (Wu et al. 2005). Importantly, primary gastric cancer tissues ablated from patients who received COXIB treatment showed significantly reduced VEGF expression and decreased MVD compared with those of the no-drug treated group (Zhou et al. 2007). These results, taken together, indicate that angiogenesis is one of the major functions of COX-2 pathway signaling in gastric tumorigenesis (Figure 20.1).

Other Effects Regulated by Cyclooxygenae-2 Pathway in Gastric Cancer

It has been demonstrated that the COX-2 pathway has an important role in suppression of apoptosis in colon cancer (Tsujii and DuBois 1996). In gastric cancer, the apoptotic index is lower in COX-2–positive cancer tissues compared with that in the COX-2–negative ones (Tatsuguchi et al. 2004). Moreover, expression of COX-2 is correlated with that of antiapoptotic factors such as Bcl-2 and Bcl-X$_L$ (Nardone et al. 2004; Tatsuguchi et al. 2004). Importantly, treatment of gastric cancer tissues with a COX-2 inhibitor resulted in increased apoptosis (Zhou et al. 2007). These results collectively indicate that COX-2 pathway signaling contributes to suppression of apoptosis in gastric cancer (Figure 20.1).

Other than induction of angiogenesis and suppression of apoptosis, the COX-2 pathway has been implicated in proliferation through activation of EGF receptor (EGFR) signaling. The PGE$_2$ signaling phosphorylates EGFR through release of TGF-α, an EGFR ligand, by activation of metalloproteinases (MMPs), which results in gastric hyperplasia (Pai et al. 2002). Accordingly, it is possible that COX-2/PGE$_2$ pathway signaling has a key role in gastric tumorigenesis through a variety of functions as described here.

Inflammation-Associated Hyperplasia and Metaplasia by Induction of Cyclooxygenae-2 and Microsomal Prostaglandin E Synthase-1 in Mouse Stomach

Hyperplasia and Mucous Metaplasia by Activation of Prostaglandin E$_2$ Pathway

K19-C2mE is a transgenic mouse line that expresses COX-2 and mPGES-1 simultaneously in the gastric epithelial cells, which mimics gastric PGE$_2$ production by *H. pylori* infection (Oshima et al. 2004). The *K19-C2mE* mice develop hyperplastic tumorous lesions in the glandular stomach (Figure 20.2A). Histologically, major cell types showing hyperplasia are mucous cells, which are similar to those found in the spasmolytic polypeptide/TFF2-expressing metaplasia (SPEM) (Oshima et al. 2005) (Figure 20.2B). SPEM is characterized by expansion of TFF2-expressing mucous cells and is associated with *H. pylori* infection and gastric adenocarcinoma, suggesting that SPEM is a preneoplastic metaplasia of gastric epithelium (Nomura et al. 2004). Accordingly, it is possible that *H. pylori* infection causes development of metaplasia called SPEM through induction of COX-2 and mPGES-1.

Notably, angiogenesis is not enhanced in the *K19-C2mE* mouse stomach, suggesting that increased PGE$_2$ signaling alone is not sufficient to drive angiogenesis *in vivo*. However, further activation of Wnt signaling in *K19-C2mE* mouse stomach resulted in marked induction of angiogenesis (see the section "Gastric Adenocarcinoma by Cooperation of Wnt and Prostaglandin E$_2$ Pathways"),

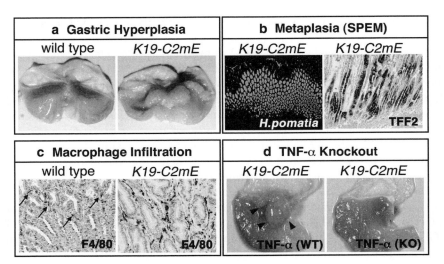

Fig. 20.2 Development of metaplastic hyperplasia by increased prostaglandin E₂ level in gastric mucosa of *K19-C2mE* mice. (**a**) Hyperplasia is found in the proximal glandular stomach of *K19-C2mE* (right) but not in wild type (left). (**b**) Histology of hyperplastic lesions of *K19-C2mE* mice with *Helix pomatia* lectin staining to detect mucous cells (left), and with TFF2 *in situ* hybridization to evaluate SPEM (right). (**c**) Immunostaining with F4/80 antibody specific for macrophages. Accumulated macrophages are found in the *K19-C2mE* stomach (right), whereas fewer macrophages are scattered in normal gastric mucosa (left). (**d**) Disruption of tumor necrosis factor (TNF)-α gene in *K19-C2mE* mice suppresses gastric lesions (right). (**a–c** reproduced from Oshima et al. 2004, with permission from Nature Publishing Group; **d** reproduced from Oshima et al. 2005, with permission from the AACR.)

suggesting that cooperation of the COX-2 pathway and Wnt signaling is important for angiogenesis in gastric tumors.

Inflammatory Response by Infectious Stimulation and Prostaglandin E₂ Induction

Heavy macrophage accumulation and their activation are found in the gastric mucosa of *K19-C2mE* mice (Figure 20.2C) (Oshima et al. 2004). Moreover, submucosal inflammatory cell infiltrations are found in the hyperplastic lesions. These inflammatory responses are suppressed by treatment of the mice with antibiotics, showing that bacterial infection is important for the induction of inflammation. Importantly, the hyperplasia and SPEM were also suppressed by antibiotic treatment, indicating that macrophage activation caused by infection is responsible for development of these gastric phenotypes. Notably, disruption of the TNF-α gene in *K19-C2mE* mice resulted in suppression of both inflammation and hyperplasia (Figure 20.2D), suggesting that TNF-α–dependent inflammation is essential for the hyperplasia and SPEM (Oshima et al. 2005). Increased PGE₂ signaling seems to be the primary cause for the macrophage accumulation, because macrophages infiltrate even in the noninflamed mucosa of *K19-C2mE* mice. A possible mechanism showing the processes from *H. pylori* infection to hyperplasia and SPEM is depicted in Figure 20.1. Although no dysplastic tumors develop in the *K19-C2mE* mice, further activation of Wnt signaling in these mice leads to gastric adenocarcinomas (see the section "Gastric Adenocarcinoma by Cooperation of Wnt and Prostaglandin E_2 Pathways"). Therefore, the PGE_2-induced inflammation, hyperplasia, and SPEM are important for gastric tumorigenesis.

Inflammatory Responses in *Helicobacter pylori*–Associated Gastric Cancer

H. pylori infection induces transdifferentiation of bone marrow derived cells (BMDCs) to gastric epithelial cells, which are considered to be an origin of gastric cancer (Houghton et al. 2004). Recently, it has also been reported that *H. pylori* infection induces expression of activation-induced cytidine deaminase (AID), an enzyme for DNA and RNA editing, which causes genetic alteration in the p53 tumor suppressor gene (Matsumoto et al. 2007). Although molecular mechanisms underlying these phenotypes have not been elucidated, it is possible that the COX-2/PGE_2 pathway induced by *H. pylori* infection contributes to recruitment of BMDCs and induction of AID in gastric mucosa. It would be interesting to examine these possibilities using *K19-C2mE* mice.

Wnt and Prostaglandin Signaling in Gastric Cancer

Activation of Wnt Signaling in Gastric Cancer Cells

Canonical Wnt Pathway in Tumorigenesis

Two Wnt signaling pathways are established: a canonical pathway (Wnt/β-catenin pathway) and a noncanonical pathway. Wnt/β-catenin transcription is now well recognized as a critical pathway in the regulation of development as well as in tumorigenesis (Taketo 2006). When Wnt signaling is in a resting state, β-catenin is phosphorylated by GSK-3β within a protein complex that contains APC and Axin (Gregorieff and Clevers 2005). Phosphorylated β-catenin is then degraded through the ubiquitin proteasome pathway (Figure 20.3). By binding of Wnt ligands to Frizzled receptors, phosphorylation of β-catenin by GSK-3β is suppressed, leading to stabilization of β-catenin. β-Catenin thus interacts with T-cell factor/lymphocyte enhancer factor to stimulate transcription of Wnt target genes, encoding important regulators of proliferation, metastasis, and angiogenesis. Accordingly, mutations in *APC*, β-catenin, or *Axin* result in stabilization of β-catenin and activation of the canonical Wnt signaling, which can cause tumor development.

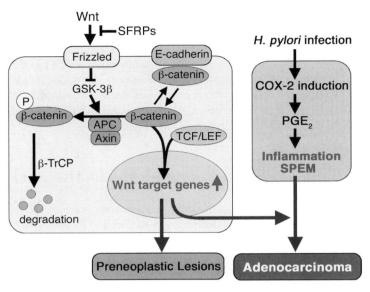

Fig. 20.3 Schematic presentation of the canonical Wnt signaling pathway in the context of gastric cancer development. In a Wnt signaling resting state, β-catenin is phosphorylated (P) and ubiquitinated by β-TrCP, resulting in degradation through ubiquitin pathway. Mutations in *APC* or β-catenin, methylation in *SFRP* genes, or downregulation of E-cadherin, or β-TrCP lead to stabilization of β-catenin in gastric epithelial cells, which activates transcription of Wnt target genes causing development of small gastric lesions with undifferentiated epithelial cells (preneoplastic lesions). However, simultaneous activation of Wnt signaling and *Helicobacter pylori*–associated cyclooxygenase-2 (COX-2)/prostaglandin E$_2$ (PGE$_2$) pathway leads to development of gastric adenocarcinoma (right side)

Accumulation and Mutation of β-Catenin in Gastric Cancer Cells

Patients with germline mutations in the *APC* gene have increased risk of gastric cancer up to 10-fold compared with those with wild-type APC (Offerhaus et al. 1992), suggesting that activation of Wnt signaling contributes to gastric carcinogenesis. Several studies have shown that nuclear and cytoplasmic accumulation of β-catenin, a hallmark of canonical Wnt activation, is found predominantly in intestinal-type gastric cancer rather than in the diffuse type (Miyazawa et al. 2000; Park et al. 1999). In contrast, other studies have indicated that β-catenin accumulation is found both in the intestinal and diffuse types (Figure 20.4) (Cheng et al. 2005; Clements et al. 2002; Oshima et al. 2006; Woo et al. 2001). It has also been reported that the incidence of β-catenin nuclear localization is 29% of gastric cancers (Clements et al. 2002). These results suggest that activation of the Wnt pathway is one of the major causes of gastric cancer development.

In colorectal cancer, mutations in the *APC* gene are detected in more than 80% of cases, and mutations in the β-catenin gene are also found in a small fraction of cancer cells that do not carry *APC* mutations. Most mutations of the β-catenin

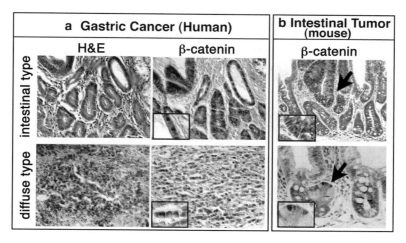

Fig. 20.4 Nuclear localization of β-catenin in human gastric cancer (**a**) and mouse intestinal adenoma (arrows) of *Apc^(Δ716)* mice (**b**). Specimens of intestinal-type and diffuse-type gastric cancer tissues are serial sections. Insets indicate higher magnifications where β-catenin is localized in the nuclei of tumor cells. (**a** reproduced from Oshima et al. 2006, with permission from Elsevier.)

gene (*CATNNB*) are detected in exon 3, and such mutant forms are no longer phosphorylated by GSK-3β. In gastric cancer, mutations in *CATNNB* exon 3 have also been reported, whereas *APC* gene mutations are rarely detected (Ebert et al. 2002; Clements et al. 2002; Park et al. 1999; Woo et al, 2001). However, *CATNNB* mutation in exon 3 was found only in 26% of Wnt-activated gastric cancers (Clements et al. 2002), suggesting that a mechanism(s) other than *APC* or *CATNNB* mutation causes Wnt signaling activation in a subpopulation of gastric cancers.

Possible Mechanisms for Wnt Activation in Gastric Cancer

E-cadherin is an intercellular adhesion molecule, and its cytoplasmic domain binds to β-catenin, linking E-cadherin to the cytoskeleton (Figure 20.3). Germline mutations of the E-cadherin gene cause hereditary diffuse-type gastric cancer. Moreover, decreased E-cadherin expression is associated with nuclear accumulation of β-cateninin gastric cancer cells (Cheng et al. 2005), suggesting that downregulation of E-cadherin leads to an increase in the cytoplasmic β-catenin resulting in activation of Wnt signaling.

β-TrCP is a ubiquitin ligase that targets phosphorylated β-catenin, and is an important negative regulator of canonical Wnt signaling (Figure 20.3). Recently, it has been reported that somatic mutations in β-TrCP are correlated with accumulation of β-catenin in gastric cancer cells (Kim et al. 2007). Therefore, mutations or downregulation of the β-TrCP gene can also lead to Wnt pathway activation.

Another important mechanism for Wnt activation is through epigenetic changes. Infection with *H. pylori* induces DNA methylation in the gastric mucosa. DNA methylation downregulates gene expression, and the level of methylation is considered to be a risk marker for gastric cancer (Nakajima et al. 2006). Recently, it has been demonstrated that *SFRP1*, *2*, and *5* genes are methylated and silenced in 91%, 96%, and 65% of primary gastric cancer tissues, respectively (Nojima et al. 2007). SFRPs (secreted frizzled-related proteins) belong to a family of secreted glycoproteins that have been identified as inhibitors of the Wnt signaling pathway (Figure 20.3). Moreover, transfection of *SFRP* genes in gastric cancer cells downregulates Wnt signaling activity, suggesting that methylation of *SFRP* genes is an important pathway for Wnt activation in gastric tumorigenesis (Nojima et al. 2007).

Gastric Adenocarcinoma by Cooperation of Wnt Signaling and Prostaglandin E₂ Pathway

Preneoplastic Lesions by Wnt Activation in Gastric Mucosa

K19-Wnt1 transgenic mice expressing *Wnt1* gene in the whole gastric epithelial cells show increased level of unphosphorylated β-catenin in the stomach (Oshima et al. 2006). The number of undifferentiated epithelial cells is significantly increased in the *K19-Wnt1* mouse stomach, suggesting that activation of the Wnt signaling suppresses differentiation of gastric epithelial cells. Importantly, multiple small lesions develop in the gastric mucosa of *K19-Wnt1* mice that consist of dysplastic epithelial cells with irregular gland branching (Figure 20.5A). Moreover, cytoplasmic and nuclear accumulation of β-catenin is found in the epithelial cells of such lesions (Figure 20.5B). However, *K19-Wnt1* mice do not develop dysplastic tumors in the stomach, indicating that activation of the Wnt signaling alone is not sufficient for gastric tumorigenesis.

Gastric Adenocarcinoma by Cooperation of Wnt and Prostaglandin E₂ Pathways

As described above, the *K19-C2mE* mice show gastric metaplastic hyperplasia, whereas *K19-Wnt1* mice develop small lesions with undifferentiated epithelial cells. Importantly, however, compound transgenic mice (*K19-Wnt1/K19-C2mE* mice), in which both Wnt and PGE₂ pathways are activated in the stomach, develop gastric adenocarcinoma with a complete penetrance (Figure 20.6A) (Oshima et al. 2006). Tumors in the compound transgenic mice consist of dysplastic epithelial cells, which invade into the smooth muscle layer (Figure 20.6B). These results indicate that simultaneous activation of the Wnt and PGE₂ pathways is responsible for the gastric adenocarcinomas. It is possible that invasion by

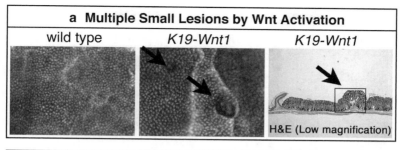

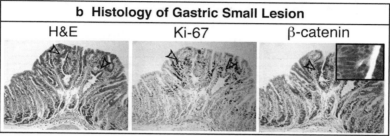

Fig. 20.5 Multiple small lesions developed in the glandular stomach of *K19-Wnt1* mice (**a, center and right**). Such lesions are not detected in wild-type mouse stomach (**a, left**). (**b**) Histology of the lesion indicated in **a**; hematoxylin and eosin staining (H&E, left), Ki-67 labeling (center), and β-catenin immunostaining (right). Arrowheads indicate undifferentiated epithelial cells with increased Ki-67 and β-catenin levels. Inset shows a higher magnification of β-catenin accumulating cells. (Reproduced from Oshima et al. 2006, with permission from Elsevier.)

tumor epithelial layers is caused by upregulation of MMP family member membrane type 3, because its expression is stimulated by Wnt signaling in gastric cancer (Lowy et al. 2006).

Although angiogenesis is not enhanced in the gastric lesions of either *K19-C2mE* or *K19-Wnt1* mice, the number of capillary vessels increases significantly in the compound *K19-Wnt1/K19-C2mE* mouse tumors (Figure 20.6B). Therefore, activation of both Wnt and PGE_2 pathways appears necessary for the induction of angiogenesis. However, it remains to be elucidated how these two pathways contribute to angiogenesis in gastric cancer.

These studies using *K19-C2mE* and *K19-Wnt1* mice suggest the following possible scenario for gastric tumorigenesis (Figures 1 and 3). *H. pylori* infection causes chronic gastritis, with activation of COX-2/PGE_2 pathway. Increased PGE_2 signaling causes inflammation and SPEM metaplasia through macrophage accumulation and activation. When Wnt signaling is activated by genetic or epigenetic alteration in the normal gastric mucosa, multiple preneoplastic lesions develop. However, Wnt activation in the PGE_2-induced inflammatory and SPEM lesions can cause gastric adenocarcinomas. Thus, cooperation of both Wnt and PGE_2 signaling can cause gastric cancer development.

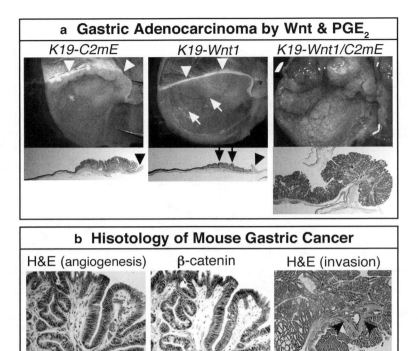

Fig. 20.6 Gastric adenocarcinomas in the *K19-Wnt1/K19-C2mE* (*K19-Wnt1/C2mE*) compound mutant mice. (**a**) Macroscopic photographs (top) and hematoxylin and esoin staining (H&E, bottom) of the glandular stomach of the *K19-C2mE* (left), *K19-Wnt1* (center), and *K19-Wnt1/C2mE* (right) mice. Arrowheads (white in top and black in bottom photos) indicate the border between the glandular stomach and forestomach. Arrows in *K19-Wnt1* point to the multiple small lesions. Note that dysplastic tumors with hyperemia develop in *K19-Wnt1/C2mE* (right). No such tumors are found in either simple transgenic strain. (**b**) Histology of tumors in the *K19-Wnt1/C2mE* mice. Tumors consist of dysplastic epithelial cells with increased number of capillary vessels (left). Nuclear accumulation of β-catenin (center) and tumor invasion into the smooth muscle layer (right) are also found in the tumor tissues. (Reproduced from Oshima et al. 2006, with permission from Elsevier.)

Transforming Growth Factor β Signaling in Gastric Cancer

Suppression of Transforming Growth Factor β Signaling in Gastric Cancer

Growth Inhibition of Epithelial Cells by Transforming Growth Factor β Pathway

TGF-β is one of a family of multifunctional cytokines that regulates proliferation, differentiation, apoptosis, and matrix accumulation. TGF-β ligands bind to the type II receptor (TGF-βRII), allowing association of the type I receptor (TGF-βRI), which causes phosphorylation of TGF-βRI by TGF-βRII (Shi and Massagué 2003) (Figure 20.7). Smad proteins are classified into three subtypes: R-Smads (Smad1, 2, 3, 5, and 8), Co-Smad (Smad4), and inhibitory I-Smad (Smad7). Smad2/3 are phosphorylated by activated TGF-βRI, whereas Smad1/5/8 are phosphorylated by bone morphogenetic protein (BMP) receptor. Phosphorylated Smad2/3 subsequently forms heteromeric complexes with Co-Smad, Smad4, resulting in transcriptional activation of the TGF-β target genes. TGF-β is a potent inducer of growth inhibition in several cell types including epithelial cells, through induction of CDK inhibitors p15 and p21 (Derynck et al. 2001). Moreover, TGF-β induces apoptosis by downregulation of Bcl-X_L and activation of caspase 3 and 8. In the intestines, differentiated epithelial cells express both TGF-β and TGF-βRII (Oshima et al. 1997). Thus, TGF-β stimulates epithelial differentiation and apoptosis possibly through autocrine and paracrine manners in the intestinal mucosa.

Mutations in Transforming Growth Factor βRII in Gastric Cancer

Genetic alterations in the TGF-βRII gene (*TGFBR2*) have been reported in colon cancer developed in hereditary nonpolyposis colon cancer (HNPCC) patients

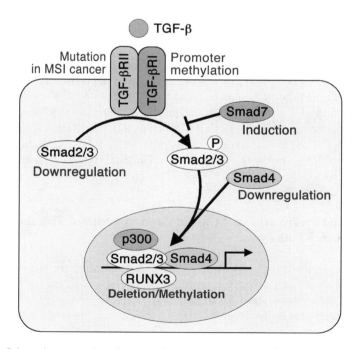

Fig. 20.7 Schematic presentation of transforming growth factor (TGF)-β signaling pathway. Genetic or epigenetic alterations reported in gastric cancer are indicated in blue next to or beneath each component. "P" indicates phosphorylation of Smad2/3. RUNX3 is indicated here because it has an important role in some functions of TGF-β pathway through binding to the Smad complex

(Markowitz et al. 1995; Parsons et al. 1995). A characteristic feature of HNPCC tumors is a defect in the DNA replication error repair system, resulting in microsatellite instability (MSI), which is termed as RER+ (replication error repair positive) phenotype. Characteristics of mutations in RER+ colon cancer cells are insertions or deletions in the 10-base pair polyadenine repeat (A10) of *TGFBR2* that causes frameshift mutations. Such mutations have also been frequently found in RER+ gastric cancer (Myeroff et al. 1995; Park et al. 1994). Transfection of the wild-type *TGFBR2* in RER+ gastric cancer cells results in reduced proliferation rate and restores TGF-β responsiveness (Chang et al. 1997; Yang et al. 1999). Moreover, such transfected cells show decreased or delayed formation of tumor xenografts, indicating that mutations in *TGFBR2* contribute to gastric tumorigenesis.

Suppression of TGF-β Receptor Type I and Smad Expression in Gastric Cancer

Downregulation of TGF-βRI expression has been found in gastric cancer cells that are resistant to TGF-β signaling but contain no detectable mutations in *TGFBR2* (Kang et al. 1999). Promoter methylation in the TGF-βRI gene (*TGFBR1*) was found in these cells, and treatment with a demethylating agent, 5-aza-2'-deoxycytidine, restored TGF-βRI expression. It has been reported that promoter methylation in *TGFBR1* is detected more frequently in gastric cancers with MSI than in those which are microsatellite stable (Pinto et al. 2003).

Reduced expression of Smad4 is also found in 75% of gastric cancers (Xiangming et al. 2001), and this is more frequent in the diffuse type (Kim et al. 2005). A reduced level of Smad3 has also been found in 37% of gastric cancers, and transfection of the Smad3 gene into gastric cancer cells that are deficient in Smad3 restores responsiveness to TGF-β, such as p21 induction and growth inhibition (Han et al. 2004). Accordingly, it is possible that downregulation of TGF-βRI or Smad3/4 expression as well as TGF-βRII mutation causes gastric carcinogenesis.

Prognosis and Predisposition of Gastric Cancer by Altered Transforming Growth Factor β Pathway

Genetic or epigenetic impairment of the TGF-β receptor genes are significantly associated with the tumor size and proliferative rate of gastric cancer cells (Pinto et al. 2003). Patients who have gastric cancer with reduced Smad4 expression show poorer clinical outcome than those with preserved expression of Smad4 (Xiangming et al. 2001). Moreover, the survival rate is significantly higher in gastric cancer patients negative for I-Smad (Smad7) expression than those positive for it (Kim et al. 2004). These results suggest that gastric cancer associated with decreased level of TGF-β signaling shows worse prognosis than those with TGF-β–intact tumors. It has also been shown that promoter polymorphisms in TGF-β1 gene

(*TGFB1*; C509T) and *TGFBR1* (G575A) are associated with decreased risk of gastric cancer (Jin et al. 2007), suggesting that the basal level of TGF-β signaling activity can affect susceptibility to gastric tumorigenesis.

Gastric Tumorigenesis by Suppression of TGF-β in Mice

Transforming Growth Factor β1 and Smad4 Knockout Mice and Gastrointestinal Tumors

Mice homozygous for a mutated *TGFB1* show wasting syndrome accompanied by a multifocal inflammatory response and necrosis, leading to organ failure (Shull et al. 1992). One of the phenotypes in *Tgfb1* (−/−) mice is hyperplasia of the glandular stomach. Hyperplasia is also found in the *Tgfb1* (+/−) mouse stomach, which is associated with inflammatory infiltrations (Boivin et al. 1996), suggesting that cooperation between TGF-β suppression and inflammation is important for gastric carcinogenesis. Consistently, *Tgfb1* (−/−), *Rag2* (−/−) compound mutant mice develop intestinal cancers associated with inflammation, whereas germ-free mice with the same genotype were free of cancer as well as inflammation (Engle et al. 2002).

Although *Smad4* (−/−) mice die *in utero*, *Smad4* (+/−) mice develop gastric polyps associated with inflammation (Takaku et al. 1999). Notably, the wild-type *Smad4* gene was lost by LOH in tumor cells, suggesting that significant suppression of the TGF-β family signaling is required for gastric tumorigenesis. Another *Smad4* (+/−) mutant mouse strain also shows gastric polyps (Xu et al. 2000). In this model, polyps progressed to carcinoma *in situ* with increasing age. Interestingly, *Smad4* LOH was found in a later stage, suggesting that haploinsufficiency of *Smad4* can initiate tumorigenesis, but LOH is required for malignant progression. Spontaneous mutant mice carrying a single nucleotide deletion in the *Smad4* gene also develop gastric adenomas in the pyloric and duodenal transition zone (Hohenstein et al. 2003), confirming the tumor suppressor effect of *Smad4* in gastric carcinogenesis.

Recently, we have demonstrated that loss of Smad4 in intestinal tumor cells induces recruitment of CCR1-positive immature myeloid cells (iMCs) to the tumor tissues, which helps invasion by secreting MMPs (Kitamura et al. 2007). Accordingly, it appears that iMCs are recruited to the TGF-β–suppressed gastric cancer epithelium, which has a key role in progression and invasion.

Other Mouse Models for Gastric Tumors

Transcription factor Stat3 is activated by gp130, the common receptor for the IL-6 cytokine family. Constitutive activation of Stat3 by gp130 mutation in mice causes

gastric adenomas (Tebbutt et al. 2002). Importantly, Stat3 activation in gastric epithelial cells results in desensitization to TGF-β through induction of inhibitory
Smad7 (Jenkins et al. 2005). These results suggest cross-talk between Stat and
Smad pathways, i.e., inflammatory cytokine and TGF-β signaling, in gastric carcinogenesis. However, transgenic mice expressing dominant negative TGF-βRII
develop gastric adenocarcinomas only when mice are infected with *H. pylori*
(Hahm et al. 2002). Accordingly, cooperation of *H. pylori* infection and suppression of TGF-β pathway is important for gastric tumorigenesis.

Considering the fact that *Tgfb1* and *Smad4* knockout mice develop inflammation-
associated tumors, these results collectively suggest that *H. pylori*–associated
inflammation together with suppression of the TGF-β signaling is critical for gastric tumorigenesis. Thus, it is also possible that activation of the PGE_2 pathway
and suppression of the TGF-β signaling cause development of gastric
adenocarcinomas.

Activation of TGF-β Pathway in Gastric Cancer Progression

Transforming Growth Factor β as a Promoter of Tumor Progression

TGF-β signaling not only induces growth inhibition of tumor cells, but also promotes cancer progression (Derynck et al. 2001). TGF-β signaling induces epithelial
to mesenchymal transition of several types of cancer cells in culture. Moreover,
activation of TGF-β induces synthesis of extracellular matrix and chemoattraction
of fibroblasts, resulting in alteration of the microenvironment. It has been reported
that TGF-β1 expression in gastric cancer is correlated with the depth of invasion
and progression stage of disease (Saito et al. 1999), and that the serum level of
TGF-β1 is significantly elevated in gastric cancer patients (Lin et al. 2006).
Moreover, patients with TGF-β1–negative gastric cancer showed significantly better prognosis. These results suggest that TGF-β signaling has an important role in
gastric cancer progression. As described above, gastric epithelial cells are sensitive
to negative growth regulation by TGF-β, but it is possible that TGF-β–resistant
cells are selected during tumorigenesis and expanded in the progression stage. Such
cell populations resistant to TGF-β–induced growth inhibition may be expanded by
TGF-β stimulation.

Effects of Transforming Growth Factor β Signaling in Gastrointestinal Cancer

It has been demonstrated that expression of TGF-β1 is associated with induction
of COX-2 expression in RER+ colon cancer and chemically induced rat colon

tumors (Shao et al. 1999), suggesting a role for TGF-β in COX-2 induction. Indeed, activated Ha-*ras* induces COX-2 in intestinal epithelial cells, which is suppressed by treatment of the cells with neutralizing antibody against TGF-β (Sheng et al. 2000). Thus, it is possible that activation of TGF-β pathway has a critical role in *ras*-induced COX-2 expression, which contributes to tumor progression (see "Induction Mechanism and Outcome Effects of Cyclooxygenae-2 Pathway in Gastric Cancer" and "Inflammation-Associated Hyperplasia and Metaplasia by Induction of Cyclooxygenae-2 and Microsomal Prostaglandin E Synthase-1 in Mouse Stomach").

Suppression of RUNX3 in Gastric Cancer

RUNX3 in Transforming Growth Factor β Pathway

RUNX proteins are a family of transcription factors that have Runt domains responsible for DNA binding (Ito and Miyazono 2003). It has been shown that RUNX3 (PEBP2αC/AML2/Cbfa3) forms a complex with Smad proteins, which regulates TGF-β signaling (Hanai et al. 1999) (Figure 20.7), suggesting that RUNX3 has a role in some of the TGF-β functions. It has been reported that RUNX3 expression is required for TGF-β–dependent induction of p21 in gastric epithelial cells (Chi et al. 2005). Moreover, RUNX3 has a role in TGF-β–induced apoptosis through induction of Bim, a proapoptotic BH3-only protein (Yano et al. 2006). These results suggest that RUNX3 contributes to TGF-β–induced gastric epithelial differentiation and apoptosis.

Suppression of RUNX3 in Mice and Human

RUNX3 is expressed in the differentiated gastric epithelial cells. Targeted disruption of the *Runx3* gene in mice causes gastric hyperplasia with stimulated proliferation and suppressed apoptosis (Li et al. 2002). Moreover, TGF-β1–induced growth inhibition is attenuated in gastric epithelial cells of *Runx3* (−/−) mice compared with those of the wild-type mice. Expression of RUNX3 is suppressed in 45%–60% of human gastric cancers, which is caused by hemizygous deletion or promoter methylation (Li et al. 2002). In another study, downregulation of RUNX3 through promoter methylation was found in 75% and 100% of primary and peritoneally metastasized gastric cancers, respectively (Sakakura et al. 2005). Moreover, stable transfection of the *RUNX3* gene in gastric cancer cells restored responsiveness to TGF-β–induced growth inhibition and apoptosis induction. These results indicate that RUNX3 is a tumor suppressor of gastric cancer, and its suppression contributes to gastric carcinogenesis.

Suppression of Bone Morphogenetic Protein Signaling in Gastric Cancer

Suppression of Bone Morphogenetic Protein in Gastrointestinal Tumors

BMP is a member of the TGF-β superfamily and binds to the BMP receptor type I and type II complex, causing phosphorylation of Smad1/5/8 (Miyazono et al. 2005). The phosphorylated Smad1/5/8 forms a complex with Smad4, leading to transcriptional activation of BMP target genes. Germline mutations in the BMP receptor type IA gene have been found in a subset of juvenile polyposis, which is an autosomal dominant gastrointestinal polyposis syndrome (Howe et al. 2001). Furthermore, transgenic expression of noggin, an endogenous antagonist of BMP signaling in mouse intestine, results in formation of ectopic crypts and development of intraepithelial hyperplasia (Haramis et al. 2004). These results indicate that suppression of the BMP pathway can cause intestinal polyposis.

BMP signaling is required for morphogenesis of the stomach during development (Narita et al. 2000), suggesting a role of BMP also in gastric tumorigenesis. Stimulation of gastric cancer cells with BMP-2 suppresses proliferation (Wen et al. 2004), and promoter methylation of the BMP-2 gene is found in gastric cancer cells (Wen et al. 2006). These results suggest that suppression of BMP signaling contributes to gastric tumorigenesis.

Suppression of Bone Morphogenetic Protein by Noggin in Mouse Stomach

K19-Noggin transgenic mice express noggin in the gastric epithelial cells (H. Oshima et al., unpublished results). One of the founder mice has developed gastric tumors on the entire surface of the glandular stomach (Figure 20.8).

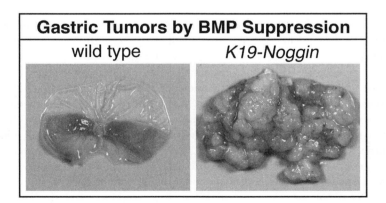

Gastric Tumors by BMP Suppression
wild type   K19-Noggin

Fig. 20.8 A gastric tumor that developed in the glandular stomach of a *K19-Noggin* mouse (right), suggesting that suppression of bone morphogenetic protein signaling contributes to gastric tumorigenesis. Such tumors are not found in the litter mate wild-type mice (left)

Histologically, tumors consist of undifferentiated epithelial cells, which is similar to those found in the *K19-Wnt1/C2mE* compound transgenic mice (Figure 20.6B). These genetic results also suggest that suppression of BMP signaling contributes to gastric tumor development.

Conclusion

Accumulating evidence has indicated that activation of Wnt signaling or suppression of TGF-β pathway is an important step in gastric cancer development. However, induction of COX-2 and activation of the PGE_2 pathway are essential for development of various types of cancer, including gastric cancer. It has been established that *H. pylori* infection is an important risk factor for gastric tumorigenesis, and *H. pylori*–associated inflammation as well as infectious stimuli induce COX-2/PGE_2 pathway signaling. Furthermore, genetic studies indicate that cooperation of the Wnt signaling and COX-2/PGE_2 pathway causes development of gastric cancer. Accordingly, it is conceivable that genetic or epigenetic alteration, which causes Wnt activation or TGF-β suppression in the *H. pylori*–infected stomach leads to gastric cancer development. Further studies using genetic mouse models will be required to verify this hypothesis.

References

Akre, K., Ekstrom, A. M., Signorello, L. B., Hansson, L. E., and Nyren, O. (2001). Aspirin and risk for gastric cancer: a population-based case-control study in Sweden. Br. J. Cancer 84:965–968.

Al-Marhoon, M. S., Nunn, S., and Soames, R. W. (2004). CagA+ Helicobacter pylori induces greater levels of prostaglandin E2 than cagA– strains. Prostaglandins Other Lipid Mediat. 73:181–189.

Boivin, G. P., Molina, J. R., Ormsby, I., Stemmermann, G., and Doetschman, T. (1996). Gastric lesions in transforming growth factor beta-1 heterozygous mice. Lab. Invest. 74:513–518.

Chang, J., Park, K., Bang, Y.-J., Kim, W. S., Kim, D., and Kim, S.-J. (1997). Expression of transforming growth factor β type II receptor reduced tumorigenicity in human gastric cancer cells. Cancer Res. 57:2856–2859.

Chang, Y.-J., Wu, M.-S., Lin, J.-T., and Chen, C.-C. (2005). Helicobacter pylori-induced invasion and angiogenesis of gastric cells is mediated by cyclooxygenase-2 induction through TLR2/TLR9 and promoter regulation. J. Immunol. 175:8242–8252.

Chang, Y. J., Wu, M. S., Lin, J. T., Sheu, B. S., Muta, T., Inoue, H., and Chen,C.-C. (2004). Induction of cyclooxygenase-2 overexpression in human gastric epithelial cells by Helicobacter pylori involves TLR2/TLR9 and c-Src-dependent nuclear factor-κB activation. Mol. Pharmacol. 66:1465–1477.

Chen, C.-N., Sung, C.-T., Lin, M.-T., Lee, P.-H., and Chang, K.-J. (2001). Clinicopathologic association of cyclooxygenase 1 and cyclooxygenase 2 expression in gastric adenocarcinoma. Ann. Surg. 233:183–188.

Cheng, X.-X., Wang, Z.-C., Chen, X.-Y., Sun, Y., Kong, Q.-Y., Liu, J., and Li, H. (2005). Correlation of Wnt-2 expression and β-catenin intracellular accumulation in Chinese gastric cancers: relevance with tumour dissemination. Cancer Lett. 223:339–347.

Chi, X.-Z., Yang, J.-O., Lee, K.-Y., Ito, K., Sakakura, C., Li, Q.-L., Kim, H.-R., Cha, E.-J., Lee, Y.-H., Kaneda, A., Ushijima, T., Kim, W.-J., Ito, Y., and Bae, S.-C. (2005). RUNX3 suppresses gastric epithelial cell growth by inducing p21$^{WAF1/CIP1}$ expression in cooperation with transforming growth factor β-activated SMAD. Mol. Cell. Biol. 25:8097–8107.

Chulada, P. C., Thompson, M. B., Mahler, J. F., Doyle, C. M., Gaul, B. W., Lee, C., Tiano, H. F., Morham, S. G., Smithies, O., and Langenbach, R. (2000). Genetic disruption of Ptgs-1, as well as Ptgs-2, reduces intestinal tumorigenesis in Min mice. Cancer Res. 60:4705–4708.

Clements, W. M., Wang, J., Sarnaik, A., Kim, O. J., MacDonald, J., Fenoglio-Preiser, C., Groden, J., and Lowy, A. M. (2002). β-Catenin mutation is a frequent cause of Wnt pathway activation in gastric cancer. Cancer Res. 62:3503–3506.

Correa, P. (2003). Helicobacter pylori infection and gastric cancer. Cancer Epidemiol. Biomarkers Prev. 12:238s–241s.

Derynck, R., Akhurst, R. J., and Balmain, A. (2001). TGF-β signaling in tumor suppression and cancer progression. Nat. Genet. 29:117–129.

Dewitt, D. L., and Smith, W. L. (1998). Primary structure of prostaglandin G/H synthase from sheep vesicular gland determined from the complementary DNA sequence. Proc. Natl. Acad. Sci. USA 85:1412–1416.

Ebert, M. P. A., Fei, G., Kahmann, S., Müller, O., Yu, J., Sung, J. J. Y., and Malfertheiner, P. (2002). Increased β-catenin mRNA levels and mutational alterations of the APC and β-catenin gene are present in intestinal-type gastric cancer. Carcinogenesis 23:87–91.

Engle, S. J., Ormsby, I., Pawlowski, S., Boivin, G. P., Croft, J., Balish, E., and Doetschman, T. (2002). Elimination of colon cancer in germ-free transforming growth factor beta 1-deficient mice. Cancer Res. 62:6362–6366.

Fletcher, B. S., Kujubu, D. A., Perrin, D. M., and Herschman, H. R. (1992). Structure of the mitogen-inducible TIS10 gene and demonstration that the TIS10-encoded protein is a functional prostaglandin G/H synthase. J. Biol. Chem. 267:4338–4344.

Fu, Y.-G., Sung, J. J. Y., Wu, K.-C., Wu, H.-P., Yu, J., Chan, M., Chan, V. Y. W., Chan, K.-K., Fan, D.-M., and Leung, W. K. (2005). Inhibition of gastric cancer-associated angiogenesis by antisense COX-2 transfectants. Cancer Lett. 224:243–252.

Futagami, S., Suzuki, K., Hiratsuka, T., Shindo,T., Hamamoto, T., Tatsuguchi, A., Ueki, N., Shinji, Y., Kusunoki, M., Wada, K., Miyake, K., Gudis, K., Tsukui,T., and Sakamoto, C. (2006). Celecoxib inhibits Cdx2 expression and prevents gastric cancer in Helicobacter pylori-infected mongolian gerbils. Digestion 74:187–198.

Gregorieff, A., and Clevers, H. (2005). Wnt signaling in the intestinal epithelium: from endoderm to cancer. Genes Dev. 19:877–890.

Gupta, R. A., and DuBois, R. N. (2001). Colorectal cancer prevention and treatment by inhibition of cyclooxygenase-2. Nat Rev. Cancer 1:11–21.

Hahm, K.-B., Lee, K. M., Kim. Y. B., Hong, W. S., Lee, W. H., Han, S. U., Kim, M. W., Ahn, B. O., Oh, T. Y., Lee, M. H., Green, J., and Kim, S. J. (2002). Conditional loss of TGF-β signaling leads to increased susceptibility to gastrointestinal carcinogenesis in mice. Aliment. Pharmacol. Ther. 16:115–127.

Han, S.-U., Kim, H.-T., Seong, D. H., Kim, Y.-S., Park, Y.-S., Bang, Y.-J., Yang, H.-K., and Kim, S.-J. (2004). Loss of the Smad3 expression increases susceptibility to tumorigenicity in human gastric cancer. Oncogene 23:1333–1341.

Hanai, J., Chen, L. F., Kanno, T., Ohtani-Fujita, N., Kim, W. Y., Guo, W.-H., Imamura, T., Ishidou, Y., Fukuchi, M., Shi, M.-J., Stavnezer, J., Kawabata, M., Miyazono, K., and Ito, Y. (1999). Interaction and functional cooperation of PEBP2/CBF with Smads. J. Biol. Chem. 44:31577–31582.

Haramis, A.-P. G., Begthel, H., van den Born, M., van Es, J., Jonkheer, S., Offerhaus, G. J. A., and Clevers, H. (2004). De novo crypt formation and juvenile polyposis on BMP inhibition in mouse intestine. Science 303:1684–1686.

Hla, T., and Neilson, K. (1992). Human cylooxygenase-2 cDNA. Proc. Natl. Acad. Sci. USA 89:7384–7388.

Hohenstein, P., Molenaar, L., Elsinga, J., Morreau, H., van der Klift, H., Struijk, A., Jagmohan-Changur, S., Smits, R., van Kranen, H., van Ommen, G.-J. B., Cornelisse, C., Devilee, P., and

Fodde, R. (2003). Serrated adenomas and mixed polyposis caused by a splice acceptor deletion in the mouse Smad4 gene. Genes Chromosomes Cancer 36:273–282.

Houghton, J., Stoicov, C., Nomura, S., Rogers, A. B., Carlson, J., Li, H., Cai, X., Fox, J. G., Goldenring, J. R., and Wang, T. C. (2004). Gastric cancer originating from bone marrow-derived cells. Science 306:1568–1571.

Howe, J. R., Bair, J. L., Sayed, M. G., Anderson, M. E., Mitros, F. A., Petersen, G. M., Velculescu, V. E., Traverso, G., and Vogelstein, B. (2001). Germline mutations of the gene encoding bone morphogenetic protein receptor 1A in juvenile polyposis. Nat. Genet. 28:184–187.

Hu, P. J., Yu, J., Zeng, Z. R., Leung, W. K., Lin, H. L., Tang, B. D., Bai, A. H. C., and Sung, J. J. Y. (2004). Chemoprevention of gastric cancer by celecoxib in rats. Gut 53:195–200.

Ito, Y., and Miyazono, K. (2003). RUNX transcription factors as key targets of TGF-β super-family signaling. Curr. Opin. Genet. Dev. 13:43–47.

Jang, T. J. (2004). Expression of proteins related to prostaglandin E_2 biosynthesis is increased in human gastric cancer and during gastric carcinogenesis. Virchows Arch. 445:564–571.

Jenkins, B. J., Graio, D., Nheu, T., Najdovska, M., Wang, B., Waring, P., Inglese, M., McLoughlin, R. M., Jones, S. A., Topley, N., Baumann, H., Judd, L. M., Giraud, A. S., Boussioutas, A., Zhu, H.-J., and Ernst, M. (2005). Hyperactivation of Stat3 in gp130 mutant mice promotes gastric hyperproliferation and desensitizes TGF-β signaling. Nat. Med. 11:845–852.

Jin, G., Wang, L., Chen, W., Hu, Z., Zhou, Y., Tan, Y., Wang, J., Hua, Z., Ding, W., Shen, J., Zhang, Z., Wang, X., Xu, Y., and Shen, H. (2007). Variant alleles of TGFB1 and TGFBR2 are associated with a decreased risk of gastric cancer in a Chinese population. Int. J. Cancer 120:1330–1335.

Joo, Y.-E., Rew, J.-S., Seo, Y.-H., Choi, S.-K., Kim, Y.-J., Park, C.-S., and Kim, S.-J. (2003). Cyclooxygenase-2 overexpression correlates with vascular endothelial growth factor expression and tumor angiogenesis in gastric cancer. J. Clin. Gastroenterol. 37:28–33.

Kang, S. H., Bang, Y.-J., Im, Y.-H., Yang, H.-K., Lee, D. A., Lee, H. Y., Lee, H. S., Kim, N. K., and Kim, S.-J. (1999). Transcriptional repression of the transforming growth factor-β type I receptor gene by DNA methylation results in the development of TGF-β resistance in human gastric cancer. Oncogene 18:7280–7286.

Kawada, M., Seno, H., Wada, M., Suzuki, K., Kanda, N., Kayahara, T., Fukui, H., Sawada, M., Kajiyama, T., Sakai, M., and Chiba, T. (2003). Cyclooxygenase-2 expression and angiogenesis in gastric hyperplastic polyp-association with polyp size. Digestion 67:20–24.

Kim, Y. H., Lee, H. S., Lee, H.-J., Hur, K., Kim, W. H., Bang, Y.-J., Kim, S.-J., Lee, K. U., Choe, K. J., and Yang, H.-K. (2004). Prognostic significance of the expression of Smad4 and Smad7 in human gastric carcinomas. Ann. Oncol. 15:574–580.

Kim, J. Y., Park, D. Y., Kim, G. H., Choi, K. U., Lee, C. H., Huh, G. Y., Sol, M. Y., Song, G. A., Jeon, T. Y., Kim, D. H., and Sim, M. S. (2005). Smad4 expression in gastric adenoma and adenocarcinoma: frequent loss of expression in diffuse type of gastric adenocarcinoma. Histol. Histopathol. 20:543–549.

Kim, C. J., Song, J. H., Cho, Y. G., Kim, Y. S., Kim, S. Y., Nam, S. W., Yoo, N. J., Lee, J. Y., and Park, W. S. (2007). Somatic mutations of the β-TrCP gene in gastric cancer. APMIS 115:127–133.

Kitamura, T., Kometani, K., Hashida, H., Matsunaga, A., Miyoshi, H., Hosogi, H., Aoki, M., Oshima, M., Hattori, M., Takabayashi, A., Minato, N., and Taketo, M. M. (2007). Smad4-deficient intestinal tumors recruit CCR1⁺ myeloid cells that promote invasion. Nat. Genet. 39:467–475.

Langman, M. J., Cheng, K. K., Gilman, E. A., and Lancashire, R. J. (2000). Effect of anti-inflammatory drugs on overall risk of common cancer: case-control study in general practice research database. BMJ 320:1642–1646.

Leung, W. K., To, K. F., Go, M. Y., Chan, K. K., Chan, F. K., Ng, E. K., Chung, S. C., and Sung, J. J. (2003). Cyclooxygenase-2 upregulates vascular endothelial growth factor expression and angiogenesis in human gastric carcinoma. Int. J. Oncol. 23:1317–1322.

Li, Q.-L., Ito, K., Sakakura, C., Fukamachi, H., Inoue, K., Chi, X.-Z., Lee, K.-Y., Nomura, S., Lee, C.-W., Han, S.-B., Kim, H.-M., Kim, W.-J., Yamamoto, H., Yamashita, N., Yano, T., Ikeda, T., Itohara, S., Inazawa, J., Abe, T., Hagiwara, T., Yamagishi, H., Ooe, A., Kaneda, A., Sugimura,

T., Ushijima, T., Bae, S.-C., and Ito, Y. (2002). Causal relationship between the loss of RUNX3 expression and gastric cancer. Cell 109:113–124.

Lim, H.Y., Joo, H. J., Choi, J. H., Yi, J. W., Yang, M. S., Cho, D. Y., Kim, H. S., Nam, D. K., Lee, K. B., and Kim, H. C. (2000). Increased expression of cyclooxygenase-2 protein in human gastric carcinoma. Clin. Cancer Res. 6:519–525.

Lim, J. W., Kim, H., and Kim, K. H. (2001). Nuclear factor-kappaB regulates cyclooxygenase-2 expression and cell proliferation in human gastric cancer cells. Lab. Invest. 81:349–360.

Lin, Y., Kikuchi, S., Obata, Y., Yagyu, K., and The Tokyo Research Group on Prevention of Gastric Cancer. (2006). Serum levels of transforming growth factor β1 are significantly correlated with venous invasion in patients with gastric cancer. J. Gastroenterol. Hepatol. 21:432–437.

Liu, F., Pan, K., Zhang, X., Zhang, Y., Zhang, L., Ma, J., Dong, C., Shen, L., Li, J., Deng, D., Lin, D., and You, W. (2006). Genetic variants in cyclooxygenase-2: expression and risk of gastric cancer and its precursors in a Chinese population. Gastroenterology 130:1975–1984.

Lowy, A. M., Clements, W. M., Bishop, J., Kong, L., Bonney, T., Sisco, K., Aronow, B., Fenoglio-Preiser, C., and Groden, J. (2006). β-Catenin/Wnt signaling regulates expression of the membrane type3 matrix metalloproteinase in gastric cancer. Cancer Res. 66:4734–4741.

Magari, H., Shimizu, Y., Inada, K., Enomoto, S., Tomeki, T., Yanaoka, K., Tamai, H., Arii, K., Nakata, H., Oka, M., Utsunomiya, H., Tsutsumi, Y., Tsukamoto,T., Tatematsu, M., and Ichinose, M. (2005). Inhibitory effects of etodolac, a selective cyclooxygenase-2 inhibitor, on stomach carcinogenesis in Helicobacter pylori-infected mongolian gerbils. Biochem. Biophys. Res. Commun. 334:606–612.

Markowitz, S., Wang, J., Myeroff, L., Parsons, R., Sun, L., Lutterbaugh, J., Fan, R. S., Zborowska, E., Kinzler, K. W., Vogelstein, B., Brattain, M., and Willson, J. K. V. (1995). Inactivation of the type II TGF-beta receptor in colon cancer cells with microsatellite instability. Science 268:1336–1338.

Matsumoto,Y., Marusawa, H., Kinoshita, K., Endo, Y., Kou, T., Morisawa, T., Azuma, T., Okazaki, I.-M., Honjo, T., and Chiba, T. (2007). Helicobacter pylori infection triggers aberrant expression of activation-induced cytidine deaminase in gastric epithelium. Nat. Med. 13:470–476.

McCarthy, C. J., Crofford, L. J., Greenson, J., and Scheiman, J. M. (1999). Cyclooxygenase-2 expression in gastric antral mucosa before and after eradication of Helicobacter pylori infection. Am. J. Gastroenterol. 94:1218–1223.

Miyazawa, K., Iwaya, K., Kuroda, M., Harada, M., Serizawa, H., Koyanagi, Y., Sato, Y., Mizokami, Y., Matsuoka, T., and Mukai, K. (2000). Nuclear accumulation of beta-catenin in intestinal-type gastric carcinoma: correlation with early tumor invasion. Virchows Arch. 437:508–513.

Miyazono, K., Maeda, S., and Imamura, T. (2005). BMP receptor signaling: transcriptional targets, regulation of signals, and signaling cross-talk. Cytokine Growth Factor Rev. 16:251–263.

Mrena, J., Wiksten, J.-P., Thiel, A., Kokkola, A., Pohjola, L., Lundin, J., Nordling, S., Ristimäki, A., and Haglund, C. (2005). Cyclooxygenase-2 is an independent prognostic factor in gastric cancer and its expression is regulated by the messenger RNA stability factor HuR. Clin. Cancer Res. 11:7362–7368.

Murakami, M., Naraba, H., Tanioka, T., Semmyo, N., Nakatani, Y., Kojima, F., Ikeda, T., Fueki, M., Ueno, A., Oh-ishi, S., and Kudo, I. (2000). Regulation of prostaglandin E$_2$ biosynthesis by inducible membrane-associated prostaglandin E$_2$ synthase that acts in concert with cyclooxygenase-2. J. Biol. Chem. 275:32783–32792.

Murata, H., Kawano, S., Tsuji, S., Tsuji, M., Sawaoka, H., Kimura, Y., Shiozaki, H., and Hori, M. (1999). Cyclooxygenase-2 overexpression enhances lymphatic invasion and metastasis in human gastric carcinoma. Am. J. Gastroenterol. 94:451–455.

Myeroff, L. L., Parsons, R., Kim, S.-J., Hedrick, L., Cho, K. R., Orth, K., Mathis, M., Kinzler, K. W., Lutterbaugh, J., Park, K., Bang, Y.-J., Lee, H. Y., Park, J.-G., Lynch, H. T., Roberts, A. B., Vogelstein, B., and Markowitz, S. D. (1995). A transforming growth factor β receptor type

II gene mutation common in colon and gastric but rare in endometrial cancers with microsatellite instability. Cancer Res. 55:5545–5547.

Nakajima, T., Maekita, T., Oda, I., Gotoda, T., Yamamoto, S., Umemura, S., Ichinose, M., Sugimura, T., Ushijima, T., and Saito, D. (2006). Higher methylation levels in gastric mucosae significantly correlate with higher risk of gastric cancers. Cancer Epidemiol. Biomarkers Prev. 15:2317–2321.

Nam, K. T., Hahm, K.-B., Oh, S.-Y., Yeo, M., Han, S.-U., Ahn, B., Kim, Y.-B., Kang, J. S., Jang, D. D., Yang, K.-H., and Kim, D.-Y. (2004). The selective cyclooxygenase-2 inhibitor nimesulide prevents Helicobacter pylori-associated gastric cancer development in a mouse model. Clin. Cancer Res. 10:8105–8113.

Nardone, G., Rocco, A., Vaira, D., Staibano, S., Budillon, A., Tatangelo, F., Sciulli, M. G., Perna, F., Salvatore, G., Di Benedetto, M., De Rosa, G., and Patrignani, P. (2004). Expression of COX-2, mPGE-synthase1, MDR-1(P-gp), and Bcl-X_L: a molecular pathway of H. pylori-related gastric carcinogenesis. J. Pathol. 202:305–312.

Narita, T., Saitoh, K., Kameda, T., Kuroiwa, A., Mizutani, M., Koike, C., Iba, H., and Yasugi, S. (2000). BMPs are necessary for stomach gland formation in the chicken embryo: a study using virally induced BMP-2 and Noggin expression. Development 127:981–988.

Nojima, M., Suzuki, H., Toyota, M., Watanabe, Y., Maruyama, R., Sasaki, S., Sasaki, Y., Mita, H., Nishikawa, N., Yamaguchi, K., Hirata, K., Itoh, F., Tokino, T., Mori, M., Imai, K., and Shinomura, Y. (2007). Frequent epigenetic inactivation of SFRP genes and constitutive activation of Wnt signaling in gastric cancer. Oncogene 26(32):4699–4713.

Nomura, S., Baxter, T., Yamaguchi, H., Leys, C., Vartapetian, A. B., Fox, J. G., Lee, J. R., Wang, T. C., and Goldenring, J. R. (2004). Spasmolytic polypeptide expressing metaplasia to preneoplasia in H. felis-infected mice. Gastroenterology 127:582–594.

Offerhaus, G. J., Giardiello, F. M., Krush, A. J., Booker, S. V., Tersmette, A. C., Kelley, N. C., and Hamilton, S. R. (1992). The risk of upper gastrointestinal cancer in familial adenomatous polyposis. Gastroenterology 102:1980–1982.

Ohno, R., Yoshinaga, K., Fujita, T., Hasegawa, K., Iseki, H., Tsunozaki, H., Ichikawa, W., Nihei, Z., and Sugihara, K. (2001). Depth of invasion parallels increased cyclooxygenase-2 levels in patients with gastric carcinoma. Cancer 91:1876–1881.

Oshima, M., Dinchuk, J. E., Kargman, S. L., Oshima, H., Hancock, B., Kwong, E., Trzaskos, J. M., Evans, J. F., and Taketo, M. M. (1996). Suppression of intestinal polyposis in Apc$^{\Delta 716}$ knockout mice by inhibition of cyclooxygenase 2 (COX-2). Cell 87:803–809.

Oshima, H., Matsunaga, A., Fujimura, T., Tsukamoto, T., Taketo, M. M. and Oshima, M. (2006). Carcinogenesis in mouse stomach by simultaneous activation of the Wnt signaling and prostaglandin E_2 pathway. Gastroenterology 131:1086–1095.

Oshima, H., Oshima, M., Inaba, K., and Taketo, M. M. (2004). Hyperplastic gastric tumors induced by activated macrophages in COX-2 /mPGES-1 transgenic mice. EMBO J. 23:1669–1678.

Oshima, M., Oshima, H., Kitagawa, K., Kobayashi, M., Itakura, C., and Taketo M. (1995). Loss of Apc heterozygosity and abnormal tissue building in nascent intestinal polyps in mice carrying a truncated Apc gene. Proc. Natl. Acad. Sci. USA 92:4482–4486.

Oshima, H., Oshima, M., Kobayashi, M., Tsutsumi, M., and Taketo, M. M. (1997). Morphological and molecular processes of polyp formation in Apc$^{\Delta 716}$ knockout mice. Cancer Res. 57:1644–1649.

Oshima, M., Oshima, H., Matsunaga, A., and Taketo, M. M. (2005). Hyperplastic gastric tumors with spasmolytic polypeptide-expressing metaplasia caused by tumor necrosis factor-α-dependent inflammation in cyclooxygenase-2/microsomal prostaglandin E synthase-1 transgenic mice. Cancer Res. 65:9147–9151.

Oshima, M., and Taketo, M. M. (2002). COX selectivity and animal models for colon cancer. Curr. Pharm. Des. 8:1021–1034.

Pai, R., Soreghan, B., Szabo, I. L., Pavelka, M., Baatar, D., and Tarnawski, A. S. (2002). Prostaglandin E_2 transactivates EGF receptor: a novel mechanism for promoting colon cancer growth and gastrointestinal hypertrophy. Nat. Med. 8:289–293.

Park, K., Kim, S.-J., Bang, Y.-J., Park, J.-G., Kim, N. K., Roberts, A, B., and Sporn, M. B. (1994). Genetic changes in the transforming growth factor β (TGF-β) type II receptor gene in human gastric cancer cells: correlation with sensitivity to growth inhibition by TGF-β. Proc. Natl. Acad. Sci. USA 91:8772–8776.

Park, W. S., Oh, R. R., Park, J. Y., Lee, S. H., Shin, M. S., Kim, Y. S., Kim, S. Y., Lee, H. K., Kim, P. J., Oh, S. T., Yoo, N, J., and Lee, J. Y. (1999). Frequent somatic mutations of the β-catenin gene in intestinal-type gastric cancer. Cancer Res. 59:4257–4260.

Parsons, R., Myeroff, L. L., Liu, B., Willson, J. K. V., Markowitz, S. D., and Kinzler, K. W. (1995). Microsatellite instability and mutations of the transforming growth factor β type II receptor gene in colorectal cancer. Cancer Res. 55:5548–5550.

Pinto, M., Oliveira, C., Cirnes, L., Machado, J. C., Ramires, M., Nogueira, A., Carneiro, F., and Seruca, R. (2003). Promoter methylation of TGFβ receptor I and mutation of TGFβ receptor II are frequent events in MSI sporadic gastric carcinomas. J. Pathol. 200:32–38.

Ristimäki, A., Honkanen, N., Jänkälä, H., Sipponen, P., and Härkönen, M. (1997). Expression of cyclooxygenase-2 in human gastric carcinoma. Cancer Res. 57:1276–1280.

Saito, H., Tsujitani, S., Oka, S., Kondo, A., Ikeguchi, M., Maeta, M., and Kaibara, N. (1999). The expression of transforming growth factor-β1 is significantly correlated with the expression of vascular endothelial growth factor and poor prognosis of patients with advanced gastric carcinoma. Cancer 86:1455–1462.

Sakakura, C., Hasegawa, K., Miyagawa, K., Nakashima, S., Yoshikawa, T., Kin, S., Nakase, Y., Yazumi, S., Yamagishi, H., Okanoue, T., Chiba, T., and Hagiwara, A. (2005). Possible involvement of RUNX3 silencing in the peritoneal metastases of gastric cancers. Clin. Cancer Res. 11:6479–6488.

Saukkonen, K., Nieminen, O., van Rees, B., Vilkki, S., Härkönen, M., Juhola, M., Mecklin, J.-P., Sipponen, P., and Ristimäki, A. (2001). Expression of cyclooxygenase-2 in dysplasia of the stomach and in intestinal-type gastric adenocarcinoma. Clin. Cancer Res. 7:1923–1931.

Saukkonen, K., Rintahaka, J., Sivula, A., Buskens, C. J., van Rees, B. P., Rio, M.-C., Haglund, C., van Lanschot, J. J. B., Offerhaus, G. J. A., and Ristimäki, A. (2003). Cyclooxygenase-2 and gastric carcinogenesis. APMIS 111:915–925.

Sawaoka, H., Kawano, S., Tsuji, S., Tsujii, M., Gunawan, E. S., Takei, Y., Nagano, K., and Hori, M. (1998). Cyclooxygenase-2 inhibitors suppress the growth of gastric cancer xenografts via induction of apoptosis in nude mice. Am. J. Physiol. 274:G1061–1067.

Schreinemachers, D. M., and Everson, R. B. (1994). Aspirin use and lung, colon, and breast cancer incidence in a prospective study. Epidemiology 5:138–146.

Seno, H., Oshima, M., Ishikawa, T., Oshima, H., Takaku, K., Chiba, T., Narumiya, S., and Taketo, M. M. (2002). Cyclooxygeanse-2 and prostaglandin E_2 receptor EP_2-dependent angiogenesis in $Apc^{\Delta716}$ mouse intestinal polyps. Cancer Res. 62:506–511.

Shao, J., Sheng, H., Aramandla, R., Pereira, M. A., Lubet, R. A., Hawk, E., Grogan, L., Kirsch, I. R., Washington, M. K., Beauchamp, R. D., and DuBois, R. N. (1999). Coordinate regulation of cyclooxygenase-2 and TGF-β1 in replication error-positive colon cancer and azoxymethane-induced rat colonic tumors. Carcinogenesis 20:185–191.

Shen, H., Sun, W.-H., Xue, Q.-P., Wu, J., Cheng, Y.-L., Ding, G.-X., Fu, H.-Y., Tsuji, S., and Kawano, S. (2006). Influences of Helicobacter pylori on cyclooxygenase-2 expression and prostaglandinE_2 synthesis in rat gastric epithelial cells in vitro. J. Gastroenterol. Hepatol. 21:754–758.

Sheng, H., Shao, J., Dixon, D. A., Williams, C. S., Prescott, S. M., DuBois, R. N., and Beauchamp, R. D. (2000). Transforming growth factor-β1 enhances Ha-ras-induced expression of cyclooxygenase-2 in intestinal epithelial cells via stabilization of mRNA. J. Biol. Chem. 275:6628–6635.

Shi, Y., and Massagué, J. (2003). Mechanisms of TGF-β signaling from cell membrane to the nucleus. Cell 113:685–700.

Shi, H., Xu, J.-M., Hu, N.-Z., and Xie, H.-J. (2003). Prognostic significance of expression of cyclooxygenase-2 and vascular endothelial growth factor in human gastric carcinoma. World J. Gastroenterol. 9:1421–1426.

Shull, M. M., Ormsby, I., Kier, A. B., Pawlowski, S., Diebold, R. J., Yin, M., Allen, R., Sidman, C., Proetzel, G., Calvin, D., Annunziata, N., and Doetschman, T. (1992). Targeted disruption of the mouse transforming growth factor-β1 gene results in multifocal inflammatory disease. Nature 359:693–699.

Smith, M. F., Jr., Mitchell, A., Li, G., Ding, S., Fitzmaurice, A. M., Ryan, K., Crowe, S., and Goldberg, J. B. (2003). Toll-like receptor (TLR) 2 and TLR5, but not TLR4, are required for Helicobacter pylori-induced NF-κB activation and chemokine expression by epithelial cells. J. Biol. Chem. 278:32552–32560.

Sonoshita, M., Takaku, K., Sasaki, N., Sugimoto, Y., Ushikubi, F., Narumiya, S., Oshima, M., and Taketo, M. M. (2001). Acceleration of intestinal polyposis through prostaglandin receptor EP2 in Apc$^{\Delta716}$ knockout mice. Nat. Med. 7:1048–1051.

Stichtenoth, D. O., Thorén, S., Bian, H., Peters-Golden, M., Jakobsson, P.-J., and Crofford, L. J. (2001). Microsomal prostaglandin E synthase is regulated by proinflammatory cytokines and glucocorticoids in primary rheumatoid synovial cells. J. Immunol. 167:469–474.

Sun, W.-H., Yu, Q., Shen, H., Ou, X.-L., Cao, D.-Z., Yu, T., Qian, C., Zhu, F., Sun, Y.-L., Fu, X.-L., and Su, H. (2004). Roles of Helicobacter pylori infection and cyclooxygenase-2 expression in gastric carcinogenesis. World J. Gastroenterol. 10:2809–2813.

Sung, J. J. Y., Leung, W. K., Go, M. Y. Y., To, K. F., Cheng, A. S. L., Ng, E. K. W., and Chan, F. K. L. (2000). Cyclooxygenase-2 expression in Helicobacter pylori-associated premalignant and malignant gastric lesions. Am. J. Pathol. 157:729–735.

Takaku, K., Miyoshi, H., Matsunaga, A., Oshima, M., Sasaki, N., and Taketo, M. M. (1999). Gastric and duodenal polyps in Smad4 (Dpc4) knockout mice. Cancer Res. 59:6113–6117.

Taketo, M. M. (2006). Wnt signaling and gastrointestinal tumorigenesis in mouse models. Oncogene 25:7522–7530.

Takeda, H., Miyoshi, H., Tamai, Y., Oshima, M., and Taketo, M. M. (2004). Simultaneous expression of COX-2 and mPGES-1 in mouse gastrointestinal hamartomas. Br. J. Cancer 90:701–704.

Tang, C., Liu, C., Zhou, X., and Wang, C. (2004). Enhanced inhibitive effects of combination of rofecoxib and octreotide on the growth of human gastric cancer. Int. J. Cancer 112:470–474.

Tatsuguchi, A., Matsui, K., Shinji, Y., Gudis, K., Tsukui, T., Kishida, T., Fukuda, Y., Sugisaki, Y., Tokunaga, A., Tajiri, T., and Sakamoto, C. (2004). Cyclooxygenase-2 expression correlates with angiogenesis and apoptosis in gastric cancer tissue. Hum. Pathol. 35:488–495.

Tebbutt, N. C., Giraud, A. S., Inglese, M., Jenkins, B., Waring, P., Clay, F. J., Malki, S., Alderman, B. M., Grail, D., Hollande, F., Heath, J. K., and Ernst, M. (2002). Reciprocal regulation of gastrointestinal homeostasis by SHP2 and STAT-mediated trefoil gene activation in gp130 mutant mice. Nat. Med. 8:1089–1097.

Thiel, A., Heinonen, M., Rintahaka, J., Hallikainen, T., Hemmes, A., Dixon, D. A., Haglund, C., and Ristimäki, A. (2006). Expression of cyclooxygenase-2 is regulated by glycogen synthase kinase-3β in gastric cancer cells. J. Biol. Chem. 281:4564–4569.

Thun, M. J., Namboodiri, M. M., Calle, E. E., Flanders, W. D., and Heath, C. W., Jr. (1993). Aspirin use and risk of fatal cancer. Cancer Res. 53:1322–1327.

Thun, M. J., Namboodiri, M. M., and Heath, C. W., Jr. (1991). Aspirin use and reduced risk of fatal colon cancer. N. Engl. J. Med. 325:1593–1596.

Tsujii, M., and DuBois, R. N. (1995). Alterations in cellular adhesion and apoptosis in epithelial cells overexpressing prostaglandin endoperoxide synthase 2. Cell 83:493–501.

Tsujii, M., Kawano, S., Tsuji, S., Sawaoka, H., Hori, M., and DuBois, R. N. (1998). Cyclooxygenase regulates angiogenesis induced colon cancer cells. Cell 93:705–716.

Uefuji, K., Ichikura, T., and Mochizuki, H. (2000). Cyclooxygenase-2 expression is related to prostaglandin biosynthesis and angiogenesis in human gastric cancer. Clin. Cancer Res. 6:135–138.

van Rees, B. P., Saukkonen, K., Ristimäki, A., Polkowski, W., Tytgat, G. N. J., Drillenburg, P., and Offerhaus, G. J. A. (2002). Cyclooxygenase-2 expression during carcinogenesis in the human stomach. J. Pathol. 196:171–179.

van Rees, B. P., Sivula, A., Thorén, S., Yokozaki, H., Jakobsson, P.-J., Offerhaus, J. A., and Ristimäki, A. (2003). Expression of microsomal prostaglandin E synthase-1 in intestinal type gastric adenocarcinoma and in gastric cancer cell lines. Int. J. Cancer 107:551–556.

Wen, X.-Z., Akiyama, Y., Baylin, S., and Yuasa, Y. (2006). Frequent epigenetic silencing of the bone morphogenetic protein 2 gene through methylation in gastric carcinomas. Oncogene 25:2666–2673.

Wen, X.-Z., Miyake, S., Akiyama, Y., and Yuasa, Y. (2004). BMP-2 modulates the proliferation and differentiation of normal and cancerous gastric cells. Biochem. Biophys. Res. Commun. 316:100–106.

Woo, D. K., Kim, H. S., Lee, H. S., Kang, Y. H., Yang, H. K., and Kim, W. H. (2001). Altered expression and mutation of β-catenin gene in gastric carcinomas and cell lines. Int. J. Cancer 95:108–113.

Wu, Y.-L., Fu, S.-L., Zhang, Y.-P., Qiao, M.-M., and Chen, Y. (2005). Cyclooxygenase-2 inhibitors suppress angiogenesis and growth of gastric cancer xenografts. Biomed. Pharmacother. 59:S289–S292.

Xiangming, C., Natsugoe, S., Takao, S., Hokita, S., Ishigami, S., Tanabe, G., Baba, M., Kuroshima, K., and Aikou, T. (2001). Preserved Smad4 expression in the transforming growth factor β signaling pathway is a favorable prognostic factor in patients with advanced gastric cancer. Clin. Cancer Res. 7:277–282.

Xiao, F., Furuta, T., Takashima, M., Shirai, N., and Hanai, H. (2001). Involvement of cyclooxygenase-2 in hyperplastic gastritis induced by Helicobacter pylori infection in C57BL/6 mice. Aliment. Pharmacol. Ther. 15:875–886.

Xie, W. L., Chipman, J. G., Robertson, D. L., Erikson, R. L., and Simmons, D. L. (1991). Expression of a mitogen-responsive gene encoding prostaglandin synthase is regulated by mRNA splicing. Proc. Natl. Acad. Sci. USA 88:2692–2696.

Xu, X., Brodie, S. G., Yang, X., Im, Y.-H., Parks, W. T., Chen, L., Zhou, Y.-X., Weinstein, M., Kim, S.-J., and Deng, C.-X. (2000). Haploid loss of the tumor suppressor Smad4/Dpc4 initiates gastric polyposis and cancer in mice. Oncogene 19:1868–1874.

Yamagata, R., Shimoyama, T., Fukuda, S., Yoshimura, T., Tanaka, M., and Munakata, A. (2002). Cyclooxygenase-2 expression is increased in early intestinal-type gastric cancer and gastric mucosa with intestinal metaplasia. Eur. J. Gastroenterol. Hepatol. 14:359–363.

Yang, H.-K., Kang, S. H., Kim, Y.-S., Won, K., Bang, Y.-J., and Kim. S.-J. (1999). Truncation of the TGF-β type II receptor gene results in insensitivity to TGF-β in human gastric cancer cells. Oncogene 18:2213–2219.

Yano, T., Ito, K., Fukamachi, H., Chi, X.-Z., Wee, H.-J., Inoue, K., Ida, H., Bouillet, P., Strasser, A., Bae, S.-C., and Ito, Y. (2006). The RUNX3 tumor suppressor upregulates Bim in gastric epithelial cells undergoing transforming growth factor β-induced apoptosis. Mol. Cell. Biol. 26:4474–4488.

Yoshimatsu, K., Golijanin, D., Paty, P. B., Soslow, R. A., Jakobsson, P.-J., DeLellis, R. A., Subbaramaiah, K., and Dannenberg, A. J. (2001). Inducible microsomal prostaglandin E synthase is overexpressed in colorectal adenomas and cancer. Clin. Cancer Res. 7:3971–3976.

Yu, H.-G., Li, J.-Y., Yang, Y.-N., Luo, H.-S., Yu, J.-P., Meier, J. J., Schrader, H., Bastian, A., Schmidt, W. E., and Schmitz, F. (2003). Increased abundance of cyclooxygenase-2 correlates with vascular endothelial growth factor-A abundance and tumor angiogenesis in gastric cancer. Cancer Lett. 195:43–51.

Zaridze, D., Borisova, E., Maximovitch, D., and Chkhikvadze, V. (1999). Aspirin protects against gastric cancer: results of a case-control study from Moscow, Russia. Int. J. Cancer 82:473–476.

Zhou, Y., Ran, J., Tang, C., Wu, J., Honghua, L., Xingwen, L., Ning, C., and Qiao, L. (2007). Effect of celecoxib on E-cadherin, VEGF, microvessel density and apoptosis in gastric cancer. Cancer Biol. Ther. 6:269–275.

Chapter 21
REG Proteins and Other Growth Factors in Gastric Cancer

Hirokazu Fukui, Yoshikazu Kinoshita, and Tsutomu Chiba

Introduction

The *regenerating gene* (*Reg*) was originally isolated from regenerating rat pancreatic islets, and its gene product was evidenced to have a trophic effect on not only islet but also gastric epithelial cells. Recently, many Reg-related genes were isolated and revealed to constitute a multigene family. Although the biologic function of Reg family proteins remains to be elucidated, accumulating evidence has suggested that Reg family genes have an important role in the pathophysiology of inflammatory diseases and their related tumor development. Moreover, we have recently shown that REG Iα protein is an important downstream molecule of STAT3 signaling and mediates the antiapoptotic effect of STAT3 by activating the Akt/Bad/Bcl-xL pathway, suggesting that REG Iα protein has a critical antiapoptotic role in gastric tumorigenesis under STAT3 hyperactivation. Accordingly, similar to other growth factors, REG Iα protein might become a therapeutic target for the prevention and treatment of gastric cancer.

REG Proteins

In 1984, Okamoto et al. found that administration of nicotinamide, a poly(ADP-ribose) synthetase inhibitor, accelerated the regeneration of pancreatic islets in 90% depancreatized rats (Yonemura et al. 1984). Subsequently, they screened the regenerating islets-derived cDNA library, and isolated a novel gene, *Reg* (i.e., regenerating gene), that encodes a 165-amino-acid protein with a 21-amino-acid signal peptide (Terazono et al. 1988). Thereafter, as suggested by the process of its discovery, *Reg* gene and its product, Reg protein were clarified to have roles in regeneration of islets by acting as a growth factor (Watanabe et al. 1994).

Interestingly, Watanabe et al. found in their early work that Reg mRNA is expressed in not only pancreas but also the gastrointestinal tract (Watanabe et al. 1990), and we subsequently demonstrated that Reg protein is expressed in enterochromaffin-like (ECL) cells in rat fundic mucosa (Asahara et al. 1996). Moreover, we found that Reg protein is mitogenic to gastric epithelial cells, suggesting that

Reg protein has an important role in regeneration of gastric mucosa as well (Fukui et al. 1998). Recently, many Reg-related genes were isolated and revealed to constitute a multigene family, the *Reg* gene family. Although the biologic functions of Reg family proteins still remain to be elucidated, various DNA array studies have accumulated evidence that Reg family genes are important for not only the pathophysiology of inflammation but also the development of cancer in the gastrointestinal tract (Dieckgraefe et al. 2000; Kim et al. 2007; Lawrance et al. 2001; Oue et al. 2004). Herein, we describe the recently advanced knowledge of Reg proteins, focusing on their roles in the gastritis–gastric cancer sequence.

Molecular Structure of the Reg Protein Family

After identification of Reg protein in the pancreas, many other proteins containing similar amino-acid sequences were isolated from human, rat, mouse, cow, and hamster (Kinoshita et al. 2004; Okamoto 1999). At present, it is well established that Reg and Reg-related proteins constitute a multiprotein family, which is classified into four subfamilies according to their primary structures (Table 21.1). Additionally, we and Okamoto et al. (1999) propose to name the Reg protein for human and other species as "REG" and "Reg," respectively. The Reg protein originally isolated from regenerating islets (Reg I) and its human homolog, REG Iα are both classified to members of the Type I group, and we mainly describe these molecules in the present chapter.

In humans, four REG family genes, i.e., REG Iα, REG Iβ, HIP/PAP, and REG III are tandemly ordered in the 140-kbp region of human chromosome 2p12 (Miyashita et al. 1995). In the mouse genome, all of the Reg family genes, i.e., *Reg I, Reg II, Reg IIIα, Reg IIIβ, Reg IIIγ, and Reg IIIδ*, were mapped to a contiguous 75-kbp region of chromosome 6C (Narushima et al. 1997). Moreover, all of these Reg family genes share a similar six-exon and five-intron structure, suggesting that the Type I, II, and III subfamilies are derived from a common ancestor gene by several duplications (Unno et al. 1993), and have reached divergency in expression and function, depend-

Table 21.1 REG gene family

| Type | Species | | |
	Human	Rat	Mouse
I	*REG Iα (PSP/Lithostathine/PTP)* *REG Iβ*	*Reg I*	*Reg I*
II			*Reg II*
III	*HIP/PAP*	*RegIII/PAP II*	*Reg IIIα*
		PAP I/Peptide 23	*Reg IIIβ*
		PAP III	*Reg IIIγ*
IV	*REG IV*		*Reg IIIδ*

Reg, regenerating gene; *PSP*, gene for pancreatic stone protein; *PTP*, gene for pancreatic thread protein; *HIP*, gene expressed in hepatocellular carcinoma, intestine and pancreas; *PAP*, gene for pancreatitis-associated protein.

ing on the different types of *Reg* genes, in the process of genetic evolution. However, *REG IV*, most recently isolated as an overexpressed gene in ulcerative colitis mucosa (Hartupee et al. 2001) and/or a chemo-drug-resistance–associated gene in colon cancer (Violette et al. 2003), is separately located on chromosome 1 and has markedly longer introns than other REG genes. Thus, although the REG IV gene has similar genomic structure with six exons and five introns, its evolution is distinct from those of other *Reg* subfamilies (Hartupee et al. 2001).

Reg family proteins usually have a 22–26 amino acid signal sequence at the N-terminus and conserve six cysteine residues (Hartupee et al. 2001; Itoh et al. 1990; Kinoshita et al. 2004; Okamoto 1999). Thus, Reg family proteins possess three disulfide bonds and form a Drickamer motif that is common in the calcium-dependent lectin (C-type lectin) superfamily (Drickamer 1988). The lectin family proteins have various biologic roles in cell recognition, migration, growth, and adhesion. Accordingly, although Reg proteins are presently suggested to act as trophic factors in the gastrointestinal tissues, additional roles for them may be elucidated in future.

REG Protein–Producing Cells and Their Physiologic Role

In 1996, Asahara et al. (1996) first found that Reg I protein is expressed in ECL cells in the rat fundic mucosa. Thereafter, we demonstrated that REG Iα protein is expressed in chief cells as well as in ECL cells in the normal human fundic mucosa (Fukui et al. 2004; Higham et al. 1999) (Figure 21.1B). Moreover, it was found that REG Iα protein is expressed in a few epithelial cells in the crypt of the small intestine or the colorectum (Figure 21.1C, D). A subpopulation of REG Iα–positive cells coexpressed chromogranin A, suggesting that some of REG Iα–positive cells

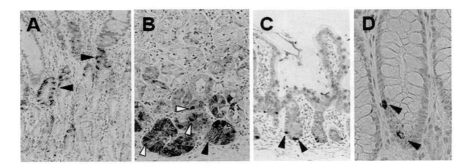

Gastric ithmus   Gastric gland   Small intestine   Colon

Fig. 21.1 Expression of REG Ia protein in the normal gastrointestinal tract in human. Arrowheads indicate REG Ia–positive cells. White arrowheads indicate REG Ia–positive endocrine cells in the gastric gland

are endocrine in nature. However, we and others have shown that not only REG Iα, but also REG IV (Oue et al. 2005) and HIP/PAP (Christa et al. 1996) are expressed in a few epithelial cells in the small intestinal or the colorectal crypt. The distribution of REG IV or HIP/PAP–positive cells is very similar to that of REG Iα–positive ones, and the former cell subpopulation are endocrine cells in the mucosa of small intestine and colorectum (Kämäräinen et al. 2003; Lasserre et al. 1999). Although distributions of different REG family proteins appear generally similar, synthesis of each REG family protein may be distinctly regulated, and each may have different biologic functions.

Interestingly, we further found that REG Iα protein is expressed in the small cells in the isthmus of the fundic mucosa (Fukui et al. 2004), where putative stem cells reside and active cell proliferation is continuously occurring (Figure 21.1A). Of note, in the small intestine and colon mucosa, REG Iα protein is expressed in cells above Paneth cells and those in the crypt base, respectively, both of which are candidate stem cells in the intestinal mucosa (Figure 21.1C, D). These findings suggest that REG Iα protein may be involved in the regeneration of epithelial cells in the gastrointestinal mucosa. In a previous study, we showed that Reg I protein has a mitogenic effect on the fraction enriched in proliferating cells prepared by counter flow elutriation *in vitro* (Fukui et al. 1998). Furthermore, Kinoshita et al. recently demonstrated that the neck zone composed of proliferating stem cells and/ or its closely related progenitor cells was greatly expanded in the gastric mucosa of Reg I–transgenic mice (Figure 21.2), and that gastric mucosal thickness was significantly greater in Reg I–transgenic mice than in control mice (Miyaoka et al. 2004). These findings strongly suggest that the growth-promoting effect of Reg I

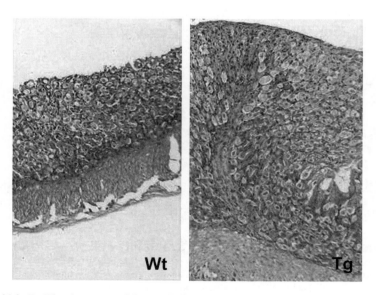

Fig. 21.2 Proliferation status of the gastric fundic mucosal cells in Reg I–transgenic mice

protein is crucial for gastric epithelial regeneration. Besides, it is also interesting that although there was no significant difference in the thickness of the foveolar (pit) layer of the fundic mucosa, the glandular layer was markedly thicker in Reg I–transgenic mice than in control mice (Figure 21.2) (Miyaoka et al. 2004). These findings suggested that Reg I protein stimulated proliferation and/or differentiation of parietal and chief cells rather than foveolar epithelial cells. In this regard, transgenic mice bearing growth factor genes such as *TGF-α* (Sharp et al. 1995) or *gastrin* (Nakajima et al. 2002) develop prominent hyperplasia of both glandular and the mucous surface cells. Although we cannot clearly differentiate the distinct actions of Reg I protein from other growth factors in the gastric mucosa at present, the distribution of Reg receptor may explain such a distinct action of Reg I protein. Indeed, *Reg receptor* is strongly expressed in the epithelial cells in the middle to deep layer of the gastric mucosa, but its expression is very weak in the cells of the surface mucous layer (Kazumori et al. 2002), suggesting the gradient efficacy of Reg I protein along the gastric epithelial axis (Figure 21.3). However, Reg I–defi-

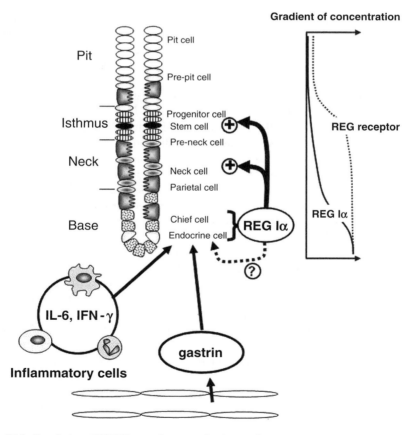

Fig. 21.3 Regulation of REG Ia protein expression in gastric mucosa

cient mice revealed that Reg I protein has a role in cell growth of the small intestinal crypt in mice, especially during the periods before and after birth (Ose et al. 2007). Moreover, in Reg I–deficient mice, the migration speed of small intestinal epithelial cells is significantly slower than that of wild-type litter mates (Ose et al. 2007). Taken together, Reg I protein is considered to be a regulator of cell growth that is required for generation and maintenance of the gastrointestinal mucosa.

REG Proteins in the Gastritis–Gastric Cancer Sequence After Helicobacter pylori Infection

Expression of REG Proteins in *Helicobacter pylori* Gastritis

We found in our early work that the *Reg I* gene was overexpressed in rat gastritis tissues induced by water-immersion restraint stress (Asahara et al. 1996), or by administration of indomethacin (Kawanami et al. 1997). In addition, others have reported that the *Reg I* gene is overexpressed in pancreatitis (Satomura et al. 1993) or inflammatory bowel disease (Sekikawa et al. 2005a). Moreover, recent comprehensive analyses using DNA array repeatedly confirmed that Reg family genes are novel genes overexpressed in inflammatory bowel disease (Dieckgraefe et al. 2000; Lawrance et al. 2001) or atrophic gastritis (Kim et al. 2007). Thus, there is no doubt that Reg family genes have an important role in the pathophysiology of inflammation in the gastrointestinal tract. As generally accepted, *Helicobacter pylori* infection is a crucial event in the development of chronic atrophic gastritis and gastroduodenal ulcer diseases, and has been implicated in the development of gastric cancer. Accordingly, we decided to investigate *Reg I* gene expression and its pathophysiologic roles in *H. pylori*–induced gastritis and gastric cancer.

We first investigated *Reg* gene expression in Mongolian gerbils infected with *H. pylori* (Fukui et al. 2003). Similar to mice, *Reg I* gene expression was detected at a very low level in the fundic mucosa of gerbils without *H. pylori* infection. However, in *H. pylori*–infected gerbils, *Reg I* mRNA became clearly detectable in the fundus 4 weeks after infection, and continued to increase up to 36 weeks. Of note, the level of *Reg I* mRNA expression in the fundic mucosa is correlated closely with severity of fundic mucosal inflammation. Conversely, it is also noteworthy that *Reg I* gene expression was diminished in accordance with the disappearance of inflammation when gerbils were eradicated of *H. pylori*.

Compatible with the data from this animal model, the enhanced level of REG Iα protein production was confirmed in the *H. pylori*–induced gastritis mucosa of humans (Yoshino et al. 2005). Furthermore, the number of REG Iα–positive cells and the level of REG Iα expression were both correlated with the severity of gastritis, and a subpopulation of REG Iα–positive cells were suggested to be ECL cells (Yoshino et al. 2005; Sekikawa et al. 2005b). These findings seem reasonable because chronic gastritis with *H. pylori* infection is generally accompanied by mild hypergastrinemia (Calam 1996) and a resulting increase of ECL cells. However, it must be noted that a considerable number of REG Iα–positive nonendocrine cells are present in *H. pylori*–induced gastritis tissues, and that metaplastic epithelial cells also express REG Iα protein (Sekikawa et al. 2005b). Further characterization of REG Iα–positive cells in gastritis or metaplastic lesions remains to be elucidated.

Mechanism for REG Protein Expression in Gastritis

Inflammation seems to be a key event for the induction of REG Iα protein expression in gastrointestinal tissues (Figure 21.3). Kazumori et al. examined the time course change of proinflammatory cytokines and *Reg I* gene expression in stress-induced gastritis and suggested that cytokine-induced neutrophil chemoattractant 2β stimulation possibly enhances *Reg I* gene expression in ECL cells in rat gastritis tissues (Kazumori et al. 2000). However, in *in vitro* studies using gastric cancer cell lines, we found that interleukin (IL)-6 and interferon (IFN)-γ stimulated *REG Iα* promoter activity and indeed enhanced *REG Iα* mRNA expression (Sekikawa et al. 2005b). In addition, others have demonstrated that HIP/PAP gene expression is enhanced in the pancreatic cell line by IL-6, IFN-γ, and TNF-α stimulation (Dusetti et al. 1996). These results suggest that various proinflammatory cytokines are involved in enhancement of REG family gene expression in gastritis. We tried to clarify the regulatory mechanism of REG Iα gene expression by IL-6 stimulation because among such candidate cytokines, IL-6 universally has important roles in inflammation and enhances *REG Iα* gene expression most strongly in gastric cancer cells.

IL-6 uses gp130 as a receptor subunit and its binding to gp130 activates the JAK/STAT signaling, which is involved in a variety of biologic responses, including the immune response, inflammation, hematopoiesis, and oncogenesis by regulating cell growth, survival, and differentiation (Bromberg 2002). We found that enhancement of REG Iα gene expression by IL-6 was accompanied by activation of STAT3 signaling; moreover, treatment with STAT3 siRNA abolished the increased *REG Iα* gene expression. Furthermore, we investigated REG Iα promoter activity by reporter assay and determined its responsible element for IL-6 stimulation, which corresponds to the consensus STAT3 binding site. Interestingly, Judd et al. recently reported that mice with STAT3 hyperactivation showed *Reg I* overexpression in the gastric mucosa and developed gastric tumors (Judd et al. 2004, 2006). These data are consistent with our *in vitro* studies and suggest that STAT3 signaling has crucial roles in *Reg I* expression in gastritis–gastric cancer sequence. Furthermore, in addition to IL-6, IFN-γ, a representative Th1 cytokine in *H. pylori*–induced gastritis, is likely to be involved in enhancement of *Reg I* expression (Sekikawa et al. 2005b).

Gastrin is also thought to be an important regulatory factor for *Reg I* gene expression in the stomach. In rat stomach, ECL cells that possess abundant gastrin receptors clearly showed enhancement of *Reg I* gene expression when stimulated by gastrin (Fukui et al. 1998). Thereafter, Dockray et al. produced *in vitro* evidence that gastrin stimulates Reg I expression via activation of PKC and RhoA, that a C-rich region (−79 to −72) of Reg I gene is critical for the response, and that

Sp-family transcription factors bind to this region of the promoter (Ashcroft et al. 2004). Importantly, gastrin is a well-known growth factor for fundic mucosal cells, and is elevated in patients with *H. pylori*–induced chronic gastritis (Calam 1996). In fact, we found that serum gastrin levels correlate significantly with *Reg I* expression in the fundus of *H. pylori*–infected gerbils. Interestingly, a recent study by Takaishi et al. indicates that endogenous gastrin and *H. felis*–induced gastritis have synergistic effects on *Reg I* expression in the stomach (Takaishi and Wang 2007). Although regulatory mechanisms for *Reg I* expression are not completely understood, it is evident that cytokines and gastrin have central roles in *Reg I* expression in *H. pylori*–induced gastritis (Figure 21.3).

Roles of REG Proteins in Gastric Carcinogenesis

Recently, much evidence has accumulated suggesting that the REG Iα protein has important roles in inflammation-associated carcinogenesis such as in gastritis–gastric cancer (Fukui et al. 2004; Sekikawa et al. 2005b; Yonemura et al. 2003) or ulcerative colitis–colitic cancer (Sekikawa et al. 2005a) sequences. In our own studies, a recombinant REG Iα protein significantly stimulated the proliferation of gastric cancer cells, which was abolished by treatment with anti-REG Iα antibody, suggesting that REG Iα protein has a trophic effect on not only normal gastric epithelial cells but also cancer cells (Figure 21.4) (Sekikawa et al. 2005b). Moreover, we also demonstrated that the REG Iα protein conferred an antiapoptotic effect on gastric cancer cells and promoted survival against H_2O_2 stimulation or serum starvation (Sekikawa et al. 2005b). Because REG Iα protein expression is sustained at high levels in chronic active gastritis, its antiapoptotic action as well as its cell proliferative effect seems to contribute to the development of gastric cancer from chronic active gastritis.

Although intracellular signaling by the REG Iα protein had remained unclear, analyses of it began when the Reg receptor was isolated from a pancreatic islet cDNA library (Kobayashi et al. 2000). *Reg receptor* gene encodes a 919-amino-acid protein, and its protein belongs to type II transmembrane protein with a long extracellular domain. REG Iα protein binds this receptor protein (Kobayashi et al. 2000) and promotes Akt phosphorylation followed by enhancement of Bcl-xL expression, exerting its antiapoptotic effect on gastric cancer cells (Sekikawa et al. 2005b). Takasawa et al. reported that Reg I protein stimulates PI3K/ATF-2/cyclin D1 signaling pathway followed by cyclin D1 activation, suggesting that Reg I protein promotes cell-cycle progression from G1 to S phase, and thus contributes to cell proliferation (Takasawa et al. 2006). Although involvement of other signal pathways including the MAPK pathway remains unknown, evidence is accumulating that Akt signaling is important in transmitting the stimulation of not only REG Iα, but also other REG family proteins (Bishnupuri et al 2006; Nishimune et al. 2000) (Figure 21.5).

It has been suggested that STAT3 activation has pivotal roles in inflammation-associated carcinogenesis by modulating its downstream gene products. Of note, Giraud et al. have clearly demonstrated that STAT3 hyperactivation caused gastric

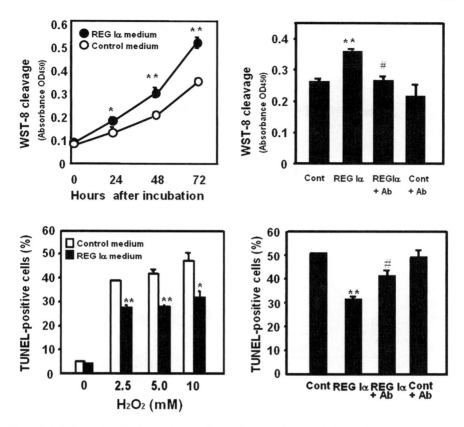

Fig. 21.4 Effects of REG Ia protein on cell growth and antiapoptosis in gastric cancer cells

tumors with chronic gastritis in mice (Judd et al. 2004, 2006). Moreover, it is interesting that in mice with STAT3 hyperactivation not only *Reg I* but also Reg IIIβ and Reg IIIγ were overexpressed in the gastric mucosa (Judd et al. 2004). Furthermore, the expression of their candidate stimulators such as IL-6 and IFN-γ were simultaneously enhanced. How then, is REG Iα protein involved in the development of gastric cancer? Although REG Iα expression is normally maintained at low level in the nonneoplastic gastric mucosa, it is markedly augmented by inflammation. Under inflammatory condition, STAT3 activation is also tightly regulated by suppressor of cytokine signaling 3 in a negative feedback loop (Levy and Darnell 2002; Starr et al. 1997), whereas in neoplastic tissues, dysregulation of this system leads to dysregulated hyperactivation of STAT3. Thus, REG Iα overexpression may reflect STAT3 hyperactivation in gastric cancer tissues. Compatible with experimental data from STAT3-hyperactivated mice, REG Iα overexpression is significantly found in gastric cancers that show STAT3 hyperactivation (unpublished data). Accordingly, REG Iα protein seems to have a critical

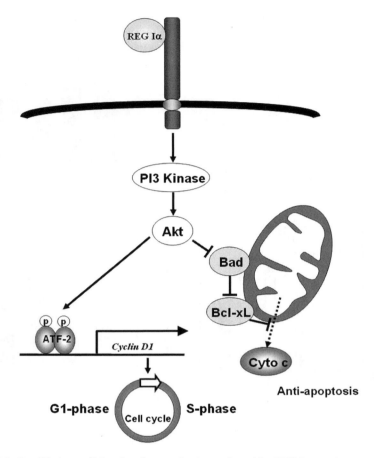

Fig. 21.5 Possible intracellular signaling mechanism activated by REG Ia protein

growth-promoting and/or antiapoptotic role in gastric tumorigenesis after STAT3 hyperactivation.

Gastrin is a growth-promoting factor for gastric epithelial cells and has roles in gastric carcinogenesis. Indeed, gastric mucosal thickness is increased in patients with hypergastrinemia in Zollinger-Ellison syndrome (Bordi et al. 1974) or in rats with hypergastrinemia evoked by proton pump inhibitors (Larsson et al. 1986). In 1998, we found that a large amount of Reg I protein is produced in ECL cells in rats with hypergastrinemia induced by proton pump inhibitors and its gene expression is directly upregulated by gastrin (Fukui et al. 1998). More importantly, we showed that Reg I protein has a trophic effect on primary cultured gastric epithelial cells (Fukui et al. 1998), and that gastric mucosal proliferation is promoted in accordance with the level of serum gastrin and *Reg I* expression in the fundus of *H. pylori*–infected gerbils (Fukui et al. 2003). Taken together, we proposed that

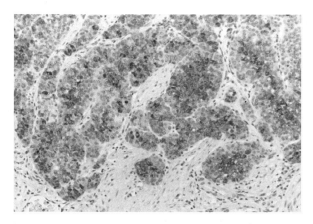

Fig. 21.6 Immunostaining of REG Ia in gastric carcinoid tumor of patients with hypergastrinemia.

gastrin stimulates gastric mucosal proliferation via two different mechanisms; directly through gastrin receptors on target cells, and indirectly through the production of Reg I protein by ECL cells. In this regard, it is interesting that REG Iα protein is expressed in gastric carcinoid tumors of patients with hypergastrinemia because of autoimmune gastritis (Figure 21.6), suggesting that the growth-promoting effect of REG Iα protein may be involved in the development of gastric carcinoid tumors. However, Reg I-overexpressing mice showed glandular cell hyperplasia but not endocrine cell hyperplasia (Miyaoka et al. 2004). In addition, Higham et al. speculated that REG Iα protein functions as an autocrine or paracrine inhibitor of ECL cells because truncation of REG Iα protein may in part contribute to the development of gastric carcinoid tumors (Higham et al. 1999). Thus, it should be elucidated whether ECL cells possess Reg receptors and whether accumulated Reg I protein in the cytoplasm of carcinoid cells has a role in tumor cell biology as an autocrine factor.

Other Growth Factors in Gastric Cancer

Receptor tyrosine kinases (RTKs) mediate the mitogenic signal of many growth factors through the phosphorylation cascade of signal transducers and nuclear proteins. When the gastric mucosa is injured by inflammation or other insults, expression of growth factors and their receptors are upregulated in epithelial or mesenchymal cells, and subsequently mucosal repair is promoted. However, RTKs/growth factors are also proposed as critical factors for tumor development and progression. Indeed, gastric cancer cells express not only a broad spectrum of growth factors including transforming growth factor (TGF)-α, epidermal growth factor

(EGF), amphiregulin, platelet-derived growth factor (PDGF), insulin-like growth factor II, and bFGF, but also various RTKs, such as the EGF receptor (EGFR) family, fibroblast growth factor receptor (FGFR) family, and vascular endothelial growth factor receptor (VEGFR) subtypes (Yokozaki et al. 1997). Recently, experimental and clinical data clearly demonstrated the usefulness of several RKT inhibitors for treatment of various tumors (Arnold et al. 2006).

Epidermal Growth Factor Receptor Family and Its Ligands

There is much evidence showing that EGFR family proteins (erb B1 to B4) and their ligands [EGF, TGF-α, amphiregulin, heparin-binding (HB)-EGF, heregulin, betacellulin] are involved in the development of gastrointestinal tumors. However, in contrast to esophageal squamous cell carcinoma, the EGFR family had been thought to have a limited role in gastric cancer development because *erb B2* overexpression was found in, at most, 12%–50% (Yonemura et al. 1991), and *erb B1* and *erb B3* overexpression in 3% and 0% of human gastric cancers, respectively (Yoshida et al. 1989). According to recent knowledge, *H. pylori* infection upregulates the expression of HB-EGF and amphiregulin in gastric cancer cells by means of bacterial production of γ-glutamyltranspeptidase, resulting in activation of PI3 kinase and/or p-38-dependent pathway (Busiello et al. 2004; Romano et al. 1998). Because cag PAI$^+$ *H. pylori* strain seems to induce greater activation of EGFR than cag PAI$^-$ strains (Keates et al. 2001), the relationship between cag PAI and EGFR signaling should be clarified in the future.

Vascular Endothelial Growth Factor and Its Receptor

VEGF family members have been identified as major regulators of lymphangiogenesis. VEGFs are also expressed in many gastric cancer cell lines and their receptors are virtually ubiquitous in human tumors. VEGF-A, a key ligand of VEGFR-2, is thought to have the most important role among the VEGF family proteins in progression of gastric cancer, including distant metastasis and peritoneal dissemination (Maeda et al. 1996). Supporting this idea, Hirakawa et al. experimentally demonstrated, using VEGF-A-overexpressing mice, that VEGF-A induced active proliferation of VEGFR-2 expressing tumor-associated lymphatic vessels, and promoted lymph node metastasis (Hirakawa et al. 2005). VEGF-C and VEGF-D also exhibit mitogenic effects for vascular and lymphatic endothelial cells and survival-promoting abilities for lymphatic endothelial cells through VEGFR-3 on such cells (Shimizu et al. 2004), suggesting that these growth factors are also involved in lymphatic invasion and lymph node metastasis. Accordingly, the VEGFR signaling pathway has been validated as a priority target for anticancer therapy. However, Caputo et al. showed that *H. pylori* infection upregulates VEGF expression in gastric mucosal cells and that this effect was dependent on *H. pylori*–derived vacA (Caputo et al. 2003; Kitadai et al. 2003). Furthermore, vacA-induced enhancement of VEGF expression is considered dependent on EGFR activation (Caputo et al. 2003), suggesting a pivotal role of EGFR signaling in gastric cancer development.

Fibroblast Growth Factor Receptor Family, Platelet-Derived Growth Factor, Hepatocyte Growth Factor, and Their Receptors

Growth factors produced from mesenchymal cells in wounded gastric tissues have major roles in the reconstitution of connective tissues, including vessel and smooth muscle cells, but also participate in epithelial reconstitution.

Of the FGF family proteins, 23 members are identified in humans that have key roles in growth and survival of stem cells during embryogenesis, tissue regeneration, and carcinogenesis (Katoh and Katoh 2006). In the stomach, FGF2 (basic FGF; bFGF) and FGF7 (keratinocyte growth factor; KGF) are major FGFs implicated in tissue regeneration (Hull et al. 1999; Kinoshita et al. 1995). bFGF and KGF are mainly expressed in mesenchymal fibroblasts and act on FGFR1 and FGFR2 on gastric mucosal cells, respectively (Katoh and Katoh 2006). In the clinical setting, bFGF expression in gastric cancer specimens is associated with higher microvascular density, tumor progression, and a worse prognosis (Zhao et al. 2005). In addition, it has been reported that KGF stimulates the proliferation of gastric cancer, especially scirrhous gastric cancer (Nakazawa et al. 2003), and indeed, scirrhous gastric cancers overexpress a variant FGFR2 that has higher affinity with KGF (Katoh and Katoh 2006). These facts suggest that KGF/FGFR2 signaling has a critical role for the development and progression of diffuse-type gastric cancer.

PDGF stimulates proliferation and chemotaxis of fibroblasts and smooth muscle cells in gastric tissue (Milani and Calabro 2001). Also, it is a potent chemoattractant for neutrophils and macrophages, and may stimulate the production and contraction of extracellular matrices. The expression of PDGF and its receptor (PDGFR) were closely related to gastric cancer growth (Chung and Antoniades 1992). Many PDGFR α and β are present in endothelial and smooth muscle cells of newly formed vessels (Milani and Calabro 2001), and therefore, PDGF/PDGFR signaling is assumed to have an angiogenic role in the progression of gastric cancers.

Hepatocyte growth factor (HGF) is a member of plasminogen-related growth factor family, which includes HGF and HGF-like/macrophage stimulating protein (Skeel et al. 1991). In the majority of tissues, expression of the HGF gene is restricted to fibroblasts or other mesenchymal cells, whereas expression of the HGF receptor, c-met, is confined to different cells that do not secrete the ligand (Sonnenberg et al. 1993). Experimental models of gastric ulceration have revealed that HGF transcripts are strongly expressed in stromal cells interposed between regenerating glands and submucosal vessels, whereas c-met mRNA expression increases in regenerating epithelial cells (Kinoshita et al. 1995; Schmassmann et al. 1997). Moreover, in damaged gastric tissues, not only the production but also the conversion of HGF to its active form is promoted (Kinoshita et al. 1997). Similar to other growth factor receptors, c-MET activation is likely to be involved in gastric cancer pathogenesis, including tumor growth, invasion, angiogenesis, and metastasis (Inoue et al. 2004). Indeed, c-MET overexpression is frequently observed in diffuse-type gastric cancers that have abundant stroma, and it has been suggested that c-MET activation is induced by HGF stimulation from the stromal component at the invasive front of such cancers.

Growth Factors Including REG Proteins as Therapeutic Targets

Several inhibitors of tyrosine kinases are currently approved by the United States Food and Drug Administration for treatment in different human malignancies (Table 21.2). However, in the case of gastric cancer, only preliminary data exist on the use of the currently available erbB1 tyrosine kinase inhibitors such as erlotinib and gefitinib (Arnold et al. 2006; Becker et al. 2006). A single study revealed that the overall response rate is 11% in the case of erlotinib treatment for gastric cancers (Dragovich et al. 2006). However, a phase II trial of gefitinib monotherapy showed better tolerance in patients with metastatic gastric adenocarcinoma, with disease control achieved in 18.3% of cases analyzed (Becker et al. 2006). Interestingly, Rojo et al. examined expression of EGFR, phosphorylated EGFR (pEGFR), pMAPK, pAkt, and Ki67 in human gastric cancer specimens before and after treatment with gefitinib (Rojo et al. 2006). After treatment, pEGFR levels were significantly reduced in gastric cancer; however, the decrease of proliferation was observed only in those with low levels of pAkt. This finding suggests that PI3K/Akt signaling is critical for gefitinib resistance, and implies that PI3K/Akt signaling should be further investigated in developing anticancer drugs of this class.

Finally, the possibility of REG proteins as a therapeutic target for gastric cancers needs to be discussed. Recently, several studies using gene-tip analysis have reported that *REG IV* is a novel gene overexpressed in gastric cancer (Mitani et al. 2007; Oue et al. 2004). *REG IV* was originally isolated as a gene associated with resistance to chemotherapeutic agents (Violette et al. 2003). Consistent with this history, REG IV protein has been shown to be a specific serum biomarker for gastric cancer patients and useful for predicting response to 5-fluorouracil (5-FU)-based chemotherapy

Table 21.2 FDA-licensed tyrosine kinase inhibitors

Agent	Target	Mode of action	Established clinical application
Imatinib	Bcr-abl	Kinase inhibitor	CML, GIST
Mesylate	PDGFR, c-kit		Hypereosinophilic syndrome
Gefitinib	erbB1/EGFR	Kinase inhibitor	Non-small-cell lung cancer
Erlotinib	erbB1/EGFR	Kinase inhibitor	Non-small-cell lung cancer
Cetuximab	erbB1/EGFR	Blocking antibody	Colorectal cancer
Trastuzmab	erbB2/HER	Blocking antibody	Breast cancer
Bevacizumab	VEGF	Neutralizing antiboty	Colorectal cancer

(Mitani et al. 2007). As for biologic functions, REG IV protein has been found to induce EGFR phosphorylation and subsequently activate the Akt/AP1 pathway in cancer cells (Bishnupuri et al. 2006), conferring a survival advantage for cell growth and apoptosis inhibition (Nanakin et al. 2007; Takehara et al. 2006). However, we demonstrated that EFGR signaling is responsible for REG IV gene expression in colon cancer cells (Nanakin et al. 2007). Thus, these findings suggest that a positive-functional loop exists between EFGR signaling and REG IV stimulation in tumors that overexpress EFGR and REG IV protein. Interestingly, Mitani et al. showed that expression of REG IV is significantly related to that of EGFR in gastric cancer, and that REG IV is a marker for prediction of resistance to 5-FU–based chemotherapy in patients with gastric cancer (Mitani et al. 2007). In this regard, the REG IV signal cascade as well as EFGR signaling is a possible target for anti-cancer therapy. Supporting this, we and others demonstrated that treatment with anti-REG protein antibodies significantly suppressed growth and inhibition of apoptosis of gastric cancer cells (Nanakin et al. 2007; Takehara et al. 2006), suggesting the feasibility of neutralizing antibody therapy targeting REG proteins. Our hope is that, in the future, the evidence obtained in this experimental work will contribute to clinical practice.

References

Arnold, D., Peinert, S., Voigt, W., Schmoll, H. J. 2006. Epidermal growth factor receptor tyrosine kinase inhibitors: present and future role in gastrointestinal cancer treatment. Oncologist 11:602–11.

Asahara, M., Mushiake, S., Shimada, S., Fukui, H., Kinoshita, Y., Kawanami, C., Watanabe, T., Tanaka, S., Ichikawa, A., Uchiyama, Y., Narushima, Y., Takasawa, S., Okamoto, H., Tohyama, M., Chiba, T. 1996. Reg gene expression is increased in rat gastric enterochromaffin-like cells following water immersion stress. Gastroenterology 111:45–55.

Ashcroft, F. J., Varro, A., Dimaline, R., Dockray, G. J. 2004. Control of expression of the lectin-like protein Reg-1 by gastrin: role of the Rho family GTPase RhoA and a C-rich promoter element. Biochem J 381:397–403.

Becker, J. C., Muller-Tidow, C., Serve, H., Domschke, W., Pohle, T. 2006. Role of receptor tyrosine kinases in gastric cancer: new targets for a selective therapy. World J Gastroenterol 12:3297–305.

Bishnupuri, K. S., Luo, Q., Murmu, N., Houchen, C. W., Anant, S., Dieckgraefe, B. K. 2006. Reg IV activates the epidermal growth factor receptor/Akt/AP-1 signaling pathway in colon adenocarcinomas. Gastroenterology 130:137–149.

Bordi, C., Cocconi, G., Togni, R., Vezzadini, P., Missale, G. 1974. Gastric endocrine cell proliferation: association with Zollinger-Ellison syndrome. Arch Pathol 98:274–8.

Bromberg J. 2002. Stat proteins and oncogenesis. J Clin Invest 109:1139–42.

Busiello, I., Acquaviva, R., Di Popolo, A., Blanchard, T. G., Ricci, V., Romano, M., Zarrilli, R. 2004. Helicobacter pylori gamma-glutamyltranspeptidase upregulates COX-2 and EGF-related peptide expression in human gastric cells. Cell Microbiol 6:255–67.

Calam J. 1996. Helicobacter pylori and hormones. Yale J Biol Med 69:39–49.

Caputo, R., Tuccillo, C., Manzo, B. A., Zarrilli, R., Tortora, G., Blanco Cdel. V., Ricci, V., Ciardiello, F., Romano, M. 2003. Helicobacter pylori VacA toxin up-regulates vascular endothelial growth factor expression in MKN 28 gastric cells through an epidermal growth factor receptor-, cyclooxygenase-2-dependent mechanism. Clin Cancer Res 9:2015–21.

Christa, L., Christa, L., Carnot, F., Simon, M. T., Levavasseur, F., Stinnakre, M. G., Lasserre, C., Thepot, D., Clement, B., Devinoy, E., Brechot, C. 1996. HIP/PAP is an adhesive protein expressed in hepatocarcinoma, normal Paneth, and pancreatic cells. Am J Physiol 271: G993–1002.

Chung, C. K., Antoniades, H. N. 1992. Expression of c-sis/platelet-derived growth factor B, insulin-like growth factor I, and transforming growth factor alpha messenger RNAs and their respective receptor messenger RNAs in primary human gastric carcinomas. Cancer Res 52:3453–9.

Dieckgraefe, B. K., Stenson, W. F., Korzenik, J. R., Swanson, P. E., Harrington, C. A. 2000. Analysis of mucosal gene expression in inflammatory bowel disease by parallel oligonucleotide arrays. Physiol Genomics 4:1–11.

Dragovich, T., McCoy, S., Fenoglio-Preiser, C. M., Wang, J., Benedetti, J. K., Baker, A. F., Hackett, C. B., Urba, S. G., Zaner, K. S., Blanke, C. D., Abbruzzese, J. L. 2006. Phase II trial of erlotinib in gastroesophageal junction and gastric adenocarcinomas: SWOG 0127. J Clin Oncol 24:4922–7.

Drickamer, K. 1988. Two distinct classes of carbohydrate-recognition domains in animal lectins. J Biol Chem 263:9557–60.

Dusetti, N. J., Mallo, G. V., Ortiz, E. M., Keim, V., Dagorn, J. C., Iovanna, J. L. 1996. Induction of lithostathine/reg mRNA expression by serum from rats with acute pancreatitis and cytokines in pancreatic acinar AR-42J cells. Arch Biochem Biophys 330:129–32.

Fukui, H., Franceschi, F., Penland, R. L., Sakai, T., Sepulveda, A. R., Fujimori, T., Terano, A., Chiba, T., Genta, R. M. 2003. Effects of Helicobacter pylori infection on the link between regenerating gene expression and serum gastrin levels in Mongolian gerbils. Lab Invest 83:1777–86.

Fukui, H., Fujii, S., Takeda, J., Kayahara, T., Sekikawa, A., Nanakin, A., Suzuki, K., Hisatsune, H., Seno, H., Sawada, M., Fujimori, T., Chiba, T. 2004. Expression of Reg Iα protein in human gastric cancers. Digestion 69:177–84.

Fukui, H., Kinoshita, Y., Maekawa, T., Okada, A., Waki, S., Hassan, M. S., Okamoto, H., Chiba, T. 1998. Regenerating gene protein may mediate gastric mucosal proliferation induced by hypergastrinemia in rats. Gastroenterology 115:1483–93.

Hartupee, J. C., Zhang, H., Bonaldo, M. F., Soares, M. B., Dieckgraefe, B. K. 2001. Isolation and characterization of a cDNA encoding a novel member of the human regenerating protein family: Reg IV. Biochim Biophys Acta 1518:287–93.

Higham, A. D., Bishop, L. A., Dimaline, R., Blackmore, C. G., Dobbins, A.C., Varro, A., Thompson, D. G., Dockray, G. J. 1999. Mutations of RegIa are associated with enterochromaffin-like cell tumor development in patients with hypergastrinemia. Gastroenterology 116:1310–8.

Hirakawa, S., Kodama, S., Kunstfeld, R., Kajiya, K., Brown, L. F., Detmar, M. 2005. VEGF-A induces tumor and sentinel lymph node lymphangiogenesis and promotes lymphatic metastasis. J Exp Med 201:1089–99.

Hull, M. A., Brough, J. L., Powe, D. G., Carter, G. I., Jenkins, D., Hawkey, C. J. 1999. Expression of cyclooxygenase 1 and 2 by human gastric endothelial cells. Gut 45:529–36.

Inoue, T., Kataoka, H., Goto, K., Nagaike, K., Igami, K., Naka, D., Kitamura, N., Miyazawa, K. 2004. Activation of c-Met (hepatocyte growth factor receptor) in human gastric cancer tissue. Cancer Sci 95:803–8.

Itoh, T., Tsuzuki, H., Katoh, T., Teraoka, H., Matsumoto, K., Yoshida, N., Terazono, K., Watanabe, T., Yonekura, H., Yamamoto, H., Okamoto, H. 1990. Isolation and characterization of human reg protein produced in Saccharomyces cerevisiae. FEBS Lett 272:85–8.

Judd, L. M., Alderman, B. M., Howlett, M., Shulkes, A., Dow, C., Moverley, J., Grail, D., Jenkins, B. J., Ernst, M., Giraud, A. S. 2004. Gastric cancer development in mice lacking the SHP2 binding site on the IL-6 family co-receptor gp130. Gastroenterology 126:196–207.

Judd, L. M., Bredin, K., Kalantzis, A., Jenkins, B. J., Ernst, M., Giraud, A. S. 2006. STAT3 activation regulates growth, inflammation, and vascularization in a mouse model of gastric tumorigenesis. Gastroenterology 131:1073–85.

Kämäräinen, M., Heiskala, K., Knuutila, S., Heiskala, M., Winqvist, O., Andersson, L. C. 2003. RELP, a novel human Reg-like protein with up-regulated expression in inflammatory and metaplastic gastrointestinal mucosa. Am J Pathol 163:11–20.

Katoh, M., Katoh, M. 2006. FGF signaling network in the gastrointestinal tract. Int J Oncol 29:163–8.

Kawanami, C., Fukui, H., Kinoshita, Y., Nakata, H., Asahara, M., Matsushima, Y., Kishi, K., Chiba, T. 1997. Regenerating gene expression in normal gastric mucosa and indomethacin-induced mucosal lesions of the rat. J Gastroenterol 32:12–8.

Kazumori, H., Ishihara, S., Fukuda, R., Kinoshita, Y. 2002. Localization of Reg receptor in rat fundic mucosa. J Lab Clin Med 139:101–8.

Kazumori, H., Ishihara, S., Hoshino, E., Kawashima, K., Moriyama, N., Suetsugu, H., Sato, H., Adachi, K., Fukuda, R., Watanabe, M., Takasawa, S., Okamoto, H., Fukui, H., Chiba, T., Kinoshita, Y. 2000. Neutrophil chemoattractant 2β regulates expression of the Reg Gene in injured gastric mucosa in rats. Gastroenterology 119:1610–22.

Keates, S., Sougioultzis, S., Keates, A. C., Zhao, D., Peek, R. M., Jr., Shaw, L. M., Kelly, C. P. 2001. cag+ Helicobacter pylori induce transactivation of the epidermal growth factor receptor in AGS gastric epithelial cells. J Biol Chem 276:48127–34.

Kim, K. R., Oh, S. Y., Park, U.C., Wang, J. H., Lee, J. D., Kweon, H. J., Kim, S. Y., Park, S. H., Choi, D. K., Kim, C. G., Choi, S. H. 2007. Gene expression profiling using oligonucleotide microarray in atrophic gastritis and intestinal metaplasia. Korean J Gastroenterol 49:209–24.

Kinoshita, Y., Ishihara, S., Kadowaki, Y., Fukui, H., Chiba, T. 2004. Peg protein is a unique growth factor of gastric mucosal cells. J Gastroenterol Hepatol 39:507–13.

Kinoshita, Y., Kishi, K., Asahara, M., Matasushima, Y., Wang, H. Y., Miyazawa, K., Kitamura, N., Chiba, T. 1997. Production and activation of hepatocyte growth factor during the healing of rat gastric ulcers. Digestion 58:225–31.

Kinoshita, Y., Nakata, H., Hassan, S., Asahara, M., Kawanami, C., Matsushima, Y., Naribayashi-Inomoto, Y., Ping, C. Y., Min, D., Nakamura, A., Chiba, T. 1995. Gene expression of kerati-nocyte and hepatocyte growth factors during the healing of rat gastric mucosal lesions. Gastroenterology 109:1068–77.

Kitadai, Y., Sasaki, A., Ito, M., Tanaka, S., Oue, N., Yasui, W., Aihara, M., Imagawa, K., Haruma, K., Chayama, K. 2003. Helicobacter pylori infection influences expression of genes related to angiogenesis and invasion in human gastric carcinoma cells. Biochem Biophys Res Commun 311:809–14.

Kobayashi, S., Akiyama, T., Nata, K., Abe, M., Tajima, M., Shervani, N. J., Unno, M., Matsuno, S., Sasaki, H., Takasawa, S., Okamoto, H. 2000. Identification of a receptor for reg (regenerat-ing gene) protein, a pancreatic beta-cell regeneration factor. J Biol Chem 275:10723–6.

Larsson, H., Carlsson, E., Mattsson, H., Lundell, L., Sundler, F., Sundell, G., Wallmark, B., Watanabe, T., Håkanson, R. 1986. Plasma gastrin and gastric enterochromaffinlike cell activa-tion and proliferation. Studies with omeprazole and ranitidine in intact and antrectomized rats. Gastroenterology 90:391–9.

Lasserre, C., Colnot, C., Bréchot, C., Poirier, F. 1999. HIP/PAP gene, encoding a C-type lectin overexpressed in primary liver cancer, is expressed in nervous system as well as in intestine and pancreas of the postimplantation mouse embryo. Am J Pathol 154:1601–10.

Lawrance, I. C., Fiocchi, C., Chakravarti, S. 2001. Ulcerative colitis and Crohn's disease: distinc-tive gene expression profiles and novel susceptibility candidate genes. Hum Mol Genet 10:445–56.

Levy, D. E., Darnell, J. E. 2002. STATs: transcriptional control and biological impact. Nat Rev Mol Cell Biol 3:651–62.

Maeda, K., Chung, Y. S., Ogawa, Y., Takatsuka, S., Kang, S. M., Ogawa, M., Sawada, T., Sowa, M. 1996. Prognostic value of vascular endothelial growth factor expression in gastric carci-noma. Cancer 77:858–63.

Milani, S., Calabro, A. 2001. Role of growth factors and their receptors in gastric ulcer healing. Microsc Res Tech 53:360–71.

Mitani, Y., Oue, N., Matsumura, .S, Yoshida, K., Noguchi, T., Ito, M., Tanaka, S., Kuniyasu, H., Kamata, N., Yasui, W. 2007. Reg IV is a serum biomarker for gastric cancer patients and predicts response to 5-fluorouracil-based chemotherapy. Oncogene 26:4383–93.

Miyaoka, Y., Kadowaki, Y., Ishihara, S., Ose, T., Fukuhara, H., Kazumori, H., Takasawa, S., Okamoto, H., Chiba, T., Kinoshita, Y. 2004. Transgenic overexpression of Reg protein caused gastric cell proliferation and differentiation along parietal cell and chief cell lineages. Oncogene 23:3572–9.

Miyashita, H., Nakagawara, K., Mori, M., Narushima, Y., Noguchi, N., Morizmi, S., Takasawa, S., Yonekura, H., Takeuchi, T., Okamoto, H. 1995. Human REG family genes are tandemly ordered in a 95-kilobase region of chromosome 2p12. FEBS Lett 377:429–33.

Nakajima, T., Konda, Y., Izumi, Y., Kanai, M., Hayashi, N., Chiba, T., Takeuchi, T. 2002. Gastrin stimulates the growth of gastric pit cell precursors by inducing its own receptors. Am J Physiol 282:G359–66.

Nakazawa, K., Yashiro, M., Hirakawa, K. 2003. Keratinocyte growth factor produced by gastric fibroblasts specifically stimulates proliferation of cancer cells from scirrhous gastric carcinoma. Cancer Res 63:8848–52.

Nanakin, A., Fukui, H., Fujii, S., Sekikawa, A., Kanda, N., Hisatsune, H., Seno, H., Konda, Y., Fujimori, T., Chiba, T. 2007. Expression of the REG IV gene in ulcerative colitis. Lab Invest 87:304–14.

Narushima, Y., Unno, M., Nakagawara, K., Mori, M., Miyashita, H., Suzuki, Y., Noguchi, N., Takasawa, S., Kumagai, T., Yonekura, H., Okamoto, H. 1997. Structure chromosomal localization and expression of mouse genes encoding type III Reg, Reg IIIα, Reg IIIβ, RegIIIγ. Gene 185:159–168.

Nishimune, H., Vasseur, S., Wiese, S., Birling, M. C., Holtmann, B., Sendtner, M., Iovanna, J. L., Henderson, C. E. 2000. Reg-2 is a motoneuron neurotrophic factor and a signalling intermediate in the CNTF survival pathway. Nat Cell Biol 2:906–14.

Okamoto, H. 1999. The Reg gene family and Reg proteins: with special attention to the regeneration of pancreatic b-cells. J Hepatobiliary Pancreat Surg 6:254–62.

Ose, T., Kadowaki, Y., Fukuhara, H., Kazumori, H., Ishihara, S., Udagawa, J., Otani, H., Takasawa, S., Okamoto, H., Kinoshita, Y. 2007. Reg I-knockout mice reveal its role in regulation of cell growth that is required in generation and maintenance of the villous structure of small intestine. Oncogene 26:349–59.

Oue, N., Hamai, Y., Mitani, Y., Matsumura, S., Oshimo, Y., Aung, P. P., Kuraoka, K., Nakayama, H., Yasui, W. 2004. Gene expression profile of gastric carcinoma: identification of genes and tags potentially involved in invasion, metastasis, and carcinogenesis by serial analysis of gene expression. Cancer Res 64:2397–405.

Oue, N., Mitani, Y., Aung, P. P., Sakakura, C., Takeshima, Y., Kaneko, M., Noguchi, T., Nakayama, H., Yasui, W. 2005. Expression and localization of Reg IV in human neoplastic and non-neoplastic tissues: Reg IV expression is associated with intestinal and neuroendocrine differentiation in gastric adenocarcinoma. J Pathol 207:185–98.

Rojo, F., Tabernero, J., Albanell, J., Van Cutsem, E., Ohtsu, A., Doi, T., Koizumi, W., Shirao, K., Takiuchi, H., Ramon, Y., Cajal, S., Baselga, J. 2006. Pharmacodynamic studies of gefitinib in tumor biopsy specimens from patients with advanced gastric carcinoma. J Clin Oncol 24:4309–16.

Romano, M., Ricci, V., Di Popolo, A., Sommi, P., Del Vecchio Blanco, C., Bruni, C. B., Ventura, U., Cover, T. L., Blaser, M. J., Coffey, R. J., Zarrilli, R. 1998. Helicobacter pylori upregulates expression of epidermal growth factor-related peptides, but inhibits their proliferative effect in MKN 28 gastric mucosal cells. J Clin Invest 101:1604–13.

Satomura, Y., Sawabu, N., Ohta, H., Watanabe, H., Yamakawa, O., Motoo, Y., Okai, T., Toya, D., Makino, H., Okamoto, H. 1993. The immunohistochemical evaluation of PSP/reg-protein in normal and diseased human pancreatic tissues. Int J Pancreatol 13:59–67.

Schmassmann, A., Stettler, C., Poulsom, R., Tarasova, N., Hirschi, C., Flogerzi, B., Matsumoto, K., Nakamura, T., Halter, F. 1997. Roles of hepatocyte growth factor and its receptor Met during gastric ulcer healing in rats. Gastroenterology 113:1858–72.

Sekikawa, A., Fukui, H., Fujii, S., Nanakin, A., Kanda, N., Uenoyama, Y., Sawabu, T., Hisatsune, H., Kusaka, T., Ueno, S., Nakase, H., Seno, H., Fujimori, T., Chiba, T. 2005a. Possible role of REG Iα protein in ulcerative colitis and colitic cancer. Gut 54:1437–44.

Sekikawa, A., Fukui, H., Fujii, S., Takeda, J., Nanakin, A., Hisatsune, H., Seno, H., Takasawa, S., Okamoto, H., Fujimori, T., Chiba, T. 2005b. REG Iα protein may function as a trophic and/or anti-apoptotic factor in the development of gastric cancer. Gastroenterology 128:642–53.

Sharp, R., Babyatsky, M. W., Takagi, H., Tågerud, S., Wang, T. C., Bockman, D. E., Brand, S. J., Merlino, G. 1995. Transforming growth factor alpha disrupts the normal program of cellular differentiation in the gastric mucosa of transgenic mice. Development 121:149–61.

Shimizu, K., Kubo, H., Yamaguchi, K., Kawashima, K., Ueda, Y., Matsuo, K., Awane, M., Shimahara, Y., Takabayashi, A., Yamaoka, Y., Satoh, S. 2004. Suppression of VEGFR-3 signaling inhibits lymph node metastasis in gastric cancer. Cancer Sci 95:328–33.

Skeel, A., Yoshimura, T., Showalter, S. D., Tanaka, S., Appella, E., Leonard, E. J. 1991. Macrophage stimulating protein: purification, partial amino acid sequence, and cellular activity. J Exp Med 173:1227–34.

Sonnenberg, E., Meyer, D., Weidner, K. M., Birchmeier, C. 1993. Scatter factor/hepatocyte growth factor and its receptor, the c-met tyrosine kinase, can mediate a signal exchange between mesenchyme and epithelia during mouse development. J Cell Biol 123:223–35.

Starr, R., Willson, T. A., Viney, E. M., Murray, L. J. L., Rayner, J. R., Jenkins, B. J., Gonda, T. J., Alexander, W. S., Metcalf, D., Nicola, N. S., Hilton, D. J. 1997. A family of cytokine-inducible inhibitors of signaling. Nature 387:917–21.

Takaishi, S., Wang, T. C. 2007. Gene expression profiling in a mouse model of Helicobacter-induced gastric cancer. Cancer Sci 98:284–93.

Takasawa, S., Ikeda, T., Akiyama, T., Nata, K., Nakagawa, K., Shervani, N. J., Noguchi, N., Murakami-Kawaguchi, S., Yamauchi, A., Takahashi, I., Tomioka-Kumagai, T., Okamoto, H. 2006. Cyclin D1 activation through ATF-2 in Reg-induced pancreatic beta-cell regeneration. FEBS Lett 580:585–91.

Takehara, A., Takehara, A., Eguchi, H., Ohigashi, H., Ishikawa, O., Kasugai, T., Hosokawa, M., Katagiri, T., Nakamura, Y., Nakagawa, H. 2006. Novel tumor marker REG4 detected in serum of patients with resectable pancreatic cancer and feasibility for antibody therapy targeting REG4. Cancer Sci 97:1191–7.

Terazono, K., Yamamoto, H., Takasawa, S., Shiga, K., Yonemura, Y., Tochino, Y., Okamoto, H. 1988. A novel gene activated in regenerating islets. J Biol Chem 263:2111–4.

Unno, M., Yonekura, H., Nakagawara, K., Watanabe, T., Miyashita, H., Morizmi, S., Okamoto, H., Itoh, T., Teraoka, H. 1993. Structure, chromosomal localization, and expression of mouse reg genes, reg I and reg II. A novel type of Reg gene, Reg II, exixts in the mouse genome. J Biol Chem 268:15974–82.

Violette, S., Festor, E., Pandrea-Vasile, I., Mitchell, V., Adida, C., Dussaulx, E., Lacorte, J. M., Chambaz, J., Lacasa, M., Lesuffleur, T. 2003. Reg IV, a new member of the regenerating gene family, is overexpressed in colorectal carcinomas. Int J Cancer 103:185–93.

Watanabe, T., Yonekura, H., Terazono, K., Yamamoto, H., Okamoto, H. 1990. Complete nucleotide sequence of human reg gene and its expression in normal and tumoral tissues: The Reg protein, pancreatic stone protein, and pancreatic thread protein are one and the same product of the gene. J Biol Chem 265:7432–9.

Watanabe, T., Yonemura, Y., Yonekura, H., Suzuki, Y., Miyashita, H., Sugiyama, K., Morizumi, S., Unno, M., Tanaka, O., Kondo, H., Bone, A. J., Takasawa, S., Okamoto, H. 1994. Pancreatic beta-cell replication and amelioration of surgical diabetes by Reg production. Proc Natl Acad Sci USA 91:3589–92.

Yokozaki, H. 1997. Molecular bases of human stomach carcinogenesis. In: Tahara, E., editor. Molecular pathology of gastroenterological cancer. Tokyo: Springer-Verlag; pp 60–3.

Yonemura, Y., Ohoyama, S., Kimura, H., Kamata, T., Matsumoto, H., Yamaguchi, A., Kosaka, T., Miwa, K., Miyazaki, I. 1991. The expression of proliferative-associated nuclear antigen p105 in gastric carcinoma. Cancer 67:2523–8.

Yonemura, Y., Sakurai, S., Yamamoto, H., Endou, Y., Kawamura, T., Bandou, E., Elnemr, A., Sugiyama, K., Sasaki, T., Akiyama, T., Takasawa, S., Okamoto, H. 2003. REG gene expression is associated with the infiltrating growth of gastric carcinoma. Cancer 98:1394–400.

Yonemura, Y., Takashima, T., Miwa, K., Miyazaki, I., Yamamoto, H., Okamoto H. 1984. Amelioration of diabetes mellitus in partially depancreatised rats by poly(ADP-ribose) synthetase inhibitors. Diabetes 33:401–4.

Yoshida, K., Tsuda, T., Matsumura, T., Tsujino, T., Hattori, T., Ito, H., Tahara, E. 1989. Amplification of epidermal growth factor receptor (EGFR) gene and oncogenes in human gastric carcinoma. Virchow Arch B Cell Pathol 57:285–90.

Yoshino, N., Ishihara, S., Rumi, M. A., Ortega-Cava, C. F., Yuki, T., Kazumori, H., Takazawa, S., Okamoto, H., Kadowaki, Y., Kinoshita, Y. 2005. Interleukin-8 regulates expression of Reg protein in Helicobacter pylori-infected gastric mucosa. Am J Gastroenterol 100:2157–66.

Zhao, Z. S., Zhou, J. L., Yao, G. Y., Ru, G. Q., Ma, J., Ruan, J. 2005. Correlative studies on bFGF mRNA and MMP-9 mRNA expressions with microvascular density, progression, and prognosis of gastric carcinomas. World J Gastroenterol 11:3227–33.

Chapter 22
Role of Bone Marrow–Derived Cells in Gastric Adenocarcinoma

JeanMarie Houghton and Timothy C. Wang

Cancer Stem Cell Hypothesis

The origin of gastric cancer has remained mysterious for many years. Waldeyer originally developed the idea that gastric cancer arose from gastric epithelial cells, and believed that almost any gastric epithelial cell could be converted to a cancer cell (Houghton et al. 2007). Our understanding of the origins of gastric cancer has undergone a considerable change in recent years, partly based on the evolving paradigm of cancer stem cells (CSCs). Although for many years physicians and scientists regarded tumors as generally homogeneous tissues, accumulating evidence supports the view that cancer is sustained by a subset of cells—cancer stem cells—that have the exclusive ability to renew and propagate a tumor.

The origin of the CSC hypothesis dates back to the discovery of the light microscope in the nineteenth century. During that time, it was recognized that tumors were composed of a heterogeneous mixture of cell types with varying levels of differentiation, similar in many respects to a normal organ. Consequently, Julius Cohnheim in 1867 proposed that tumors were derived not from normal adult tissues but from "embryonal cell rests" which represented residual embryonic cells that were left behind during the development of the adult organ (Rather 1978). Nevertheless, during most of the twentieth century, it was believed that most cells in a tumor should be competent for tumor formation. Cohnheim's theory was largely ignored for 100 years until the 1970s when additional experimental evidence came to light. Studies in leukemia and multiple myeloma showed that only a small subset of cells was capable of extensive proliferation and could give rise to colony-forming units *in vitro* or splenic colonies in mice (Houghton et al. 2007). Similarly, work in solid tumors also showed that fewer than 1 in 1,000 cells were clonogenic *in vitro* or *in vivo*. Based on his studies of mouse teratocarcinoma, Pierce (1974) speculated that tumors contain a very small number of malignant cells that sustain the tumor and gave rise to daughter cells with varying degrees of differentiation and function.

The actual existence of CSCs was first directly demonstrated by John Dick in 1994, when his group proved the hypothesis to be largely true for acute myelogenous leukemia (AML) (Lapidot et al. 1994). Only a small subset of human leukemic

AML cells were capable of reproducing the disease in NOD/SCID mice, and these cells were quite similar, if not identical, to hematopoietic stem cells (HSCs) with the CD34+CD38– phenotype. These AML CSCs could be passed from animal to animal consistent with the property of self-renewal. This approach has been used to demonstrate the existence of CSCs in a number of other tumors, and the ability of a subset of cells, enriched by sorting for specific cell-surface markers, to reca-pitulate tumors in NOD/SCID mice through serial transplantation. This has become the "gold standard" for defining a CSC. For example, breast CSCs were shown to be CD44+CD24low/– (Al-Hajj et al. 2003), whereas brain cancer (glioblastoma) stem cells were enriched by sorting for CD133+ cells (Singh et al. 2004). CSCs have also been demonstrated for prostate cancer (Collins et al. 2005), melanoma (Fang et al. 2005), head and neck cancer (Prince et al. 2007), colon cancer (Dalerba et al. 2007; O'Brien et al. 2007; Ricci-Vitiani et al. 2007), hepatocellular cancer (Ma 2007), and pancreatic cancer (Herman et al. 2007; Li et al. 2007). No CSCs have been demonstrated for gastric cancer. In these studies, the population of cells enriched for the CSC is shown to have better growth in soft agar and in immuno-deficient mice compared with the non-CSC population. In some of these studies, as few as 100 cells can give rise to tumors in immunodeficient mice; in other studies, much larger numbers of cells are required. These data suggest that the putative stem cell markers used to date enrich for a CSC population, but we do not yet have a surface-marker signature that will allow isolation of the CSC at the single-cell level. Based on these data, an American Association for Cancer Research workshop agreed on the definition of cancer stem cells as "cells within a tumor that possess the capacity for self-renewal and that can cause the heterogeneous lineages of can-cer cells that constitute the tumor" (Clarke et al. 2006).

The majority of data examining human tumor cells transplanted back into the original host in an immunodeficient mouse model (Houghton et al. 2007) or studies examining mouse tumor cells in a mouse model demonstrate the rare nature of CSCs in solid and blood-borne tumors. However, recent data challenging the notion of the CSC as a rare cell type suggest that in some models (the authors' report on preB/B lymphoma cells from three independent Eµ-myc transgenic mice) the majority of cells within a tumor may retain the CSC phenotype (Kelly et al. 2007).

Over the past decade, the "cancer stem cell" hypothesis has achieved support and increased acceptance (Jordan et al. 2006), although the number of cells within a tumor possessing this phenotype is debated. The data supporting the hypothesis are incomplete, in that many of the markers used to purify the CSCs (e.g., CD44, CD133, side populations, BrdU retention) are not highly specific and thus CSCs have mostly been enriched and not completely purified. The assays used to define CSCs may simply test for the ability of cells to grow under specific conditions (e.g., immunodeficient mice), and the demonstration that ablation of a specific cellular subset prevents tumor survival has for the most part not been achieved. Despite these limitations, a growing consensus suggests that cancer can now be viewed much like differentiated tissues and organs, a heterogeneous collection of cells sustained by a small number of progenitors that may represent 1 in 10 to 1 in 1,000,000 cells, depending on the model system and tumor type analyzed. In the

CSC model, it is thought that only these CSCs are able to propagate the tumors and give rise to invasive lesions and metastases. The model has tremendous implications for cancer therapy because currently most of our therapies are successful in eradicating non-CSCs but not the CSCs, and consequently most solid tumors that shrink in response to therapies recur within a few months to a few years (Reya et al. 2001).

Gastric Cancer Arises from Gastritis Through a Series of Pathologic Stages

If we accept that gastric cancer, similar to most other malignancies, is sustained by CSCs, then it is important to begin to determine how and where these CSCs arise. Indeed, as has been pointed out in other reviews, the CSCs that sustain advanced tumors may not be precisely the same as the cancer cells that initiate the tumor. Gastric cancer is strongly linked to *Helicobacter pylori* infection and chronic inflammation. The association between inflammation and cancer has long been recognized but poorly understood, and credit for the association is generally given to Virchow, who in the nineteenth century first reported the frequent presence of leukocytes in neoplastic tissues (Balkwill and Mantovani 2001). Chronic inflammation of the antrum and body of the stomach, also known as chronic type B gastritis, represents the first step in the Correa cascade (see Chapter 1) leading to intestinal-type gastric cancer. Most gastric adenocarcinomas progress from chronic gastritis to atrophic gastritis, intestinal metaplasia, dysplasia, and then to gastric cancer, and this entire process is dependent on continued chronic inflammation. In addition to classical intestinal metaplasia, numerous studies have also pointed to pseudopyloric metaplasia or SPEM, a lesion positive for TFF2 expression, as a precursor of gastric cancer (see Chapter 13). The preneoplastic lesions that precede the development of diffuse gastric cancer are less well understood. Metaplasia is a particularly interesting entity, because it is a fairly permanent alteration that suggests a marked genetic and epigenetic program of the local stem cell.

In the classical model, inflammation is thought to have a role primarily in tumor promotion, acting on tumor cells that have already been initiated to promote their growth and further neoplastic progression. Today, more information is known regarding the induction of chronic inflammation by *H. pylori* (see Chapter 17). Inflammation leads to both increased oxidative stress—caused by the release of oxygen radicals and tissue hypoxia—and increased tissue turnover caused by repeated cycles of tissue destruction and regeneration. The latter is thought to result in an increased rate of proliferation, which might lead to more frequent mitotic error and an increased rate of mutagenesis, although this relationship has not been formally established. Although oxidative stress is certainly of some importance in carcinogenesis, work in recent years has highlighted important roles for specific immune subsets (e.g., macrophages and T cells) and proinflammatory cytokines. It has become clear that the immune system can enhance tumor progression and metastatic spread, possibly contributing to cancer initiation. However, given that cancer arises from CSCs, the mechanism by which chronic inflammation initiates or promotes these stem cells needs to be addressed.

One area of increasing interest is the potential effect of chronic inflammation on the stem cell niche. Recent work has demonstrated that the behavior and lineage-specific differentiation patterns of stem cells are governed primarily by the local microenvironment, also known as the stem cell niche, a concept first proposed by Schofield (1978). This niche consists of a variety of cell types, including stromal cells (fibroblasts and myofibroblasts), endothelial cells, macrophages, and many others. Stromal cells are believed to generate instructive regulatory signals constituting a microenvironment that governs stem cell behavior. Studies in *Drosophila*, particularly with *Drosophila* germline cells, have shown the importance of the niche signals in lineage determination. Indeed, these studies indicate that stem cells may be somewhat interchangeable, and germline stem cells that are lost can be efficiently replaced by secondary cell types, with the niche (e.g., cap cells) possessing the crucial role in the ultimate behavior of the stem cell that occupies the niche (Xie and Spradling 2000). This concept of the empty niche, and the cell types that can replace resident tissue stem cells and occupy the niche, may be fundamental to understanding the pathogenesis of CSCs. Changes to the stem cell niche that occur in the setting of inflammation in the gastrointestinal (GI) tract (Powell et al. 2005; Pull et al. 2005) may also lead to changes in stem cell behavior and are covered in more detail in Chapter 23.

The Gastric Stem Cell as a Potential Origin of Gastric Cancer Stem Cells

Tissue stem cells have in recent years been viewed as the best candidate for tumorigenic progenitors or CSCs. If a tumor can be regarded as an aberrant organ, then the simplest model might be one in which the tumor arises from a conversion of the resident tissue stem cell. Abundant research has implicated the existence of resident tissue stem cells in many tissues, particularly the skin and the gut. These tissue stem cells are thought to be slowly cycling cells that undergo asymmetric divisions giving rise to a daughter stem cell to sustain the stem cell pool, and to a more rapidly dividing cell (transit amplifying progenitor) that then gives rise to terminally differentiated cells to populate epithelial tissues. Both normal resident tissue stem cells and CSCs are long-lived and have extensive proliferate potential and self-renewal abilities, and can give rise to heterogeneous cell types. In addition to tissue stem cells, another source of CSCs seems to be transit-amplifying progenitor cells that acquire self-renewing capability through genetic mutation. Progenitor cells are derived from stem cells, have extensive proliferative ability, but more restricted self-renewal capacity and often a more limited range of differentiation.

The mouse stomach is lined by an epithelium that is constantly being renewed and is likely sustained by a population of multipotential stem cells. The stomach can be divided into two types of epithelium: the oxyntic mucosa located in the body and corpus that contains parietal and zymogenic (chief) cells (zymogenic units), and the antral units, which contain mucous cells, endocrine (e.g., gastrin) cells, and only rarely parietal cells. In each case, these gastric units are composed of a planar surface, tubular invaginations (called pits), and tubular extensions (called glands). In the mouse, zymogenic units are composed of approximately 200 cells. The stem/progenitor cells reside at a midpoint location and give rise, via bidirectional migration patterns, to three predominant lineages that include pit, parietal, and zymogenic cells.

Through a number of different lines of investigation, the multipotent stem cell has been localized to a region in the gastric unit that lies between the gastric pits and glands, just above the neck region in an area known as the gastric isthmus. The first characterization of these stem cells was performed by Karam and LeBlond (1992, 1993) using tritiated thymidine labeling combined with electron microscopy. On the basis of these studies, an undifferentiated granule-free cell was identified that appeared to have the most primitive phenotype and share some morphologic features of proliferating cells located in the stem cell zone of the small intestine, and seemed to give rise to other types of progenitor cells:pre-pit cell precursor, pre-neck cell precursor, and pre-parietal cells. Of note, however, was the very high rate of labeling (e.g., 30%) 30 minutes after a 3H-thymidine injection, suggesting more rapid cycling than one would expect of a true stem cell and suggesting that these cells may actually be a TA daughter cell, likely adjacent to the true stem cell.

Support for the existence of multipotent stem cells in the adult mouse gastric epithelium was provided by studies with chimeric animals by Bjerknes and Cheng (2002). These investigators used chemical mutagenesis to label random epithelial cells by loss of transgene function in adult hemizygous ROSA26 mice, a mouse strain that expresses the *lacZ* transgene (and β-galactosidase) in all tissues. Their work clearly demonstrated that many glands showed loss of transgene function in all major epithelial cell types, consistent with clonal inheritance of a single mutation, thereby demonstrating the existence of multipotent stem cells. Additional glands showed loss of transgene function in a single lineage, indicating the presence of long-lived progenitors in the stomach.

Gastric epithelial cell progenitors have been characterized in a molecular manner through analysis of an adult transgenic mouse with genetically engineered, mutant diphtheria toxin A fragment (tox176)-mediated ablation of parietal cells. In contrast to the wild-type adult mouse, in which progenitors make up <3% of the total epithelial cell population, tox176 mice have >20% progenitor cells in their gastric epithelium. By comparing the gene expression profile of the gastric mucosa from WT mice to that from tox176 mice and embryonic day 18 mice (which show >80 progenitors), Mills et al. (2002) were able to perform a "dissection-free" analysis of transcripts that were enriched in gastric epithelial progenitors (GEPs). These studies identified 147 transcripts enriched in GEPs, and showed a prominent enrichment of growth factor response pathways (e.g., insulinlike growth factor), and products required for mRNA processing and cytoplasmic localization (Mills et al. 2002).

This molecular characterization was extended further through the sequencing of cDNA libraries prepared from laser capture microdissected adult GEPs, taking advantage of the gnotobiotic tox176 transgenic mouse model in which these cells are represented in increased numbers (Giannakis et al. 2006). Four thousand thirty-one unique transcripts were obtained and compared using Gene Ontology (GO) terms with cDNA libraries from intact adult stomach. Wnt/β-catenin, G-protein-coupled receptor signaling, tyrosine metabolism, serotonin receptor signaling, +-aminobutyric acid receptor signaling, and amino acid metabolic pathways were prominently represented (Giannakis et al. 2006). Of particular interest, however, was the finding that antibodies to Dcamkl1, a microtubule-associated kinase that is the product of a GO-enriched transcript identified in the comparison, mark single cells in the isthmal stem cell niche of gastric units. These solitary Dcamkl1-positive cells did not express any biomarkers associated with differentiated cell types, and are adjacent to rapidly cycling BrdU-positive progenitors but are not labeled with BrdU after a 1.5-hour exposure. Consequently, Dcamkl1 has been proposed as a candidate marker of adult gastric stem cells (Giannakis et al. 2006). In addition, gastric stem cells are marked by a 12.4-Kb villin promoter/enhancer fragment. Using the villin promoter upstream of *lacZ*, EGFP and Cre, rare cells in the gastric glands, are labeled by the transgene and shown to give rise to all gastric lineages in the gland (Qiao et al. 2007).

Thus, although resident gastric stem cells, or early gastric progenitors, remain good candidates for gastric CSCs, other candidates for CSCs have recently come under consideration. These other candidates—circulating progenitor cells—were identified during the course of studies focused on mouse models of *Helicobacter*-dependent gastric cancer.

Different Types of Adult Stem Cells

Overview

The previous sections discussed the concept of CSCs, and touched on their potential relationship with stem cells within the GI tract. To fully explore the role of stem cells in cancer, we will first need to compare and contrast the different types of stem cells recognized. Stem cells, regardless of their source, are defined as cells with the following properties: self-renewal, primitive or nondifferentiated phenotype, and the ability to give rise to specialized cells. There are many recognized types of stem cells with varying degrees of differentiation potential. The most primitive is the embryonic stem cell, derived from early preimplantation embryos. These cells give rise to all of the cell types of the body, and as such are the most versatile stem cells recognized, carrying the greatest differentiation potential. Adult stem cells, also capable of self renewal, are derived from tissues of the adult animal and are responsible for tissue maintenance and repair.

Adult Stem Cells

Research on adult stem cells began in earnest in the 1960s, when it was discovered that the bone marrow contains at least two stem cell types: hematopoietic and mesenchymal. The HSCs are responsible for all the formed elements of the blood, and the mesenchymal stem cells (MSCs) or stromal cells were initially recognized as support cells for hematopoiesis. Later, MSCs were recognized to also be the cell or origin for bone, cartilage, fat, and fibrous connective tissue. Tissue stem cells reside within the peripheral tissues and give rise to the differentiated cells of the organ. A putative surface-marker expression pattern and location of the stem cell niche has been proposed for different organs, to our knowledge more complete for some niches than for others. We are, however, a long way from completely defining the peripheral stem cell and our knowledge base lags behind that for the hematopoietic stem cell.

In general, adult stem cells have traditionally been thought to possess a more limited repertoire of differentiation fates than the embryonic stem cell. Although they are widely recognized to produce the specialized cell types of the organ they reside in, they have traditionally been thought to be restricted in their ability to differentiate across lineage boundaries.

The Changing Paradigm of Adult Stem Cells

In recent years, the theory of stem cell lineage restriction has been challenged with emerging evidence from multiple laboratories that adult stem cells possess a greater degree of plasticity than previously imagined. For example, neural stem cells have been shown to differentiate into several cells of the hematopoietic lineages (Bjornson et al. 1999), and epithelial cells within the GI tract, muscle, lung, and skin (Clarke et al. 2000). Cells originating in the bone marrow have been isolated from virtually every organ of the body as differentiated cells and seem to not only take on the phenotypic characteristics of peripheral cells, but also to function appropriately in their new location (Harris et al. 2006; Vassilopoulos et al. 2003).

Additional Adult Stem Cell Types Recognized

More recently, additional adult stem cell types have been isolated. Cord blood derived from unrestricted somatic stem cells (USSCs) (Kogler et al. 2004) have attracted much attention for their potential use in regenerative medicine applications as their differentiation ability and growth properties are believed to lie between that of adult stem cells and embryonic cells. USSCs may therefore offer greater differentiation potential than adult cells, without the ethical issues surrounding the use of embryonic stem cells. USSCs can potentially be banked and stored for the patients future use, or used transplanted into nonrelated patients where they carry the immune caveats associated with conventional organ transplantation. Marrow-derived endothelial precursor cells (EPCs) (Urbich 2004; Sata 2006) have also been recognized to participate in wound healing, tissue repair, and scar formation. An additional stem cell type, termed the multipotent adult progenitor cell (MAPC) has been isolated from long-term MSC cultures (Jiang et al. 2002). MAPCs are capable of differentiating down all cell lineages including the hematopoietic cell lines; however, it is still not clear if MAPCs are a culture artifact or if culture conditions enrich for a rare primitive cell type with pluripotent potential. Interestingly, in independent experiments, a more primitive MSC population expressing stage-specific embryonic antigen-1 (SSEA-1) was isolated from adult mouse bone marrow. This population appears to contain the most primitive MSC population found to date. Studies addressing the multipotential differentiation capacity of these cells relied on culture expansion techniques, however, so the question remains if the phenotype of this cell type is real or artificially induced by culture. (Anjos-Alfonso 2007).

Stems Cells: A Continuum of Potential?

The stem cells residing within the bone marrow are not a stationary population; rather, they are in a dynamic relationship with bone, blood, and peripheral tissues. Hematopoietic stem cells enter the circulation and home back to areas of the bone undergoing remodeling to reestablish foci for hematopoiesis. Integrins such as VLA-4 (Burger et al. 2003; Carstanjen et al. 2005; Tong 2003) and adhesion receptors such as CD44, CXCR4, and its ligand stromal derived factor-1 (Peled et al. 2004), c-kit, and other receptor ligand interactions are essential for the coordinated mobilization and eventual homing of cells back to the marrow, and under pathologic conditions, may have a role in the retention of cells in the periphery. Mobilization into the circulation is increased by injury and stress, which act in part by release of cytokines and chemokines from the injured tissue; this results not only in increased circulation of cells, but also the potential for engraftment of these cells in peripheral sites (Bowie et al. 2006; Wojakowski et al. 2004). Indeed, mouse models of disease and data from humans having undergone bone marrow or solid organ transplantation demonstrate clearly that the level of bone marrow–derived cell (BMDC) engraftment into peripheral tissues is increased with injury and inflammation. Furthermore, tissue damage and ongoing inflammation may drive BMDCs to engraft into the peripheral stem cell niche ensuring retention of these cells long term (Houghton et al. 2004).

Bone Marrow–Derived Cells and Cancer

As discussed in the first section, cancer is increasingly recognized as a stem cell disease. Bone marrow–derived stem cells possess the plasticity for transdifferentiation, reside in peripheral organs in increasing numbers as a result of inflammation and damage, and depend heavily on environmental cues and cell–cell signaling for differentiation decisions and growth regulation. It seems logical then, to entertain the notion that BMDCs might contribute to cancer either directly or indirectly in situations such as chronic tissue damage and inflammation. Indeed, there are ample data supporting the contributions of BMDCs to tumor formation (Table 22.1). EPCs have been shown to contribute directly to angiogenesis in forming tumors, participating in a crucial step for tumor growth and metastasis (Kopp et al. 2006; Li 2006; Spring et al. 2005). The level of participation, however, varies widely among tumor types and the model used for study. Cancer-associated fibroblasts are in part derived from bone marrow cells, and depending on the model studied and the tissue type involved, BMDCs may form a minority of cells—or the majority of activated fibroblasts within preneoplastic and neoplastic tissues (Direkze et al. 2004; Iwano et al. 2002). (The role of BMDCs in the tumor stroma is covered in detail in Chapter 23.) Additionally, BMDCs may contribute to established cancers through cell mimicry, cell fusion, or may initiate cancer directly. Once residing in a stem cell niche of a damaged, inflamed organ, the bone marrow–derived stem cell is poised for abnormal differentiation, gene mutation, and transformation. To fully investigate

Table 22.1 Evidence for bone marrow–derived stem cell contribution to tumors

Study	Cell source	Tumor model	Phenotype of BMDC	Fusion
Rodents				
Davidoff (2001)	Total BM	Heterotopic neuroblastoma	Tumor vasculature	NT
Ishii (2003)	Total BM	Pancreatic CA	Myofibroblasts	NT
Houghton (2004)	Total BM	Chronic Hf infection model	Adenocarcinoma	N
Direkze (2004)	Total BM	Pancreatic CA	Myofibroblasts fibroblasts	NT
Spring (2005)	Total BM	Insulinoma HCC	Endothelium	NT
Li (2006)	Total BM	Sarcoma	Endothelium	NT
Human				
Barozzi (2003)	Kidney	Kaposi sarcoma	Sarcoma	NT
Chakraborty (2004)	Liver and BM	Renal cell	Renal cell	Suggested
Peters (2005)	BM	Various	Endothelium	NT
Aractingi (2005)	Kidney	Basal cell CA	Basal cell	
		Squamous cell	Stromal cells	
		Basal cell CA	Stromal cells	
		Actinic keratosis	Stromal cells	NT
		Keratoacanthoma	Stromal cells	
		Various benign lesions		
Cogle (2007)	BM	Colon adenoma	Colonocyte	
		Lung cancer	Lung cancer cell	N
Avital (2007)	BM	Lung		
		Laryngeal	Tumor cell	NT
		Kaposi sarcoma		
		Glioblastoma		

Abbreviations: BMDC: bone marrow–derived cell; BM: bone marrow; CA: cancer; NT: not tested; N: negative; HF: Helicobacter felis infection; HCC: hepatocellular carcinoma.

this possibility, we must first understand the animal models used to study this phenomenon, the strengths and shortcomings of the methods for tracking cells *in vivo*, and finally, the evidence from human disease.

Methods to Track Bone Marrow Cells in Murine Studies

To investigate a role for BMDCs in peripheral tissues, one must utilize a model that allows analysis of several key aspects of stem cell transdifferentiation biology. Their small size, repertoire for molecular, genetic, and immunology reagents available, and spectrum of accepted disease models make the mouse a valuable model for evaluating BMDC biology. Various cellular markers have been used to track BMDCs within the mouse, and each of these markers has strengths and limitations. Ideally, both a nuclear and a cytoplasmic tag permanently mark the cells of bone marrow origin, allowing accurate identification of these cells within peripheral tissues, and allowing analysis at a single cell level. We will discuss those markers most widely used, and methods of introducing tagged cells for analysis.

Cytoplasmic Markers

β-Galactosidase

A widely used marker is β-galactosidase (Schmidt et al. 1998). The *Escherichia coli lacZ* gene can be placed under a variety of promoters. In the most common model, the ROSA26 mouse, random insertion of retroviral gene trap vector ROSA26 places the β-galactosidase transgene under an unknown endogenous promoter. *LacZ* is expressed in all tissues of the developing embryo and in most tissues, including all hematopoietic cells of the adult transgenic mouse (Monastersky and Robl 1995); however, expression can be patchy and unpredictable because of gene inactivation in adult tissues. β-Galactosidase catalyzes the cleavage of lactose to galactose and glucose. Incubation of cells, whole mount tissue sections, or embryos with the β-galactosidase substrate, X-gal, produces an insoluble blue dye which allows identification of cells with *lacZ* activity by their intense blue color. This provides a useful and powerful model to study the engraftment of single cells into peripheral tissues. The bacterial *lacZ* functions optimally at a pH of 8–9, whereas mammalian endogenous β-galactosidase activity is rarely detected after incubation at pH greater than 7.5 (Weiss et al. 1999). Substrate pH can therefore be used to optimize detection; however, caution must still be used in interpreting β-galactosidase activity in tissues that contain high endogenous β-galactosidase activity such as the GI tract. Additionally, antibody directed against the bacterial protein allows immunohistochemical detection as a further means of verification.

Green Fluorescent Protein, or Enhanced Green Fluorescent Protein

Multiple fluorescent proteins are available for *in vivo* cell detection. The most widely used is green fluorescent protein (GFP), and a modified EGFP (enhanced GFP), derived from the jellyfish *Aequorea victoria*. GFP fluoresces when exposed to a 488-nm light allowing cells to be detected by direct fluorescence, fluorescence-activated cell sorter, or immunohistochemistry using antibodies directed toward GFP. Multiple transgenic mice are available commercially where GFP is driven off a host of ubiquitous or tissue-specific promoters, allowing precise tailoring of gene expression to suit experimental needs. Despite widespread use and strengths of the model system, interpretation of results must be done cautiously. Tissue autofluorescence must be carefully controlled for, and differences in tissue-specific promoter activity driving GFP may lead to underestimation of engraftment as GFP signal may be attenuated once a BMDC transdifferentiates in the peripheral organs. As with β-galactosidase detection, multiple detection methods minimize error and provide additional confidence. All transgenes can be detected in tissue samples by polymerase chain reaction where analysis provides global (+) or (–) information for the tissue, but does not assign a signal to a particular cell type or after laser capture microscopy of multiple positive cells which allows correlation with histology.

Nuclear Markers

The most frequently used nuclear markers are the X and Y chromosomes. Cells transplanted from gender-mismatched animals are detected by fluorescent *in situ* hybridization (FISH) for sex chromosomes. X and Y chromosomes are reliably detected in tissue sections and single-cell preparations. Single-cell preparations offer a better detection yield because whole nuclei are preserved whereas tissue sections typically between 4- and 10-μm thick will exclude the target chromosome in 20%–50% of cells, thus underestimating the presence of cells. Combining nuclear marker detection with a cytoplasmic marker such as GFP or β-galactosidase further increases confidence of the cell's origin. Additionally, combining Y-FISH with tissue-specific markers confirms transdifferentiation of a marrow derived cell. (For a comprehensive review on optimizing techniques for tracking transplanted stem cells *in vivo*, see review by Brazelton and Blau 2005.)

Preparing Mice for Transplants

Early studies examining the fate of transplanted cells used standard irradiation or chemotherapy-based marrow ablation protocols before cell infusion. Marrow ablation results in high levels of engraftment, but its use is complicated by collateral damage to peripheral tissues. It has been argued that subsequent engraftment of

BMDCs into the periphery may be artifactually increased. Injection of cells into nonirradiated hosts bypasses this issue, at the expense of decreased engraftment of cells. An additional method to evaluate the fate of bone marrow cells without peripheral tissue injury, yet with abundant cells for study is to use a parabiosis model whereby two animals are joined surgically, sharing circulation (Bailey et al. 2006). Marrow cells from one animal can freely circulate into the other providing a continual source of cells. Peripheral injury is minimal; however, stress hormones may be elevated and local inflammatory cytokines may be altered. When comparing studies, it is important that the method of cell transfer as well as the method for identification of cells within the peripheral tissues be taken into consideration.

The Role of Bone Marrow–Derived Cells During Acute Injury and Tissue Repair

Under normal physiologic conditions, multiple types of epithelial cells have been derived from bone marrow cells including epithelium of the lung, GI tract, and skin (Jiang et al. 2002; Krause et al. 2001). Experiments using transplantation of a single bone marrow–derived stem cell demonstrate diverse epithelial lineages supporting the existence of a single pluripotent stem cell rather than multiple committed progenitor cells as the cell of origin (Krause et al. 2001). Within the uninjured GI tract, isolated bone marrow–derived epithelial cells in the gastric pits of the stomach, the small intestinal villi, the colonic crypt, and rarely in the esophagus, appear as single differentiated epithelial cells, and do not appear to engraft into the stem cell niche. These cells are long lived, as they can be recovered months after transplantation, but their role within the peripheral tissue is not clear. Studies in human females receiving bone marrow transplant from male donors suggests that BMDCs are found in increased numbers within the inflamed epithelium, and the level of engraftment correlates somewhat with the level of graft-versus-host disease present. However, cells are found as terminally differentiated cells, questioning their long-term fate (Matsumoto et al. 2005; Okamoto 2002). Several questions arise from these findings. For example, does a BMDC contribute to healing of the GI tract by transdifferentiation to an epithelial phenotype, thus helping to acutely restore mucosal integrity—or is there a long-term function of these cells? If BMDCs are recruited to a wound, do they remain once healing is complete?

To begin addressing some of these questions, we used a model of gastric mucosal injury to evaluate the role of BMDCs in wound healing of the stomach. To track the BMDCs *in vivo*, C57BL/6J mice were irradiated to achieve myloablation and transplanted with gender-mismatched bone marrow from mice that express a nonmammalian β-galactosidase enzyme [C57BL/*6JGtrosa26* (ROSA26)], mice that expressed GFP [C57BL/6J-*beta-actin-EGFP* (GFP)], or control C57BL/6J litter mates. The mice were allowed to recover for 4 weeks before any experiments.

Engraftment of BMDCs was determined by several different methods. ROSA26 BMDCs were detected by analyzing enzyme activity and specific β-galactosidase immunohistochemistry (two cytoplasmic markers). In those mice transplanted with GFP marrow, GFP was detected by fluorescence-activated cell sorting of cytokeratin-positive, single-cell preparations, and GFP immunohistochemistry of tissue sections. X and Y chromosome FISH was used as an additional means to detect BMDCs in gender-mismatched transplants (Houghton et al. 2004). Ulceration of the gastric mucosa by cryoinjury or direct application of acetic acid (Houghton et al. 2004) causes an acute self-limited ulcer which heals in 10–14 days without sequela. Examination of ulcers in the BMDC-transplanted mice demonstrated a brisk influx of BMDCs into the area, within areas of inflammation. Early in the course of ulcer healing, labeled cells were lymphocytes, neutrophils, and occasionally fibroblasts within granulation tissue and underlying regenerating mucosa. At later time points, after complete healing of the ulcer, bone marrow–derived leukocytes were rarely recovered. Rare bone marrow–derived fibroblasts and endothelial cells incorporated into the healed mucosa. We were unable however, to detect any epithelial cells that were bone marrow derived, suggesting that normal healing of the gastric mucosa does not require bone marrow cells (Houghton et al. 2004).

Next, we analyzed if specific cell lineages could be repopulated by the bone marrow. We used targeted chemical ablation of parietal cells and evaluated acute and long-term repopulation of the parietal cell mass. Similar to restoration of the mucosa during ulceration, the parietal cell mass is restored without evidence for direct BMDC participation.

The models of ulcer and parietal cell ablation are short-term, self-limited models and as such have a different tissue microenvironment than is found in chronic conditions leading to mucosal damage. To begin to evaluate BMDC activity in an inflammatory model, we used the well-described *H. felis*/C57BL/6 mouse model of gastric inflammation and injury (Cai et al. 2005; Wang et al. 1999). This mouse model provides a continuum of disease from the initial acute inflammation with minimal mucosal damage, through atrophy, metaplasia culminating in gastric adenocarcinoma. C57BL/6 mice transplanted with marked bone marrow were infected with *H. felis*, and evaluated at various time points for the presence of BMDCs within the gastric mucosa. As expected, acute *Helicobacter* infection was associated with an influx of bone marrow–derived inflammatory cells into the tissue. However, at early time points, these cells were neutrophils, lymphocytes, and occasional fibroblasts, and we did not detect any engraftment or differentiation of BMDCs to an epithelial cell phenotype. This is consistent with a similar model of colon inflammation, whereby BMDCs were seen to differentiate as activated fibroblasts within inflammatory lesions of the colon (Direkze et al. 2004) intercalated with colonic glands, without evidence for direct differentiation as epithelial cells. Based on these findings, acute inflammation of the GI tract, specifically the stomach, does not lead to significant engraftment of BMDCs as epithelium. It remains a possibility that bone marrow cells could appear transiently or if injected into the stomach could aid ulcer healing, but these conditions are inefficient in giving rise to engraftment of BMDCs in the stomach.

Bone Marrow–Derived Cells Engraft with Chronic Inflammation and Give Rise to Metaplasia and Dysplasia

In addition to providing a model of acute injury and inflammation, the *Helicobacter*–infected mouse model is ideal for evaluating the effects of chronic inflammation on bone marrow cell migration and engraftment in the stomach. In this model, inflammation is maximal at about 8 weeks of infection, and then continues at a more modest level for the life of the mouse. Between 8 and 20 weeks of infection, there is loss of the oxyntic glands, and a restructuring of the gastric architecture to include metaplastic cell lineages, reflecting the effects of an abnormal tissue milieu on rapidly proliferating cells (Cai et al. 2005). After several months, mucosal atrophy, parietal cell loss, and mucus cell metaplasia are prominent. Interestingly, engraftment of BMDCs first becomes evident at about 20 weeks of infection, and corresponds with the appearance of these metaplastic cell lineages. With time, the number of bone marrow–derived glands increases dramatically, suggesting a threshold for recruitment needs to be reached (Houghton et al. 2004). Based on these studies, it seems that longstanding inflammation and inflammatory-mediated damage to the epithelium are required for BMDC engraftment within the gastric epithelium, an environment strongly linked to the development of cancer in many settings.

We know from the previous data that the chronic inflammatory environment is sufficient for BMDC engraftment within the gastric mucosa. We reasoned that bone marrow–derived stem cells, as the ultimate uncommitted adult stem cell, might represent the ideal candidate for transformation if placed in a favorable environment. The C57BL/6 model of *H. felis*–induced gastric cancer is optimal for studying the role of stem cells in inflammatory-mediated cancers because C57BL/6 mice do not develop gastric cancer under control conditions, and reliably develop cancer with infection. With *Helicobacter* infection, the gastric mucosa progresses through a series of changes including metaplasia and dysplasia (Wang et al. 1999), culminating in GI intraepithelial neoplasia (Boivin et al. 2003) by 12–15 months of infection (Cai et al. 2005; Wang et al. 1999). This reiterates human disease, where gastric cancer in the absence of *Helicobacter* infection is unusual, and longstanding infection carries a significant (up to 1%–3%) risk of gastric cancer (Nomura et al. 1991).

Progressive parietal and chief cell loss is a hallmark of chronic *Helicobacter* infection. Of the few parietal or chief cells that we isolated from the infected mice, none were bone marrow derived, strongly suggesting that marrow cells do not differentiate toward the parietal or chief cell phenotype under the experimental conditions we used. At 30 weeks of infection, antralized glands and metaplastic cells at the squamocolumnar junction were replaced by marrow-derived cells. The severity of intraepithelial dysplasia increased over time, and by 1 year of infection, most mice developed invasive neoplastic glands. All of the intraepithelial neoplasia in mice infected for 12–16 months arose from donor marrow cells, strongly suggesting an inherent vulnerability of this population of cells to transformation.

In addition to epithelial cells within the tumor, BMDCs also comprise a subset of cells within the tumor stroma and within seemingly uninvolved epithelium and subepithelial spaces adjacent to the tumors. We have recovered adipocytes, fibroblasts, endothelial cells, and myofibroblasts derived from bone marrow precursors in areas adjacent to dysplasia and neoplasia (Houghton et al. 2004; Li 2006).

Fusion

It is apparent from animal studies and correlative human data that BMDCs have the ability to transdifferentiate to epithelial cells within the GI tract. The recovery of bone marrow–derived epithelial cells is greater under conditions of inflammation and mucosal repair and BMDCs may be found as both differentiated cells, and within the stem cell niche. The mechanism by which the marrow-derived cells acquire phenotype of peripheral cells is not known. Two possibilities are direct transdifferentiation whereby the cell responds to environmental cues to adopt the phenotype of the surrounding cells, or through fusion of a BMDC with a peripheral stem cell or a differentiated cell whereby the fusion partner provides instruction via cytoplasmic or nuclear signals. Furthermore, if fusion was to occur, several scenarios are plausible. Stable fusion with a differentiated cell would result in a tetraploid cell that could either maintain separate nuclei, or undergo nuclear fusion, combining genetic material. In either of these scenarios, reprogramming of either cell may occur. Alternately, fusion of cells (with or without nuclear fusion) may be followed by a cytoreductive division, returning the cell to diploid status. Both nuclear reprogramming, and resorting of chromosomes has the potential to confer stable alterations to the cells.

There is compelling evidence both for and against direct differentiation and fusion with a peripheral cell as a mechanism for bone marrow–derived stem cells adopting a peripheral cell phenotype. (See Stoicov et al. 2005 for a summary of studies addressing fusion, and their findings.) Although there are conflicting data reported, it may be that both possibilities are correct, and the manner in which the bone marrow stem cell engrafts and differentiates depends on what type of tissue is involved and the mechanism of injury. For example, early studies that demonstrate fusion as a common event evaluated liver (Vassilopoulos et al. 2003; Wang 2003.) and muscle (Daging 2004), two tissue types that may be uniquely able to support multinucleate cells. Additionally, in these organs, severe injury was required, a condition that normally supports cell fusion in liver and muscle.

BMDCs engraft within the gastric mucosa and differentiate as metaplastic and dysplastic gastric glands during ongoing *H. felis* infection (Houghton et al. 2004). Bone marrow–derived gastric adenocarcinoma contains both a Y chromosome (donor marker) and either β-gal or GFP (donor marker) (Houghton, unpublished data) in greater than 60% of the cells examined. Analysis of Y chromosome in gastric tissue samples from male mice detects between 60%–65% of the Y chromosome, suggesting that the gastric tissue in the experimental group is derived from direct BMDC transdifferentiation, and not through fusion (where one would expect a discordant X-, Y-FISH and cytoplasmic markers). Multinucleate cells were not detected in this tumor model, further supporting a lack of stable fusion (Houghton et al. 2004).

More recent studies evaluating the role of fusion in the luminal GI tract do support a role for fusion. Engraftment of BMDCs within the intestinal tract of mice is an infrequent event, and may depend on irradiation-induced damage to the stem cell compartment. In a study by Rizvi et al. (2006), $0.25\% \pm 0.05\%$ of intestinal crypts in irradiated mice were marrow derived. The authors were unable to recover BMDCs in the intestinal epithelium of the nonirradiated mouse. Using a combination of EGFP and Y-chromosome identification, up to 60% of bone marrow–derived small intestinal cells resulted from fusion, presumably with an intestinal stem cell or long-lived progenitor cell as all cell lineages were clonally represented. Using a combination of EGFP and β-gal expression suggested fusion might contribute to a greater degree than estimated by EGFP and Y-chromosome colocalization. Results of this study suggest that fusion occurs between a BMDC and tissue-specific stem cell under conditions of stem cell damage, and in the APC-tumor model, with tumor cells as well.

These seemingly contradictory results must be interpreted cautiously because extrapolating results from one area of the GI tract to another area might not be valid. Also, the behavior of wild-type BMDCs in a genetically tumor-prone APC-mouse model might not mimic findings in other model systems, such as the inflammation-induced gastric cancer model.

Further studies addressing the role of fusion are ongoing. One method to address this issue is to use transplanted cells containing a dual reporter within a lox-construct such as those derived from the ZEG mouse (Novak et al. 2000). Fusion can be determined when the ZEG cells, or cells containing a similar construct are introduced into a cre-recombinase–expressing host. Cre can be ubiquitously expressed, or driven off a tissue-specific promoter allowing determination of fusion even in the presence of a reductive division, and allow identification of the host fused cell.

Which Bone Marrow–Derived Cell Gives Rise to Dysplasia?

BMDCs give rise to dysplasia and adenocarcinoma in the mouse model of gastric cancer, raising the important question, Which cell within the marrow is responsible for this phenotype? For our studies addressing BMDC engraftment within the gastric mucosa of *Helicobacter*-infected mice, and studies from other laboratories addressing engraftment of BMDCs within the GI tract, whole bone marrow was used. Whole bone marrow contains HSCs, MSCs, and a heterogeneous group of precursor cells with varying degrees of differentiation potential, precluding accurate identification of the responsible cell. Studies addressing the contributions of HSC and MSC populations to peripheral tissue architecture have suggested that MSCs are a likely candidate for transdifferentiation and MSCs have several unique properties that make them attractive candidates for transformation. Mesenchymal cells from the marrow have shown plasticity, and have been recovered as differenti-ated cells from virtually every organ system studied (Pittenger et al. 1999). *In vitro* data support the notion that MSCs are more mutagenic than other cell types, and may transform more easily than other cell types (Rubio et al. 2005; Serakinci et al. 2004). Also, studies suggest that MSCs may be home to components of the inflam-matory environment (Kassem 2004; Shimizu et al. 2001), offering a mechanism by which these cells might enter the tissue during infection and inflammation. Inherently mutagenic MSCs exposed to the conflicting growth-signaling environ-ment of local tissue inflammation may allow transformation and formation of peripheral cancers.

Although our data suggest that MSCs can give rise to gastric cancer, this does not preclude other cell types such as a gastric stem cell from initiating cancer. The cell type involved might depend on the stimulus initiating the tumor, and on the local environment within the tissue. Further investigation in this area will likely prove exciting and interesting as the mechanisms of cancer initiation are further unraveled.

Human Studies and Bone Marrow–Derived Cells in Cancer

Until recently, there have been few studies that have examined the possible cellular origins of human cancers. Most of these studies have used FISH for X and Y chromo-somes on cancers arising in patients that have undergone gender-mismatched bone marrow transplantation (BMT). Several early published reports on this topic have resulted in negative studies. The first published study from Johns Hopkins involved six individuals who developed cancers after BMT and FISH revealed that BMDCs had contributed to 4.9% of the tumor endothelium, with no evidence of direct contribution of the donor cells to the tumor itself (Peters et al. 2005). The second tumor was diagnosed 1–15 years (mean 6.7 years) after BMT, and included diverse neoplasias (spindle cell sarcoma, Hodgkin's lymphoma, submandibular carci-noma, thyroid carcinoma, osteogenic sarcoma, glossal carcinoma). A separate report the following year from The Netherlands described five women who devel-oped breast neoplasia 1.5–10 years after six mismatched stem cell transplants and again none of the nonhematologic malignancies were of donor origin (Smith et al. 2006). However, as pointed out in the response to this last report (Avital et al. 2006), it is extremely difficult to prove that bone marrow stem cells do not participate in the formation of epithelial malignancies, primarily because of the incomplete nature of bone marrow reconstitution and the likely long lead time between CSC engraftment and the clinical diagnosis of cancer. The Gompertzian model of human breast cancer would predict that an average of 8 years of growth is required for one cell to achieve a clinically detectable tumor (Norton et al. 1988), and BMDC engraftment may in fact occur many years before cancer growth commences. Thus, in many of these cases, it is likely that the cancers had actually been initiated years before the stem cell or whole bone marrow transplants were performed.

In contrast to these negative studies, a number of reports have provided supporting evidence that epithelial malignancies can arise from donor cells or BMDCs (Table 22.1). The earliest publications were single case reports. A report in 2004 described a child who, after allogeneic liver and bone marrow transplants, developed a *de novo* metastatic renal cell carcinoma (Chakraborty et al. 2004). Although the donor and recipient were both males, the bone marrow donor was blood group A+ whereas the recipient was O+. Donor DNA (based on the presence of the A allele) was detected in laser microdissected tumor cells in 16 of 21 metastatic tumor DNA samples. In 2005, a second report appeared of a female patient that developed a primary renal cell carcinoma 2 years after a BMT from a donor male (her son). Y-FISH revealed the presence of donor Y chromosomes in the tumor cells. In both of these cases, tumor cell hybridization with donor bone marrow cells was suggested as a possible explanation, rather than differentiation of BMDCs into tumor cells.

The first large series of donor-derived cancers was described in patients who had undergone not bone marrow but kidney transplants (Table 22.1). Aractingi et al. (2005) reported a series of 48 malignant and benign cutaneous tumors that developed in 14 females that had received a male kidney. Male cells were detected in 5 of 15 squamous cell carcinomas and 3 of 5 basal cell carcinomas, as well as in many benign skin lesions (actinic keratosis, keratoacanthoma, benign cutaneous lesions). In one patient, a female patient transplanted with a male kidney, a basal cell carcinoma was largely donor derived, with Y-chromosome-positive, cytokeratin-positive cells found in the tumor nest but not in the normal epidermis, suggesting clonal expansion of a single donor cell. In addition, CD45+ donor cells were noted in some skin sections. Kidneys are known to harbor mesenchymal progenitor cells that show some degree of multipotency and multilineage differentiation (Plotkin and Goligorsky 2006). In addition, hematopoietic microchimerism has been shown to occur frequently in recipients of solid organ transplants, with circulating CD45+ donor cells within the recipient's peripheral blood lymphocytes (Elwood et al. 1997). Thus, the donor cells giving rise to the transformed keratinocytes likely came from the graft kidney but the specific type of cell responsible (HSC, MSC, etc.) is unknown.

More comprehensive studies on the potential role of bone marrow stem cells in human cancers were reported in 2007 (Table 22.1). Cogle et al. (2007) identified colonic neoplasias in two women who were diagnosed shortly (less than 2 months) after hematopoietic cell transplantation from male donors. The patients had undergone colonoscopic evaluations for diarrhea, and colonic adenomas were identified and resected. Y-FISH analysis revealed that the adenomas contained donor-derived colonocytes in the adenoma epithelia with no evidence for cell fusion. Of 1,000 adenoma epithelial cells studied, 1%–4% originated from donor BMDCs. Another woman who underwent gender-mismatched stem cell transplants was diagnosed with lung cancer 4 years posttransplant, and the lung cancer demonstrated a donor origin as evidenced by male cells (Y-chromosome positive). Approximately 20% of

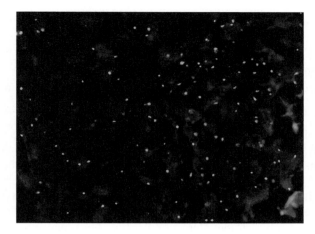

Fig. 22.1 Female tumor cells are present in solid tumors in male patients. The photomicrograph shows a laryngeal squamous cell carcinoma from a male patient that had undergone a bone marrow transplant 7 years earlier from a female donor. The tissue was stained for Y chromosome (red dots) and X chromosomes (green dots) by fluorescent *in situ* hybridization. Nuclei were counterstained in blue with DAPI stain. The tumor appears to be polyclonal, with the red circle on the left indicating a region that is largely XY and of recipient (male) origin, whereas the green circle on the left shows an area that is largely XX and of donor (female) origin. (This figure was provided courtesy of I. Avital and R. Downey, MSKCC.) (*See Color Plates*)

the lung cancer cells were of donor marrow origin, with no evidence for fusion. Analysis of skin cancer—both squamous cell and basal cell—revealed no evidence for donor contribution. Overall, however, this study provides strong support that BMDCs can contribute directly to the neoplastic lineage.

Finally, the most recent study from Memorial Sloan Kettering Cancer Center investigated four male patients who developed solid organ cancers after total body irradiation and bone marrow reconstitution from female donors (Avital et al. 2007). The patients were identified from a prospective registry of 12,000 patients with transplants, and the cancers were diverse and included lung adenocarcinoma, laryngeal squamous cell carcinoma, glioblastoma, and Kaposi's sarcoma. The cancers were diagnosed 1–7 years (mean 3.5 years) after the BMT. X- and Y-FISH were performed to investigate the origin of the neoplastic cells, and donor-derived cancer cells were observed in each instance, ranging from 2.5% to 6% of the tumor cellularity. Thus, these tumors were a mixture of donor and recipient cells, but with regions that individually were clearly clonal in origin (Figure 22.1), consistent with the emerging view that many tumors begin as polyclonal lesions.

Taken together, evidence is emerging that bone marrow–derived stem cells directly contribute to several human tumors, and have a role in solid organ carcinogenesis.

Summary and Future Directions

Accumulated evidence suggests that chronic inflammation promotes many types of cancers, and that this occurs in part through the mobilization of BMDCs. BMDCs can clearly give rise to tumor cells, as demonstrated in murine models (Houghton et al. 2004) and in human studies (Avital et al. 2007; Cogle et al. 2007). Thus, a new model of carcinogenesis has developed that involves the recruitment of BMDCs to the epithelial compartment, where it gives rise to metaplasia and dysplasia (Figure 22.2). A number of studies have demonstrated that chronic inflammation can also lead to recruitment of BMDCs to both the liver and the colon (Brittan et al. 2005; Forbes et al. 2004), where they can contribute to activated myofibroblasts as well as endothelial cells, two cell types that are critical to the stem cell niche and tissue microenvironment (see Chapter 23). Therefore, it is likely that the interaction between bone marrow–derived epithelial cells and bone marrow–derived stromal and nonepithelial cells is central to tumor development.

In addition, BMDCs have contributed to the metastatic niche. In recent reports, bone marrow–derived vascular endothelial growth factor receptor 1positive hematopoietic progenitor cells were shown to home to tumor-specific premetastatic sites before the arrival of tumor cells (Kaplan et al. 2005). Thus, BMDCs seem to be contributing to both tumor cells and the tumor niche, and have roles in both cancer initiation and cancer progression.

Given the novelty of these recent discoveries, work in other models of cancer is needed to address these mechanisms, and to clarify the importance of these pathways in human cancer. Characterizing and developing markers for the BMDC subpopulations that can contribute to solid tumors should be given priority, not only to facilitate clinical studies and early detection in patients but also for the development of targeted therapies. It seems likely that some solid tumors are derived from resident tissue stem cells—rather than circulating BMDCs—making it necessary for the overall importance of the two paradigms to be understood. The possible role of cell fusion by BMDCs has yet to be excluded as a possible mechanism. The role of an aberrant stem cell niche in contributing the CSC development from BMDCs, and the role of prevention and/or reversibility of preneoplastic lesions (by eradication of *H. pylori*) are clearly worth addressing further in the *Helicobacter* mouse model.

Finally, the precise relationship of chronic inflammation to BMDCs needs further investigation. One suspects that these studies will reveal the involvement of many cell types and specific factors—including cytokines, chemokines, hypoxia, and oxidative stress—in the mobilization, recruitment, engraftment, and progression of stem cells toward cancer.

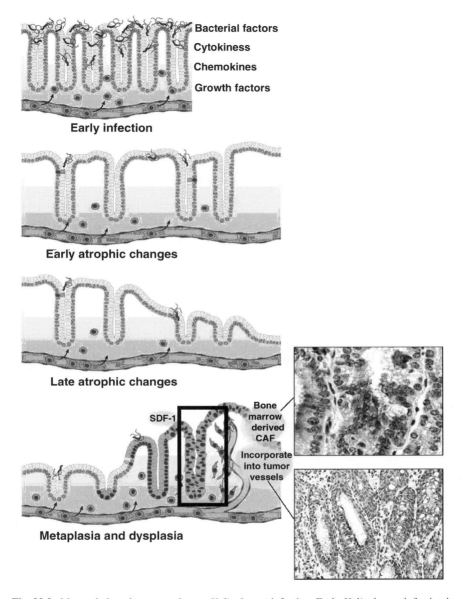

Fig. 22.2 Mucosal alterations secondary to *Helicobacter* infection. Early *Helicobacter* infection is associated with inflammatory cell infiltrate within the gastric mucosa and submucosa, release of cytokines and chemokines and growth factors (first panel). With time, atrophy of the mucosa becomes prominent with loss of specialized cells. At this stage, rare bone marrow–derived cells are found as terminally differentiated cells, as fibroblasts, and occasionally as endothelial cells within the submucosa (second and third panels). Long-term infection is associated with both metaplastic and dysplastic cell changes and the appearance of intraepithelial neoplasia (fourth panel). An increase in bone marrow-derived, cancer-associated fibroblasts (CAF) and endothelial cells are also seen associated with the appearance of intraepithelial neoplasia. Long-term infection in a mouse model utilizing marked bone marrow shows that the metaplastic, dysplastic, and neoplastic cells are derived from bone marrow progenitor cells. (Micrographs of panel 4 courtesy of Arlin B. Rogers, MIT.) Reproduced from Correa/Houghton 2007, with permission from Elsevier

References

Al-Hajj M., Wicha M.S., Benito-Hernandez A., Morrison S.J., and Clarke M.F. 2003. Prospective identification of tumorigenic breast cancer cells. Proc Natl Acad Sci USA 100(7):3983–3988.

Anjos-Alfonso F., and Bonnet D. 2007. Nonhematopoietic/endothelial SSEA-1+ cells define the most primitive progenitors in the adult murine bone marrow mesenchymal compartment. Blood 109(3):1298–1306.

Aractingi S., Kanitakis J., Euvrard S., Le Danff C., Peguillet I., Khosrotehrani K., Lantz O., and Carosella E.D.2005. Skin carcinoma arising from donor cells in a kidney transplant recipient. Cancer Res 65(5):1755–1760.

Avital I., Moreira A., and Downey R.J. 2006. The origin of epithelial neoplasms after allogeneic stem cell transplantation. Haematologica 91:283–284.

Avital I., Moreira A., Klimstra D., Leversha M., Papadopolous E., Brennan M., and Downey R.J. 2007. Donor derived human bone marrow stem cells potentially contribute to solid organ cancers. Stem Cells 25(11):2903–2909.

Bailey A., Willenbring H., Jiang S., Anderson D., Schroeder D., Wong M., Grompe M., and Fleming W. 2006. Myeloid lineage progenitors give rise to vascular endothelium. PNAS 103(35):13156–13161.

Balkwill F., and Mantovani A. 2001. Inflammation and cancer: back to Virchow? Lancet 357(9255):539–545.

Barozzi P., Luppi M., Facchetti F., Mecucci C., Alu M., Sarid R., Rasini V., Ravazzini L., Rossi E., Festa S., Crescenzi B., Wolf D.G., Schulz T.F., and Torelli G. 2003. Erratum to "Post-transplant Kaposi sarcoma originates from the seeding of donor-derived progenitors." Nat Med 9(7):975.

Bjerknes M., and Cheng H. 2002. Multipotential stem cells in adult mouse gastric epithelium. Am J Physiol Gastrointest Liver Physiol 283(3):G767–G777.

Bjornson C., Reitze R., Reynolds B., Magli M., and Vescovi A. 1999 Turning brain into blood: a hematopoietic fate adopted by neural stem cells in vivo. Science 283:534–537.

Boivin G.P., Washington K., Yang K., Ward J.M., Pretlow T.P., Russell R., Besselsen D.G., Godfrey V.L., Doetschman T., Dove W.F., Pitot H.C., Halberg R.B., Itzkowitz S.H., Groden J., and Coffey R.J. 2003. Pathology of mouse models of intestinal cancer: consensus report and recommendations. Gastroenterology 124:762–777.

Bowie M.B., McKnight K.D., Kent D.G., McCaffrey L., Hoodless P.A., and Eaves C.J. 2006. Hematopoietic stem cells proliferate until after birth and show a reversible phase-specific engraftment defect. J Clin Invest 116(10):2808–2816.

Brittan M., Chance V., Elia G., Poulsom R., Alison M.R., MacDonald T.T., and Wright N.A. 2005. A regenerative role for bone marrow following experimental colitis: contribution to neovasculogenesis and myofibroblasts. Gastroenterology 128(7):1984–1995.

Burger J.A., Spoo A., Dwenger A., Burger M., and Behringer D. 2003. CXCR4 chemokine receptors (CD184) and alpha4beta1 integrins mediate spontaneous migration of human CD34+ progenitors and acute myeloid leukaemia cells beneath marrow stromal cells (pseudoemperipolesis). Br J Haematol 122(4):579–589.

Cai X., Carlson J., Stoicov C., Li H., Wang T.C., and Houghton J. 2005. Helicobacter felis eradication restores normal architecture and inhibits gastric cancer progression in C57BL/6 mice. Gastroenterology 128:1937–1952.

Carstanjen D., Gross A., Kosova N., Fichtner I., and Salama A. 2005. The alpha4beta1 and alpha-5beta1 integrins mediate engraftment of granulocyte-colony-stimulating factor-mobilized human hematopoietic progenitor cells. Transfusion 45(7):1192–2000.

Chakraborty A., Lazova R., Davies S., Backvall H., Ponten F., Brash D., and Pawelek J. 2004. Donor DNA in a renal cell carcinoma metastasis from a bone marrow transplant recipient. Bone Marrow Transplant 34(2):183–186.

Clarke M.F., Dick J.E., Dirks P.B., Eaves C.J., Jamieson C.H., Jones D.L., Visvader J., Weissman I.L., and Wahl G.M. 2006. Cancer stem cells—perspectives on current status and future directions: AACR workshop on cancer stem cells. Cancer Res 66(19):9339–9344.

Clarke D., Johansson C., Wilbertz J., Veress B., Nilsson E., Karlstrom H., Lendahl U., and Frisen J. 2000. Generalized potential of adult neural stem cells. Science 288:1660–1663.

Cogle C.R., Theise N.D., Fu D., Ucar D., Lee S., Guthrie S.M., Lonergan J., Rybka W., Krause D.S., and Scott E.W. 2007. Bone marrow contributes to epithelial cancers in mice and humans as developmental mimicry. Stem Cells 25(8):1881–1887.

Collins A.T., Berry P.A., Hyde C., Stower M.J., and Maitland N.J. 2005. Prospective identification of tumorigenic prostate cancer stem cells. Cancer Res 65(23):10946–10951.

Correa P. 1992. Human gastric carcinogenesis: a multistep and multifactorial process—First American Cancer Society Award Lecture on Cancer Epidemiology and Prevention. Cancer Res 52:6735–6740.

Dalerba P., Dylla S.J., Park I.K., Liu R., Wang X., Cho R.W., Hoey T., Gurney A., Huang E.H., Simeone D.M., Shelton A.A., Parmiani G., Castelli C., and Clarke M.F. 2007. Phenotypic characterization of human colorectal cancer stem cells. Proc Natl Acad Sci USA 104(24):10158–10163.

Davidoff A.M., Ng C.Y., Brown P., Leary M.A., Spurbeck W.W., Zhou J., Horwitz E., Vanin E.F., and Nienhuis A.W. 2001. Bone marrow-derived cells contribute to tumor neovasculature and, when modified to express an angiogenesis inhibitor, can restrict tumor growth in mice. Clin Cancer Res 7(9):2870–2879.

Direkze N.C., Hodivala-Dilke K., Jeffery R., Hunt T., Poulsom R., Oukrif D., Alison M.R., and Wright N.A. 2004. Bone marrow contribution to tumor associated myofibroblasts and fibroblasts. Cancer Res 64:8492–8495.

Elwood E.T., Larsen C.P., Maurer D.H., Routenberg K.L., Neylan J.F., Whelchel J.D., O'Brien D.P., and Pearson T.C. 1997. Microchimerism and rejection in clinical transplantation. Lancet 349(9062):1358–1360.

Fang D., Nguyen T.K., Leishear K., Finko R., Kulp A.N., Hotz S., Van Belle P.A., Xu X., Elder D.E., and Herlyn M. 2005. A tumorigenic subpopulation with stem cell properties in melanomas. Cancer Res 65(20):9328–9337.

Forbes S.J., Russo F.P., Rey V., Burra P., Rugge M., Wright N.A., and Alison M.R. 2004. A significant proportion of myofibroblasts are of bone marrow origin in human liver fibrosis. Gastroenterology 126(4):955–963.

Forman D., Newell D.G., Fullerton F., Yarnell J.W., Stacey A.R., Wald N., and Sitas F. 1991. Association between infection with Helicobacter pylori and risk of gastric cancer: evidence from a prospective investigation. BMJ 302:1302–1305.

Giannakis M., Stappenbeck T.S., Mills J.C., Leip D.G., Lovett M., Clifton S.W., Ippolito J.E., Glasscock J.I., Arumugam M., Brent M.R., and Gordon J.I. 2006. Molecular properties of adult mouse gastric and intestinal epithelial progenitors in their niches. J Biol Chem 281(16):11292–11300.

Harris M., Brown G.A.J., Jorgensen M., Kaushal S., Ellis E.A., Grant M.B., and Scott E.W. 2006. Bone marrow-derived cells home to and regenerate retinal pigment epithelium after injury. Invest Ophthalmol Vis Sci 47(5):2108–2113.

Hermann P.C. Huber S.L., Herrler R., Aicher A., Ellwart J.W, Guba M., Bruns C.J., and Hesschen C. 2007. Distinct populations of cancer stem cells determines tumor growth and metastatic activity in human pancreatic cancer. Cell Stem Cell 1:313–323.

Houghton J.M., Morozov A., Smirnova I., and Wang T.C. 2007. Stem cells and cancer. Semin Cancer Biol 17(3):191–203.

Houghton J.M., Stoicov C., Nomura S., Rogers A.B., Carlson J., Li H., Cai X., Fox J.G., Goldenring J.R., and Wang T.C. 2004. Gastric cancer originating from bone marrow-derived cells. Science 306:1568–1571.

Ishii G., Sangai T., Oda T., Aoyagi Y., Hasebe T., Kanomata N., Endoh Y., Okumura C., Okuhara Y., Magae J., Emura M., Ochiya T., and Ochiai A. 2003. Bone-marrow-derived myofibroblasts contribute to the cancer-induced stromal reaction. Biochem Biophys Res Commun 309(1):232–240.

Iwano M., Plieth D., Danoff T.M., Xue C., Okada H., and Neilson E.G. 2002. Evidence that fibroblasts derive from epithelium during tissue fibrosis. J Clin Invest 110(3):341–350.

Jiang Y., Jahagirdar B.N., Reinhardt R.L., Schwartz R.E., Keene C.D., Ortiz-Gonzalez X.R., Reyes M., Lenvik T., Lund T., Blackstad M., Du J., Aldrich S., Lisberg A., Low W.C., Largaespada D.A., and Verfaillie C.M. 2002. Pluripotency of mesenchymal stem cells derived from adult marrow. Nature 418:41–49.

Jordan C.T., Guzman M.L., and Noble M. 2006. Cancer stem cells. N Engl J Med 355(12):1253–1261.

Kaplan R.N., Riba R.D., Zacharoulis S., Bramley A.H., Vincent L., Costa C., MacDonald D.D., Jin D.K., Shido K., Kerns S.A., Zhu Z., Hicklin D., Wu Y., Port J.L., Altorki N., Port E.R., Ruggero D., Shmelkov S.V., Jensen K.K., Rafii S., and Lyden D. 2005. VEGFR1-positive haematopoietic bone marrow progenitors initiate the pre-metastatic niche. Nature 438(7069):820–827.

Karam S.M., and Leblond C.P. 1992. Identifying and counting epithelial cell types in the "corpus" of the mouse stomach. Anat Rec 232(2):231–246.

Karam S.M., and Leblond C.P. 1993.Dynamics of epithelial cells in the corpus of the mouse stomach. I. Identification of proliferative cell types and pinpointing of the stem cell. Anat Rec 236(2):259–279.

Kassem M. 2004. Mesenchymal stem cells: biological characteristics and potential clinical applications. Cloning Stem Cells 6:369–374.

Kelly P.N., Dakic A., Adams J.M., Nutt S.L., and Strasser A. 2007. Tumor growth need not be driven by rare cancer stem cells. Science 317(20):337.

Kögler G., Sensken S., Airey J.A., Trapp T., Müschen M., Feldhahn N., Rüdiger S., Sorg V., Fischer J., Rosenbaum C., Greschat S., Knipper A., Bender J., Degistirici Ö., Gao J., Caplan A.I., Colletti E.J., Almeida-Porada G., Müller H.W., and Zanjani E., Wernet P. 2004. A new human somatic stem cell from placental cord blood with intrinsic pluripotent differentiation potential. J Exp Med 200(2):123–135.

Kopp H.G., Ramos C.A., and Rafii S. 2006. Contribution of endothelial progenitors and proangiogenic hematopoietic cells to vascularization of tumor and ischemic tissue. Curr Opin Hematol 13(3):175–181.

Krause D.S., Theise N.D., Collector M.I., Henegariu O., Hwang S., Gardner R., Neutzel S., and Sharkis S.J. 2001. Multi-organ, multi-lineage engraftment by a single bone marrow-derived stem cell. Cell 105:369–377.

Lapidot T., Sirard C., Vormoor J., Murdoch B., Hoang T., Caceres-Cortes J., Minden M., Paterson B., Caligiuri M.A., and Dick J.E. 1994.A cell initiating human acute myeloid leukaemia after transplantation into SCID mice. Nature 367(6464):645–648.

Li C., Heidt D.G., Dalerba P., Burant C.F., Zhang L., Adsay V., Wicha M., Clarke M.F., and Simeone D.M. 2007. Identification of pancreatic cancer stem cells. Cancer Res 67(3):1030–1037.

Li B., Sharpe E.E., Maupin A.B., Teleron A.A., Pyle A.L., Carmeliet P., and Young P.P. 2006. VEGF and PlGF promote adult vasculogenesis by enhancing EPC recruitment and vessel formation at the site of tumor neovascularization. FASEB J 20:1495–1497.

Li H.C., Stoicov C., Rogers A.B., and Houghton J.M. 2006. Stem cells and cancer: evidence for bone marrow stem cells in epithelial cancers. World J Gastroenterol 12(3):363–371.

Ma S., Chan K.W., Hu L., Lee T.K., Wo J.Y., Ng I.O., Zheng B.J., and Guan X.Y. 2006. Identification and characterization of tumorigenic liver cancer stem/progenitor cells. Gastroenterology. 132(7):2542–2556.

Matsumoto T., Okamoto R., Yajima T., Mori T., Okamoto S., Ikeda Y., Mukai M., Yamazaki M., Oshima S., Tsuchiya K., Nakamura T., Kanai T., Olano H., Inazawa J., Hibi T., and Watanabe M. 2005. Increase of bone marrow derived secretory lineage epithelial cells during regeneration in the human intestine. Gastroenterology 128:1851–1867.

Mills J.C., Andersson N., Hong C.V., Stappenbeck T.S., and Gordon J.I. 2002. Molecular characterization of mouse gastric epithelial progenitor cells. Proc Natl Acad Sci USA 99(23):14819–14824.

Monastersky G.M., and Robl J.M., editors. 1995. Strategies in transgenic animal science. Washington, D.C.: ASM Press.

Nomura A., Stemmermann G.N., Chyou P.H., Kato I., Perez-Perez G.I., and Blaser M.J. 1991. Helicobacter pylori infection and gastric carcinoma among Japanese Americans in Hawaii. N Engl J Med 325:1132–1136.

Norton L. 1988. A Gompertzian model of human breast cancer growth. Cancer Res 48(24):7067–7071.

Novak A., Guo C., Yang W., Nagy A., and Lobe C.G. 2000. Z/EG, a double reporter mouse line that expresses enhanced green fluorescent protein upon Cre-mediated excision. Genesis 28(3–4):147–155.

O'Brien C.A., Pollett A., Gallinger S., and Dick J.E. 2007. A human colon cancer cell capable of initiating tumour growth in immunodeficient mice. Nature 445(7123):106–110.

Okamoto R., Yajima T., Yamazaki M., Kanai T., Mukai M., Okamoto S., Hibi T., Inazawa J., and Wantanabe M. 2002. Damaged epithelia regenerated by bone marrow derived cells in the human gastrointestinal tract. Nat Med 8(9):1011–1017.

Parsonnet J., Friedman G.D., Vandersteen D.P., Chang Y., Vogelman J.H., Orentreich N., and Sibley R.K. 1991. Helicobacter pylori infection and the risk of gastric carcinoma. N Engl J Med 325:1127–1131.

Peled A., Kollet O., Ponomaryov T., Petit I., Franitza S., Grabovsky V., Slav M.M., Nagler A., Lider O., Alon R., Zipori D., and Lapidot T. 2000. The chemokine SDF-1 activates the integrins LFA-1, VLA-4, and VLA-5 on immature human CD34(+) cells: role in transendothelial/stromal migration and engraftment of NOD/SCID mice. Blood 95(11):3289–3296.

Peters B.A., Diaz L.A., Polyak K., Meszler L., Romans K., Guinan E.C., Antin J.H., Myerson D., Hamilton S.R., Vogelstein B., Kinzler K.W., and Lengauer C. 2005. Contribution of bone marrow-derived endothelial cells to human tumor vasculature. Nat Med 11(3):261–262.

Pierce G.B. 1974. Neoplasms, differentiations and mutations. Am J Pathol 77(1):103–118.

Pittenger M.F., Mackay A.M., Beck S.C., Jaiswal R.K., Douglas R., Mosca J.D., Moorman M.A., Simonetti D.W., Craig S., and Marshak D.R. 1999. Multilineage potential of adult human mesenchymal stem cells. Science 284(5411):143–147.

Plotkin M.D., and Goligorsky M.S. 2006. Mesenchymal cells from adult kidney support angiogenesis and differentiate into multiple interstitial cell types including erythropoietin-producing fibroblasts. Am J Physiol Renal Physiol 291(4):F902–912.

Powell D.W., Adegboyega P.A., Di Mari J.F., and Mifflin R.C. 2005. Epithelial cells and their neighbors. I. Role of intestinal myofibroblasts in development, repair, and cancer. Am J Physiol Gastrointest Liver Physiol 289(1):G2–7.

Prince M.E., Sivanandan R., Kaczorowski A., Wold G.T., Kaplan M.J., Dalerba P., Weissman I.L., Clarke M.F., and Ailles L. E. Identification of a subpopulation of cells with cancer stem cell properties in head and neck squamous cell carcinoma. Proc Natl Acad Sci USA 104(3):973–978.

Pull S.L., Doherty J.M., Mills J.C., Gordon J.I., and Stappenbeck T.S. 2005. Activated macrophages are an adaptive element of the colonic epithelial progenitor niche necessary for regenerative responses to injury. Proc Natl Acad Sci USA 102(1):99–104.

Qiao X.T., Ziel J.W., Madison B.B., McKimpson W., Todisco A., Merchant J.L., Samuelson L.C., and Gumucio D.L. 2007. Prospective identification of a multi-lineage progenitor in murine stomach epithelium. Gastroenterology (in press).

Rather L.J. 1978. The genesis of cancer: a study in the history of ideas. Baltimore: The Johns Hopkins University Press.

Reya T., Morrison S.J., Clarke M.F., and Weissman I.L. 2001. Stem cells, cancer, and cancer stem cells. Nature 414(6859):105–111.

Ricci-Vitiani L., Lombardi D.G., Pilozzi E., Biffoni M., Todaro M., Peschle C., and De Maria R. 2007. Identification and expansion of human colon-cancer-initiating cells. Nature 445(7123):111–115.

Rizvi A.Z., Swain J.R., Davies P.S., Bailey A.S., Decker A.D., Willenbring H., Grompe M., Fleming W.H., and Wong M.H. 2006. Bone marrow-derived cells fuse with normal and transformed intestinal stem cells. PNAS 103:6321–6325.

Rubio D., Garcia-Castro J., Martin M., de la Fuente R., Cigudosa J., Lloyd A.C., and Bernad A. 2005. Spontaneous human adult stem cell transformation. Cancer Res 65:3035–3039.

Sata M. 2006. Role of circulating vascular progenitors in angiogenesis vascular healing and pulmonary hypertension: lessons from animal models. Arterioscler Thromb Vasc Biol 26:1008–1014.

Schmidt A., Tief K., Foletti A., Hunziker A., Penna D., Hummler E., and Beermann F. 1998. lacZ transgenic mice to monitor gene expression in embryo and adult. Brain Res Brain Res Protoc 3(1):54–60.

Schofield R. 1978. The relationship between the spleen colony-forming cell and the haemopoietic stem cell. Blood Cells 4(1–2):7–25.

Serakinci N., Guldberg P., Burns J.S., Abdallah B., Schrodder H., Jensen T., and Kassem M. 2004. Adult human mesenchymal stem cell as a target for neoplastic transformation. Oncogene 23:5095–5098.

Shi D., Reinecke H., Murry C.E., and Torok-Storb B. 2004. Myogenic fusion of human bone marrow stromal cells, but not hematopoietic cells. Blood 104:290–294.

Shimizu K., Sugiyama S., Aikawa M., Fukumoto Y., Rabkin E, Libby P., and Mitchell R. 2001. Host bone-marrow cells are a source of donor intimal smooth-muscle-like cells in murine aortic transplant arteriopathy. Nat Med 7:738–741.

Singh S.K., Hawkins C., Clarke I.D., Squire J.A., Bayani J., Hide T., Henkelman R.M., Cusimano M.D., and Dirks P.B. 2004. Identification of human brain tumour initiating cells. Nature 432(7015):396–401.

Smith M.J., van Cleef P.H., Schattenberg A.V., and van Krieken J.H. 2006. The origin of epithelial neoplasms after allogeneic stem cell transplantation. Haematologica 91(2):283–284.

Spring H., Schüler T., Arnold B., Hämmerling G.J., and Ganss R. 2005. Chemokines direct endothelial progenitors into tumor neovessels. Proc Natl Acad Sci USA 102(50):18111–18116.

Stoicov C., Li H., Carlson J., and Houghton J. 2005. Bone marrow cells as the origin of stomach cancer. Future Oncol 1(6):851–862.

Tong C., and Xie Y. 2003. Correlation between VLA-4 integrin and hematopoietic cell migration. J Exp Hematol 11(3):230–234.

Uemura N., Okamoto S., Yamamoto S., Matsumura N., Yamaguchi S., Yamakido M., Taniyama K., Sasaki N., and Schlemper R.J. 2001. Helicobacter pylori infection and the development of gastric cancer. N Engl J Med 345:784–789.

Vassilopoulos G., Wang P.R., and Russell D.W. 2003. Transplanted bone marrow regenerates liver by cell fusion. Nature 422(6934):901–904.

Wang T.C., Goldenring J.R., Dangler C., Ito S., Mueller A., Jeon W.K., Koh T.J., and Fox J.G. 1999. Mice lacking secretory phospholipase A2 show altered apoptosis and differentiation with Helicobacter felis infection. Gastroenterology 114:675–689.

Wang X., Willenbring H., Akkari Y., Torimaru Y., Foster M., Al-Dhalimy M., Lagasse E., Finegold M., Olson S., Grompe M. 2003. Cell fusion is the principal source of bone-marrow-derived hepatocytes. Nature 422:897–901.

Weiss D.J., Liggitt D., and Clark J.G. 1999. Histochemical discrimination of endogenous mammalian beta-galactosidase activity from that resulting from lac-Z gene expression. Histochem J 31(4):231–236.

Wojakowski W., Tendera M., Michalowska A., Majka M., Kucia M., Maslankiewicz K., Wyderka R., Ochala A., and Ratajczak M.Z. 2004. Mobilization of CD34/CXCR4+, CD34/CD117+, c-met+ stem cells, and mononuclear cells expressing early cardiac, muscle, and endothelial markers into peripheral blood in patients with acute myocardial infarction. Circulation 110(20):3213–3220.

Wong M.H., Saam J.R., Stappenbeck T.S., Rexer C.H., and Gordon J.I. 2000. Genetic mosaic analysis based on Cre recombinase and navigated laser capture microdissection. Proc Natl Acad Sci USA 97(23):12601–12606.

Xie T., and Spradling A.C. 2000. A niche maintaining germ line stem cells in the Drosophila ovary. Science 290(5490):328–330.

Chapter 23
Stromal Cells and Tumor Microenvironment

Andrea Varro

Introduction

Cancer initiation is attributable to genetic mutations that confer dominance on cancer-producing stem cells over their normal counterparts (1). Mutations in epithelial cells that lead to cancer initiation and progression have been widely studied in the past decade or so (2). It has also been clear for some time that tumor survival and progression (including metastasis) depends on a microenvironment that is capable of supporting increased cell proliferation, migration and invasion, and decreased apoptosis, that together give advantage to transformed cancer cells compared with their nontransformed counterparts (1, 3–8). The relevant microenvironment is provided by the stroma, and consists of host cells that continually interact with cancer cells producing growth factors, proteases, protease inhibitors, cytokines, and extracellular matrix (ECM) proteins (9–12).

At present, rather little is known of the differences that might occur between the stromal components of different epithelial carcinomas. In the stomach, 95% of all malignant epithelial tumors are adenocarcinomas; other gastric cancers include adenosquamous, squamous, small-cell or neuroendocrine, hepatoid, chorio, embryonal, and medullary or lymphoepithelioma-like carcinomas. The extent to which stroma differs in these tumors is difficult to assess at the present time, but it is worth noting that medullary or lymphoepithelioma-like tumors in which lymphocytic infiltrates are prominent may differ from other gastric tumors (13). Moreover, gastrointestinal stromal-derived cancers (see Chapter 5), which are often characterized by activating c-kit or platelet-derived growth factor receptor A (PDGFR-A) mutations, are also a special case (14, 15).

It has been estimated that 60%–90% of the mass of gastrointestinal tumors is occupied by stroma (16). The ECM proteins of the latter act as regulators of stromal cell migration, invasion, adhesion, and since some of the paracrine mediators released by the host cells are associated with ECM proteins, it may also function as a pool for growth factors and other chemical mediators of stromal cell interactions (17). One of the best studied (cellular) components of stroma is the myofibroblast, which is a smooth muscle-like fibroblast; other important host cells include vascular cells that have a role in supporting increased angiogenesis [i.e., endothelial cells, pericytes, vascular smooth muscle cells (vSMCs)], and inflammatory cells including mast cells, lymphocytes, and macrophages (Table 23.1).

Although there is growing realization that stroma has a crucial role in determining the progression of cancer, it is not yet clear whether stroma should be considered an expanded version of an existing normal microenvironment assembled, for example, during chronic inflammation or injury or whether it should be considered an alter-

Table 23.1 Key cell types in stroma and some of their protein products

	Myofibroblasts/ fibroblasts*	Endothelial cells†	Pericytes‡	Inflammatory cells§
Growth factors	TGF-β, PDGF, IGF, HGF, KGF, VEGF, FGF, EGF, endothelins	PDGF, VEGF, ANG2,	bFGF, Ang-1 VEGF, gremlin ANG1, 2	VEGF, TGF-β, PDGF, IGF, TGF-α, bFGF
Receptors	PDGFR, c-kit, IGF-1R, c-Met, FGFR, uPAR, TGF-β, EGF, PARs, IL1R, TLR	HGFR, VEGR, IGF-R, Notch, PDGFR, N-cad UNC5B, CCR7, CXCR3, EphB4, FGFR, PAR	PDGFR, TLR4, IL1R, C-Met, TNF-αR	CCR9, CCR5, CXCR3, IL-2R, CCR4, IL-12R, CCL2R, TGF-βR, PDGFR, EGFR
Proteases	uPA, MMPs	MMPs, uPA	MMP-2	MMPs, uPA,
Protease inhibitors	TIMPs, PAI-1, serpins	TIMPs	TIMPs	TIMPs, PAI-2
Cytokines	IL-1, IL-6, IL-10, TNF-α, LIF	LC4	IL-1β, IL-6, TNF-α,	Leukotrienes, IFNS, TNF-α, IL-10, IL-1
Chemokines	IL-8, SDF-1,	IL-8	SDF-1	IL-8, IL-4, CLC20
ECM molecules	Collagen, laminin, fibronectin	Vitronectin	Fibronectin	Fibronectin
Adhesion molecules	ICAM, VCAM	PECAM-1, VCAM,	MAdCAM-1,	L-selectin,
Others	Prostaglandins, Wnt, natriuretic peptides	CD31	Calponin, H-caldesmon	NO, histamine

Protein products (growth factors, surface receptors, protease systems, chemokines, cytokines, and molecules) characteristic of the cells comprising stroma.

*(10, 40, 44, 47, 68, 70, 87); (1, 3–5, 16, 41, 42, 48–52, 55, 60–62, 74, 86, 88–96).

†(5, 22–24, 61, 70, 76, 97, 98).

‡(22, 25, 89, 99, 100).

§(3, 28, 33, 61, 70, 101).

native microenvironment deregulated by the cancer cells. Early work highlighted the importance of the ECM in both breast and ovarian carcinogenesis and common mechanisms in cancer initiation that are determined by the organ-specific microenvironment (18–21). It remains unclear, however, whether (or how) microenvironments in primary versus local and distant metastatic sites might differ. This chapter provides, first, an overview of the characteristics of host cells contributing to stroma in cancer. Second, the biology of the myofibroblast is considered in more detail because it is considered to be the resident niche cell in normal tissues, and it is one of the most abundant cell types in stroma that is increased early in carcinogenesis. In the context of gastric cancer, the chapter also highlights the importance of signaling between epithelial cells and myofibroblasts initiated by *Helicobacter pylori* infections that can progress into preneoplastic changes leading to gastric cancer. Finally, therapeutic opportunities provided by the stromal components in cancer are considered.

Composition of Stroma

Cells Comprising the Vasculature

Angiogenic expansion, i.e., blood and lymphatic vessel growth, is an important requirement for cancer progression and metastasis (22), and depends on the activity of several cell types including endothelial cells, pericytes, and vSMCs. Endothelial cells can be divided into two distinct groups depending whether they are constituents of vascular or lymphatic vessels. They differ in the type of vascular endothelial growth factor receptor (VEGFR) expressed: vascular endothelial cells mostly express and signal through VEGFR-2 receptors, whereas lymphatic endothelial cells, especially in tumor-induced lymphangiogenesis, mostly signal through VEGFR-3 (23). Both vascular and lymphatic endothelial cells respond to VEGF, but lymphatic endothelium preferentially responds to VEGF-C and -D. Lymphatic vessels that are VEGFR-3 positive are significantly correlated with the extent of lymph node involvement in gastric cancer (24).

Pericytes are located within the basement membrane of arterioles, capillaries, and postcapillary venules. They envelop endothelial cells and are considered to be important in stabilizing vessel walls and regulating blood flow in the microcirculation. An increasing body of experimental work suggests a role for them in influencing endothelial cell maturation, proliferation, survival, migration, and permeability. They are typically characterized by expression of desmin and α-smooth muscle actin (SMA), but there are organ- and disease-specific differences in the expression of these markers. Pericyte development requires intact PDGFR-β signaling because mice null for this receptor, or for PDGF-B, have no pericytes (25, 26). There are differences in shape and the extent of colocalization of desmin and α-SMA between normal and cancer-derived pericytes. In general, cancer-associated pericytes uniformly coexpress desmin and α-SMA, have irregular shapes including cytoplasmic processes that project toward the tumor parenchyma, and associate loosely with endothelial cells compared with their normal counterparts; the latter might explain the leakiness and sensitivity of tumor vessels to VEGF. Several reports (22, 25) have established that pericytes proliferate at the beginning of angiogenesis in mouse cancer models and they are most abundant on blood vessels at the growing front of tumors where angiogenesis is essential for tumor progression. The origin of pericytes in cancer is not yet clear; transdifferentiation from smooth muscle cells or myofibroblasts remains a possibility, as does homing from the bone marrow and differentiation via the mesenchymal stem cell lineage (Figure 23.1).

Vascular smooth muscle cells are essential components of the vasculature. Whereas pericytes surround capillaries and immature vessels, vSMCs provide the covering in all larger and mature vessels and have a key role in controlling blood flow (22). Differentiation of vSMCs from progenitor cells is stimulated by transforming growth factor (TGF)-β (27); vSMC development also requires intact PDGFR-β signaling because mice lacking PDGF-B or PDGFR-β are deficient in vSMCs, and develop microaneurysms and vessel leakiness (25).

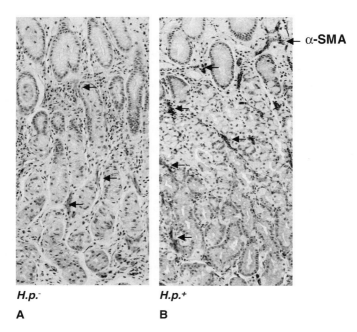

H.p.⁻ *H.p.⁺*

A   **B**

Fig. 23.1 Increased gastric α-smooth muscle actin (SMA) in patients infected with *Helicobacter pylori* (*H.p.*). (**A**) Representative image showing immunohistochemical localization of α-SMA (black arrows) in the gastric corpus of *H.p.* negative and (**B**) positive patients. Note increased number of α-SMA–positive cells in the proliferating region in the patient infected with *H.p.* [Reproduced with permission of Gastroenterology from McCaig et al. (53).]

Inflammatory Cells

Resident inflammatory cells, i.e. leukocytes, mast cells, lymphocytes, and macrophages are recruited initially by chemotactic signals after inflammation and injury; for example, in the case of gastric cancer, these events are evoked in response to *H. pylori* infection (28–30). These cells can migrate out from the lamina propria after epithelial loss caused by mucosa damage (31). They may also influence cancer initiation and cancer progression by secreting soluble factors that influence the stem cell niche and may generate reactive oxygen species that cause DNA damage (32, 33). Within this class of stromal cell, activated macrophages that differentiate from monocytes in response to cytokines such as interleukin (IL)-6 and leukemia inhibitory factor (LIF) (34), are the main source of growth factors and cytokines, whereas mast cells are also important for production, storage, and release of inflammatory mediators such as histamine (33). Moreover, selective recruitment of different subsets of T lymphocytes in cancer could contribute to impaired local antitumor responses (28, 35). In gastric cancer, a decrease in CXCR3-expressing lymphocytes in the tumor mucosa has been reported compared with the tumor-free counterpart (28). Activated macrophages are often called tumor-associated macrophages (TAM); they express the surface markers CD11B and F4/80 and recent evidence suggests that they are probably a particularly important inflammatory cell type in the neoplastic microenvironment because they coordinate inflammatory networks (34, 36–39).

Fibroblasts

Fibroblasts *in vivo* are normally slowly dividing, quiescent cells with low motility and characterized by expression of vimentin but not α-SMA. They function normally as components of connective tissues. However, because of lack of rigorous identification (other than their morphology), the term *fibroblast* is often used synonymously for myofibroblast (see below). Nevertheless, fibroblasts can transdifferentiate to myofibroblasts, characterized by α-SMA expression, upon stimulation by TGF-β and PDGF (40), and during carcinogenesis (41, 42).

Myofibroblasts

The term *myofibroblast* was first used to describe a subset of fibroblasts displaying smooth muscle–like features (43). These cells can also be defined by their localization, morphologic appearance, and expression of certain cytoskeleton markers: α-SMA, vimentin, and desmin. Originally, myofibroblasts expressing only vimentin were classified as V type, whereas if also expressing myosin as VM type; those expressing vimentin, desmin, and α-SMA were named VAD type, but those that expressed only vimentin and α-SMA were termed VA type (44–46). Their localization within a given organ can be diverse, ranging from close proximity to epithelial cells, or to neighboring blood vessels or the smooth muscle layers of the gut (16, 41, 44, 47–50). For these reasons, there is still some confusion in the literature over the use of the term myofibroblast, and for clarity in this chapter, the term is applied to subepithelial, VA/VAM-type myofibroblasts. The close proximity of some myofibroblasts to epithelial cells is thought to be functionally relevant and is the basis of the use of the term "niche cell" (48, 50–54). Gastric and intestinal myofibroblast cultures were first established from fibroblast-like cells migrating out of the lamina propria from mucosal samples denuded of epithelial cells through discrete pores in the subepithelial basement membrane (48, 55). Phenotypical characterization confirmed that they expressed cyclooxygenase (COX)-1 and COX-2 and released prostaglandin E_2 (48, 55). The origin of these cells is still unclear. There are proposals that they may originate from fibroblasts (43) and/or smooth muscle cells (20, 56). However, recent studies in mice and humans involving bone marrow transplants from a male donor to a female recipient suggest that myofibroblasts could derive from circulating bone marrow–derived mesenchymal stem cells via transdifferentiation (47, 57). Taken as a whole, the evidence points to the existence of multiple, different subclasses of myofibroblast, but further work is needed to develop reliable cell-surface markers for different populations.

Myofibroblasts are required for normal epithelial differentiation (44, 58, 59), and they enhance barrier function and modulate electrogenic chloride secretion of intestinal epithelial cells (47, 52). Crucially, in the present context, they have an active part in transepithelial signaling initiated by luminal challenges including bacterial infection (53, 60) and injury (16, 44, 51, 61–63). In the stomach, infection with *H. pylori* (or *H. felis* in mice) increases the number of myofibroblasts in the submucosa (53). The increase in the myofibroblast population has been attributed to increased expression and secretion of epithelial paracrine mediators that in turn stimulate the release of factors from the myofibroblast themselves that stimulate their proliferation and migration. One significant example is the increased expression of a protease, matrix metalloproteinase (MMP)-7 from the epithelium that stimulates release from myofibroblasts of insulin-like growth factor (IGF)-II and its binding protein, IGFBP-5. Cleavage of IGFBP-5 by MMP-7 results in the increased bioavailability of IGF-II that subsequently stimulates proliferation of both epithelial cells and myofibroblasts (53, 64, 65). In addition, MMP-7 activates other MMPs such as MMP-3 and MMP-8, which may also contribute to increased bioavailability of IGF-II through similar actions (Figure 23.2). Other relevant mechanisms include stimulation by the bacterial product, lipopolysaccharide, which, acting via toll-like receptor (TLR), is reported to increase expression and release of IL-8 from myofibroblasts (60).

The mechanisms described above seem to be activated during injury, damage, or infection and to be part of the mechanisms of tissue repair. But, in addition, there is growing evidence that myofibroblasts have an active role in cancer initiation by secreting growth factors, proteases, cytokines, and chemokines (16, 41, 50) that facilitate the recruitment of other cell types, such as macrophages, pericytes, and endothelial cells. Moreover, myofibroblasts are in close proximity with cancer cells

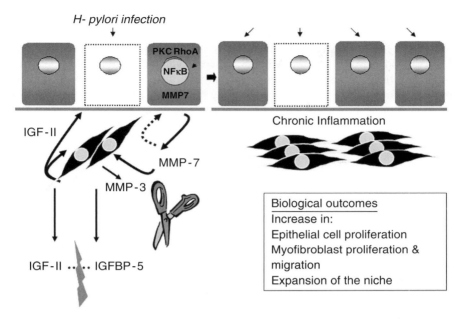

Fig. 23.2 Schematic representation of *Helicobacter pylori* stimulated transepithelial signaling via matrix metalloproteinase (MMP)-7 secretion leading to myofibroblast activation in the gastric corpus. *H. pylori* infection increases MMP-7 expression and secretion, which in turn increases the bioavailable pool of insulin-like growth factor (IGF)-II via stimulating the release of MMP-3 and cleavage of its binding partner, IGFBP-5, from myofibroblasts, thereby stimulating both epithelial cell proliferation and myofibroblast migration and proliferation. The outcome of chronic activation of the system is led to an expansion of the stem cell niche

and a prominent role in cancer progression including metastasis has started to emerge in the last few years. Thus, myofibroblasts are a source of growth factors stimulating proliferation such as hepatocyte growth factor (HGF) and members of the IGF and fibroblast growth factor (FGF) families. In addition, myofibroblasts produce a wide range of proteases, including urokinase plasminogen activator (uPA) and MMPs; these should not be thought to function in isolation, and, for example, collaboration between uPA and the MMP system is described in several cancers (1, 5, 40, 66, 67). In addition, in the breast stromal-derived factor (SDF)-1 has a crucial role in promoting cancer progression by not only stimulating tumor growth but also stimulating neoangiogenesis (68).

Interactions Within the Stroma

Interactions among host cells in the stroma provide all the necessary conditions for supporting cancer progression. For example, myofibroblasts and inflammatory cells secrete PDGF to stimulate angiogenesis and lymphangiogenesis, including pericyte maturation (69). In addition, by secreting proteases, such as MMPs, they participate in the remodeling of the ECM (1, 5). Stromal cells express specific MMPs in cell type–restricted manner; for example, gastric myofibroblasts express MMP-1 and MMP-3 but not MMP-8, whereas colonic myofibroblasts express MMP-1, MMP-3, and MMP-8 and macrophages also express MMP-8, thereby providing the capacity to define the shedding of growth factors and degradation of ECM and other proteins in patterns characteristic of different tumors (53, 64, 70).

Interactions Between Cancer Cells and Their Microenvironment

There is increasing evidence that the behavior of carcinomas is strongly influenced by the interaction between the cancer and host cells (4, 7, 10, 12, 68, 71–73). Cancer cells produce a wide variety of biologically active substances, including proteases, cytokines, chemokines, and growth factors that act as mitogens and chemoattractants for myofibroblasts, and inflammatory cells through mechanisms similar to those described above. In response, host cells also produce and secrete proteases, cytokines, chemokines, and growth factors that further stimulate cancer cell proliferation, epithelial–mesenchymal transition, and inhibit apoptosis, thereby providing survival advantage to cancer cells not only at the primary but at metastatic sites too (Figure 23.3). In addition, there is increased evidence that invasion is a consequence of a cross-talk between cancer cells and the different host cell types, especially myofibroblasts, endothelial cells, and leukocytes, which are themselves invasive (3, 9, 11, 74). Major players that have attracted attention recently include the TGF-β and HGF pathways; in particular, there is strong evidence for complex interactions between TGF-β and HGF in regulating proliferation and invasion of cancer cells (4, 5, 11).

New Therapeutic Tools in Targeting Stromal Cells

Because cancer initiation and progression are highly influenced by the microenvironment, therapy targeted on the stroma might be used effectively as a viable option for both prevention and inhibition of progression of cancer in combination with existing chemotherapeutic agents targeting the cancer cells themselves (11, 69, 74–80). In addition to the inhibition of signal transfer from host cells to cancer cells, such as the growth factor receptor inhibitors already in clinical use, and the inhibition of angiogenesis, taking advantage of the opportunities to target stromal cells could therefore also contribute to the objective of reducing growth-factor drive to tumor cells (81–85). Because myofibroblasts are a key determinant of the cancer microenvironment, they could represent a specific, novel, therapeutic target by selective inhibition of organ-specific myofibroblast-derived targets (5, 7, 12, 40, 53, 64, 77, 82, 82, 86). In addition, therapeutic interventions of TAM generation might enhance existing cancer therapies (34, 36, 37).

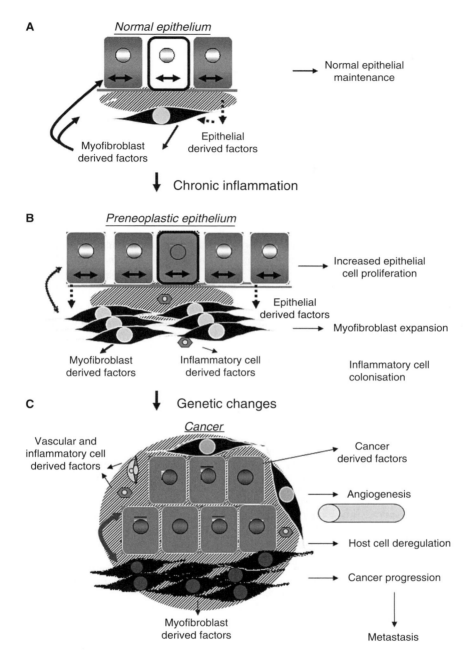

Fig. 23.3 Schematic representation of host and epithelial cell interactions leading to cancer initiation on a background of chronic inflammation. **(top)** Normal epithelial maintenance is regulated by the niche influenced by epithelial cells and the resident niche cell, the myofibroblast. **(middle)** In chronic inflammation, there is colonization by inflammatory cells and increased myofibroblast cell numbers leading to preneoplastic changes in the epithelium. **(bottom)** Because of the influence of the expanded microenvironment, there is increased chance of genetic mutations leading to cancer initiation. New blood vessel formation stimulated by both cancer and host cell factors leads to cancer progression. Host cells adapt (deregulation) to altered microenvironment by changes in their gene expression and intercellular signaling

References

1. Bhowmick, N.A., Neilson, E.G. and Moses, H.L. Stromal fibroblasts in cancer initiation and progression. Nature 432: 332–337, 2004.
2. Christofori, G. Cancer: division of labour. Nature 446: 735–736, 2007.
3. Mareel, M. and Leroy, A. Clinical, cellular, and molecular aspects of cancer invasion. Physiol. Rev. 83: 337–376, 2003.
4. Bhowmick, N.A., Chytil, A., Plieth, D., Gorska, A.E., Dumont, N., Shappell, S., Washington, M.K., Neilson, E.G. and Moses, H.L. TGF-beta signaling in fibroblasts modulates the oncogenic potential of adjacent epithelia. Science 303: 848–851, 2004.
5. Kalluri, R. and Zeisberg, M. Fibroblasts in cancer. Nat. Rev. Cancer 6: 392–401, 2006.
6. van Kempen, L.C., Rhee, J.S., Dehne, K., Lee, J., Edwards, D.R. and Coussens, L.M. Epithelial carcinogenesis: dynamic interplay between neoplastic cells and their microenvironment. Differentiation 70: 610–623, 2002.
7. Jodele, S., Blavier, L., Yoon, J.M. and DeClerck, Y.A. Modifying the soil to affect the seed: role of stromal-derived matrix metalloproteinases in cancer progression. Cancer Metastasis Rev. 25: 35–43, 2006.
8. Liotta, L.A. and Kohn, E.C. The microenvironment of the tumour-host interface. Nature 411: 375–379, 2001.
9. Lippert, E., Falk, W., Bataille, F., Kaehne, T., Naumann, M., Goeke, M., Herfarth, H., Schoelmerich, J. and Rogler, G. Soluble galectin-3 is a strong, colonic epithelial-cell-derived, lamina propria fibroblast-stimulating factor. Gut 56: 43–51, 2007.
10. Orimo, A., Gupta, P.B., Sgroi, D.C., renzana-Seisdedos, F., Delaunay, T., Naeem, R., Carey, V.J., Richardson, A.L. and Weinberg, R.A. Stromal fibroblasts present in invasive human breast carcinomas promote tumor growth and angiogenesis through elevated SDF-1/CXCL12 secretion. Cell 121: 335–348, 2005.
11. Bierie, B. and Moses, H.L. Tumour microenvironment: TGFbeta—the molecular Jekyll and Hyde of cancer. Nat. Rev. Cancer 6: 506–520, 2006.
12. Balkwill, F. Cancer and the chemokine network. Nat. Rev. Cancer 4: 540–550, 2004.
13. Lauwers, G.Y. and Shimizu, M. Pathology of gastric cancer. In: Rustgi, A.K., editor. Gastrointestinal cancers. Saunders: Edinburgh; pp. 321–330, 2003.
14. Hirota, S., Isozaki, K., Moriyama, Y., Hashimoto, K., Nishida, T., Ishiguro, S., Kawano, K., Hanada, M., Kurata, A., Takeda, M., Muhammad, T.G., Matsuzawa, Y., Kanakura, Y., Shinomura, Y. and Kitamura, Y. Gain-of-function mutations of c-kit in human gastrointestinal stromal tumors. Science 279: 577–580, 1998.
15. Heinrich, M.C., Corless, C.L., Duensing, A., McGreevey, L., Chen, C.J., Joseph, N., Singer, S., Griffith, D.J., Haley, A., Town, A., Demetri, G.D., Fletcher, C.D. and Fletcher, J.A. PDGFRA activating mutations in gastrointestinal stromal tumors. Science 299: 708–710, 2003.
16. Powell, D.W., Adegboyega, P.A., Di Mari, J.F. and Mifflin, R.C. Epithelial cells and their neighbors. I. Role of intestinal myofibroblasts in development, repair, and cancer. Am. J. Physiol. Gastrointest. Liver Physiol. 289: G2–G7, 2005.
17. Schmidt, D.R. and Kao, W.J. The interrelated role of fibronectin and interleukin-1 in biomaterial-modulated macrophage function. Biomaterials 28: 371–382, 2007.
18. Bissell, M.J., Radisky, D.C., Rizki, A., Weaver, V.M. and Petersen, O.W. The organizing principle: microenvironmental influences in the normal and malignant breast. Differentiation 70: 537–546, 2002.
19. Radisky, D., Muschler, J. and Bissell, M.J. Order and disorder: the role of extracellular matrix in epithelial cancer. Cancer Invest. 20: 139–153, 2002.
20. Ronnov-Jessen, L., Petersen, O.W., Koteliansky, V.E. and Bissell, M.J. The origin of the myofibroblasts in breast cancer. Recapitulation of tumor environment in culture unravels diversity and implicates converted fibroblasts and recruited smooth muscle cells. J. Clin. Invest. 95: 859–873, 1995.

21. Roskelley, C.D. and Bissell, M.J. The dominance of the microenvironment in breast and ovarian cancer. Semin. Cancer Biol. 12: 97–104, 2002.

22. Adams, R.H. and Alitalo, K. Molecular regulation of angiogenesis and lymphangiogenesis. Nat. Rev. Mol. Cell Biol. 8: 464–478, 2007.

23. Plate, K. From angiogenesis to lymphangiogenesis. Nat. Med. 7: 151–152, 2001.

24. Yonemura, Y., Fushida, S., Bando, E., Kinoshita, K., Miwa, K., Endo, Y., Sugiyama, K., Partanen, T., Yamamoto, H. and Sasaki, T. Lymphangiogenesis and the vascular endothelial growth factor receptor (VEGFR)-3 in gastric cancer. Eur. J. Cancer 37: 918–923, 2001.

25. Morikawa, S., Baluk, P., Kaidoh, T., Haskell, A., Jain, R.K. and McDonald, D.M. Abnormalities in pericytes on blood vessels and endothelial sprouts in tumors. Am. J. Pathol. 160: 985–1000, 2002.

26. Lindahl, P., Johansson, B.R., Leveen, P. and Betsholtz, C. Pericyte loss and microaneurysm formation in PDGF-B-deficient mice. Science 277: 242–245, 1997.

27. Chen, S. and Lechleider, R.J. Transforming growth factor-beta-induced differentiation of smooth muscle from a neural crest stem cell line. Circ. Res. 94: 1195–1202, 2004.

28. Enarsson, K., Johnsson, E., Lindholm, C., Lundgren, A., Pan-Hammarstrom, Q., Stromberg, E., Bergin, P., Baunge, E.L., Svennerholm, A.M. and Quiding-Jarbrink, M. Differential mechanisms for T lymphocyte recruitment in normal and neoplastic human gastric mucosa. Clin. Immunol. 118: 24–34, 2006.

29. Guruge, J.L., Falk, P.G., Lorenz, R.G., Dans, M., Wirth, H.P., Blaser, M.J., Berg, D.E. and Gordon, J.I. Epithelial attachment alters the outcome of Helicobacter pylori infection. Proc. Natl. Acad. Sci. U.S.A. 95: 3925–3930, 1998.

30. Bodger, K., Ahmed, S., Michael, A., Khan, A.L., Pazmany, L., Pritchard, D.M., Dimaline, R., Dockray, G.J. and Varro, A. Altered gastric corpus expression of tissue inhibitors of metalloproteinases in human and murine Helicobacter infection. J. Clin. Pathol. 61: 72–28, 2007.

31. Mahida, Y.R., Galvin, A.M., Gray, T., Makh, S., McAlindon, M.E., Sewell, H.F. and Podolsky, D.K. Migration of human intestinal lamina propria lymphocytes, macrophages and eosinophils following the loss of surface epithelial cells. Clin. Exp. Immunol. 109: 377–386, 1997.

32. Fukuda, Y., Ishizaki, M., Kudoh, S., Kitaichi, M. and Yamanaka, N. Localization of matrix metalloproteinases-1, -2, and -9 and tissue inhibitor of metalloproteinase-2 in interstitial lung diseases. Lab. Invest. 78: 687–698, 1998.

33. Sala, A. and Folco, G. Neutrophils, endothelial cells, and cysteinyl leukotrienes: a new approach to neutrophil-dependent inflammation? Biochem. Biophys. Res. Commun. 283: 1003–1006, 2001.

34. Duluc, D., Delneste, Y., Tan, F., Moles, M.P., Grimaud, L., Lenoir, J., Preisser, L., Anegon, I., Catala, L., Ifrah, N., Descamps, P., Gamelin, E., Gascan, H., Hebbar, M. and Jeannin, P. Tumor-associated leukemia inhibitory factor and IL-6 skew monocyte differentiation into tumor-associated-macrophage-like cells. Blood 110: 4319–4330, 2007.

35. Gerard, C. and Rollins, B.J. Chemokines and disease. Nat. Immunol. 2: 108–115, 2001.

36. Porta, C., Subhra, K.B., Larghi, P., Rubino, L., Mancino, A. and Sica, A. Tumor promotion by tumor-associated macrophages. Adv. Exp. Med. Biol. 604: 67–86, 2007.

37. Sica, A., Rubino, L., Mancino, A., Larghi, P., Porta, C., Rimoldi, M., Solinas, G., Locati, M., Allavena, P. and Mantovani, A. Targeting tumour-associated macrophages. Expert. Opin. Ther. Targets 11: 1219–1229, 2007.

38. Nakagawa, J., Saio, M., Tamakawa, N., Suwa, T., Frey, A.B., Nonaka, K., Umemura, N., Imai, H., Ouyang, G.F., Ohe, N., Yano, H., Yoshimura, S., Iwama, T. and Takami, T. TNF expressed by tumor-associated macrophages, but not microglia, can eliminate glioma. Int. J. Oncol. 30: 803–811, 2007.

39. Nazareth, M.R., Broderick, L., Simpson-Abelson, M.R., Kelleher, R.J., Jr., Yokota, S.J. and Bankert, R.B. Characterization of human lung tumor-associated fibroblasts and their ability to modulate the activation of tumor-associated T cells. J. Immunol. 178: 5552–5562, 2007.

40. De, W.O. and Mareel, M. Role of myofibroblasts at the invasion front. Biol. Chem. 383: 55–67, 2002.

41. Adegboyega, P.A., Mifflin, R.C., DiMari, J.F., Saada, J.I. and Powell, D.W. Immunohistochemical study of myofibroblasts in normal colonic mucosa, hyperplastic polyps, and adenomatous colorectal polyps. Arch. Pathol. Lab. Med. 126: 829–836, 2002.

42. Desmouliere, A., Darby, I.A. and Gabbiani, G. Normal and pathologic soft tissue remodeling: role of the myofibroblast, with special emphasis on liver and kidney fibrosis. Lab. Invest. 83: 1689–1707, 2003.

43. Gabbiani, G. The cellular derivation and the life span of the myofibroblast. Pathol. Res. Pract. 192: 708–711, 1996.

44. Powell, D.W., Mifflin, R.C., Valentich, J.D., Crowe, S.E., Saada, J.I. and West, A.B. Myofibroblasts. I. Paracrine cells important in health and disease. Am. J. Physiol. 277: C1–C9, 1999.

45. Eyden, B. The myofibroblast: a study of normal, reactive and neoplastic tissues, with an emphasis on ultrastructure. Part 1—normal and reactive cells. J. Submicrosc. Cytol. Pathol. 37: 109–204, 2005.

46. Eyden, B. The myofibroblast: a study of normal, reactive and neoplastic tissues, with an emphasis on ultrastructure. Part 2—tumours and tumour-like lesions. J. Submicrosc. Cytol. Pathol. 37: 231–296, 2005.

47. Powell, D.W., Mifflin, R.C., Valentich, J.D., Crowe, S.E., Saada, J.I. and West, A.B. Myofibroblasts. II. Intestinal subepithelial myofibroblasts. Am. J. Physiol. 277: C183–C201, 1999.

48. Wu, K.C., Jackson, L.M., Galvin, A.M., Gray, T., Hawkey, C.J. and Mahida, Y.R. Phenotypic and functional characterisation of myofibroblasts, macrophages, and lymphocytes migrating out of the human gastric lamina propria following the loss of epithelial cells. Gut 44: 323–330, 1999.

49. Valentich, J.D., Popov, V., Saada, J.I. and Powell, D.W. Phenotypic characterization of an intestinal subepithelial myofibroblast cell line. Am. J. Physiol. 272: C1513–C1524, 1997.

50. Mutoh, H., Sakurai, S., Satoh, K., Osawa, H., Tomiyama, T., Kita, H., Yoshida, T., Tamada, K., Yamamoto, H., Isoda, N., Ido, K. and Sugano, K. Pericryptal fibroblast sheath in intestinal metaplasia and gastric carcinoma. Gut 54: 33–39, 2005.

51. McKaig, B.C., Makh, S.S., Hawkey, C.J., Podolsky, D.K. and Mahida, Y.R. Normal human colonic subepithelial myofibroblasts enhance epithelial migration (restitution) via TGF-beta3. Am. J. Physiol. 276: G1087–G1093, 1999.

52. Beltinger, J., McKaig, B.C., Makh, S., Stack, W.A., Hawkey, C.J. and Mahida, Y.R. Human colonic subepithelial myofibroblasts modulate transepithelial resistance and secretory response. Am. J. Physiol. 277: C271–C279, 1999.

53. McCaig, C., Duval, C., Hemers, E., Steele, I., Pritchard, D.M., Przemeck, S., Dimaline, R., Ahmed, S., Bodger, K., Kerrigan, D.D., Wang, T.C., Dockray, G.J. and Varro, A. The role of matrix metalloproteinase-7 in redefining the gastric microenvironment in response to Helicobacter pylori. Gastroenterology 130: 1754–1763, 2006.

54. Andoh, A., Bamba, S., Brittan, M., Fujiyama, Y. and Wright, N.A. Role of intestinal subepithelial myofibroblasts in inflammation and regenerative response in the gut. Pharmacol. Ther. 114: 94–106, 2007.

55. Mahida, Y.R., Beltinger, J., Makh, S., Goke, M., Gray, T., Podolsky, D.K. and Hawkey, C.J. Adult human colonic subepithelial myofibroblasts express extracellular matrix proteins and cyclooxygenase-1 and -2. Am. J. Physiol. 273: G1341–G1348, 1997.

56. Rajkumar, V.S., Howell, K., Csiszar, K., Denton, C.P., Black, C.M. and Abraham, D.J. Shared expression of phenotypic markers in systemic sclerosis indicates a convergence of pericytes and fibroblasts to a myofibroblast lineage in fibrosis. Arthritis Res. Ther. 7: R1113–R1123, 2005.

57. Direkze, N.C., Forbes, S.J., Brittan, M., Hunt, T., Jeffery, R., Preston, S.L., Poulsom, R., Hodivala-Dilke, K., Alison, M.R. and Wright, N.A. Multiple organ engraftment by bone-marrow-derived myofibroblasts and fibroblasts in bone-marrow-transplanted mice. Stem Cells 21: 514–520, 2003.

58. Marsh, M.N. and Trier, J.S. Morphology and cell proliferation of subepithelial fibroblasts in adult mouse jejunum. I. Structural features. Gastroenterology 67: 622–635, 1974.

59. Nishida, T., Tsuji, S., Kimura, A., Tsujii, M., Ishii, S., Yoshio, T., Shinzaki, S., Egawa, S., Irie, T., Yasumaru, M., Iijima, H., Murata, H., Kawano, S. and Hayashi, N. Endothelin-1, an ulcer

inducer, promotes gastric ulcer healing via mobilizing gastric myofibroblasts and stimulates production of stroma-derived factors. Am. J. Physiol. Gastrointest. Liver Physiol. 290: G1041–G1050, 2006.

60. Otte, J.M., Rosenberg, I.M. and Podolsky, D.K. Intestinal myofibroblasts in innate immune responses of the intestine. Gastroenterology 124: 1866–1878, 2003.

61. Jackson, L.M., Wu, K.C., Mahida, Y.R., Jenkins, D. and Hawkey, C.J. Cyclooxygenase (COX) 1 and 2 in normal, inflamed, and ulcerated human gastric mucosa. Gut 47: 762–770, 2000.

62. Seymour, M.L., Zaidi, N.F., Hollenberg, M.D. and MacNaughton, W.K. PAR1-dependent and independent increases in COX-2 and PGE2 in human colonic myofibroblasts stimulated by thrombin. Am. J. Physiol. Cell Physiol. 284: C1185–C1192, 2003.

63. Andoh, A., Bamba, S., Brittan, M., Fujiyama, Y. and Wright, N.A. Role of intestinal subepithelial myofibroblasts in inflammation and regenerative response in the gut. Pharmacol. Ther. 114: 94–106, 2007.

64. Hemers, E., Duval, C., McCaig, C., Handley, M., Dockray, G.J. and Varro, A. Insulin-like growth factor binding protein-5 is a target of matrix metalloproteinase-7: implications for epithelial-mesenchymal signaling. Cancer Res. 65: 7363–7369, 2005.

65. Varro, A., Kenny, S., Hemers, E., McCaig, C., Przemeck, S., Wang, T.C., Bodger, K. and Pritchard, D.M. Increased gastric expression of MMP-7 in hypergastrinemia and significance for epithelial-mesenchymal signaling. Am. J. Physiol. Gastrointest. Liver Physiol. 292: G1133–G1140, 2007.

66. Anastasiadis, P.Z., Moon, S.Y., Thoreson, M.A., Mariner, D.J., Crawford, H.C., Zheng, Y. and Reynolds, A.B. Inhibition of RhoA by p120 catenin. Nat. Cell Biol. 2: 637–644, 2000.

67. Nakayama, H., Enzan, H., Miyazaki, E. and Toi, M. Alpha smooth muscle actin positive stromal cells in gastric carcinoma. J. Clin. Pathol. 55: 741–744, 2002.

68. Orimo, A. and Weinberg, R.A. Stromal fibroblasts in cancer: a novel tumor-promoting cell type. Cell Cycle 5: 1597–1601, 2006.

69. Aghi, M., Cohen, K.S., Klein, R.J., Scadden, D.T. and Chiocca, E.A. Tumor stromal-derived factor-1 recruits vascular progenitors to mitotic neovasculature, where microenvironment influences their differentiated phenotypes. Cancer Res. 66: 9054–9064, 2006.

70. Coussens, L.M. and Werb, Z. Inflammation and cancer. Nature 420: 860–867, 2002.

71. Stoeltzing, O., McCarty, M.F., Wey, J.S., Fan, F., Liu, W., Belcheva, A., Bucana, C.D., Semenza, G.L. and Ellis, L.M. Role of hypoxia-inducible factor 1alpha in gastric cancer cell growth, angiogenesis, and vessel maturation. J. Natl. Cancer Inst., 96: 946–956, 2004.

72. Krtolica, A., Parrinello, S., Lockett, S., Desprez, P.Y. and Campisi, J. Senescent fibroblasts promote epithelial cell growth and tumorigenesis: a link between cancer and aging. Proc. Natl. Acad. Sci. U.S.A. 98: 12072–12077, 2001.

73. Balkwill, F. and Coussens, L.M. Cancer: an inflammatory link. Nature 431: 405–406, 2004.

74. Yashiro, M., Nakazawa, K., Tendo, M., Kosaka, K., Shinto, O. and Hirakawa, K. Selective cyclooxygenase-2 inhibitor downregulates the paracrine epithelial-mesenchymal interactions of growth in scirrhous gastric carcinoma. Int. J. Cancer 120: 686–693, 2007.

75. Overall, C.M. and Dean, R.A. Degradomics: systems biology of the protease web. Pleiotropic roles of MMPs in cancer. Cancer Metastasis Rev. 25: 69–75, 2006.

76. Gupta, G.P., Nguyen, D.X., Chiang, A.C., Bos, P.D., Kim, J.Y., Nadal, C., Gomis, R.R., Manova-Todorova, K. and Massague, J. Mediators of vascular remodelling co-opted for sequential steps in lung metastasis. Nature 446: 765–770, 2007.

77. Albini, A. and Sporn, M.B. The tumour microenvironment as a target for chemoprevention. Nat. Rev. Cancer 7: 139–147, 2007.

78. Barcellos-Hoff, M.H., Park, C. and Wright, E.G. Radiation and the microenvironment—tumorigenesis and therapy. Nat. Rev. Cancer 5: 867–875, 2005.

79. Coussens, L.M., Fingleton, B. and Matrisian, L.M. Matrix metalloproteinase inhibitors and cancer: trials and tribulations. Science 295: 2387–2392, 2002.

80. Rhee, J.S. and Coussens, L.M. RECKing MMP function: implications for cancer development. Trends Cell Biol. 12: 209–211, 2002.

81. Overall, C.M. and Kleifeld, O. Tumour microenvironment—opinion: validating matrix met-alloproteinases as drug targets and anti-targets for cancer therapy. Nat. Rev. Cancer 6: 227–239, 2006.

82. Mareel, M. and Madani, I. Tumour-associated host cells participating at invasion and metastasis : targets for therapy? Acta Chir. Belg. 106: 635–640, 2006.

83. Lazzeri, E. and Romagnani, P. CXCR3-binding chemokines: novel multifunctional therapeutic targets. Curr. Drug Targets Immune Endocr. Metab. Disord. 5: 109–118, 2005.

84. Furuya, M., Nishiyama, M., Kasuya, Y., Kimura, S. and Ishikura, H. Pathophysiology of tumor neovascularization. Vasc. Health Risk Manag. 1: 277–290, 2005.

85. Baselga, J. Targeting tyrosine kinases in cancer: the second wave. Science 312: 1175–1178, 2006.

86. Offersen, B.V., Nielsen, B.S., Hoyer-Hansen, G., Rank, F., Hamilton-Dutoit, S., Overgaard, J. and Andreasen, P.A. The myofibroblast is the predominant plasminogen activator inhibitor-1-expressing cell type in human breast carcinomas. Am. J. Pathol. 163: 1887–1899, 2003.

87. Selman, M., Ruiz, V., Cabrera, S., Segura, L., Ramirez, R., Barrios, R. and Pardo, A. TIMP-1, -2, -3, and -4 in idiopathic pulmonary fibrosis. A prevailing nondegradative lung microenvironment? Am. J. Physiol. Lung Cell Mol. Physiol. 279: L562–L574, 2000.

88. Powers, C.J., McLeskey, S.W. and Wellstein, A. Fibroblast growth factors, their receptors and signaling. Endocr. Relat. Cancer 7: 165–197, 2000.

89. Bonner, J.C. Regulation of PDGF and its receptors in fibrotic diseases. Cytokine Growth Factor Rev. 15: 255–273, 2004.

90. Seymour, M.L., Binion, D.G., Compton, S.J., Hollenberg, M.D. and MacNaughton, W.K. Expression of proteinase-activated receptor 2 on human primary gastrointestinal myofibroblasts and stimulation of prostaglandin synthesis. Can. J. Physiol. Pharmacol. 83: 605–616, 2005.

91. Mifflin, R.C., Saada, J.I., Di Mari, J.F., Adegboyega, P.A., Valentich, J.D. and Powell, D.W. Regulation of COX-2 expression in human intestinal myofibroblasts: mechanisms of IL-1-mediated induction. Am. J. Physiol. Cell Physiol. 282: C824–C834, 2002.

92. McKaig, B.C., McWilliams, D., Watson, S.A. and Mahida, Y.R. Expression and regulation of tissue inhibitor of metalloproteinase-1 and matrix metalloproteinases by intestinal myofibroblasts in inflammatory bowel disease. Am. J. Pathol. 162: 1355–1360, 2003.

93. Flemstrom, G. and Sjoblom, M. Epithelial cells and their neighbors. II. New perspectives on efferent signaling between brain, neuroendocrine cells, and gut epithelial cells. Am. J. Physiol. Gastrointest. Liver Physiol. 289: G377–G380, 2005.

94. Varga, J. and Abraham, D. Systemic sclerosis: a prototypic multisystem fibrotic disorder. J. Clin. Invest. 117: 557–567, 2007.

95. Iredale, J.P. Models of liver fibrosis: exploring the dynamic nature of inflammation and repair in a solid organ. J. Clin. Invest. 117: 539–548, 2007.

96. Lawson, W.E., Polosukhin, V.V., Zoia, O., Stathopoulos, G.T., Han, W., Plieth, D., Loyd, J.E., Neilson, E.G. and Blackwell, T.S. Characterization of fibroblast-specific protein 1 in pulmonary fibrosis. Am. J. Respir. Crit. Care Med. 171: 899–907, 2005.

97. Kuijper, S., Turner, C.J. and Adams, R.H. Regulation of angiogenesis by eph-ephrin interactions. Trends Cardiovasc. Med. 17: 145–151, 2007.

98. Arora, P., Ricks, T.K. and Trejo, J. Protease-activated receptor signalling, endocytic sorting and dysregulation in cancer. J. Cell Sci. 120: 921–928, 2007.

99. Edelman, D.A., Jiang, Y., Tyburski, J.G., Wilson, R.F. and Steffes, C.P. Cytokine production in lipopolysaccharide-exposed rat lung pericytes. J. Trauma 62: 89–93, 2007.

100. Kane, R., Stevenson, L., Godson, C., Stitt, A.W. and O'Brien, C. Gremlin gene expression in bovine retinal pericytes exposed to elevated glucose. Br. J. Ophthalmol. 89: 1638–1642, 2005.

101. Marino, A.P., da, S.A., dos, S.P., Pinto, L.M., Gazzinelli, R.T., Teixeira, M.M. and Lannes-Vieira, J. Regulated on activation, normal T cell expressed and secreted (RANTES) antagonist (Met-RANTES) controls the early phase of trypanosoma cruzi-elicited myocarditis. Circulation 110: 1443–1449, 2004.

Chapter 24
Future Prospects for *Helicobacter pylori* Vaccination

Dominique Velin and Pierre Michetti

Current Therapies of *Helicobacter* Infection

Helicobacter pylori eradication therapies have revolutionized the natural course of peptic ulcer disease. Antibiotic treatment of *H. pylori* infection is relatively success-ful—the organism is eradicated in approximately 80% of patients. Numerous studies have suggested that the yearly relapse rate of 80% for duodenal ulcer and 60% for gastric ulcer is reduced to less than 5% after successful *H. pylori* eradication (Nervi et al. 2006). Eradication of *H. pylori* is strongly recommended in duodenal and gastric ulcer disease, mucosa-associated lymphoid tissue lymphoma, atrophic gastritis, postgastric cancer resection, and in first-degree relatives of gastric cancer patients (Malfertheiner et al. 2007). Triple therapy, consisting of two antibiotics—clarithromycin and amoxicillin or metronidazole—in combination with a proton pump inhibitor, has become the first-line option for infection with *H. pylori* and has been recommended at several consensus conferences. Because of antibiotic resist-ance, bismuth-based quadruple therapy has also become a first-line regimen in areas with exceedingly high rates of clarithromycin and metronidazole resistance, and is the preferred second-line option otherwise. Triple therapies based on levofloxacin and/or rifabutin mainly with combination of amoxicillin are options if multiple eradication failure occurs (Wolle and Malfertheiner 2007). In clinical practice, *H. pylori* eradication with the recommended treatment regimens will fail in approxi-mately 20% of patients. Antimicrobial resistance is the primary cause of treatment failure and mainly concerns two of the major antibiotic components in current antibiotic regimens: macrolides and nitroimidazoles. Clarithromycin resistance is well understood, its mechanism has been defined, and its clinical relevance has defi-nitely been proven. The prevalence of clarithromycin resistance is well documented, with important differences noted between northern and southern Europe, where resistance rates in adults were <5% and 20%, respectively (Megraud, 2004).

In addition to generating antibiotic resistance, current therapies are relatively expensive and thus not fully accessible for people living in less-developed coun-tries. Therefore, new therapies still need to be developed (Kabir 2007; Wolle and Malfertheiner 2007). One of therapeutic approaches that needs to be further evalu-ated is the development of preventive and/or therapeutic vaccination strategies.

Helicobacter pylori Subverts the Mechanisms of Defense

A hallmark feature of infection with *H. pylori* is a chronic, active inflammatory response with an inability of the host to clear the infection, resulting in tissue damage. In addition, *H. pylori* infection leads to the generation of adaptive immune responses, which are also ineffective at eradicating the pathogen from the stomach.

Acidic pH of the Stomach: Role of Urease

With food and saliva intake, humans ingest thousands of microorganisms each day, most of which are unable to colonize the stomach. Under fasting conditions, the human gastric luminal pH is <2; this low pH is incompatible with the proliferation of most bacteria within the gastric lumen. A major virulence factor of *H. pylori* is urease, which is expressed by all strains. Urease is composed of two subunits, α, which is approximately 24 kDa, and β, which is approximately 68 kDa. *H. pylori* produces a large amount of urease, representing 5%–10% of the total protein content of the bacteria. The most important role urease has is to hydrolyze urea into CO_2 and NH_3, which aids in buffering the low pH of the stomach. An *H. pylori* null mutant defective in production of urease is unable to colonize animal models (Eaton and Krakowka 1994), suggesting that acid resistance conferred by urease is essential for *H. pylori* to overcome a major antibacterial natural barrier of the stomach.

Helicobacter pylori *Is a Flagellated Bacterium Living in the Stomach Lumen*

After entering the stomach, *H. pylori* penetrates the mucus gastric layer (Schreiber et al. 2004) but does not traverse the epithelial barrier (Hazell et al. 1986), and therefore it is considered a noninvasive pathogen. Most of *H. pylori* organisms are free living in the mucus layer, but some organisms attach to the apical surface of gastric epithelial cells (Hazell et al. 1986) and small amounts have been shown to invade epithelial cells (Semino-Mora et al. 2003). Flagella permit bacterial motility, which allows bacterial penetration of the mucus layer. *H. pylori* null mutant defective in production of flagella is unable to colonize gnotobiotic piglets (Eaton et al. 1996). After mucus entry, several outer membrane proteins, including BabA, SabA, AlpA, AlpB, and HopZ can mediate bacterial adherence to gastric epithelial cells. Attachment of *H. pylori* to gastric epithelial cells results in activation of numerous signaling pathways and permits efficient delivery of toxins or other effector molecules into the cells (Guruge et al. 1998). Flagella and the expression of adhesins are major virulence factors allowing *H. pylori* colonization and persistence, and are important in subverting the antibacterial function of the gastric mucus layer.

Acute Helicobacter pylori *Infection Leads to the Development of Innate and Adaptive Immune Responses*

Recently, 20 human volunteers were experimentally infected with *H. pylori* (Graham et al. 2004). Gastric biopsies performed 2 weeks after infection showed infiltration of lymphocytes and monocytes, along with significantly increased expression of interleukin (IL)-1β, IL-8, and IL-6 in the gastric antrum (Graham et al. 2004). Anti-*H. pylori* immunoglobulin (Ig)M and IgG responses were detected in the serum of infected individuals. In addition, 4 weeks after infection, the numbers of gastric CD4+ and CD8+ T cells were increased compared with pre-infection levels (Nurgalieva et al. 2005). These data provide evidence that gastric inflammation develops within a short period of time after *H. pylori* infection.

Chronic Helicobacter pylori *Infection Is Characterized by the Presence of a Systemic and Local Anti-Helicobacter* pylori *Immune Response*

Gastric mucosal biopsies from humans persistently infected with *H. pylori* reveal an increased infiltration of various types of leukocytes compared with biopsies from uninfected humans (Dixon et al. 1996). Lymphocytes (T and B cells), macrophages, neutrophils, mast cells, and dendritic cells are usually present (Dixon et al. 1996; Suzuki T et al. 2002). B cells and CD4+ T cells together with dendritic cells sometimes organize into lymphoid follicles (Terres and Pajares 1998) reflecting ongoing antigen presentation and chronic immune responses. *H. pylori*–specific CD4+ T cells are detectable in the gastric mucosa and peripheral blood of infected but not uninfected humans (Di Tommaso et al. 1995). Levels of cytokines (interferon-γ, tumor necrosis factor α, IL-1β, IL-6, IL-7 IL-8, IL-10, and IL-18) are increased in the stomach of *H. pylori*–infected humans compared with those uninfected (Lindholm et al. 1998). IL-4 has not been detected in the gastric mucosa of most *H. pylori*–infected individuals (Lindholm et al. 1998). Therefore, it has been concluded that *H. pylori* infection leads to a Th1-polarized response. Associated with cellular responses, a humoral immune response is elicited in nearly all *H. pylori*–infected humans (Perez-Perez et al.1988). Serum IgA and IgG antibodies in chronically infected persons are directed toward many different *H. pylori* antigens (Perez-Perez et al. 1988). A local antibody response directed toward *H. pylori* antigens is also detectable with chronic *H. pylori* infection. These subjects have remarkably higher frequencies of total IgA- and IgM-secreting cells than the noninfected subjects, whereas the frequencies of IgG-secreting cells were virtually the same in the different groups. In addition, most infected subjects have IgA antibody–secreting cells reacting with *H. pylori* membrane proteins, flagellin, and urease, whereas none of the noninfected subjects have any detectable *H. pylori*–reactive antibody-secreting cells (Mattsson et al. 1998). Secretory IgA antibodies to *H. pylori* antigens (such as urease α and β subunits and the 66-kDa heat shock protein) are detectable in gastric juice, suggesting that potentially anti-*H. pylori* secretory antibodies might have access to the bacteria living in the lumen of stomach (Hayashi et al. 1998).

Taken together, these observations indicate that in *H. pylori* infection, although eliciting an immune response, the host remains unable to clear the bacteria.

Preclinical Development of Vaccines Directed Against *Helicobacter pylori*

Animal Models of Helicobacter pylori Infection

Most of our knowledge in the immune responses directed toward *H. pylori* derives from studies in subjects with chronic infection. Consequently, such information may have limited utility in developing strategies for a successful therapeutic and/or prophylactic vaccine. It is not known from observing natural infection what type of response should be elicited to induce clearance of the bacteria, given that this does not occur in the natural course of *H. pylori* infection. The design of vaccines have therefore concentrated on the use of animal models of acute and chronic *Helicobacter* infections.

The seminal work reported in 1990 by Lee et al. demonstrated the feasibility of studying different aspects of the pathology and the immune response induced by *Helicobacter* species in mice. These investigators using germ-free mice and *H. felis*, a bacteria that naturally infects cats and dogs, achieved successful long-term colonization and associated gastritis in these mice. This model became very popular and a large number of immunization studies were performed in *H. felis*–infected mice. This was made possible by the fact that vaccine candidate antigens are shared between *H. felis* and *H. pylori* species (i.e., urease and heat shock proteins). Thereafter, *H. pylori* strains have been adapted to the mouse stomach and this experimental model reproduces several aspects of the human infection (Lee et al. 1997; Marchetti et al. 1995). *H. pylori*–infected mice have been used to investigate the role of *H. pylori*–specific virulence factors such as VacA, CagA, and urease in gastric colonization and inflammation (Ghiara et al. 1995). Successful colonization with *H. pylori* has been reported in rats, guinea pigs, Mongolian gerbils, gnotobiotic pigs, cats, and beagle dogs (Del Giudice et al. 2001). However, because of the limitation of specific immune reagents in these other animals, and the development of transgenic mice, most of them have not been exploited as research models for *H. pylori* vaccines. *H. pylori* naturally infects some species of nonhuman primates, and the pathologic changes in the stomach resulting from *H. pylori* infection are very similar to those observed in humans (Dubois et al. 1994). This animal model is therefore the animal model of choice when evaluating the final phases of preclinical development of a vaccine candidate. However, the cost of nonhuman primates and the high rates of natural infections with *H. pylori* as well as *H. heilmanii* limit the use of this model in large-scale studies.

Curative and Therapeutic Vaccines

Virtually all vaccines in use today are prophylactic, designed to prevent acquisition of infection. There is growing interest, however, in therapeutic vaccines that might be used to treat preexisting infections. Because of the large numbers of people who are already infected with *H. pylori*, therapeutic vaccination has merit. However, prophylactic immunization could also be used to complement therapeutic vaccination to more efficiently limit *H. pylori* transmission. *H. pylori* is acquired by oral ingestion and is mainly transmitted within families in early childhood (Feldman 2001; Rowland et al. 1999). It seems likely that in industrialized countries, direct transmission usually occurs from person to person by vomitus, saliva, or feces; additional transmission routes, such as water, may be important in developing countries (Goodman et al. 1996; Parsonnet et al. 1999). Therefore, therapeutic vaccination should be useful to cure the disease in adults whereas prophylactic vaccination presumably should be more readily adapted to children. Using vaccines, *H. pylori* infections and/or gastritis could be substantially prevented, reduced, or eliminated by prophylactic and therapeutic vaccinations (Corthesy-Theulaz et al. 1995; Czinn et al. 1993; Doidge et al. 1994; Ikewaki et al. 2000). It has been suggested that a 10-year prophylactic vaccination program in the United States starting in 2010 could reduce *H. pylori* prevalence to 0.07% by the end of the twenty-first century (Rupnow et al. 2001).

Source of Helicobacter pylori–*Derived Antigens*

Whole-Cell Vaccines

The first murine vaccine studies used *H. felis* bacterial lysates or chemically inactivated whole-cell bacteria delivered to mice orogastrically along with cholera toxin (CT) or *Escherichia coli* heat-labile enterotoxin (LT) as mucosal adjuvants (Czinn et al. 1993; Doidge et al. 1994). Protection using crude *Helicobacter* antigen preparations is very efficient, often reaching 100%. However, the development of such vaccines would encounter quality-control and regulatory issues related to the quality and reproducibility of vaccine preparations. In addition, the relatively undefined nature of whole-cell vaccines and their potential for reactogenicity [lipopolysaccharide (LPS)] and autoimmunity (Lewis blood group and proton pump epitopes) make it difficult to envision their use in humans.

Antigen-Based Vaccines

For the reasons described above, the most promising approach is thus the development of vaccines against *H. pylori* based on defined antigens.

Urease. Urease is an abundant protein of *H. pylori*, essential for colonization and pathogenic events; thus, this antigen is an ideal vaccine candidate. In addition, proteomic analysis identified urease as a dominant immunogenic protein (McAtee et al. 1998). Urease is conserved among the different gastric species of the *Helicobacter* genus (Ferrero and Labigne 1993). Different routes of immunization (oral, intranasal, intrarectal, intramuscular) with urease (ureB subunit or inactive form of holoenzyme) associated with different adjuvants (CT, LT, alum, QS21) have been shown to protect or cure *H. felis* or *H. pylori* infections in mice and ferrets (Del Giudice et al. 2001). A significant reduction in the number of bacteria colonizing the stomach was obtained in rhesus monkeys primed by oral immunization and then boosted by intramuscular immunization (Lee et al. 1999a, 1999b; Solnick et al. 2000).

VacA, CagA, and NAP. Virulence factors are well-known protective antigens for vaccine development. For example, in the acellular pertussis vaccine, one of the antigens is a detoxified form of the pertussis toxin which is a major virulence factor of *Bordetella pertussis*. Therefore, vaccine formulations based on antigens involved in the pathogenesis are potentially promising, and in the case of *H. pylori* infection, antigens such as VacA, CagA, or NAP could be suitable candidates. These three antigens have been shown to cure and/or prevent *H. pylori* colonization of the mouse stomach when used in vaccine preparations (Ghiara et al. 1997; Satin et al. 2000). Although this approach is promising, many strains do not express CagA and VacA, limiting their potential as vaccine candidates if used alone.

Other antigens. Heat shock proteins were shown to protect mice against *Helicobacter* infections (Ferrero et al. 1995; Yamaguchi et al. 2000). Although these antigens are promising, it has to be noted that heat shock proteins are highly conserved phylogenetically and might have the potential to prime immune response against human heat shock proteins. Catalase (Radcliff et al. 1997), Lpp20 (Keenan et al. 2000), 50/59 Kda (Dunkley et al. 1999), and different fusion proteins (Kang et al. 2005; Kim et al. 2001) also confer protection to mice against *Helicobacter* spp. infection.

Vaccine Formulations

Our understanding of *H. pylori*–induced pathogenesis, the study of the *H. pylori*–induced immune response in infected individuals, the availability of animal models of *Helicobacter* spp. infections, and the identification of different protective antigens led to the generation of vaccine candidates for clinical phases of development. Because *H. pylori* lives in close association with the epithelial mucous membrane of the stomach, the goal of clinical *Helicobacter* spp. vaccine development was initially to assess induction of gastric immune responses. Therefore, most of the initial vaccine preparations contained mucosal adjuvants or were engineered in mucosal live vector systems.

Mucosal adjuvants. To achieve significant levels of protection in animal models, strong mucosal adjuvants were required, such as CT, LT, the fully nontoxic LT mutant LTK63 or CTA1-DD adjuvant (Akhiani et al. 2006; Corthesy-Theulaz et al.

1995; Marchetti et al. 1998; Michetti et al. 1994). Other mucosal adjuvants [CpG (Shi et al. 2005) and chitosan (Moschos et al. 2004)] have also been evaluated and some of them—mixed with *H. pylori* antigens—have been shown to be effective in limiting *H. pylori* infection in mice.

Live vectors. Considering known effective human vaccines, those consisting of a living organism have often proven to be the most effective. The live attenuated strain of poliovirus induces solid protection against the spread of the wild-type virus (Ghendon and Robertson 1994). The vaccine used for immunization against typhoid also consists of an attenuated strain of the causative pathogen, *Salmonella typhi* (Cryz 1993). Attenuated strains usually possess a mutation in a particular virulence factor that renders it harmless. It has thus been suggested that attenuated *S. typhi* expressing *H. pylori* antigens might constitute a bivalent vaccine against *S. typhi* and *H. pylori* infections. A recombinant strain of *S. typhimurium*, which infects mice, has been modified to express the A and B subunits of urease from *H. pylori* (Corthesy-Theulaz et al. 1998; Gomez-Duarte et al. 1998). Nasal and oral immunization with this recombinant strain demonstrated a significant reduction in *Helicobacter* colonization in mice (Corthesy-Theulaz et al. 1998; Gomez-Duarte et al. 1998).

In addition to the use of attenuated bacterial vector, a study conducted by Corthesy et al. (2005) reported the use of the nonpathogenic, noninvasive lactic acid bacterial strain. These carriers do not induce pronounced proinflammatory responses, which renders them best suited for immunocompromised subjects, infants, and elderly individuals. *Lactobacillus plantarum* producing *H. pylori* UreB was used as oral-delivery vehicles in a mouse model of *H. felis* infection. The recombinant strain induced serum UreB–specific antibody to levels similar to those elicited by intragastric administration of recombinant UreB/CT and afforded partial protected against *H. felis* infection (Corthesy et al. 2005).

Recently, a vaccine vector has been generated that used poliovirus genomes, in which capsid genes were replaced with the gene encoding the β subunit of *H. pylori* urease (UreB replicon). Systemic injection of UreB replicon in mice bearing the gene for poliovirus receptor showed prophylactic and therapeutic efficacy against *H. pylori* (Smythies et al. 2005). Although this study needs confirmation, this approach is interesting and additional studies should be conducted.

DNA vaccination. DNA vaccines consist of the gene encoding the desired antigen inserted into a bacterial plasmid containing appropriate eukaryotic promoters, thus enabling target mammalian host cells to express the vaccine antigen. The plasmid is delivered either as naked DNA or by an attenuated carrier microorganism such as *Salmonella* sp., which can invade host cells. The advantage of bacteria is that they naturally contain unmethylated CpG oligonucleotides, which have adjuvant activity. A DNA vaccine based on the urease B gene administrated intranasally was reported to induce a small but significant decrease in *H. pylori* stomach colonization in mice. However, the vector without the antigen also showed a trend toward reduction in bacterial colonization density (Hatzifoti et al. 2006). Immunized mice with DNA plasmids encoding heat shock proteins and catalase have induced suppression of *Helicobacter* spp. colonization but not a complete protection (Miyashita et al. 2002; Todokori et al. 2000). DNA vaccination may prove to be a feasible approach to eradication of *H. pylori* in the future.

Mechanisms of Action of the Vaccine-Induced *Helicobacter* Clearance

Based on the localization of *H. pylori* infection on the surface of the stomach mucosa, it may be deduced that a predominant secretory IgA antibody would be effective for protection (Brandtzaeg 2007). However, numerous studies in animals suggested that T cells and innate immune responses are of prime importance for protection. Indeed, vaccination-induced protection against *H. pylori* in mice requires major histocompatibility complex class II–restricted CD4$^+$ T cells (Del Giudice et al. 2001; Ermak et al. 1998; Pappo et al. 1999), and both Th-1 and Th-2 CD4$^+$ T cell responses mediate protection (Aebischer et al. 2001; Sawai et al. 1999). B cells (antibodies), however, are not required for protection (Ermak et al. 1998), although they may be beneficial (Lee et al. 1995). $\alpha4\beta7$ integrin–mediated homing processes are also critical for host protection (Michetti et al. 2000). Even if CD4$^+$ T cells appear as key players in the *Helicobacter* spp. clearance (Ermak et al. 1998), their mechanisms of action remain unclear. Indeed, although CD4$^+$ T cells can migrate in the stomach mucosa (Michetti et al. 2000), they do not pass through the mucosa and therefore will never be in direct contact with the bacteria. Therefore, the role of CD4$^+$ T cells in the bacterial eradication is certainly indirect and the final mechanisms leading to *Helicobacter* spp. clearance after vaccination remain undefined.

We recently reported a major role of mast cells in vaccine-induced *Helicobacter* clearance (Velin et al. 2005). Previously, Berg et al. showed the central role for neutrophils and of complement activation in *Helicobacter* spp. eradication in the IL-10$^{-/-}$ mouse model (Ismail et al. 2003a, 2003b), suggesting that inflammatory processes are key players in bacterial clearance. In addition, Garhart et al. (2002) showed that reduction in bacteria counts in immunized mice was predicted by the CD4$^+$ T cell response and correlated with the development of gastriti. Together, these observations suggest that CD4$^+$ T cells and other elements of the inflammatory cascade are essential for clearance of *Helicobacter* spp. from the stomach mucosa. Because most of the mechanisms of action of vaccine-induced *Helicobacter* clearance described above have been studied with the urease-based vaccine, the generalization of these data to a nonurease-based vaccine is not possible. Indeed, it can be envisaged that nonurease-based vaccines possibly prime different immune mechanisms responsible for eradication of *Helicobacter* spp.

Human Clinical Trials of Anti-*Helicobacter pylori* Vaccine

Urease-Based Vaccine

To date, four randomized, double-blind, placebo-controlled clinical trials have been conducted with mucosally administered recombinant urease in healthy asymptomatic volunteers. The first two were conducted orally in subjects with *H. pylori* infection and chronic gastritis, the third was conducted in uninfected subjects, and the last was conducted with rectally administered urease, together with LT. In the first trial, urease was orally administered alone (without a mucosal adjuvant). This study showed that urease was safe and well tolerated, but in the absence of a mucosal adjuvant did not alter the course of the infection (Kreiss et al. 1996). The design of the second clinical trial was similar to the first, except that urease was administered with LT as an adjuvant. There were no serious adverse events, no clinically significant laboratory abnormalities, and no subject discontinued the trial for safety-related reasons. At 1 month after the final vaccination, all volunteers remained infected with *H. pylori*. However, those receiving urease + LT experienced, on average, a larger decrease in *H. pylori* colonization levels in the gastric mucosa from baseline than did those volunteers receiving LT alone or placebo (Michetti et al. 1999). Although the study had small sample sizes per group and was not powered to detect significant differences between treatment groups, it did provide preliminary evidence suggesting that therapeutic immunization is achievable in humans. The third and fourth trials were conducted in volunteers without detectable preexisting exposure to *H. pylori*. In the third trial, a dose of 60 mg of urease was administered weekly for 5 consecutive weeks by the oral route as either a liquid formulation or in enteric-coated capsules, all with low-dose LT (Banerjee et al. 2002). A fourth trial was conducted with the objective to evaluate the safety and immunogenicity of urease when administered intrarectally in combination with LT adjuvant to healthy volunteers (Sougioultzis et al. 2002). Although these last two trials confirmed the safety of urease, the immunogenicity of the vaccine remained weak.

Live Vector Vaccines

Three *Salmonella*-based recombinant vaccines expressing urease have been orally administered to human volunteers. In a first trial, a pho/phoQ deleted *Salmonella enterica* serovar *typhi* strain expressing both urease subunits was administered to *H. pylori*–negative persons (DiPetrillo et al. 1999). None of the orally immunized volunteers developed serum antiurease antibodies. In a subsequent study, volunteers were immunized with an *S. enterica* serovar *typhimurium* phoP/phoQ-negative strain expressing *H. pylori* urease. Weak humoral responses against urease were detected in 3 of 6 immunized volunteers (Angelakopoulos and Hohmann 2000). Using the known vaccine strain *S. enterica* var. Ty21 expressing *H. pylori* urease, orally immunized volunteers did not develop detectable humoral antiurease response, but some of them mounted weak T-cell responses to urease (Bumann et al. 2001; Metzger et al. 2004). In a study taking advantage of the model of *H. pylori* challenge in humans developed by Graham (Graham et al. 2004), Aebischer et al. documented that 3 of 9 volunteers vaccinated orally with this urease-expressing Ty21a-based construct cleared the infection after subsequent challenge with *H. pylori* (Aebischer et al. 2005). This study awaits full publication, but illustrates that the proper immune correlates of vaccine-induced immunity against *H. pylori* requires further study.

Inactivated Whole Killed Cells

Formalin-killed *H. pylori* cells with varying doses of LT_{R192G} were administrated orally to a group of both *H. pylori*–uninfected and –infected volunteers. The adjuvant is a modified form of *E. coli* LT, with a glycine residue substituted for the arginine at position 192 from the amino terminus of the A1 subunit of the molecule. Removal of this arginine residue renders the molecule trypsin-insensitive, thereby interfering with its activation to an enterotoxic form (Kotloff et al. 2001). However, some side effects such as diarrhea, fever, and vomiting were observed in some of the volunteers. Significant anti-*H. pylori*–specific anti-IgA in fecal and salivary specimens was detected in some volunteers. Remarkably, this antibody response was associated with the detection of *H. pylori*–specific antibody-secreting cells in the gastric tissues of uninfected volunteers (Losonsky et al. 2003). Although these data clearly established that this vaccine preparation primed mucosal immune response in human to *H. pylori* antigens, the vaccine regimen was unable to eradicate preexisting infection (Kotloff et al. 2001).

Bottleneck of Vaccine Development

Lack of Major Investments

In the United States, it has been estimated that there will be a "natural" decrease in *H. pylori* prevalence from 12.0% in 2010 to 4.2% in 2100. With the introduction of a prophylactic vaccine targeting all infants beginning in 2010, prevalence would decrease to 0.7% by year 2100. In the same period, incidence of *H. pylori*–attributable gastric cancer would decrease from 4.5 to 0.4 per 100,000 with vaccine (compared with 1.3 per 100,000 without vaccine). Incidence of *H. pylori*–attributable duodenal ulcer would decrease from 33.3 to 2.5 per 100,000 with vaccine (compared with 12.2 per 100,000 without vaccine). With continuous vaccination in developing countries, *H. pylori*–attributable gastric cancer would only decrease from 31.8 to 5.8 per 100,000 by 2100 (Rupnow et al. 2001). These data clearly support a positive impact of the introduction of a prophylactic vaccine in developing countries to control *H. pylori* infection. The situation is less obvious for the industrialized countries where *H. pylori* prevalence is declining naturally. To our knowledge, most of the pharmaceutical companies that were previously involved in the development of anti-*H. pylori* vaccines stopped their efforts. Therefore, the major bottleneck of the anti-*H. pylori* vaccine is the lack of major investments in this field.

Oral Tolerance

It is well known that most individuals develop lifelong clinical and immunologic tolerance to both food antigens and indigenous gut flora (Strobel 2001). Several mechanisms have been proposed for the development of oral tolerance, ranging from the deletion of antigen-specific T cells, to immune deviation, induction of anergy, and suppression by regulatory T cells (Strobel 2001). Oral tolerance to food antigens and gut flora is mainly acquired during early childhood (Strobel 2001). During this period, the immature mucosal immune system is educated to tolerate novel diet and colonization by novel strains and species of bacteria. Because *H. pylori* infection is often acquired during this time of oral tolerance development, it can be postulated that our mucosal immune system is programmed to tolerate *H. pylori* colonization (Feldman 2001; Rowland et al. 1999; Strobel 2001). Therefore, it is possible that *H. pylori* is recognized by the host's immune system as a component of the gut flora allowing for its persistence in the stomach.

Helicobacter Immune Escape and/or Helicobacter Immune Control?

Although *H. pylori* infection elicits innate and adaptive immune responses, the bacteria elude these protective mechanisms and persist for decades in our stomach. Several mechanisms of immune evasion used by *H. pylori* have been described. For instance, contrary to the LPS and flagellins of other gram-negative bacteria, the LPS and flagellins of *H. pylori* do not adequately activate the antigen-presenting cells via the Toll-like receptors (Gewirtz et al. 2004; Suda et al. 2001). Studies also indicate that *H. pylori* is capable of evading the host immune responses by interfering directly with T-cell proliferation. Using Jurkat cells, a human T-cell line, it was demonstrated that VacA interfered with calcium-signaling events inside the cell and prevented activation of the calcium-dependent phosphatase calcineurin (Boncristiano et al. 2003; Gebert et al. 2003). The subsequent dephosphorylation of nuclear factor of activated T cells, a transcription factor that regulates immune responses, was suppressed resulting in inhibition of IL-2 expression and proliferation of T cells. In addition to VacA, *H. pylori* arginase can impair T-cell function during infection. Using Jurkat T cells and human normal lymphocytes, it was found that a wild-type *H. pylori* strain, but not an arginase mutant strain, inhibited T-cell proliferation, depleted L-arginine, and reduced the expression of the CD3ζ chain of the T-cell receptor (Zabaleta et al. 2004). In addition, Akhiani et al. recently demonstrated that bacteria were completely cleared from B-cell–deficient mutant mice within the context of severe gastric inflammation, although initial colonization was comparable to the one observed in wild-type mice (Akhiani et al. 2004). This suggests that the presence of antibodies directed toward the bacteria results in less-severe inflammation and facilitates the chronic infection.

Furthermore, *H. pylori* infection also leads to the generation of regulatory T cells (Treg). Indeed, it has been documented that *H. pylori*–induced T-cell response is actively downregulated partly by immunosuppressive CD25+ T cells or Treg (Lindholm et al. 1998; Lundgren et al. 2003; Raghavan et al. 2003). Lundgren et al. (2003) have shown that the memory T-cell responses to *H. pylori* antigens in the peripheral blood are under the control of Treg in *H. pylori*–infected asymptomatic individuals. Removal of Treg specifically from the memory T-cell population increased the proliferative responses to *H. pylori* antigens and, importantly, the addition of Treg back to the memory T cells suppressed the *H. pylori*–specific responses but failed to suppress responses to unrelated antigens. Moreover, CD4+CD25[high] T cells (putative Treg) isolated from the gastric and duodenal mucosa of *H. pylori*–infected asymptomatic carriers express the specific Treg marker *FOXP3* (Lundgren et al. 2005). Although widely acknowledged to have a role in the maintenance of self-tolerance, recent studies indicate that Treg can be activated and expanded against bacterial, viral, and parasite antigens *in vivo* (Belkaid et al, 2006). Such pathogen-specific Treg can prevent infection-induced immunopathology but may also increase the load of infection and favor pathogen persistence by suppressing protective immune responses. Therefore, it can be anticipated that these *H. pylori*–specific Treg cells maintain a balance between bacterial infection and development of tissue damage affecting the gastric mucosa, favoring chronicity of the infection.

Therefore, the natural immune response to *H. pylori* observed in infected individuals seems permissive for the bacteria to persist and for the host to protect itself from inflammation-mediated pathogenesis. We can postulate that *H. pylori* has evolved to control and/or to escape host defenses, and that these defenses are adapted to tolerate *H. pylori*. However, the virulence factors expressed by *H. pylori*, the genetic background of the host, and environmental factors can affect this equilibrium and lead to incidental spontaneous *H. pylori* eradication or to *H. pylori*–induced pathogenesis such as ulcers and gastric cancer.

The Adjuvants

To date, oral and nasal vaccinations in human clinical use are mainly composed of live attenuated organisms. From the adaptation of *H. pylori* to its human host, it can be predicted that oral administration of an attenuated strain of *H. pylori* will be well tolerated but still unable to generate a protective immune response. Oral delivery of nonreplicating vaccines is difficult because of poor stability of immunogenic proteins, peptides, and DNA in the acidic and enzyme-rich gastrointestinal environment. In animal models, powerful mucosal adjuvants associated with *H. pylori* antigens were able to overcome these conditions and have been shown to generate a protective immune response. Unfortunately, although nontoxic in mice, CT and LT, even at very low doses (e.g., 1–5 µg) may induce severe diarrhea in humans (Levine et al. 1983). In addition, when LT has been nasally administrated in humans as part of a commercial vaccine (Glück et al. 2000), cases of facial paresis were noted that led to withdrawal of product (the Swiss inactivated influenza vaccine Nasalflu) (Mutsch et al. 2004). Therefore, it is important to develop genetically detoxified mutants of mucosal adjuvants, tolerated by humans, before embarking on new clinical trials of nasal LT-based vaccinations. As potential alternatives, several non-CT non-LT–based mucosal adjuvants are currently in preclinical development [e.g., CpG (Shi et al. 2005) and chitosan (Moschos et al. 2004)]. However, these adjuvants need to be evaluated in humans.

Immune Correlates

To successfully navigate a vaccine through the early stages of clinical development into efficacy studies, the immune response that the vaccine preparation has to induce should be defined. Without this knowledge, testing vaccine efficacy against the infectious agent is compromised. Preclinical studies of vaccine preparations have to be performed in different animal models in order to define as best as possible these immune correlates required to clear the infection. To date, with urease-based vaccine, CD4$^+$ T-cell responses and inflammatory tissue responses are necessary to eradicate or to prevent *H. pylori* infection. In human and animal models, the CD4$^+$ T-cell responses are divided in three arms: Th1, Th2, and Th17 (Reiner 2007). These distinct CD4$^+$ T-cell responses vary among different vaccination protocols. Future experiments are needed to define which type of T-helper subsets are responsible for vaccine-induced bacterial clearance.

Future of Vaccine Development

Starting from its discovery by Barry Marshall and Robin Warren in 1984 (Marshall and Warren 1984), phenomenal knowledge on the natural disease progression and pathogenesis of *H. pylori* has been documented. Although *H. pylori* seems to mimic a commensal bacteria, its persistence can lead to alterations of the gastric mucosa and to disease (Ernst and Gold 2000; Uemura et al. 2001; Zucca et al. 1998). Hence, gastroenterologists use a combination of antisecretory and antimicrobial agents to eradicate *H. pylori* (Wolle and Malfertheiner 2007). Similar to other antimicrobial treatments, the therapy may select resistant *H. pylori* strains (Wolle and Malfertheiner 2007). Therefore, there is a need to develop alternative therapies to eradicate *H. pylori* infection. The development of a vaccine against *H. pylori* is one of these alternatives. Although the goal is complex, as usual in the development of a new therapy, this option clearly provides an opportunity to control *H. pylori*–induced pathogenesis on a large scale, required in developing countries (Rupnow et al. 2001). In addition, in developed countries, individuals infected with antibiotic-resistant *H. pylori* strains will possibly need alternative therapies.

In animal models, considerable insight has been generated regarding the immune mechanisms leading to vaccine-induced *Helicobacter* clearance. The immune correlates defined in animals can certainly be used in human clinical trials. Protective *H. pylori*–derived antigens such as urease, VacA, CagA, NAP, and others can be easily produced in prokaryote expression systems at reasonable costs. Most of these antigens already have been administrated to human and shown to be safe (Banerjee et al. 2002; Kreiss et al. 1996; Malfertheiner et al. 2002; Michetti et al. 1999; Sougioultzis et al. 2002). A recent demonstration in animal models that parenteral vaccination was as effective as mucosal vaccination to eradicate *H. pylori* infection (Gottwein et al. 2001) suggests that intramuscular vaccination strategy might also be possible. This is an important step, because the selection of appropriate adjuvants, able to generate adapted T-helper responses capable of clearing *Helicobacter* infection, is well established. Other options such as recombinant live bacteria (*Salmonella*, noninvasive lactic acid bacteria, and poliovirus) might also constitute alternatives to develop an efficient vaccine against *H. pylori*.

Our point of view is that the development of an anti-*H. pylori* vaccine should be pursued in humans. Financial support is certainly required if this option is pursued. Given the magnitude of the target population, major funding agencies, such as the National Institutes of Health or the Bill & Melinda Gates Foundation, should be encouraged to support the development of a safe, effective vaccine.

References

Aebischer, T., Bumann, D., Epple, H. D., Graham, D. Y., Metzger, W., Schneider, T., Stolte, M., Zeitz, M., and Meyer, T. F. (2005). Development of a vaccine against Helicobacter pylori [abstract]. Helicobacter. 10:547.

Aebischer, T., Laforsch, S., Hurwitz, R., Brombacher, F., and Meyer, T. F. (2001). Immunity against Helicobacter pylori: significance of interleukin-4 receptor alpha chain status and gender of infected mice. Infect Immun. 69:556–558.

Akhiani, A. A., Schon, K., Franzen, L. E., Pappo, J., and Lycke, N. (2004). Helicobacter pylori-specific antibodies impair the development of gastritis, facilitate bacterial colonization, and counteract resistance against infection. J Immunol. 172:5024–5033.

Akhiani, A. A., Stensson, A., Schon, K., and Lycke, N. (2006). The nontoxic CTA1-DD adjuvant enhances protective immunity against Helicobacter pylori infection following mucosal immunization. Scand J Immunol. 63:97–105.

Angelakopoulos, H., and Hohmann, E. L. (2000). Pilot study of phoP/phoQ-deleted Salmonella enterica serovar typhimurium expressing Helicobacter pylori urease in adult volunteers. Infect Immun. 68:2135–2141.

Banerjee, S., Medina-Fatimi, A., Nichols, R., Tendler, D., Michetti, M., Simon, J., Kelly, C. P., Monath, T. P., and Michetti, P. (2002). Safety and efficacy of low dose Escherichia coli enterotoxin adjuvant for urease based oral immunisation against Helicobacter pylori in healthy volunteers. Gut. 51:634–640.

Belkaid, Y., Blank, R. B., and Suffia I. (2006). Natural regulatory T cells and parasites: a common quest for host homeostasis. Immunol Rev. 212:287–300.

Boncristiano, M., Paccani, S. R., Barone, S., Ulivieri, C., Patrussi, L., Ilver, D., Amedei, A., D'Elios, M. M., Telford, J. L., and Baldari, C. T. (2003). The Helicobacter pylori vacuolating toxin inhibits T cell activation by two independent mechanisms. J Exp Med. 198:1887–1897.

Brandtzaeg, P. (2007). Induction of secretory immunity and memory at mucosal surfaces. Vaccine. 25(30):5467–5484.

Bumann, D., Metzger, W. G., Mansouri, E., Palme, O., Wendland, M., Hurwitz, R., Haas, G., Aebischer, T., von Specht, B. U., and Meyer, T. F. (2001). Safety and immunogenicity of live recombinant Salmonella enterica serovar Typhi Ty21a expressing urease A and B from Helicobacter pylori in human volunteers. Vaccine. 20:845–852.

Corthesy, B., Boris, S., Isler, P., Grangette, C., and Mercenier, A. (2005). Oral immunization of mice with lactic acid bacteria producing Helicobacter pylori urease B subunit partially protects against challenge with Helicobacter felis. J Infect Dis. 192:1441–1449.

Corthesy-Theulaz, I. E., Hopkins, S., Bachmann, D., Saldinger, P. F., Porta, N., Haas, R., Zheng-Xin, Y., Meyer, T., Bouzourene, H., Blum, A. L., and Kraehenbuhl, J. P. (1998). Mice are protected from Helicobacter pylori infection by nasal immunization with attenuated Salmonella typhimurium phoPc expressing urease A and B subunits. Infect Immun. 66:581–586.

Corthesy-Theulaz, I., Porta, N., Glauser, M., Saraga, E., Vaney, A. C., Haas, R., Kraehenbuhl, J. P., Blum, A. L., and Michetti, P. (1995). Oral immunization with Helicobacter pylori urease B subunit as a treatment against Helicobacter infection in mice. Gastroenterology. 109:115–21.

Cryz, S. J. (1993). Post-marketing experience with live oral Ty21a vaccine. Lancet. 341:49–50.

Czinn, S. J., Cai, A., and Nedrud, J. G. (1993). Protection of germ-free mice from infection by Helicobacter felis after active oral or passive IgA immunization. Vaccine. 11:637–642.

Del Giudice, G., Covacci, A., Telford, J. L., Montecucco, C., and Rappuoli, R. (2001). The design of vaccines against Helicobacter pylori and their development. Annu Rev Immunol. 19:523–563.

DiPetrillo, M. D., Tibbetts, T., Kleanthous, H., Killeen, K. P., and Hohmann, E. L. (1999). Safety and immunogenicity of phoP/phoQ-deleted Salmonella typhi expressing Helicobacter pylori urease in adult volunteers. Vaccine. 18:449–459.

Di Tommaso, A., Xiang, Z., Bugnoli, M., Pileri, P., Figura, N., Bayeli, P. F., Rappuoli, R., Abrignani, S., and De Magistris, M. T. (1995). Helicobacter pylori-specific CD4+ T-cell clones from peripheral blood and gastric biopsies. Infect Immun. 63:1102–1106.

Dixon, M. F., Genta, R. M., Yardley, J. H., and Correa, P. (1996). Classification and grading of gastritis. The updated Sydney System. International Workshop on the Histopathology of Gastritis, Houston 1994. Am J Surg Pathol. 20:1161–1181.

Doidge, C., Crust, I., Lee, A., Buck, F., Hazell, S., and Manne, U. (1994). Therapeutic immunisation against Helicobacter infection. Lancet. 343:914–915.

Dubois, A., Fiala, N., Heman-Ackah, L. M., Drazek, E. S., Tarnawski, A., Fishbein, W. N., Perez-Perez, G. I., and Blaser, M.J. (1994). Natural gastric infection with Helicobacter pylori in monkeys: a model for spiral bacteria infection in humans. Gastroenterology. 106:1405–1417.

Dunkley, M. L., Harris, S. J., McCoy, R. J., Musicka, M. J., Eyers, F. M., Beagley, L. G., Lumley, P. J., Beagley, K. W., and Clancy, R. L. (1999). Protection against Helicobacter pylori infection by intestinal immunisation with a 50/52-kDa subunit protein. FEMS Immunol Med Microbiol. 24:221–225.

Eaton, K. A., and Krakowka S. (1994). Effect of gastric pH on urease-dependent colonization of gnotobiotic piglets by Helicobacter pylori. Infect Immun. 62:3604–3607.

Eaton, K. A., Suerbaum, S., Josenhans, C., and Krakowka, S. (1996). Colonization of gnotobiotic piglets by Helicobacter pylori deficient in two flagellin genes. Infect Immun. 64:2445–2448.

Ermak, T. H., Giannasca, P. J., Nichols, R., Myers, G. A., Nedrud, J., Weltzin, R., Lee, C. K., Kleanthous, H., and Monath, T. P. (1998). Immunization of mice with urease vaccine affords protection against Helicobacter pylori infection in the absence of antibodies and is mediated by MHC class II-restricted responses. J Exp Med. 188:2277–2288.

Ernst, P. B., and Gold, B. D. (2000). The disease spectrum of Helicobacter pylori: the immunopathogenesis of gastroduodenal ulcer and gastric cancer. Ann Rev Microbiol. 54:615–640.

Feldman, R. A. (2001). Epidemiologic observations and open questions about disease and infection caused by Helicobacter pylori. In: Achtman, M., and Suerbaum, S., editors. Helicobacter pylori: molecular and cellular biology. Wymondham: Horizon Scientific Press; pp. 29–51.

Ferrero, R. L., and Labigne, A. (1993). Cloning, expression and sequencing of Helicobacter felis urease genes. Mol Microbiol. 9:323–333.

Ferrero, R. L., Thiberge, J. M., Kansau, I., Wuscher, N., Huerre, M., and Labigne, A. (1995). The GroES homolog of Helicobacter pylori confers protective immunity against mucosal infection in mice. Proc Natl Acad Sci USA. 92:6499–6503.

Garhart, C. A., Redline, R. W., Nedrud, J. G., and Czinn, S. J. (2002). Clearance of Helicobacter pylori infection and resolution of postimmunization gastritis in a kinetic study of prophylactically immunized mice. Infect Immun. 70:3529–3538.

Gebert, B., Fischer, W., Weiss, E., Hoffmann, R., and Haas, R. (2003). Helicobacter pylori vacu-olating cytotoxin inhibits T lymphocyte activation. Science. 301:1099–1102.

Gewirtz, A. T., Yu, Y., Krishna, U. S., Israel, D. A., Lyons, S. L., and Peek, R. M. (2004). Helicobacter pylori flagellin evades Toll-like receptor 5-mediated innate immunity. J Infect Dis 189:1914–1920.

Ghendon, Y., and Robertson, S. E. (1994). Interrupting the transmission of wild polioviruses with vaccines: immunological considerations. Bull World Health Organ. 72:973–983.

Ghiara, P., Marchetti, M., Blaser, M. J., Tummuru, M. K., Cover, T. L., Segal, E. D., Tompkins, L. S., and Rappuoli, R. (1995). Role of the Helicobacter pylori virulence factors vacuolating cytotoxin, CagA, and urease in a mouse model of disease. Infect Immun. 63:4154–4160.

Ghiara, P., Rossi, M., Marchetti, M., Di Tommaso, A., Vindigni, C., Ciampolini, F., Covacci, A., Telford, J. L., De Magistris, M. T., Pizza, M., Rappuoli, R., and Del Giudice, G. (1997). Therapeutic intragastric vaccination against Helicobacter pylori in mice eradicates an other-wise chronic infection and confers protection against reinfection. Infect Immun. 65:4996–5002.

Gluck, R., Mischler, R., Durrer, P., Furer, E., Lang, A. B., Herzog, C., and Cryz, S. J. (2000). Safety and immunogenicity of intranasally administered inactivated trivalent virosome-formu-lated influenza vaccine containing Escherichia coli heat-labile toxin as a mucosal adjuvant. J Infect Dis. 181:1129–1132.

Goodman, K. J., Correa, P., Tengana Aux, H. J., Ramirez, H., DeLany, J. P., Guerrero, P. O., Lopez, Q. M., and Collazos, P. T. (1996). Helicobacter pylori infection in the Colombian Andes: a population-based study of transmission pathways. Am J Epidemiol. 144:290–299.

Gomez-Duarte, O. G., Lucas, B., Yan, Z. X., Panthel, K., Haas, R., and Meyer, T. F. (1998). Protection of mice against gastric colonization by Helicobacter pylori by single oral dose immunization with attenuated Salmonella typhimurium producing urease subunits A and B. Vaccine. 16:460–471.

Gottwein, J. M., Blanchard, T. G., Targoni, O. S., Eisenberg, J. C., Zagorski, B. M., Redline, R. W., Nedrud, J. G., Tary-Lehmann, M., Lehmann, P. V., and Czinn, S. J. (2001). Protective anti-Helicobacter immunity is induced with aluminum hydroxide or complete Freund's adjuvant by systemic immunization. J Infect Dis. 184:308–314.

Graham, D. Y., Opekun, A. R., Osato, M. S., El-Zimaity, H. M., Lee, C. K., Yamaoka, Y., Qureshi, W. A., Cadoz, M., and Monath, T. P. (2004). Challenge model for Helicobacter pylori infection in human volunteers. Gut. 53:1235–1243.

Guruge, J. L., Falk, P. G., Lorenz, R. G., Dans, M., Wirth, H. P., Blaser, M. J., Berg, D. E., and Gordon, J. I. (1998). Epithelial attachment alters the outcome of Helicobacter pylori infection. Proc Natl Acad Sci USA. 95:3925–3930.

Hatzifoti, C., Roussel, Y., Harris, A. G., Wren, B. W., Morrow, J. W., and Bajaj-Elliott, M. (20066). Mucosal immunization with a urease B DNA vaccine induces innate and cellular immune responses against Helicobacter pylori. Helicobacter. 11:113–122.

Hayashi, S., Sugiyama, T., Yokota, K., Isogai, H., Isogai, E., Oguma, K., Asaka, M., Fujii, N., and Hirai, Y. (1998). Analysis of immunoglobulin A antibodies to Helicobacter pylori in serum and gastric juice in relation to mucosal inflammation. Clin Diagn Lab Immunol. 5:617–621.

Hazell, S. L., Lee, A., Brady, L., and Hennessy, W. (1986). Campylobacter pyloridis and gastritis: association with intercellular spaces and adaptation to an environment of mucus as important factors in colonization of the gastric epithelium. J Infect Dis. 153:658–663.

Ikewaki, J., Nishizono, A., Goto, T., Fujioka, T., and Mifune, K. (2000). Therapeutic oral vaccina-tion induces mucosal immune response sufficient to eliminate long-term Helicobacter pylori infection. Microbiol Immunol. 44:29–39.

Ismail, H. F., Fick, P., Zhang, J., Lynch, R. G., and Berg, D. J. (2003a). Depletion of neutrophils in IL-10(–/–) mice delays clearance of gastric Helicobacter infection and decreases the Th1 immune response to Helicobacter. J Immunol. 170:3782–3789.

Ismail, H. F., Zhang, J., Lynch, R. G., Wang, Y., and Berg, D. J. (2003b). Role for complement in development of Helicobacter-induced gastritis in interleukin-10-deficient mice. Infect Immun. 71:7140–7148.

Kabir, S. (2007). The current status of Helicobacter pylori vaccines: a review. Helicobacter. 12:89–102.

Kang, Q. Z., Duan, G. C., Fan, Q. T., and Xi, Y. L. (2005). Fusion expression of Helicobacter pylori neutrophil-activating protein in E. coli. World J Gastroenterol. 11:454–456.

Keenan, J., Oliaro, J., Domigan, N., Potter, H., Aitken, G., Allardyce, R., and Roake, J. (2000). Immune response to an 18-kilodalton outer membrane antigen identifies lipoprotein 20 as a Helicobacter pylori vaccine candidate. Infect Immun. 68:3337–3343.

Kim, B. O., Shin, S. S., Yoo, Y. H., and Pyo, S. (2001). Peroral immunization with Helicobacter pylori adhesin protein genetically linked to cholera toxin A2B subunits. Clin Sci (Lond). 100:291–298.

Kotloff, K. L., Sztein, M. B., Wasserman, S. S., Losonsky, G. A., DiLorenzo, S. C., and Walker, R.I. (2001). Safety and immunogenicity of oral inactivated whole-cell Helicobacter pylori vaccine with adjuvant among volunteers with or without subclinical infection. Infect Immun. 69:3581–3590.

Kreiss, C., Buclin, T., Cosma, M., and Michetti, P. (1996). Safety of oral immunization with recombinant urease in patients with Helicobacter pylori infection. Lancet. 347:1630–1631.

Lee, A., Fox, J. G., Otto, G., and Murphy, J. (1990). A small animal model of human Helicobacter pylori active chronic gastritis. Gastroenterology. 99:1315–1323.

Lee, A., O'Rourke, J., De Ungria, M. C., Robertson, B., Daskalopoulos, G., and Dixon, M. F. (1997). A standardized mouse model of Helicobacter pylori infection: introducing the Sydney strain. Gastroenterology. 112:1386–1397.

Lee, C. K., Soike, K., Giannasca, P., Hill, J., Weltzin, R., Kleanthous, H., Blanchard, J., and Monath, T. P. (1999a). Immunization of rhesus monkeys with a mucosal prime, parenteral boost strategy protects against infection with Helicobacter pylori. Vaccine. 17:3072–3082.

Lee, C. K., Soike, K., Hill, J., Georgakopoulos, K., Tibbitts, T., Ingrassia, J., Gray, H., Boden, J., Kleanthous, H., Giannasca, P., Ermak, T., Weltzin, R., Blanchard, J., and Monath, T. P. (1999b). Immunization with recombinant Helicobacter pylori urease decreases colonization levels following experimental infection of rhesus monkeys. Vaccine. 17:1493–1505.

Lee, C. K., Weltzin, R., Thomas, W., Jr., Kleanthous, H., Ermak, T. H., Soman, G., Hill, J. E., Ackerman, S. K., and Monath, T. P. (1995). Oral immunization with recombinant Helicobacter pylori urease induces secretory IgA antibodies and protects mice from challenge with Helicobacter felis. J Infect Dis. 172:161–172.

Levine, M. M., Kaper, J. B., Black, R. E., and Clements, M. L. (1983). New knowledge on pathogenesis of bacterial enteric infections as applied to vaccine development. Microbiol Rev. 47:510–550.

Lindholm, C., Quiding-Jarbrink, M., Lonroth, H., Hamlet, A., and Svennerholm, A. M. (1998). Local cytokine response in Helicobacter pylori-infected subjects. Infect Immun. 66: 5964–5971.

Losonsky, G. A., Kotloff, K. L., and Walker, R. I. (2003). B cell responses in gastric antrum and duodenum following oral inactivated Helicobacter pylori whole cell (HWC) vaccine and LT(R192G) in H pylori seronegative individuals. Vaccine. 21:562–565.

Lundgren, A., Stromberg, E., Sjoling, A., Lindholm, C., Enarsson, K., Edebo, A., Johnsson, E., Suri-Payer, E., Larsson, P., Rudin, A., Svennerholm, A. M., and Lundin, B. S. (2005). Mucosal FOXP3-expressing CD4+ CD25high regulatory T cells in Helicobacter pylori-infected patients. Infect Immun. 73:523–531.

Lundgren, A., Suri-Payer, E., Enarsson, K., Svennerholm, A. M., and Lundin, B. S. (2003). Helicobacter pylori-specific CD4+ CD25high regulatory T cells suppress memory T-cell responses to H. pylori in infected individuals. Infect Immun. 71:1755–1762.

Malfertheiner, P., Megraud, F., O'morain, C., Bazzoli, F., El-Omar, E., Graham, D., Hunt, R., Rokkas, T., Vakil, N., and Kuipers, E. J. (2007). Current concepts in the management of Helicobacter pylori infection: the Maastricht III Consensus Report. Gut. 56:772–781.

Malfertheiner, P., Schultze, V., Del Giudice, G., Rosenkranz, B., Kaufmann, S. H. E.,. Winau, F, Ulrichs, T., Theophil, E., Jue, C. P., Novicki, D., Norelli, F., Contorni, M., Berti, D., Lin, J. S., Schwenke, C., Goldman, M., Tornese, D., Ganju, J., Palla, E., Rappuoli, R., and Scharschmidt, B.

(2002). Phase I safety and immunogenicity of a three-component H. pylori vaccine. Gastroenterology 122 Suppl A-585, Abstr. W1195.

Marchetti, M., Arico, B., Burroni, D., Figura, N., Rappuoli, R., and Ghiara P. (1995). Development of a mouse model of Helicobacter pylori infection that mimics human disease. Science. 267:1655–1658.

Marchetti, M., Rossi, M., Giannelli, V., Giuliani, M. M., Pizza, M., Censini, S., Covacci, A., Massari, P., Pagliaccia, C., Manetti, R., Telford, J. L., Douce, G., Dougan, G., Rappuoli, R., and Ghiara, P. (1998). Protection against Helicobacter pylori infection in mice by intragastric vaccination with H. pylori antigens is achieved using a non-toxic mutant of E. coli heat-labile enterotoxin (LT) as adjuvant. Vaccine. 16:33–37.

Marshall, B., and Warren, J. R. (1984). Unidentified curved bacilli in the stomach of patients with gastritis and peptic ulceration. Lancet. 323:1311–1315.

Mattsson, A., Quiding-Jarbrink, M., Lonroth, H., Hamlet, A., Ahlstedt, I., and Svennerholm, A. (1998). Antibody-secreting cells in the stomachs of symptomatic and asymptomatic Helicobacter pylori-infected subjects. Infect Immun. 66:2705–2712.

McAtee, C. P., Lim, M. Y., Fung, K., Velligan, M., Fry, K., Chow, T., and Berg, D. E. (1998). Identification of potential diagnostic and vaccine candidates of Helicobacter pylori by two-dimensional gel electrophoresis, sequence analysis, and serum profiling. Clin Diagn Lab Immunol. 5:537–542.

Megraud, F. (2004). H pylori antibiotic resistance prevalence, importance, and advances in testing. Gut. 53:1374–1384.

Metzger, W. G., Mansouri, E., Kronawitter, M., Diescher, S., Soerensen, M., Hurwitz, R., Bumann, D., Aebischer, T., Von Specht, B. U., and Meyer, T. F. (2004). Impact of vector-priming on the immunogenicity of a live recombinant Salmonella enterica serovar typhi Ty21a vaccine expressing urease A and B from Helicobacter pylori in human volunteers. Vaccine. 22:2273–2277.

Michetti, P., Corthesy-Theulaz, I., Davin, C., Haas, R., Vaney, A. C., Heitz, M., Bille, J., Kraehenbuhl, J. P., Saraga, E., and Blum, A. L. (1994). Immunization of BALB/c mice against Helicobacter felis infection with Helicobacter pylori urease. Gastroenterology. 107:1002–1011.

Michetti, M., Kelly, C. P., Kraehenbuhl, J. P., Bouzourene, H., and Michetti, P. (2000). Gastric mucosal alpha(4)beta(7)-integrin-positive CD4 T lymphocytes and immune protection against Helicobacter infection in mice. Gastroenterology. 119:109–118.

Michetti, P., Kreiss, C., Kotloff, K. L., Porta, N., Blanco, J. L., Bachmann, D., Herranz, M., Saldinger, P. F., Corthesy-Theulaz, I., Losonsky, G., Nichols, R., Simon, J., Stolte, M., Ackerman, S., Monath, T. P., and Blum, A. L. (1999). Oral immunization with urease and Escherichia coli heat-labile enterotoxin is safe and immunogenic in Helicobacter pylori-infected adults. Gastroenterology. 116:804–812.

Miyashita, M., Joh, T., Watanabe, K., Todoroki, I., Seno, K., Ohara, H., Nomura, T., Miyata, M., Kasugai, K., Tochikubo, K., Itoh, M., and Nitta, M. (2002). Immune responses in mice to intranasal and intracutaneous administration of a DNA vaccine encoding Helicobacter pylori-catalase. Vaccine. 20:2336–2342.

Moschos, S. A., Bramwell, V. W., Somavarapu, S., and Alpar, H. O. (2004). Adjuvant synergy: the effects of nasal coadministration of adjuvants. Immunol Cell Biol. 82:628–637.

Mutsch, M., Zhou, W., Rhodes, P., Bopp, M., Chen, R. T., Linder, T., Spyr, C., and Steffen, R. (2004). Use of the inactivated intranasal influenza vaccine and the risk of Bell's palsy in Switzerland. N Engl J Med. 350:896–903.

Nervi, G., Liatopoulou, S., Cavallaro, L. G., Gnocchi, A., Dal-Bo, N., Rugge, M., Iori, V., Cavestro, G. M., Maino, M., Colla, G., Franze, A., and Di Mario. F. (2006). Does Helicobacter pylori infection eradication modify peptic ulcer prevalence? A 10 years' endoscopical survey. World J Gastroenterol. 12:2398–2401.

Nurgalieva, Z. Z., Conner, M. E., Opekun, A. R., Zheng, C. Q., Elliott, S. N., Ernst, P. B., Osato, M., Estes, M. K., and Graham, D. Y. (2005). B-cell and T-cell immune responses to experimental Helicobacter pylori infection in humans. Infect Immun. 73:2999–3006.

Pappo, J., Torrey, D., Castriotta, L., Savinainen, A., Kabok, Z., and Ibraghimov, A. (1999). Helicobacter pylori infection in immunized mice lacking major histocompatibility complex class I and class II functions. Infect Immun. 67:337–341.

Parsonnet, J., Shmuely, H., and Haggerty, T. (1999). Fecal and oral shedding of Helicobacter pylori from healthy infected adults. JAMA. 282:2240–2245.

Perez-Perez, G. I., Dworkin, B. M., Chodos, J. E., and Blaser, M. J. (1988). Campylobacter pylori antibodies in humans. Ann Intern Med. 109:11–17.

Radcliff, F. J., Hazell, S. L., Kolesnikow, T., Doidge, C., and Lee, A. (1997). Catalase, a novel antigen for Helicobacter pylori vaccination. Infect Immun. 65:4668–4674.

Raghavan, S., Fredriksson, M., Svennerholm, A. M., Holmgren, J., and Suri-Payer, E. (2003). Absence of CD4+CD25+ regulatory T cells is associated with a loss of regulation leading to increased pathology in Helicobacter pylori-infected mice. Clin Exp Immunol. 132:393–400.

Reiner, S. L. (2007). Development in motion: helper T cells at work. Cell. 129:33–36.

Rowland, M., Kumar, D., Daly, L., O'Connor, P., Vaughan, D., and Drumm B. (1999). Low rates of Helicobacter pylori reinfection in children. Gastroenterology. 117:336–341.

Rupnow, M. F., Shachter, R. D., Owens, D. K., and Parsonnet, J. (2001). Quantifying the population impact of a prophylactic Helicobacter pylori vaccine. Vaccine. 20:879–885.

Satin, B., Del Giudice, G., Della Bianca, V., Dusi, S., Laudanna, C., Tonello, F., Kelleher, D., Rappuoli, R., Montecucco, C., and Rossi, F. (2000). The neutrophil-activating protein (HP-NAP) of Helicobacter pylori is a protective antigen and a major virulence factor. J Exp Med. 191:1467–1476.

Sawai, N., Kita, M., Kodama, T., Tanahashi, T., Yamaoka, Y., Tagawa, Y., Iwakura, Y., and Imanishi, J. (1999). Role of gamma interferon in Helicobacter pylori-induced gastric inflammatory responses in a mouse model. Infect Immun. 67:279–285.

Schreiber, S., Konradt, M., Groll, C., Scheid, P., Hanauer, G., Werling, H. O., Josenhans C., and Suerbaum, S. (2004). The spatial orientation of Helicobacter pylori in the gastric mucus. Proc Natl Acad Sci USA. 101:5024–5029.

Semino-Mora, C., Doi, S. Q., Marty, A., Simko, V., Carlstedt, I., and Dubois, A. (2003). Intracellular and interstitial expression of Helicobacter pylori virulence genes in gastric precancerous intestinal metaplasia and adenocarcinoma. J Infect Dis. 187:1165–1177.

Shi, T., Liu, W. Z., Gao, F., Shi, G. Y., and Xiao, S. D. (2005). Intranasal CpG-oligodeoxynucleotide is a potent adjuvant of vaccine against Helicobacter pylori, and T helper 1 type response and interferon-gamma correlate with the protection. Helicobacter. 10:71–79.

Smythies, L. E., Novak, M. J., Waites, K. B., Lindsey, J. R., Morrow, C. D., and Smith, P. D. (2005). Poliovirus replicons encoding the B subunit of Helicobacter pylori urease protect mice against H. pylori infection. Vaccine. 23:901–909.

Solnick, J. V., Canfield, D. R., Hansen, L. M., and Torabian, S. Z. (2000). Immunization with recombinant Helicobacter pylori urease in specific-pathogen-free rhesus monkeys (Macaca mulatta). Infect Immun. 68:2560–2565.

Sougioultzis, S., Lee, C. K., Alsahli, M., Banerjee, S., Cadoz, M., Schrader, R., Guy, B., Bedford, P., Monath, T. P., Kelly, C. P., and Michetti, P. (2002). Safety and efficacy of E coli enterotoxin adjuvant for urease-based rectal immunization against Helicobacter pylori. Vaccine. 21:194–201.

Strobel S. (2001). Immunity induced after a feed of antigen during early life: oral tolerance v. sensitisation. Proc Nutr Soc 60:437–442.

Suda, Y., Kim, Y. M., Ogawa, T., Yasui, N., Hasegawa, Y., Kashihara, W., Shimoyama, T., Aoyama, K., Nagata, K., Tamura, T., and Kusumoto, S. (2001). Chemical structure and biological activity of a lipid A component from Helicobacter pylori strain 206. J Endotoxin Res. 7:95–104.

Suzuki, T., Kato, K., Ohara, S., Noguchi, K., Sekine, H., Nagura, H., and Shimosegawa, T. (2002). Localization of antigen-presenting cells in Helicobacter pylori-infected gastric mucosa. Pathol Int. 52:265–271.

Terres, A. M., and Pajares, J. M. (1998). An increased number of follicles containing activated CD69+ helper T cells and proliferating CD71+ B cells are found in H. pylori-infected gastric mucosa. Am J Gastroenterol. 93:579–583.

Todoroki, I., Joh, T., Watanabe, K., Miyashita, M., Seno, K., Nomura, T., Ohara, H., Yokoyama, Y., Tochikubo, K., and Itoh, M. (2000). Suppressive effects of DNA vaccines encoding heat shock protein on Helicobacter pylori-induced gastritis in mice. Biochem Biophys Res Commun. 277:159–163.

Uemura, N., Okamoto, S., Yamamoto, S., Matsumura, N., Yamaguchi, S., Yamakido, M., Taniyama, K., Sasaki, N., and Schlemper, R. J. (2001) Helicobacter pylori infection and the development of gastric cancer. N Engl J Med. 345:784–789.

Velin, D., Bachmann, D., Bouzourene, H., and Michetti, P. (2005). Mast cells are critical mediators of vaccine-induced Helicobacter clearance in the mouse model. Gastroenterology. 129:142–155.

Wolle, K., and Malfertheiner, P. (2007). Treatment of Helicobacter pylori. Best Pract Res Clin Gastroenterol. 21:315–324.

Yamaguchi, H., Osaki, T., Kai, M., Taguchi, H., and Kamiya S. (2000). Immune response against a cross-reactive epitope on the heat shock protein 60 homologue of Helicobacter pylori. Infect Immun. 68:3448–3454.

Zabaleta, J., McGee, D. J., Zea, A. H., Hernandez, C. P., Rodriguez, P. C., Sierra, R. A., Correa, P., and Ochoa, A.C. (2004). Helicobacter pylori arginase inhibits T cell proliferation and reduces the expression of the TCR zeta-chain (CD3zeta). J Immunol. 173:586–593.

Zucca, E., Bertoni, F., Roggero, E., Bosshard, G., Cazzaniga, G., Pedrinis, E., Biondi, A., and Cavalli, F. (1998). Molecular analysis of the progression from Helicobacter pylori-associated chronic gastritis to mucosa-associated lymphoid-tissue lymphoma of the stomach. N Engl J Med. 338:804–810.

Index